Reproductive Biology and Phylogeny of Chondrichthyes

Sharks, Batoids and Chimaeras

Reproductive Biology and Phylogeny of Chondrichthyes

Sharks, Batoids and Chimaeras

Volume edited by
WILLIAM C. HAMLETT
Professor
Department of Anatomy and Cell Biology
Indiana University School of Medicine
Notre Dame, Indiana
USA

Volume 3 of Series:
Reproductive Biology and Phylogeny

Series edited by
BARRIE G.M. JAMIESON
School of Integrative Biology
University of Queensland
St. Lucia, Queensland
Australia

THE UNIVERSITY OF QUEENSLAND
AUSTRALIA

CRC Press
Taylor & Francis Group
Boca Raton London New York

CRC Press is an imprint of the
Taylor & Francis Group, an **informa** business
A SCIENCE PUBLISHERS BOOK

First published 2005 by Science Publishers, Inc.

Published 2022 by CRC Press
Taylor & Francis Group
6000 Broken Sound Parkway NW, Suite 300
Boca Raton, FL 33487-2742

ISBN-13: 978-1-57808-271-1 (set)
ISBN-13: 978-1-57808-314-5 (hbk) (vol.3)

Visit the Taylor & Francis Web site at
http://www.taylorandfrancis.com

and the CRC Press Web site at
http://www.crcpress.com

Library of Congress Cataloging-in-Publication Data

Reproductive biology and phylogeny of Chondrichthyes: sharks, batoids, and chimaeras/
volume edited by William C. Hamlett.
p. cm. --(Reproductive biology and phylogeny; v. 3)
Includes bibliographical references and index.
ISBN 1-57808-314-1
1. Chondrichthyes—Reproduction. I. Hamlett, William C. II. Series.

QL638.6.R467 2005
597.3--dc22 2004065324

Preface to the Series

This series was founded by the present series editor, Barrie Jamieson, in consultation with Science Publishers, Inc., in 1991. The series bears the title 'Reproductive Biology and Phylogeny' and this title is followed in each volume with the name of the taxonomic group which is the subject of the volume. Each publication has one or more invited volume editors and a large number of authors of international repute. The level of the taxonomic group which is the subject of each volume varies according, largely, to the amount of information available on the group, the advice of proposed volume editors, and the interest expressed by the zoological community in the proposed work. The order of publication reflects these concerns, and the availability of authors for the various chapters, and it is not proposed to proceed serially through the animal kingdom in a presumed phylogenetic sequence. Nevertheless, a second aspect of the series is coverage of the phylogeny and classification of the group, as a necessary framework for an understanding of reproductive biology. Evidence for relationships from molecular studies is an important aspect of the phylogenetic sections. Apart from the chapter(s) on phylogeny classification, chapters do not necessorily have a phylogenetic theme.

It is not claimed that a single volume can, in fact, cover the entire gamut of reproductive topics for a given group but it is believed that the series gives an unsurpassed coverage of reproduction and provides a general text rather than being a mere collection of research papers on the subject. Coverage in different volumes will be vary in terms of topics, though it is clear from the first volumes that the standard of the contributions by the authors will be uniformly high. The stress will vary from group to group; for instance, modes of external fertilization or vocalization, important in one group, might be inapplicable in another.

The first two volumes, on Urodela, edited by Professor David Sever, and on Anura, edited by myself, reflected the above criteria and the interests of certain research teams. This, the third volume, has resulted from our good fortune in the acceptance by Professor William Hamlett of an invitation to edit a volume on Chondrichthyes. Will is an outstanding authority, renowned for the breadth and depth of his studies on chondrichthyans. In contributions to this volume he has demonstrated, again, the highest standards of scholarship and research and an exquisite ability to illustrate his subject with line drawings and photographically. His enthusiasm, combined with scientific rigor, for the study of these often maligned animals is apparent in all the chapters to which he has contributed and in his design of the contents of the volume. In his choice of topics for the sixteen chapters he has been able to draw on a distinguished group of authors, many of them notable authorities, others highly competent younger workers who demonstrate their already established expertise in these pages.

Other volumes in preparation are on Gymnophiona (J.-M. Exbrayat), Annelida (G. Rouse and F. Pleijel), Cetacea (D. Miller) and Birds (B.G.M. Jamieson).

While volume editing is by invitation, reproductive biologists who consider that a given taxonomic group should be included in the series and may wish to undertake the task of editing a volume should not hesitate to make their views known to the series editor.

Special thanks are due to Professor Mick McManus, Executive Dean, Faculty of Biological and Chemical Science, Universiy of Queensland, for the encouragement that he has given me in the preparation of these volumes.

I am grateful to the publishers for their support in producing this series. Sincere thanks must be given the volume editors and the authors, who have freely contributed their chapters, in very full schedules. The editors and publishers are confident that the enthusiasm and expertise of these contributors will be reflected by the reception of the series by our readers.

THE UNIVERSITY
OF QUEENSLAND
AUSTRALIA

7 June 2004

Barrie Jamieson
School of Integrative Biology
University of Queensland
Brisbane

Preface to this Volume

As I sit here in northern Indiana with -25° C windchill outside writing about chondrichthyan reproduction my thoughts go back to July 1988 Rome, Italy when I met the original "Godfather" of chondrichthyan reproductive biology, Professor Silvio Ranzi. I was a young assistant professor when I organized a symposium "Evolutionary and contemporary biology of elasmobranchs" and Professor Ranzi was gracious enough to attend and speak on his seminal work from the 1930s when he was at the Stazione Zoologica di Napoli. All of us who study chondrichthyans, in particular reproductive and evolutionary biology, owe a debt of gratitude for his insightful and pioneering work. It has inspired and guided me since 1977 when I began studies for my Ph.D. Throughout my career I focussed on bringing modern techniques such as electron microscopy, immunohistochemistry, biochemical analyses to further elucidate mechanisms that govern maternal-fetal relations in chondrichthyes and to expand on the Master's work.

Professor Ranzi's research was primarily published in Italian and his work was brought to a broader audience in 1942 with publication of Joseph Needham's volume *Biochemistry and Morphogenesis*. Two of Professor Ranzi's monographs from 1932 and 1934 numbering several hundred pages are still the most commonly cited works regarding nutrient sources in chondrichthyes. I am fortunate to own these papers that are profusely illustrated with masterful color pen and ink drawings. Nothing can match the joy I have when I thumb through the pages. It is more akin to appreciating a fine art book than a mere scientific study. Had his career only generated these epoch making works his esteemed place in Italian and international science would be established but he continued his illustrious career for many more years and moved on to new areas of inquiry.

Much of what follows is excerpted from Professor Ranzi's obituary written by Professor Baccio Baccetti in Accademia Nazionale dei Lincei. 1999.

Professor Ranzi was born in Rome October 16, 1902 and died April 16, 1996. He received his Doctor of Natural Science degree in 1924 from the University of Rome. He published his first paper in 1923 at age 21 and his last at age 84 in 1986. He spent 1926-1938 at the Stazione Zoologica di Napoli with return periods in 1943-1945. He spent 1938 at the University of Perugia before moving to the University of Milano from 1939-1996. He married

Contents

CHAPTER 1

Phylogenetic Relationships among the Major Lineages of Modern Elasmobranchs

Gavin J. P. Naylor[1], *J. A. Ryburn*[1], *O. Fedrigo*[2] *and J. A. López*[3]

1.1 INTRODUCTION

Modern elasmobranchs (sharks, skates and rays) are the extant survivors of one of the earliest offshoots of the vertebrate evolutionary tree. Their basal placement on the tree has drawn attention from researchers interested in understanding features characterizing the ancestral vertebrate condition. While their basal placement has long been appreciated, most studies have adopted the stance that a single token elasmobranch is sufficient to represent the basal vertebrate condition. Usually, the exemplar chosen is the spiny dogfish *Squalus acanthias*, the stalwart of comparative anatomy classes. However, the use of a single taxon is valid only to the extent that it is representative of the basal condition. This is unlikely to be the case for all traits. Instead, it is more likely that only a few of the character states seen in the exemplar will actually reflect the basal condition while the majority will be uniquely derived for the exemplar itself. A quick survey of the unique and atypical traits seen in the model organisms *Drosophila*, *Caenorhabditis*, *Danio* and *Mus* attests to the fact that single representative "model" organisms are often more "atypical" than they are "representative".

To estimate the basal condition for any group we must use a phylogenetic tree. Trees allow us to distinguish traits that are likely common to the ancestral condition from those that are unique to particular lineages. In order to do this, it is important: (a) that the phylogenetic tree used be accurate, and (b) that the taxa compared provide an even, balanced and representative coverage of the evolutionary trajectory (Yang *et al.* 1995). In this paper we set out to

[1]School of Computational Science, Florida State University, Tallahassee, Florida, 32306, USA.
[2]Duke University, Department of Biology, 139 Biological Science Building, Science Drive, Durham, NC 27708 USA
[3]Florida Museum of Natural History, Dickinson Hall, Ichthyology Deptartment, University of Florida, Gainesville, Florida 32611, USA.

establish a phylogenetic tree based on DNA sequence comparisons for the major lineages of extant elasmobranchs. We base our estimates on comparisons of DNA sequence data derived from four genes (three mitochondrial and one nuclear totaling 5,811 nucleotide base pairs) and employ a taxon-sampling scheme which, by design, includes divergent representatives of each of the currently recognized eight orders of sharks and three of the five orders of batoids.

While our primary goal is to understand the pattern of diversification that gave rise to modern elasmobranchs, we anticipate that the phylogeny we present will provide a framework that can be used to better estimate the suite of traits that characterize the ancestral vertebrate condition. We caution, however, that this goal is possible only to the extent that the sampled diversity of modern elasmobranchs arises from evenly spaced cladogenesis from the vertebrate ancestral condition to the present day. To the extent that cladogenesis is "clumped" over time, or worse yet, is restricted to a relatively recent radiation near the tips of the tree, it may be impossible to accurately determine the vertebrate ancestral condition.

1.1.2 A Brief History of Elasmobranch Higher Systematics

Despite 150 years of formal study, the phylogenetic relationships among different groups of elasmobranchs are not yet well established. In this section we review prior work on elasmobranch systematics and phylogeny. Our coverage is far from exhaustive but gives some context and background to the problem.

Early interest in the systematics and diversity among elasmobranchs dates back to descriptions of families of sharks and rays by Muller and Henle (1841), Duméril (1865) and Günther (1870). Various putative classifications were subsequently forwarded by Regan (1906), Garman (1913), Jordan (1923), White (1936, 1937), Moy-Thomas (1939), Bigelow and Schroeder (1948, 1953, 1957), Arambourg and Bertin (1958), Compagno (1973, 1977), Jarvik (1977), Schaeffer and Williams (1977), Maisey (1984a, 1984b). While there has been reasonably broad based acceptance of the family level groupings proposed, many of the inter-relationships have been debated, including the notion that modern elasmobranchs may not form a monophyletic group. For a review the reader is directed to Maisey (1984a).

It is now generally accepted that the "modern level" or Neoselachian elasmobranchs *sensu* Moy-Thomas (1939) and Schaeffer and Williams (1977), are a monophyletic group comprising all extant forms of sharks and rays and a scattering of extinct, but unambiguously modern level forms such as *Palaeospinax*, *Synechodus* and a few ray-like forms (Schaeffer and Williams 1977; Compagno 1977; Schaeffer 1981; Maisey 1982; 1984a, 1984b; Thies 1983). Furthermore it has been established that this monophyletic group is distinct from the vast majority of fossil sharks including the hybodonts, ctenacanths and xenacanths (Maisey 1984a).

While the monophyly of modern level elasmobranchs is well established, the same cannot be said for the inter-relationships among them. One particularly long-standing dispute concerns the placement of the batoids (skates and rays) within modern elasmobranchs. The traditional view places the batoids as the sister group to all recent sharks (Bigelow and Schroeder 1948, 1953, 1957). However recent cladistic analyses of morphological data have suggested that batoids may be a highly derived group nested within sharks (Shirai 1992). Another controversy concerns whether or not the four superordinal groupings forwarded by Compagno (1977) represent valid monophyletic groups. The four superorders are the Galeomorphii (comprising the Carcharhiniformes, Lamniformes, Orectolobiformes and Heterodontiformes), the Squalomorphii (Squaliformes, Hexanchiformes and Pristiophoriformes), the Squatinomorphii (*Squatina*) and the Rajomorphii (comprising all skates and rays). Maisey suggested a modification to this arrangement and proposed that Compagno's squalomorphii be combined with the squatinomorphii to form what he termed the "Orbitostylic group" of sharks, a grouping that Maisey proposed based on the presence of a distinctive articulation between the orbital process of the palatoquadrate and the brain case close to the interorbital wall.

Shirai (1992) published a comprehensive anatomical survey and rigorous cladistic analysis based on skeletal and myological variation of 46 genera of elamobranchs. While he focused on variation within squaloids (*sensu* Compagno 1984) he also included several batoid taxa and representatives of all the major orders of extant elasmobranchs. Shirai concluded from his analysis that the batoids were nested deeply within a superordinal group he termed "Squalea" comprising *Chlamydoselachus*, hexanchoids, "squaloids", *Squatina* and pristiophorids. Within the Squalea he proposed that squaloids constituted a paraphyletic group comprising several sequential sister taxa to a monophyletic hypnosqualean group containing *Squatina*, pristiophorids and all batoids. In 1996 Shirai expanded his morphological survey to include a greater diversity of taxa. This expanded survey corroborated the conclusions presented in his 1992 work. In 1996, in reviewing Shirai's work, de Carvalho questioned the independence of some of Shirai's characters but endorsed the major components of Shirai's phylogenetic arrangement. The phylogenetic hypotheses of de Carvalho and Shirai differ in their treatment of Hexanchiformes and Squaliformes; de Carvalho (1996) suggests that *Chlamydoselachus, Notorhynchus, Hexanchus* and *Heptranchias* form a monophyletic group. Shirai, by contrast suggests that *Chlamydoselachus* is the sister taxon to a clade comprising the remaining hexanchoids and his hypnosqualean group (squaloids, Squatiniformes, Pristiophoriformes and batoids). Furthermore, de Carvalho suggests that Squaliformes are monophyletic when *Echinorhinus* is excluded whereas Shirai's scheme has more extensive paraphyly within Squaliformes. We present trees for competing hypotheses forwarded by various authors in Fig. 1.1.

If one thing stands out from the morphologically based systematic work on extant elasmobranchs it is the extent of character homoplasy (de Carvalho 1996). Indeed it is the homoplasy in conjunction with the difficulties in asserting homology across widely varying structures that has led to much of the current phylogenetic uncertainty. Sharks are an ancient group whose

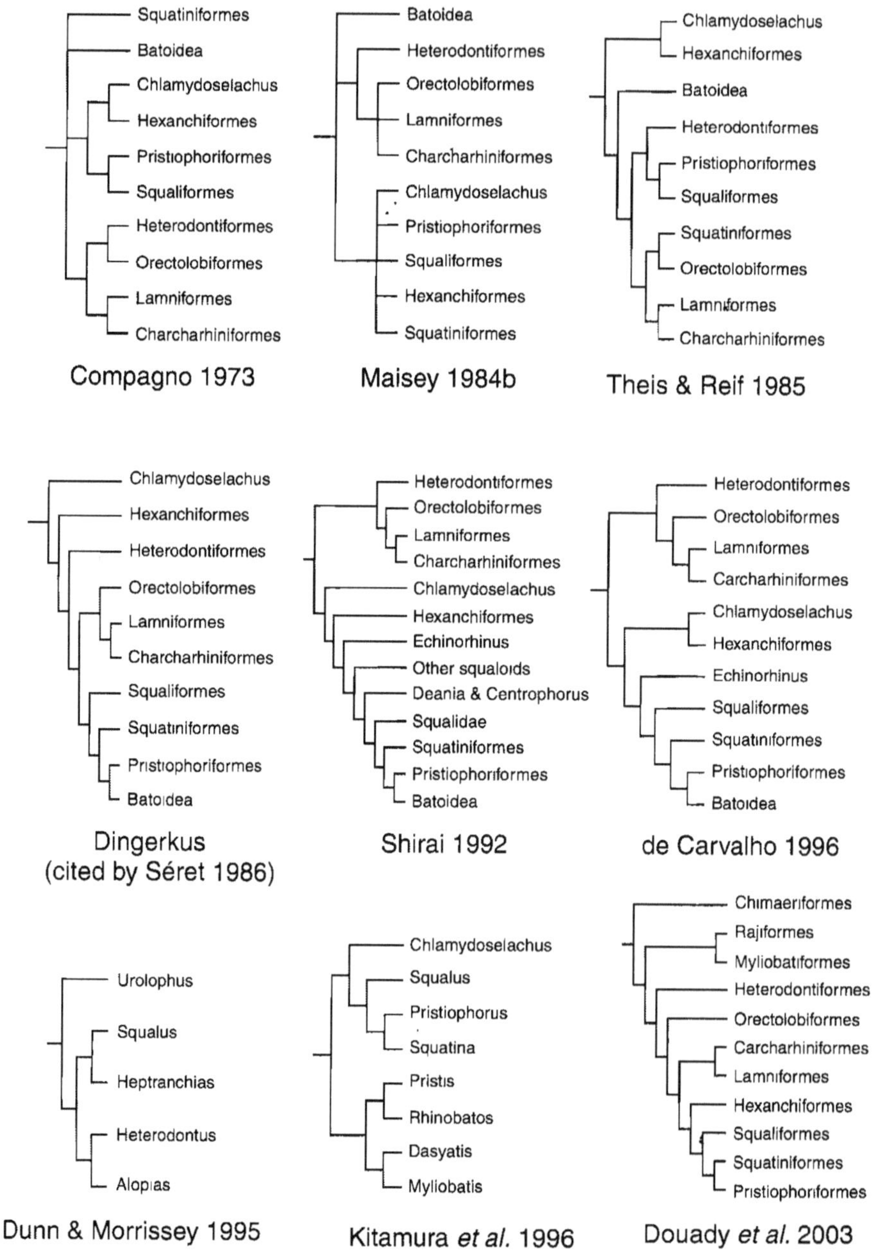

Fig. 1.1 Nine hypotheses of relationship among extant elasmobranches forwarded by various authors. The last 3 are based on molecular sequence data

morphology is conserved relative to other groups of similar age. It appears that many of the character suites in elasmobranchs have been constrained over the course of their evolution. Those that have been free to change have often flipped backward and forward between alternating character states over the course of their evolution. Such tight constraints have seemingly led to character state distributions that are highly homoplasious.

Given the homoplasy and the inherently ambiguous nature of the phylogenetic signal in the morphological data several researchers have sought to investigate the problems of elasmobranch phylogeny with molecular sequence comparisons. Dunn and Morrissey (1995) used a 303 bp fragment of 12s rRNA for five representative elasmobranch taxa (*Squalus, Heptranchias, Heterodontus, Alopias and Urolophus*). Kitamura *et al.* (1996) used 732bp nucleotide sites of the mitochondrial gene cyt-B for eight taxa (*Squalus, Squatina, Rhinobatos, Pristis, Pristiophorus, Myliobatis, Dasyatis and Chlamydoselachus*) in conjunction with three distant outgroups (carp, sturgeon and lamprey) taken from Gen Bank. Douady *et al.* (2003) used a 2.4 kb segment of mitochondrial DNA comprising predominantly 12s and 16s rRNA for 19 elasmobranch taxa with a taxon sampling scheme weighted toward lamniform and carcharhiniform taxa (comprising nine out of the 19 taxa).

It is now widely appreciated that different genes can yield different phylogenies (Kim 1996; Naylor and Brown 1998) and that taxon sampling can have a strong effect on phylogenetic inference (Hillis 1996; Kim 1996; Rannala *et al.* 1998; Graybeal 1998; Poe 1998). Given this, we set out to (a) sequence four different protein-coding genes from representative taxa from diverse lineages within each of the putatively monophyletic orders of sharks and (b) to explore the variation in phylogenetic signal among genes.

We had four specific goals:

(1) To evaluate the molecular evidence for elasmobranch phylogeny based on a balanced taxon-sampling scheme across extant orders of sharks and rays using multiple genes.
(2) To assess the degree of phylogenetic signal concordance between the nuclear and mitochondrial genes.
(3) To explore the effects, if any, of three factors known to compromise phylogenetic signal in sequence data (a) saturation due to multiple hits (b) non-stationarity in base composition and (c) variation in codon usage bias among taxa.
(4) To evaluate the fit of morphological hypotheses proposed to date to the hypotheses yielded by the different molecular data sets.

1.2 MATERIALS AND METHODS

1.2.1 Choice of Taxa

We chose the two most divergent representatives of each elasmobranch order where possible. We have assumed that currently recognized orders (*sensu* Compagno 1984) are monophyletic (See updated checklist of living Chondrichthyes in Chapter **16** of this volume). This contrasts with the schemes of Shirai (1992, 1996) that present a paraphyletic Squaloidea, and that of de Carvalho (1996) which removes *Echinorhinus* from an otherwise monophyletic Squaloidea. The taxa used for the current study are shown in Table 1.1.

Table 1.1. Taxa used to represent elasmobranch diversity in the current study. Samples were chosen to represent distant relatives within each order. The ID column corresponds to the assignments in tissue data base of the senior author.

Order	*Family*	*Representative*	*I.D.*	*Locality*	*Collector*
Carcharhiniformes	Carcharhinidae	*Carcharhinus limbatus*	1340	Kota Kinabalu, Malaysia	Rachel Cavanagh
	Scyliorhinidae	*Apristurus macrorhynchus*	1015	Tahsi,Taiwan	Gavin Naylor
Heterodontiformes	Heterodontidae	*Heterodontus francisci*	1110	Baja Calif, Mexico	Janine Caira
Hexanchiformes	Chlamydoselachidae	*Chlamydoselachus anguineus*	1481	North of Scotland	Pamela McGarr
	Hexanchidae	*Heptranchias perlo*	977	Tahsi, Taiwan	Gavin Naylor
Lamniformes	Odontaspididae	*Odontaspis ferox*	1085	São Miguel Island, Azores	José Azevedo
	Mitsukurinidae	*Mitsukurina owstoni*	1057	Tasmania, Australia	John Stevens
Myliobatiformes	Myliobatidae	*Aetobatus narinari*	1602	Darwin, Australia	Gavin Naylor
	Dasyatidae	*Dasyatis annotata*	1588	Thailand	Janine Caira
Orectolobiformes	Orectolobidae	*Orectolobus ornatus*	1491	South West Australia	Colin Simpfendorfer
	Stegostomatidae	*Stegostoma fasciatum*	1295	Kota Belud, Malaysia	Rachel Cavanagh
Pristiformes	Pristidae	*Pristis clavata*	1610	Darwin, Australia	Gavin Naylor
Pristiophoriformes	Pristiophoridae	*Pristiophorus japonicus*	1049	Japan	Kazunari Yano
Rajiformes	Rajidae	*Raja rhina*	1615	Gulf of Alaska, USA	Gaines Tyler
Squaliformes	Oxynotidae	*Oxynotus paradoxus*	1850	West Coast Scotland	Marianne Du Buit
	Squalidae	*Squalus acanthias*	919	Ocean City, Md, USA	Gavin Naylor
Squatiniformes	Squatinidae	*Squatina tergocellatoides*	1005	Tahsi, Taiwan	Gavin Naylor
Chimaeriformes	Chimaeridae	*Chimaera phantasma*	971	Tahsi, Taiwan	Gavin Naylor

1.2.2 Choice of Genes

Both empirical and simulation studies indicate that certain properties of DNA sequences predispose them to yield accurate phylogenies under parsimony. The percentage of sites free to vary (Shoemaker and Fitch 1989; Palumbi 1989), degree of among-site-rate-variation (ASRV) (Yang 1994, 1996; Sullivan *et al.* 1995, 1996), base composition (Sidow and Wilson 1990; Collins *et al.* 1994), stationarity (Saccone *et al.* 1990; Lockhart *et al.* 1992; Steel 1994), and length of gene (Comeron *et al.* 1999) have all been shown to affect phylogenetic accuracy. Different methods of phylogenetic reconstruction make different assumptions. When estimating phylogeny from sequences we should strive to choose genes whose evolutionary dynamics meet the assumptions of the inference method to be used. For parsimony, simulation work suggests that an ideal gene should be single copy with respect to the taxa being investigated and should exhibit the following properties:

(1) An overall base composition that is even (25% A:G:C:T) and stationary across taxa
(2) A codon usage that is unbiased and stationary across taxa
(3) A low degree of among site rate variation

We identified RAG-1 as potentially promising based on these criteria, and because it has no introns in elasmobranchs and could be readily amplified from genomic DNA. It might be argued that the criteria we present for selecting phylogenetically useful genes implicitly assumes parsimony to be the method of choice because the properties deemed "*desirable*" are identified within a parsimony framework. Indeed the case could be made that because maximum likelihood approaches can be tailored to accommodate features that cause problems for parsimony that choosing data is *less important* than choosing the appropriate inference model with which to analyze it. From a theoretical standpoint this is correct. However, when faced with empirical data, it is not always easy to identify which model most accurately describes the data. Likelihood ratio tests (Goldman 1993; Posada and Crandall 1998) merely allow us to determine that one model appears to fit better than another. Finding which of a collection of homogeneous stationary models best fits a heterogeneous non-stationary data set is akin to trying to find out which of various rectangular pegs best fits into an eccentrically shaped hole—one will always be better than the alternatives—but the possibility exists that they might *all* be very poor fits. This is rarely considered. Given this, we resort to a conservative strategy that is less prone to violate assumptions *over a broad range of models*—namely using relatively slowly-evolving genes that exhibit even and stationary base compositions and codon usage profiles.

We included sequence from three mitochondrial genes that are routinely used in molecular phylogenetic studies: Cytochrome b (Cyt-b), NADH-dehydrogenase 2 (NADH-2) and NADH-dehydrogenase 4 (NADH-4). The three mitochondrial genes were included to provide a frame of reference against which the signal and patterns of evolution exhibited by the RAG-1 sequences could be compared.

1.2.3 Laboratory Protocols

The DNA sequences reported here were obtained from PCR products from genomic DNA preparations. Genomic DNA's were extracted with the High Pure PCR Template Preparation Kit (Boehringer-Manheim) following the manufacturer's recommendations. The isolated genomic DNA was diluted ten-fold and the dilutions were used as PCR templates without quantitation

Table 1.2 shows the sequences and location of all the primers used in amplification and sequencing of the genes included in the analyses. Different specimens required the use of different sets of these primers for amplification and sequencing. The RAG-1 sequences were obtained from

Table 1.2 Set of primers used in amplification and sequencing of the four genes used in the study. Location is given relative to the first base of the first amino acid coding codon. Negative numbers indicate primer is upstream of start codon.

Target Gene	*Primer Name*	*Sequence*	*Location**
RAG-1			
	Rag-1For	5'- ATG GAA ATG GAG AGG CCT CTG CAA ATG -3'	1 - 27
	Rag-1For5	5'- CCC ACC CAC TTC TGC TAC AAC TGC TGG -3'	523 - 549
	Rag-1For7	5'- AAY GAR CAC AGR CAA GCW GAT GA -3'	1354 - 1376
	Rag-1For9	5'- CCY GTY GAT GAT GTC ACA CAA G -3'	404 - 424
	Rag-1For11	5'- TGC CTT TGG ACA AGA AGA CTT TGG A -3'	197 - 221
	Rag-1For15	5'-ATC TCA AAY KTC AGG TKA AGG C-3'	1256 - 1277
	Rag-1For20	5'-GCA GTC TGC CTT GCC ATT AGA-3'	1420 - 1440
	Rag-1For22	5'-CTG CGA GAT CTG AAA CTT CAG GTG-3'	1249 - 1271
	Rag-1Rev3	5'- GAA GAC ATT TTG AAG CAT C -3'	2983 - 3001
	Rag-1Rev4	5'- GTT TGA GAC CTT TNC GTT TGC GAG T -3'	670 - 694
	Rag-1Rev7	5'- CCT CTG TAA ATA TTT TGA AGT GTA CAG CC -3'	3005 - 3033
	Rag-1Rev10	5'- GCR TGY AAD GGC TGR AAG ATC TG -3'	1510 - 1532
	Rag-1Rev11	5'- AAT GGC TGG AAG ATC TGC TTC C -3'	1505 - 1526
	Rag-1Rev12	5'- CTG GCT TCA GCC CTG ATC CAT G -3'	1399 - 1421
	Rag-1Rev13	5'- AGT GGT GCT TCA GAA TAT CTT C -3'	2983 - 3004
	Rag-1Rev15	5'- TGG ACT GCC TTG CRT TCA T -3'	2950 - 2968
NADH-2			
	Ile-Mustelus	5'- AAG GAC CAC TTT GAT AGA GT -3'	-194 - -175
	Asn-Mustelus	5'- AAC GCT TAG CTG TTA ATT AA -3'	1217-1237
	Galeus-IntF	5'- CCA ACC TCT GCC ACA CT -3'	274 - 291
	Leptocharias-IntF	5'- CCA ACA TCT GCC ACA CT -3'	274 - 291
	ND2-442-IF	5'- CCA ACC TCC GCC ACA CT -3'	274 - 291
	ND2-batoids-IFA	5'- CAC TTY TGA CTW CCA GAA GT -3'	334 - 353
	Aeto-IF	5'- CAA CCA AGT ATC CAT CAC ACT -3'	270 - 291
NADH-4			
	ND4	5'- TGA CTA CCA AAA GCT CAT GTA GAA GC -3'	645-670
	Leu-Scyliorhinus	5'- CAT AAC TCT TGC TTG GAG TTG CAC CA -3'	1566-1591
CYT-B			
	GluDG	5'- TGA CTT GAA RAA CCA YCG TTG -3'	-42 - -22
	C61121H	5'- CTC CAG TCT TCG RCT TAC AAG -3'	1175-1195

three contiguous, overlapping amplicons while the mitochondrial gene sequences were obtained from a single amplicon. An internal sequencing primer was needed to determine the full length sequence of the NADH-2 amplicon. Optimal temperature cycling conditions and reagent concentrations for the PCR were determined with the aid of a thermocycler equipped with a temperature gradient block. Denaturation temperature and duration were 94°C and 30 secs. respectively. Annealing was set for 30 secs. at temperatures ranging from 56°C to 48°C varying with primer pair and template. Extension was carried out at 72°C for between 60 and 90 secs. depending on fragment length. The amplification of most of the RAG-1 fragments required the use of touchdown temperature profiles. The DNA Synthesis and Sequencing Facility at Iowa State University carried out the sequencing reactions using reagents (Dye-Terminators and Big-Dye Terminators) and equipment (Automated Sequencers) from Perkin-Elmer according to manufacturer's recommendations.

1.2.4 Analysis

1.2.4.1 Parsimony signal

Uniformly weighted parsimony bootstrap analyses were carried out separately for the mitochondrial and nuclear data sets to evaluate the topological congruence between the two data sets. Bootstrap analyses were carried out for four types of representation—nucleotide, purine/pyrimidine, codon, and amino acid—for both mitochondrial and nuclear data sets.

1.2.4.2 Partition homogeneity tests

A series of partition homogeneity tests were carried out to assess the difference in signal between the mitochondrial and nuclear partitions for each of the four types of representation. We also conducted partition homogeneity tests to assess differences in signal between the different representations of the data. These tests were carried out on the combined data set (nuclear + mitochondrial).

1.2.4.3 Saturation analysis

Most parsimonious trees (MPTs) were computed for the nuclear and mitochondrial data sets separately for each of the four types of representation (a total of eight data sets). Patristic distances were computed for each of the resulting MPTs. Pair-wise distance was plotted against the corresponding patristic distance to provide a graphical indication of the relative extent of saturation.

1.2.4.4 Base composition

Base compositional evenness was measured and base compositional stationarity among taxa computed for each data set and evaluated using a *Chi* squared test (four tests were carried out: two data representations: nucleotide (A,G,C,T) and transversions (R,Y), for the nuclear and mitochondrial data sets, separately).

1.2.4.5 Codon usage bias

Codon usage was contrasted across taxa. Patterns of bias were evaluated to determine if they might account for any differences in topology seen among analyses.

1.2.4.6 Combined analysis

The nuclear and mitochondrial data sets were pooled and subjected to a combined parsimony analysis. This was done for each of the four types of representation: nucleotides, purine/pyrimidine, codons and amino acids.

1.2.4.7 Likelihood approaches incorporating rate variation

Data sets were subjected to likelihood analysis that incorporated among site rate variation (ASRV). This was carried out to determine if differences in ASRV might be responsible for signal discordance among data sets.

1.3 RESULTS

1.3.1 Parsimony Analysis

The mitochondrial data set comprised 2940 nucleotides from Cytochrome b (Cyt-b), NADH-dehydrogenase 2 (NADH-2) and NADH-dehydrogenase 4 (NADH-4). The nuclear data set comprised 2871 nucleotides all from RAG-1. The RAG-1 nucleotide, transversion, and codon data sets all yielded exactly the same most parsimonious tree topology (Fig. 1.2). The RAG-1 amino acid data set and the four different representations of the mitochondrial data set all yielded different topologies under parsimony. We have not depicted these topologies as they are not robust to re-sampling. Instead, we show the tree topologies resulting from parsimony bootstrap analyses for each of the eight data sets together with their corresponding saturation plots (Fig. 1.3).

The tree topologies are consistent in five out of the eight bootstrap trees and with the most parsimonious tree resulting from the RAG-1 nucleotide, purine/pyrimidine, and codon analyses shown in Fig. 1.2. Topologies that are inconsistent with this prevailing pattern occur in: (A) the RAG-1 amino acids, which place the batoids as the sister group to the galeomorph clade (*Heterodontus*, ((*Orectolobus*, *Stegostoma*), (*Misukurina*, *Odontaspis*), (*Apristurus*, *Carcharhinus*))); (B) the bootstrap analysis of the mitochondrial nucleotides which suggests that the hexanchiform clade (*Chlaydoselachus*, *Heptranchias*) is the sister group to the squaloid clade (*Oxynotus*, *Squalus*) to the exclusion of *Squatina* and *Pristiophorus*, and that *Squatina* and *Pristiophorus* no longer comprise a monophyletic clade as they do in the majority of the analyses; and (C) the bootstrap analysis of the mitochondrial codons, which yields a tree that is topologically distinct from all of the other trees in placing the two orectolobiform taxa *Stegostoma* and *Orectolobus* as the most basal representatives of extant modern sharks, implying that the galeomorph taxa do not form a monophyletic group (see Fig. 1.3).

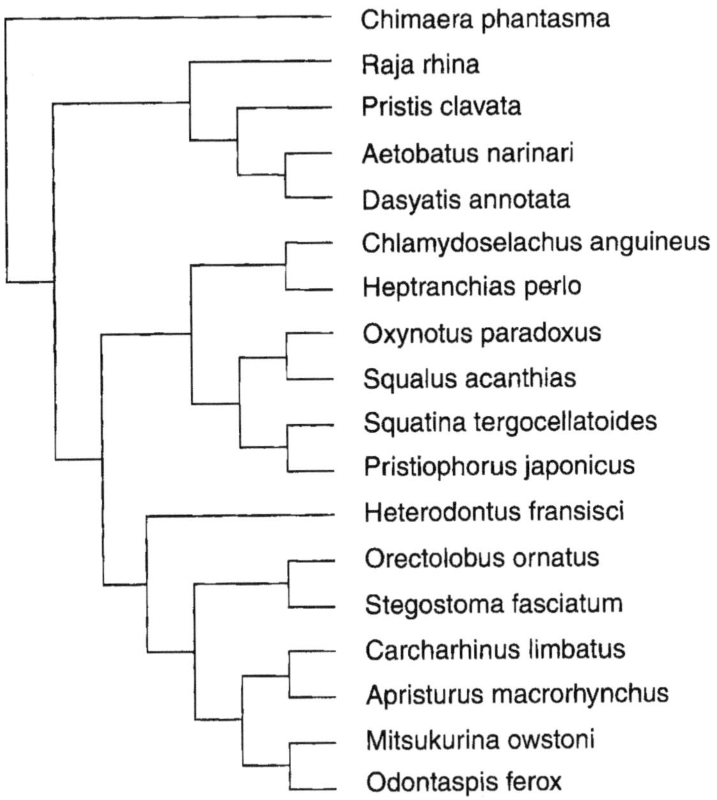

Fig. 1.2 The prevailing phylogenetic signal resulting from the majority of analyses.

1.3.2. Partition Homogeneity Tests

Partition homogeneity tests indicate that the signal between nuclear and mitochondrial partition is compatible at both the Amino acid (AA) and purine/pyrimidine (RY) levels but is significantly different at the Codon (Cod) ($p = 0.04$) and Nucleotide (Nuc) ($p = 0.006$) levels (see Table 1.3).

Table 1.3 Partition homogeneity test p values contrasting mitochondrial and nuclear signals at different levels of representation.

	AAs	*Codons*	*RYs*	*Nucs*
Mito/Nuc PH test	0.162	0.004	0.318	0.006

Tests contrasting the signal between different representations of the same data set (all four genes combined) indicate that the signal is compatible between all representations at the $p = 0.05$ level. However the contrast between codons and transversions and that between codons and amino acids are close to being significantly different (Table 1.4).

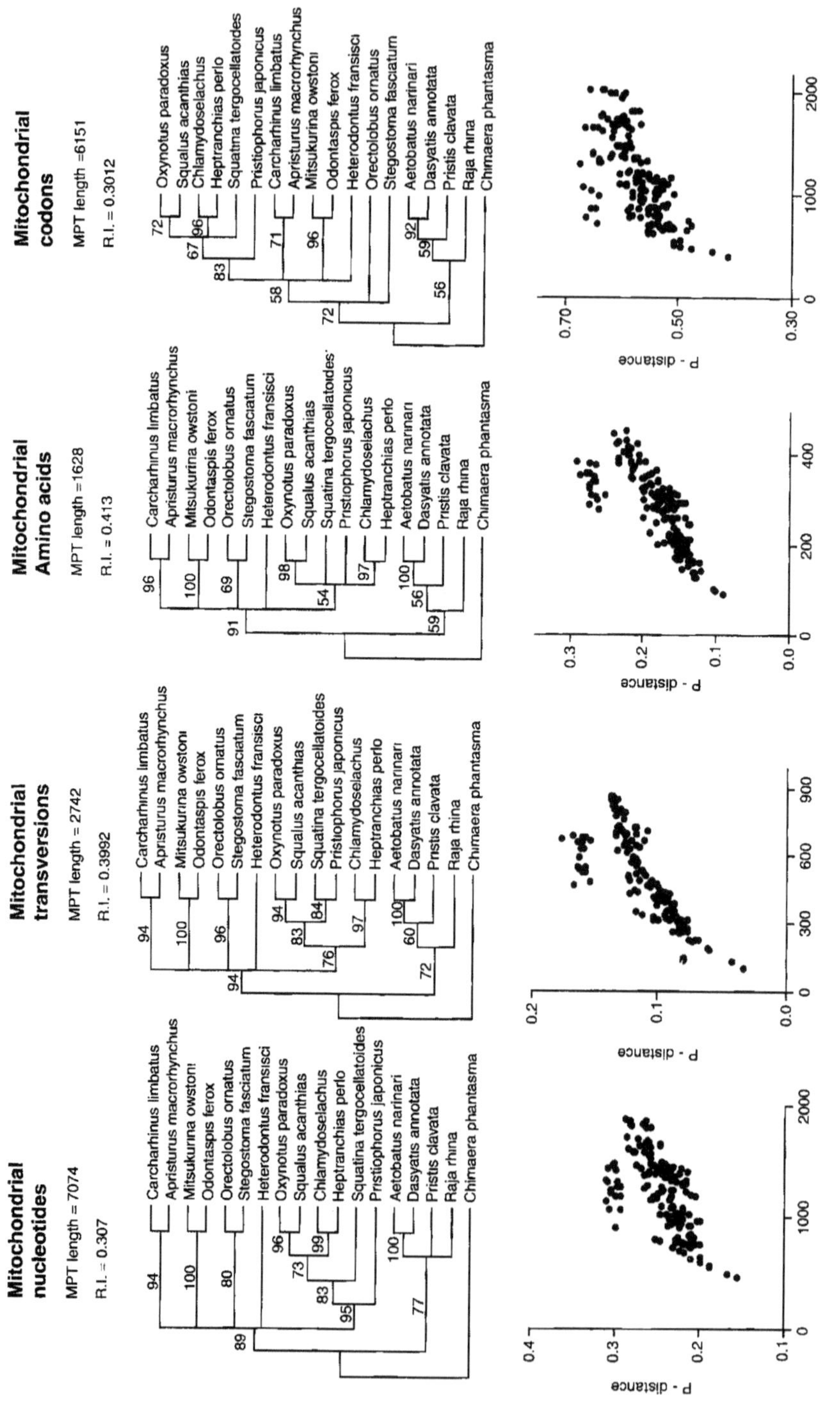

Mitochondrial codons
MPT length =6151
R.I. = 0.3012
Mitochondrial Amino acids
MPT length =1628
R.I. = 0.413
Mitochondrial transversions
MPT length = 2742
R.I. = 0.3992
Mitochondrial nucleotides
MPT length = 7074
R.I. = 0.307
Carcharhinus limbatus
Apristurus macrorhynchus
Mitsukurina owstoni
Odontaspis ferox
Orectolobus ornatus
Stegostoma fasciatum
Heterodontus fransisci
Oxynotus paradoxus
Squalus acanthias
Chlamydoselachus
Heptranchias perlo
Squatina tergocellatoides
Pristiophorus japonicus
Aetobatus narinari
Dasyatis annotata
Pristis clavata
Raja rhina
Chimaera phantasma
P - distance
Patristic distance

(Fig. 1.3 Contd. ...)

(Fig. 1.3 Contd. ...)

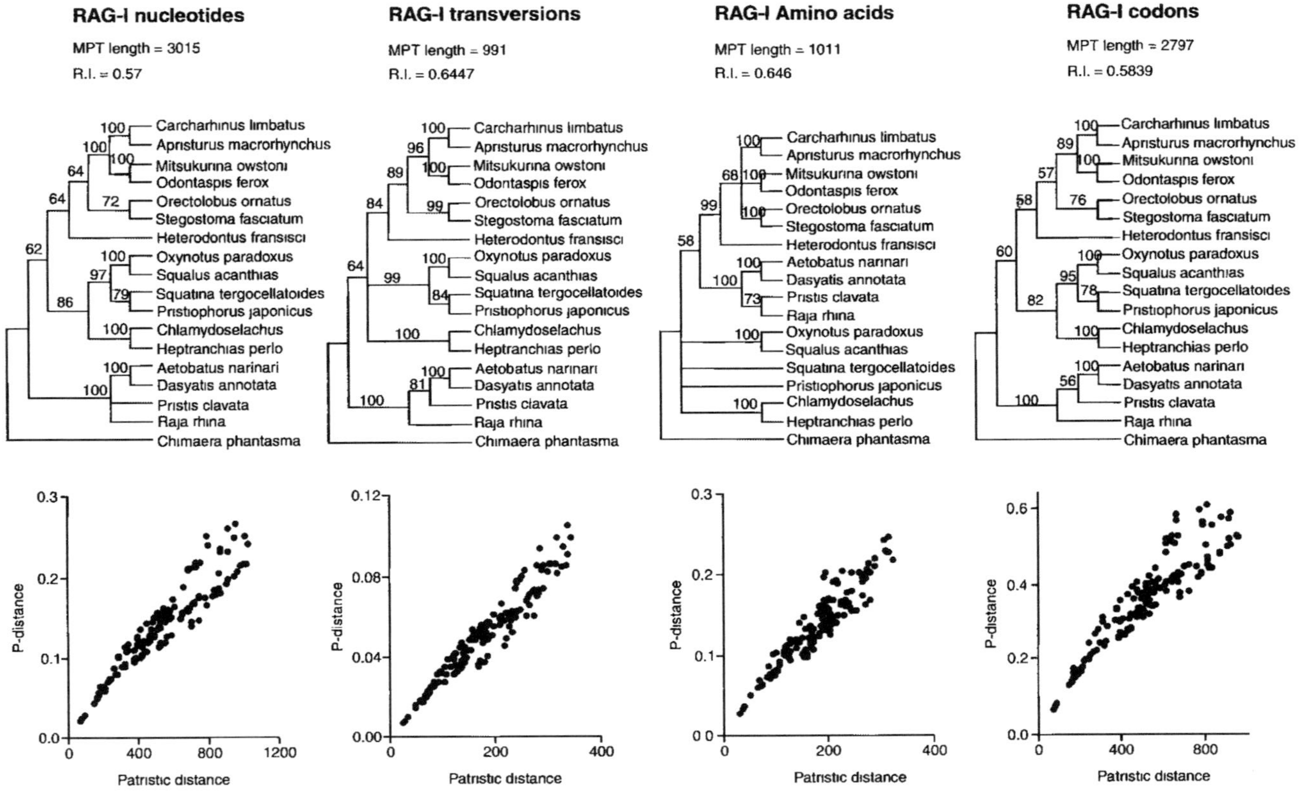

Fig. 1.3 Parsimony bootstrap analyses for eight analyses shown with the corresponding saturation plot. The eight analyses comprised 2 data sets (nuclear and mitochondrial) analyzed at four levels of representation (nucleotides, transversions, codons and amino acids).

Table 1.4 Partition homogeneity p values contrasting different representations of the combined mitochondrial and nuclear data set.

	Nuc / RY	*Nuc / AA*	*RY / AA*	*Cod / Nuc*	*Cod / RY*	*Cod / AA*
PH test	0.68	0.67	0.748	0.51	0.077	0.065

1.3.3. Saturation Analyses

Saturation plots associated with each of the eight data sets (nuclear and mitochondrial for four different representations of the same data) are shown in Fig. 1.3. The two data sets exhibiting the highest degree of saturation, the mitochondrial nucleotide and the mitochondrial codon data sets, also yield the most anomalous trees relative to the prevailing signal depicted in Fig. 1.2. Not surprisingly, these are the two data sets identified by the partition homogeneity tests as being most distinct.

1.3.4. Base Composition

Overall base compositional stationarity is shown in tables 1.5 (RAG-1) and 1.6 (mitochondrial genes). *Chi* squared tests indicate that both the nuclear and the mitochondrial data sets exhibit statistically significant deviation from stationarity at the nucleotide level (Table 1.7). Neither showed any such deviation at the purine/pyrimidine, level.

Table 1.5 Base composition of RAG-1 for the taxa examined in this study.

Taxon	*A*	*C*	*G*	*T*	*# sites*
Carcharhinus	0.31375	0.19832	0.23199	0.25594	2868
Apristurus	0.31469	0.19685	0.23077	0.25769	2861
Mitsukurina	0.31647	0.19845	0.2255	0.25957	2847
Odontaspis	0.31473	0.20202	0.23108	0.25218	2871
Orectolobus	0.30825	0.20655	0.23232	0.25287	2871
Stegostoma	0.29278	0.21685	0.24369	0.24668	2853
Heterodontus	0.32321	0.18988	0.22269	0.26422	2865
Oxynotus	0.31916	0.19686	0.22822	0.25575	2870
Squalus	0.32027	0.19791	0.22433	0.25749	2871
Squatina	0.31905	0.20078	0.22814	0.25202	2871
Pristiophorus	0.31727	0.19944	0.22803	0.25527	2849
Chlamydoselachus	0.32729	0.19083	0.21997	0.2619	2814
Heptranchias	0.32224	0.19492	0.22472	0.25812	2833
Aetobatus	0.32323	0.19854	0.21978	0.25845	2871
Dasyatis	0.31121	0.20696	0.2297	0.25213	2701
Pristis	0.31996	0.19658	0.22891	0.25455	2692
Raja	0.31675	0.20498	0.22911	0.24916	2693
Chimaera	0.28131	0.24679	0.24165	0.23026	2723
Mean	0.3146	0.2023	0.22889	0.2542	2823.56

Table 1.6 Base composition for the mitochondrial genes for the taxa examined in this study.

Taxon	*A*	*C*	*G*	*T*	*# sites*
Carcharhinus	0.30034	0.28322	0.10753	0.3089	2921
Apristurus	0.28689	0.27694	0.13109	0.30508	2915
Mitsukurina	0.29743	0.29365	0.11286	0.29605	2915
Odontaspis	0.30775	0.25545	0.1108	0.326	2879
Orectolobus	0.30943	0.26708	0.11202	0.31148	2928
Stegastoma	0.3355	0.28855	0.10281	0.27313	2918
Heterodontus	0.30499	0.30806	0.10485	0.2821	2928
Oxynotus	0.29733	0.262	0.11523	0.32545	2916
Squalus	0.29414	0.25677	0.12204	0.32705	2917
Squatina	0.28796	0.26573	0.12004	0.32627	2924
Pristiophorus	0.28919	0.295	0.12697	0.28884	2922
Chlamydoselachus	0.31216	0.24522	0.10724	0.33538	2928
Heptranchias	0.30764	0.27372	0.10312	0.31552	2919
Aetobatus	0.29174	0.33048	0.11279	0.265	2917
Dasyatis	0.30511	0.29619	0.1073	0.2914	2917
Pristis	0.31179	0.29949	0.10769	0.28103	2925
Raja	0.284	0.31175	0.11853	0.28571	2919
Chimaera	0.29814	0.28157	0.12257	0.29772	2415
Mean	0.30122	0.28286	0.11355	0.30237	2890.17

Table 1.7 Chi square test of stationarity in base composition among the taxa examined separately for each of the mitochondrial and nuclear partitions at the level of both nucleotides and transversions. The statistical significance of the non-stationarity for the nucleotide representation of the nuclear data partition (p value 0.0181) is nearly all due to the base composition of chimaera. When chimaera is removed and the test repeated, the p value is 0.99. By contrast, the statistical significance of the non-stationarity for the nucleotide representation of the mitochondrial data partition (p value 0.00001) is nearly all due to deviation from base compositional stationarity at the 3rd codon position. When 3rd codon positions are eliminated from analysis the new p value is 0.7302.

	Mitochondrial	*Nuclear*
Nucleotides	0.00001	0.0181
Transversions	0.705	0.987

1.3.5. Codon Usage

The orectolobiform taxa had a different pattern of codon usage in the mitochondrial data set relative to the other taxa. Differences in codon usage were found to be restricted to the following amino acids Glycine (G), Isoleucine (I), Leucine (L), Asparagine (N), Proline (P) and Serine (S). Hypothesizing that this difference in codon usage might be responsible for the topology resulting from the codon level analysis of the mitochondrial data, we reassigned codon character states such that each of the amino acids G, I, L, N, P, S was scored as using the same codon across all taxa. This allowed us to retain information associated with transformation from one amino acid to another while eliminating transformations from codon to codon within amino acids exhibiting non-stationary codon usage. We re-ran the bootstrap analysis on this adjusted data set and obtained a tree, which, while considerably less resolved, was consistent with the predominant signal seen in the majority of the data sets (Fig. 1.4). We conclude that differences in codon usage among taxa in the mitochondrial data set are responsible for the anomalous topology seen in the codon level analysis.

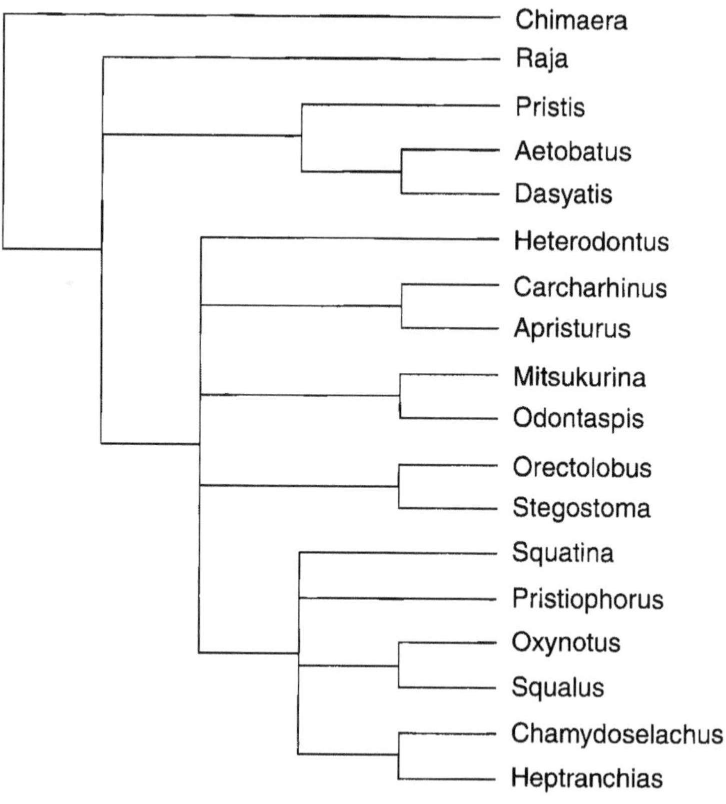

Fig. 1.4 Bootstrap analysis of mitochondrial codon data set modified such that amino acids with non-stationary codon usage across taxa (i.e. G, I, L, N, P, S) were scored as using the same codon across all taxa.

1.3.6. Combined Analyses

A parsimony analysis of the combined nuclear and mitochondrial data sets yielded a different most parsimonious tree topology for each of the four different representations (nucleotides, transversions, codons and amino acids). The trees resulting from these analyses are depicted in Fig. 1.5. Only the tree resulting from the parsimony analysis of the transversion, data is identical to that seen in Fig. 1.2.

1.3.7. Likelihood Approaches

There are conditions under which maximum likelihood recovers the correct tree where parsimony fails (Felsenstein 1978). One such set of conditions is when there is extensive rate variation among sites and lineages. To explore this possibility we ran analyses of the mitochondrial data set under likelihood using the HKY (Hasegawa *et al* 1985) model, with an estimated proportion of sites assumed to be invariant, and with the rates of the

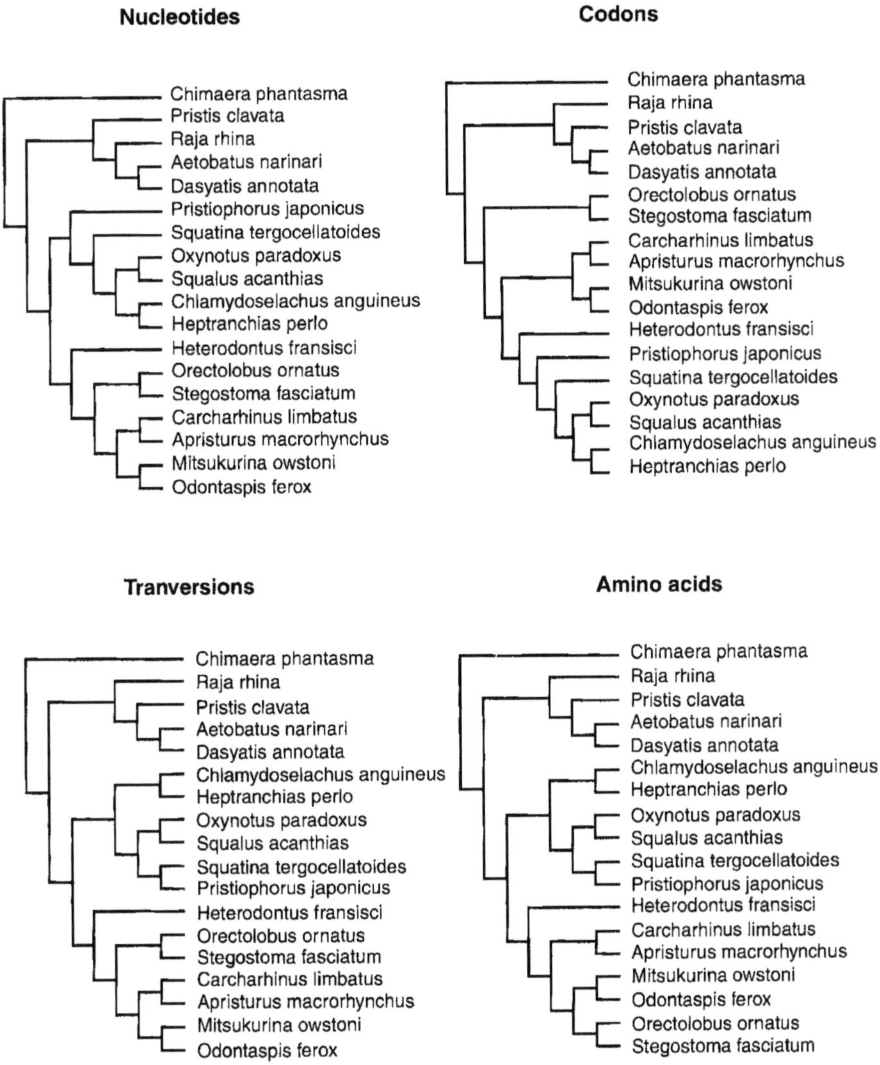

Fig. 1.5 Parsimony analyses of the combined nuclear and mitochondrial data set analyzed at four levels of representation (nucleotides, transversions, codons and amino acids).

remaining assumed to be distributed according to a gamma distribution with an alpha value of 0.5.

The tree that resulted from the likelihood analysis of the mitochondrial data set (Fig. 1.6) was markedly different from that obtained from the same data set using parsimony but was almost identical to the tree yielded by parsimony analysis of the RAG-1 nucleotide, transversion, and codon analyses (Fig. 1.2). The only difference being the placement of the carcharhiniform taxa as the sister group to the orectolobiform taxa rather than as the sister group to the Lamniformes. As discussed previously, this is likely due to the non-stationary codon usage profiles of the Orectolobiformes. Likelihood analysis of the mitochondrial transversion data yields the same topology as that of the majority of the analyses shown in Fig. 1.2 (likelihood of the combined nucleotide data set of all mitochondrial genes and RAG-1 also yielded the topology seen in Fig. 1.2).

In summary five out of the eight bootstrap analyses yield a topology that is identical to that shown in Fig. 1.2. Two of the three data sets that yield

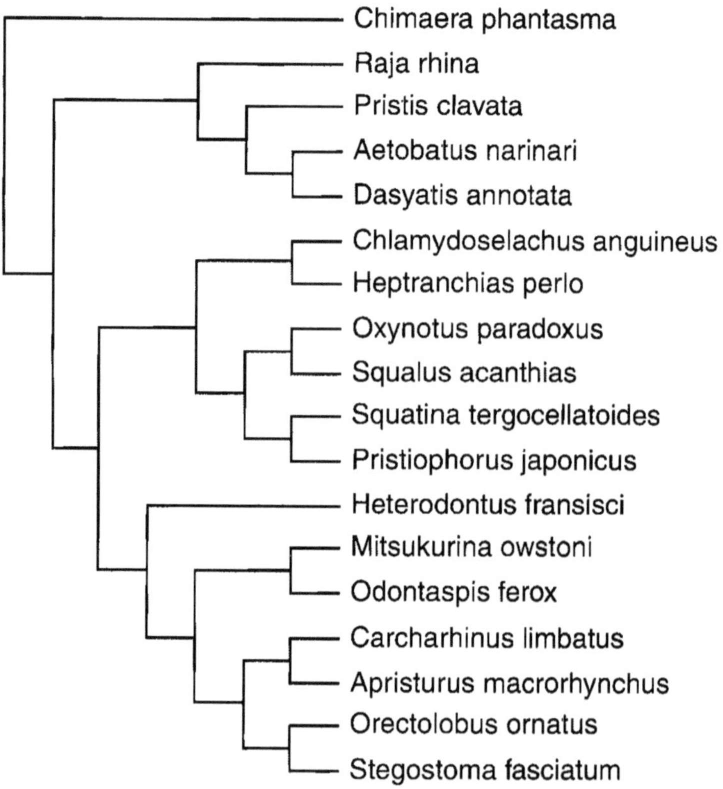

Fig 1.6 Likelihood analysis of the combined mitochondrial nucleotide data set using the HKY+1+G model of sequence evolution. The topology is almost identical to that seen in Fig. 1.2 except for the sister group relationship between the Orectolobiformes and the Carcharhiniformes.

different topologies (the mitochondrial nucleotide and codon data sets) exhibit features known to be problematic for phylogenetic inference (saturation in conjunction with non-stationary codon usage).

1.4 DISCUSSION

The mitochondrial genes exhibit saturation at the level of nucleotides and codons. Both saturated data sets yield trees that are distinct from the prevailing signal reflected in the topology of Fig. 1.2. The degree of saturation in the mitochondrial data set is ameliorated when the data are recoded as transversions or amino acids.

By contrast, the RAG-1 sequences show little saturation in any of the representations and appear well suited to estimating phylogeny for this set of taxa. However, while RAG-1 does not appear saturated in any of the four representations (nucleotides, transversion, codons and amino acids) a parsimony bootstrap of the RAG-1 amino acid data set yielded a tree that was topologically different from that seen in the majority of analyses in placing batoids as the sister group to galeomorph sharks. This placement has weak (58%) bootstrap support and requires only two extra steps to fit the data set to the topology shown in Fig. 1.2. Interestingly, eight characters supported the monophyly of sharks with a retention index of 1.00. One of these (character #148: changing from a Histidine to Tyrosine) was unambiguous over all most parsimonious reconstructions. By contrast, only one character (#6) supported the galeomorph/batoid grouping with a retention index of 1.00 on the most parsimonious tree. This character change was not unambiguous over all reconstructions. Thus while there are several characters with a high retention index that favor the branch depicting shark monophyly, this signal is overturned by homoplasious changes occurring further toward the tips of the tree. An analysis of branch lengths suggests that the inferred sister group relationship between batoids and galeomorph sharks might be an artifact due to long branch attraction. In contrast to the RAG-1 results, the amino acid representation of the mitochondrial data set yields a tree that is topologically consistent with the signal seen in the majority of analyses Fig 1.2 However, the codon representation of the same data set yields a topology in which orectolobiformes fall as the sister group to all other sharks. This begs the question "How is it that the codon representation can appear well behaved in the RAG-1 data set but not in the mitochondrial data set while the amino acid representation shows the opposite pattern?" The answer may lie in the mapping of codons to amino acids. In the mitochondrial data set, there is considerable variation in the codon usage patterns among taxa. This variation reflects non-stationarity in the evolutionary process and results in a misleading mitochondrial inference at the codon level. In the RAG-1 data set the codon usage patterns are stationary among taxa. The anomalous signal resulting from the analysis of the RAG-1 amino acid data is harder to explain but may be a simple consequence of stochastic homoplasy. It is worth noting that the anomalous inferences associated with both the mitochondrial codon and the RAG-1 amino acid data sets exhibit low bootstrap support values (58%).

1.4.1 Signal Heterogeneity Across Different Levels of Representation of the Combined Mitochondrial and Nuclear Data Set

The fact that combined analyses at the level of both nucleotides and codons yield topologies that are different from most of the other analyses is unsettling. It underscores concerns voiced in the literature that it may be poor practice to simply *assume* that combining data sets will result in accurate phylogenetic estimates (Barret *et al.* 1991; de Queiroz 1993). The fact that combined parsimony analyses yields different solutions for a data set comprising nearly 6000 base pairs should be a matter for concern.

Rather than pooling data sets and assuming that homoplasy will become randomly distributed with respect to the historical signal, it would be preferable to identify potentially misleading influences from the outset and incorporate them into an explicit evolutionary model. Alternatively one might choose to remove characters that demonstrably violate the assumptions of the inference method *a priori*. In the current study, we recognized that there was non-stationarity in mitochondrial codon usage in a subset of amino acids for a subset of the taxa. When we controlled for this we obtained a topology consistent with that of the nuclear sequence data. We wish to underscore that this is not an endorsement for arbitrarily removing characters that do not fit a cherished hypothesis. If characters are to be removed, their removal should be contingent on their *a priori* failure to meet an explicit requirement of the inference model (such as non-stationarity in codon usage across taxa).

1.4.2 Alternative ways of "explaining" Anomalous Patterns in the Data

We were able to reconcile the mitochondrial codon data set to the tree topology yielded by the nuclear data by accommodating deviation in codon usage for a subset of amino acids. We were also able to reconcile topological inconsistencies by representing ASRV as a discrete approximation to a gamma distribution in a likelihood framework. The fact that two approaches, each emphasizing different aspects of the data, can reconcile the mitochondrial data to a topology consistent with that of the nuclear data begs the question "Which explanation is correct?" Is it the non-stationarity of codon usage bias that is causing the anomalous inference, or is the rate variation? Or both? Or neither? Clearly there are multiple views of what is relevant. In our case, we regard non-stationarity in codon usage to be a more satisfying explanation than the failure to accommodate ASRV simply because it has a more proximal connection to the underlying biology. Modelling ASRV as a gamma distribution merely provides an efficient means of improving the fit between model and data. However, it does not provide us with any clues as to which of several possible biological forces might be responsible for shaping the distribution of rates. The observed distribution could be consistent with a particular set of structural constraints, a particular history of functional adaptation, nucleotide mutability or codon usage. By

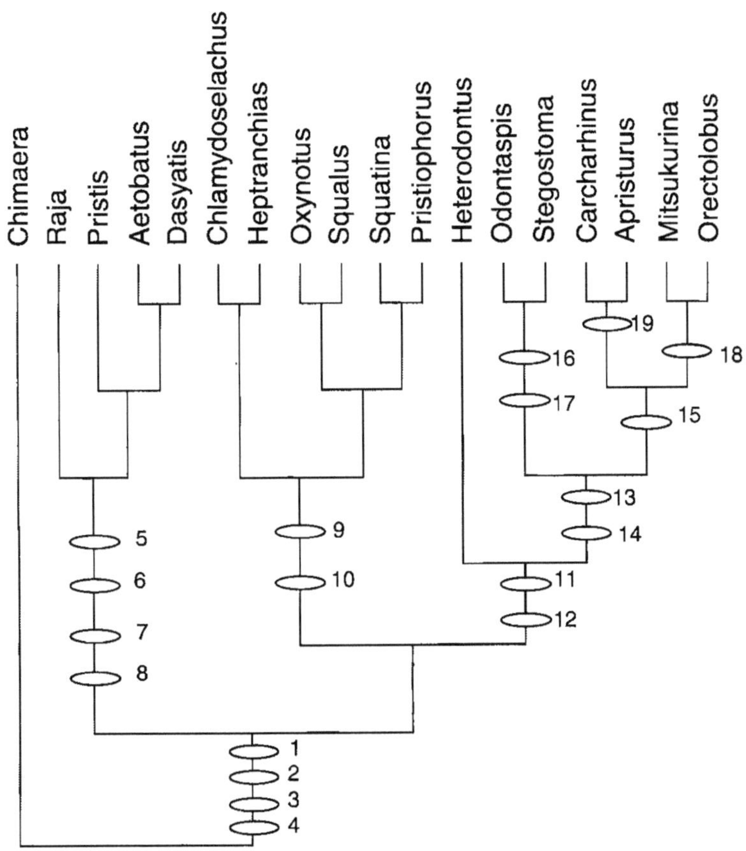

1 Bell-shaped basicranial vascular circuit
2 Adult hypophyseal duct closed externally
3 Segmented, calcified notochordal sheath
4 Occipital condyles
5 Cranio-palatine articulation absent
6 Ceratobranchial 5 attached to anterior margin of coracoid
7 Scapular process fused to synarcual
8 First cervical synarcual present
9 Basitrabecular process present
10 Cranio-palatine articulation of "orbitostylic" type
11 Pharyngo-branchial blade present
12 Hyomandibular fossa immediately behind orbit (i.e. Not separate from orbit)
13 Superficial opthalmic passing through separate foramen distinct from pro-oticum
14 Clasper: ventral margin of cartilage elongated
15 Tripodial rostrum
16 Pectoral Mesopterygium & Metapterygium expanded distally
17 Nasal barbel innervated by branch of opthalmicus superficialis
18 Nictatating lower eyelid
19 Spiral valve

Fig 1.7 Morphological characters taken from Shirai (1992) mapped on to the predominant phylogenetic signal observed in the current study.

contrast, explaining the discrepancy as an outcome of differential patterns of codon usage among taxa, provides more than a statistical improvement in fit between the model and the data—it narrows down the suite of viable explanations to a tangible *cause* for the difference between the mitochondrial and nuclear signals.

1.4.3 Contrasting the Molecular Phylogeny with Previous Hypotheses

Of the morphologically based hypotheses depicted in Fig. 1.1 the topology forwarded by Maisey (1984b) is the most compatible with that suggested by the molecular data (Fig. 1.2). However, it is also the least resolved and therefore most predisposed to fit any hypothesis. Maisey's groupings are based on the idea that neoselachians fall into 3 monophyletic groups: First those with an orbitostylic jaw suspension (Hexanchiformes, Squaliformes, Pristiophoriformes and Squatiniformes); second the galeomorphs (Heterodontiformes, Orectolobiformes, Lamniformes and Carcharhiniformes; and third the Batoids (skates and rays). There is considerable molecular support for this scheme. Within the galeomorph sharks the molecular data strongly support the topology suggested by Shirai (1992, 1996) and de Carvalho (1996) (Heterodontiformes, (Orectolobiformes, (Lamniformes, Carcharhiniformes). We mapped morphological character data from Shirai (1996) and Maisey (pers. comm.) on to the tree using parsimony. Diagnostic characters whose placement on the tree was unambiguous (i.e. not optimization dependent) are shown in Fig. 1.7. Interestingly, while there are several morphological characters that support the tree topology presented, characters that diagnose the monophyly of sharks as distinct from rays are conspicuously absent. If the tree we present is correct, then the lack of morphological characters that unambiguously diagnose sharks as a monophyletic group would go a long way to explain why Shirai (1996) and de Carvalho (1996) obtained the tree topologies they did. The phylogram of the likelihood tree also provides a possible explanation for the lack of diagnostic characters for the monophyly of sharks. It seems that the early cladogenetic events that gave rise to the orbitostylic sharks, the batoids and the galeomorph occurred over a relatively short space of time, as evidenced by the short internodal branch lengths at the base of the inferred tree and the almost simultaneous appearance in the fossil record of representatives of each of the three groups. We conjecture that if batoids are *truly* the sister group to a monophyletic shark clade as suggested by our data, there may not have been sufficient time between cladogenetic events to fix many morphological character states in the lineage that would give rise to sharks.

1.5 ACKNOWLEDGEMENTS

We thank John Marchalonis and Sam Schluter for providing us with the RAG-1 sequence of *Carcharhinus plumbeus* from which we were able to design primers. We thank John Maisey for comments on an earlier version of the manuscript, and Vicente Faria for editorial improvements.

1.6 LITERATURE CITED

Arambourg, C., and Bertin. L. 1958a. Classe des Chondrichthyens (Chondrichthyes). In P.-P. Grassé (ed.). *Traité de Zoologie. Anatomie, Systématique, Biologie*, Vol. 13 3: 2010-2015, figs. 1410-1411.

Arambourg, C., and Bertin. L. 1958b. Sous-classe des Sélaciens (Selachii). In P.-P. Grassé (ed.). *Traité de Zoologie. Anatomie, Systématique, Biologie*, Vol.13 3: 2016-2056, figs. 1412-1449.

Barret, M., Donoghue M. J., and Sober E. 1991. Against consensus. Systematic Zoology 40: 486-493.

Bigelow, H. B. and Schroeder, W. C. 1948. Sharks. Pp. 59-546. In Tee-Van, J. *et al.* (eds), *Fishes of the Western North Atlantic*. Memoir of the Sears Foundation, Marine Research 1(1). Yale University, New Haven.

Bigelow, H. B. and Schroeder, W. C. 1953. Part 2. Sawfishes, Guitarfishes, Skates and Rays; Chimaeroids In *Fishes of the Western North Atlantic*. Sears Foundation for Marine Research, Yale University, New Haven.

Bigelow, H. B. and Schroeder, W. C. 1957. *A study of the sharks of the suborder Squaloidea*. Bulletin of the Museum of Comparative Zoology, Harvard 117: 1-150.

de Carvalho, M. R. 1996. Higher-level elasmobranch phylogeny, basal squaleans, and paraphyly. Pp. 35-62. In M. L. J. Stiassny, L. R. Parenti, and D. Johnson (eds), *Interrelationships of fishes*. Academic Press, San Diego.

Collins, T. M., Wimberger, P. H. and Naylor, G. J. P. 1994. Compositional bias, character-state bias, and character-state reconstruction using parsimony. Systematic Biology 43(4): 482-496.

Comeron, J. M., Kreitman, M. and Aguadé, M. 1999. Natural selection on synonymous sites is correlated with gene length and recombination in *Drosophila*. Genetics 151: 239-249.

Compagno L. J. V. 1973. Interrelationship of living elasmobranchs. Pp. 15-61. In P.H. Greenwood *et al.* (eds), *Interrelationships of fishes*. Academic Press, New York.

Compagno L. J. V. 1977 Phyletic relationships of living sharks and rays. American Zoologist 17: 303-322.

Compagno, L. J. V., 1984. *Sharks of the world. An annotated and illustrated catalogue of sharks species known to date. FAO Fisheries Synopsis* Nº 125, 4 (1 and 2): 655 pp.

Douady, C. J., Dosay, M., Shivji, M. S. and Stanhope, M. J. 2003. Molecular phylogenetic evidence refuting the hypothesis of Batoidea (rays and skates) as derived sharks. Molecular Phylogeny and Evolution 26: 215-221.

Duméril, A. H. A. 1865. *Histoire naturelle des poissons ou ichthyologie générale*. Tome premier, Élasmobranches, Plagiostomes et Holocéphales ou Chimères. pp. 1-720.

Dunn, K. A. and Morrissey J. F. 1995. Molecular phylogeny of elasmobranchs. Copeia 1995: 526-531.

Felsenstein, J. 1978. Cases in which parsimony or compatibility methods will be positively misleading. Systematic Zoology 27: 401-416.

Garman, S. 1913. The Plagiostomia (Sharks, skates and rays). Memoirs of the Museum of Comparative Zoology XXXVI i-xiv+1-515 pages, 77 plates.

Goldman, N. 1993. Statistical tests of models of DNA substitution. Journal of Molecular Evolution 36: 182-198.

Graybeal A. 1998. Is it better to add taxa or characters to a difficult phylogenetic problem? Systematic Biology 47(1): 9-17.

Günther, A. C. 1870. *Catalogue of the fishes of the British Museum*. Vol. viii, pp. i-xxv, 1.

Hasegawa, M., H. Kishino and T. Yano., 1985. Dating of the human-ape splitting by a molecular clock of mitochondrial DNA. Journal of Molecular Evolution 21:160-174.

Hillis, D. M. 1996. Inferring complex phylogenies. Nature 383: 130-131.

Jarvik, E. 1977. The systematic position of acanthodian fishes. Pp. 199-225. In S. M. Andrews, R. S. Miles and A. D. Walker (eds), *Problems in vertebrate evolution*. Academic Press, London.

Jordan, D. S. 1923. A classification of fishes, including families, and genera as far as known. Stanford University Publications University Series in Biological Sciences III (2) 79-243 + i-x.

Kim, J. 1996. General inconsistency conditions for maximum parsimony: Effects of branch lengths and increasing numbers of taxa. Systematic Biology 45: 363–374.

Kitamura, T., Takemura, A., Watabe, S., Taniuchi, T. and Shimizu, M. 1996. Molecular phylogeny of the sharks and rays of superorder squalea based on mitochondrial Cytochrome b gene. Fisheries Science 62(3): 340-343.

Lockhart, P. J., Howe, C. J., Bryant, D. A., Beanland, T. J. and Larkum, A. W. D. 1992. Substitutional bias confound inference of cyanelle origins from sequence data. Journal of Molecular Evolution 34: 153-162.

Maisey, J. G. 1982. The anatomy and interrelationships of Mesozoic hybodont sharks. American Museum Novitates 2724: 1-48.

Maisey, J. G. 1984a. Higher elasmobranch phylogeny and biostratigraphy. Zoological Journal of the Linnean Society 82: 33-54.

Maisey, J. G. 1984b. Chondrichthyan phylogeny: A look at the evidence. Journal of Vertebrate Paleontology 4: 359-371.

Moy-Thomas, J. A. 1939. The early evolution and relationships of the elasmobranchs. Biological Reviews 14: 1-26, 12 figs.

Muller, J. and Henle, F. G. J. 1841. Systematische beschreibung der Plagiostomen. Berlin. Plagiostomeni-xxii.

Naylor, G. J. P. and Brown, W. M. 1998. Amphioxus mitochondrial DNA, chordate phylogeny, and the limits of inference based on comparison of sequences. Systematic Biology 47(1): 61-76.

Poe, S. 1998. Sensitivity of phylogeny estimation to taxon sampling. Systematic Biology 47(1): 18-31.

Palumbi, S. R. 1989. Rates of molecular evolution and the proportion of nucleotide positions free to vary. Journal of Molecular Evolution 29: 180-187.

Posada, D. and Crandall, K. A. 1998. Modeltest: Testing the model of DNA substitution. Bioinformatics 14: 817-818.

de Queiroz, A. 1993. For consensus (sometimes). Systematic Biology 42: 368-372.

Rannala, B., Huelsenbeck, J. P., Yang, Z. and Nielsen, R. 1998. Taxon sampling and the accuracy of large phylogenies. Systematic Biology 47(4): 702–710.

Regan, C. T. 1906. A classification of the selachian fishes. Proceedings of the Zoological Society of London 1906 (2): 722-758, text-figs. 115-124.

Saccone, C. C., Lanavé, C., Pesole, G. and Preparata, G. 1990. Influence of base composition on quantitative estimates of gene evolution. Methods in Enzymology 183: 570-583.

Schaeffer, B. 1981. The xenacanth shark neurocranium with comments on elasmobranch monophyly. Bulletin of the American Museum of Natural History 169, pp. 66.

Schaeffer, B. and Williams M. E. 1977. Relationships of fossil and living elasmobranchs. American Zoologist 17: 293-302.

Schoemaker, J. S. and Fitch, W. M. 1989. Evidence from nuclear sequences that invariable sites should be considered when sequence divergence is calculated. Molecular Biology and Evolution 6: 270-289.

Séret, B. 1986. Classification et phylogénèse des chondrichthyens. Océanis, 12: 161-180. Paris.

Shirai, S. 1992. Squalean phylogeny: a new framework of "squaloid" sharks and related taxa. Sapporo, Hokkaido University Press. 151 pp.
Shirai, S. 1992b. Phylogenetic relationships of the Angel sharks with comments of elasmobranch phylogeny (Chondrichthyes, Squatinidae) Copeia 505-518.
Shirai, S. 1996. Phylogenetic interrelationships of neoselachians (Chondrichthyes: Euselachii). Pp. 9-34. In M. L. J. Stiassny, L. R. Parenti and G. D. Johnson (eds). *Interrelationships of fishes*. Academic Press San Diego.
Sidow, A. and Wilson, A. C. 1990. Compositional statistics: An improvement of evolutionary parsimony and its application to deep branches in the tree of life. Journal of Molecular Evolution. 31(1): 51-68.
Steel, M. 1994. Recovering a tree from the leaf colourations it generates under a Markov model. Applied Mathematics Letters 7: 19-23.
Sullivan, J., Holsinger, K. E. and Simon, C. 1995. Among site rate variation and phylogenetic analysis of 12S rRNA data in sigmodontine rodents. Molecular Biology and Evolution 12: 988-1001.
Sullivan, J., Holsinger, K. E. and Simon, C. 1996. The effect of topology on estimates of among site rate variation. Journal of Molecular Evolution 42: 308-312.
Thies, D. 1983. Jurazeitliche Neoselachier aus Deutschland und S-England. [Jurassic Neoselachians from Germany and [Southern]-England.] [in German, with English summ.]. CFS, Courier Forschungsinst. Senckenberg 58, Pp. 116.
Thies, D. and Reif, W. E. 1985. Phylogeny and evolutionary ecology of Mesozoic, Neoselachii. News Jahrbuch für Geologie und Palaontologie Monatshefte, Abh. 169: 331-361, Stuttgart.
White, E. G. 1936. A classification and phylogeny of the elasmobranch fishes. American Museum Novitates 837: 1-16.
White, E. G. 1937. Interrelationships of the elasmobranchs with a key to the order Galea. Bulletin of the American Museum of Natural History 74: 25-138, 66 figs., 51 pls.
Yang, Z. 1994. Maximum likelihood phylogenetic estimation from DNA sequences with variable rates over sites: Approximate methods. Journal of Molecular Evolution 39: 306-314.
Yang, Z., Kumar, S. and Nei, M. 1995. A new method of inference of ancestral nucleotide and amino acid sequences. Genetics 141: 1641-1650.
Yang, Z. 1996. Among site rate variation and its impact on phylogenetic analyses. Trends in Ecology and Evolution. 11: 367-372.

CHAPTER 2

Population and Reproductive Genetics in Chondrichthyes

Edward J. Heist

2.1 MOLECULAR MARKERS AND STOCK STRUCTURE IN FISHES

Since the development of allozyme electrophoresis in the 1960's, molecular genetics has been broadly applied to the identification of discrete genetic stocks of fishes (Utter 1991). Genetic traits useful for detecting stock structure are those that are polymorphic within populations and whose various forms (alleles or haplotypes) are presumed to be selectively equivalent (i.e. neutral). Such traits include allozymes (Murphy *et al.* 1996; May 2003), mitochondrial DNA (mtDNA) restriction patterns (Lansman *et al.* 1981; Billington 2003) or sequences (Bernatchez and Danzmann 1993; Keeney *et al.* 2004), and more recently microsatellite DNA (Ashley and Dow 1994; O'Connell and Wright 1997). Essentially, differences in the frequencies of variant forms of molecular markers (alleles for allozymes and microsatellites, haplotypes in mtDNA) among geographic groups are used to determine the presence and degree of reproductive isolation and hence stock structure.

While there are many definitions of what constitutes isolated fish stocks in the published literature (reviewed in Carvalho and Hauser, 1994), stocks of fishes are essentially units that are reproductively isolated to the extent that each stock exhibits independent recruitment dynamics. Gene frequencies of neutral markers change due to genetic drift such that over time isolated stocks assume different gene frequencies. Thus if two or more stocks of fishes have been isolated for a sufficient number of generations, scoring of molecular genetic traits will detect these significant differences in allele frequencies. Migration among stocks either in the form of a small number of individuals each generation or large scale episodic movements preclude divergence of

Fisheries and Illinois Aquaculture Center, Southern Illinois University at Carbondale, Carbondale, IL 62901-6511.

gene frequencies such that stocks that are essentially reproductively isolated will exhibit genetic homogeneity (Waples 1998). Thus the failure to detect stock structure using polymorphic genetic characters does not prove that reproductively isolated stocks do not currently exist, however the demonstration of gene frequency differences at presumably neutral loci is sound evidence for some degree of reproductive isolation and hence stock structure.

The amount of recurrent migration among stocks that is sufficient to prevent genetic drift from producing measurable differences in gene frequency is very small. Under the island model of migration, in which a species is divided into a large number of stocks with equal size and equal migration among stocks, the equilibrium relationship between migration and drift is

$$F_{ST} = \frac{1}{4N_e m + 1}$$

where F_{ST} is Wright's (1969) fixation index (a measure of the magnitude of genetic divergence among stocks), N_e is the effective population size of each stock, and m is the migration rate. The product $N_e m$ can be taken as the number of breeders in a stock that are migrants from other stocks. Following this relationship a migration of greater than 10 individuals per generation reduces F_{ST} to below 0.02. A genetic study would need to employ large sample sizes in order to detect such a small number of migrants. Furthermore, the island model is unrealistic for most species. If stocks are more likely to receive migrants from stocks that are geographically more proximate a stepping stone model may be more representative. Because migrants from proximate populations will likely exhibit similar allele frequencies a single migrant under a stepping stone model is worth 1.2 to 2 migrants in an island model (Crow 1986). The lack of barriers to migration and gene flow throughout the oceans results in less stock structure across similar geographic distances in the marine relative to freshwater fishes including Chondrichthyes (Ward et al. 1994; Waples 1998).

2.2 ALLOZYMES

Allozymes are allelic forms of typically enzymatic proteins that differ in their rate of migration in an electric field through some separatory medium such as starch gels (Aebersold *et al.* 1987), cellulose acetate plates (Easteal and Boussy 1987; Hebert and Beaton 1989), or polyacrylamide (Hames and Rickwood 1990). Allozymes were the first molecular technique to be widely applied to identification of stock structure in fishes (Utter 1991; Ward and Grewe 1994) and have been broadly applied to marine, freshwater, and anadromous species. The first examination of allozyme diversity in chondrichthyans was that of Smith (1986) who reported heterozygosity in seven species including shallow water demersal species (*Squalus acanthias, Mustelus lenticulatus, Galeorhinus australis* (now *G. galeorhinus*)), deepwater demersal species (*Deania calcea, Centroscymnus crepidator, Etmopterus baxteri*), and a pelagic species (*Prionace glauca*). Allozyme heterozygosity was low for all species ranging from 0.001 in *M. lenticulatus* to 0.037 in *P. glauca*.

In a review of published studies on stock structure in marine, freshwater, and anadromous fishes Ward *et al.* (1994) concluded that while total allozyme heterozygosity was similar in all three groups of fishes, subpopulation heterozygosity was much less in the marine environment indicating that stocks were far less structured. The only chondrichthyan data in that study were from *Carcharhinus sorrah* and *C. tilstoni* (data originally in Lavery and Shaklee 1989), both of which exhibited very low heterozygosities (0.035 and 0.037 respectively) and values for G_{ST}, a multilocus estimator of F_{ST}, were low (0.007 and 0.009 respectively) and did not differ significantly from zero. Since the publication of Ward *et al.* (1994), low heterozygosities and negligible values of F_{ST} have based on allozyme studies have been reported for *Carcharhinus plumbeus* (mean heterozygosity = 0.005) (Heist *et al.* 1995) and *Mustelus antarcticus* (mean heterozygosity = 0.006) (MacDonald 1988). Subsequent to the findings of MacDonald (1988), Gardner and Ward (1998) found significant structure (mean G_{ST} = 0.113 over seven loci) in *M. antarcticus* in a study that included greater geographic sampling and detected much larger heterozygosity (mean heterozygosity = 0.099) within geographic samples. The contradiction between the studies of MacDonald (1988) and that of Gardner and Ward (1998) illustrate one of the limitations of allozymes, namely that laboratories often differ in the resolution of alleles (DNA-based methods have the potential to be much more objective). Gardner and Ward (2002) detected mean heterozygosities of 0.000 to 0.100 in four species of *Mustelus* from Australia and New Zealand. Gaida (1997) detected considerable amounts of genetic variation (mean heterozygosity = 0.056) and significant genetic structure (mean F_{ST} = 0.085 over six loci) in *Squatina californica* collected from different islands in southern California's Channel Island chains. In that study, *S. californica* from nearby Santa Rosa and Santa Cruz Islands did not exhibit significant differences in allele frequencies while frequencies from both islands differed in *S. californica* from the more distant San Clemente Island. The most likely explanation why significant stock structure was detected in *Mustelus* and *Squatina* but not in either of the three species of *Carcharhinus* listed above is the vagility of the species. *S. californica* is a demersal shallow water species (maximum depth less than 200 m (Compagno 1984) and those that inhabit the offshore channel islands are presumably reproductively isolated owing to the deep (>500 m) channels between San Clemente and islands to the north (Gaida 1997).

With the development of polymerase chain reaction (PCR) based DNA technologies allozymes have declined in importance as a research tool for identifying stock structure of fishes. During the late 1980's and 90's, PCR-based studies of mitochondrial DNA restriction fragment analysis (RFLP) and sequences (Billington 2003), nuclear DNA fingerprinting (Franck *et al.* 1991; Wright 1993) and microsatellite DNA (Franck *et al.* 1991; Wright, 1993, O'Connell and Wright 1997) have largely replaced allozyme analysis. It appears unlikely that there will be a resurgence of allozyme studies for population genetics in Chondrichthyes. Allozyme heterozygosity in Chondrichthyes is low (Smith 1986; Heist 1999) and thus very large sample sizes are necessary to achieve a reasonable level of statistical power given the expected low levels of genetic divergence among chondrichthyan stocks (Heist 2004). Selective neutrality is an assumption of polymorphic genetic markers for determining stock structure in fishes (Ward and Grewe 1994) and allozyme alleles are less likely to be selectively neutral than the variation scored by many DNA-based methods. Finally the allozyme technique requires the collection of very fresh tissues, often from multiple tissue types, followed by storage at very cold temperatures (-20° or colder) prior to electrophoresis. This makes shipboard sampling, especially nonlethal shipboard sampling, very inconvenient or impossible (Heist 1999).

2.3 MITOCHONDRIAL DNA

Mitochondrial DNA is a haploid, organellar genome that typically exists as a single haplotype found in all of an individual's cells (Attardi 1985). Some individuals and some species exhibit more than one form of mtDNA. This condition, known as heteroplasmy, has been reported in several species of cartilaginous (Brown *et al.* 1992; Ludwig *et al.* 2000; Grunwald *et al.* 2002) and bony fishes (Magoulas and Zouros 1993; Broughton and Dowling 1997; Ravago *et al.* 2002). Heteroplasmy may be related to paternal leakage in which the mechanisms that prevent inheritance of the paternal mtDNA break down and has been demonstrated in bony fishes (Magoulas and Zouros 1993), birds (Kvist *et al.* 2003), and mammals (Gyllensten *et al.* 1991) including humans (Schwartz and Vissing 2002). While it is generally believed that mtDNA does not undergo recombination, a feature which makes it attractive for phylogenetic and phylogeographic work (Avise 2000), recent studies (Hoarau *et al.* 2002; Rokas *et al.* 2003) indicate that heteroplasmy, perhaps following paternal leakage, may permit recombination among different mtDNA haplotypes within a single individual. Nevertheless most individuals of most vertebrate species possess a single form of mtDNA that is inherited intact from the female parent. The combination of haploidy and maternal inheritance results in an approximately fourfold reduction in effective population size for mtDNA relative to nuclear DNA (Birky *et al.* 1989).

Mitochondrial DNA of Chondrichthyes is similar to that of other vertebrates in that it consists of a single loop of double stranded DNA comprising 13 protein coding genes, two genes for ribosomal RNA, and 22 transfer RNAs (Meyer 1993). The genome is very compact with no introns and little DNA between genes with the exception of a single non-coding control region that contains the origin of mtDNA heavy strand replication and typically about 1kb of non-coding sequence. Mitochondrial genome sizes of elasmobranchs range from 16,707 to 16,783 base pairs (Heist 2004 and references cited therein). The only mtDNA genome completely sequenced in a holocephalan (*Chimaera monstrosa*) is also the largest mtDNA genome (18,580 base pairs) sequenced in any vertebrate (Arnason *et al.* 2001).

The first studies of mtDNA variation in fishes including Chondrichthyes employed restriction fragment length polymorphism (RFLP) analysis of whole molecule mtDNA. Later studies employed direct sequencing of segments of mtDNA amplified using PCR (Mullis and Faloona 1987; Saiki *et al.* 1988). RFLP analysis of whole molecule mtDNA is performed by first isolating intact mtDNA from nuclear DNA and other cellular components typically using cesium chloride density gradient ultracentrifugation (Lansman *et al.* 1981). Aliquots of purified mtDNA are digested with type II endonucleases that recognize specific 4- to 6-base motifs and cleave the circular DNA into one or more linear strands of DNA that can be resolved on agarose or polyacrylamide gels and visualized using ethidium bromide staining, silver staining, or auto-radiography of radio labeled fragments (Dowling *et al.* 1996; Billington 2003). Differences in the numbers and sizes of fragments can be used to estimate the fraction of nucleotide sites that differ among any two haplotypes and to estimate the amount of genetic variation within and among populations (Nei 1987).

Heist (1994) scored whole-molecule mtDNA variation in three species of sharks: *Rhizoprionodon terraenovae, Carcharhinus plumbeus,* and *Isurus oxyrinchus. C. plumbeus* exhibited very low levels of intra-population variation. Using twelve restriction enzymes in 95 *C. plumbeus* from the mid-Atlantic Bight and the west coast of Florida, Heist *et al.* (1995) detected only five unique haplotypes, one of which was found in 87 of 95 sharks. Nucleotide sequence divergence (nsd) was 0.036%, indicating that on average less than one tenth of one percent of nucleotide sites differed among sharks. *R. terraenovae* collected from the same locations and an additional site in Veracruz, Mexico, were slightly more variable (nsd = 0.13%) (Heist *et al.* 1996b). Neither *C. plumbeus* nor *R. terraenovae* exhibited significant differences in haplotype frequencies among sites, consistent with a single stock of each species. These findings are consistent with traditional tag/recapture data in sandbar sharks which shows a great deal of migration among the mid-Atlantic Bight and Gulf of Mexico (Kohler *et al.* 1998). The results from *R. Terraenovae* need to be viewed with caution since sample sizes were relatively low (Heist *et al.* 1996b). The null hypothesis in any study of genetic stock structure is that gene frequencies are identical across samples and failure to reject the null hypothesis should never be taken as proof that the null hypothesis is true (Dizon *et al.* 1995). Given the relatively little movement in tagged *R. terraenovae* (Kohler *et al.* 1998), there may be multiple fishery stocks that may or may not be detected employing larger sample sizes and perhaps a more sensitive methodology (Heist *et al.* 1996b). Nine of the restriction enzymes used to score *C. plumbeus* from the western North Atlantic were applied to mtDNA from 16 *C. plumbeus* from the eastern Indian Ocean (Western Australia). The use of only nine restriction enzymes reduced the number of Atlantic haplotypes to only two however three distinct haplotypes were present in the eastern Indian Ocean, all of which differed from those in the Atlantic by a minimum of four restriction sites (Fig. 1). The low levels of mtDNA nucleotide diversity in both *C. plumbeus* and *R. terraenovae* may not be surprising given the low mutation rate of shark mtDNA compared to that of mammals (Martin *et al.* 1992). However, nucleotide variation in *I. oxyrinchus* (Heist *et al.* 1996a) was much higher (nsd = 0.347%) indicating a higher mutation rate, a higher long term effective population size, or both in this pelagic species. Haplotype frequencies in *I. oxyrinchus* collected from the South Atlantic, South Pacific, and North Pacific were not significantly different from one another, while haplotype diversity in the North Atlantic was slightly lower and thus haplotype frequencies were significantly different from those in other basins (Heist *et al.* 1996a). A reanalysis of the data using a more powerful statistical approach showed that haplotype frequencies between the South Atlantic and North Pacific were also significantly different (Schrey and Heist 2003). The results of this study demonstrate that while it seems unlikely that shortfin mako comprises

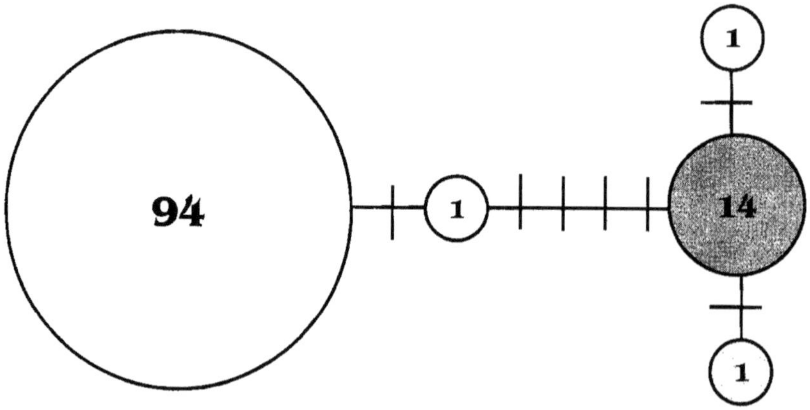

Fig. 2.1 Distribution of *Carcharhinus plumbeus* mtDNA haplotypes in specimens from Virginia (unshaded) and Western Australia (shaded). Numbers inside circles are sample sizes. Slashes indicate restriction site differences among haplotypes. From Heist, E. J. 1994. Population Genetics of Selected Species of Sharks. Ph.D. Dissertation. College of William and Mary, Gloucester Point, VA.

a single worldwide population in terms of fishery stocks, there has been sufficient female-mediated migration among ocean basins to retard divergence of mtDNA haplotype frequencies. Estimate of nsd for three species of *Mustelus* from Australia and New Zealand ranged from 0.00% to 0.21% (Gardner and Ward 2002) indicating that some species of this coastal genus exhibit considerable stock structure.

2.4 DNA MICROSATELLITES

DNA microsatellites (also called simple sequence repeats) are short (1-6 base pair) motifs that are repeated in tandem (e.g. GA_n and $GACA_n$ where n refers to the number of times the unit is repeated) in the genome of eukaryotic organisms (Ashley and Dow 1994; Heist *et al.* 1996b; O'Connell and Wright 1997). These motifs, which occur more commonly in the genome of vertebrates than would be expected by point mutations (Tautz *et al.* 1986) have a high rate of mutation for copy number, presumably due a process known as slipped-strand mispairing (Levinson and Gutman 1987). The high mutation rate results in a great deal of intraspecific variation, with some loci possessing fifty or more alleles and heterozygosities approaching 100%. Microsatellite loci are among the most variable genetic elements known to date and this high variation makes them useful for studies across a variety of scales, including comparisons among species, genetic stock structure within species, and calculations involving relatedness (e.g. paternity) within families (MacDonald and Potts 1997).

Microsatellites are scored by amplifying the tandem repeat using PCR primers complementary to the sequences flanking the repeat. PCR products are resolved on a separatory medium, typically a polyacrylamide gel. The products are visualized most often by labeling one of the PCR primers prior to amplification with either a radionuclide (^{32}P or ^{33}P) and scoring the gel via autoradiography (Fig. 2) or by employing a fluorescently tagged primer and resolving the fragments using an automated DNA sequencer/analyzer. Microsatellites are codominant data, meaning that both alleles present in heterozygotes can be scored and thus analyses are similar to those performed for allozymes.

One drawback of the microsatellite technique is that a library of DNA fragments must be constructed and screened to identify microsatellite repeat motifs and their flanking regions. This is a costly and time-consuming operation. To date polymorphic microsatellite loci have been developed in *Carcharhinus plumbeus* (Heist and Gold 1999b), *Carcharodon carcharias* (Pardini *et al.* 2000), *Negaprion brevirostris* (Feldheim *et al.* 2001a; Feldheim *et al.* 2001b), *Isurus oxyrinchus* (Schrey and Heist 2002), *Carcharhinus limbatus* (Keeney and Heist 2003) and *Ginglymostoma cirratum* (Heist *et al.* 2003).

Primer sets often exhibit cross-species utility, although there is often a decrease in variation and reliability of amplification the more distant

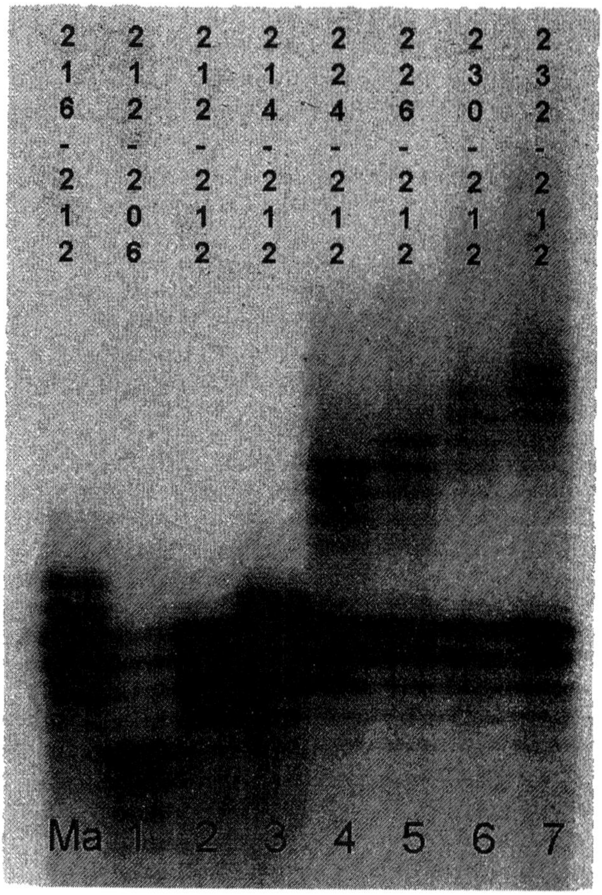

Fig. 2.2 Demonstration of multiple paternity in litter of *Ginglymostoma cirratum* using a single microsatellite locus. Lanes represent a mature female *G. cirratum* (Ma) and seven of her offspring. Numbers at top refer to genotypes (e.g. Ma is a 216/212 heterozygote). All offspring possess allele 212 found in Ma but there are a total of seven paternal alleles (216, 206, 212, 214, 224, 226, 230, and 232) indicating that at least four fathers sired the litter.

the phylogenetic relationship between the species from which the primers were developed and the species in which they are being used. Keeney and Heist (2003) characterized cross-species amplification of sixteen microsatellite loci developed in *Carcharhinus limbatus*. Eleven of the 16 primer pairs reliably amplified polymorphic loci in *C. melanopterus* while the numbers of reliable polymorphic loci amplified in 10 other species of *Carcharhinus* ranged from one in *C. isodon* to 10 in *C. altimus*. Numbers of polymorphic loci amplified in other genera included eight loci in *Galeocerdo cuvier*, six in *Prionace glauca*, four in *Negaprion brevirostris*, one to three in two species of *Rhizoprionodon*, and four to seven in two species of *Sphyrna*. Of the five polymorphic microsatellite loci developed in *I oxyrinchus* by Schrey and Heist (2002) all were polymorphic in *Lamna nasus* and *L. ditropis*, and two were polymorphic in *Carcharodon carcharias* and *Alopias vulpinus*.

Microsatellite based studies in chondrichthyans to date have detected very low levels of stock structure. Heist and Gold (1999b) found no evidence of differences in allele frequencies among *Carcharhinus plumbeus* from the Gulf of Mexico and Mid-Atlantic Bight, consistent with a previous study that employed allozymes and mtDNA (Heist *et al.* 1995). Pardini *et al.* (2001) reported that microsatellite based estimates of F_{ST} among *C. charcharodon* from Australia/New Zealand and South Africa were all nonsignificant. Estimates of F_{ST} among *Negaprion brevirostris* from Florida/Bahamas and Brazil were statistically significant ($P < 0.05$ and $P < 0.0001$ for two different estimators) however Feldheim *et al.* (2001b) argued that because the magnitude of the estimators were so small (0.016 and 0.026 respectively) the values were not biologically significant. Schrey and Heist (2003) found very little evidence of variation in microsatellite allele frequency among *I. oxyrinchus* collected from the North Atlantic, South Atlantic (Brazil), South Africa, South Pacific (Australia), and North Pacific. Pairwise estimates of F_{ST} ranged from several negative values (typically assumed to be zero) to a high value of 0.011 between the North Atlantic and the North Pacific, the two most distant locations in terms of mako dispersal potential. This highest value was the only statistically significant pairwise value although one estimator of F_{ST} across all samples was also significant. Schrey and Heist (2003) concluded that this study does not provide evidence for genetic stock structure in *I. oxyrinchus*. Hueter *et al.* (2004) likewise found very low and non-significant F_{ST} values between *Carcharhinus limbatus* neonates and young of the year collected from nursery areas in the western North Atlantic, Gulf of Mexico and Caribbean Sea.

2.5 FORENSICS AND IDENTIFICATION OF CRYPTIC SPECIES

The morphological similarity of many species of chondrichthyans makes identification of specimens and recognizing unnamed species challenging. Genetic markers provide additional means of identifying and verifying species identity by demonstrating reproductive isolation among sympatric forms or by providing estimates of the amount of genetic divergence among allopatric forms. Allozymes have been used to demonstrate the presence of three species of *Squatina* where previous morphology-based studies assumed the presence of only a single species (Sole-Cava *et al.* 1983; Sole-Cava and Levy 1987). Lavery and Shaklee (1991) demonstrated that two forms of "blacktip shark" that were previously assumed to be *Carcharhinus limbatus* and possessed very slight differences in pigmentation were reproductively isolated based on nearly fixed differences at two allozyme loci and significant frequency differences at two other loci. Lavery and Shaklee concluded that the rare form was *C. limbatus* based on comparisons with specimens from the type locality and that the common form was *C. tilstoni*. Eitner (1995) suggested the presence of an additional species of *Alopias* based on an allozyme study but unfortunately no specimens were retained and the conclusions of that study remain unverified. Identification of specimens that have been processed by fishers are especially difficult. Smith and Benson (2001) showed that 40% of shark fillets labeled *M. lenticulatus* in New Zealand were actually from other species and in some cases from prohibited species. Protocols for genetically identifying carcasses in commercial catches of sharks have been developed using mitochondrial (Heist and Gold 1999a; Chan *et al.* 2003; Chapman *et al.* 2003) and nuclear DNA (Pank *et al.* 2001; Shivji *et al.* 2002; Chapman *et al.* 2003) markers.

2.6 PHILOPATRY AND SEX-BIASED DISPERSAL

Studies that employ both mtDNA and microsatellites sometimes reveal disparate levels of genetic stock structure. Explanations for this phenomenon include sex-biased dispersal, differences in genetic effective population size of the two markers, and differences in the rate at which the two markers approach equilibrium with respect to migration and genetic drift (Buonaccorsi *et al.* 2001). Because mtDNA is maternally inherited it often shows a strong phylogeographic signal in species in which females exhibit natal philopatry such as sea turtles (Karl *et al.* 1992; Allard *et al.* 1994; Bowen and Karl 1997) and marine mammals (Gladden *et al.* 1997; Lyrholm *et al.* 1999; Escorza-Trevino and Dizon 2000). Microsatellites are biparentally inherited nuclear markers and in species in which males move more than females, low F_{ST} values will occur as a result of male-mediated gene flow. Thus in species that exhibit sexual segregation or differences in movements between the sexes, like many chondrichthyans, differences between maternally and biparentally inherited markers are often taken as an indication of sex-biased dispersal. However, it is important to remember that large discrepancies between mtDNA and microsatellite-based estimates of F_{ST} can also result from differences in the rates of genetic drift and mutation in the two markers (Buonaccorsi *et al.* 2001). A combination of genetic and nongenetic (tracking and tagging) data indicate that many sharks exhibit philopatry, which may take the form of females returning to their natal nursery areas to pup (Hueter *et al.* 2004).

The most striking difference between mtDNA and microsatellites in chondrichthyes come from a study by Pardini *et al.* (2001) which showed that *Carcharodon carcharias* mtDNA haplotypes from Australia/New Zealand and South Africa were almost entirely distinct, with only one individual from Australia/New Zealand exhibiting a haplotype more similar to South African haplotypes yet there were no significant differences in microsatellite allele frequencies. The authors attributed the discrepancy between the two data types to a greater degree of female fidelity against a backdrop of male roaming. There was a smaller discrepancy between the signal present in mtDNA and nuclear microsatellites in shortfin mako which may be explained by female philopatry and male-mediated gene flow (Schrey and Heist 2003). In that study a power analysis was used to demonstrate that the differences between the signals provided by mtDNA and microsatellites were significant in that if the mtDNA signal was taken to be an accurate representation of gene flow the microsatellite-based estimates should have been much higher.

Genetic studies of *Carcharhinus limbatus* employing mitochondrial DNA (Keeney *et al.* 2003) and microsatellites (Hueter *et al.* 2004) demonstrate large and significant differences in female-mediated mitochondrial DNA haplotype frequencies with much smaller and non-significant differences in microsatellite allele frequencies among young of the year blacktip sharks collected from continental nurseries in the South Atlantic Bight (Georgia and South Carolina), the west coast of Florida, Texas, Quintana Roo, and Belize. Tagging data (Kohler *et al.* 1998) exhibit considerable movement in blacktip sharks between the Atlantic and Gulf and between the eastern and western Gulf of Mexico. If such movement is indicative of the distance between birth and reproduction of blacktip sharks, frequencies at both nuclear and mitochondrial markers would be homogeneous. Yet the strong signal in the mtDNA data (Hueter *et al.* 2004; Keeney *et al.* 2004) indicates that females tend to return to their natal regions to deliver their pups. Thus, movement does not necessarily coincide with gene flow if members of either sex are philopatric in regards to reproduction. Because many oceanic chondrichthyans utilize continental nursery areas for parturition and development (Pratt and Carrier 2001), differentiation of genetic stocks may be possible even when these stocks sometimes overlap. One unique feature that likely contributed to the strong signal in the mtDNA data was that only neonate and young of the year sharks collected during the spring and summer months were used, and thus were presumably collected near their site of parturition. Because mtDNA is inherited intact from the female parent (see above), scoring mtDNA variation in young sharks is equivalent to sampling their mothers. Movement by male blacktip sharks appears to be greater since little if any genetic structure was detected in the bi-parentally inherited microsatellite markers (Hueter *et al.* 2004).

Microsatellites and mtDNA are powerful at resolving population structure at differing levels of isolation. The availability of multiple independent loci with sufficient variation makes microsatellite loci very good markers for estimating current levels of gene flow (Shaw *et al.* 1999). However the slipped-strand mutation process of microsatellites (Levinson and Gutman 1987) tends to produce a great deal of homplasy (alleles that are identical in state but not of shared common ancestry) and thus microsatellites are not as powerful for higher levels of divergence (Shaw *et al.* 1999). Microsatellites, which can possess fifty or more alleles and heterozygosities approaching 100%, can also be too polymorphic for studies of stock structure as the maximum value that F_{ST} can obtain is equal to homozygosity (Hedrick 1999). Thus estimates based on loci that are too polymorphic may underestimate F_{ST}. The presumed lack of recombination (but see Rokas *et al.*2003) allows mtDNA to evolve with far less homplasy, thus the amount of divergence among mtDNA haplotypes is an indication of the time to shared common ancestry (Avise 2000). However, because mtDNA represents a single genetic locus subject to random genetic drift it does not provide as sensitive an estimate to low levels of gene flow as do multiple microsatellite loci (Buonaccorsi *et al.* 2001).

2.7 GENETICS AND REPRODUCTIVE ECOLOGY IN CHONDRICHTHYES

The development of highly polymorphic loci (especially microsatellite DNA) has fostered the use of molecular genetics to study reproductive biology of organisms in new ways. Highly polymorphic markers can be used to accurately genetically assign parentage to offspring (Gerber *et al.* 2000; Avise *et al.* 2002) and determine the degree of genetic relatedness among any pair of individuals (Blouin *et al.* 1996; Goodnight and Queller 1999). Multiple paternity in chondrichthyes was first demonstrated in nurse sharks by Ohta *et al.* (2000) and almost simultaneously by Saville *et al.* (2002). Both of these studies utilized major histocompatibility (MHC) loci, a component of the jawed vertebrate immune system, and showed that nurse shark litters contain a minimum of three and four fathers respectively. Multiple paternity in *Negaprion brevirostris* was demonstrated by Feldheim *et al.* (2001a) who showed that a minimum of three sires fathered one litter using multiple microsatellite loci. Multiple paternity was also confirmed in one litter of *Prionace glauca* and is uncommon but does occur in *Sphyrna tiburo* (M. Shivji, personal communication). Current research in my laboratory indicates that some large nurse shark litters possess a minimum of six sires based on the number of multilocus genotypes (E. J. H. unpublished data) (Fig. 2.2).

The most detailed study of molecular ecology of a chondrichthyan to date is that of Feldheim *et al.* (2002) as part of a long term study of *Negaprion brevirostris* ecology in Bimini Lagoon. Attempts were made to collect all juvenile and neonate *N. brevirostris* over successive years and to collect as many adults as possible. Feldheim *et al.* (2002) were able to identify the female parent of 119 of 897 free-living juvenile and young of the year *N. brevirostris* could be genetically assigned to one of five sampled female adults. Some offspring of the same female parent were born in multiple seasons spaced two years apart, consistent with the presumed two-year reproductive cycle in *N. brevirostris*. By reconstructing litters of offspring assigned to the same female parent it was possible to demonstrate additional cases of multiple paternity and to infer partial genotypes of the sires. Similar studies are currently underway on a population of *Ginglymostoma cirratum* in the Dry Tortugas, FL (E. J. H. unpublished data). While many species of Chondrichtyes are wide ranging, the high degree of site-fidelity of *N. brevirostris* (Feldheim *et al.* 2002) and *G. cirratum* (Carrier and Pratt 1998) make them ideal candidates for future studies of reproductive ecology in Chondrichthyes.

2.8 LITERATURE CITED

Aebersold, P. B., Winans, G. A., Teel, D. J., Milner, G. B., and Utter, F. 1987. Manual for starch gel electrophoresis: A method for the detection of genetic variation, 19 pp. NOAA, Washington, D.C.

Allard, M. W., Miyamoto, M. M., Bjorndal, K. A., Bolten, A. B., and Bowen, B. W. 1994. Support for Natal Homing in Green Turtles from Mitochondrial-DNA Sequences. Copeia: 34-41.

Arnason, U., Gullberg, A., and Janke, A. 2001. Molecular phylogenetics of gnathostomous (jawed) fishes: Old bones, new cartilage. Zoologica Scripta. 30: 249-255.

Ashley, M. V., and Dow, B. D. 1994. The use of microsatellite analysis in population biology: Background, methods and potential applications, Pp. 185-201. In B. Schierwater, B. Streit, and R. DeSalle (eds), *Molecular Ecology and Evolution: Approaches and Applications*. Birkhauser Verlag, Basel, Switzerland.

Attardi, G. 1985. Animal mitochondrial DNA: An extreme example of genetic economy. International Review of Cytology. 93: 93-145.

Avise, J. C. 2000. Phylogeography: The History and Formation of Species. Harvard University Press, Cambridge, MA.

Avise, J. C., Jones, A. G., Walker, D., and Dewoody, J. A. 2002. Genetic mating systems and reproductive natural histories of fishes: Lessons for ecology and evolution. Annual Review of Genetics. 36: 19-45.

Bernatchez, L., and Danzmann, R. G. 1993. Congruence in control-region sequence and restriction site variation in mitochondrial DNA of brook charr (*Salvelinus fontinalis* Mitchill). Molecular Biology and Evolution. 10: 1002-1014.

Billington, N. 2003. Mitochondrial DNA, Pp. 59-100. In E. M. Hallerman (ed.), *Population Genetics: Principles and Applications for Fisheries Scientists*. American Fisheries Society, Bethesda, MD.

Birky, C. W. J., Fuerst, P., and Marayama, T. 1989. Organelle gene diversity under migration, mutation, and drift: Equilibrium expectations, approach to equilibrium, effects of heteroplasmic cellsm and comparison to nuclear genes. Genetics. 121: 613-627.

Blouin, M. S., Parsons, M., Lacaille, V., and Lotz, S. 1996. Use of microsatellite loci to classify individuals by relatedness. Molecular Ecology. 5: 393-401.

Bowen, B. W., and Karl, S. A. 1997. Population genetics, phylogeography, and molecular evolution, Pp. 29-50. In P. L. Lutz and J. A. Musick (eds), *The Biology of Sea Turtles*. CRC Press, New York.

Broughton, R. E., and Dowling, T. E. 1997. Evolutionary dynamics of tandem repeats in the mitochondrial DNA control region of the minnow *Cyprinella spiloptera*. Molecular Biology and Evolution. 14: 1187-1196.

Brown, J. R., Beckenbach, A. T., and Smith, M. J. 1992. Mitochondrial-DNA length variation and heteroplasmy in populations of white sturgeon (*Acipenser transmontanus*). Genetics. 132: 221-228.

Buonaccorsi, V. P., Mcdowell, J. R., and Graves, J. E. 2001. Reconciling patterns of inter-ocean molecular variance from four classes of molecular markers in blue marlin (*Makaira nigricans*). Molecular Ecology. 10: 1179-1196.

Carrier, J. C., and Pratt, H. L. 1998. Habitat management and closure of a nurse shark breeding and nursery ground. Fisheries Research. 39: 209-213.

Carvalho, G. R., and Hauser, L. 1994. Molecular genetic and the stock concept in fisheries. Reviews in Fish Biology & Fisheries. 4: 326-350.

Chan, R. W. K., Dixon, P. I., Pepperell, J. G., and Reid, D. D. 2003. Application of DNA-based techniques for the identification of whaler sharks (*Carcharhinus spp.*) caught in protective beach meshing and by recreational fisheries off the coast of New South Wales. Fishery Bulletin. 101: 910-914.

Chapman, D. D., Abercrombie, D. L., Douady, J., Pikitch, E. K., Stanhope, M., and Shivji, M. 2003. A streamlined bi-organelle, multiplex PCR approach to species identification: Application to global conservation and trade monitoring of the great white shark, *Carcharodon carcharias*. Conservation Genetics. 4: 415-425.

Compagno, L. J. V. 1984. FAO species catalogue. Vol. 4. Sharks of the world. An annotated and illustrated catalogue of shark species known to date. Part 1. Hexanchiformes to Lamniformes. FAO Fish. Synop. (125) Vol. 4, Pt. 1 Pp. 249.

Crow, J. F. 1986. Basic concepts in population, quantitative, and evolutionary genetics. W. H. Freeman, New York.

Dizon, A. E., Taylor, B. L., and O'corry-Crowe, G. M. 1995. Why statistical power is necessary to link analyses of molecular variation to decisions about population structure, Pp. 288-294. In J. L. Neilson and D. A. Powers (eds), *Evolution and the Aquatic Ecosystem*. American Fisheries Society, Bethesda, MD.

Dowling, T. E., Moritz, C., Palmer, J. D., and Riesenberg, L. H. 1996. Nucleic Acids III: Analysis of fragments and restriction sites, Pp. 249-320. In D. M. Hillis, C. Moritz, and B. K. Mable (eds), *Molecular Systematics*. Sinauer Associates, Inc., Sunderland, MA.

Easteal, S., and Boussy, I. A. 1987. A sensitive and efficient isoenzyme technique or small arthropods and other invertebrates. Bulletin of Entomological Research. 77: 407-415.

Eitner, B. J. 1995. Systematics of the genus *Alopias* (Lamniformes: Alopiidae) with evidence for the existence of an unrecognized species. Copeia 1995: 562-571.

Escorza-Trevino, S., and Dizon, A. E. 2000. Phylogeography, intraspecific structure and sex-biased dispersal of Dall's porpoise, *Phocoenoides dalli*, revealed by mitochondrial and microsatellite DNA analyses. Molecular Ecology. 9: 1049-1060.

Feldheim, K. A., Gruber, S. H., and Ashley, M. V. 2001a. Multiple paternity of a lemon shark litter (Chondrichthyes : Carcharhinidae). Copeia 2001: 781-786.

Feldheim, K. A., Gruber, S. H., and Ashley, M. V. 2001b. Population genetic structure of the lemon shark (*Negaprion brevirostris*) in the western Atlantic: DNA microsatellite variation. Molecular Ecology. 10: 295-303.

Feldheim, K. A., Gruber, S. H., and Ashley, M. V. 2002. The breeding biology of lemon sharks at a tropical nursery lagoon. Proceedings of the Royal Society of London Series B-Biological Sciences. 269: 1655-1661.

Franck, J. P. C., Harris, A. S., Bentzen, P., Denovan-Wright, M., and Wright, J. M. 1991. Organization and evolution of satellite, minisatellite and microsatellite DNAs in teleost fishes. In N. MacLean (ed.), *Oxford Surveys on Eukaryotic Genes*. Oxford University Press, Oxford.

Gaida, I. H. 1997. Population structure of the Pacific angel shark, *Squatina californica* (Squatiniformes : Squatinidae), around the California Channel Islands. Copeia 1997: 738-744.

Gardner, M. G., and Ward, R. D. 1998. Population structure of the Australian gummy shark (*Mustelus antarcticus* Gunther) inferred from allozymes, mitochondrial DNA and vertebrae counts. Marine and Freshwater Research. 49: 733-745.

Gardner, M. G., and Ward, R. D. 2002. Taxonomic affinities within Australian and New Zealand *Mustelus* sharks (Chondrichthyes: Triakidae) inferred from allozymes, mitochondrial DNA and precaudal vertebrae counts. Copeia 2002: 356-363.

Gerber, S., Mariette, S., Streiff, R., Bodenes, C., and Kremer, A. 2000. Comparison of microsatellites and amplified fragment length polymorphism markers for parentage analysis. Molecular Ecology. 9: 1037-1048.

Gladden, J. G. B., Ferguson, M. M., and Clayton, J. W. 1997. Matriarchal genetic population structure of North American beluga whales *Delphinapterus leucas* (Cetacea: Monodontidae). Molecular Ecology. 6: 1033-1046.

Goodnight, K. F., and Queller, D. C. 1999. Computer software for performing likelihood tests of pedigree relationship using genetic markers. Molecular Ecology. 8: 1231-1234.

Grunwald, C., Stabile, J., Waldman, J. R., Gross, R., and Wirgin, I. 2002. Population genetics of shortnose sturgeon *Acipenser brevirostrum* based on mitochondrial DNA control region sequences. Molecular Ecology. 11: 1885-1898.

Gyllensten, U., Wharton, D., Josefsson, A., and Wilson, A. C. 1991. Paternal Inheritance of Mitochondrial-DNA in Mice. Nature. 352: 255-257.

Hames, B. D., and Rickwood, D. 1990. Gel Electrophoresis of Proteins: A Pratical Approach. IRL Press, New York.

Hebert, P. D. N., and Beaton, M. J. 1989. Methodologes for allozyme analysis using cellulose acetate electrophoresis. Helena Laboratories, Beaumont Texas.

Hedrick, P. W. 1999. Highly variable loci and their interpretation in evolution and conservation. Evolution. 53: 313-318.

Heist, E. J. 1994. Population Genetics of Selected Species of Sharks, Pp. 88. In *Marine Science*. College of William and Mary, Gloucester Point, VA.

Heist, E. J. 1999. A review of population genetics in sharks, Pp. 161-168. In J. A. Musick (ed.), *Life in the Slow Lane*. American Fisheries Society Symposium 23, Bethesda, MD.

Heist, E. J. 2004. Genetics of Sharks, Skates, and Rays. In J. C. Carrier, J. A. Musick, and M. R. Heithaus (eds), *Biology of Sharks, Skates, and Rays*. CRC Press, Boca Raton, FL.

Heist, E. J., and Gold, J. R. 1999a. Genetic identification of sharks in the US Atlantic large coastal shark fishery. Fishery Bulletin. 97: 53-61.

Heist, E. J., and Gold, J. R. 1999b. Microsatellite DNA variation in sandbar sharks (*Carcharhinus plumbeus*) from the Gulf of Mexico and mid-Atlantic Bight. Copeia. 1999: 182-186.

Heist, E. J., Graves, J. E., and Musick, J. A. 1995. Population genetics of the sandbar shark (*Carcharhinus plumbeus*) in the Gulf of Mexico and Mid-Atlantic Bight. Copeia. 1995: 555-562.

Heist, E. J., Jenkot, J. L., Keeney, D. B., Lane, R. L., Moyer, G. R., Reading, B. J., and Smith, N. L. 2003. Isolation and characterization of polymorphic microsatellite loci in nurse shark (*Ginglymostoma cirratum*). Molecular Ecology Notes. 3: 59-61.

Heist, E. J., Musick, J. A., and Graves, J. E. 1996a. Genetic population structure of the shortfin mako (*Isurus oxyrinchus*) inferred from restriction fragment length polymorphism analysis of mitochondrial DNA. Canadian Journal of Fisheries & Aquatic Sciences. 53: 583-588.

Heist, E. J., Musick, J. A., and Graves, J. E. 1996b. Mitochondrial DNA diversity and divergence among sharpnose sharks, *Rhizoprionodon terraenovae*, from the Gulf of Mexico and Mid-Atlantic Bight. Fishery Bulletin. 94: 664-668.

Hoarau, G., Holla, S., Lescasse, R., Stam, W. T., and Olsen, J. L. 2002. Heteroplasmy and evidence for recombination in the mitochondrial control region of the flatfish *Platichthys flesus*. Molecular Biology and Evolution. 19: 2261-2264.

Hueter, R. E., Heupel, M. R., Heist, E. J., and Keeney, D. B. 2004. The implications of philopatry in sharks for the management of shark fisheries. Journal of Northwest Atlantic Fishery Science. in press.

Karl, S. A., Bowen, B. W., and Avise, J. C. 1992. Global population genetic-structure and male-mediated gene flow in the green turtle (*Chelonia mydas*) - RFLP analyses of anonymous nuclear loci. Genetics. 131: 163-173.

Keeney, D. B., and Heist, E. J. 2003. Characterization of microsatellite loci isolated from the blacktip shark and their utility in requiem and hammerhead sharks. Molecular Ecology Notes. 3: 501-504.

Keeney, D. B., Heupel, M. R., Hueter, R. E., and Heist, E. J. 2004. Genetic heterogeneity among blacktip shark, *Carcharhinus limbatus*, continental nurseries along the U.S. Atlantic and Gulf of Mexico. Marine Biology. 3: 1039-1046.

Kohler, N. E., Casey, J. G., and Turner, P. A. 1998. NMFS cooperative shark tagging program 1962-93: An atlas of shark tag and recapture data. Marine Fisheries Review. 60: 1-87.

Kvist, L., Martens, J., Nazarenko, A. A., and Orell, M. 2003. Paternal leakage of mitochondrial DNA in the great tit (*Parus major*). Molecular Biology and Evolution. 20: 243-247.

Lansman, R. A., Shade, R. O., Shapira, J. F., and Avise, J. C. 1981. The use of restriction endonucleases to measure mitochondrial DNA sequence relatedness in natural populations. Journal of Molecular Evolution. 17: 214-226.

Lavery, S., and Shaklee, J. B. 1989. Population genetics of two tropical sharks *Carcharhinus tilstoni* and *C. sorah*, in Northern Australia. Australian Journal of Marine and Freshwater Research. 40: 541-557.

Lavery, S., and Shaklee, J. B. 1991. Genetic evidence for separation of two sharks, *Carcharhinus limbatus* and *C. tilstoni*, from Northern Australia. Marine Biology. 108: 1-4.

Levinson, G., and Gutman, G. A. 1987. Slipped-strand mispairing: a major mechanism for DNA sequence evolution. Molecular Biology & Evolution. 4: 203-221.

Ludwig, A., May, B., Debus, L., and Jenneckens, I. 2000. Heteroplasmy in the mtDNA control region of sturgeon (*Acipenser*, *Huso* and *Scaphirhynchus*). Genetics. 156: 1933-1947.

Lyrholm, T., Leimar, O., Johanneson, B., and Gyllensten, U. 1999. Sex-biased dispersal in sperm whales: contrasting mitochondrial and nuclear genetic structure of global populations. Proceedings of the Royal Society of London Series β-Biological Sciences. 266: 347-354.

Macdonald, C. M. 1988. Genetic variation, breeding structure and taxonomic status of the gummy shark *Mustelus antarcticus* in southern Australian waters. Australian Journal of Marine and Freshwater Research. 39: 641-648.

Macdonald, D. B., and Potts, W. K. 1997. DNA microsatellites as genetic markers at several scales. In D. P. Mindell (ed.), *Avian molecular evolution and systematics*. Academic Press, Sand Diego.

Magoulas, A., and Zouros, E. 1993. Restriction-Site Heteroplasmy in Anchovy (*Engraulis encrasicolus*) Indicates Incidental Biparental Inheritance of Mitochondrial-DNA. Molecular Biology and Evolution. 10: 319-325.

Martin, A. P., Naylor, G. J. P., and Palumbi, S. R. 1992. Rates of mitochondrial DNA evolution in sharks are slow compared with mammals. Nature. 357: 153-155.

May, B. 2003. Allozymes. In E. M. Hallerman (ed.), *Population Genetics: Principles and Applications for Fisheries Scientists*. American Fisheries Society, Bethesda, MD.

Meyer, A. 1993. Evolution of mitochondrial DNA in Fishes, Pp. 1-38. In P. W. Hochachka and T. P. Mommsen (eds), *Biochemistry and Molecular Biology of Fishes*. Vol. 2. Elsevier, New York.

Mullis, K. B., and Faloona, F. A. 1987. Specific synthesis of DNA in vitro via a polymerase-catalyzed chain reaction. Methods in Enzymology. 155: 335-350.

Murphy, R. W., Sites, J. M. J., Buth, D. G., and Haufler, C. H. 1996. Proteins: Isozyme Electrophoresis, Pp. 51-120. In D. M. Hillis, C. Moritz, and B. K. Mable (eds), *Molecular Systematics*. Sinauer Associates, Inc., Sunderland, MA.

Nei, M. 1987. Molecular Evolutionary Genetics. Columbia University Press, New York.

O'connell, M., and Wright, J. M. 1997. Microsatellite DNA in fishes. Reviews in Fish Biology & Fisheries. 7: 331-363.

Ohta, Y., Okamura, K., Mckinney, E. C., Bartl, S., Hashimoto, K., and Flajnik, M. F. 2000. Primitive synteny of vertebrate major histocompatibility complex class I and class II genes. Proceedings of the National Academy of Sciences of the United States of America. 97: 4712-4717.

Pank, M., Stanhope, M., Natanson, L., Kohler, N., and Shivji, M. 2001. Rapid and simultaneous identification of body parts from the morphologically similar sharks *Carcharhinus obscurus* and *Carcharhinus plumbeus* (Carcharhinidae) using multiplex PCR. Marine Biotechnology. 3: 231-240.

Pardini, A. T., Jones, C. S., Noble, L. R., Kreiser, B., Malcolm, H., Bruce, B. D., Stevens, J. D., Cliff, G., Scholl, M. C., Francis, M., Duffy, C. A. J., and Martin, A. P. 2001. Sex-biased dispersal of great white sharks—In some respects, these sharks behave more like whales and dolphins than other fish. Nature. 412: 139-140.

Pardini, A. T., Jones, C. S., Scholl, M. C., and Noble, L. R. 2000. Isolation and characterization of dinucleotide microsatellite loci in the Great White Shark, *Carcharodon carcharias*. Molecular Ecology. 9: 1176-1178.

Pratt, H. L., and Carrier, J. C. 2001. A review of elasmobranch reproductive behavior with a case study on the nurse shark, *Ginglymostoma cirratum*. Environmental Biology of Fishes. 60: 157-188.

Ravago, R. G., Monje, V. D., and Juinio-Menez, M. A. 2002. Length and sequence variability in mitochondrial control region of the milkfish, *Chanos chanos*. Marine Biotechnology. 4: 40-50.

Rokas, A., Ladoukakis, E., and Zouros, E. 2003. Animal mitochondrial DNA recombination revisited. Trends in Ecology & Evolution. 18: 411-417.

Saiki, R. K., Gelfand, D. H., Stoffel, S., Scharf, S. J., Higuchi, R., Horn, G. T., Mullis, K. B., and Erlich, H. A. 1988. Primer-directed enzymatic amplification of DNA with a thermostable DNA polymerase. Science. 239: 487-491.

Saville, K. J., Lindley, A. M., Maries, E. G., Carrier, J. C., and Pratt, H. L. 2002. Multiple paternity in the nurse shark, *Ginglymostoma cirratum*. Environmental Biology of Fishes. 63: 347-351.

Schrey, A. W., and Heist, E. J. 2002. Microsatellite markers for the shortfin mako and cross-species amplification in Lamniformes. Conservation Genetics. 3: 459-461.

Schrey, A. W., and Heist, E. J. 2003. Microsatellite analysis of population structure in the shortfin mako (*Isurus oxyrinchus*). Canadian Journal of Fisheries & Aquatic Sciences. 60: 670-675.

Schwartz, M., and Vissing, J. 2002. Paternal inheritance of mitochondrial DNA. New England Journal of Medicine. 347: 576-580.

Shaw, P. W., Turan, C., Wright, J. M., O'connell, M., and Carvalho, G. R. 1999. Microsatellite DNA analysis of population structure in Atlantic herring (*Clupea harengus*), with direct comparison to allozyme and mtDNA RFLP analyses. Heredity. 83: 490-499.

Shivji, M., Clarke, S., Pank, M., Natanson, L., Kohler, N., and Stanhope, M. 2002. Genetic identification of pelagic shark body parts for conservation and trade monitoring. Conservation Biology. 16: 1036-1047.

Smith, P. J. 1986. Low Genetic Variation in Sharks (Chondrichthyes). Copeia. 1986: 202-207.

Smith, P. J., and Benson, P. G. 2001. Biochemical identification of shark fins and fillets from the coastal fisheries in New Zealand. Fishery Bulletin. 99: 351-355.

Sole-Cava, A. M., and Levy, J. A. 1987. Biochemical evidence for a third species of angel shark off the east coast of South America. Biochemical Systematics and Ecology. 15: 139-144.

Sole-Cava, A. M., Voreen, C. M., and Levy, J. A. 1983. Isozymic Differentiation of Two Sibling Species of Squatina (Chondrichthyes) in South Brazil. Comparative Biochemistry and Physiology. 75B: 355-358.

Tautz, D., Trick, M., and Dover, G. A. 1986. Cryptic simplicity in DNA is a major source of genetic variation. Nature. 322: 652-656.

Utter, F. M. 1991. Biochemical genetics and fishery management—An historical perspective. Journal of Fish Biology. 39: 1-20.

Waples, R. S. 1998. Separating the wheat from the chaff—Patterns of genetic differentiation in high gene flow species. Journal of Heredity. 89: 438-450.

Ward, R. D., and Grewe, P. M. 1994. Appraisal of molecular-genetic techniques in fisheries. Reviews in Fish Biology and Fisheries. 4: 300-325.

Ward, R. D., Woodwark, M., and Skibinski, D. O. F. 1994. A comparison of genetic diversity levels in marine, freshwater, and anadromous fishes. Journal of Fish Biology. 44: 213-232.

Wright, J. M. 1993. DNA Fingerprinting of Fishes, Pp. 58-91. In T. P. Mommsen and P. W. Hochachka (eds), *Molecular Biology Frontiers*. Vol. 2. Elsevier, New York.

Wright, S. 1969. Evolution and the Genetics of Populations. Volume 2, The Theory of Gene Frequencies. University of Chicago Press, Chicago. 511 pp.

CHAPTER 3

Reproductive Evolution of Chondrichthyans

John A. Musick and Julia K. Ellis

3.1 INTRODUCTION

3.1.1 Chondrichthyan Reproduction

The living Chondrichthyes are comprised of about 1100 species of neoselachian elasmobranchs (sharks and rays) and more than 30 species of holocephalans (chimaeras) (Compagno 1990, 2002). Although the number of living chondrichthyans is small relative to some other vertebrate groups, a diversity of reproductive modes has evolved (Wourms 1977). Wourms (1981) pointed out that these modes could be divided into two major categories based on fetal nutrition: lecithotrophy, where the entire development of the embryo is supported solely by the yolk; and matrotrophy, where at least part of the fetal development is augmented by additional maternal input of nutrients. In addition, chondrichthyan reproductive modes may be further divided by whether embryonic development is external to the mother's body (oviparity), or internal (viviparity).

3.1.2 Oviparity

Oviparity is obviously a lecithotrophic mode of reproduction. All chondrichthyan eggs deposited externally have leathery, structurally complex and remarkably durable shells (Hamlett and Koob 1999). Oviparity may be divided into two types: single (= external) oviparity and multiple (= retained) oviparity (Nakaya 1975; Compagno 1990). The former is the only type of reproduction in the Heterdontiformes and the batoid family Rajidae and occurs along with various forms of viviparity in the Orectolobiformes and the carcharhiniform family Scylorhinidae. In this type of oviparity one egg is

Virginia Institute of Marine Science, College of William and Mary, P.O. Box 1346, Gloucester Point, Virginia 23062 deposited at a time from each oviduct, usually in pairs; tens of eggs (but perhaps hundreds for a few species) may be deposited over the course of a spawning season. Multiple oviparity occurs only in a small number of scylorhinid species (and perhaps an orectolobiform) and entails the retention of a small number of eggs (usually = 10) in the oviduct during most of development before deposition and hatching on the seabed.

3.1.3 Yolk-sac Viviparity

Viviparity includes both lecithotrophic and a variety of matrotrophic modes of reproduction (Table 3.1) (Wourms 1977, 1981; Compagno 1990; Wourms and Lombardi 1992). Yolk-sac viviparity involves retention of fertilized eggs throughout development within the uterus with no additional maternal nutritional input beyond the yolk. This form of lecithotrophic reproduction is the most widespread among elasmobranchs and occurs in all living orders except the Heterodontiformes (which is oviparous) and the Lamniformes which has more advanced forms of viviparity (Compagno 1990). Yolk-sac viviparity was formerly called "ovoviviparity", a term widely used, understood, and accepted in the biological community. The term has been abandoned by most recent authors (Wourms 1977, 1981; Compagno 1990; Hamlett 1999) as suggested by Budker (1958) and Hoar (1969). Ranzi (1932, 1934) showed that although some "ovoviviparous" elasmobranchs, including some Torpediniformes and Squaliformes, lost 23-46 percent organic weight during gestation, another "ovoviviparous" squalid actually gained 1 percent and three triakids gained 11-369 percent. In true yolk-sac viviparity, substantial (= 20-25%) weight loss is expected because the organic material in the egg must provide not only material for structural development of the embryo but also for energetic costs of development (Chapter **13** of this volume). Thus, weight loss less than about 20% or weight gain during development would require some sort of matrotrophic contribution. In the cases cited above, this contribution appeared to be from a mucoid secretion or histotroph from the uterus that could be ingested or absorbed by the developing embryo. Thus, in some groups, the line between yolk-sac viviparity and limited histotrophy may be difficult to discern without data on the organic content of the eggs and term embryos. Consequently, the term "ovoviviparity" was abandoned and replaced by the unfortunate term "aplacental viviparity", which includes three major modes of elasmobranch reproduction: yolk-sac viviparity, histotrophy, and oophagy (see below). The term "aplacental viviparity" obfuscates the true diversity of elasmobranch reproduction and through implication elevates the importance of placental viviparity, which is restricted to a small number of families at the terminal nodes of the Carcharhiniformes. In addition, "aplacental viviparity" describes a mode by what it is not instead of what it is, and is uninformative. The term "aplacental viviparity" would best be abandoned, and the four modes of chondrichthyan viviparity recognized above should be used instead (Table 3.1).

Table 3.1 Chondrichthyan modes of reproduction.

		Lecithotrophic	*Matrotrophic*
Oviparity			
	Single	+	
	Multiple	+	
Viviparity			
	Yolk-sac	+	
	Limited Histotrophy		+
	Lipid Histotrophy		+
	Carcharhinid Oophagy		+
	Lamnid Oophagy		+
	Placental		+

3.1.4 Histotrophy

Histotrophy reaches its zenith in the batoid Myliobatiformes, which produce a protein- and lipid-rich histotroph from highly developed trophonemata. Embryos in this group unequivocally obtain matrotrophic nutrition and exhibit an increase in organic content of 1680-4900 percent (Needham 1942). Lipid histotrophy is clearly different from the limited "mucoid" histotrophy cited above and results in term embryos that may have gained one to two orders of magnitude more in mass than embryos of limited histotrophs. It is useful to recognize these modes separately (Table 3.1) in order to gain greater insights into the reproductive ecology and evolution of elasmobranchs.

3.1.5 Oophagy

Oophagy is a form of matrotrophic viviparity where, after initial yolk-sac nutrition, developing embryos ingest unfertilized eggs to support further development. Oophagy may result in very large (> 100 cm TL) neonates in some species (Hamlett and Koob 1999). Oophagy is the mode of reproduction in all members of the Lamniformes, and has evolved in one small family of carcharhiniform sharks, the Pseudotriakidae (Yano 1992, 1993). The mechanisms of oophagy are different in the two groups: the lamniforms, throughout most of their pregnancy, continuously produce unfertilized eggs which the developing embryos ingest and store in a large bulging yolk-stomach; the carcharhiniforms include a multitude of apparently unfertilized ova within the same egg envelope as the developing embryo, which then ingests this self-contained food source and stores it in the external yolk sac. Adelphophagy is a form of lamniform oophagy in which the largest developing embryo in each uterus consumes all the smaller embryos then relies on maternal production of unfertilized eggs for the duration of development. This reproductive mode is definitively known for only one species, *Carcharias taurus* (Gilmore *et al.* 1983; Gilmore 1991; Hamlett and Koob 1999).

3.1.6 Placental Viviparity

Placental viviparity has evolved only in five families of higher carcharhiniform sharks (Compagno 1988). In the vast majority of placental sharks, early development is supported by the yolk. The timing of placentation varies among species, occurring later in some than in others. In addition, limited histotrophy may function to support embryonic growth before and perhaps even after placentation (Hamlett 1989; Hamlett and Hysell 1998, Hamlett and Koob 1999; Chapter **15** of this volume.)

3.1.7 The Plesiomorphic Reproductive State

In virtually all previous analyses of the evolution of reproduction in modern elasmobranchs, oviparity has been assumed to represent the plesiomorphic state (Wourms 1977; Compagno 1990; Wourms and Lombardi 1992; Callard *et al.* 1995; Dulvy and Reynolds 1997). However, no empirical evidence has been offered to support this dogmatic assumption. Dulvy and Reynolds (1997) concluded from a cladistic analysis that oviparity was the plesiomorphic reproductive mode in modern elasmobranchs, but their use of the Holocephali as an outgroup in their phylogenetic analysis pre-ordained their conclusion. All the living holocephalans for which information is available are oviparous. However, the living holocephalans are a relic of a once diverse and dynamic group of Paleozoic chondrichthyans (Grogan 1993; Grogan and Lund 2000) with reproductive modes that included viviparity (Lund 1990). In addition, recently Grogan and Lund (2004) have argued that viviparity was the dominant mode of reproduction in most of the chondrichthyans (both elasmobranch and holocephalan) in the well-known Mississippian Bear Gulch deposit of Montana (USA). (This site includes a wide diversity of very well preserved chondrichthyan fossils and is one of the most intensely studied in the world.) Therefore, although the living holocephalans are oviparous, the Paleozoic chondrichthyans from which they evolved, and also the distant ancestors of the neoselachians, already included viviparous forms. The present paper examines the hypothesis that yolk-sac viviparity, not oviparity, is the plesiomorphic mode of reproduction in the Neoselachii and perhaps for the Chondrichthyes as a whole.

3.2 PHYLOGENETIC PATTERNS

3.2.1 Neoselachii

All living elasmobranchs are considered to be monophyletic and within the sub-class Neoselachii (Compagno 1977; Maisey *et al.* 2004). This group also includes a scattering of extinct but modern level fossils from the Mesozoic and perhaps a small number of Paleozoic forms (Maisey *et al.* 2004). The sister group of neoselachians are the hybodonts, which arose during the Paleozoic, radiated widely with a diversity of ecomorphotypes in the Mesozoic and became extinct in the Cretaceous (Maisey *et al.* 2004). Extant clades of neoselachians have historically been separated into two cohorts, batoids (Batoidea) and sharks (Selachii) (Bigelow and Schroeder 1948, 1953). However, morphological analyses during the 1990s suggested that the batoids were a terminal group among the squalean sharks, and they were included in the clade Hypnosqualea along with the Squatiniformes and Pristiophoriformes (Shirai 1992, 1996; Carvalho 1996). Recent molecular analyses including both nuclear and mitochondrial genes (Douady *et al.* 2003; Maisey *et al.* 2004; Chapter **1** of this volume) contradict this phylogeny and recognize the traditional arrangement where the batoids are the sister group of the sharks, which in turn are comprised of two major superorders, the Galeomorphii and Squalomorphii (Maisey *et al.* 2004; Chapter **1** of this volume) (Fig. 3.1). This arrangement is supported by the paleontological data which show that the batoids were already separated from the other neoselachians by the early Jurassic if not earlier (Thies 1983; Maisey *et al.* 2004). In this section,0 I will revisit the patterns of the major modes of reproduction among the Batoidea, the Squalamorphii, and the Galeomorphii using the most recent phylogenetic information for each group and including paleontological information.

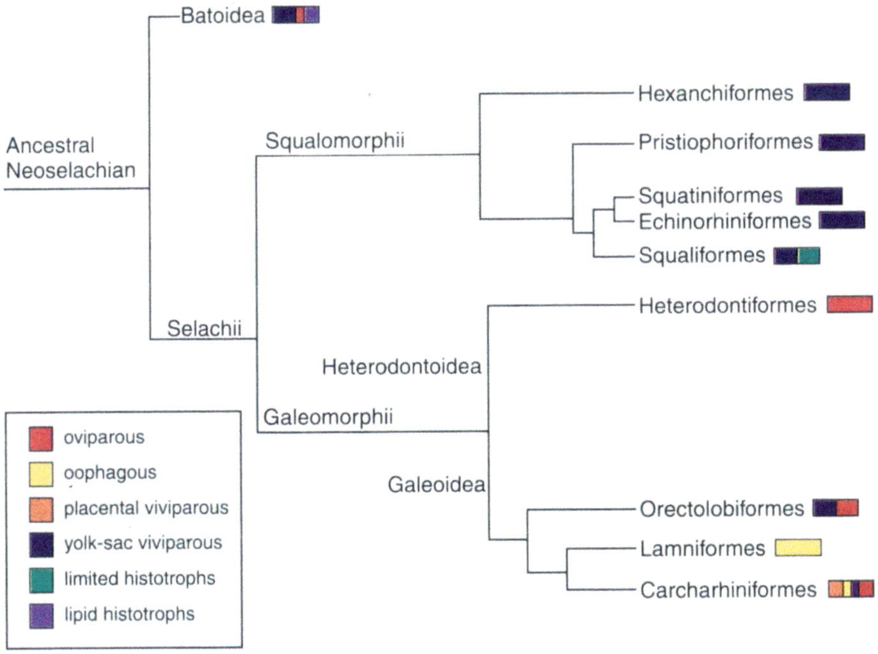

Fig. 3.1 Phylogeny of the elasmobranchs with reproductive modes. Modified after Musick *et al.* 2004.

3.2.2 Cohort Batoidea

The following discussion is based on the recent batoid phylogeny by McEachran and Aschliman (2004) who found that the Torpediniformes are basal to the rest of the living batoids followed by the Pristiformes (Fig. 3.2).

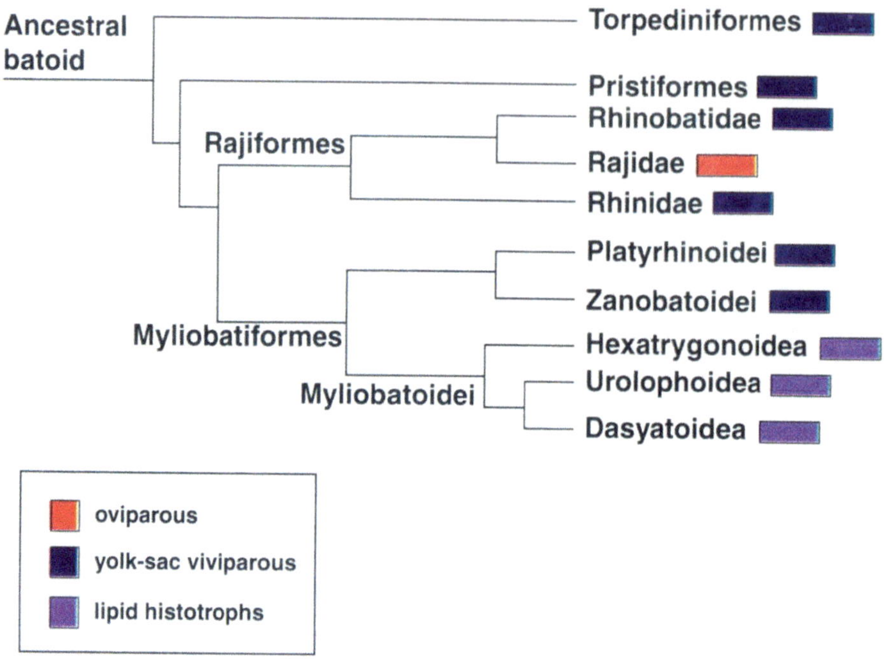

Fig. 3.2 Phylogeny of the Batoidea with reproductive modes. Modified after McEachran and Aschliman 2004.

The torpedoes exhibit yolk-sac viviparity (Ranzi 1932, 1934). The Pristiformes also exhibit yolk-sac viviparity (Thorson *et al.* 1983; Compagno 1990) and perhaps limited histotrophy. Observations by Setna and Sarangdhar (1949) of a "milky secretion" in the uterus of *Pristis cuspidatus* should not be misconstrued to mean that *Pristis* is histotrophic in the same way as the Myliobatiformes, which produce a histotroph rich in lipids. However, limited histotrophy, which involves production of mucoproteins in the uterus, is widespread among viviparous elasmobranchs (Chapter **13** of this volume) and may occur in the Pristiformes.

The next node in the batoid classification leads to two orders, one, the Rajiformes, with the Rajidae (skates) at its terminus and the other, the Myliobatiformes, with the Myliobatoidei (stingrays) as most derived (Fig. 3.2). Thus the depressed disc-shaped morphology in these two taxa evolved through separate ancestral taxa, the rhinobatoids and platyrhinids, respectively (McEachran and Aschliman 2004). Both of the latter two taxa had been placed formerly in the guitarfish order Rhinobatiformes (Compagno 1999), and both have yolk-sac viviparity as their mode of reproduction (Compagno 1990; Ebert 2003; Chapter **13** of this volume). The earliest known batoid fossils are rhinobatoids from the lower Jurassic (Cappetta *et al.* 1993). The Rajidae have single oviparity and deposit large numbers of leathery eggs. The Myliobatoidei produce a lipid-rich histotroph and bear a small number of large young (Hamlett and Koob 1999). The organic content of developing embryos in this group increases up to 4900% and is higher than that in most placental sharks (Chapter **13** of this volume). Lipid histotrophy is apparently limited to the Myliobatoidea, although limited histotrophy may be widespread among batoids as in other groups. All of the basal clades within the Batoidea, including the oldest, have yolk-sac viviparity, and the plesiomorphic reproductive mode in the cohort Batoidea is unequivocally yolk-sac viviparity.

3.2.3 Superorder Squalomorphii

The Squalomorphii comprise five extant orders (Fig. 3.1): the Hexanchiformes, Pristiophoriformes, Squatiniformes, Echinorhiniformes, and Squaliformes. All of these orders, except the Squaliformes, are depauperate with few lower taxa. The Hexanchiformes is basal and also is the oldest order dating at least back to the lower Jurassic. All squalomorphs exhibit yolk-sac viviparity with limited histotrophy present in many species, particularly among the Squaliformes. Oviparity is unknown in this entire superorder and yolk-sac viviparity is obviously the plesiomorphic reproductive mode.

3.2.4 Superoder Galeomorphii

The galeomorphs are a morphologically diverse group of sharks that consists of four extant orders: Heterodontiformes, Orectolobiformes, Lamniformes and Carcharhiniformes. The Heterodontiformes had been placed close to hybodont sharks by early workers (Smith 1942), but both recent morphological (Maisey 1984; de Carvalho 1996; Shirai 1996) and molecular (Maisey *et al.* 2004; Chapter 1 of this volume) evidence agree that the heterodontiforms are most closely allied with the galeomorphs, if distantly. The separation between the superorder Heterodontoidea and the Galeoidea—which comprises the Orectolobiformes, Lamniformes and Carcharhiniformes (de Carvalho 1996)—dates back to the lower Jurassic at least (Cappetta *et al.* 1993). The extant Heterodontiformes are a very small group of small benthic species all of which are oviparous (Compagno 2001).

The Orectolobiformes are basal to the Galeoidea (Fig. 3.1). Recent molecular (Maisey *et al.* 2004) and morphological (Goto 2001) cladistic analyses of the orectolobiforms concur (Fig. 3.3) and suggest that the order may be subdivided into two suborders, the Parascylloidei and Orectoloboidei. The parascylloids include only one family of small benthic oviparous sharks (Compagno 2001). The Orectoloboidei includes two superfamilies, the Orectoloboideia and Ginglymostoidea. The superfamily Orectoloboidea contains two families, the Orectolobidae and Brachaeluridae, both of which have a form of yolk-sac viviparity (Compagno 2001). The Ginglymostoidea includes the Hemiscylliidae, a group of small benthic oviparous sharks, and a second clade, including the Ginglymostomidae, Rhincodontidae and Stegostomatidae (Compagno 2001). All of the ginglymostomids and *Rhincodon* have yolk-sac viviparity, whereas *Stegostoma fasciatus* is a large oviparous species (Compagno 2001, 2002). The oldest fossil orectolobiforms are within

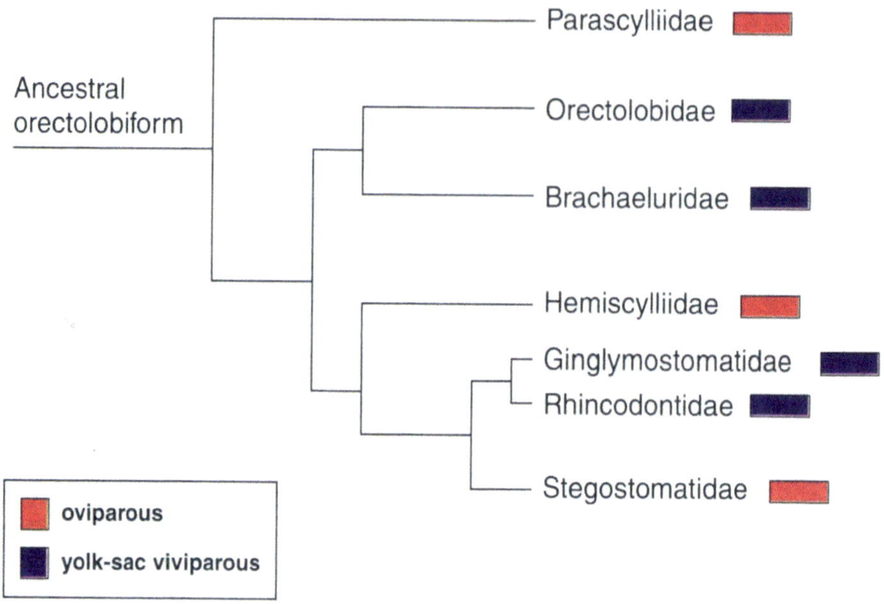

Fig. 3.3 Phylogeny of the Orectolobiformes with reproductive modes. Modified after Compagno 1988 and Goto 2001.

the Brachaeluridae (lower Jurassic, 180 mya), and Orectolobidae (middle Jurassic, 160 mya) (Cappetta *et al.* 1992), families with yolk-sac viviparity. The oviparous parascylliids and hemiscylliids did not appear until the middle Cretaceous (125 mya), although cladistic analysis suggests the parascylliids may be older. The oldest Orectolobiformes were contem-poraneous with the oldest Heterodontiformes.

The relationships of the remaining two orders of galeoid sharks, the Lamniformes and the Carcharhiniformes, have been debated for many years. White (1937) considered the Lamniformes to be more closely related to the Orectolobiformes than the Carcharhiniformes, and Applegate (1974) believed both Lamniformes and Carcharhiniformes were derived from Orectolobiformes. More recent studies, both morphological and molecular, recognize the Lamniformes and Carcharhiniformes to be sister groups (Maisey 1984; de Carvalho 1996; Shirai 1996; Maisey *et al.* 2004).

All of the Lamniformes for which reproductive modes are known are viviparous with oophagy. Most recent classifications place *Mitsukurina* and *Carcharias* as the two most primitive clades within the order (Shirai 1996; Martin and Naylor 1997). Nothing is known about reproduction in *Mitsukurina*, but *Carcharias taurus* appears to be unique among elasmobranchs in that it exhibits adelphophagy (see above) (Gilmore *et al.* 1983; Gilmore 1991; Chapter **14** of this volume). Adelphophagy results in two very large (= 100 cm) neonates and represents the extreme in the alternative reproductive strategy of investing in large young with high survivorship (versus a large number of small young with low survivorship) (Stearns 1992; Cortés 2004).

It is unclear whether adelphophagy is a plesiomorphic stage in the evolution of pure oophagy or an autapomorphic condition confined to *C. taurus*. Information on reproduction of *Mitsukurina* should shed light on this question. In oophagous species the initial stages of embryonic development are supported solely by the yolk-sac, and oophagy most probably evolved from simple yolk-sac viviparity.

Compagno (1988) divided the Carcharhiniformes into two suborders: the Scyliorhinoidei, containing the families Scyliorhinidae, Proscylliidae and Pseudotriakidae and the Carcharhinoidei, including the Leptochariidae, Triakidae, Hemigaleidae and Carcharhinidae (here including the Sphyrnidae). Recent molecular analysis (Maisey *et al.* 2004) placed the Pseudotriakidae closer to the Carcharhinidae (Fig. 3.4) but did not include any proscylliids in the study. (They used *Gollum*, previously classified as a proscylliid but now included in the Pseudotriakidae (Compagno 1999)). The Scyliorhinidae have been considered to be the most primitive carcharhiniforms (White 1937) because of their posteriorly placed dorsal fins and reduced vertebral calcifications. However, posterior dorsals are typical of benthic morphotypes (Compagno 1988, 1990), and reduced calcification is widespread among several orders of elasmobranchs which are found primarily in bathyal habitats (as are most scyliorhinids) (Compagno 1984). Compagno (1988) concluded "if lamnoids are the immediate sister group of carcharhinoids [as recent studies have concluded]... the proscylliid or even triakid habitus with the first dorsal forwards might be primitive for carcharhinoids and scyliorhinoids derived...."

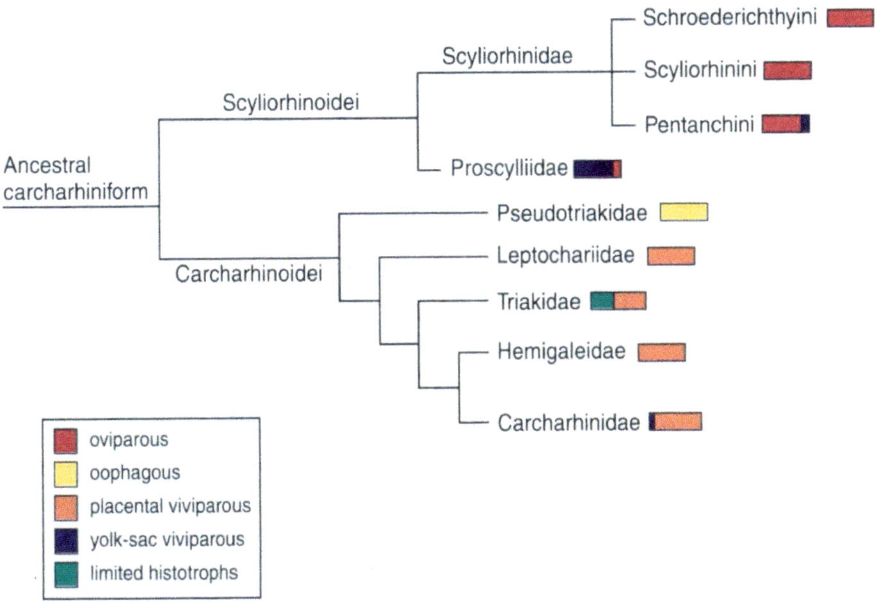

Fig. 3.4 Phylogeny of the Carcharhiniformes with reproductive modes. Modified after Compagno 1988 and Maisey *et al.* 2004.

The earliest proposed scyliorhinid fossil is *Macrourogaleus hassei* from the upper Jurassic of Europe (Cappetta *et al.* 1992). However, this specimen is in very poor condition, lacks its dentition, and consists of only a vague impression of its body outline (Cappetta 1987). The next earliest fossil scyliorhinid is *Scyliorhinus destombedii* from the lower Cretaceous of northern France (Cappetta 1987). No fewer than 18 species of *Scyliorhinus* have been recognized from early Cretaceous to Eocene deposits, most based on teeth. Cappetta (1987) contended that nearly all scyliorhinid fossils have been placed in the genus *Scyliorhinus* out of ignorance of the dentition of modern genera, and that "undoubtedly several fossil genera exist." He continued to note that "the genus *Scyliorhinus* as used by paleontologists is heterogeneous." Compagno (1988) suggested that some of the early fossil "scyliorhinids" may actually be proscylliids (which have similar dentition). Regardless, an upper Jurassic origin for the Carcharhiniformes (with the appearance of the Scyliorhinoidea) fits well with their phylogenetic position as sister group to the Lamniformes. The oldest lamniform fossil appears to be *Paleocarcharias* from the upper Jurassic of Europe (Duffin 1988).

Following Compagno's (1988) conclusions that the proscylliids are the primitive sister group of the scyliorhinids and thus the most primitive of living carcharhiniforms (Fig. 3.4), their modes of reproduction may provide particular insight into the plesiomorphic state in the order. Of the three genera of proscylliids, *Eridacnis* and *Ctenacis* both have yolk-sac viviparity, whereas *Proscyllium* is oviparous. Compagno (1988) pointed out that *Eridacnis* and *Ctenacis* were more closely related to each other than either was to *Proscyllium*, and that of the three genera, *Proscyllium* was the closest to the Scyliorhinidae, particularly the genus *Schroederichthys*. Given *Proscyllium*'s position close to the Scyliorhinidae, the characters which ally it to *Ctenacis* and *Eridacnis* should be more closely examined to determine whether they are principally plesiomorphic. If so, *Proscyllium* should be allied with the Scyliorhinidae as its most primitive member, thus clearly defining yolk-sac viviparity in the *Ctenacis-Eridacnis* clade as plesiomorphic relative to the rest of the Carcharhiniformes (Fig. 3.5). Regardless, oviparity in the Scyliorhinidae is derived. Yolk-sac viviparity is indicated as the plesiomorphic state in carcharhiniforms not only by its presence in *Ctenacis* and *Eridacnis*, but also by the sister group relationship between Carcharhiniformes and Lamniformes in which the plesiomorphic state is unambiguously yolk-sac viviparity.

All of the scyliorhinids are small benthic sharks and most have single oviparity (Compagno 1988). Multiple oviparity is present in the five species of *Halaelurus* (Nakaya 1975; Compagno 1988; Francis pers. comm.). However, in the closely related genus *Bythaelurus*, species are either single oviparous or yolk-sac viviparous with only two young (Compagno 1988; Francis pers. comm.). The appearance of yolk-sac viviparity in a group with single oviparity (Compagno 1988) contradicts the suggestion that yolk-sac viviparity has evolved from single oviparity through an intermediate stage of multiple oviparity (Nakaya 1975; Wourms *et al.* 1988; Compagno 1990). Multiple oviparity has also evolved in *Galeus melastomus*. The genus *Galeus* also includes

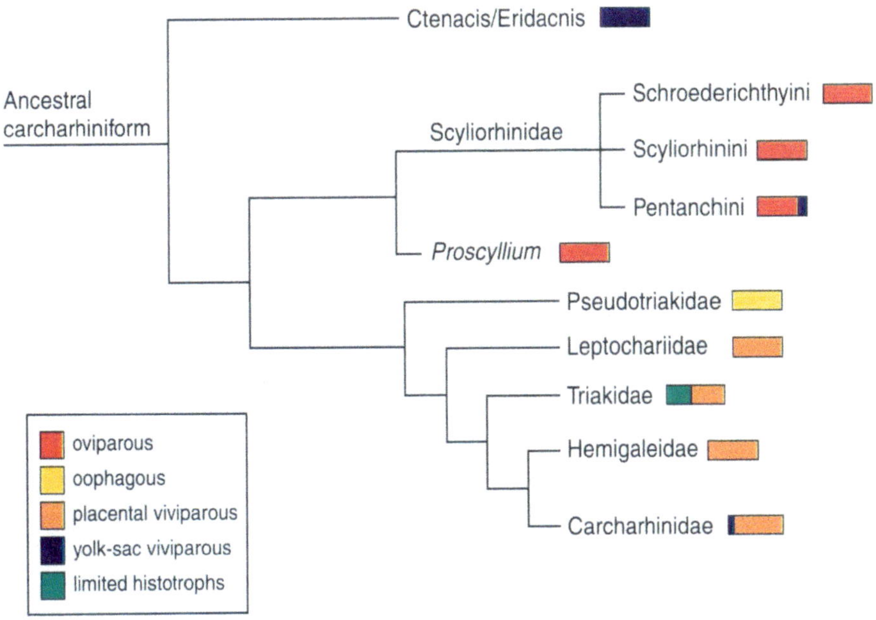

Fig. 3.5 Alternate phylogeny of the Carcharhiniformes with reproductive modes.

five apparently single oviparous species and two species with yolk-sac viviparity. All of the genera that have yolk-sac viviparity (*Bythaelurus*, *Galeus*, and *Cephalurus*) are closely related and within the subtribe Galeini (Compagno 1988). Apparently the mode of reproduction in the subtribe has remained evolutionarily labile. Springer (1979) suggested within the subspecies of *Galeus arae*, *G. arae arae* was yolk-sac viviparous, whereas *Galeus arae antillensis* was oviparous. Although the interrelationships among the sub-families of scyliorhinids are unresolved, the appearance of yolk-sac viviparity among the Galeini probably represents an evolutionary reversal in an oviparous family (Scyliorhinidae) that is an offshoot from the main line of carcharhiniform evolution.

The Pseudotriakidae fall somewhere between the Scyliorhinoidei and higher Carcharhinoidea (Compagno 1988; Maisey *et al*. 2004). The two genera in this family, *Gollum* and *Pseudotriakis*, exhibit a unique form of oophagy quite different from that in the Lamniformes (see above). As in other viviparous groups, early development of embryos is supported by the yolk sac and oophagy commences as development proceeds.

The Leptochariidae appears to be an ancient carcharhiniform clade (Compagno 1988), and has been classified as the sister group to a clade that includes the Triakidae and the Hemigaleidae and Carcharhinidae (including Sphyrnidae) together (Fig. 3.5). Alternatively, it might also be placed within the Triakidae as the sister group to all other triakids (Compagno 1988). Placental viviparity first appears in the Leptochariidae, is present along with limited histotrophy in the triakids and is found in all hemigaleids and carcharhinids (except *Galeocerdo cuvier*, a primitive carcharhinid which is yolk-sac viviparous). The absence of placental viviparity in some triakids may represent a loss and evolutionary reversal, or Leptochariidae may have evolved its unique globular placenta (Compagno 1988) independently. Triakids without a placenta still retain the uterine compartments and persistent egg envelope (Storrie 2004) that are the hallmarks of all placental species. In addition, non-placental triakids produce a copious mucoid histotroph and may exhibit embryonic mass increases that approach those of some placental species (Needham 1942; Hamlett and Koob 1999; Storrie 2004).

3.3 MORPHOLOGICAL AND PHYSIOLOGICAL PATTERNS

3.3.1 Oviducal Gland

The ovidical gland (= nidimental or shell gland) is a complex structure located just below the anterior oviduct and above the uterus in virtually all living chondrichthyans (Hamlett *et al.* 1998; Hamlett and Koob 1999). Fertilization takes place in the oviducal gland or just anterior to it. Histologically, four distinct zones can be discerned within this structure, a proximal club zone, papillary zone, baffle zone, and terminal zone (Hamlett *et al.* 1998; Chapter 10 of this volume). The club and papillary zones produce the various jelly coats that surround and protect the egg and developing embryos (Koob and Straus 1998). The baffle zone forms the egg envelope, capsule or membrane that encloses egg and jelly. In oviparous species, the baffle zone produces the leathery shell. The terminal zone is where sperm storage may occur in many species. Although the basic four zone structure of oviducal glands seems to be nearly universal, among most elasmobranch reproductive modes, the size of the gland is considerably larger in oviparous species (Hamlett and Koob 1999) and more elaborate at least in the Heterodontiformes (Hamlett pers. comm.). Hamlett *et al.* (1998) have characterized the oviducal gland of oviparous species as "specialized".

3.3.2 Uterus

The uterus in all elasmobranchs is a complex structure that may provide many roles in protecting and supporting the developing embryos, most importantly structural accommodation of the eggs and embryos, supplying oxygen to the uterine lumen and biosynthesis and secretion of structural or nutritional materials (Hamlett and Koob 1998). In oviparous species the uterus harbors the egg capsule during capsule sclerotization and thereafter until oviposition (up to several days) (Hamlett and Hysell 1998). Regardless of earlier characterization of the oviparous uterus as a simple conduit to the outside (Wourms *et al.* 1988), it is very sophisticated (Koob and Hamlett 1998) with structural specializations. In rajids the uterus has vascularized longitudinal folds lined with cilia and microvilli and with branched tubular glands. In scyliorhinids the intrauterine mucosa is folded, vascularized, and highly secretory in structure (Otake 1990). The oviparous uterus contributes to capsule surface structure and chemistry, and may facilitate biochemical processes associated with capsule polymerization, including provision of oxygen and absorption of water (Koob and Hamlett 1998; Hamlett and Koob 1999).

In yolk-sac viviparous species, the uterus specializes in regulating the intrauterine milieu, including supplying oxygen, water, and minerals (but not organic material) for the developing embryo, and regulating wastes (Hamlett and Koob 1999). The uterine wall in yolk-sac viviparous species is vascularized and folded with non-secretory villi. This arrangement with minor variations is similar in Squaliformes, Squatiniformes, Pristiophoriformes, primitive Rajiformes and primitive Carcharhiniformes (Ranzi 1932, 1934; Needham 1942; Compagno 1988; Otake 1990; Chapter **13** of this volume). Fine structure of the uterus in the Hexanchiformes has not been described but is probably similar to that in the Squaliformes. Limited histotrophy is a natural progression from yolk-sac viviparity and there is a thin line between the two (see above). This progression involves a proliferation of secretory cells that produce a nutritive mucous, and perhaps other organic substances, that may be ingested or absorbed by the developing embryo. Limited histotrophy has been reported in the Squaliformes, Rajiformes, and among the Carcharhiniformes in the families Pseudotriakidae (where limited histotrophy may support oophagy, (Yano 1992, 1993) and Triakidae. The uterus in the latter group is quite unlike that in the other taxa with limited histotrophy and has uterine compartments similar to those in the placental members of the family and in all other placental carcharhiniforms (Otake 1990). Uterine compartments isolate each embryo from its siblings and greatly increase the surface area available for metabolic exchange between the mother and fetus (Hamlett 1989). All placental species pass through a histotrophic stage after absorption of the yolk sac and before placental implantation (Hamlett and Koob 1999; Chapter **15** of this volume).

In the stingrays (Myliobatoidei), all of which have lipid histotrophy, the uterus develops large villous projections termed trophonemata (Hamlett *et al.* 1996a, b; Hamlett and Hysell 1998), which increase the surface area for histotrophic secretions and respiratory exchange. The oophagous lamniforms initially have a uterus with a smooth epithelium, but as embryos grow and require more oxygen, the uterus forms highly vascularized longitudinal folds. There is no provision for uterine secretion (Hamlett and Hysell 1998).

The evolutionary patterns of uterine structure suggest that species with yolk-sac viviparity and oophagy have the simplest condition with some folding and vascularization and minimal development of non-secretory villi. In the oviparous state, large secretory crypts are present along with cilia, both absent in the yolk-sac viviparous state. In the histotrophic species there is a progression in the development of secretory structures from modest development in limited histotrophs, culminating in the trophonemata found in the lipid histotrophs. In placental species, the development of uterine compartments was probably a necessary stage before placentation evolved.

3.3.3 Claspers

Claspers (mixopterygii) are paired, grooved extensions of the posterior base of the pelvic fins and are supported by an endoskeleton. They serve as intromittent organs to introduce sperm into the female's reproductive system thus facilitating internal fertilization (Compagno 1999a). The evolution of claspers has involved the coordinated development of the muscles required to pump sperm and to maneuver the claspers during copulation. Claspers are one of two principal synapomorphies which tie the Chondrichthyes together as a monophyletic group (Grogan and Lund 2004). All male chondrichthyans have claspers despite arguments to the contrary based on upper Devonian fossils of *Cladoselache*. Grogan and Lund (2004) have pointed out that these fossils were likely to be female. They base their conclusion on the well-known habit of extant elasmobranchs to be sexually segregated temporally and geographically. Also, other upper Devonian elasmobranchs such as *Diademodus* from the same deposit as *Cladoselache* had pelvic claspers, and all other male members of the cladodont group had claspers. Therefore, *Cladoselache* cannot be used as evidence that the plesiomorphic state within male chondrichthyans was unmodified pelvic fins (Dulvy and Reynolds 1997). Thus, claspers and internal fertilization probably have been defining features of all Chondrichthyes since the earliest evolution of the group. With internal fertilization comes the strong potential if not the probability of viviparity.

3.3.4 Urea Retention

All Chondrichthyes retain urea while in sea water so that they can be in approximate osmotic equilibrium with the environment and at the same time can maintain characteristic low vertebrate ion levels (Smith 1953). Ureosmotic regulation was thought to be unique to Chondrichthyes before its discovery in the living coelacanth, *Latimeria chalumnae* (Pickford and Grant 1967). Urea is mostly generated by the ornithine-urea cycle when used as a significant osmolyte and as the principal form for excreting nitrogenous waste (Griffith 1991). A complete ornithine-urea cycle has been shown in representatives of all gnathostome classes except birds in which it has been lost. As an osmotic regulator urea retention has now been confirmed not only in elasmobranchs and coelacanths but also in some marine adapted amphibians and reptiles and in some other marine and freshwater fishes (Griffith 1991). Extrapolating from living fishes to the Devonian and before, by which time the ureogenic elasmobranchs, coelocanths and other major vertebrate groups had diverged, Griffith (1991) proposed a hypothesis for the evolution of ureosmotic regulation:

1. A functional ornithine-urea cycle was absent in early agnathans (as with extant agnathans), but all component enzymes were present.
2. A complete ornithine-urea cycle evolved in early gnathostomes as a means for detoxifying ammonia during early embryogenesis.

Depeche *et al*. (1979) found high levels of urea in the developing embryos of the viviparous teleost, *Poecilia reticulata* (the guppy). Griffith (1991) concluded that urea synthesis was important in internal embryonic development where there was a restricted opportunity to exchange ammonia with the environment combined with high protein catabolism (of ovovitelline from the yolk). Following this logic, we would suggest that urea retention in early chondrichthyans evolved along with internal fertilization and yolk-sac viviparity initially as an embryonic adaptation to avoid ammonia toxicity. Urea retention into the adult stage would involve simple paedomorphosis (Griffith, 1991) and would allow early chondrichthyans to osmoregulate more efficiently in the marine environment, thus increasing their ability to occupy a broad diversity of niches.

3.4 EVOLUTIONARY IMPLICATIONS

3.4.1 Oviparity

Single oviparity has evolved in taxonomic groups whose members are mostly of small body size (< 100 cm TL) (Callard *et al.* 1995) and therefore would

Table 3.2 Available data on shark fecundity for species of ≤100 cm TL with single oviparous and viviparous modes of reproduction.

Species	*Size*	*Fecundity*	*References*
Single Oviparous	Total Length (cm)	Eggs/Year	
Hemiscyllium ocellatum	100	22	Bennett and Kyne 2003
Scyliorhinus canicula	100	29-190	Mellinger.1983; Compagno 1984; Capape *et al.* 1991; Ellis and Shackley 1997
Scyliorhinus retifer	50	44-53	Castro *et al.* 1988
Average Fecundity		60.0	
Viviparous		Litter Size	
Aculeola nigra	60	≧3	Compagno 1984
Centrophorus moluccensis	98	2	Compagno 1984
Centrophorus uyato	100	1	Compagno 1984
Centroscymnus crepidater	90	4-8	Cox and Francis 1997
Deania profundurum	76	5-7	Compagno *et al.* 1989
Etmopterus brachyurus	22.7	2	Compagno *et al.* 1989
Etmopterus granulosus	38	10-13	Last and Stevens 1994
Etmopterus hillanus	50	4-5	Compagno 1984
Euprotomicrus bispinatus	27	8	Compagno 1984
Isistius brasilliensis	50	6-12	Ebert 2003
Oxynotus bruniensis	72	7	Compagno 1984
Squalus blainville	95	3-4	Compagno 1984
Squalus japonicus	95	4.08	Chen *et al.* 1981
Squalus megalops	71	2-4	Compagno 1984; Last and Stevens 1994
Squalus rancureli	77	3	Compagno 1984
Average Fecundity		4.6	

Table 3.3 Available data on batoid fecundity for species with single oviparous and viviparous modes of reproduction. Parentheses indicate average fecundity.

Species	*Size*	*Fecundity*	*References*
Oviparous	Total Length (cm)	Eggs/Year	
Amblyraja radiata	102	2-88	del Rio Iglesias 2001
Dipturus batis	250	40	du Buit 1977; Walker and Hislop 1998
Leucoraja erinacea	54	30	Johnson 1979
Leucoraja naevus	70	90	du Buit 1976
Raja asterias	70	34-112	Capape 1977
Raja brachyura	120	40-90	Holden *et al.* 1971; Walker and Hislop 1998
Raja clavata	90	60-140	Holden 1975; Ryland and Ajayi 1984
Raja eglanteria	79	60	Luer and Gilbert 1985
Raja miraletus	60	32-90	Abd El Aziz *et al.* 1987
Raja montagui	80	25-60	Holden *et al.* 1971
Raja polystigma	53	20-62	Capape 1978
Average Fecundity		58.9	
Viviparous (histotroph)	Disc Width (cm)	Litter Size	
Dasyatis americana	200	2-10 (4.2)	Henningsen 2000
Dasyatis centroura	220	2-6	Capape 1993
Dasyatis dipterura	88	1-4	Ebert 2003
Dasyatis longus	156	1-3	Villavicencio Garayzar *et al.* 1994
Dasyatis marmorata	440	2-4	Capape and Zaouali 1995
Dasyatis pastinaca	60	6	Capape 1983
Dasyatis sabina	37	1-4 (2.6)	Snelson *et al.* 1988
Dasyatis sayi	73	1-6	Snelson *et al.* 1987
Dasyatis tortonesei	80	(4)	Capape 1978
Potamotrygon circularis	59.5	4-11 (5.8)	Thorson *et al.* 1983
Potamotrygon motoro	46	(6.3)	Thorson *et al.* 1983
Pteroplatytrygon violacea	80	4-13	Ebert 2003
Average Fecundity		4.4	
Viviparous (yolk-sac)	Total Length (cm)	Litter Size	
Platyrhinoides triseriata	91	1-15	Ebert 2003
Rhinobatos cemiculus	230	5-12 (7.5)	Capape and Zaouali 1994
Rhinobatos granulatus	280 SL	3-5	Prasad 1951
Rhinobatos horkelii	130	4-12	Lessa *et al.* 1986
Rhinobatos hynnicephalus	44	2-9 (4.6)	Wenbin and Shuyan 1993
Rhinobatos lentiginosus	75	6	Bigelow and Schroeder 1953
Rhinobatos productus	170	6-28 (9-11)	Ebert 2003
Rhinobatos rhinobatos	162	6-8	Capape *et al.* 1997
Rhyncobatus djiddensis	310	3-5	Prasad 1951; Compagno *et al.* 1989
Zapteryx exasperata	97	4-11	Ebert 2003
Average Fecundity		6.7	

have very limited fecundity if viviparous. Therefore, oviparity appears to be an adaptation in small species to increase fecundity (Holden 1973) contrary to the assertion of Wourms and Lombardi (1992). They claimed that brood sizes were similar in oviparous and viviparous species and attempted to prove their point by comparing the fecundity in *Prionace glauca* and *Hexanchus griseus*, two very large (> 300 cm TL) viviparous species, with scyliorhinids and rajids, most of which are small (< 100 cm TL) (Musick *et al.* 2004; Appendix 3.1). When small oviparous species are compared to small viviparous species, the differences are striking, with fecundity in oviparous forms averaging at least an order of magnitude higher than that in viviparous forms (Tables 3.2 and 3.3). The average annual fecundity in the scyliorhinids is 60.0 (eggs/year) compared to 4.6 (pups/year) in small squaliforms (Table 3.2), and the average fecundity in the rajids is 58.9 (eggs/year) compared to 5.5 (pups/year) in the myliobatiforms and rhinobatiforms (Table 3.3).

The disk-shaped batoid morphology appears to closely restrict the coelomic space and thus further restrict uterine capacity. The average *annual* fecundity for some species in the viviparous groups may be even smaller because they may not breed every year (Dodd 1983). Another selective advantage accrues to small species of oviparous sharks and rays through "bet hedging" (Stearns 1992). Small individuals are subject to proportionately higher predation than larger individuals (Peterson and Wroblewski 1984; Chen and Watanabe 1989; Cortés 2004), and if a pregnant viviparous shark is eaten, her evolutionary fitness equals zero. Species with simple oviparity avoid that problem, and even with egg predation rates of 20-60 percent (Frisk *et al.* 2002), their evolutionary fitness may be insured. These predation rates on cleidoic elasmobranch eggs are far lower than on non-cleidoic Actinopterygian eggs (Winemiller and Rose 1993). Multiple oviparity, where a moderate number of eggs are retained in the mother's uterus for a substantial portion of the developmental period before deposition (Nakaya 1975), has probably evolved from single oviparity, where and when egg predation rates may be very high particularly during the early stages of development. Likewise, the reversal to yolk-sac viviparity in two species of small scyliorhinids of the genus *Bythaelurus* also may have been selected for because of high egg predation rates. The evolution of cleidoic oviparity among chondrichthyans may have appeared in some taxa as early as the Paleozoic, but the evidence is sparse (Grogan and Lund 2004).

3.4.2 Parsimony

Past studies of the evolution of reproductive modes in modern elasmobranchs have been predicated on oviparity as the plesiomorphic reproductive state (Wourms 1977; Wourms and Lombardi 1992; Dulvy and Reynolds 2002). Wourms and Lombardi (1992) estimated that viviparity evolved from oviparity 18-20 times. Dulvy and Reynolds' analysis suggested that there were 9 to 10 transitions from oviparity to viviparity, and two "reversals" back to oviparity in the Rajidae and the orectolobiform *Stegostoma*. In contrast, hypothesizing that yolk-sac viviparity is the plesiomorphic state in living elasmobranchs requires the evolution of oviparity once each in the Heterodontiformes, Rajidae, and Scyliorhinidae (including *Proscyllium*) and three times in the Orectolobiformes for a total of six transitions, and reversals to viviparity in some species of the Galeini among the oviparous scyliorhinids. Thus plesiomorphic yolk-sac viviparity is more parsimonious because it requires three to four fewer transitions and fewer reversals than in the alternate hypothesis (Table 3.4).

Table 3.4 Hypotheses of plesiomorphic and apomorphic elasmobranch reproductive states with numbers of transitions and reversals.

State				
Plesiomorphic?	*Apomorphic*	*Transitions*	*Reversals*	*Source*
Oviparity	Viviparity	18-20	No reversals cited	Wourms, 1977
Oviparity	Viviparity	9-10	2	Dulvy and Reynolds 2002
Viviparity	Oviparity	6	1	This paper

Yolk-sac viviparity is clearly the plesiomorphic state in all orders of Batoidea and all squalomorphs. The situation in the galeomorphs may be a bit more equivocal because the Heterodontiformes are an old oviparous group that is the sister group of the remainder of the galeomorphs. However, the fossil record shows that the oldest Heterodontiformes were concurrent with the oldest yolk-sac viviparous Orectolobiformes (Brachaeluridae). In addition, the sister group relationship between the viviparous batoids and the selachians, of which the viviparous squalomorph Hexanchiformes are the oldest clade, would dictate that the ancestral neoselachians also had yolk-sac viviparity. Limited histotrophy has evolved from yolk-sac viviparity in several lineages and might be expected in virtually all major taxa in which yolk-sac viviparity is found. Lipid histotrophy has evolved once in the myliobatoid stingrays. Likewise, placental viviparity has evolved once in the higher carcharhiniforms. There are two independently derived forms of oophagy, lamniform oophagy and carcharhiniform oophagy, each of which has evolved one time. Recognizing that yolk-sac viviparity is plesiomorphic simplifies the pattern of reproductive evolution in living elasmobranchs and provides a straightforward sequence leading to other modes of reproduction.

The Chondrichthyes appear to be the oldest gnathostome group (Miller *et al.* 2003; Kikugawa *et al.* 2004) and may have evolved from some thelodont agnath ancestor in the Silurian (Marss *et al.* 2002). Early gnathostome evolution was apparently rapid with divergence into chondrichthyan and placoderm, and osteichthyan clades. Chondrichthyans and placoderms (Miles 1967) apparently evolved intromittent organs and internal fertilization and viviparity early on probably in response to high egg predation by the newly evolved gnathostomes. Viviparity is widespread among invertebrate groups (Marshall *et al.* 2003), including the ascidians, a chordate group basal to the vertebrates (Young, 1950). Sarcopterygian reproductive evolution is equivocal, with the living Dipnoi having benthic non-cleidoic eggs and the living coelacanth having yolk-sac viviparity. However, in contrast to Chondrichthyes, actinopterygian reproduction evolved in another direction predicated on non-cleidoic eggs. Within that evolutionary trajectory several adaptations have evolved to decrease egg predation or to increase fitness in spite of predation. These include nest building and parental protection on one hand, and the production of very large numbers of small pelagic eggs on the other. It is significant that these adaptations never evolved in the Chondrichthyes, probably because they had already evolved a successful strategy (viviparity) to avoid egg predation.

3.5 ACKNOWLEDGMENTS

The authors wish to thank Eileen Grogan and Dick Lund for stimulating the ideas presented here as well as discussions with other colleagues, including Will Hamlett, José Castro, and Greg Cailliet. Thanks also to colleagues who assisted with references, including Ian Callard and Matthias Stehmann. This is VIMS contribution #2595 and a contribution from the National Shark Research Consortium.

3.6 REFERENCES

Abdel-Aziz, S. H., Ezzat, A. and Hussein, M. 1987. Sexuality, reproduction and fecundity of *Raja miraletus* (L) from the Mediterranean waters off Alexandria. Bulletin of the Institute of Oceanography and Fisheries, Cairo 13(1): 119-132.

Abdel-Aziz, S. H., Khalil, A. N. and Abdel-Maguid, S. A. 1993. Reproductive cycle of the common guitarfish, *Rhinobatos rhinobatos* (Linnaeus, 1758), in Alexandria waters, Mediterranean Sea. Australian Journal of Marine and Freshwater Research 44(3): 507-517.

Applegate, S.P. 1974. A revision of the higher taxa of Orectoloboids. Journal of the Marine Biological Association of India 14(2): 743-751.

Bennett, M. B. and Kyne, P. M. 2003. Epaulette shark, *Hemiscyllium ocellatum* (Bonaterre, 1788). Pp. 58. In R. D. Cavanagh, P. M. Kyne, S. L. Fowler, J. A. Musick and M. B. Bennett (eds), *The Conservation Status of Australasian Chondrichthyans: Report of the IUCN Shark Specialist Group Australia and Oceania Regional Red List Workshop*. University of Queensland, School of Biomedical Sciences, Brisbane, Australia.

Bigelow, H. B. Schroeder, W. C. 1948. Sharks. Pp. 59-546. In J. Tee-Van, C. M. Breder, A. E. Parr, W. C. Schroeder, and L. P. Schultz (eds), *Fishes of the Western North Atlantic*. Memoirs of the Sears Foundation for Marine Research. Yale University, New Haven, Connecticut.

Bigelow, H. B. and Schroeder, W. C. 1953. *Fishes of the Western North Atlantic, Part II, Sawfishes, Guitarfishes, Skates and Rays; Chimaeroids*. Memoirs of the Sears Foundation for Marine Research, Yale University, New Haven, Connecticut. 588 pp.

Budker, P. 1958. La viviparite chez les selaciens. Pp. 1755-1790. In P. Grasse (ed.), *Traite de Zoologie*. Masson, Paris.

Callard, I. P., Putz, O., Paolucci, M. and Koob, T. J. 1995. Elasmobranch reproductive life-histories: Endocrine correlates and evolution. Pp. 204-208. In F. Goetz and P. Thomas (eds), *Proceedings of the Fifth International Symposium on the Reproductive Physiology of Fish*, Austin, Texas. Fish Symposium 95.

Capape, C. 1976. Contribution to the biology of the Dasyatidae of the Tunisian coasts. I. *Dasyatis pastinaca* (Linne, 1758). Geographic and bathymetric distribution, sexuality, reproduction, fecundity. Annali del Museo Civico di Storia Naturale di Genova 81: 22-32.

Capape, C. 1977. Contribution to the biology of Tunisian coastal Rajidae. 4. *Raja asterias* Delaroche, 1809: Geographical and bathymetric distributions, sexuality, reproduction and fecundity. Bulletin du Musee d'Histoire Naturelle de Paris (3e Serie) (Zoologie.) 435: 305-326.

Capape, C. 1978. Contribution to the biology of the Dasyatidae off Tunisian coasts. 3. *Dasyatis tortonesei* Capape, 1975. Geographical and bathymetric repartition, sexuality, reproduction, fecundity. Bulletin de l'Institut National Scientifique et Technique d'Oceanographie et de Peche 5(1-4): 97-110.

Capape, C. 1993. New data on the reproductive biology of the thorny stingray, *Dasyatis centroura* (Pisces: Dasyatidae) from off the Tunisian coasts. Environmental Biology of Fishes 38(1-3): 73-80.

Capape, C., Ben Brahim, R., and Zaouali, J. 1997. Aspects of the reproductive biology of the common guitarfish (*Rhinobatos rhinobatos*) in Tunisian waters. Ichtyophysiologica Acta 20: 113-127.

Capape, C. and Zaouali, J. 1994. Distribution and reproductive biology of the blackchin guitarfish, *Rhinobatos cemiculus* (Pisces: Rhinobatidae), in Tunisian waters (central Mediterranean). Australian Journal of Marine and Freshwater Research 45(4): 551-561.

Capape, C. and Zaouali, J. 1995. Reproductive biology of the marbled stingray, *Dasyatis marmorata* (Steindachner, 1892) (Pisces: Dasyatidae) in Tunisian waters (Central Mediterranean). Journal of Aquariculture and Aquatic Sciences 7: 108-119.

Capape, C., Seck, A. A., Gueye-Ndiaye, A., Diatta, Y. and Diop, M. 2002. Reproductive biology of the smoothback angel shark, *Squatina oculata* (Elasmobranchii: Squatinidae), from the coast of Senegal (eastern tropical Atlantic). Journal of the Marine Biological Association of the United Kingdom 82: 635-640.

Capape, C., Tomasini, J. A. and Bouchereau, J. L. 1991. Observations sur la biologie de la reproduction de la petite roussette *Scyliorhinus canicula* (Linnaeus, 1758) (Pisces, Scyliorhinidae) du Golfe du Lion (France meridionale). Ichtyophsiologica Acta 14: 87-109.

Cappetta, H. 1987. Chondrichthyes II, Mesozoic and Cenozoic Elasmobranchii. Pp. 1-193. In H. P. Schultze (ed.), *Handbook of Paleoichthyology*, vol. 3B. Gustav Fischer Verlag, Stuttgart.

Cappetta, H. 1992. Carcharhiniformes nouveaux (Chondrichthyes, Neoselachii) de l'Yprésien du Bassin de Paris. Geobios 25(5): 639-646.

Cappetta, H., Duffin, C. J. and Zidek, J. 1993. Chondrichthyes. Pp. 593-609. In M. J. Benton (ed.), *The Fossil Record*. Chapman and Hall, London.

Carpenter, K. E., Krupp, F., Jones, D. A. and Zajonz, U. 1997. *The Living Marine Resources of Kuwait, Eastern Saudi Arabia, Bahrain, Qatar, and the United Arab Emirates*. FAO Species Identification Guide for Fishery Purposes. FAO, Rome. 293 pp.

Castro, J. I., Bubucis, P. M. and Overstrom, N. A. 1988. The reproductive biology of the chain dogfish, *Scyliorhinus retifer*. Copeia 3: 740-746.

Chen, C., Taniuchi, T. and Nose, Y. 1981. Some aspects of reproduction in the pointed-snout dogfish *Squalus japonicus* taken off Nagasaki and Choshi. Bulletin of the Japanese Society of Scientific Fisheries 47(9): 1157-1164.

Chen, S. and Watanabe, S. 1989. Age dependence of natural mortality coefficient in fish population dynamics. Nippon Suisan Gakkaishi/Bulletin of the Japanese Society of Scientific Fisheries 55(2): 205-208.

Chirichigno, N. and Cornejo, R. 2001. *Catálogo comentado de los peces marinos del Perú*. Publicación Especial. Instituto del Mar del Perú, Calllao. 314 pp.
Compagno, L. J. V. 1977. Phyletic relationships of living sharks and rays. American Zoology 17(2): 303-322.
Compagno, L. J. V. 1984. Sharks of the World: An Annotated and Illustrated Catalogue of Shark Species Known to Date, Part 2, Carcharhiniformes. FAO Fisheries Synopsis No. 125 4(2). FAO, Rome. 655 pp.
Compagno, L. J. V. 1988. *Sharks of the Order Carcharhiniformes*. Princeton University Press, Princeton, New Jersey. 572 pp.
Compagno, L. J. V. 1990. Alternate life history styles of cartilaginous fishes in time and space. Environmental Biology of Fishes 28: 33-75.
Compagno, L. J. V. 1999. Systematics and body form. Pp. 1-42. In W. C. Hamlett (ed.), *Sharks, Skates, and Rays: The Biology of Elasmobranch Fish*. Johns Hopkins University Press, Baltimore and London.
Compagno, L. J. V. 2001. Sharks of the World: An Annotated and Illustrated Catalogue of Shark Species Known to Date, Volume 2. Bullhead, mackerel and carpet sharks (Heterodontiformes, Lamniformes and Orectolobiformes). FAO Species Catalogue for Fishery Purposes 1(2). FAO, Rome. 269 pp.
Compagno, L. J. V., Ebert, D. A. and Smale, M. J. 1989. *Guide to Sharks and Rays of Southern Africa*. Struik Publishers, Cape Town, South Africa. 160 pp.
Cortés, E. 2004. Life-history patterns, demography, and population dynamics Ch. 15, Pp. 449-470. In J. Carrier, J. Musick and M. Heithaus (eds), *The Biology of Sharks and Their Relatives*. CRC Press, Boca Raton, FL.
Cox, G. and Francis, M. 1997. *Sharks and Rays of New Zealand*. Canterbury Univ. Press, Canterbury. 68 pp.
de Carvalho, M. R. and De Figueiredo, J. L. 1994. *Psammobatis extenta* (Garman, 1913): a senior synonym of *Psammobatis glansdissimilis* McEachran, 1983 (Chondrichthyes, Rajidae). Copeia 1994(4): 1029-1033.
de Carvalho, M. 1996. Higher-level elasmobranch phylogeny, basal squaleans, and paraphyly. Pp. 35-62. In M. Stiassny, L. Parenti and G. Johnson (eds), *Interrelationships of Fishes*. Academic Press, San Diego, London.
del Rio Iglesias, J. 2001. Some aspects of the thorny skate (*Raja radiata* Donovan, 1808) reproductive biology in NAFO Division 3N Regulatory Area. Northwest Atlantic Fisheries Organization Scientific Council Research Document 02/118: 13.
Deng, S.-M., Xiong, G.-Q. and Zhan, H.-X. 1983. Description of three new species of elasmobranchiate fishes from deep waters of the east China Sea. Oceanologia et Limnologia Sinica 14(1): 64-70.
Depeche, J., Gilles, R., Daufresne, S. and Chiapello, H. 1979. Urea content and urea production via the ornithine-urea cycle pathway during the ontogenic development of two teleost fishes. Comparative Biochemistry and Physiology 63A(1): 51-56.
Devadoss, P. 1998. Observations on the breeding and development in some batoid fishes. Indian Journal of Fisheries 45(3): 271-283.
Dodd, J. M. 1983. Reproduction in cartilaginous fishes (Chondrichthyes). Pp. 31-95. In W. S. Hoar, D. J. Randall and E. M. Donaldson (eds), *Fish Physiology*, Volume 9. Academic Press, New York.
Douady, C. J., Dosay, M., Shivji, M. S. and Stanhope, M. J. 2003. Molecular phylogenetic evidence refuting the hypothesis of Batoidea (rays and skates) as derived sharks. Molecular Phylogeny and Evolution 26: 215-221.
Du Buit, M. H. 1976. The ovarian cycle of the cuckoo ray, *Raja naevus* (Mueller and Henle), in the Celtic Sea. Journal of Fish Biology 8(3): 199-201.

Du Buit, M. H. 1977. Age et croissance de *Raja batis* et de *Raja naevus* en Mer Celtique. Journal du Conseil International pour l'Exploration de la Mer 37(3): 261-275.

Duffin, C. J. 1988. The upper Jurassic selachian *Palaeocarcharias* de Beaumont (1960). Zoological Journal of the Linnean Society 94(3): 271-286.

Dulvy, N. K.and Reynolds, J. D. 1997. Evolutionary transitions among egg-laying, live-bearing and maternal inputs in sharks and rays. Proceedings of the Royal Society of London, Series B: Biological Sciences. 264(1386): 1309-1315.

Dulvy, N.K. and Reynolds, J. D. 2002. Predicting extinction vulnerability in skates. Conservation Biology 16(2): 440-450.

Ebert, D. A. 2003. *Sharks, Rays, and Chimaeras of California*. University of California Press, Berkeley and Los Angeles. 284 pp.

Ellis, J. R. and Shackley, S.E. 1997. The reproductive biology of *Scyliorhinus canicula* in the Bristol Channel, U.K. Journal of Fish Biology 51: 361-372.

Frisk, M. G., Miller, T. J. and Fogarty, M. J. 2002. The population dynamics of little skate *Leucoraja erinacea*, winter skate *Leucoraja ocellata*, and barndoor skate *Dipturus laevis*: Predicting exploitation limits using matrix analyses. ICES Journal of Marine Science 59: 576-586.

Gilmore, R. G. 1991. The reproductive biology of lamnoid sharks. Underwater Naturalist 19: 64-67.

Gilmore, R. G., Dodrill, J. W. and Linley, P. A. 1983. Reproduction and embryonic development of the sand tiger shark, *Odontaspis taurus* (Rafinesque). Fishery Bulletin 81(2): 201-226.

Goto, T. 2001. Comparative anatomy, phylogeny and cladistic classification of the order Orectolobiformes (Chondrichthyes, Elasmobranchii). Memoirs of the Graduate School of Fisheries Sciences, Hokkaido University 48(1): 1-100.

Griffith, R. W. 1991. Guppies, toadfish, lungfish, coelacanths and frogs: A scenario for the evolution of urea retention in fishes. Pp. 199-218. In J. A. Musick, M. N. Bruton and E. K. Balon (eds), *The Biology of Latimeria chalumnae and the Evolution of Coelacanths*. Environmental Biology of Fishes (32)1-4, The Hague.

Grogan, E. D. 1993. The structure of the holocephalan head and the relationships of the Chondrichthyes. Ph.D. Dissertation, School of Marine Science, College of William and Mary, Williamsburg, Virginia.

Grogan, E. D. and Lund, R. 2000. *Debeerius ellefseni* (Fam. Nov., Gen. Nov., Spec. Nov.), an autodiastylic chondrichthyan from the Mississippian Bear Gulch Limestone of Montana (USA), the relationships of the chondrichthyes, and comments on gnathostome evolution. Journal of Morphology 243(3): 219-245.

Grogan, E. D. and Lund, R. 2004. Origin and relationships of early Chondrichthyes. Pp. 3-31. In J. C. Carrier, J. A. Musick, and M. R. Heithaus (eds), *Biology of Sharks and Their Relatives*. CRC Press, Boca Raton, Florida.

Grove, J. S. and Lavenberg, R. J. 1997. *The Fishes of the Galápagos Isands*. Stanford University Press, Stanford, California. 863 pp.

Hamlett, W. C. 1989. Evolution and morphogenesis of the placenta in sharks. Pp. 35-52. In W. C. Hamlett and B. Tota (eds), *Eighth International Symposium on Morphological Sciences*, Rome, Italy. Journal of Experimental Zoology, supplement 2: 35-52.

Hamlett, W. C., Musick, J.A., Eulitt, A. M., Jarrell, R. L. and Kelly, M. A. 1996a. Ultrastructure of uterine trophonemata, accommodation for Uterolactation, and gas exchange in the southern stingray, *Dasyatis americana*. Canadian Journal of Zoology 74: 1417-1430.

Hamlett, W. C., Musick, J. A., Eulitt, A. M., Jarrell, R. L. and Kelly, M. A. 1996b. Ultrastructure of fetal alimentary organs: Stomach and spiral intestine in the southern stingray, *Dasyatis americana*. Canadian Journal of Zoology 74: 1431-1443.

Hamlett, W. C., Knight, D. P., Koob, T. J., Jezior, M., Loung, T. Rozycki, T., Brunette, N. and Hysell, M. K. 1998. Survey of oviducal gland structure and function in elasmobranchs. Journal of Experimental Zoology 282: 399-420.

Hamlett, W. C. and Hysell, M. K. 1998. Uterine specializations in elasmobranchs. Journal of Experimental Zoology 282(4-5): 438-459.

Hamlett, W. C. and Koob, T. 1999. Female reproductive system. Pp. 398-443. In W. C. Hamlett (ed.), *Sharks, Skates and Rays: The Biology of Elasmobranchs Fishes*. Johns Hopkins University Press, Baltimore, Maryland.

Hatooka, K., Yamada, U. and Aizawa, M. 2002. Rajidae. Pp. 164-174. In T. Nakabo (ed.), *Fishes of Japan, English Edition, Volume 1*. Tokai University Press, Tokyo.

Henningsen, A. D. 2000. Notes on reproduction in the southern stingray, *Dasyatis americana* (Chondrichthyes: Dasyatidae), in a captive environment. Copeia 2000(3): 826-828.

Hoar, W. S. 1969. Reproduction. Pp. 1-72. In W. S. Hoar and D. J. Randall (eds), *Fish Physiology, Volume III, Reproduction and Growth: Bioluminescence, Pigments and Poisons*. Academic Press, New York and London.

Holden, M. J. 1973. Are long-term sustainable fisheries for elasmobranchs possible? Rapports et Proces-Verbaux des Reunions du Conseil International pour l'Exploration de la Mer 164: 360-367.

Holden, M. J. 1975. The fecundity of *Raja clavata* in British waters. Journal du Conseil International pour l'Exploration de la Mer 36(2): 110-118.

Holden, M. J., Rout, D. W. and Humphreys, C. N. 1971. The rate of egg laying by three species of ray. Journal du Conseil International pour l'Exploration de la Mer 33(3): 335-339.

Ishiyama, R. 1967. *Fauna Japonica, Rajidae (Pisces)*. Biogeographical Society of Japan, Tokyo. 82 pp., 32 plates.

Johnson, G. G. 1979. The biology of the little skate, *Raja erinacea* Michill 1825, in Block Island Sound, Rhode Island: University of Rhode Island. 118 pp.

Kikugawa, K., Katoh, K., Kuraku, S., Sakurai, H., Ishida, O., Iwabe, N. and Miyata, T. 2004. Basal jawed vertebrate phylogeny inferred from multiple nuclear DNA-coded genes. BioMed Central Biology 2: 3.

Koob, T. J. and Hamlett, W. C. 1998. Microscopic structure of the gravid uterus in the little skate, *Raja erinacea*. Journal of Experimental Zoology 282: 421-437.

Koob, T. J. and Straus, J. W. 1998. On the role of egg jelly in *Raja erinacea* egg capsule. Bulletin of the Mount Desert Island Biological Laboratory 37: 117-119.

Lamilla, J. F. 2003. Life History of Deepsea Chilean Chondrichthyes. Pp. 21. In *Conservation and Management of Deepsea Chondrichthyan Fishes*, Abstracts. FAO and IUCN Shark Specialist Group Pre-Conference Meeting, Deepsea 2003, University of Otago, Portobello Marine Laboratory, South Island, New Zealand.

Last, P. R. and Compagno, L. J. V. 1999a. Anacanthobatidae. Legskates. Pp. 1462-1466. In K. E. Carpenter and V. H. Niem (eds), *FAO Identification Guide for Fishery Purposes, The Living Marine Resources of the Western Central Pacific*. FAO, Rome.

Last, P. R. and Compagno, L. J. V. 1999b. Arhyncobatidae. Softnose skates. Pp. 1457-1461. In K. E. Carpenter and V. H. Niem (eds), *FAO Identification Guide for Fishery Purposes, The Living Marine Resources of the Western Central Pacific*. FAO, Rome.

Last, P. R. and Compagno, L. J. V. 1999c. Rajidae. Hardnose skates. Pp. 1452-1456. In K. E. Carpenter and V. H. Niem (eds), *FAO Identification Guide for Fishery Purposes, The Living Marine Resources of the Western Central Pacific*. FAO, Rome.

Last, P. R. and Stevens., J. D. 1994. *Sharks and Rays of Australia*. CSIRO Australia. 513 pp.

Lessa, R., Vooren, C. M. and Lahaye, J. 1986. Development, sex cycle, migrations and fecundity of females of the guitar fish *Rhinobatos horkelii* (Muller and Henle, 1841) from southern Brazil. Atlantica. Rio Grande 8: 5-34.

Lloris, D. and Rucabado, J. 1991. Ichthyofauna of the Beagle Channel (Tierra del Fuego), ecological aspects and biogeographical analysis. Publicaciones Especiales. Instituto Español de Oceanografía, no. 8. 182 pp.

Luer, C. A. and Gilbert, P. W. 1985. Mating behavior, egg deposition, incubation period, and hatching in the clearnose skate, *Raja eglanteria*. Environmental Biology of Fishes 13(3): 161-171.

Lund, R. 1990. Chondrichthyan life history styles as revealed by the 320 million years old Mississippian of Montana. Environmental Biology of Fishes 27(1): 1-19.

Maisey, J. G. 1984. Higher elasmobranch phylogeny and biostratigraphy. Zoological Journal of the Linnean Society 82(1-2): 33-54.

Maisey, J. G., Naylor, G. J. P. and Ward, D. 2004. Mesozoic elasmobranchs, neoselachian phylogeny, and the rise of modern neoselachian diversity. (in press). In G. Arratia and A. Tintori (eds) *Mesozoic Fishes III. Systematics, Paleoenvironments and Biodiversity*. Verlag Pfeil, Munich.

Marshall, D. J., Bolton, T. F. and Keough, M. J. 2003. Offspring size affects the post-metamorphic performance of a colonial marine invertebrate. Ecology 84(12): 3131-3137.

Marss, T., Wilson, M. V. H. and Thorsteinsson, R. 2002. New thelodont (Agnatha) and possible chondrichthyan (Gnathostomata) taxa established in the Silurian and Lower Devonian of the Canadian Arctic Archipelago. Proc. Estonian Acad. Sci. Geol. 51(2): 88-120.

Martin, A. P. and Naylor, G. J. P. 1997. Independent origins of filter feeding in megamouth and basking sharks (Order Lamniformes) inferred from phylogenetic analysis of cytochrome b gene sequences. Pp. 39-50. In K. Yano, J. F. Morrissey, Y. Yabumoto, and K. Nakaya (eds), *Biology of the Megamouth Shark*. Tokai University Press, Tokyo.

McEachran, J. D. 1982. Revision of the South American skate genus *Sympterygia* (Elasmobranchii: Rajiformes). Copeia 1982(4):867-890.

McEachran, J. D. 1995. Rajidae. Pp. 773-777. In W. Fischer, F. Krupp, W. Schneider, C. Sommer, K. E. Carpenter and V. H. Niem (eds), *Guia FAO Para Identificacion de Especies para los Fines de la Pesca, Pacifico Centro-Oriental, Volumen II*. FAO, Rome.

McEachran, J. D. and Aschliman, N. 2004. Phylogeny of Batoidea. Pp. 79-114. In J. C. Carrier, J. A. Musick, and M. R. Heithaus (eds), *Biology of Sharks and Their Relatives*. CRC Press, Boca Raton, Florida.

McEachran, J. D. and de Carvalho, M. R. 2002. Rajidae. Pp. 531-561. In K.E. Carpenter (ed.), *The Living Marine Resources of the Western Central Atlantic*, Volume 1, Introduction, Molluscs, Crustaceans, Hagfishes, Sharks, Batoid Fishes and Chimaeras. FAO, Rome.

McEachran, J. D. and Fechhelm, J. D. 1998. *Fishes of the Gulf of Mexico, Volume 1, Myxiniformes to Gasterosteiformes*. University of Texas Press, Austin. 1112 pp.

McEachran, J. D. and Miyake, T. 1984. Comments on the skates of the tropical eastern Pacific: one new species and three new records (Elasmobranchii: Rajiformes). Proceedings of the Biological Society of Washington 97(4): 773-787.

Mecklenburg, C. A., Mecklenburg, T. W. and Thorsteinson, L. K. 2002. *Fishes of Alaska*. American Fisheries Society, Bethesda, Maryland. 1037 pp.

Mellinger, J. 1983. Egg-case diversity among dogfish, *Scyliorhinus canicula* (L.): a study of egg laying rate and nidamental gland secretory activity. Journal of Fish Biology 22: 83-90.

Miles, R. S. 1967. Observations on the ptyctodont fish, *Rhamphodopsis* Watson. Journal of the Linnean Society London, Zoology 47: 99-120.

Miller, R. F. 2003. The oldest articulated chondrichthyan from the Early Devonian period. Nature 425: 501-504.

Miller, R. G. 1993. A History and Atlas of the Fishes of the Antarctic Ocean. Foresta Institute, Nevada.

Misra, K. S. *The fauna of India and the adjacent countries. Pisces.* Issued by the Zoological Survey of India, Calcutta. Manager of Publications, Delhi. 256 pp.

Musick, J. A., Harbin, M. M and Compagno, L. J. V. 2004. Historical Zoogeography of the Selachii. Pp. 33-77. In J. C. Carrier, J. A. Musick, and M. R. Heithaus (eds), *Biology of Sharks and Their Relatives.* CRC Press, Boca Raton, Florida.

Muus, B. J. and Nielsen, J. G. 1999. *Sea Fish.* Scandanavian Fishing Year Book, Hedehusene, Denmark. 340 pp.

Nakaya, K. 1975. Taxonomy, comparative anatomy and phylogeny of Japanese catsharks, Scyliorhinidae. Memoirs of the Faculty of Fisheries, Hokkaido University 23(1): 1-94.

Needham, J. 1942. *Biochemistry and Morphogenesis.* Cambridge University Press, Cambridge, U.K. 768 pp.

Otake, T. 1990. Classification of reproductive modes in sharks with comments on female reproductive tissues and structures. Pp. 111-130. In H. L Pratt, S. H. Gruber, and T. Taniuchi (eds), *Elasmobranchs as Living Resources: Advances in the Biology, Ecology, Systematics, and the Status of the Fisheries.* National Oceanic and Atmospheric Administration Technical Report, National Marine Fisheries Service 90.

Peterson, I. and Wroblewski, J. S. 1984. Mortality rate of fishes in the pelagic ecosystem. Canadian Journal of Fisheries and Aquatic Sciences 41: 1117-1120.

Pickford, G. E. and Grant, F. B. 1967. Serum osmolarity in the coelocanth, *Latimeria chalumnae*: urea retention and ion regulation. Science 155: 568-570.

Prasad, R. 1951. Observations on the egg-cases of some oviparous and viviparous elasmobranchs, with a note on the formation of the elasmobranch egg-case. Journal of the Bombay Natural History Society 49: 755-762.

Ranzi, S. 1932. Le basi fisio-morfologische dello sviluppo embrionale dei Selaci—Parti I. Pubb. Staz. Zool. Napoli 13: 209-240.

Ranzi, S. 1934. Le basi fisio-morfologische dello sviluppo embrionale dei Selaci—Parti II and III. Pubb. Staz. Zool. Napoli 13: 331-437.

Ryland, J. S. and Ajayi, T. O. 1984. Growth and population dynamics of three *Raja* species in Carmarthen Bay, British Isles. Journal du Conseil International pour l'Exploration de la Mer 4: 111-120.

Scott, W. B. and Scott, M. G. 1988. *Atlantic Fishes of Canada.* Canadian Bulletin of Fisheries and Aquatic Sciences 219. 731 pp.

Setna, P. B. and Sarangdhar, P. N. 1950. Breeding habits of Bombay elasmobranchs. Rec. Indian Mus. 47: 107-124.

Shirai, S. 1992. Squalean Phylogeny. A New Framework of "Squaloid" Sharks and Related Taxa. Hokkaido University Press, Sapporo. 151 pp.

Shirai, S. 1996. Phylogenetic interrelationships of neoselachians (Chondrichthyes, Euselachii). Pp. 9-34. In M. L. J. Stiassny, L. R. Parenti and G. D. Johnson (eds), *Interrelationships of Fishes.* Academic Press, San Diego, London.

Smith, B. 1942. The heterodontid sharks: Their natural history and the external development of *Heterodontus japonicus* based on the notes and drawings by Bashford Dean. Pp. 151-802. In *The Bashford Dean Memorial Volume Archaic Fishes, Article VIII.* American Museum of Natural History, New York.

Smith, H. W. 1953. *From Fish to Philosopher.* Little, Brown and Co., Boston. 264 pp.

Snelson, F. F., Jr., Williams-Hooper, S. E. and Schmid, T. H. 1988. Reproduction and ecology of the Atlantic stingray, *Dasyatis sabina*, in Florida coastal lagoons. Copeia 1988(3): 729-739.

Snelson, F. F., Jr., Williams-Hooper, S. E. and Schmid, T. H. 1989. Biology of the stingray, *Dasyatis sayi*, in Florida coastal lagoons. Bulletin of Marine Science 45: 15-25.

Springer, S. 1979. A revision of the catsharks, family Scyliorhinidae. National Oceanic and Atmospheric Administration, National Marine Fisheries Service, NOAA Technical Report 422: 1-152.

Stearns, S. C. 1992. *The Evolution of Life Histories*. Oxford University Press, Oxford. 249 pp.

Stehmann, M. 1990. Rajidae. Pp. 29-50. In J. C. Quero, J. C. Hureau, C. Karrer, A. Post and L. Saldanha (eds), *Check-list of the Fishes of the Eastern Tropical Atlantic*. Junta Nacional de Investigaçao Cientifica e Tecnológica, Lisbon, Portugal.

Stehmann, M. 1995. First and new records of skates (Chondrichthyes, Rajiformes, Rajidae) from the West African continental slope (Morocco to South Africa), with descriptions of two new species. Archive of Fishery and Marine Research 43(1): 1-119.

Stehmann, M. and Bürkel, D. L. 1984. Rajidae. Pp. 163-196. In P. J. P. Whitehead, M. -L. Bauchot, J. -C. Hureau, J. Nielsen and E. Tortonese (eds), Rajidae. United Nations Educational, Scientific and Cultural Organization (UNESCO), Paris.

Stehmann, M. and Bürkel, D. L. 1990. Rajidae. Pp. 86-97. In O. Gon and P. C. Heemstra (eds), *Fishes of the Southern Ocean*. J. L. B. Smith Institute of Ichthyology, Grahamstown.

Storrie, M. 2004. Microstructural modifications of the reproductive tissues of the gummy shark (*Mustelus antarcticus*) during maturation and gestation. Ph.D. Dissertation, School of Ecology and Environment, Faculty of Science and Technology, Deakin University, Warrnambool, Victoria, Australia.

Thies, D. 1983. Jurazeitlicher Neoselachier aus Deutschland und S. England. Cour. Forschungsinst. Senckenberg 58: 1-116.

Thorson, T. B., Langhammer, J. K. and Oetinger, M. I. 1983. Reproduction and development of the South American freshwater stingrays, *Potamotrygon circularis* and *P. motoro*. Environmental Biology of Fishes 9(1): 3-24.

Villavicencio Garayzar, C. J., Hoffmann, C. D. and Melendez, E. M. 1994. Size and reproduction of the ray *Dasyatis longus* (Pisces: Dasyatidae), in Almejas Bay, Baja California Sur, Mexico. Revista de Biología Tropical, San Jose. 42(1-2): 375-377.

Walker, P. A. and Hislop, J. R. G. 1998. Sensitive skates or resilient rays? Spatial and temporal shifts in ray species composition in the central and north-western North Sea between 1930 and the present day. ICES Journal of Marine Science 55: 392-402.

Wenbin, Z. and Shuyuan, Q. 1993. Reproductive biology of the guitarfish, *Rhinobatos hynnicephalus*. Environmental Biology of Fishes 38: 81-93.

White, E. G. 1937. Interrelationships of the elasmobranchs with a key to the order Galea. Bulletin of the American Museum of Natural History 74: 25-138.

Winemiller, K. O. and Rose, K. A. 1993. Why do most fish produce so many tiny offspring? American Naturalist 142: 585-603.

Wourms, J. P. 1977. Reproduction and development in chondrichthyan fishes. American Zoologist 17: 379-410.

Wourms, J. P. 1981. Viviparity: The maternal-fetal relationship in fishes. American Zoologist 21(2): 473-515.

Wourms, J. P. and Lombardi, J. 1992. Reflections on the evolution of piscine viviparity. American Zoologist 32: 276-293.

Wourms, J. P., Grove, B. D. and Lombardi, J. 1988. The maternal-embryonic relationship in viviparous fishes. Pp. 1-134. In W. S. Hoar and D. J. Randall (eds), *Fish Physiology, Volume 11*. Academic Press, San Diego.

Yano, K. 1992. Comments on the reproductive mode of the false cat shark *Pseudotriakis microdon*. Copeia 1992(2): 460-468.
Yano, K. 1993. Reproductive biology of the slender smoothhound, *Gollum attenuatus*, collected from New Zealand waters. Environmental Biology of Fishes 38: 59-71.
Young, J. Z. 1950. *The Life of Vertebrates*. Claredon Press, Oxford. 767 pp.

APPENDIX 3.1

Skate total length (mean = 71.6 ± 41.9), habitat and FAO region for 230 species of skates. FAO Regions are designated as follows: ANE = Indian Ocean, Antarctic, ANW = Pacific Ocean, Antarctic, ANC = Atlantic Ocean, Antarctic; EIO = Eastern Indian Ocean, WIO = Western Indian Ocean. For remaining FAO Region codes, first letter indicates western (W) or eastern (E); second letter indicates central (C), north (N) or south (S); and third letter indicates Atlantic (A) or Pacific (P).

Species	*Total Length (cm)*	*Habitat*	*FAO Regions*	*Source*
Anacanthobatidae				
Anacanthobatis americanus	38	183-915 m	WCA	McEachran and de Carvalho 2002
Anacanthobatis borneenisis	38	600-1700 m	WNP	Hatooka *et al.* 2002
Anacanthobatis donghaiensis	44	200-1000 m	WNP	Deng *et al.* 1983
Anacanthobatis folirostris	62	300-512 m	WCA	McEachran and Fechhelm 1998; McEachran and de Carvalho 2002
Anacanthobatis longirostris	75	520-1052 m	WCA	McEachran and de Carvalho 2002
Anacanthobatis marmoratus	25	230-322 m	ESA, WIO	Compagno *et al.* 1989
Anacanthobatis melanosoma	59	900-1100 m	WNP, WCP	Last and Compagno 1999a
Anacanthobatis ori	21	1000-1725 m	WIO	Compagno *et al.* 1989
Anacanthobatis sp. A	54	420-1120 m	EIO	Last and Stevens 1994
Anacanthobatis sp. B	57	680-880 m	WEP	Last and Stevens 1994
Cruriraja andamanica	21	510 m	WIO	Misra 1969
Cruriraja atlantis	34	512-777 m	WCA	McEachran and de Carvalho 2002
Cruriraja cadenati	38	457-896 m	WCA	McEachran and de Carvalho 2002
Cruriraja durbanensis	31	859 m	ESA	Compagno *et al.* 1989
Cruriraja parcomaculata	55	195-620 m	ESA	Compagno *et al.* 1989
Cruriraja poeyi	34	366-870 m	WCA	McEachran and de Carvalho 2002
Cruriraja rugosa	49	366-1007 m	WCA	McEachran and de Carvalho 2002
Cruriraja triangularis	41	220-675 m	WIO	Compagno *et al.* 1989
Arhynchobatidae				
Arynchobatis asperrimus	75	90-1070 m	WSP	Cox and Francis 1997
Bathyraja abyssicola	157	362-2906 m	WNP, ENP	Hatooka *et al.* 2002; Mecklenburg *et al.* 2002; Ebert 2003
Bathyraja aleutica	154	148-900 m	WNP	Ishiyama 1967; Hatooka *et al.* 2002; Ebert 2003
Bathyraja andriashevi	120	1390-1480 m	WNP	Hatooka *et al.* 2002
Bathyraja bergi	95	100-500 m	WNP	Hatooka *et al.* 2002

Bathyraja brachyurops	64	81-313 m	ESP	Lloris and Rucabado 1991
Bathyraja diplotaenia	85	300-1000 m	WNP	Hatooka *et al.* 2002
Bathyraja eatonii	100	15-800 m	ANE, ANW, ANE	Stehmann and Burkel 1990
Bathyraja fedorovi	73	1370-1550 m	WNP	Hatooka *et al.* 2002
Bathyraja griseocauda	49	94-585 m	ESP	Miller 1993
Bathyraja hesperafricana	34.2	750-2000	ECA	Stehmann 1995
Bathyraja irrasa	120	300-1200 m	ANE	Stehmann and Burkel 1990
Bathyraja isotrachys	75	100-1480 m	WNP	Ishiyama 1967; Hatooka *et al.* 2002
Bathyraja kincaidi	56	200-500 m	ECP, ENP	Ebert 2003
Bathyraja lindbergi	93	120-950 (possibly 2000) m	WNP, ENP	Hatooka *et al.* 2002; Mecklenburg *et al.* 2002
Bathyraja longicauda	80	605-735 m	ESP	McEachran and Miyake 1984; Chirichigno Fonseca 2001
Bathyraja maccaini	120	to 500 m	ANC, ANE	Stehmann and Burkel 1990
Bathyraja maculata	120	73-1110 (usually 100-650) m	WNP, ENP	Mecklenburg *et al.* 2002
Bathyraja matsubarai	120	200-1205 m	WNP	Hatooka *et al.* 2002
Bathyraja meriodionalis	120	760-800 m	ANC	Stehmann and Burkel 1990
Bathyraja microtrachys	70	1995-2900 m	ECP, ENP	Ebert 2003
Bathyraja minispinosa	83	150-1420 (usually 200-800) m	WNP, ENP	Mecklenburg *et al.* 2002
Bathyraja pallida	160	2400-2950 m	ENA	Stehmann and Burkel 1984
Bathyraja parmifera	150	15-1602 m	WNP, ENP	Mecklenburg *et al.* 2002
Bathyraja radiata	105	735-1060 m	ESP	McEachran and Miyake 1984
Bathyraja richardsoni	186	0-2500 m	WSP, WNA, ENA, ESP	McEachran and Miyake 1984; Scott and Scott 1988; Stehmann and Burkel 1984; Cox and Francis 1997
Bathyraja shuntovi	140	300-1470 m	WSP	Cox and Francis 1997
Bathyraja simoterus	94	300- m	WNP	Ishiyama 1967; Hatooka *et al.* 2002
Bathyraja smirnovi	100	100-950 m	WNP	Hatooka *et al.* 2002
Bathyraja smithii	120	440-1020 m	ESA	Compagno *et al.* 1989
Bathyraja sp. A	120	2300 m		Last and Stevens 1994
Bathyraja spinicauda	170	140-800 m	WNA, ENA	Stehmann and Burkel 1984; Scott and Scott 1988
Bathyraja spinosissima	150	800-2938 m	ECP	Ebert 2003
Bathyraja trachouros	90		WNP	Hatooka *et al.* 2002
Bathyraja trachura	89	400-2550 m	WNP, ENP, ECP	Mecklenburg *et al.* 2002; Ebert 2003
Bathyraja tzinovskii	71	2500 m	WNP	Hatooka *et al.* 2002
Bathyraja violacea	73	20-1100 m (usually 100-800 m)	WNP, ENP	Mecklenburg *et al.* 2002
Irolita sp. A	42	150-200 m	EIO	Last and Stevens 1994

Irolita waitei	52	50-200 m		Last and Stevens 1994
Notoraja asperula	51	200-1300 m	WSP	Cox and Francis 1997
Notoraja ochroderma	36	400-465 m	WCP	Last and Compagno 1999b
Notoraja sp. A	60	840-1120 m		Last and Stevens 1994
Notoraja sp. B	36	400-465 m	WEP	Last and Stevens 1994
Notoraja sp. C	45	590-760 m	EIO	Last and Stevens 1994
Notoraja sp. D	53	820-930 m	EIO	Last and Stevens 1994
Notoraja spinefera	80	170-1460 m	WSP	Cox and Francis 1997
Notoraja tobitukai	50	300-1000 m	WNP	Hatooka *et al.* 2002
Pavoraja alleni	35	200-460 m	EIO	Last and Stevens 1994
Pavoraja nitida	35	30-390 m	WSP	Last and Stevens 1994
Pavoraja sp. A	57	800-880 m	WEP	Last and Stevens 1994
Pavoraja sp. B	54	610-1200 m	EIO	Last and Stevens 1994
Pavoraja sp. C	33	200-520 m	EIO	Last and Stevens 1994
Pavoraja sp. D	30	300-400 m	WEP	Last and Stevens 1994
Pavoraja sp. E	37	210-500 m	WEP	Last and Stevens 1994
Pavoraja sp. F	37	360-739 m	WEP, WSP	Last and Stevens 1994
Psammobatis extenta	24.5	shelves	ESP, WSA	de Carvalho and de Figueiredo 1994
Pseudoraja fischeri	58	412-576 m	WCA	McEachran and de Carvalho 2002
Rhinoraja albomaculata	14.4	130-434 m	ESP, WSA	Lloris and Rucabado 1991
Rhinoraja interrupta	86	55-1372 m	WNP, ENP, ECP	Mecklenburg *et al.* 2002
Rhinoraja kujiensis	100	600-800 m	WNP	Ishiyama 1967
Rhinoraja longi	70	300-980 m	WNP	Hatooka *et al.* 2002
Rhinoraja longicauda	70	549-914 m	WNP	Ishiyama 1967
Rhinoraja murrayi	60	30-650 m	ANE	Stehmann and Burkel 1990
Rhinoraja odai	60	330-350 m	WNP	Ishiyama 1967; Hatooka *et al.* 2002
Rhinoraja taranetzi	70	15-550 m	WNP, ENP	Mecklenburg *et al.* 2002
Sympterygia acuta	42.3	shelves	ESP	McEachran 1982
Sympterygia bonapartei	61	shelves	ESP	McEachran 1982
Sympterygia brevicaudata	47	shelves	ESP	McEachran 1982
Sympterygia lima	53.7	shelves	ESP	McEachran 1982
Rajidae				
Amblyraja badia	100	1100-2300 m	WNP, ENP, ECP	Hatooka *et al.* 2002
Amblyraja frerichsi	120	800-2500 m	ESP	Lamilla 2003
Amblyraja georgiana	100	20-250, 660, 1130 m	ANC, ANW	Stehmann and Burkel 1990
Amblyraja hyperborea	106	300-1500 m	EIO, WSP, WEP, WNP, ENA	Stehmann and Burkel 1984; Last and Stevens 1994; Cox and Francis 1997
Amblyraja jenseni	85	1907 m	WNA	Bigelow and Schroeder 1953
Amblyraja radiata	62	20-1000 m	ESA, WIO	Compagno *et al.* 1989
Amblyraja radiata	102	18-1000 m	WNA, WCA	Stehmann and Burkel 1984; Scott and Scott 1998; McEachran and de Carvalho 2002

Amblyraja reversa	60	1499 m	WIO	Misra 1969
Amblyraja robertsi	77	1350 m	ESA, WIO	Compagno *et al.* 1989
Amblyraja taaf	90	150-600 m	ANE	Stehmann and Burkel 1990
Breviraja claramaculata	29	293-896 m	WCA	McEachran and de Carvalho 2002
Breviraja colesi	40	220-415 m	WCA	McEachran and de Carvalho 2002
Breviraja marklei	45.1	443-988 m		McEachran and Miyake 1987
Breviraja mouldi	41	353-776 m	WCA	McEachran and de Carvalho 2002
Breviraja nigriventralis	44	546-776 m	WCA, WSA	McEachran and de Carvalho 2002
Breviraja spinosa	33	366-671 m	WCA	McEachran and de Carvalho 2002
Dactylobatus armatus	32	338-685 m	WCA	McEachran and de Carvalho 2002
Dactylobatus clarki	75	366-915 m	WCA	McEachran and de Carvalho 2002
Dipturus batis	250	100-1000 m	ECA, ENA	Stehmann and Burkel 1984; Stehmann 1990
Dipturus bullisi	77	183-549 m	WCA	McEachran and de Carvalho 2002
Dipturus campbelli	66	137-403 m	ESA, WIO	Compagno *et al.* 1989
Dipturus doutrei	100	450-600 m	ESA, WIO	Compagno *et al.* 1989
Dipturus garricki	107	275-476 m	WCA	McEachran and de Carvalho 2002
Dipturus gigas	140	300-400 m	WNP	Ishiyama 1967; Hatooka *et al.* 2002
Dipturus gudgeri	140	160-700 m	EIO, WSP	Last and Stevens 1994
Dipturus innominatus	240	15-1310 m	WSP	Cox and Francis 1997
Dipturus johannisdavesi	26.3	220-549 m	WIO	Misra 1969
Dipturus kwangtungensis	65	20-80 m	WNP	Hatooka *et al.* 2002
Dipturus laevis	152	0-750 m	WNA	Bigelow and Schroeder 1953; Scott and Scott 1988
Dipturus lancerostratus	82	430-439 m	WIO	Compagno *et al.* 1989
Dipturus linteus	112	55-1371 m	WNA, ENA	Stehmann and Burkel 1984; Scott and Scott 1988
Dipturus macrocaudus	120	300-400 m	WNP	Hatooka *et al.* 2002
Dipturus nasulus	118	10-1500 m	WSP	Cox and Francis 1997
Dipturus nidarosiensis	200	200-1000 m	ENA	Stehmann and Burkel 1984; Stehmann 1990
Dipturus olseni	57	55-384 m	WCA	McEachran and de Carvalho 2002
Dipturus oregoni	144	475-1079 m	WCA	McEachran and de Carvalho 2002
Dipturus oxyrhynchus	150	90-900 m	ENA, MED	Stehmann and Burkel 1984; Stehmann 1990
Dipturus pullopunctatus	130	50-457 m	ESA	Compagno *et al.* 1989
Dipturus springeri	160	88-740 m	ESA, WIO	Compagno *et al.* 1989
Dipturus stenorhyncus	90	625-741 m	WIO	Compagno *et al.* 1989

Dipturus teevani	84	311-732 m	WCA	McEachran and de Carvalho 2002
Dipturus tengu	100	60-150 m	WNP	Ishiyama 1967; Hatooka *et al.* 2002
Fenestraja atripinna	29	366-951 m	WCA	McEachran and de Carvalho 2002
Fenestraja cubensis	23	440-869 m	WCA	McEachran and de Carvalho 2002
Fenestraja ishiyamai	36	503-950 m	WCA	McEachran and de Carvalho 2002
Fenestraja mamillidens	29.2	1091 m	WIO	Misra 1969
Fenestraja plutonia	27	293-1024 m	WCA	McEachran and de Carvalho 2002
Fenestraja sibogae	31	290 m	WCP	Last and Compagno 1999c
Fenestraja sinusmexicanus	36	56-1096 m	WCA	McEachran and de Carvalho 2002
Genus A (Formerly Raja) binoculata	244	3-800 m	ENP, ECP	Mecklenburg *et al.* 2002
Genus A (Formerly Raja) cortezensis	35.8	to 80 m	ECP	McEachran 1995
Genus A (Formerly Raja) inornata	76	17-67	ECP, ENP	Ebert 2003
Genus A (Formerly Raja) pulchra	100	50-100 m	WNP	Hatooka *et al.* 2002
Genus A (Formerly Raja) rhina	137	20-1000 m	ENP, ECP	Mecklenburg *et al.* 2002; Ebert 2003
Genus A (Formerly Raja) stellulata	76	to 732 m (usually <100 m)	ECP, ENP	Ebert 2003
Genus B (Formerly Raja) ackleyi	41	32-384 m	WCA	McEachran and de Carvalho 2002
Genus B (Formerly Raja) bahamensis	54	366-411 m	WCA	McEachran and de Carvalho 2002
Genus B (Formerly Raja) cervigoni	51	37-174 m	WCA	McEachran and de Carvalho 2002
Genus B (Formerly Raja) eglanteria	79	0-111 m	WNA, WCA	Bigelow and Schroeder 1953; McEachran and de Carvalho 2002
Genus B (Formerly Raja) equitorialis	50	shelfs	ECP, ESP	McEachran 1995
Genus B (Formerly Raja) texana	53.7	0-91 m	WCA	McEachran and de Carvalho 2002
Genus B (Formerly Raja) velezi	75.6	35-140 m	ECP, ESP	McEachran 1995
Gurgesiella atlantica	49	247-960 m	WCA	McEachran and de Carvalho 2002
Gurgesiella dorsalifera	53	500-800 m	WSA	McEachran and de Carvalho 2002
Gurgesiella furvescens	52	slopes	ESP	Grove and Lavenberg 1997
Leucoraja circularis	120	70-300 m	ENA, MED	Stehmann and Burkel 1984; Stehmann 1990

Leucoraja compagnoi	29.2	550 m	ESA	Stehmann 1995
Leucoraja erinacea	54	26-111 m; 329 m (Scott and Scott 1988)	WNA	Bigelow and Schroeder 1953; Scott and Scott 1988
Leucoraja fullonica	100	30-600 m	ECA, MED, ENA	Stehmann and Burkel 1984; Stehmann 1990
Leucoraja garmani	44	37-366 m	WNA, WCA	McEachran and de Carvalho 2002
Leucoraja lentiginosa	44	53-588 m	WCA	Bigelow and Schroeder 1953; McEachran and de Carvalho 2002
Leucoraja leucosticta	80	70-600 m	ECA	Stehmann 1990
Leucoraja melitensis	50	60-600 m	MED	Stehmann and Burkel 1984
Leucoraja naevus	70	20-250 m, 400 m	ENA, ECA	Stehmann and Burkel 1984; Stehmann 1990
Leucoraja ocellata	109	0-73 m	WNA, WCA	Bigelow and Schroeder 1953; Frisk *et al.* 2002
Leucoraja wallacei	92	95-432 m	ESA, WIO	Compagno *et al.* 1989
Leucoraja yucatanensis	30	192-457 m	WCA	McEachran and de Carvalho 2002
Malacoraja kreffti	70	1200 m	ENA	Stehmann and Burkel 1984
Malacoraja senta	61	46-874 m	WNA	McEachran and de Carvalho 2002
Malacoraja spinacidermis	70	864-1350 m (juvies); Adults >1500 m	ECA, ESA, ENA	Stehmann and Burkel 1984; Compagno *et al.* 1989; Stehmann 1990
Neoraja africana	30	900-1030 m	ECA	Stehmann 1990
Neoraja caerulea	30	600-1260 m	ENA	Stehmann and Burkel 1984
Neoraja carolinensis	29	695-1010 m	WCA	McEachran and de Carvalho 2002
Neoraja stehmanni	35	292-1025 m	ESA	Compagno *et al.* 1989
Okamejei acutispina	45	50-100 m	WNP	Ishiyama 1967; Hatooka *et al.* 2002
Okamejei australis	50	50-180 cm	WEP, WSP	Last and Stevens 1994
Okamejei boesmani	55	70-90 m	WNP	Hatooka *et al.* 2002
Okamejei cerva	60	20-470 m	EIO, WSP	Last and Stevens 1994
Okamejei heemstrai	51.5	500 m	WIO	McEachran and Fechhelm 1982
Okamejei hollandi	55	60-90 m	WNP	Ishiyama 1967
Okamejei kenojei	66	30-100 m	WNP	Ishiyama 1967; Hatooka *et al.* 2002
Okamejei lemprieri	52	0-170 m	EIO, WSP	Last and Stevens 1994
Okamejei meerdervoorti	33	80-90 cm	WNP	Hatooka *et al.* 2002
Okamejei pita	46	shallow	WIO	Carpenter *et al.* 1997
Okamejei powelli	36	122-237 m	WIO	Misra 1969
Okamejei schmidti	50	20-50 m	WNP	Hatooka *et al.* 2002
Raja africana	80	50-400 m	ESA, MED	Stehmann and Burkel 1984; Stehmann 1990

Raja asterias	70	to 170 m	ECA, MED	Stehmann and Burkel 1984; Stehmann 1990
Raja brachyura	120	to 100 m	ENA, ECA, MED	Stehmann and Burkel 1984; Stehmann 1990
Raja clavata	90	to 300 m	ENA, ECA, MED, ESA, WIO	Stehmann 1990
Raja herwigi	50	55-102 m	ECA	Stehmann 1990
Raja maderensis	80	to 150 m	ENA, ECA	Stehmann and Burkel 1984; Stehmann 1990
Raja microocellata	80	to 100 m	ENA, ECA	Stehmann and Burkel 1984; Stehmann 1990
Raja miraletus	60	17-306 m	ENA, ECA, MED, ESA, WIO	Stehmann and Burkel 1984; Compagno *et al.* 1989; Stehmann 1990
Raja montagui	80	to 100 m	ENA, MED, ECA	Stehmann and Burkel 1984; Stehmann 1990
Raja polystigma	53	100-400 m	MED	Capape 1978; Stehmann and Burkel 1984
Raja radula	70	to 300 m	MED	Stehmann and Burkel 1984; Stehmann 1990
Raja rondeleti	50	moderate depths		MED Stehmann and Burkel 1984
Raja sp. A	70	40-250 m	EIO, WSP	Last and Stevens 1994
Raja sp. B	90	450-600 m	EIO, WSP	Last and Stevens 1994
Raja sp. C	63	70-450 m	WSP	Last and Stevens 1994
Raja sp. D	48	20-200 m	EIO	Last and Stevens 1994
Raja sp. E	58	200-250 m	EIO	Last and Stevens 1994
Raja sp. F	72	200-440 m	EIO	Last and Stevens 1994
Raja sp. G	77	225-550 m	WEP	Last and Stevens 1994
Raja sp. H	76	240-650 m	WEP, WSP	Last and Stevens 1994
Raja sp. I	115	400-1030 m	EIO, WEP, WSP	Last and Stevens 1994
Raja sp. J	133	800-1400 m	EIO, WSP	Last and Stevens 1994
Raja sp. K	76	440-650 m	WEP	Last and Stevens 1994
Raja sp. L	67	5 m	EIO	Last and Stevens 1994
Raja sp. M	36	20-35 m	EIO	Last and Stevens 1994
Raja sp. N	56	400-735 m	EIO	Last and Stevens 1994
Raja sp. O	40	350-420 m	EIO	Last and Stevens 1994
Raja sp. P	55	860-1500 m	EIO, WSP	Last and Stevens 1994
Raja straeleni	91	0-690 m	ECA, ESA	Compagno *et al.* 1989
Raja undulata	100	to 200 m	ENA, ECA, MED	Stehmann and Burkel 1984; Stehmann 1990
Rajella annandalei	33	400-830 m	WCP	Last and Compagno 1999c
Rajella barnardi	68	170-913 m	ESA	Compagno *et al.* 1989 (entry for Rajella confundens)
Rajella bathyphila	90	600-2173 m	WNA, ENA, ECA	Bigelow and Schroeder 1948; Stehmann and Burkel 1984; Stehmann 1990; Stehmann 1995

Rajella bigelowi	55	650-4156 m	WNA, WCA, ENA, ECA	Stehmann and Burkel 1984; Stehmann 1990
Rajella caudaspinosa	58	310-718 m	ESA	Compagno *et al.* 1989
Rajella dissimilis	70	719-1016 m (1620 m, Stehmann)	ECA, ESA	Compagno *et al.* 1989; Stehmann 1990
Rajella fuliginea	45	731-1280 m	WCA	McEachran and de Carvalho 2002
Rajella fyllae	60	170-2050 m	WNA, ENA	Bigelow and Schroeder 1953; Stehmann 1990; Muus *et al.* 1999
Rajella leopardus	95	300-923 m; 170-1920 m (Stehmann)	ECA, ESA	Compagno *et al.* 1989; Stehmann 1990
Rajella nigerrima	45.7	590-1000	ESP	McEachran and Miyake 1984; Lamilla 2003
Rajella purpuriventralis	51	732-2010 m	WCA	McEachran and de Carvalho 2002
Rajella ravidula	67	496-1016 m	ECA, ESA	Compagno *et al.* 1989, Stehamann 1995
Rajella sadowskyii	75	1200 m	ESP	Lamilla 2003
RostroRaja alba	230	30-366 m	ENA, ECA, MED, ESA, WIO	Compagno *et al.* 1989; Stehmann 1990
Western Pacific species (Formerly Raja) koreana	74	30-120 m	WNP	Hatooka *et al.* 2002
Western Pacific species (Formerly Raja) polyommata	36	140-310 m	WEP	Last and Stevens 1994
Western Pacific species (Formerly Raja) whitleyi	170	0-170 m	EIO, WSP	Last and Stevens 1994

CHAPTER 4

Reproduction in Fisheries Science

Terence I. Walker

4.1 INTRODUCTION

Assessment of populations of chondrichthyan species requires a quantitative approach to the study of reproduction. Measures of reproductive rate, together with mortality rate, are required for stock assessment of species harvested by fisheries and for ecological risk assessment of bycatch species (Walker 2004). Such measures are also required for species assessment against criteria established under each of several risk of extinction categories developed by the Species Survival Commission of the International Union for Conservation of Nature and Natural Resources (IUCN) (Hilton-Taylor 2000).

Population demography is based on simple models for constructing life tables from reproductive and mortality parameters or on age-based models using Leslie matrices. Stock assessment of harvested species for fishery management require more complex model and data requirements to account for density-dependent mechanisms for population regulation (Wood *et al.* 1979; Walker 1992), somatic growth and trends in abundance of the animals, and history of extracted catches by the fishery (Punt and Walker 1998). Some fishery assessment models also account for complex interactions between the animals and the fishing gear, movement of the animals, and stock structuring (Punt *et al.* 2000).

All these types of assessment, irrespective of model complexity, require the same specific parameters for representing three key components of reproduction. The first of these components is the sex ratio at birth, which can be determined by counting embryos or neonates of each sex. The second component is the relationship between the annual number of offspring and maternal age or size of the animals (litter size

Primary Industries Research Victoria, P.O. Box 114, Queenscliff, Victoria 3225, Australia curve). This can be determined by ageing or measuring a sample of animals in maternal condition and counting the number of their offspring. The third is the relationship between the proportion of the female population contributing to annual recruitment and the age or size of animals (maternity ogive). The third is the most complex component to determine, as it inevitably requires information on the periodicity of each of the ovarian cycle and gestation before estimating values of the parameters of a maternity ogive.

For fisheries assessment, a fourth component is required if the management objectives for a fishery are expressed in terms of mature population number or mature biomass. The fourth requires the relationship between the proportion of the female population in mature condition at any time and the age or size of the animals (maturity ogive); an explicit definition of maturity is essential for this purpose. A maturity ogive might also be required for the males. Maturity ogives are otherwise not needed for determining the dynamics of a population. The female maturity ogive is often used as an approximation to the maternity ogive, if the maternity ogive is not available. However, the model outputs for any species, as demonstrated later in this chapter, is highly dependent on the adopted definition of maturity.

Each of the litter size curve, maternity ogive and maturity ogive are best determined as a function of age. However, they can be expressed as a function of size provided the relationship between size and age of the animals is known. For chondrichthyan species, this is usually required for each sex separately because of differences in their growth. For fisheries assessment, it is also an advantage to have the relationship between size and mass of the animals. Because this relationship can vary depending on the sex and reproductive condition of the animals, methods for determining these relationships are described. Although the relationships between size and age usually vary between females and males, their determination can be complex and are beyond the scope of this volume. The reader is referred to texts on age and growth for determining these relationships.

Chondrichthyan species are renowned for their wide range of reproductive strategies. They are usually categorised as oviparous species or viviparous species, where viviparous species are further categorised as placental and aplacental with several aplacental species exhibiting oophagy and one species exhibiting intra-uterine cannibalism. A simpler categorisation between lecithotrophic species and matrotrophic species has recently been proposed. This dichotomy is based on relative contributions to the mass gain of embryos during gestation either from the initial egg yolk or from nutrients provided by the mother via a placenta, trophonemata, sibling embryos, or continuous ovulation during pregnancy (see Chapter **13** of this volume).

The variation in reproductive mode and differences between species or between populations within species, in the period between successive birth events, or laying of egg clutches, prevents prescriptive procedures for determining maternity ogives. Whereas parturition and ovulation in many species are annual, in other species they are biennial, triennial or possibly longer, and while synchrony of gestation and of the ovarian cycle occur in the populations of many species, in other species they appear to be asynchronous. The periods of gestation and ovarian cycles are particularly difficult to determine for species inhabiting the deep cold waters of the continental slopes and abyssal plains of the world because their duration are of several years. They may also be asynchronous. Determining sex ratio of embryos or period of gestation in oviparous species requires observing eggs in the wild or in captivity.

The methods described in this chapter are demonstrated by presenting the results from analyses of reproductive data available from southern Australia for *Galeorhinus galeus*, a species known to form six separate populations in widely separated regions of the world. This species is selected because of the long period of the ovarian cycle and complexity associated with sampling to determine its maternity and maturity ogives.

Before describing the methods and results for *G. galeus*, a brief description of relevant components of its life history is presented. Also, the three terms mature condition, pregnant condition and maternal condition central to population dynamics are explicitly defined. Brief descriptions of the structure and function of the typical reproductive systems of female and male chondrichthyan animals are presented, adopting preferred terminology, to simplify description of the methods.

Methods for estimating the sex ratio at birth and the parameters of the litter size curve are explained first. Methods are then described for determining the ovarian and gestation periods, which are required to derive the frequency of parturition or egg laying. Next, methods for estimating the parameters of maturity ogives of females and males separately and maternity ogives are described. Finally, methods are described for estimating mass-size relationships for different breeding conditions.

4.2 TERMINOLOGY AND DESCRIPTION OF REPRODUCTIVE SYSTEM

4.2.1 Female Reproductive System

Chondrichthyan fishes typically have paired or single ovaries and paired oviducts. Each oviduct is differentiated into a funnel-shaped ostium, anterior oviduct, oviducal gland with, in some species, an isthmus leading to the uterus, cervix and the urogenital sinus, which is common to both reproductive tracts.

The chondrichthyan ovary is attached to the dorsal wall of the body cavity by the mesovarium mesentery and has three main functions: generate germ cells, accumulate yolk and synthesise and secrete hormones. In mature animals, the ovary consists of small follicles, developing follicles of various sizes, preovulatory follicles undergoing atresia, and corpora lutea, which are all embedded together in a dense stroma of connective tissue. A follicle consists of an oocyte surrounded by granulosa cells and delimited by the basal lamina, the size of which for many depends on the stage of the reproductive cycle. The structure has an oocyte plasmalemma, zona pellicida, follicular epithelium and connective tissue theca. The follicles are small with little or no yolk in juveniles (Hamlett and Koob 1999) but enlarge initially through folliculogenesis as the follicular cells begin to mature. With further development, follicles enlarge as the oocytes accumulate yolk through the process of vitellogenesis, whereby phosvitin and lipovitellin of hepatic origin are deposited in the oocyte (Storrie 2004). Ovulation of the largest oocytes occurs when they attain a particular size. Atretic follicles form by degeneration of preovulatory follicles and resorption of yolk from the oocytes. Atresia of a follicle can occur at any size to form a corpus atretica. Following ovulation, the follicle wall collapses and forms the corpus luteum, consisting of a lipid-filled cell derived from the granulosa cells (Hamlett and Koob 1999).

Following ovulation, ova (ovulated oocytes), moved by cilia in the peritoneal cavity to a single ostium (or paired ostia) that bifurcates into the left and right oviducts. Ova move through the oviducts to the oviducal gland where they are fertilized and encapsulated with egg jelly (Hamlett and Koob 1999). In *Galeorhinus galeus*, initially spherical ova are compressed to an ellipsoid shape by the time they are encapsulated in a brownish-yellow, transparent, flexible egg-case membrane, the free ends of which are spirally twisted, and deposited in the uterus (personal observation).

The uterine walls are thick, spongy and vascular during early gestation but as gestation advances, they become thin, semi-transparent, and further vascularized. As described for *Mustelus manazo* (Teshima and Koga 1973), the external yolk sac of *Galeorhinus galeus* is large during the early embryonic stages but, as the embryo grows, the external yolk sac becomes progressively smaller as the yolk is consumed. As parturition approaches, the contracted yolk-sac and the short stalk by which it is attached appear to be incorporated into the body of the embryo when the yolk is completely absorbed. The egg case membrane remains intact, and unfolds and stretches to accommodate the developing embryo and increasing amounts of enclosed clear fluid (personal observation).

4.2.2 Male Reproductive System

Chondrichthyan fishes typically have external paired claspers (mostly ventrally grooved copulatory appendages), which are extensions of the posterior bases of the pelvic fins. These are calcified and articulate at their bases in mature animals. The internal organs include the testes, genital ducts, Leydig gland and the alkaline gland. The paired genital ducts cover the elongate kidneys embedded in the dorsal abdominal wall, consist of the efferent ductules, epididymis, ductus deferens and seminal vesicle, and are covered by the peritoneum. There are two equally developed elongate testes, each embedded in the anterior portion of the long irregular epigonal gland, which is a lymphomyeloid organ supporting the testis.

The testis has two functions: spermatogenesis (germ cell generation) and steroidogenesis (secretion of steroid hormones). It is packed with spherical spermatocysts consisting of many spermatoblasts (each with a single Sertoli cell and its synchronously developing isogenetic clone germ cells) bounded by a basement membrane. In a particular spermatocyst, all germ cells pass through spermatogenesis as synchronously developing clone cells. Spermatocyst development commences with a single germ cell and a single Sertoli cell and terminates at sperm release when the Sertoli cells fragment to release the spermatozoa and Sertoli cell remnants into the lumen of the efferent ductules. Initially, spermatogonia undergo repeated mitotic divisions to produce 16 germ cells per Sertoli cell. Subsequent meiosis results in 64 spermatozoa per Sertoli cell (Hamlett 1999). Viewed in a transverse histological section, *Galeorhinus galeus* testes are diametric where the germinal zone consists of a strip along the distolateral surface. Development of the spermatocysts occurs diametrically (across the width of the testis towards the efferent ductules located medially). Spermatocyst development was initially described as passing through 18 different stages for the small-spotted catshark (*Scyliorhinus canicula*) (Mellinger 1965) but this was subsequently reduced to 7 stages for the bonnethead shark (*Sphyrna tiburo*) (Parsons and Grier 1992). These stages cover the various mitotic divisions of diploid primary spermatocytes to produce primary spermatocytes, which in turn divide through meiosis to produce haploid secondary spermatocytes. The secondary spermatocytes develop into spermatotids with emerging flagella and then into tightly bundled mature spermatozoa, which are shed into the interstitial spaces of the testis as the spermatocyst disintegrates before passing through the efferent ductules into the epididymis.

Spermatozoa occur throughout the lumina of the epididymis, ductus deferens and seminal vesicles, along with secretions from the genital ducts and from the Leydig gland. The process of spermiogenesis, where spermatids mature into motile sperm, occurs in these genital ducts. Secretions from the Leydig gland facilitate maturation of spermatozoa and matrical material form in the lumena of the seminal vesicles and associate with individual and previously bundled sperm. The sperm are held together by a sticky matrix as either spermatophores (sperm encapsulated) or spermatozeugmata (sperm not encapsulated but tails of peripheral sperm protruding). Ciliated epithelial columnar cells lining the lumen convey the spermatozoa through the genital ducts; only the seminal vesicles have a muscular wall. At copulation and ejaculation, sperm transfer from the seminal vesicle through the urogenital papilla to the dorsal groove of each clasper. Spermatozoa acquire the potential for modest motility while in the terminal regions of the genital ducts but acquire active, robust motility at ejaculation (Hamlett 1999).

4.2.3 Maturity and Ovarian Cycle

Application of demographic models to chondrichthyan populations, originally developed for mammals and other vertebrate groups, without recognising the peculiarities of the reproductive cycles of chondrichthyan species, can lead to bias through implicit assumptions that are not correct. In many reported demographic analyses, female mature condition, pregnant condition and maternal condition are often incorrectly equated. Investigation of the dynamics of a chondrichthyan population requires clear definitions for distinguishing between these three breeding conditions, which relate to the timing and periodicity of ovulation and parturition.

The period from the onset of maturity to the start of first pregnancy for females of most vertebrate species is comparatively short and the reproductive cycle is annual or less, with a few exceptions, particularly among large mammals. For chondrichthyan species, this period can also be annual or less but in other species it is several years. In these other species, the period from the onset of maturity of an animal to first ovulation or the period between successive ovulation cycles can be 2, 3 or, as possibly in deep-sea squalids, more years. Similarly, the period from fertilization to parturition is often more than one year in these other species. Fertilization occurs during the relatively short period following ovulation when the ovum passes through the oviduct and oviducal gland before encapsulation by an egg envelop before entering into the uterus. Storage of sperm in the terminal zone of the oviducal gland, evident in several species (Hamlett *et al.* 2002; Smith *et al.* 2004), ensures a supply of sperm for progressive fertilization of ova released by ovulation over a period of several weeks, or possibly months as in *Galeorhinus galeus*. Sperm storage is probably a mechanism for avoiding the problem of accumulating *in utero* eggs obstructing or retarding fresh sperm transiting the uteri to the oviducal glands.

For an animal experiencing first pregnancy, the period from when an oocyte begins yolking to ovulation, together with the period of gestation for the first pregnancy, is more than one year for most species. However, for subsequent pregnancies, vitellogenesis can proceed concurrently with gestation so parturition can be annual or, as argued for some species, more frequently. For species producing large-sized oocytes, such as *G. galeus* and squalids from coastal waters (~40 mm diameter) (Hanchet 1988) and deep-sea continental slopes (60–87 mm diameter) (Yano and Tanaka 1988; Girard and Du Buit 1999; Guallart and Vincent 2001) the period for vitellogenesis is 2, 3 or possibly more years. For our purpose here, the period of the ovarian cycle is defined as the period from completion of one ovulation to completion of the next, and the period of gestation is defined as the period from fertilization to parturition.

Female maturity in chondrichthyan species is not unambiguously defined but animals are mostly assumed to be mature when the ovarian follicles are enlarged. However, this is arbitrary, subjective and likely to provide unrepeatable results. Less ambiguous criteria that might be adopted for defining maturity relate to levels of selected hormones in the blood, the onset of vitellogenesis, first mating, first sperm storage in the oviducal gland or first ovulation. However, the maturity ogives derived for any species would vary depending on the criterion adopted. For population studies requiring large-scale sampling, the onset of vitellogenesis can be rapidly assessed by macroscopic inspection of the ovaries.

Female maturity was assumed to coincide with the onset of vitellogenesis for *G. galeus*. Animals where the diameter of the largest ovarian follicle was >3 mm were assumed to be vitellogenic animals and in mature condition. Animals where the diameter of the largest follicle was 0–3 mm were assumed to be non-vitellogenic and in immature condition.

4.2.4 Pregnancy, Maternity and Parturition Frequency

The size of a total population depends on birth rate and death rate; immigration and emigration rates can affect the sizes of sub-populations interacting within the total population. Birth rate relates to the number of births for viviparous species or to the number of eggs layed for oviparous species. The number of births or eggs layed can be calculated from the number of females in the population and two mathematical equations. One equation expresses litter size as a function of age (or size) of animal (litter size curve). The other equation expresses the proportion of the female population giving birth or laying eggs by the end of the year to contribute to annual recruitment (0+ year-old cohort) at the beginning of the following year. This is referred to here as the proportion of females in maternal condition, and is expressed as a function of age (or size) of animal. The relationship is referred to as the maternity ogive and year applies to the 12-month period prior to the date (or similar date) for completion of parturition.

In *Galeorhinus galeus* for example, as will be demonstrated later in this chapter, parturition frequency is triennial, which implies about one-third of the female population having reached first ovulation (or maternity) gives birth at the end of any year and contributes to recruitment at the beginning of the following year. As parturition occurs during November–January, the birth dates for all offspring can conveniently be fixed at 1 January. Hence, with respect to maternity, a female observed at any time of the year is in maternal condition if it is in pregnant condition and expected to give birth prior to 1 January or if it is in post-partum condition and recently gave birth prior to 1 January. Any other female observed is in non-maternal condition. Pregnant condition of a female is defined by the presence of *in utero* eggs or embryos; non-pregnant condition of a female is defined by the absence of *in utero* eggs and embryos.

4.3 LIFE HISTORY OF *GALEORHINUS GALEUS*

The triakid *Galeorhinus galeus* occurs as six genetically distinct populations (Ward and Gardner 1997) off western North America, eastern South America, southern Africa, southern Australia and New Zealand, and in the eastern North Atlantic Ocean (Compagno 1984). There is some mixing by a small proportion of the large animals undertaking migrations between New Zealand and Australia (Hurst *et al.* 1999) but genetic studies indicate that there is no interbreeding between these populations. There may be genetic or behavioural sub-structuring within the populations but this remains uncertain.

The coastal semi-pelagic species is presently harvested for its meat, cartilage and fins. The species is particularly long-lived, caught by many fishing methods, and demonstrates how such a low-productivity species can be severely depleted if not adequately managed. A fishery based on *Galeorhinus galeus* for its liver oil in California collapsed during the 1940s (Walker 1999). Recent assessments list the species as endangered off eastern South America, vulnerable off southern Australia and southern Africa, and near threatened off New Zealand. The maximum size and mass of *G. galeus* varies between the six populations suggesting that the reproductive parameters required for population assessments will probably vary between the populations. Information on TL-at-maturity (not maturity ogives), litter size and sex ratio among embryos are reported for California (Ripley 1946), eastern South America (Peres and Vooren 1991; Lucifora 2003), and the eastern North Atlantic (Capapé and Mellinger 1988). Only in Australia (Punt and Walker 1998) are all the required parameters for quantitative assessment reported.

A theory of movement of *Galeorhinus galeus* off southern Australia related to reproductive cycles was first developed from tag data and fishing information. According to this theory, pregnant sharks move into shallow nursery areas in Tasmania and Victoria to give birth and then move to deeper waters. The adults tend to move inshore during summer and offshore, or north to the warmer waters of New South Wales and South Australia, during early winter, before returning south during spring. The neonates and young juveniles tend to remain in the nursery areas before moving to eastern Bass Strait. Older juveniles distribute more widely across southern Australia (Olsen 1954). Archival tagging experiments demonstrating deep-water diurnal feeding patterns off the continental shelf (West and Stevens 2001), information on catch composition from the fishery when it expanded from Bass Strait through South Australia to include the Great Australian Bight, and reproductive data presented in this chapter, generally support this theory. Pregnant females at most stages of gestation are observed in waters of the Great Australian Bight, including the eastern region on the south coast of Western Australia, where they occur for much of the period of gestation, before returning to eastern Bass Strait and Tasmania to give birth (Fig. 4.1).

As will be discussed further, spatial and temporal variations in size, breeding condition and sex ratio of each population and the use of gillnets for harvesting these populations can cause biases in parameter estimates caused by non-representative sampling and possibly length-selective fishing mortality.

4.4 SAMPLING *GALEORHINUS GALEUS*

4.4.1 Capture of Sharks

Sampling to investigate the population biology of *Galeorhinus galeus* was undertaken during three separate periods (1973–76, 1986–87, and 1992–01) mainly in Bass Strait (BS) and waters off South Australia (SA) (Fig. 4.1).

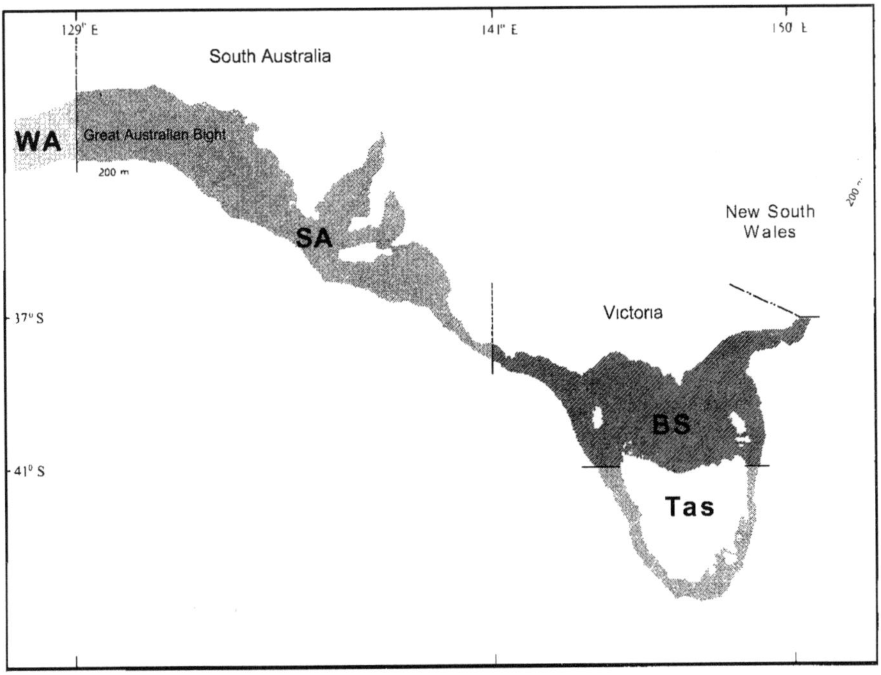

Fig. 4.1 Definition of adopted regions for *Galeorhinus galeus*
WA, Western Australia; SA, South Australia; BS, Bass Strait, and Tas, Tasmania

During 1973–76, the animals were caught using experimental gillnets of mesh-size ranging 2–9 inches (51–229 mm), in steps of 1 inch (25 mm), and hooks attached to sinking longlines. The animals were captured at 162 sites mainly in BS (126 sites) but also off eastern Tasmania south of latitude 41° South (20 sites) (grouped with BS samples) and SA (16 sites). During 1986–1987, the animals were caught in experimental gillnets of mesh-size ranging 5–8 inches (127–203 mm), in steps of 1 inch. The animals were captured at 144 sites (60 sites in BS and 84 sites in SA). During 1992–01, the sampling was opportunistic with most animals caught during 1998–01 at 153 fishing sites (91 sites in BS and 62 sites in SA) by gillnets of 6-inch (152 mm) or 6½-inch (165 mm) mesh-size on board commercial fishing vessels.

4.4.2 Biological Sampling

Specimens of *Galeorhinus galeus* were dissected to investigate their reproduction and other aspects of their biology. They were measured to the nearest millimetre as total length (TL); the tail of each animal was first allowed to take a natural position and the upper caudal lobe was then placed parallel to the body axis. Recorded data included sex, TL, fullness of the stomach and several reproductive indices for each animal. Also recorded, when the sea conditions permitted (mostly on a research vessel during 1973–76), were the wet mass of total body, liver, ovary of females and left testis of males.

For females, at-sea macroscopic inspection of the condition of the paired uteri, paired oviducal glands, and single ovaries was undertaken to investigate breeding condition, litter size, period of gestation and the growth of embryos. Records were made of the diameters of the three largest oocytes and the presence of *corpora atretica* or *corpora lutea* in the ovary and, for pregnant animals, the number of *in utero* eggs and embryos in each uterus. In addition, the TL, sex, uterus (left or right) and wet mass (with and without yolk sac) of each embryo and uterus and wet mass of each *in utero* egg were recorded for many of the pregnant sharks. Indices were adopted for recording the condition of the ovary, oviducal gland, and uteri from rapid visual inspection. Ovary index (O) was based size and colour of the follicles (O = 1–4). Oviducal gland index (G) was based on shape and size of the gland (G = 1–3). Uterus index (U) was based on appearance, size and contents of the uteri (U = 1–6) (Table 4.1).

For males, at-sea macroscopic inspection of condition of the testes, seminal vesicles, and claspers was undertaken to investigate maturity by adopting three indices of breeding condition. Testis index (T) was based on shape, size and relative predominance of testis tissue to epigonal gland tissue (T = 1–3). Seminal vesicle index (V) was based on appearance, thickness of the wall, and presence or absence of seminal fluid (V = 1–3). The length of left clasper was measured from the basipterygium to the distal end and clasper index (C) was based on appearance and rigidity (C = 1–3) (Table 4.1). A histological approach to determining male maturity

Table 4.1 Indices adopted for staging reproductive condition.
Assumptions on mature and immature conditions made for *Galeorhinus galeus* are also listed.

Organ	*Index*	*Description*	*Maturity assumption*
Female			
Ovary	O = 1	Largest follicles white and of diameter <2 mm	Immature
	O = 2	Largest oocytes yolking and of diameter 2–3 mm	Immature
	O = 3	Largest oocytes with yellowish yolk and of diameter >3 mm	Mature
	O = 4	Yolked oocytes of diameter >3 mm and extensive corpora atretica present	Mature
Oviducal gland	G = 1	Indistinct from anterior oviduct	Immature
	G = 2	Distinct but only partly formed	Immature
	G = 3	Enlarged and heart-shaped	Mature
Uterus	U = 1	Uniformly thin tubular structure	Immature
	U = 2	Thin tubular structure partly enlarged posteriorly	Uncertain
	U = 3	Uniformly enlarged tubular structure	Uncertain
	U = 4	In utero eggs present without macroscopically visible embryos present	Mature
	U = 5	In utero embryos macroscopically visible	Mature
	U = 6	Enlarged tubular structure distended	Mature
Male			
Testis	T = 1	Thin tissue strip with epigonal gland predominant	Immature
	T = 2	Thickened strip with epigonal gland tissue extensive	Immature
	T = 3	Enlarged and predominant with epigonal gland tissue negligible	Mature
Seminal vescicle	V = 1	Thin translucent walls and seminal fluids absent	Immature
	V = 2	Thickened opaque walls and seminal fluids present	Mature
	V = 3	Thickened opaque walls and seminal fluids absent	Mature
Clasper[A]	C = 1	Pliable with no calcification	Immature
	C = 2	Partly calcified	Immature
	C = 3	Rigid and fully calcified	Mature

[A]Adopted for periods 1986–1987 and 1992–01, but not for period 1973–76.

was adopted during 1973–76. Two or three pieces of testis tissue (4–8 mm thick) were removed at-sea by transverse section from the left testis and stored temporarily in Bouin's solution. The Bouin's solution was renewed every 12 h for about 36 h and then replaced with 10% neutralised formalin, for subsequent laboratory processing.

4.5 STATISTICAL METHODS

The equations for the litter size curve and the maturity and maternity ogives are expressed as a function of TL rather than age for the statistical analyses undertaken for *Galeorhinus galeus* in Australia. This allows for more robust parameter estimates for *G. galeus* because sample sizes of available data are much larger with TL than with age. Relationships equating TL and age for population assessment of *G. galeus* in Australia are available (Grant *et al.* 1979; Moulton *et al.* 1992). Alternative parameters for the same or similar equations presented as a function of TL in the following can be also expressed as a function of age.

4.5.1 Determining Litter Size and Sex Ratio of Embryos

Simple mathematical equations can be selected to represent the relationships between the number of macroscopically visible *in utero* embryos, p, and maternal TL, l. The linear relationship (Conrath and Musick 2002; Jones and Ugland 2001) between p and l is given by

$$p = a + bl$$

where a and b are parameters estimated by linear regression. One example of a curvilinear relationship between p and maternal l is given by

$$p = ce^{a-bl}$$

where a and b are parameters estimated by linear regression after reformulation to $\ln(p) = a + bl$, and c is a constant to correct for bias caused by logarithmic transformation of p for the regression (Beauchamp and Olson 1973). The linear relationship of *in utero* embryos against TL was adopted for *Galeorhinus galeus* in the present study and a curvilinear relationship was adopted for *Mustelus antarcticus* in southern Australia (Walker 1983; Lenanton *et al.* 1990) from inspection of scattergrams of the data.

For *Galeorhinus galeus*, the effects of factors region and period, the region × period interaction term, and covariate TL on the linear relationship between p and l were statistically tested by analysis of covariance (ANCOVA) through initially including all three terms in the generalized linear model

$$p = \text{Region Period Region} \times \text{Period TL}$$

and then sequentially excluding the term with the highest statistically non-significant P value ($P > 0.05$), until only significant terms were present (stepwise backward elimination). The ANCOVAs were undertaken using the general linear modelling procedure (Proc GLM) of the computer statistical package SAS (SAS Institute, NC, USA).

Paired t-tests were applied to pregnant females with *in utero* eggs (U = 4) or macroscopically visible embryos (U = 5 animals) to test two hypotheses. (1) The sex ratio of *in utero* embryos is 1 : 1. (2) The number of *in utero* embryos and infertile eggs in the left uterus equals the number in the right uterus. The SAS means procedure (Proc Means) was used to compute the required statistics. The statistics were the differences in the means, the standard errors of the difference in the means, the values of the Student t statistic, and the probabilities that the differences are statistically significant (Cody and Smith 1997).

4.5.2 Determining Period of Gestation and Growth of Embryos

The period of gestation and growth of embryos can be determined by plotting mean TL of embryos observed in pregnant females (U = 5 animals) and mean TL values of 0 for *in utero* eggs observed in pregnant females (U = 4 animals) against month and then evaluating the seasonal pattern. In addition, for *Galeorhinus galeus*, mass gain or loss from egg to full-term embryo during gestation was investigated for a sample of pregnant females (U = 5 animals). This was undertaken by separately plotting each of four variables against mean embryo TL for U = 5 animals. These variables were the mean wet mass of embryos, the mean wet mass of external yolks, the sum of these two quantities and the mean external yolk wet mass expressed as a proportion of the sum of the two quantities.

4.5.3 Determining Period of Ovarian Cycle

The ovarian cycle can be investigated by examining the ovary and measuring the diameters of the largest follicles in animals caught throughout the year. The diameters of the largest follicles vary widely between individual animals and vary depending on uterus condition, so seasonal patterns of follicle growth for each of the six uterus conditions defined in Table 4.1 need to be examined separately.

In *Galeorhinus galeus*, pregnant females with macroscopically visible *in utero* embryos (U = 5 animals) provide the least ambiguous basis for determining seasonal growth rates of follicles and for distinguishing between annual, biennial and longer ovarian cycles. The earliest observations of small *in utero* embryos were during late February and the latest observations of near-term embryos were during January. This provides a period, close to one full year, that can be adopted for measuring annual rate of follicle growth.

None of the other five uterus conditions provided such clear information on annual rate of follicle growth. The data indicate that females with uterus conditions U = 1 and U = 2 exhibit little or no change in follicle diameter over the 12-month period from January to December. Females with uterus conditions U = 4 were few in number and occurred for only several months at the end of the year or very early in the year, and therefore provide no information on annual growth of follicles. These animals, however, do provide information on the timing of ovulation and on follicle diameter at the time of ovulation. Similarly, females with uterus condition U = 6 were few in number, displayed little variation in size, and therefore provided no information on annual growth rate of follicles. Nevertheless, these animals did provide information on follicle diameter at the time of parturition and during the period immediately following parturition. That the uterus condition U = 6 was not commonly observed at other times of the year suggests that after parturition the distended uterus contracts to resemble the uterus condition U = 3. This implies that animals recorded with uterus condition U = 3 might be a mixture of animals approaching first pregnancy (all U = 3) and animals between pregnancies (U = 6 changing to resemble U = 3). Unlike animals with uterus conditions U = 5, U = 4, and U = 6, which can be related to the timing of ovulation and parturition, animals with uterus condition U = 3 cannot be so reliably related to either of these events. Hence, these animals are possibly less reliable for determining annual growth rate of follicles.

Pooling data from BS and SA, annual growth rate for animals of uterus condition U = 5 was determined by the linear relationship between follicle diameter, o, and Julian day, t, given by

$$o = a'' + b''t$$

where a'' and b'' are parameters estimated by linear regression. For each of BS and SA separately, the regression line for U = 5 animals was then compared with the scattergrams of follicle diameter against Julian day for each of the U = 3, U = 4 (ovulating), and U = 6 animals separately. A similar regression was undertaken for the U = 3, U = 4 (ovulating), and U = 6 animals combined and compared with that for the U = 5 animals and the scattergrams. These comparisons provided a basis for considering whether the ovarian cycle is annual, biennial, triennial or longer.

4.5.4 Determining Size-at-maturity and Size-at-maternity

The proportion of the population of animals mature at any TL can be determined by classing each animal as in mature condition or immature condition and applying logistic regression for females (Mollet *et al.* 2000; Conrath and Musick 2002) and males separately. Similarly, for females, the proportion of the population of animals in maternal condition at any TL can be determined by classing each animal as in maternal condition or non-maternal condition and applying logistic regression.

For *Galeorhinus galeus*, a female was classed as in mature condition if the largest ovarian follicle was >3 mm in diameter (size at first yolking); otherwise it was classed as in immature condition. Given uncertainty of the best indicator of maturity of males, the results from methods based on alternative criteria for assuming the mature condition and the immature condition are compared. Males were classed by testis condition as mature if T = 3 and immature if T = 1 or T = 2. Similarly, they were classed by seminal vesicle condition as mature if V = 2 or V = 3 and immature if V = 1 and by clasper condition as mature if C = 3 and immature if C = 1 or C = 2 (Table 4.1). There were too few data available to apply a fourth method based on stages of spermatogenesis.

A female was classed as in maternal condition at the time of dissection, if it would have given birth to young before or soon after the following 1 January. To implement this criterion, females were classed as in maternal condition if they met any one of three criteria. These criteria were pregnant with visible embryos (U = 5) during February–December, pregnant with *in utero* eggs (U = 4) during January–May, non-pregnant, or in post-partum condition with distended uteri (U = 6) during November–December. All other females were classed as in non-maternal condition; that is U = 1, U = 2, U = 3, U = 4 during June–December, U = 5 during January with full-term embryos, or U = 6 during January–October.

Logistic regression was adopted to determine the proportion of females in mature condition, the proportion of males in mature condition, and the proportion of females in maternal condition as a function of TL. Females or males in mature condition were assigned a maturity condition value of 1, whereas those in immature condition were assigned a maturity condition value of 0. Similarly, females in maternal condition were assigned a maternal condition value of 1, whereas females in non-maternal condition were assigned a maternal condition value of 0.

These proportions are given by P as a function of TL, l, where P is determined by logistic regression analysis. P is given by a random dichotomous variable y taking the value of 1 with a probability of φ for the mature or maternal condition and the value of 0 with a probability of $1-\varphi$ for the immature or non-maternal condition. This point-binomial variable has a probability distribution such that

$$P = P^y(y;\varphi) = \varphi^y(1-\varphi)^{(1-y)}$$

where $y = 0{,}1$

The likelihood function, $L(y;\varphi)$, takes the form

$$L(y;\varphi) = \prod_{j=1}^{N} \{\varphi_j y_j (1-\varphi_j)^{(1-y_j)}\}$$

where φ_j represents the probability that individual j in a random sample of N animals from the shark population which is judged to be in mature condition or, alternatively, maternal condition at the time of sampling. The logistic equation adopted to express P as a function of l is given by

$$P = \frac{c'''}{(1 + e^{-(a''' + b''' l)})}$$

where a''', b''' and c''' are parameters (Walker 1994) but to provide parameters that are more biologically meaningful, the equation is reformulated as

$$P = P_{\max}\left(1 + e^{-\ln(19)\left(\frac{l - l_{50}}{l_{95} - l_{50}}\right)}\right)^{-1}$$

where P_{max} is the maximum proportion of animals in mature condition or maternal condition (equivalent to c'''), l_{50} and l_{95} and are the lengths at which 50% and 95% of the maximum proportion of animals in mature condition or maternal condition (Punt and Walker 1998).

The parameters a''', b''' and c''', P_{max}, l_{50} and l_{50}, with 95% confidence intervals, were estimated by the method of maximum likelihood using the probit procedure (Proc Probit) of the computer statistical package SAS (SAS Institute, Cary, North Carolina, USA). This applies a modified Newton-Raphson algorithm for estimation.

The standard error for any length, l, is given by

$$se_l = P_l(1-P_l)/\mathrm{N}$$

For *Galeorhinus galeus*, the effects of factors region and period and the region x period interaction term on the logistic relationships between P and l were statistically tested by initially including all three terms in the logistic regression model

$$P=\left(\frac{n}{N}\right)=\text{Region Period Region x Period TL}$$

where n is the number of mature or maternal animals, N is the total number of animals included by SAS is the analysis. SAS assigns the data to length-classes automatically, unless specified by the user, and discards data where there is more than one length-class with values of only 0 and more than one length-class with only values of 1.

For *Galeorhinus galeus*, the model was then run repeatedly and the least statistically non-significant factor or interaction term was sequentially deleted by stepwise backward elimination until only statistically significant terms ($P < 0.05$) remained in the model. The logistic relationships between P and l were tested for the effects of the two regions BS and SA, the three sampling periods 1973–76, 1986–87, and 1998–01, and the region x period interactions. These tests were undertaken by applying the logistic procedure (Proc Logistic) of the computer statistical package SAS. The terms were tested by the χ^2 likelihood-ratio test (Rao 1973; Silvey 1975).

The SAS probit procedure sets $1-c''' = 0.000$. This is appropriate for the maturity ogive where all large-sized animals in the population are in mature condition, and hence the proportion of large-sized animals in the population mature is 1.000. Similarly, this is appropriate for the maternity ogive where all of the large-sized animals in the population are in maternal condition, and hence the proportion of large-sized animals in the population in maternal condition is 1.000; parturition frequency is annual. However, this is not appropriate where parturition frequency is biennial, triennial or some other period.

Application of the SAS probit procedure is more complex to apply to any parturition frequency, γ, other than 1 year. For example, if parturition is biennial where half the population gives birth each year then $\gamma = 0.50$ or if parturition is triennial where one-third the population gives birth each year then $\gamma = 0.33$. This was undertaken for *G. galeus* by categorising the number of animals in maternal condition and the number of observations into 100-mm length-classes. For parturition frequency, where the ratio of number in maternal condition / number of observations exceeds γ within a 100-mm length-class, the number in mature condition needs to be adjusted to produce the ratio γ. For SAS probit analysis, the number of observations in each 100-mm length-class (or some other selected range) is multiplied by γ. A weight statement was used to weight the values in each length-class by the original number of observations in that length-class. The ogive relationships, with 95% CI, produced by the SAS probit procedure can then be divided by γ to give the required parameter values of the maternity ogive, with 95% CI.

4.5.5 Determining Total Body Mass at Size

The relationship between total body mass, w and TL, l, can be determined using the power curve

$$w = a'''' \, c'''' l^{b''''}$$

adopted commonly for sharks (Olsen 1954) and bony fishes (Ricker 1958) without the constant c'''', where a'''' and b'''' are parameters determined by linear regression of $ln(w)$ against $ln(l)$, and c'''' is a factor for correcting for biases caused by natural logarithmic transformation (Beauchamp and Olson 1973).

For *Galeorhinus galeus*, linear regression relationships were determined separately for males, non-pregnant females, pregnant females with *in utero* eggs, and pregnant females with *in utero* embryos. These four relationships were determined separately because females grow larger than males and because pregnant animals weigh more than non-pregnant animals at any length. The Student *t*-test was applied to test for differences between the slopes (parallelism) and intercepts (elevation) for selected pairs of straight lines determined from the four $ln(w)$–$ln(l)$ regression fits (Kleinbaum *et al.* 1988).

4.6 APPLICATION OF METHODS TO *GALEORHINUS GALEUS*

4.6.1 Litter Size and Sex Ratio of Embryos

Macroscopically visible *in utero* embryos were examined in 63 pregnant *G. galeus* (U = 5 animals); there were 22 U = 5 animals from BS and 41 from SA. ANCOVA testing for effects of factors region and period, region x period interaction, and covariate maternal TL by stepwise backward elimination indicated that only maternal TL was statistically significant (Table 4.2). This allowed data from the two regions (BS and SA) and three periods (1973–76, 1986–87 and 1998–01) to be pooled to provide a single relationship between the number of *in utero* embryos and maternal TL determined from linear regression. These animals ranged in size 1429–1680 mm TL and carried 15–43 *in utero* embryos (Fig. 4.2).

Table 4.2 Hypothesis testing for females with in utero embryos

Analyses of covariance (ANCOVA) testing for the effects of region (Bass Strait and South Australia), period (1973–76, 1986–87 and 1998–01), and region x period interaction on the number of in utero embyos against total length for pregnant females with macroscopically visible embryos (U = 5) by stepwise backward elimination of non-significant factors. d.f., degrees of freedom; M.S., mean square; ns, not significant; P is probability of statistical significance (*P < 0.05; **P < 0.01; ***P < 0.001).

Source of variation		*d.f.*	*M.S.*	*P*	
Step 1					
	Region	1	20.814	0.6108	ns
	Period	2	5.604	0.8749	ns
	Region x period	2	88.443	0.1516	ns
	Maternal length	1	529.230	0.0008	***
	Residual	56	41.847		
Step 2					
	Region	1	71.307	0.1971	ns
	Period	2	56.334	0.2751	ns
	Maternal length	1	530.626	0.0008	***
	Residual	57	42.664		
Step 3					
	Region	1	106.569	0.0932	ns
	Maternal length	1	603.990	0.0004	***
	Residual	59	43.128		

The overall mean number (±s.e.) of *in utero* embryos and infertile eggs in the 63 pregnant females with macroscopically visible embryos was 31.24 (±0.88) where 30.18 (±0.92) were embryos and 1.06 (±0.20) were infertile eggs indicating 96.6% of *in utero* eggs develop as embryos. The mean number of embryos in the left uterus (14.71±0.53) was not significantly different from the mean number of embryos in the right uterus (15.46±0.46). Similarly, the mean number of infertile eggs in the left uterus (0.60±0.12) was not different from the mean number of infertile eggs in the right uterus (0.46±0.13). The mean number of male embryos (14.56±0.62) was not significantly different from the mean number of female embryos (14.76±0.62); 2.8% of the embryos were too small to confidently identify their sex (0.86±0.54) (Tables 4.3 and 4.4). Of the 63 U = 5 animals, 28 animals had zero infertile eggs (44.4% of animals), 19 had one (30.2%), 11 had two (17.4%), two had three (3.2%), one had four (1.6%), one had six (1.6%), and 1 had ten (1.6%) infertile eggs. In addition to counting the

Table 4.3 Mean number (±se) of *in utero* embryos and eggs in pregnant females
s.e., standard error for 63 pregnant females with macroscopically visible embryos (U = 5 animals).

Mean (±s.e.) of number of embryos and infertile eggs											
Left uterus			Right uterus			Embryos			Total		
Embryos	Eggs	Total	Embryos	Eggs	Total	Male	Female	Unknown	Embryos	Eggs	Total
14.71±0.53	0.60±0.12	15.32±0.51	15.46±0.46	0.46±0.13	15.92±0.44	14.56±0.62	14.76±0.62	0.86±0.54	30.18±0.92	1.06±0.20	31.24±0.88

Table 4.4 Testing sex ratio and number of *in utero* and infertile eggs embryos between left and right uteri
Statistical paired t-tests comparing the number of *in utero* embryos and eggs between the left and right uteri and between male and female embryos in 63 pregnant females with macroscopically visible embryos (U = 5 animals). ns, not significant; P is the probability of statistical significance (*P < 0.05; **P < 0.01; ***P < 0.001).

Comparison	*Mean difference*	*s.e. of difference*	*t value*	*P*	
Left uterus embryos & eggs – right uterus embryos & eggs	–0.603	0.355	1.70	0.0946	ns
Left uterus embryos – right uterus embryos	–0.746	0.375	1.99	0.0512	ns
Left uterus eggs – right uterus eggs	0.143	0.139	1.03	0.3088	ns
Male embryos – female embryos[A]	–0.210	0.719	0.29	0.7715	ns

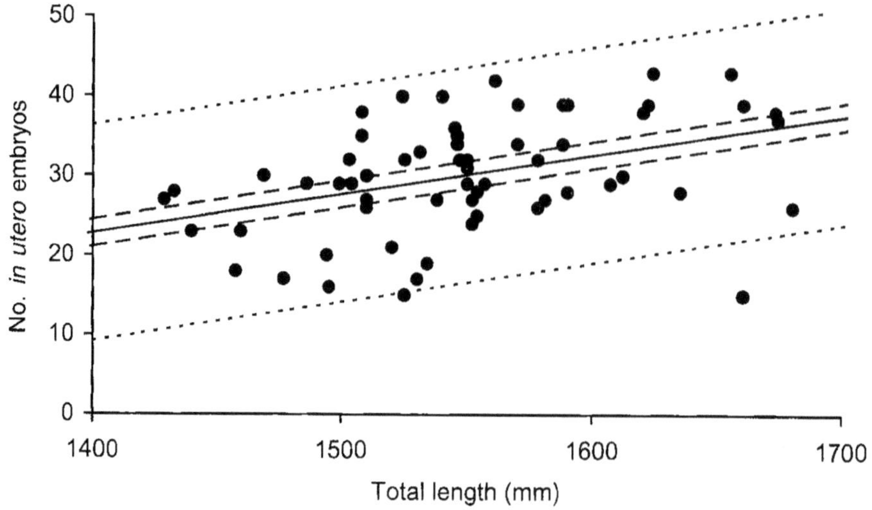

Fig. 4.2 Number of *in utero* embryos against maternal total length
Mean number of embryos (—), 95% confidence limits (– –), 95% prediction intervals (- - -), and raw data (•) are plotted against maternal total length of pregnant females with macroscopically visible embryos (U = 5). Values of parameters and statistical quantitites for the equation $p = a' + b'l$ are given in the following tabulation:

a' (±s.e.)	b' (±s.e.)	n	r^2	*rmse*	*P*
−46.0 (±22.2)	0.0491 (±0.0143)	63	0.148	6.724	***

where l is maternal total length measured in millimetres, p is number of in utero embryos, a' and b' are parameters, n is sample size, r^2 is square of regression correlation coefficient, *rmse* is root mean square error, and *P* is the probability of statistical significance (*P < 0.05; **P < 0.01; ***P < 0.001) for linear regression.

embryos and infertile eggs in these 63 U = 5 animals, eggs were counted in four pregnant females with only *in utero* eggs (U = 4 animals). Based of the simultaneous presence of large follicles in the ovary and eggs *in utero*, these animals were all judged to be in the process of ovulation (Table 4.5).

4.6.2 Period of Gestation and Growth of Embryos

Mean TL of embryos (with standard error) measured in each of 54 pregnant females (U = 5 animals) and an assigned mean TL value of 0 mm for *in utero* eggs observed in 65 pregnant females (U = 4 animals) are plotted against month (Fig. 4.3). *In utero* embryos were observed during the eight-month period May–January and *in utero* eggs were observed during the seven-month period June–January. *In utero* embryos were not observed during the three-month period during February–April; pregnant females with early-stage embryos (U = 5 animals) occur in the Great Australian Bight and more easterly regions in SA (anecdotal information from fishers), and possibly in oceanic waters away from the continental shelf. These regions were not sampled at that time of the year. The data for the period May–January indicate a high degree of synchrony in gestation between

Table 4.5 Number of *in utero* eggs in ovulating pregnant females
TL, total length; LFD, largest follicle diameter in four U = 4 animals.

Animal	Date	TL (mm)	LFD (mm)	No. in utero eggs		
				Left uterus	Right uterus	Total
1	Oct-92	1580	52	4	4	8
2	Dec-92	1600	53	22	22	44
3	Nov-95	1480	47	5	5	10
4	Nov-98	1570	34	6	6	12

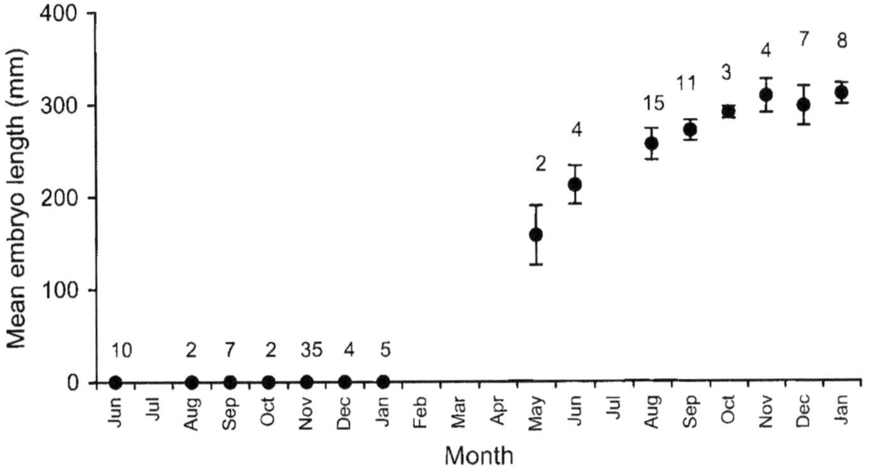

Fig. 4.3 Mean embryo length against month
Derived from the mean embryo length of the litter from each of 54 pregnant animals with macroscopically visible embryos and 65 pregnant animals with only *in utero* eggs; •, overall mean; bars, standard deviation .

U = 5 animals. This synchrony, together with the long period for the presence of *in utero* eggs, is evidence that the period of gestation exceeds one year and the frequency of parturition cannot be annual.

The highest mean embryo wet mass of 149 g towards the end of gestation observed in one pregnant animal when mean TL of its embryos exceeded 300 mm was about double the highest mean egg wet mass of 80 g at the beginning of gestation in another animal (Fig. 4.4). Given *Galeorhinus galeus* is aplacental, this approximately 100% mass gain from egg to full-term embryo suggests *G. galeus* provides nutrients to the embryos by way of intra-uterine nutrients (histotroph), although the mass gain might be from hydration. There is little or no mass gain in the combined mass of embryo and external yolk sac for the first half of gestation but this increases progressively towards the end of gestation (Fig. 4.4).

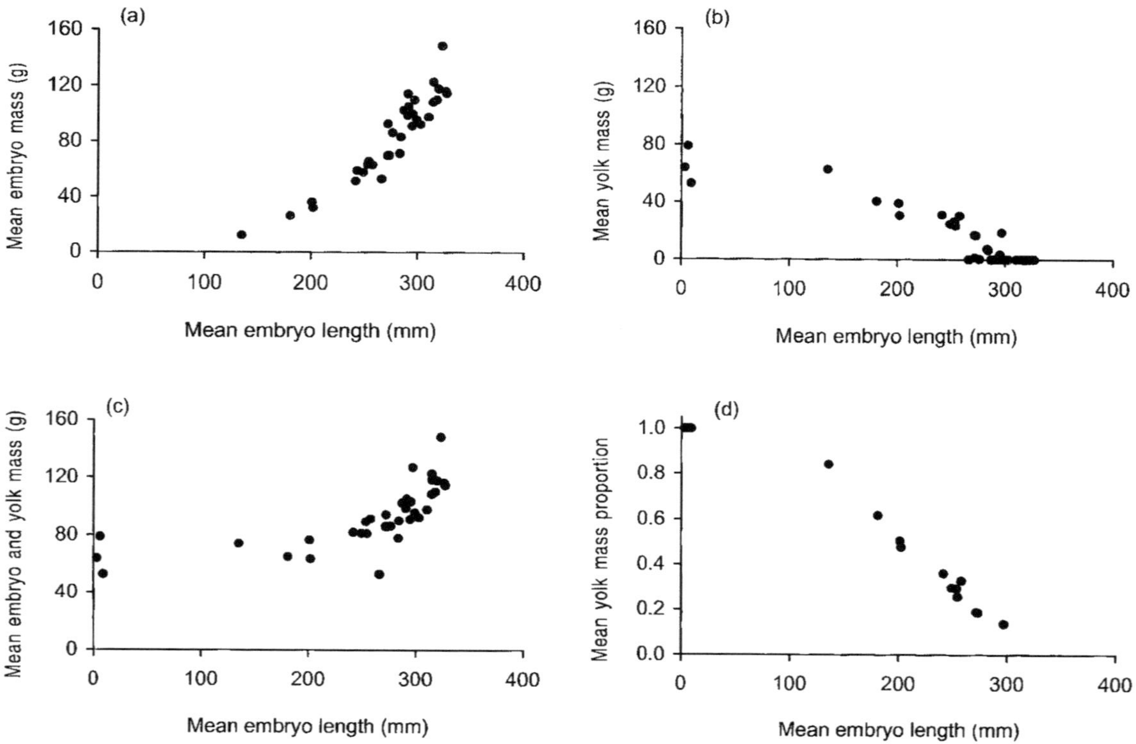

Fig. 4.4 Mass gain of embryos and mass loss from yolk sac during embryonic development
Mean mass of embryos (a), mean mass of yolk sacs (b), mean mass of embryo and yolk (c) and yolk sac as a proportion of sum yolk sac and embryo mass (d) against mean embryo length. Each data point is derived from the mean embryo mass, mean yolk mass and mean embryo length determined for the litter of each of 36 pregnant animals with macroscopically visible embryos. Yolk mass proportion is yolk sac mass/(embryo mass + yolk sac mass).

4.6.3 Ovarian Cycle

The diameter of the largest follicle recorded in the ovary from each of 715 females (230 from BS and 485 from SA) ranged 1–56 mm. There was little difference between the diameters of the three largest follicles (Table 4.6), so all statistical analyses of follicle data were undertaken using only the first measured follicle, which was judged visually to be the largest when measured.

Table 4.6 Comparison of diameters of three largest ovarian follicles for each uterus condition n, sample size; s.e., standard error.

Uterus condition[A]	*n*	*Mean diameter (±s.e.) of three largest ovarian follicles (mm)*		
		Follicle 1	*Follicle 2*	*Follicle 3*
U = 1	226	1.40±0.06	1.38±0.06	1.36±0.06
U = 2	238	2.08±0.12	2.08±0.12	2.07±0.12
U = 3	94	14.68±1.26	14.39±1.24	14.22±1.23
U = 4	16	34.06±3.95	34.25±4.07	31.44±4.20
U = 5	72	9.63±0.36	9.33±0.33	9.15±0.32
U = 6	24	13.58±0.65	13.83±0.79	13.50±0.66

[A]Defined in Table 4.1

Patterns in plots of largest follicle diameter (LFD) against Julian day were not evident where all the data were pooled; patterns in LFD against Julian day were evident only when the animals were considered for each uterus condition separately. LFD was consistently small for animals with uterus conditions U = 1–2, a clear indication that these animals were immature or at early stages of folliculogenesis or vitellogenesis. There was little variation in LFD among U = 1 animals (n = 232, mean 1.4 mm, s.d. 1.0 mm, range 1–9 mm) or among U = 2 animals (n = 244, mean 2.1 mm, s.d. 1.8 mm, range 1–15 mm) (Fig. 4.5). LFD varied widely among animals with uterus conditions U = 3 (1–56 mm), U = 4 (7–53 mm), U = 5 (3–19 mm) and U = 6 (4–19 mm). From uterus condition alone, these animals can be classed as pregnant or post-partum (U = 4–6) or as mature and approaching ovulation (U = 3). The patterns in plots of LFD against Julian day are generally similar between BS and SA, apart from the lack of U = 4 and U = 5 animals in BS prior to Julian day 274 (October) (Fig. 4.6).

For pregnant females with macroscopically visible embryos (U = 5 animals), linear regression of LFD against Julian day for animals from BS and SA pooled indicated that annual growth of LFD is 10.2 mm y^{-1}. At the end of the year when gestation is complete or approaching completion, the ovarian follicles are much too small for ovulation. The predicted mean LFD increased from 1.9 to 12.1 mm during one year, evidence that the ovarian cycle, and hence frequency of parturition, exceeds one year (Figs 4.5, 4.6).

The U = 3, U = 4 and U = 6 animals were then examined to assess whether they conformed to the hypothesis of a two-year, three-year or longer ovarian

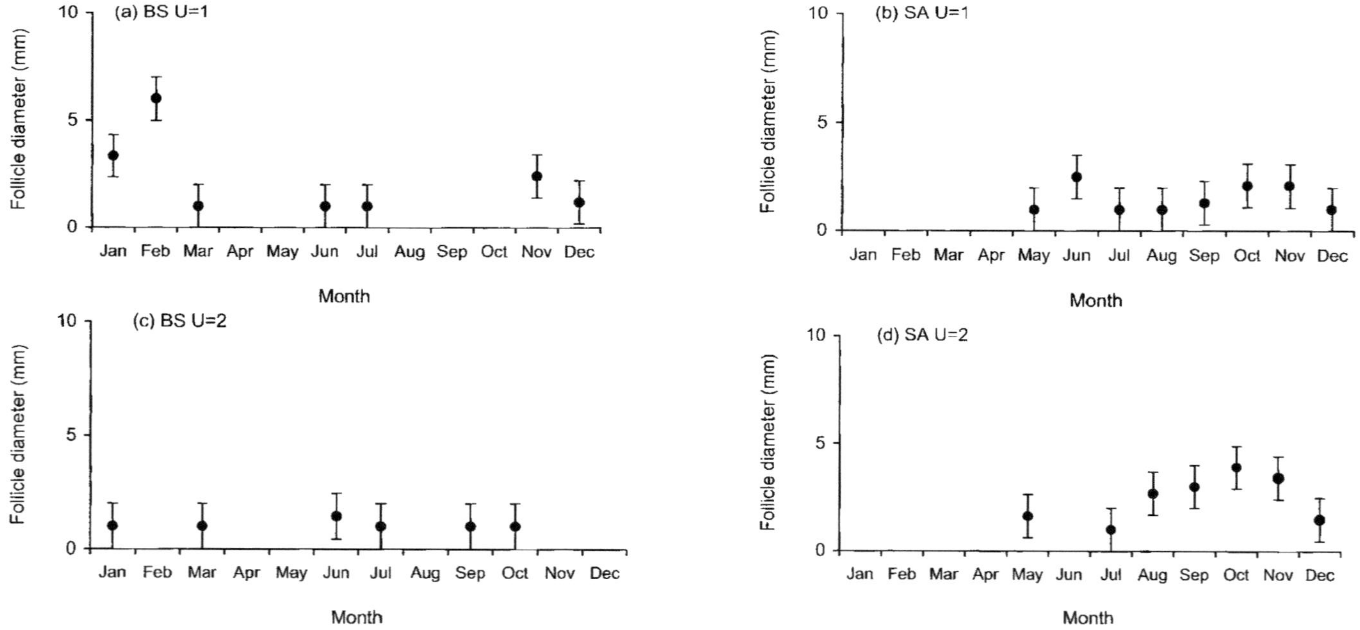

Fig. 4.5 Ovarian follicle diameter against month by region for uterus conditions U = 1–2
Mean oocyte diameter (±s.d.) plotted against month for each of the two uterus conditions U=1 (a and b) and U=2 (c and d) from Bass Strait (BS) (left) and South Australia (SA) (right) during 1973–76, 1986–87 and 1998–01. Sample size (n) and annual mean follicle diameter are given in the following tablulation:

Fig.	*U*	*Region*	*n*	*Mean (±s.d.) (mm)*	*Range (mm)*
(a)	1	BS	113	1.2 (1.0)	1– 7
(b)	1	SA	119	1.6 (1.0)	1– 9
(c)	2	BS	65	1.1 (0.5)	1– 3
(d)	2	SA	179	2.5 (1.9)	1–15

cycle. For BS and SA separately, scattergrams of LFD plotted against Julian day for animals in each of these three uterus conditions showed that LFD values were in three broad clusters (<18 mm, 18–33 mm and 34–56 mm). This provides evidence in favour of the three-year ovarian cycle, consistent with the three-year ovarian cycle hypothesised for the stock of *G. galeus* occurring off eastern South America (Peres and Vooren 1991). Growth of the largest ovarian follicles was described by determining the linear relationship between LFD and adjusted Julian day for the 3-year period. Adjusted Julian day for the three-year period was calculated by adding to Julian day 365 days if LFD ranged 18–33 mm or 730 days if LFD ranged 34–56 mm; Julian day was unadjusted if LFD <18 mm. The predicted mean LFD trajectory for the three-year ovarian cycle was determined by linear regression pooling data from BS and SA and then plotted on each of the three scattergrams of uterus condition for each of BS and SA separately. The U = 4 animals where ovulation was complete (LFD <30 mm) were excluded from the regression analysis. The LFD trajectory was extrapolated through a second and third year but displayed for a 1 year period by presenting the trajectory as three parallel lines, one for each year. The linear relationship for U = 3, U = 4 (ovulating only), and U = 6 animals and the linear relationship for U = 5 animals were not significantly different (t-test, $t = 1.906$, d.f. = 228 and $P > 0.05$ for comparison of slopes and $t = 1.045$, d.f. = 228, and $P > 0.05$ for comparison of elevations) (Figs 4.6, 4.7).

Among most U = 3 animals, the individual LFD values were consistent with the hypothesis for a three-year ovarian cycle in both BS and SA (Fig. 4.6). The animals are distinct between the second and third years but are less distinct between the first and second years.

Among U = 4 animals, the individual LFD values were also consistent with the hypotheses for a three-year ovarian cycle, particularly in SA. These data provide reliable information on the timing of ovulation and on magnitude of LFD at the time of ovulation. The animals were classed as ovulating or ovulated on the basis of LFD, which were variously clustered in each of BS (ranging 8–30 mm LFD) and SA (ranging 7–11 mm, one 34 mm, and 42–53 mm LFD). Animals were classed as ovulating (in process of ovulation) if they contained eggs *in utero* and ≥34-mm LFD; animals were classed as ovulated (ovulation complete) if they contained eggs *in utero* and <34-mm LFD. No U = 4 animals were found ovulating in BS but four ovulated animals were captured during the period from 13 October (Julian day 286) to 14 November (Julian day 318). With the exception of one ovulated animal captured on 21 June (Julian day 172), all U = 4 animals captured in SA during the period from 21 June (Julian day 172) to 4 December (Julian day 338) were ovulating (Fig. 4.6).

Among most U = 6 animals, the individual LFD values have a similar distribution to those of the U = 5 animals; they are clustered near the trajectory for the first year. The single LFD value near the mean LFD trajectory at the end of the first year in each of BS and SA were likely to

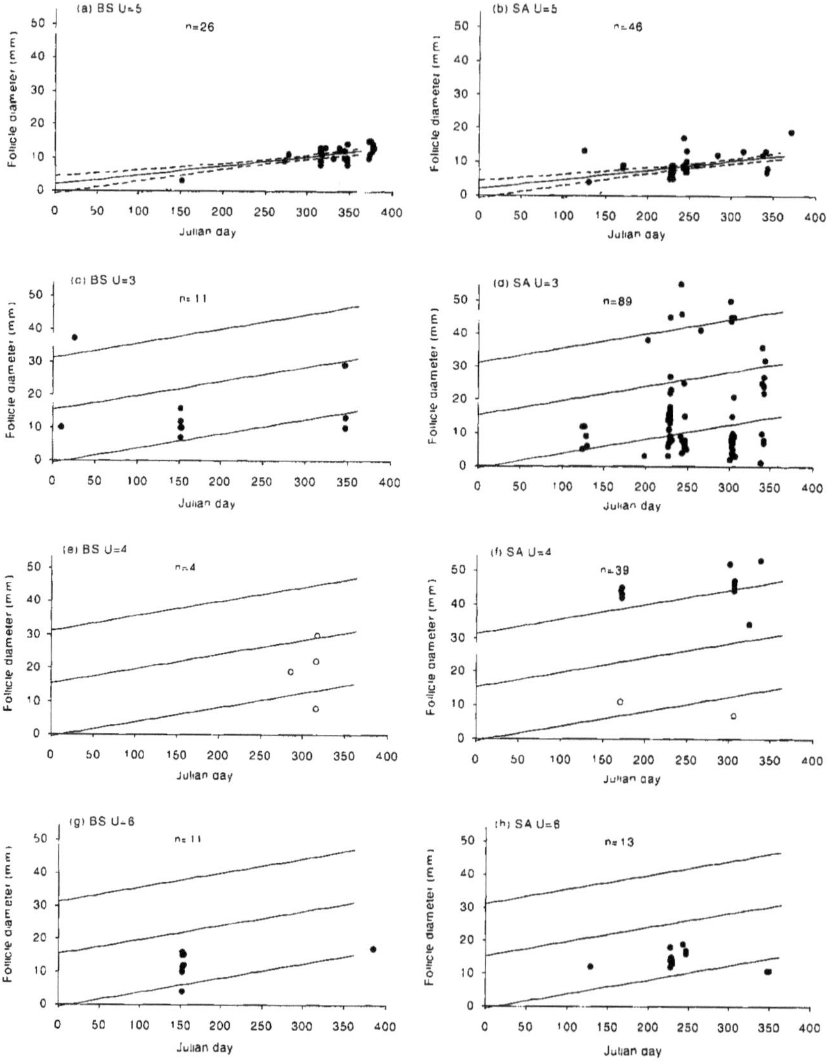

Fig. 4.6 Largest follicle diameter against Julian day by region for uterus conditions U = 3–6
Follicle diameter against Julian day for females from Bass Strait (BS) (left) and South Australia (SA) (right) during 1973–76, 1986–87 and 1998–01 combined for each of the of four uterus conditions (U = 3–6). Mean follicle diameter (—) with 95% confidence limits (– –) and 95% prediction intervals (- - -) are presented for pregnant females with *in utero* embryos (U = 5) (a and b), non-pregnant animals (U = 3) (c and d), pregnant animals with *in utero* eggs (U = 4, ovulating (•) and ovulated (○)) (e and f), and postpartum females (U = 6) (g and h). Values of parameters and statistical quantities for the regression equation o = a'+$b't$ for U = 5 animals and for U = 3, U = 4 (ovulating) and U = 5 animals pooled from BS and SA are given in the following tablulation:

U	*a' (±se)*	*b'(±se)*	*n*	r^2	*rmse*	*P*
5	1.93 (1.25)	0.0280 (±0.0044)	72	0.354	2.467	***
3, 4 (ovulating), 6	–0.56 (0.73)	0.0436 (±0.0011)	160	0.903	5.166	***

Fig. 4.6 *Contd. ...*

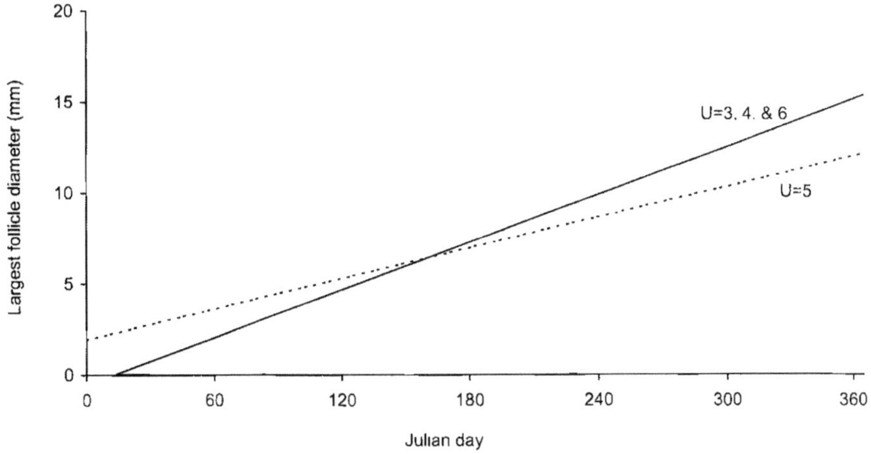

Fig. 4.7 Comparison of LFD against Julian days for U = 5 condition with U = 3, 4 (ovulating) and 6 conditions

The regression line of largest follicle diameter (LFD) against Julian day for females of uterus condition U = 5 (——) is compared with the regression line for uterus conditions U = 3, 4 (ovulating only), and 6 (- - -). The data are pooled from animals collected from Bass Strait and South Australia during 1973–76, 1986–87 and 1998–01.

be animals that had recently given birth. The other animals clustered near the trajectory for the first year during earlier months are inconsistent with the three-year ovarian cycle in BS and SA (Fig. 4.6); they are most likely U = 3 animals incorrectly classed as U = 6.

Ovary mass was available for only 43 animals. Ovary mass was available for U = 1 animals (n = 11, mean 13.0 g, s.d. 11.2 g, range 1–32 g), U = 3 animals (n = 2, mean 70.5 g, s.d. 3.5 g, range 68–73 g), U=4 animals (n = 7, mean 568.3 g, s.d. 660.0 g, range 48–1880 g), and U = 5 animals (n = 23, mean 169.1 g, s.d. 244.1 g, range 44–910 g). Ovary mass was not taken for U = 2 or U = 6 animals.

Percentage GSI was available for 41 animals. GSI was available for U = 1 animals (n = 11, mean 0.15, s.d. 0.09, range 0.02–0.28), U = 3 animals (n = 2, mean 0.42, s.d. 0.01, range 0.41–0.42), U = 4 animals (n = 7, mean 2.58, s.d. 2.76, range 0.28–7.23), and U = 5 animals (n = 21, mean 0.70, s.d. 0.21, range 0.22–3.88). Percentage GSI was not taken for U = 2 or U = 6 animals.

Percentage HSI was available for 64 animals. Percentage HSI was available for U = 1 animals (n = 25, mean 5.47, s.d. 1.35, range 3.32–9.00), U = 3 animals (n = 3, mean 6.46, s.d. 5.05, range 2.75–12.22), U = 4 animals (n = 12, mean 10.48, s.d. 2.52, range 5.69–14.15), and U = 5 animals (n = 24,

---

Fig. 4.6 *Contd.* ...

where *t* is Julian day (for up to 3 years), *o* is follicle diameter, *a'* and *b'* are parameters, *n* is sample size, r^2 is square of regression correlation coefficient, and rmse is root mean square error for the regression, and *P* is probability of statistical significance (*P < 0.1; **P < 0.01; ***P < 0.001).

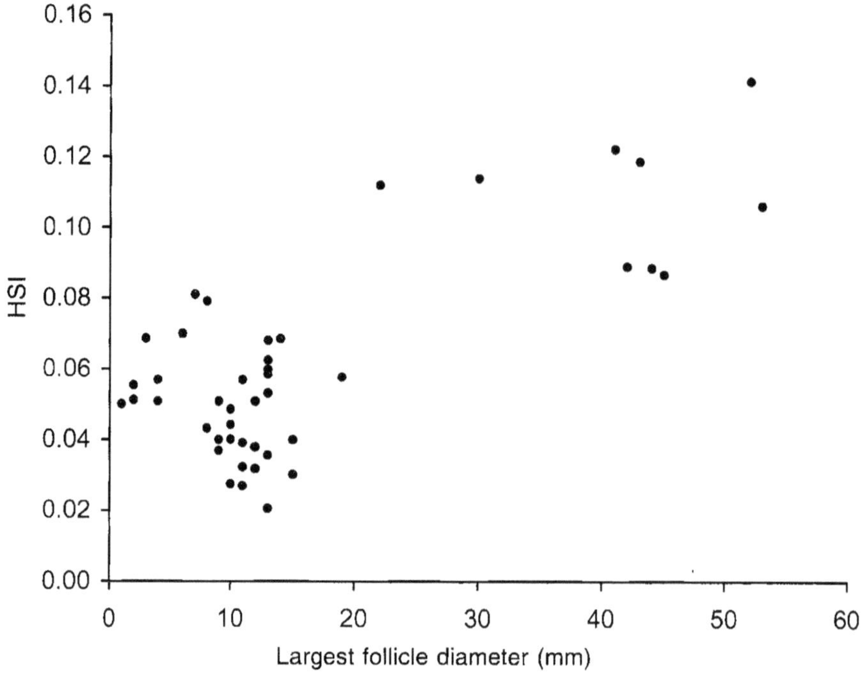

Fig. 4.8 HSI against largest ovarian follicle diameter
Spearman correlation between hepatic somatic index (HSI) and ovarian largest follicle diameter is 0.397 (P = 0.0076 **) for U = 1–6 animals where *P* is probability of statistical significance (*P < 0.1; **P < 0.01; ***P < 0.001).

mean 4.65, s.d. 1.51, range 2.06–8.24). Percentage HSI was not taken for U = 2 or U = 6 animals. HSI tends to rise with LFD (Fig. 4.8.), suggesting the liver mass increases concurrently with the process of vitellogenesis.

4.6.4 Female Maturity Ogives

For *Galeorhinus galeus*, hypothesis testing for the effects of region, period and region x period interaction on logistic regression models was undertaken by stepwise backward elimination of statistically non-significant terms by log-likelihood ratio tests. For the full model, only the effect of period (P = 0.0040) was statistically significant; the effects of region and region x period interaction were not significant. Hence, further hypothesis testing was undertaken for only the effect of period; the two regions BS and SA were pooled. The effect of period between 1973–76 and 1986–87 (P = 0.0006) and between 1973–76 and 1998–01 (P=0.0032) was highly significant but was not significant between 1986–87 and 1998–01 (P = 0.2415) (Table 4.7). Hence, the data were pooled for the periods 1986–87 and 1998–01 for subsequent probit analysis (Table 4.8). Separate figures are therefore presented for the 1973–76 period and for the 1986–87 and 1998–01 periods combined (Fig. 4.9). The large difference in sample size between 1973–76 (56 mature and 104 immature animals) and the combined periods

Table 4.7. Hypothesis testing for female maturity based on largest follicle diameter

Testing for the effects of region (Bass Strait and South Australia), period (1973–76, 1986–87, and 1998–01) and region x period interaction on logistic regression of the proportion of the animals in the population mature against TL. Only final model in stepwise backward elimination to show effect of period is presented. An animal is classed as mature where the largest follicle diameter is >3 mm. χ^2, *Chi* square; ns, not significant; P is probability of statistical significance (*P < 0.05; **P < 0.01; ***P < 0.001).

Source of variation	*Number of animals*		χ^2	*P*	
	Mature	*Immature*			
Region x Period	375	491			
Intercept			148.557	<0.0001	***
Period			8.274	0.0040	**
TL			153.393	<0.0001	***
Between 1973–76 and 1986–87	222	423			
Intercept			86.070	<0.0001	***
Period			11.784	0.0006	***
Total length			102.721	<0.0001	***
Between 1973–76 and 1998–01	212	172			
Intercept			62.411	<0.0001	***
Period			8.678	0.0032	**
Total length			64.105	<0.0001	***
Between 1986–87 and 1998–01	316	387			
Intercept			125.823	<0.0001	***
Period			1.372	0.2415	ns
Total length			128.154	<0.0001	***
Between 1973–76 and 1986–87, 1998–01	375	491			
Intercept			0.010	0.9201	ns
Period			13.362	0.0003	***
Total length			154.284	<0.0001	***

1986–87 and 1998–01 (245 mature and 458 immature animals) accounts for the wide difference between the magnitude of the 95% confidence intervals for these two periods. The l_{50} value (with 95% CI) of 1349 (1339, 1358) mm for mature condition during 1986–87 and 1998–01 combined is larger than the value of 1244 (1184, 1304) mm for 1973–76 (Fig. 4.9), suggesting an actual or apparent increase in the size-at-maturity. An apparent change in size-at-maturity could occur from the effects of sampling different age classes between the two periods or from the effects of length-selective fishing mortality. As demonstrated for *Mustelus antarcticus* (Walker *et al.* 1998), gillnets of 6–7-inch mesh-size used in the intensive shark fishery of southern Australia have the effects of selectively removing large young animals and small old animals from the population. This can give rise to the "phenomenon of apparent change of growth rate" reflected in apparent rather than actual changes in the shape of growth curves (Lee 1920; Ricker 1969). Such changes in a population could also affect maturity ogives.

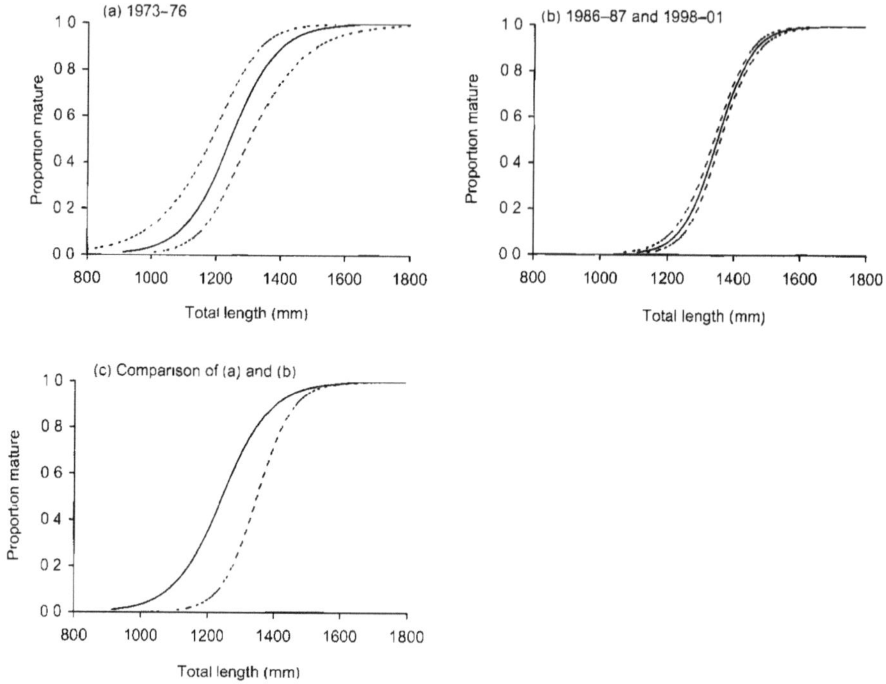

Fig. 4.9 Female length-at-maturity ogives
Proportion of population mature against TL (——), with 95% confidence intervals (- - -), for females sampled during 1973–76 (a), 1986–87 and 1990–01 combined (b), and comparison of the mean ogives for 1973–76 (——) and 1986–87 and 1998–01 (- - -). An animal is defined as mature if the largest ovarian follicle diameter is >3 mm. Values of parameters and statistical quantities for the equation $P = P_{max}(1+e^{-\ln(19)(l-l_{50})/(l_{95}-l_{50})})^{-1}$ determined from probit analysis are given in the following tabulation:

Period	l_{50} *(CI)*	l_{95} *(CI)*	P_{max}	*n*	*N*	*ML*	*P*
1973–76	1244 (1184, 1304)	1458 (1385, 1587)	1.000	59	163	−22.622	***
1986–87, 1998-01	1349 (1339, 1358)	1502 (1487, 1521)	1.000	316	703	−443.147	***

where *l* is total length measured in millimetres, P is proportion of animals at TL *l*, l_{50} and l_{95} are parameters, P_{max} is an asymptotic constant, *n* is the total number of animals classed as mature, and *N* is the total number of animals selected in statistical procedure, *ML* is maximum likelihood, and *P* is probability of statistical significance (*P < 0.1; **P < 0.01; ***P < 0.001).

4.6.5 Female Maternity Ogives

Pregnant female *Galeorhinus galeus* tend to aggregate in discrete regions at certain times of the year, so it was important to have wide spatial and temporal sampling for maternal condition. Because the maturity ogive for 1973–76 period was statistically different from the maturity ogive for the 1986–87 and 1998–01 periods combined and because the sample size for maternity was small during 1973–76, only data for the combined periods were used for determining a maternity ogive.

Table 4.8 Number of females sampled in each length-class for maturity condition An animal is classed as mature where the largest follicle diameter is >3 mm.

Period	*Length-class (mm)*	*No. of animals*			*Proportion mature*
		Immature	*Mature*	*Total*	
1973–76	<700	46		46	0.000
	700–799	9		9	0.000
	800–899	17		17	0.000
	900–999	12		12	0.000
	1000–1099	10		10	0.000
	1100–1199	7	1	8	0.125
	1200–1299	3	2	5	0.400
	1300–1399		4	4	1.000
	1400–1499		10	10	1.000
	1500–1599		26	26	1.000
	1600–1699		13	13	1.000
	Total	104	56	160	
	<700	9		9	0.000
	700–799	46		46	0.000
	800–899	60		60	0.000
	900–999	54		54	0.000
	1000–1099	52		52	0.000
1986–87 and	1100–1199	43		43	0.000
1998–01	1200–1299	71	3	74	0.000
combined	1300–1399	67	35	102	0.343
	1400–1499	53	97	150	0.647
	1500–1599	3	86	89	0.966
	1600–1699		24	24	1.000
	Total	458	245	703	

In preparation for probit analysis the females were classed as in either maternal or non-maternal condition for each 100-mm TL-class, which indicated that about one-third of the females in the largest TL-classes were in maternal condition. These data are consistent with the results for the ovarian cycle, which indicate that most animals ovulate every third year. Before weighting the total number in each TL-class by a factor of 3, the total number of animals in the 1500–1599 mm TL-class needed to be adjusted from 90 to 93, so as to adjust the proportion in maternal condition from 0.344 to 0.333 (Table 4.9). This is because each value after weighting for probit analysis must be 1.000 or less.

The probit analysis produced a curve and 95% confidence intervals (CI), which had maximum values of 1.000. These curves were then divided by 3 for the purpose of readjusting the curve back to a maximum value of 0.333. After adjustment, TL at which 50% (l_{50}) and 95% (l_{95}) of the animals were in maternal condition, with 95% CI, and P_{max} derived from the ogives presented in Fig. 4.10 are tabulated as follows.

l_{50} (95% CI) (mm)	l_{95} (95% CI) (mm)	P_{max}
1421 (1423, 1423)	1488 (1484, 1492)	0.333

Table 4.9 Number of females samples in each length-class for maternal and non maternal condition during 1986-87 and 1998-01

Animals were classed as in maternal condition if U = 4 (Jan–May). U = 5 (Feb–Dec). or U = 6 (Nov–Dec); otherwise they were classed as in non-maternal condition. For probit analysis the observed value of 90[A] animals in the 1500–1599 mm length-class was increased to 93 animals to prevent the ratio of the number of maternal animals / total number of animals within that length-class from exceeding 0.333.

Length-class	*No. of females for each uterus index*[B]						*No. of females for each condition*			*Proportion in*
(mm)	*1*	*2*	*3*	*4*	*5*	*6*	*Non-maternal*	*Maternal*	*Total*	*maternal condition*
<700	25	7					32	0	32	0.000
700–799	25	47					72	0	72	0.000
800–899	67	13					80	0	80	0.000
900–999	34	32					66	0	66	0.000
1000–1099	42	29					71	0	71	0.000
1100–1199	17	26					43	0	43	0.000
1200–1299	40	33					73	0	73	0.000
1300–1399	32	59	10		1	1	102	1	103	0.010
1400–1499	9	53	57	11	11	9	137	13	150	0.087
1500–1599		2	30	16	29	13	59	31	90[A]	0.344
≥1600			6	6	8	4	16	8	24	0.333
Total	291	301	103	33	49	27	751	53	804	

[B]Defined in Table 4.1

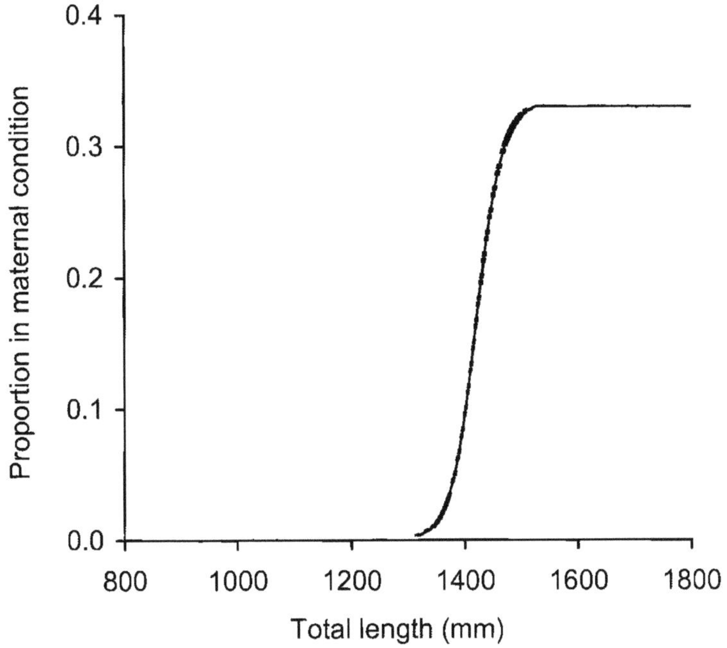

Fig. 4.10 Proportion of female population in maternal condition versus TL
Proportion of female population in maternal condition against TL (——) with 95% con-fidence intervals (- - -) for females during 1986–87 and 1998–01 combined. Values of parameters and statistical quantities for the equation $P_l = P_{max}(1+e^{-\ln(19)(l-l_{50})/(l_{95}-l_{50})})^{-1}$ determined from probit analysis are given in the following tabulation:

l_{50} (*CI*)	l_{95} (*CI*)	P_{max}	*n*	*N*	*ML*	*P*
1421 (1420, 1423)	1488 (1484, 1492)	0.333	53	269	–4953.89	***

where *l* is total length measured in millimetres, P_l is proportion of animals at TL, *l*. l_{50} and l_{95} are parameters, P_{max} is an asymptotic constant, n is the total number of animals classed as in maternal condition, and *N* is the total number of animals selected in statistical procedure, *ML* is maximum likelihood, and *P* is probability of statistical significance (*P < 0.1; **P < 0.01; ***P < 0.001). Animals were classed as being in maternal condition if U = 4 (Jan–May), U = 5 (Feb–Dec), or U = 6 (Nov–Dec).

The TL-at-maternity is considerably larger than TL-at-maturity (Fig. 4.11). If the common practice of weighting the maturity ogive by the frequency of parturition (0.333 for *Galeorhinus galeus*) was adopted the proportion of animals in maternal condition (Fig. 4.12) would be markedly over-estimated. This in turn would create a major bias if applied in a population model by overestimating the number of births. For *G. galeus*, there is a three-year lag between first maturity defined as LFD >3mm and maternal condition, during which time the animals would have undertaken considerable growth. It follows that the greater the time period between maturity and maternity the greater the bias caused by growth of the animals. For this reason it is necessary to estimate the maternity ogive completely independently of the maturity ogive.

Table 4.10 Hypothesis testing for proportion of female population in maternal condition

Logistic regression testing for the effects of period (1973–76, 1986–87 and 1998–01) within the Bass Strait region and effects of region (Bass Strait and South Australia) during 1998–01. Animals were classed as in maternal condition if U = 4 (Jan–May), U = 5 (Feb–Jan) or U = 6 (Nov–Dec); otherwise they were classed as in non-maternal condition. χ^2, chi square; ns, not significant; *P < 0.05; **P < 0.01; ***P < 0.001.

Source of variation	*Number of animals*		χ^2	*P*	
	Maternal	*Non-maternal*			
Between 1973–76, 1986–87, and 1998–01	85	770			
Intercept			1405.916	<0.0001	***
Period			585.460	<0.0001	***
Total length			1725.837	<0.0001	***
Between 1973–76 and 1986–87	72	573			
Intercept			862.998	<0.0001	***
Period			97.850	<0.0001	***
Total length			1293.626	<0.0001	***
Between 1973–76 and 1998–01	42	328			
Intercept			263.358	<0.0001	***
Period			394.760	<0.0001	***
Total length			400.119	<0.0001	***
Between 1986–87 and 1998–01	56	639			
Intercept			1350.949	<0.0001	***
Period			389.728	<0.0001	***
Total length			1619.825	<0.0001	***

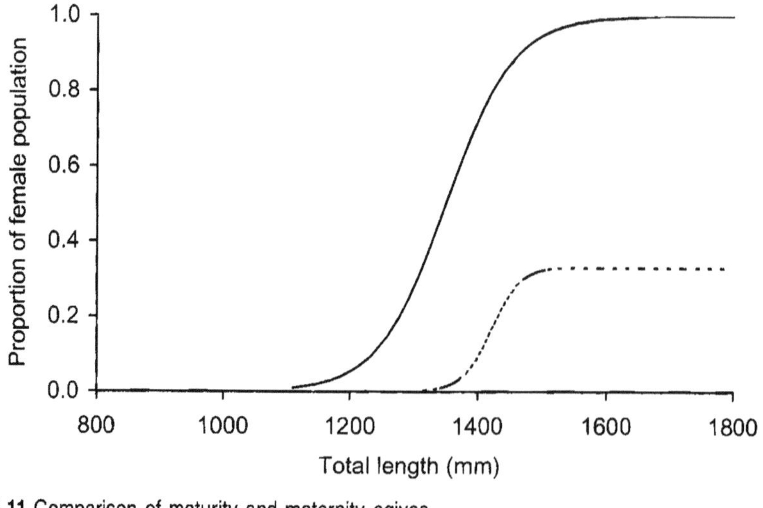

Fig. 4.11 Comparison of maturity and maternity ogives
Proportion of female population in mature condition (——) or maternal condition (- - -) against TL for females from southern Australia during 1986–87 and 1998–01.

4.6.6 Male Maturity Ogives

The results from indices of maturity based on testis, seminal vesicle and clasper condition were compared. Testing for spatial and temporal effects was avoided and the data were pooled between BS and SA and between 1973–76, 1986–87 and 1998–01 (Table 4.11). The reason for pooling the data across all three sampling periods is that the maturity indices require subjective judgement by the field observers and different observers collected the data between the three periods. The reason for pooling the data between the regions is that males are highly migratory and form part of the one population irrespective of where they are captured.

Based on TL at which 50% (l_{50}) and 95% (l_{95}) of the animals were in mature condition, with 95% CI, presented in Fig. 4.13, the testis condition and seminal vesicle condition are in reasonable agreement. The value of l_{50} is 1291 (1276, 1306) mm for testis condition and 1260 (1244, 1276) mm for seminal vesicle condition. The value of l_{50} of 1312 (1301, 1322) mm for clasper condition is larger than the values for the other two methods and the ogive has a different shape (Fig. 4.13). All three methods require subjective judgement but clasper condition is considered the most subjective and the least reliable indicator of maturity for *G. galeus*.

4.6.7 Total Body Mass at Size

The largest female *Galeorhinus galeus* sampled during field operations (1745 mm TL) was longer than the largest male sampled (1628 mm). The highest total body mass of a female recorded (32.3 kg) was more than 50% above the highest mass of a male recorded (21.0 kg).

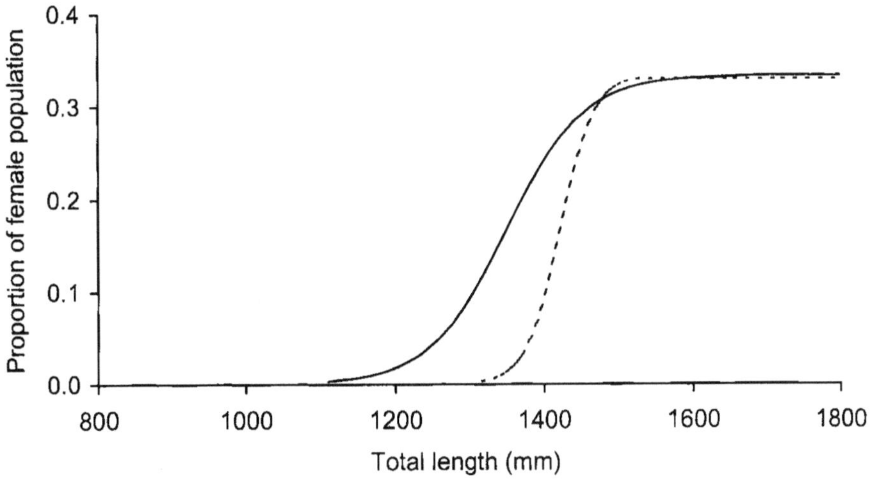

Fig. 4.12 Bias in one-third maturity ogive compared with maternity ogive
Proportion of female population in maternal condition (- - -) and one-third of female population in mature condition (——) against TL for females from southern Australia during 1986–87 and 1998–01.

Table 4.11 Number of males sampled in each length-class by method for maturity
Males were classed as immature for T = 1 or T = 2 and mature for T = 3 testis condition, as immature for V = 1 and mature for V = 2 or V = 3 seminal vesicle condition, and immature for C = 1 or C = 2 and mature for C = 3 clasper condition.

Method	*Length-class (mm)*	*No. of animals for each index*[A]			*No. of animals*			*Proportion mature*
		1	*2*	*3*	*Immature*	*Mature*	*Total*	
Testis condition	<600	35	0	0	35	0	35	0.000
	600–699	39	0	0	39	0	39	0.000
	700–799	59	0	0	59	0	59	0.000
	800–899	74	0	0	74	0	74	0.000
	900–999	65	0	1	65	1	66	0.015
	1000–1099	60	2	1	62	1	63	0.016
	1100–1199	44	14	2	58	2	60	0.033
	1200–1299	18	26	23	44	23	67	0.343
	1300–1399	3	15	51	18	51	69	0.739
	1400–1499	0	1	76	1	76	77	0.987
	1500–1599	0	1	41	1	41	42	0.976
	1600–1699	0	0	3	0	3	3	1.000
	Total	397	59	198	456	198	654	
Seminal vescicle condition	<600	36	0	0	36	0	36	0.000
	600–699	43	0	0	43	0	43	0.000
	700–799	65	0	0	65	0	65	0.000
	800–899	76	0	0	76	0	76	0.000
	900–999	67	1	0	67	1	68	0.015
	1000–1099	62	2	0	62	2	64	0.031
	1100–1199	60	1	1	60	2	62	0.032
	1200–1299	34	25	1	34	26	60	0.433
	1300–1399	6	52	8	6	60	66	0.909
	1400–1499	2	71	10	2	81	83	0.976
	1500–1599	1	42	5	1	47	48	0.979

	1600–1699	0	4	0	0	4	4	1.000
	Total	452	198	25	452	223	675	
Clasper condition	<600	1	0	0	1	0	1	0.000
	600–699	6	0	0	6	0	6	0.000
	700–799	34	13	0	47	0	47	0.000
	800–899	28	1	0	29	0	29	0.000
	900–999	45	3	0	48	0	48	0.000
	1000–1099	37	2	0	39	0	39	0.000
	1100–1199	44	5	0	49	0	49	0.000
	1200–1299	25	19	8	44	8	52	0.154
	1300–1399	5	13	49	18	49	67	0.731
	1400–1499	0	0	59	0	59	59	1.000
	1500–1599	0	0	21	0	21	21	1.000
	1600–1699	0	0	2	0	2	2	1.000
	Total	225	56	139	281	139	420	

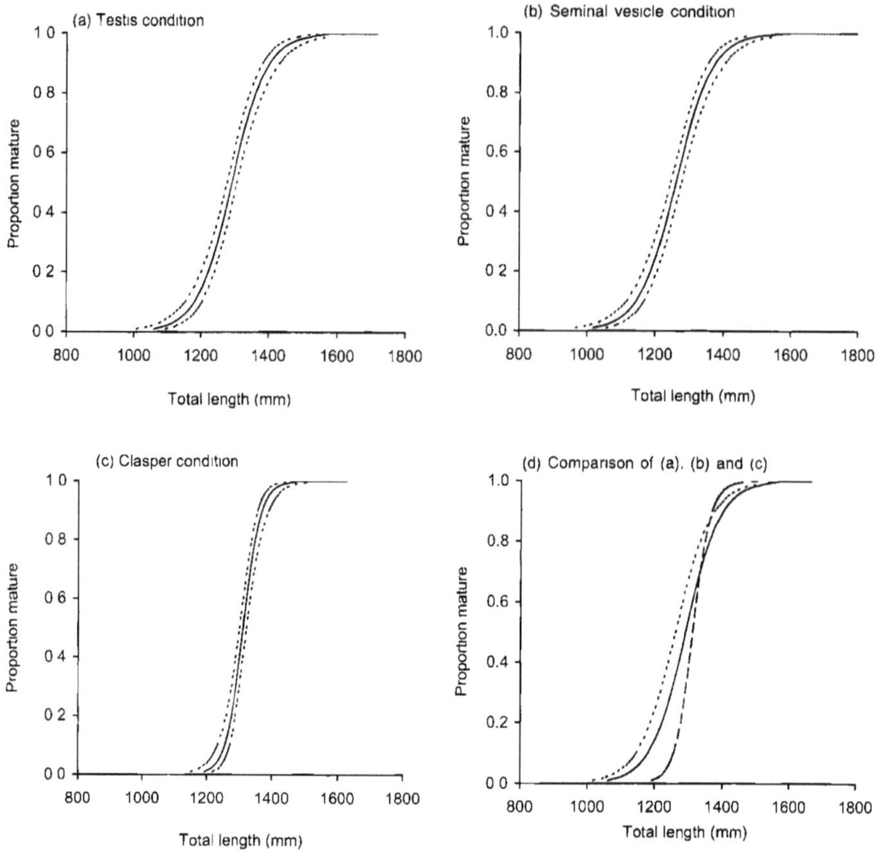

Fig. 4.13 Maturity of males based on testis, seminal vesicle and clasper condition
Proportion of population mature against TL (——) with 95% confidence intervals (- - -) for males determined from testis condition (a), seminal vesicle condition (b), clasper condition (c) and comparison of mean ogives for testis (——), seminal vesicle (- - -) and clasper condition (– – –) (d). Males were classed as immature for T = 1 or T = 2 and mature for T = 3 testis condition, as immature for V = 1 and mature for V = 2 or V = 3 seminal vesicle condition, and immature for C = 1 or C = 2 and mature for C = 3 clasper condition. Values of parameters and statistical quantities for the equation $P = P_{max}(1+e^{-\ln(19)(l-l_{50})/(l_{95}-l_{50})})^{-1}$ determined from probit analysis are given in the following tabulation:

Method	l_{50} (*CI*)	l_{95} (*CI*)	P_{max}	*n*	*N*	*ML*	*P*
Testis	1291 (1276, 1306)	1439 (1413, 1473)	1.000	198	566	–146.841	***
Seminal vesicle	1260 (1244, 1276)	1415 (1389, 1449)	1.000	223	585	–141.998	***
Clasper	1312 (1301, 1322)	1388 (1372, 1413)	1.000	139	357	–80.727	***

where l is total length measured in millimetres, P is proportion of animals at TL l, l_{50} and l_{95} are parameters, P_{max} is an asymptotic constant, n is the total number of animals classed as mature, and N is the total number of animals selected in statistical procedure, ML is maximum likelihood, and P is probability of statistical significance (*$P < 0.1$; **$P < 0.01$; ***$P < 0.001$).

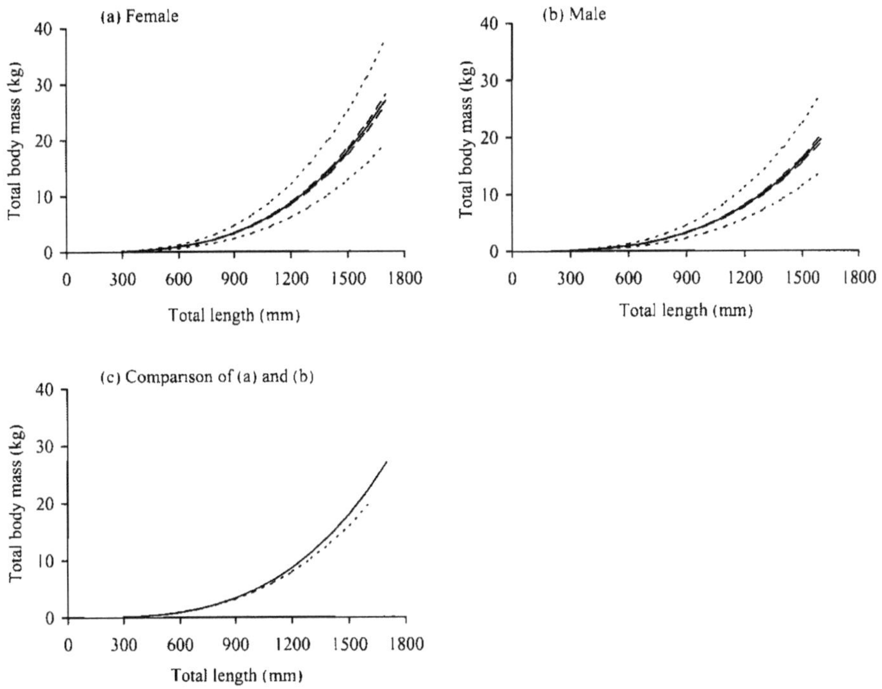

Fig. 4.14 Relationships between total body mass and total length.
Plots of mean total mass against TL (——), with 95% confidence limits (– – –) and 95% prediction intervals (- - -), for females (a), males (b), and comparison of the mean curves for females (——) and males (- - -) (c) in southern Australia during the periods 1973–76, 1986–87 and 1995–01 combined. Values for parameters and statistical quantities from linear regression analysis to derive the equation $w = acl^b$ are given in the following tablulation:

Sex	*a (s.e. range)* x 10^{-9}	*b(se)*	*c*	*n*	r^2	*rmse*	*P*
Females	0.699 (0.592–0.825)	3.276 (0.024)	1.015	401	0.979	0.172	***
Males	1.810 (1.530–2.150)	3.129 (0.024)	1.015	355	0.979	0.173	***

where *w* is total body mass, *l* is total length, *a* and *b* are parameters, *c* is the Beauchamp and Olson (1973) correction factor, *n* is sample size, r^2 is square of correlation coefficient, *rmse* is root mean square error, and P is probability of statistical significance (*P < 0.1; **P < 0.01; ***P<0.001) for the regression equation $\ln(w) = a + b\ln(l)$.

Statistical comparison of the slopes and intercepts by the Student *t*-test for selected pairs of straight line ln(*w*)–ln(*l*) relationships determined from linear regression fits indicate that two of the relationships tested were significantly different. The ln(*w*)–ln(*l*) relationship for pregnant females carrying *in utero* eggs (U = 4) and the relationship for pregnant females carrying *in utero* embryos (U = 5) were not significantly different (t-test, $t = 1.410$, d.f. = 58, and $P > 0.05$ for comparison of slopes and $t = 1.405$, d.f. = 58, and $P > 0.05$ for comparison of elevations). Similarly, the ln(*w*)–

ln(l) relationship for pregnant females (U = 4 and U = 5 pooled) and the relationship for non-pregnant females (U = 1, U = 2, U = 3, and U = 6 pooled) were not significantly different (t-test, t = 0.958, d.f. = 399, and P > 0.05 for comparison of slopes and t = 0.987, d.f. = 399, and P > 0.05 for comparison of elevations). The ln(w)–ln(l) relationship for females and the relationship for males were highly significantly different (t-test, t=4.285, d.f. = 754, and P < 0.001 for comparison of slopes, and t = 4.030, d.f. = 754, and P < 0.001 for comparison of elevations). Hence, the relationships of total body mass against TL, with 95% confidence limits on the mean curves and 95% prediction intervals are presented separately for females and males. These curves indicate that for a given length, the mean body mass is higher for females than for males (Fig. 4.14).

It is common among chondrichthyan species for females to attain higher TL and total body mass than males (Klimley 1987; Jakobsdóttir 2001; Simpfendorfer *et al.* 2001). For *Galeorhinus galeus*, the curves for these relationships coincide for low TL but diverge with increasing TL. In some species, for a given TL in the size range for mature animals, the mean body mass is higher for pregnant females than for non-pregnant females and is higher for non-pregnant females than for males (unpublished data). This is consistent with increasing mass of the ovaries and perhaps liver in mature females and the presence of *in utero* eggs and embryos in pregnant females.

4.7 CONCLUSIONS FOR *GALEORHINUS GALEUS*

Several important conclusions can be drawn from application of the methods described in this chapter to *Galeorhinus galeus*.

1. The results obtained for maturity and maternity are dependent on explicit definitions of terms. A definition of female maturity based on the diameter of the largest ovarian follicle provides an objective criterion unlikely to be affected by the field observer. Similarly, a definition of maternity based on uterus condition provides objective criteria unlikely to be affected by the field observer, other than occasional uncertainty distinguishing between the U = 3 and U = 6 conditions. A definition of male maturity is more problematic; maturity based on seminal vesicle, testis, and clasper condition all require a degree of judgement.
2. Pregnant condition of a female is defined by the presence of *in utero* eggs or embryos; non-pregnant condition of a female is defined by the absence of *in utero* eggs and embryos. A female observed at any time of the year is in maternal condition if it is in pregnant condition and expected to give birth prior to 1 January or if it is in post-partum condition and recently gave birth prior to 1 January (November–December). Any other female observed is in non-maternal condition (Fig. 4.15).

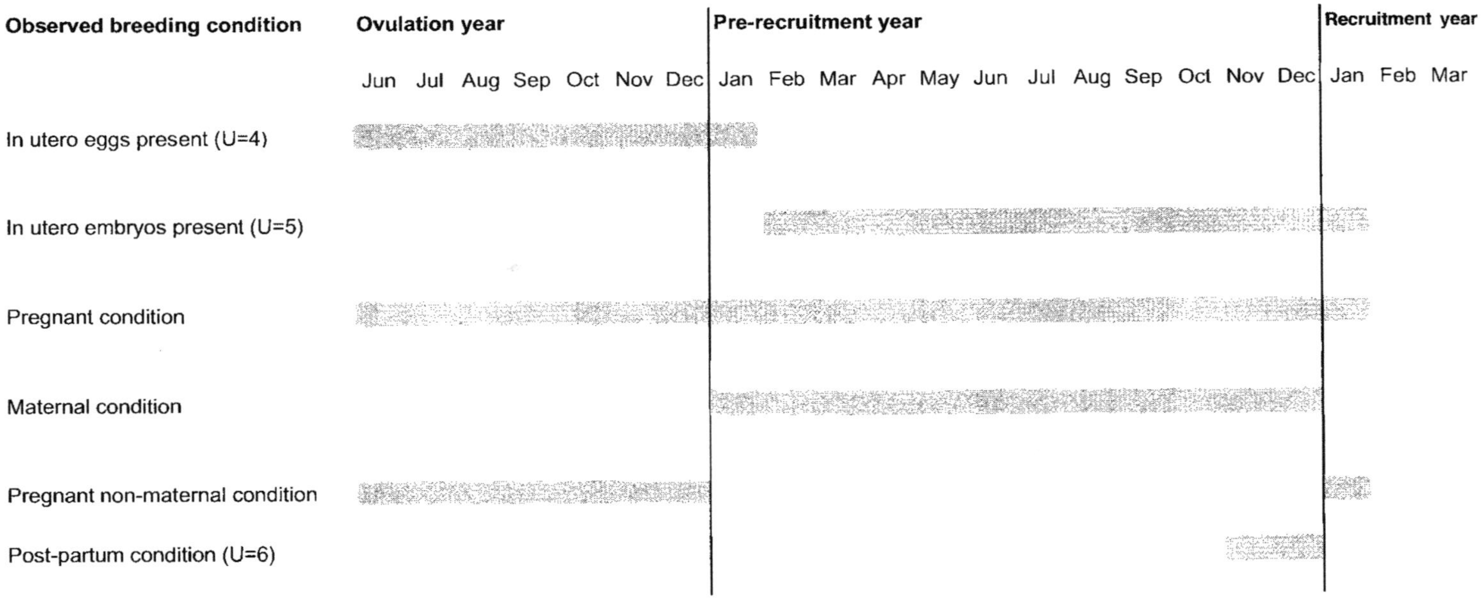

Fig. 4.15 Synchrony of breeding condition in *Galeorhinus galeus*
The U = 6 condition can occur at any time but only Nov–Dec is presented when they are in maternal condition.

3. Females attain a higher TL (1745 mm) and body mass (32.3 kg) than do males (1628 mm, 21.0 kg) and, for a given TL, the mean body mass is higher for females than for males; mass is not significantly affected by breeding condition.
4. Ovarian follicles have diameters ranging 1–57 mm and ovary mass can reach 1880 g in a pregnant animal during the early stages of ovulation and 910 g in a pregnant animal with close to full-term embryos. Ovary mass is mostly less than 73 g. A correlation between hepatic somatic index (liver mass/total body mass) (HSI) and largest follicle diameter indicates liver mass increases during the process of vitellogenesis. The highest HSI occurs in ovulating animals, which suggests liver mass and presumably its lipid content increase prior to pregnancy.
5. The ovarian cycle in most mature females has a period of three years. It is expected that the ovarian cycle of an individual is synchronous with one-third of the population of mature females; another one-third of the population is out of phase and ahead by one year and the other one-third is out of phase and ahead by two years (Fig. 4.16).
6. *In utero* eggs without macroscopically visible embryos are present during the eight-month period June–January; most animals observed in this condition are ovulating, suggesting that the process of ovulation in an individual is several months duration.
7. Macroscopically visible embryos are present during the 12-month period February–January and most attain a size of more than 300 mm TL at full-term; wet mass gain from egg to full-term is about 100%.
8. The number of *in utero* embryos range 15–43 and increases linearly with maternal length; this relationship is not affected by sampling region or sampling period.
9. The sex ratio of embryos is 1:1, the number *in utero* embryos in the left uterus is not significantly different from the number in the right uterus, and 3.4% of oocytes ovulated remain infertile in the uterus during pregnancy.
10. Parturition frequency for an individual female is mostly triennial. It is expected that an individual is synchronous with about one-third of the female population having ovulated at least once (having reached maternity); another one-third of the population is out of phase by one year and the other one-third is out of phase by two years (Fig. 4.16). Natural mortality can be expected to marginally increase the proportion in the population in maternal condition.
11. Female TL-at-maturity determined from diameter of the largest ovarian follicle appears to have increased between 1973–76 and 1986–87 with no change between 1986–87 and 1998–01.
12. The TL-at-maternity (l_{50} = 1421 mm, l_{95} = 1488 mm, and P_{max} = 0.333) is considerably larger than TL-at-maturity (l_{50} = 1349 mm, l_{95} = 1502 mm, and P_{max} = 1.000) for the periods 1986–87 and 1998–01 combined.

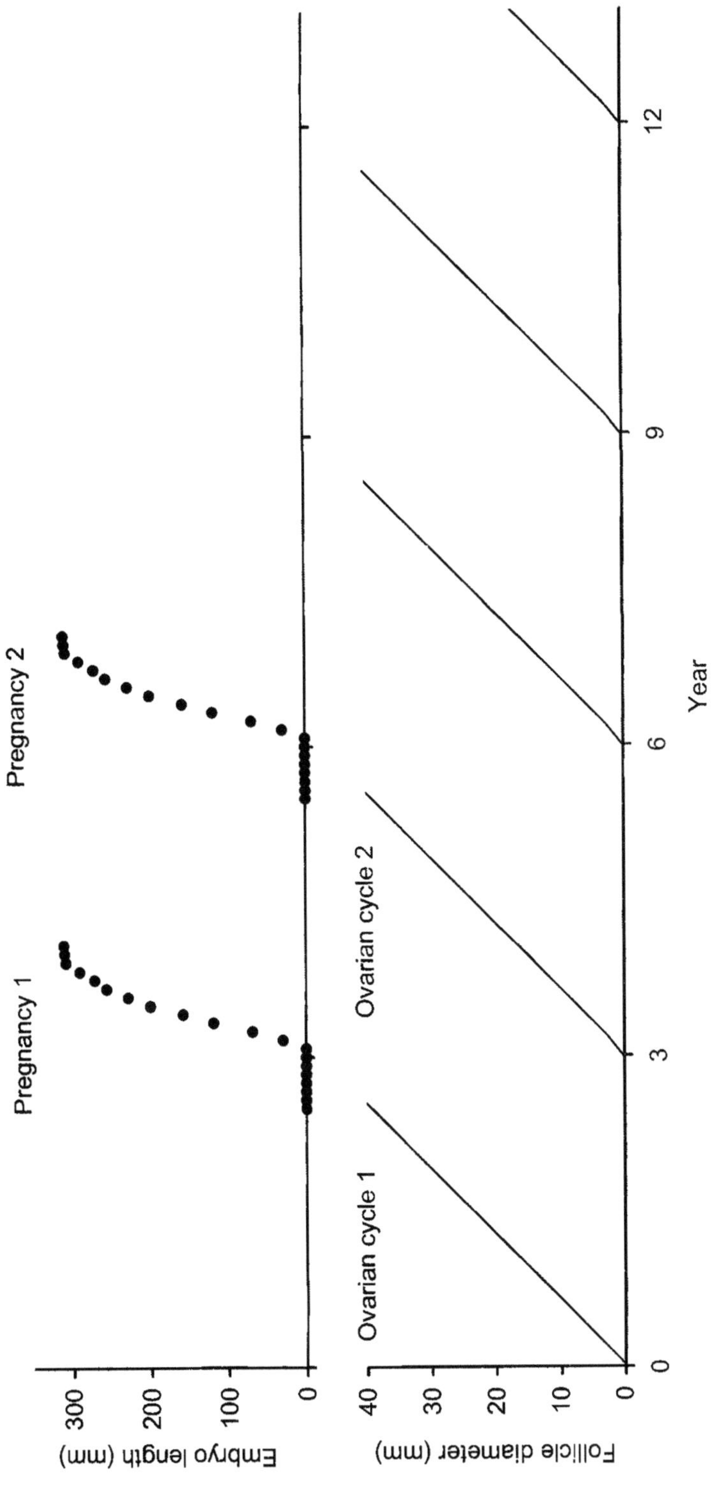

Fig. 4.16 Periodicity of the ovarian cycle and gestation for a mature animal of *Galeorhinus galeus*
The ovarian cycle and frequency of parturition are triennial: it is 4 years from the beginning of vitellogenesis for a particular follicle to full-term embryo.

Weighting the maturity ogive by the frequency of parturition (0.333 for *G. galeus*), as is commonly practised, would markedly over-estimate recruitment in any population dynamics model. It is essential to estimate the maternity ogive completely independently of the maturity ogive.

13. Male TL-at-maturity determined from indices of maturity based on macroscopic inspection of testis condition, seminal vesicle condition and clasper condition is highest for clasper condition (l_{50} = 1312 mm, l_{95} = 1388 mm, and P_{max} = 1.000). The shape of the TL-at-maturity ogives for testis condition (l_{50} = 1291 mm, l_{95} = 1439 mm, and P_{max} = 1.000) is similar but higher than for seminal vesicle condition (l_{50} = 1260 mm, l_{95} = 1415 mm, and P_{max} = 1.000). The data are pooled across regions and sampling period because males are highly migratory, often aggregate by size and breeding condition, and methods of determining maturity are subjective and likely to vary between observers.

4.8 ACKNOWLEDGMENTS

Acknowledgment is due to Lauren P. Brown, Russell J. Hudson, Dr Ian A. Knuckey, Peter L. Moulton, and Stephen R. Saddlier, formerly or presently of Primary Industries Research Victoria (PIRVic), who participated in field sampling of sharks and laboratory processing. Anne S. Gason of PIRVic provided advice on statistical analysis of the data and application of the computer statistical package SAS. Professor William C. Hamlett of the Indiana University School of Medicine is thanked for valuable discussions. The Fisheries Research and Development Corporation in Australia contributed most of the funding through several projects.

4.9 LITERATURE CITED

Beauchamp, J. J. and Olson, J. S. 1973. Corrections for bias in regression estimates after logarithmic transformation. Ecology 54: 1403–1407.

Capapé, C. and Mellinger, J. 1988. Nouvelles données sur la biologie de la reproduction du milandre, *Galeorhinus galeus* (Linné, 1778), (Pisces, Triakidae) des côtes tunisiennes. Cahiers de Biologie Marine 29: 135–146.

Cody, R. P. and Smith, J. K. 1997. *Applied Statistics and the SAS Programming Language.* Prentice Hall, Upper Saddle River, NJ. 445 pp.

Compagno, L. J. V. 1984. FAO species catalogue. Vol. 4. Sharks of the world. An annotated and illustrated catalogue of shark species known to date. Part 2. Carcharhiniformes. FAO Fisheries Synopsis 125: 251–655.

Conrath, C. L. and Musick, J. A. 2002. Reproductive biology of the smooth dogfish, *Mustelus canis*, in the northwest Atlantic Ocean. Environment Biology of Fishes 64: 367–377.

Girard, M. and Du Buit, M.-H. 1999. Reproductive biology of two deep-water sharks from the British Isles, *Centroscymnus coelolepis* and *Centrophorus squamosus*

(Chondrichthyes: Squalidae). Journal of the Marine Biological Association of the United Kingdom 79: 923–931.

Grant, C. J., Sandland, R. L. and Olsen, A. M. 1979. Estimation of growth, mortality and yield per recruit of the Australian school sharks, *Galeorhinus australis* (Macleay), from tag recoveries. Australian Journal of Marine and Freshwater Research 30: 625–637.

Guallart, J. and Vincent, J. J. 2001. Changes in composition during embryo development of the gulper shark, *Centrophorus granulosus* (Elasmobranchii, *Centrophoridae*): An assessment of maternal-embryonic nutritional relationships. Environmental Biology of Fishes 61: 135–150.

Hamlett, W. C. 1999. Male reproductive system. Pp. 444–470. In W. C. Hamlett (ed.), *Sharks, Skates and Rays: The Biology of Elasmobranch Fishes*. The Johns Hopkins University Press, Baltimore, MD.

Hamlett, W. C. and Koob, T. J. 1999. Female reproductive system. Pp. 398–443. In W. C. Hamlett (ed.), *Sharks, Skates and Rays: The Biology of Elasmobranch Fishes*. The Johns Hopkins University Press, Baltimore, MD.

Hamlett, W. C., Musick, J. A., Hysell, C. K. and Sever, D. M. 2002. Uterine epithelial-sperm interaction, endometrial cycle and sperm storage in the terminal zone of the ovicucal gland in the placental smoothhound, *Mustelus canis*. Journal of Experimental Zoology 292: 129–144.

Hanchet, S. 1988. Reproductive biology of *Squalus acanthias* from the east coast, South Island, New Zealand. New Zealand Journal of Marine and Freshwater Research 22: 537–549.

Hilton-Taylor, C. 2000. *2000 IUCN Red List of Threatened Species*. IUCN The World Conservation Union, Gland, Switzerland. 61 pp.

Hurst, R. J., Bagley, N. W., McGregor, G. A. and Francis, M. P. 1999. Movement of the New Zealand school shark, *Galeorhinus galeus*, from tag returns. New Zealand Journal of Marine and Freshwater Research 33: 29–48.

Jakobsdóttir, K. B. 2001. Biological aspects of two deep-water squalid sharks: *Centroscyllium fabricii* (Reinhardt, 1825) and *Etmopterus princeps* (Collett, 1904) in Icelandic waters. Fisheries Research 51: 247–265.

Jones, T. S. and Ugland, K. I. 2001. Reproduction of female spiny dogfish, *Squalus acanthias*, in the Oslofjord. Fishery Bulletin 99: 685–690.

Kleinbaum, D. G., Kupper, L. L. and Muller, K. E. 1988. *Applied regression analysis and other multivariable methods*. PWS–Kent Publishing Company, Boston, MA. 718 pp.

Klimley, A. P. 1987. The determinants of sexual segregation in the scalloped hammerhead shark, *Sphyrna lewini*. Environmental Biology of Fishes 18: 27–40.

Lee, R. M. 1920. A review of the methods of age and growth determination in fishes by means of scales. Fishery Investigations Ser. 2 , 4: 1–32.

Lenanton, R. C. J., Heald, D. I., Platell, M., Cliff, M. and Shaw, J. 1990. Aspects of the reproductive biology of the gummy shark, *Mustelus antarcticus* Günther, from waters off the south coast of Western Australia. Australian Journal of Marine and Freshwater Research 41: 807–822.

Lucifora, L. O. 2003. Ecología y conservación de los grandes tiburones costeros de Bahía Anegarda, Provincia de Buenos Aires, Argentina. Ph.D. Thesis, Universidad Nacional de Mar del Plata, Mar del Plata, Argentina. 406 pp.

Mellinger, J. 1965. Stades de la spermatogenese cher *Scyliorhinus caniculus* (L.): description, donnees histochimigues, variations normales et experimentales. Zeitschrift fur Zellforschung 67: 653–673.

Mollet, H. F., Cliff, G., Pratt Jnr, H. L. and Stevens, J. D. 2000. Reproductive biology of the female shortfin mako, *Isurus oxyrinchus* Rafinesque, 1810, with comments on the embryonic development of lamnoids. Fishery Bulletin 98: 299–318.

Moulton, P. M., Walker, T. I. and Saddlier, S. R. 1992. Age and growth studies of gummy shark, *Mustelus antarcticus* Günther, and school shark, *Galeorhinus galeus* (Linnaeus), from southern-Australian waters. Australian Journal of Marine and Freshwater Research 43: 1241–1267.

Olsen, A. M. 1954. The biology, migration, and growth rate of the school shark, *Galeorhinus australis* (Macleay) (Carcharhinidae) in south-eastern Australian waters. Australian Journal of Marine and Freshwater Research 5: 353–410.

Parsons, G. R. and Grier, H. J. 1992. Seasonal changes in shark testicular structure and spermatogenesis. Journal of Experimental Zoology 261: 173–184.

Peres, M. B. and Vooren, C. M. 1991. Sexual development, reproductive cycle, and fecundity of the school shark *Galeorhinus galeus* off southern Brazil. Fishery Bulletin 89: 655–667.

Punt, A. E., Pribac, F., Walker, T. I., Taylor, B. L. and Prince, J. D. 2000. Stock assessment of school shark, *Galeorhinus galeus*, based on a spatially-explicit population dynamics model. Marine and Freshwater Research 51: 205–220.

Punt, A. E. and Walker, T. I. 1998. Stock assessment and risk analysis for the school shark (*Galeorhinus galeus*) off southern Australia. Marine and Freshwater Research 49: 719–731.

Rao, C. R. 1973. *Linear Statistical Inference and Its Applications*. John Wiley and Sons, New York, NY. 265 pp.

Ricker, W. E. 1958. Handbook of computations for biological statistics of fish populations. Bulletin of Fisheries Research Board of Canada 119: 1–191.

Ricker, W. E. 1969. Effects of size-selective mortality and sampling bias on estimates of growth, mortality, production, and yield. Journal of the Fisheries Research Board of Canada 26: 479–541.

Ripley, W. E. 1946. The soupfin shark and the fishery. California Division of Fish and Game Fish Bulletin 64: No. 64, 7–37.

Silvey, S. D. 1975. *Statistical Inference*. Chapman and Hall, London, UK. 191 pp.

Simpfendorfer, C. A., Goodreid, A. B. and McAuley, R. B. 2001. Size, sex and geographic variation in the diet of the tiger shark, *Galeocerdo cuvier*, from Western Australian waters. Environmental Biology of Fishes 61: 37–46.

Smith, R. M., Walker, T. I. and Hamlett, W. C. 2004. Microscopic organization of the oviducal gland of the holocephalan elephant fish, *Callorhynchus milii*. Marine and Freshwater Research 55: 155–164.

Storrie, M. T. 2004. Microscopic modifications of the reproductive tissues of the female gummy shark (*Mustelus antarcticus*) throughout maturation and gestation. Ph.D. Thesis, Deakin University, Warrnambool, Victoria, Australia. 153 pp.

Teshima, K. and Koga, S. 1973. Studies on sharks. V. Taxonomic characteristics of reproductive organs in Japanese *Mustelus*. Marine Biology 23: 337–341.

Walker, T. I. 1983. Investigations of the gummy shark, *Mustelus antarcticus* Günther, from south-eastern Australian waters, Report to Fishing Industry Research Committee. June 1983. Pp. 1–94. In A. E. Caton (ed.), *Proceedings of the Shark Assessment Workshop, South East Fisheries Committee Shark Research Group*. 7–10 March 1983. Melbourne. Department of Primary Industry, Canberra, ACT, Australia.

Walker, T. I. 1992. A fishery simulation model for sharks applied to the gummy shark, *Mustelus antarcticus* Günther, from southern Australian waters. Australian Journal of Marine and Freshwater Research 43: 195–212.

Walker, T. I. 1994. Fishery model of gummy shark, *Mustelus antarcticus*, for Bass Strait. Pp. 422–438. In I. Bishop (ed.), *Resource Technology '94 New Opportunities Best Practice*. 26–30 September 1994. University of Melbourne, Melbourne. The Centre for Geographic Information Systems & Modelling, The University of Melbourne, Melbourne, Australia.

Walker, T. I. 1999. *Galeorhinus galeus* fisheries of the world. In 'Case studies of management of elasmobranch fisheries'. FAO Fisheries Technical Paper 378/2: 728–773.

Walker, T. I. 2004. Chapter 13. Management measures. Pp. 285–321. In J. A. Musicki and R. Bonfil (eds), *Technical Manual for the Management of Elasmobranchs.* Asia Pacific Economic Cooperation Secretariat, Singapore.

Walker, T. I., Taylor, B. L., Hudson, R. J. and Cottier, J. P. 1998. The phenomenon of apparent change of growth rate in gummy shark (*Mustelus antarcticus*) harvested off southern Australia. Fisheries Research 39: 139–163.

Ward, R. D. and Gardner, M. G. 1997. Stock structure and species identification of school and gummy sharks in Australasian waters. Project FRRF 93/11 and FRDC 93/64. February 1997. CSIRO Marine Research, Hobart, Tasmania, Australia. 92 pp.

West, G. J. and Stevens, J. D. 2001. Archival tagging of school shark, *Galeorhinus galeus*, in Australia: Initial results. Environmental Biology of Fishes 60: 283–298.

Wood, C. C., Ketchen, K. S. and Beamish, R. J. 1979. Population dynamics of spiny dogfish (*Squalus acanthias*) in British Columbia waters. Journal of the Fisheries Research Board of Canada 36: 647–656.

Yano, K. and Tanaka, S. 1988. Size at maturity, reproductive cycle, fecundity and depth segregation of the deep sea squaloid sharks *Centroscymnus owstoni* and *C. coelolepsis* in Surga Bay, Japan. Nippon Suisan Gakkaishi 54: 167–174.

CHAPTER 5

Elasmobranch Courtship and Mating Behavior

Harold L. Pratt, Jr.[1] *and Jeffrey C. Carrier*[2]

5.1 INTRODUCTION

Elasmobranchs have elaborate organ systems for reproduction and many have complex reproductive life histories, but we are just beginning to understand that elasmobranchs may also exhibit involved pre-copulatory and copulatory behaviors (Pratt and Carrier 2001). Shark, skate and ray behavior has been generally dismissed as unsophisticated by both biologists and non-biologists. They have been cast as primitive animals rather than the ancient and highly evolved fishes that we have come to know. For example, many sharks and rays have relatively large brains overlapping some mammals and birds in brain-body ratios (Northcutt 1977, 1978).

Like mammals, all elasmobranchs have internal fertilization and most have long gestation periods, up to 22 months in the spiny dogfish, *Squalus acanthias*, and as long as 3.5 yr in the frilled shark (*Chlamydoselachus anguineus*) (Tanaka *et al.* 1990). In elasmobranchs, extended development may produce a central nervous system capable of mediating complex social and sexual behavior (Demski 1990b). A high degree of sophistication has been revealed by investigations into interactions of large groupings of scalloped hammerheads, *Sphyrna lewini* (Klimley 1987), dominance hierarchies in sand tigers, *Carcharias taurus* (Gordon 1993), and complex pair and group mating behaviors in nurse sharks, *Ginglymostoma cirratum* (Carrier *et al.* 1994).

Much is known about elasmobranch reproduction and development (Wourms 1977; Wourms 1981; Dodd 1983; Gilmore *et al.* 1983; Hamlett and Koob 1999; Carrier *et al.* 2004); but the quantitative study of elasmobranch

[1] Mote Center for Shark Research, Summerland Key, 24244 Overseas Highway, Summerland Key, FL 33042
[2] Department of Biology, Albion College, Albion, MI 49224

reproductive behavior is a relatively new endeavor. These studies have lagged behind other life history and fisheries research largely because of the lack of direct observations. Elasmobranch reproductive behavior was briefly reviewed by Demski (1990b) in his work on neuroendocrine mechanisms. Bres (1993) reviewed shark behavior, including reproduction and reported that few species of sharks have been directly observed in mating activities. Pratt and Carrier (2001) provided a comprehensive review and added the perspective of their own investigations of nurse sharks in the Dry Tortugas archipelago, Florida.

As we shall see, elasmobranch reproductive behavior has been implied from the examination of freshly caught animals, from laboratory studies of reproductive structures and function, and determined from direct observations of captive or free swimming wild animals. The understanding of elasmobranch reproduction and reproductive behavior has become increasingly important as a consequence of their recent exploitation. Slow growth rates, late age at first maturity and low reproductive rates have allowed shark populations in particular to be reduced by directed and incidental fisheries world wide (Musick 1999). The reproductive behavior, habits, breeding areas, sexual segregation of females and nursery grounds of sharks must be thoroughly understood for successful conservation and management.

The terms courtship and mating are used interchangeably in this report as they are in the literature (Klimley 1980, Tricas 1980) to describe both pre-copulatory (usually courtship) and copulatory (usually mating) behaviors of male and female elasmobranchs. 'Courtship,' an activity for mate identification and assessment, may seem an unwarranted term for the apparently rough and direct pre-copulatory behavior of elasmobranchs, but detailed observations of the nurse shark are showing that 'courtship' may eventually prove to be accurate even in the popular sense of females being 'choosy' and selective through avoidance behaviors and males competing for the cooperation of a specific female during the prelude to mating. (Appendix 5.1 summarizes and consolidates the terminology related to studies of reproduction in elasmobranchs.)

5.2 REPRODUCTIVE STRUCTURES AND BEHAVIOR

Elasmobranchs are a diverse taxonomic group but share similar organ systems (claspers, siphon sacs, ovaries with large follicles, oviduct with specialized tissues) and often-complex reproductive behaviors. Successful mating requires the male to physically hold the female and to introduce one, or perhaps both (Leigh-Sharpe 1920), claspers into the cloaca and common vagina. Repeated copulations may be necessary in some species for successful fertilization. Sperm viability, the extent to which males and females of different species may store and maintain sperm and the role of mating behavior in sperm competition has yet to be established. Recent research has also revealed the role of the electrosensory system

(Tricas *et al.* 1995) and the neuroendocrine system (Dodd 1983; Rasmussen and Gruber 1990; Tricas *et al.* 2000) in courtship and mating. A brief assessment of important reproductive adaptations is provided in the following sections to add background to understanding behaviors.

5.2.1 Indicators of Male Sexual Activity

5.2.1.1 Sperm and sperm delivery

Many elasmobranchs have relatively large testes and during the mating season carry profuse sperm in the upper and lower epididymis (Pratt 1979). Copious sperm production is usually an indicator of sperm competition (Eberhard 1998), which may be a facet of many shark mating systems. In the mature sexually active male shark, sperm packets are stored in the ampulla epididymis (Jones and Jones 1982) as encapsulated spermatophores in *Cetorhinus maximus* and lamnids (Matthews 1950) or with naked exposed tails as spermatozeugmata in carcharhinids (Pratt and Tanaka 1994, Hamlett 1999).

During or immediately before copulation, sperm exit the paired ampullae to the urogenital sinus and then are injected into the apopyle and clasper groove of one forward-rotated and copulating clasper. From the clasper groove, the sperm are propelled into the female's vagina and uterus either through the hydraulic action of the corresponding muscular subdermal siphon sac (Gilbert and Heath 1972) or by the action of smooth muscles in the walls of the ampullae epididymides or both. This transfer may be further augmented by changes in hydrostatic pressure in the coelomic cavity generated by the male's undulations and thrusting during copulation.

5.2.1.2 Claspers

In copulation, at least one of the two paired intermittent organs, the claspers, rotates forward and receives sperm from the single urogenital papilla through an opened apopyle on the dorsal surface of the clasper (Leigh-Sharpe 1920). The terminal cartilages variously bear sharpened ridges, hooks or barbs that expand out when flexed to lock the clasper in the vagina and often leave abrasions and lesions in the vaginal wall. In species that bear clasper spines or hooks, the relative bluntness of the spine can serve as an index of a male's mating activity. Mature virgin males of some species posses a needle-sharp spine sometimes invested in a tunic of epidermis. Larger adults have naked, blunted yellowed spurs. These can be difficult but not impossible to observe in live sharks. If the clasper is used in a splayed position (that is, not under the urogenital papilla) then sperm must be transferred to the clasper groove prior to copulation.

It is possible that in some or all species of sharks, the role of the siphon sacs is not sperm propulsion as suggested by Gilbert and Heath, but to assist sperm competition and wash any rival sperm from the female's vagina before copulation (Eberhard 1985). Whitney *et al.* (unpublished) have examined detailed video records of clasper and siphon sac function in *Triaenodon obesus*. They conclude that the siphon sac function is one of sperm propulsion. In this species the sac is divided into three zones that aid to fill and isolate seawater and transport sperm to the female. No evidence was found for washing rival sperm in this species.

During *Ginglymostoma cirratum* mating, only one clasper is used during copulation and in copulation attempts. For example, if a female is approached from her left side, her left pectoral fin is grasped and the male's left clasper crosses his cloaca and is inserted (Carrier *et al.* 1994). It is not clear for other species whether one or both claspers are used. Most detailed accounts of copulation report the use of only a single clasper (Klimley 1980, Tricas and Lefeuvre 1985, Carrier *et al.* 1994). Gordon (1993) reported that the claspers of *Carcharias taurus* may be 'splayed' laterally at ninety degrees or crossed at a strong angle, seemingly at will while swimming. Dral (1980) reported that *Chiloscyllium griseum*, can also cross and splay its claspers. It is not known which other species have this ability nor its significance to reproductive behavior.

5.2.1.3 Sexual biting and holding

Successful elasmobranch mating always requires male 'biting' or holding behavior. Some courtship bites are preliminary and may serve to stop the female or signal male intent. These bites are less tenacious than feeding bites and usually do not appear to employ full force or full closure of the jaw. Springer (1967) noted: 'Among the larger species, some cooperation on the part of the female seems necessary.' In most species in which mating behavior has been observed, the male bites and often holds the female, possibly as a pre-copulatory releasing mechanism to invoke female acquiescence, to facilitate insertion of the clasper and to maintain the proper position and proximity until sperm transfer is complete (Fig. 5.1).

During the mating season, female elasmobranchs and some males will usually bear some combination of courtship marks on their fins, flanks and elsewhere on their bodies. Some scaring patterns are species specific, depending on male biting, 'grasping' and mating behaviors and the presence and employment of dentition, horns, spines and denticles. Most commonly found are tooth cuts and abrasions on the female pectoral fins, but marks on males of some species are not uncommon. Gordon (1993) reported several types of biting behavior in captive *Carcharias taurus*. Males bit females to mate, and also snapped at smaller species in the tank, but never at each other. The mature females also bit the mating males as noted in wild populations by Gilmore *et al.* (1983). These observations indicate that there may be sexual hierarchies that are maintained through aggressive behavior (Gordon 1993). These hierarchies may serve to synchronize mating sequence in some sharks.

Social hierarchies and interactions may precede or follow active mating, and carry over into non-reproductive periods, particularly in

Fig. 5.1 Male nurse shark, *Ginglymostoma cirratum*, (upper) of the Dry Tortugas study population bites and holds the entire left pectoral fin of a female while copulating. The female with cupped pelvic fins remains motionless as her head is pressed into the algae covered coral rubble bottom of the shallow study area. Sharks are 2.3 to 2.4 m in total length. Original.

species which remain together or repeatedly use particular mating sites over several annual cycles as in *Carcharias taurus* (Gilmore *et al.* 1983) and *Ginglymostoma cirratum* (Carrier *et al.* 1994).

It should be cautioned that not all bite marks result from reproductive activities. Bites around the head and gills of large female shortfin makos, *Isurus oxyrinchus*, and white sharks, *Carcharodon carcharias* dissected by Pratt *et al.* (1982 and unpublished), are not usually accompanied by signs of recent activity around the cloaca, clasper or by recently introduced sperm. All of the *I. oxyrinchus* surveyed by Pratt in New England waters with these marks have been in a reproductively resting condition and we concluded that some of these head and gill area bites, slashes and resultant scars, while possibly involving hierarchies, usually result from agonistic interactions over food rather than mating activity. While bites and scars on female pectoral fins in the Dry Tortugas *G. cirratum* population are common, their dorsal fins are typically unmarked. However the first and second dorsal fins of the adult mating males frequently show missing arcs the size of nurse shark jaws and ragged fin edges are common.

Differences in dentition have been shown between genders and some seasonal changes in dentition may reflect changes in behaviors related to mating. Taniuchi and Shimizu (1993) showed dentition differences in adult stingrays (*Dasyatis akajei*) and noted that the teeth of adult males are more pointed than those of females. Food preferences were not sufficiently different between genders to account for changes in dentition. Rather they suggested that dentition changes in males at sexual maturity as a mechanism to assist with establishing adequate grips during mating. Kajiura and Tricas (2000) showed similar patterns in *Dasyatis sabina*, and further showed a periodic shift in male dental morphology correlated with the time of mating. They further noted that replacement cycles did not differ between the sexes so that explanations of differences could not therefore be attributed to differences in the wearing of teeth.

5.2.2 Indicators of Female Sexual Activity

5.2.2.1 Sperm and ova

Some female elasmobranchs may store inseminated sperm for varying periods (Metten 1941; Prasad 1944, 1945; Pratt 1993; Conrath and Musick 2002; Hamlett *et al.* 2002). Histological analysis of freshly bitten female *Prionace glauca* caught and dissected a short time after copulation reveals large volumes of fresh sperm in their reproductive tracts, especially in the lumen and diverticula of the oviducal glands (Pratt 1979).

Some females may delay fertilization and store sperm for many months in the oviducal gland (Pratt 1993). Other species immediately fertilize ova following copulation (Gilmore *et al.* 1983). Ova are apparently released sequentially, usually alternating between right and left uterine tracts. Fertilization may occur in the oviducal gland for elasmobranchs that store sperm and in the upper oviduct above the oviducal gland or in its vicinity for those that do not (Gilmore *et al.* 1983). Sperm storage and mating behavior may have evolved together to complement a species mating system. Repeated seasonal copulations may be necessary to fertilize all ova in both oviducts if sperm is not stored.

5.2.2.2 Oviducts and fertilization

All known species of sharks have two functional oviducts, but gravid females often have developing embryos in one oviduct and infertile ova in the other. This is rare in smaller species but frequent among larger ones (Springer 1967) and may be the result of limited matings where mating is only from one side and consequently, only one oviduct is filled (unilateral fertilization). In *Carcharias taurus*, Gilmore and his colleagues (1983) have shown that both oviducts are always in complete synchrony with regard to physical condition, oviducal gland activity, fertilization and capsulation of ova and embryos as well as uterine condition (Gilmore *et al.* 1983, Gilmore 1993). Some rays (for example, stingrays) have only one functional oviduct and one copulation by one male may suffice to fertilize all of the eggs. Pectoral fin choice by the mating male may dictate the clasper he uses and influence which oviduct ultimately receives his sperm. Repeated symmetrical insemination may especially be needed in species with a minimum of body flexibility.

5.2.2.3 Vagina, cloaca and mating

In the mature female, the common vagina bifurcates to join the paired uteri. The vagina of *Prionace glauca* is thick-walled to receive the sharp terminal edges of the clasper during mating (Pratt 1979). Abrasions, lesions and dark bruises from claspers are common on vaginal walls during and after the mating season and are a good index of female sexual activity in *P. glauca* and other species. In *Ginglymostoma cirratum*, after numerous mating encounters, the walls of the cloaca, and the skin of the pelvic fins around the cloaca and even the abdominal epidermis of the female can be hematose. The vagina and surrounding tissue will also appear red and hematose during and after parturition and sometimes as a result of fishing and capture technique. Conclusions about mating activity should be made with respect to season and circumstance.

5.2.3 Elasmobranch Mating Systems

5.2.3.1 Electrosensory system

The electrosensory system serves an important function in the social and reproductive behavior of round stingrays, *Urolophus halleri*, (Tricas *et al.* 1995). Like other elasmobranchs, *U. halleri* produce a weak electric field that is rhythmically modulated by the movements of the spiracles and gills during ventilation. This signal is detected by adult conspecifics via the ampullae of Lorenzini, which are found in pores located on the dorsal and ventral cranial surfaces of most elasmobranchs. In *U. halleri*, this sensory system is used by reproductively active males to locate mates and by females to locate buried consexuals. This system is probably employed by other elasmobranchs as well, particularly those that live in the more two-dimensional world of the benthic batoids.

5.2.3.2 Neuroendocrine control of mating

Mating behavior usually occurs seasonally following the recrudescence of the gonads and animal migration to mating areas. Temperature and photoperiod may be the major environmental regulators of the neuroendocrine mechanisms that drive sexual development and behavior. They have been shown to control the endocrine regulation of spermatogenesis in *Scyliorhinus canicula* (Dobson and Dodd 1977). Lunar and solar influences may play a role in some cases. Dodd (1983) provides a comprehensive overview of reproduction and endocrine control in cartilaginous fishes.

Hormonal control of sexual development and behavior is well documented in several elasmobranch species (Sumpter and Dodd 1979; Rasmussen *et al.* 1992; Wright and Demski 1993; Tricas *et al.* 2000). Demski (1990b) found that gonadotropin-releasing hormones mediate brain control of testicular and ovarian development via the brain-pituitary axis and speculates that the 'following' behavior noted by Myrberg and Gruber (1974) and Klimley (1980) may be triggered by female pheromones that trigger sexual behavior in males. Rasmussen and Gruber (1990) found elevated levels of the steroid hormones estradiol and testosterone in the serum of free ranging, actively courting female *Negaprion brevirostris*. Tricas *et al.* (2000) established in their study population of male and female *Dasyatis sabina* that social aggression during mating is mediated by production of gonadal steroids over a long season of seven months.

5.2.3.3 Energetics of mating sharks

Reproductive behavior may be revealed or influenced by the feeding habits and activity of sharks during courtship and mating. Springer (1967) found that immature and female sharks of the coastal Florida-Gulf and Mexico-Caribbean regions had large livers containing a high proportion of oil. In contrast, during the same season, sexually mature males had smaller, thin livers with low oil content. He postulated that this might be due to prolonged fasting during courtship. Alternatively, thin livers may also result from extreme male activity during mating such as increased 'patrolling', male-male aggression establishing hierarchies and the competitive pre-copulatory behavior required to grasp and mate with females. Other factors affecting male condition may include the reproductive costs associated with production of vast quantities of sperm in those species that engage in sperm competition. Male reproductive costs may be found to be as high in some species as for females.

5.2.3.4 Polyandry, polygyny and mating systems

The tactics of polyandry, polygyny sperm storage, and sperm competition are poorly understood in elasmobranchs and are just beginning to be revealed as mating systems come under scrutiny. Polyandry and polygyny are characteristic of many vertebrates and invertebrates and will probably be found to be common in sharks and batoids (Table 5.1). Their interplay has to affect mating behavior. For example, multiple matings may be needed to insure the fertilization of eggs in both uteri or the insemination of multiple clutches of eggs. When polyandry is common, male-male competition will be keen for access to females. In time, behavioral strategies may evolve that help males maximize their exposure to females (group behaviors, blocking, forced copulation). Females may develop counter-strategies to limit male access (refuging and avoidance behaviors such as shielding, arching, and shoaling, see Glossary). Multiple paternities are direct evidence of polyandry. The recent spate of work in this field (Ohta *et al.* 2000; Feldheim *et al.* 2001; Saville *et al.* 2002) has revealed a great deal about the behavior of the subject species.

5.3 REPRODUCTIVE BEHAVIOR

Because of the elasmobranch trait of internal fertilization, their mating behaviors differ markedly from that of other fishes. It more often resembles the joining of mammals than the broadcast spawning typical of most teleosts. Consequently the mating of sharks came to human

Table 5.1 Summary of observed courtship and mating behaviors in elasmobranch fishes.

General behavior and species	*Descriptions / Notes*	*References*
Sharks		
PRECOPULATORY AND COURTSHIP		
Following		
Carcharhinus melanopterus	'Close follow' near female's vent possibly olfactory-mediated.	2
Ginglymostoma cirratum	Male and female swim parallel and synchronously side-by-side.	3
Negaprion brevirostris	Swimming with body axes in parallel.	4
Female avoidance		
Carcharias taurus	Female 'shields' with pelvics close to substrate.	
Ginglymostoma cirratum	'Lying on back' the female rests motionless and rigid.	1
	Female 'pivots and rolls' on her back when a male bitesher pectoral fin.	3
Triaenodon obesus	Ventral surface to substrate.	24
Female acceptance		
Carcharias taurus	'Submissive' body, 'cupping' and 'flaring' of pelvic fins.	1
Ginglymostoma cirratum	Female arches body toward male, 'cups' pelvic fins.	5
Biting		
Heterodontus francisci	Male bites and wraps female pectoral fin body, tail, gills.	7
Scyliorhinus retifer	Male bites and wraps female pectoral fin body, tail, gills.	8
Scyliorhinus torazame	Male bites and wraps female pectoral fin body, tail, gills.	9
Ginglymostoma cirratum	Male bites and holds female's pectoral fin.	3, 5
Carcharhinus sp.	Male bites and holds female's pectoral fin.	10
Triaenodon obesus	Male bites and holds female's pectoral fin.	9, 13
Positioning and alignment		
Ginglymostoma cirratum	'Nudging' female into position with head.	3
Ginglymostoma cirratum	After 'pectoral bite' male rolls female, then aligns for insertion.	5
Sphyrna lewini	'Torso thrust' with 'clasper flexion' possibly filling siphon sacs.	6
Carcharias taurus	'Crossing' or 'splaying' claspers as position requires.	1
Group		
Ginglymostoma cirratum	Multiple males compete or cooperate for a mate. A cooperative Behavior or a single male 'blocking' a mating pair.	
Triaenodon obesus	Multiple males compete in mid-water for one female.	24
Insertion and copulation	Insertion of one or more claspers into the cloaca leading toejaculation.	5
COPULATORY		

Table 5.1 *Contd.* ...

Table 5.1 *Contd.* ...

General behavior and species	*Descriptions / Notes*	*References*
Male bites female while at rest		
Heterodontus francisci	Male wraps around females body.	7
Scyliorhinus retifer	Male wraps around females body.	8
smaller shark sp.	Male wraps around females body.	7, 8, 11, 12
Triaenodon obesus	Heads to substrate, sharks undulate to keep tails elevated.	13
Ginglymostoma cirratum	'Lying parallel on substrate' less than two pectoral widths apart during bouts of 'parallel swimming'.	3
Ginglymostoma cirratum	Heads to substrate, tails elevated or lying parallel. 'Copulation' sometimes in groups of many males.	5
Heterodontus francisci	Male crosses female's body, rhythmic motion for up to 35 m.	7
Parallel swimming 'in copula'		
Negaprion brevirostris	Coordinated pair swimming while copulating.	4
Carcharodon carcharias	Possible coordinated pair swimming while copulating.	14
Polygyny		
Ginglymostoma cirratum	Males will mate with many females over several weeks.	21
Polyandry		
Ginglymostoma cirratum	Females will mate with many males over several weeks.	21
POST COPULATORY	Pair remains together or departs rapidly.	5
Stalking		
Carcharias taurus	Male aggression toward other species in a captive environment.	1
Batoids		
PRECOPULATORY AND COURTSHIP		
Following		
Aetobatus narinari	Rapid 'chase', close to tail of female.	9
Manta birostris	Rapid 'chase', close to tail of female.	15
Myliobatis californica	Male ventral to female with wing beats synchronized.	16
Myliobatis californica	Males 'follow' females.	17
Dasyatis americana	Multiple males follow females.	22, 23
Female avoidance		
Urolophus halleri	Females bury in sand to 'avoid' males.	18
Aetobatus narinari	Females raise back out of water and slap wings on surface in response to male nipping.	16
Urolophus halleri	Females spine males with caudal spine.	19
Female acceptance		
Raja eglanteria	'Back arching', 'pectoral fin undulations' to attract males.	20

Table 5.1 *Contd.* ...

Table 5.1 *Contd.* ...

General behavior and species	*Descriptions / Notes*	*References*
Biting		
Aetobatus narinari	'Gouging', bites on female dorsal surface.	9
Rhinoptera bonasus	'Gouging', bites on female dorsal surface.	16
Rhinoptera javanica	'Gouging', bites on female dorsal surface.	9
Manta birostris	Male grasps pectoral fin tips (nipping).	15
Dasyatis americana	Male grasps pectoral fin tips (biting).	22, 23
Group		
Dasyatid and myliobatid rays	Common for multiple males to 'follow' single females.	9, 16, 17
Rhinoptera javanica	Many captive males overwhelmed a female for multiple matings. Mortality sometimes resulted from wounds and exhaustion.	9
Dasyatis americana	Seven to eight males follow, two attach at once.	22, 23
Other behaviors		
Aetobatus narinari	Males 'bob' and 'sway' while 'following' 'avoiding' females.	16
Dasyatis americana	Rapid succession of quick copulations.	22, 23
COPULATORY		
While reposed on bottom		
Raja eglanteria	Copulate for one to four hours while at rest on bottom. Male holds trailing edge of female's pectoral fin, swings tail beneath hers and inserts one clasper.	20
While swimming		
Manta birostris	Copulation near the surface, abdomen to abdomen.	15
Aetobatus narinari	Mating abdomen to abdomen in the mid-depths of the tank. Insertion time was 0.5 to 1.5 min.	9
Rhinoptera javanica	Starts at the surface or mid-depth, abdomen to abdomen, continues on the bottom	9
Rhinoptera bonasus	Starts at the surface or mid-depth, abdomen to abdomen,continues on the bottom.	9
Dasyatis americana	Starts over bottom, may continue or move to substrate.	22, 23
Polyandry		
Aetobatus narinari	A captive female mated many times in successionwith 3 to 4 males in 1 h.	9
Rhinoptera javanica	Multiple matings common.	9
Dasyatis americana	Multiple matings common.	22, 23
POST COPULATORY		
Manta birostris	Male remains attached to pectoral fin tip briefly.	15

Reference key: 1- Gordon (1993), 2- Johnson and Nelson (1978), 3- Klimley (1980), 4- Clark (1963), 5- Carrier *et al.* (1994), 6- Klimley (1985), 7- Dempster and Herald (1961), 8- Castro *et al.* (1988), 9- Uchida *et al.* (1990), 10- Clark (1975), 11- Gilbert and Heath (1972), 12- Dral (1980), 13- Tricas and LeFeuvre (1985), 14- Francis (1996), 15- Yano *et al.* (1999), 16- Tricas (1980), 17- Feder, (1974), 18- Tricas *et al.* (1995), 19- Michael (1993), 20- Luer and Gilbert (1985), 21- Pratt and Carrier (2001), 22- DeLoach (1999), 23- Chapman *et al.* (pers. comm), 24- Whitney *et al.* (pers. comm.).

attention hundreds of years ago. According to Louis Agassiz (1871), Aristotle noted what must be the first record of shark mating behavior: 'The cartilaginous fishes in copulation hang together after the fashion of dogs, the long tailed ones mounting the others, unless the latter have a thick tail preventing this, when they will come together belly to belly.'

In the last ten years, the development of inexpensive durable instruments such as electronic tags and video cameras has permitted us to extend our observations underwater. We can record and gather data effectively in both aquaria and in the ocean itself on the intimate lives of large reclusive elasmobranchs. Early reproductive observations by pioneers like Don Nelson, Scott Johnson and Eugenie Clark set the groundwork for techniques that can now be augmented with remote transmitters, monitors, cameras and satellite tags.

In the following account, key observations and records are presented first for captive animals then for animals mating in the oceans. Except for newer literature and research updates, summaries are brief. Table 5.1 and Appendix 5.1 summarize and clarify key findings and terminology. For a more detailed record see Pratt and Carrier (2001).

Records of the mating of small sharks in aquaria were the first insights into elasmobranch behavior, albeit a special circumstance. Though more difficult to obtain, observations in natural settings are ultimately more valid and, except for reactions to the researchers' presence, reflect more closely the true condition of wild elasmobranch reproductive behavior.

5.3.1 Mating in Captive Elasmobranchs

Other than Aristotle's contribution, the first published records of shark copulation came from several observations of catsharks, usually *Scyliorhinus canicula*, held in relatively small aquaria. Catsharks mature at a small size; 39 to 44 cm in *S. canicula* (Compagno 1984), so mating in a small aquarium is feasible. There is a common theme from the earliest reports. Males of these small slender flexible species of sharks initially bite the female as part of pre-copulatory behavior and entwine their bodies around the female while inserting one clasper for 20 min or longer (Bolau 1881; Schensky 1914; Hardy 1959; Gilbert 1981).

The best of these descriptions was the more recent observation of Castro *et al.* (1988) for the chain dogfish, *Scyliorhinus retifer.* A male and female circled tightly for an hour at the bottom of a large tank at the Mystic Marine Life Aquarium. The male bit the female in the gill region, flanks or tail. Eventually the male bit the female's tail and would not release it. The female struggled violently, and then became listless. The male moved its bite up the left ventral flank until it reached the left pectoral axilla, wrapped its body around the female and flexed its left clasper to the right, toward the midline, and copulated nearly motionless for 30 sec. The male released its bite but remained coiled around the female. The female then broke away and both swam to separate areas of the tank. Biting behavior was observed on the following days indicating that mating may occur repeatedly in this species.

Similar observations exist for horn sharks, *Heterodontus francisci*, in the Steinhart Aquarium (Dempster and Herald 1961), the male epaulette shark, *Hemiscyllium ocellatum* (West and Carter 1990; Michael 1993) and the grey bamboo shark, *Chiloscyllium griseum* (Dral 1980).

Clark (1963) reported an observed copulation of two *Negaprion brevirostris* in the Cape Haze Marine Lab's seaside shark pen. The sharks were side by side, heads slightly apart but the posterior half of their bodies in such close contact and the swimming movements so perfectly synchronized that they gave the appearance of a single individual with two heads, as they swam for at least one half hour.

Klimley (1980) was the first to detail courtship and copulation of *Ginglymostoma cirratum* from his observations in the shark channel exhibit at the Miami Seaquarium. He based his findings on 15 courtship attempts and one copulation in the channel and from interpretations of photographs taken from land and boats of wild matings at the Dry Tortugas.

The importance of captive work in modern large aquaria was shown by Uchida *et al.* (1990). Mating was observed in two shark species at the Okinawa Expo Aquarium. The cloudy catshark, *Scyliorhinus torazame* was observed mating twice and *Triaenodon obesus* was also observed mating twice. The mating *Triaenodon obesus* were born in the tank and were from the same litter. The male bit the female on her right pectoral fin and used his medially flexed right clasper in a copulation pattern that is frequent among the larger sharks.

Gordon (1993) reported three instances of pre-copulatory and copulatory behavior of two adult males and one adult female *C. taurus* held at Oceanworld, Manly, Sydney, Australia. Several distinctive behaviors related to mating and dominance were identified, some of which help clarify field observations. Both sexes were observed to perform 'courtship bites' to facilitate mating or as a reaction by females to a persistent male. Gordon (1993) was first to document a reproductive dominance hierarchy. The alpha male swam in increasingly larger circles and began splaying its claspers, then approached the female and exhibited tailing and nosing. Copulation occurred as the male bit into the right flank and trailing edge of the pectoral fin of the female. The male swam side by side with the female, copulating with the right clasper for one to two min. After copulation, the male showed little interest in the female. Gordon (1993) speculates that a chemical stimulant (pheromone) attracted the males in his study. In contrast to Klimley's (1980) findings for *Ginglymostoma cirratum*, he reports lighter colored males and darker females in precopulatory sand tiger sharks, *Carcharias taurus*.

Mating in skates and rays is complex and can take several forms (Table 5.1). Some small skates mate with ventral surfaces opposed

(Wourms 1977). Some larger rays can apparently mate with the female upright, above or below the male (Tricas 1980; Uchida *et al.* 1990; Yano *et al.* 1999). Demski (1990b) remarks that the variety of mating positions observed in elasmobranchs is probably related to the great differences in body form and swimming habits found in these fishes. Libby and Gilbert (1960) and Luer and Gilbert (1985) discovered that male and female clearnose skates, *Raja eglanteria* mate side by side with the pair in an upright position. The male bites the caudal margin of the female's pectoral and inserts one clasper, flexed medially 90 degrees, into her oviduct, In a rare record of female precopulatory attractive behavior, they note that the female skate often draws the attention of males by 'back arching' and 'pectoral fin undulations.' The alar and head spines of the male may pierce the skin of the female to obtain purchase during a copulation that lasts up to four hours. In a captive setting, Uchida *et al.* (1990) observed mating and copulation once in the spotted eagle ray, *Aetobatus narinari*, and six times in the cownose ray, *Rhinoptera javanica*. At the Okinawa Expo Aquarium, the mating *A. narinari* have enough space for what Uchida *et al.* (1990) call 'typical patterns of mating behavior.' Captive mating and related behavioral observations are very valuable because they can point the way to understanding patterns of activities in the wild, but these results must also be interpreted with caution. Confined animals, even in the largest aquaria, may still be restricted in ways we cannot understand or appreciate. Avoidance and mate selection are difficult. They may have unique uses for different substrates and habitat that may be in short supply or completely lacking because of the engineering restrictions of the tank or pool.

5.3.2 Mating in Wild Elasmobranchs

Elasmobranch copulation must be a very common seasonal occurrence in the sea, however it is rarely observed and, consequently, poorly documented. Dissection of sharks caught in sport and commercial fisheries has revealed that mating activities are indeed very common at certain times. However, mating activities are seen serendipitously in nature and hardly ever by researchers. Mating elasmobranchs are reclusive; they are usually not seen mating near human activity or fishing operations (*Ginglymostoma cirratum* is an exception). They live in a concealing medium and even if mating is witnessed, it is seldom that the entire event is observed and recorded. This is especially true for the larger sharks and rays.

5.3.2.1 Sexual segregation

In many species of elasmobranchs, members of the same size and sex gather or travel in groups. Sharks that have been reported to segregate by sex include school sharks, *Galeorhinus galeus* (Olsen 1954), *Carcharhinus plumbeus* (Springer 1960), bonnetheads, *Sphyrna tiburo* (Myrberg and Gruber

1974), *Prionace glauca* (Pratt 1979), *Carcharias taurus* (Gilmore *et al.* 1983), grey reef sharks *Carcharhinus amblyrhynchos* (McKibben and Nelson 1986), *Sphyrna lewini*, (Klimley 1985, Klimley 1987) and *Squalus acanthias*, (Callard *et al.* 1988). The movements, migrations and distribution of sharks are influenced both directly and indirectly by water temperatures (Casey and Kohler 1992). Consequently some species migrate seasonally, apparently following temperature isotherms, currents and probably seasonal movements of prey species. In some elasmobranchs, reproductive condition and probably the need to accentuate, enhance or limit reproductive success is a factor in segregation and movements

Migrations often take sharks to specific places where reproductive events may occur (Springer 1967). Economakis and Lobel (1998) observed aggregations of adult female *Carcharhinus amblyrhynchos* in warm shallow waters at Johnston Atoll and suggest that this behavior may hasten embryonic development. They did not notice recent mating bites on the females or male sharks in the area, so aggregation in this instance is probably not for mating activity. Taylor (1993) reports that it is not unusual for female *C. amblyrhynchos* to gather in shallow channels and lagoons in the Northwestern Hawaiian Islands. They seemed to favor warmer shallow water reef areas during times of maximum sunlight, between 11 a.m. and 3 p.m. May through August. He speculated that the resultant increase in body temperature, an estimated 1.0 to 1.5°C in this case, may hasten embryonic development. Pratt and Carrier (unpublished) have observed large gatherings of rotund female *Ginglymostoma cirratum* in similar, warm shallow waters at the Dry Tortugas three to four months after mating and one to two months before parturition (Fig. 5.2).

Demski (1990a) noted that in those elasmobranchs that form schools, the sight and sounds of a critical mass of individuals may be necessary to trigger mating behavior as in some birds and mammals. Klimley (1985) suggested that *Sphyrna lewini* females compete for position near the center of the female school. This action could be similar to Hamilton's (1971) 'selfish herd' premise, a behavior used by schooling and herding species to escape predation. Female segregation in *S. lewini* and other species of sharks may spread out male courtship advances and could aid in female choice in mating with or 'avoiding' a particular male.

Frequent mating may have a high cost for females due to blood loss or possible infection from bites or manipulations and also to males from possible female defensive maneuvers (Springer 1967, Gordon 1993). Nearly every female *Prionace glauca* on the southern New England continental shelf in summer bears fresh, deep dermal lacerations from male tooth cuts. Dissection of these females typically shows vaginal cuts and edema and reproductive tracts that have large volumes of fresh sperm in the lumen of the uteri and oviducal gland (Pratt 1979). If sexual selection is regarded as competition between the sexes (Krebs and Davies 1993), then females segregating away from the males when not engaged

Fig. 5.2 Female *Ginglymostoma cirratum* sharks in warm shallow waters at the Dry Tortugas three to four months after mating. Original.

in mating activity may be the best behavior for both breeding success and survival from repeated and prolonged mating attempts.

5.3.2.2 Wild Mating: Observations from above the sea's surface

The most numerous observations of wild mating come from inferences resulting from dissection of landed animals. With some exceptions, most of the records reported here are from sharks. Batoids do not usually come under the close scrutiny of reproductive biologists. This perhaps reflects commercial capture and the use on research trips of commercial gear that targets sharks.

Stuart Springer worked with commercial fisheries in the 1940's and 50's and made many accurate deductions from shipboard dissections. In his 1967 paper he wrote, 'Among the larger carcharhinid sharks males harass the females persistently and violently to induce essential cooperation in mating because the jaw armature of the males is such that the male could not bite and hold the female without producing a very severe injury and because the relatively inflexible male carcharhinid trunk region cannot by itself hold the female for mating as do the more supple catsharks and dogfish.' Springer (1960) speculated that this harassment in sharks of itself represents a form of courtship.

Courtship behavior has been inferred from necropsies of *Cetorhinus maximus* (Matthews 1950), *Carcharhinus plumbeus* (Springer 1960), *Prionace glauca* (Stevens 1974; Pratt 1979), *Carcharias taurus* (Gilmore *et al.* 1983), the blacktip shark, *Carcharhinus limbatus* (Castro 1996) smooth dogfish, *Mustelus canis* (Conrath and Musick 2002) and many others. Examinations of these species have revealed tooth cuts, abrasions and punctures on the female's body and, in some cases, wounds to the cloaca and vaginal walls, presumably caused by mating activities and by clasper terminal edges or spurs after insertion. The presence of fresh sperm in the uteri or oviducal gland in sharks with fresh bite marks was reported by Pratt (1979) as noted above for *P. glauca* and for the finetooth shark, *Carcharhinus isodon* and *C. limbatus* by Castro (1993, 1996). Conrath and Musick also used the amount of sperm stored in the shell gland by month to determine mating season (Conrath and Musick 2002).

White sharks are uncommon throughout their range. Even in areas where they are frequently seen, they can be unpredictably absent. Their biology is incomplete and records of their reproductive condition are few and sketchy. Evidence of suspected mating by *Carcharodon carcharias* in Australian and New Zealand waters is presented by Francis (1996). Most records are of bite marks on females or observations of seminal fluid oozing from claspers or the genital papillae of adult males. He includes three field reports of probable copulation in free-living white sharks seen swimming side-by-side. Like Clark's (1963) record, they were seen from above, a limited viewpoint. There was an initial grasp and then a motionless but obvious copulation that lasted 40 minutes before the sharks separated.

Ebert (1996) reported on copulation of two sevengill sharks, *Notorynchus cepedianus* in Luderitz Bay, South Africa. The sharks were swimming slowly at the surface. The male's jaws were clamped on the female's flank, just behind the left pectoral fin. The bodies were side to side with the males sagging below the female's. The number of claspers inserted was not discernible.

As in *Carcharodon carcharias*, mating behavior in *Cetorhinus maximus* is proving difficult to detail. Harvey-Clark *et al.* (1999) interpreted five minutes of aerial video footage and still photographs taken of 13 *C. maximus* in close circling formation and overlapping interactions as being of mating behavior. Copulation could not be confirmed. Sims *et al.* (2000) report 25 separate interactions of possible reproductive significance while tracking *C. maximus* during a five-year study in the Western Basin of the English Channel. Again, pre-copulatory activities (grasp) and copulation was not witnessed, however they observed 'nose-to-tail following', 'flank approach' and 'parallel swimming' behaviors. It was difficult to determine sex consistently from their vessel, but based on comparison with the observations of Carrier *et al.* (1994), they suggest that the observed behaviors may constitute part of a courtship behavior.

5.3.2.3 Wild Mating: Observations from below the sea's surface

It is difficult to discern reproductive processes or behavior based on examinations of dead animals or on brief encounters from above the sea's surface. Because they usually occur under less than optimal conditions, results can be misleading and even lead to errors. Evidence of recent mating from these sources is very valuable to establish the time and place and other particulars of mating; however, underwater field observations, though at times troublesome, are necessary to obtain details of mating behavior. Larger species are often reclusive, challenging to find and sometimes dangerous to approach when reproductively active. Successful observation requires that the presence and actions of the observers do not modify the natural behaviors under study.

Underwater reproductive behavioral observations began when Eugenie Clark (1975) became the first to document courtship biting by sharks in the open sea. Johnson and Nelson (1978) reported pre-copulatory behavior in the blacktip reef shark, *Carcharhinus melanopterus* and *Triaenodon obesus*. In both species the male followed very closely behind the female, with its snout less than 30 cm behind the lead shark's tail. The female held its tail up in an unnaturally erect posture that permitted the male to orient its snout to her vent. Johnson and Nelson (1978) hypothesized olfaction-mediated recognition and pairing after observing males intercepting females when visual and sound cues were absent. Two mating *T. obesus* were observed and photographed copulating off Molokini in the Hawaiian Islands (Tricas and Lefeuvre 1985). Pair contact was maintained by the male's tight oral grasp of the female's entire left pectoral fin and by his left clasper anchored one-third of its length into the female's cloaca. Observations on batoids show similar patterns to shark courtship, with chasing, grasping and biting behavior. Working off Eniwetok Atoll, Tricas (1980) observed what appeared to be complex courtship activity in *Aetobatus narinari*. Tricas (1980) also noted a brief observation of dorsal clasper flexion in a bat ray *Myliobatis californica*, and a possible insertion attempt as the male swam in position just under and slightly behind a larger female. Reed and Gilmore (1981) found that male roughtail stingrays, *Dasyatis centroura*, also bite the pelvic fins of the females, inflicting scars during mating behavior.

In yellow stingrays, *Urolophus jamaicensis*, the males and females simultaneously grasp the leading edge of each other's pectoral fins. The ray grasps its partner's fin edge in its mouth and flips over at a slight angle to its mate (Dugger 1987). Male *U. halleri* also bite female pectoral fins (Nordell 1994). Tricas *et al.* (1995) discovered that reproductively active male *U. halleri* use their electrosensory system to locate females during courtship. Females also used their electrosense to locate buried consexuals (Tricas *et al.* 1995).

Manta birostris mating behavior was observed by Yano *et al.* (1999) in the waters off the Ogasawara Islands, Japan. They recorded the abdomen–to–abdomen courtship behavior of two males with one female. Yano *et al.* divided the mating behavior of *M. birostris* into: (1) 'chasing', the male rapidly follows behind the tail of the female and assails her several times, (2) 'nipping', the male nips the tip of the pectoral fin of the female and then moves to the ventral surface of the female, (3) 'copulating', the male inserts a clasper into the cloaca of the female and copulates abdomen–to–abdomen, (4) 'post-copulating', and (5) 'separating'.

DeLoach (1999) provided an account of mating in the southern stingray, *Dasyatis americana*, in which one female was chased by seven or eight smaller males, forcibly held to the sea floor and mated by two of them in quick succession.

Fowler (1906) described mating *G. cirratum* around the Marquesas Islands, Florida. Gudger (1912) made early observations of mating of free-living *G. cirratum* in the Dry Tortugas, Florida. Rouse (1985) provided fortuitous underwater photographs of *G. cirratum* mating on the bottom in 35 m of water off the East Coast of Florida. The authors and our collaborators working from 1992 to the present have added detailed underwater observations of free-living *G. cirratum* reproduction to those of Fowler, Gudger, Klimley and Rouse (Carrier *et al.* 1994, Pratt and Carrier 1995, 2001). Four stages of mating were identified: 'precoupling' which included the 'following' behavior similar to Klimley's 'parallel swimming'; 'coupling', in which at least one female pectoral fin was grasped; 'positioning and alignment' where rolling and body position were similar to Klimley's 'nudging' and a prelude to 'insertion and copulation' (Klimley 1980). Sharks copulated in right and left, side-to-side and ventral-to-ventral positions. Ten of the sixteen events observed in 1992 involved multiple males. (*see* Appendix 5.1 for a summary of reproductive terminology.)

Chapman *et al.* (pers. comm.) provide the first detailed, complete sequence of mating in free-living *Dasyatis americana*. Their observations are augmented with a video record. Four mating events were observed with five copulations. Events followed the pattern of five behaviors identified by Yano *et al.* (1999) and clarified by Pratt and Carrier (2001). The event was characterized as 1) close-following, 2) pre-copulatory biting, 3) insertion/copulation, 4) resting and 5) separation. One event involved two males, one biting the trailing edge of each pectoral, each mated ventral-to-ventral with the female, one right after the other. The clasper of the first male may have remained inserted while the second male copulated (Fig. 5.3). The five copulations lasted between 10 and 33 s. Another two copulations occurred 35 m to 2 h after a witnessed parturition. Henningson (2000) has noted that the interval between parturition and the next mating season may be very short in this species. As noted above for *Dasyatis sabina*, Maruska *et al.* (1996) suggested that copulations many months before ovulation may be a stimulus to initiate the female reproductive cycle.

Despite much excellent work, few investigations have had the opportunity for long-term, systematic studies of mating in an elasmobranch population. Most species are too inaccessible and too intractable to be followed for any period of time. Not many elasmobranch field research endeavors last over a decade. The authors have constructed a research setting in which to evaluate the process of mating, mate

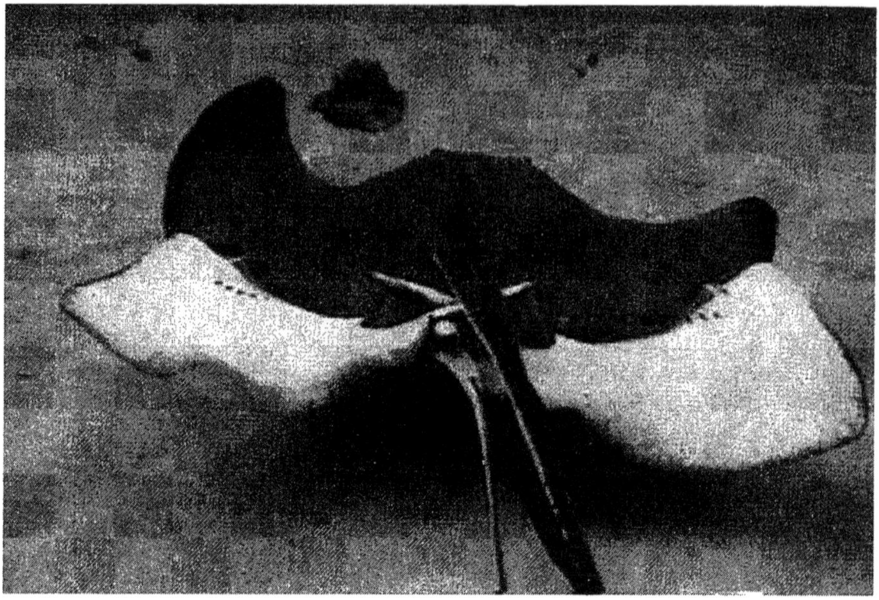

Fig. 5.3 Mating *Dasyatis americana* two males copulate in quick succession with female (center) (from Chapman *et al.* pers. comm.).

selection and possible social structure in *Ginglymostoma cirratum* over the long term. The following is an account of some current efforts.

5.4 RECENT TECHNIQUES AND RESULTS IN STUDYING *GINGLYMOSTOMA CIRRATUM* MATING BEHAVIOR

As diving methodology, underwater image recording and telemetry instruments improve, they allow the observation and recording of elasmobranch reproductive behaviors with more precision. Videography and telemetry provide a durable unbiased record, a detailed time series for analysis and quantification of underwater behaviors not previously possible. It must be noted, though, that in behavioral work, sometimes the most basic approaches are the most profitable. A few observers using quiet methods and simple equipment can often be the least disruptive and the most effective. Often a combination of simplicity and technology are the most valuable.

The authors have conducted studies on a population of *G. cirratum* in the Dry Tortugas National Park, Florida, since 1991. Our recent investigations include: 1) determining juvenile and adult distribution in the local reefs and entire archipelago using tagging; 2) defining elements of social structure with DNA fingerprinting; 3) elucidating behaviors, such as dominance hierarchies, sperm competition, male rivalry, male cooperation and possible kin interactions using direct field observations and sampling of tagged adult, juvenile, and neonate *G. cirratum* and 4) characterization of the archipelago as a critical habitat for mating and nursery grounds, and management of the site to ensure reproductive success (Carrier and Pratt 1998).

5.4.1 Less-invasive Sampling and Tagging

5.4.1.1 Tagging - spear gun to hoop net

In the study's early years, our inability to identify individual adult sharks hampered our understanding of their courtship and mating. We began by using a modified spear gun to deliver diver-identifiable NMFS dart tags near the first dorsal fin of mature male and female *G. cirratum* after mating events. Standard sampling using baited hooks was rejected to prevent adding an unnecessary stimulus to the water and the possibility of unnaturally attracting other shark species to the study area. The tags work well on many species (Kohler *et al.* 1998), but are relatively short lived on nurse sharks. The tag's stainless steel anchors are permanent, but the monofilament streamer wears thin after four to five years depending on site of attachment on the body. Marine bio-fouling also obliterates tag numbers and covers the coded beads we used for identification. Spear techniques were discontinued in 1999 when a hoop net capture technique was developed. Both males and females may be captured on the shallow water mating grounds during and just after the courtship and mating attempts in the reinforced net. Sharks were usually caught at the end of the day at the conclusion of behavioral observations. The captured sharks became quiescent after five minutes in the net and were then moved to shallow water for measuring and tagging. We have been able to capture 48 large (219-275 cm TL) reproductively active sharks from our study population using these nets. Ten of the 48 adults captured, sampled, retagged and released between 2000 and 2003 were originally tagged by the authors with spear-placed tags in 1994, 1995, and 1996.

To provide a durable marker, all captured G. cirratum were tagged with passive integrated transponder (P. I. T.) tags, under the dermis at the base of the first dorsal fin. One or two colored-coded nylon cattle ear, Dalton® 'Jumbotags' and/or 'rototags' were placed in the first dorsal fin of adult females and second dorsal fin of males. The two varying colors allow sharks to be identified by color combination alone. Fouling is still a problem, but if any part of the tag is visible, color makes shark identification more practical than our previous techniques.

5.4.1.2 Results 1991 – 2003 tagging

Diver identifiable tags are vital in determining the identity, frequency and time period in which individuals participate in mating. Since 1993, we have tagged 215 nurse sharks (86 adults and 129 juveniles) in the Dry Tortugas study population. Of the identified adults, 34 were subsequently sighted (visually recaptured or netted), at least once, and some more frequently. One individual has been seen in various mating activities 65 times over the course of 12 years and several juveniles have been physically recaptured five or more times. Observations of tags and sharks with natural markings indicate that many adult males visit the study site to participate in reproduction faithfully every year, with three dominant males and one 'alpha' male consistently observed since 1992. Results from our tag studies continue to support our hypothesis that adult females visit the study area to mate in alternate years. Females that have actively mated in one year have never returned the following year, but some females return after an absence of two or more years. This biennial mating pattern is common in elasmobranch females (Wourms 1981; Pratt and Casey 1990; Gilmore 1993) and probably allows post-partum females the time to rebuild reproductive reserves before mating again.

5.4.2 Video recording and 'Crittercam'

A challenge in underwater observation is to unobtrusively view animal interactions that, because of depth, or swimming speed are beyond the scope of human senses and abilities. The shallow water courtship and mating we have seen and followed for years in *G. cirratum* has yielded rich insights to shark reproduction. Video is invaluable for the long recording times and ease of use it affords, but it has limitations. We have used free diving and video techniques to record mating on the shallow mating grounds, but possible mating activity outside the lagoon in deeper water among the reefs has been enigmatic. Females travel from these deeper areas to seek refuge in the shallows and males return to deeper waters after brief sorties in the shallows. To better investigate the question of deep water mating, we sought the expertise of National Geographic's remote imaging team and 'Crittercam' (Fig. 5.4) an animal-borne, self-contained video and telemetry system.

Attaching a large housing with camera, antennae and sensors to an animal brings with it the uncertainty that the natural behavior patterns will overcome the handling and the intrusive addition. In the case of nurse sharks, the animals were large relative to the size of the unit. The resultant images taken from the animal's point of view showed that the males immediately resumed mating. Preliminary results from just a few deployments reveal that when males that mate in shallow water return to the deeper reefs (5-10 m), they tend to rest for long periods in sandy gullies on the sea floor. No courtship and mating has yet been observed in deeper water, though fast pursuit of females and attempts by males carrying Crittercam to secure a grip have been observed several times in shallow water (< 2 m).

5.4.3 Telemetry and Movements During Mating Season

Ultrasonic telemetry is a powerful tool to determine the fine scale daily activities of sharks and is being used by the authors to track male and

Fig. 5.4 'Crittercam' clamped to the first dorsal fin of a 248 cm male nurse shark, *Ginglymostoma*. Original.

female adult sharks when they leave the study lagoon to determine their home ranges and activity cycles. Selected male and female adults were tagged with Vemco (Shad Bay, Nova Scotia) or Sonotronics (Tucson, AR) ultrasonic transmitters for telemetry. Vemco VR2 sea bottom monitors with built-in hydrophones and recorders were deployed around the study area to track tagged sharks and continue to provide data that support these earlier observations. Transmitters were always accompanied by separate conventional tags designed to remain after the transmitter breaks away.

Males tagged with ultrasonic transmitters showed a great deal of active swimming in their local patrolling behaviors supporting Springer's (1967) remarks of increased activity of males during the mating season. Monitor data and active tracking with a moving hydrophone show males repeatedly patrolled the coast of the islands that bracket the study site in 1-12 m of water, resting at intervals near reef structure. This confirms to some extent the Crittercam records of deeper water resting.

5.5 THE AUTHORS' CURRENT RESEARCH INTO SHARK MATING BEHAVIOR

5.5.1 Male Helping Behaviors

The evolution of cooperation among animals has been vexing to biologists since Darwin (1859). If 'survival of the fittest' operates at an individual level, then how do cooperative behaviors evolve? There are many theories (See Dugatkin (1997) for an overview and recent perspective from a fish oriented writer.) Cooperative behaviors are well known from a wide variety of animals engaged in hunting prey, nest building and territorial defense. Some of the most complex non-human social structures (cetaceans) are seen in the sea (Connor *et al*. 2001). Many birds and mammals put themselves at risk to make alarm calls. At first these actions seem to be altruistic: actions that increase the recipient's lifetime number of offspring at a cost to its own survival and reproduction (Trivers 1971). But recent research in birds and mammals has shown that there can also be a genetic basis to some seemly altruistic actions, especially where kin can be recognized (Krebs and Davies 1993). In some cases the altruism is apparently reciprocal, as in males procuring females for other males in the olive baboon (Packer 1977). While helping behaviors are known in fish (Taborski 1984), they were not noticed in sharks until Carrier *et al*. (1994) first reported them as part of the group behaviors in the Dry Tortugas mating population of *G. cirratum*.

Although participants in *Ginglymostoma cirratum* courtship bouts are often single males selecting a 'refuging' female, our work shows that most successful encounters are between a group of males and a lone female. For instance, four to six males may surround the female in shallow water, competing for an effective bite on either pectoral fin (Fig. 5.5), but it is clear to an observer that from the start in most events with multiple males, there is one male that does not compete for a fin: the designated 'blocker.' During competition, before and during clasper insertion, and throughout the ensuing copulation, this male's role is to 'block' the forward motion of the interacting group and then the mating couple by laying against their rising and falling heads and moving his body with the mating pair. The 'helping' male holds his body approximately at right angles to the mating in a firm arch, absorbing the copulating males

Fig. 5.5 Five males surround a female *Ginglymostoma cirratum* that was refuging in shallow water. Two are competing for an effective bite, one on each pectoral fin. Original.

Fig. 5.6 A 'helping' cooperative male *Ginglymostoma cirratum* (lower left, note clasper) positions its body at right angles to the copulating pair. The female is in the upper center of photo; male is upside-down and arched on the right with a grasp on the female's left pectoral fin. All sharks are about 2.5 m long. Original.

forward thrusting motion and keeping both sharks from sliding across the sea floor (Fig. 5.6).

Why do some males help other males to mate? The authors and their cooperators are now utilizing tagging and DNA analysis to define male roles and investigate possible kinship to determine if this male helping behavior is a mutualism or reciprocal altruism (Trivers 1971) and to further explore and investigate this unique act in their ongoing study of *G. cirratum*.

5.5.2 *Triaenodon obesus*

From a recent video documentation, three mating events involving nine whitetip reef sharks, *Triaenodon obesus* were filmed in the Cocos Islands, Costa Rica. Mating behavior and siphon sac function were observed and analyzed from video records (Fig. 5.7).

Whitney *et al.* (unpublished) describe several mating behaviors including 'arch', 'avoid', 'following', 'group behavior', 'pectoral grasp', 'positioning and alignment', 'clasper flexion', 'insertion and copulation' and 'separating', for the first time in this species. They also report four new mating behaviors, 'siphon isthmus constriction', 'reverse thrusting', 'post-mating gaping' and 'non-copulatory ejaculation' that have never been seen in any elasmobranch. This work also presents the first hypothesis of

Fig. 5.7 *Triaenodon obesus* male (right) with siphon sac partially inflated attempts clasper insertion on the sea floor (from Whitney *et al.* unpublished, Fig. 5.4).

siphon sac function to be based on observations of mating sharks. The authors believe that the siphon sacs in *Triaenodon obesus* are used as mechanisms of sperm propulsion, not for flushing the female's reproductive tract of sperm from previous males.

5.6 THE EMERGING BEHAVIOR TEMPLATE

Some of the reproductive activities described in the previous sections reveal behaviors that seem common to most elasmobranchs investigated to date. The following section provides a revised and updated overview based on our older review (Pratt and Carrier 2001). References to specific authors are omitted for clarity as all are mentioned in the preceding section.

Elasmobranch courtship begins when one animal signals a potential mate that it is reproductively receptive to 'copulation'. In most instances, the female probably initiates mating, perhaps involuntarily, through some combination of secretions and behavioral cues, but work on urolophid stingrays and captive *Carcharias taurus* shows that the male may sometimes initiate mating. The precopulatory cues are probably principally olfactory, with females employing chemical cues (pheromones) that may be combined with motor displays. These include: female 'back arching', 'submissive' body tilting, 'pectoral fin undulations', pelvic fin 'cupping' and 'flaring' and perhaps other, as yet undescribed physical cues.

Males may or may not cease feeding, become aggressive toward other species and may compete for dominance with other males of the same species. One male may approach the female either alone or in a 'group' and attempt to 'follow', perhaps 'nosing' her cloaca. The female may raise her tail, or incline her body. After preliminary interactions, the male's mouth is used for first contact in most elasmobranchs. Typically the male moves up the female's body to 'bite' her pectoral fin. Biting may act as a releaser for female mating behaviors and precopulatory biting may stimulate ovulation and other physiological changes. Batoids may also bite and 'gouge' and hold females with spines and denticles. Female urolophid stingrays may bite the males in return. Males of some larger shark species (*G. cirratum*, most known carcharhinids) concentrate on one of the pectoral fins. More flexible species such as *Prionace glauca*, catsharks and horn sharks, may initially 'grasp' anywhere on the female's body, usually just behind the eyes, including the pectoral fins. Smaller and more limber sharks are sometimes able to copulate from a body bite or even a body 'wrap' without using a 'courtship bite', but the rule in larger, less flexible sharks seems to necessitate an unshakable oral grasp of the pectoral fin.

After attaining a hold on the female, actions vary with species and female cooperation. Male *Ginglymostoma cirratum*, if in shallow water (< 1 m), will lift and 'carry' the female to deeper water (1.5-2 m) or may attempt to do so. In, *Triaenodon obesus* and possibly hammerheads, the courting sharks may meet in mid-water, sometimes in groups (Fig. 5.8) and fall to or seek the sea floor as the male grasps the female. Here, the female's rostrum becomes positioned against the substrate, perhaps intentionally by the male to help balance his thrusts. The male then catches the female's tail between his tail and more the distant (crossed) clasper and begins copulation. Some species, such as *Heterodontus francisci*, reportedly can use either clasper. Male *T. obesus* and *G. cirratum* use a crossed clasper on the opposite side from the grasped female fin. Some elasmobranchs may have the option of using either clasper in either direction, 'crossed' or 'splayed'. Skates and some rays copulate on the sea floor; other pelagic or more powerful animals, such as *Aetobatus narinari* and *M. birostris*, perform a two-part ballet in mid-water.

During or immediately before copulation, sperm exit the paired ampullae to the urogenital sinus and then are injected into the apopyle and clasper groove of one forward-rotated and copulating clasper. From the clasper groove, the sperm are propelled into the female's vagina and uterus either through the hydraulic action of the corresponding muscular subdermal siphon sac or by the action of smooth muscles in the walls of the ampullae epididymides or both. This transfer may be further augmented by changes in hydrostatic pressure within the coelomic cavity generated by the male's thrusting behavior during copulation. In *G. cirratum* thrusting may be aided by the male cooperative behavior known as blocking (see section 5.5.1).

In response to the male's maneuvers, an 'accepting' female *Ginglymostoma cirratum* 'arches' her body toward the male, 'cups' her

Fig. 5.8 A group of five courting *Triaenodon obesus*, meet in mid-water off the Cocos Islands. The group then swam and fell to the sea floor where it reduced to three, then two sharks (from Whitney *et al.* (unpublished, Fig. 5.6).

pelvic fins upward, becomes immobile and usually becomes docile and receptive. The female *Aetobatus narinari* stops in flight to permit a grasp. The clasper may be partly or totally inserted into the cloaca and presumably into the common vagina and perhaps one uterus. It is locked in place as the terminal clasper structures take the form of an expanded reverse funnel. If the clasper rhipidion bears sharpened ridges, hooks and spurs, they find purchase in the female's thickened vaginal wall at this time and may assist in anchoring the clasper during movements accompanying copulation.

At any given encounter, the female often chooses not to mate. Perhaps the timing of ovulation is not auspicious, or perhaps she finds the current male undesirable, too young or weak. An unaccepting, 'avoiding' female will retreat from the approaching aggressive male. A female *Carcharias taurus* may 'shield' her cloaca by swimming near the substrate. A female *G. cirratum* caught in an unwanted 'pectoral bite' will first 'arch' her body away from the male, bend towards him, then twist out of his grasp with a fast 'pivot and roll' to escape. Some male *G. cirratum* will let go of the female pectoral fin spontaneously after four to five minutes of waiting, perhaps in exhaustion or as a consequence of oxygen deprivation due to the blockage of the buccal cavity by the large pectoral fin of the female. *Triaenodon obesus* males swim away with their mouths agape as if seeking oxygen.

If the male is accepted or endured, copulation ensues and the female shark becomes relatively immobile in those species so far observed (Table 5.1) (*Ginglymostoma cirratum, Scyliorhinus retifer, S. canicula, Triaenodon obesus, Aetobatus narinari, Manta birostris* and anecdotally, *Negaprion brevirostris, Carcharodon carcharias* and *Negaprion cepedianus*). The male thrusts and rotates his tail, probably accompanied by internal contractions of the ampullae epididymides and perhaps the siphon sacs for several seconds to a few minutes in larger species of sharks to many minutes and even hours in certain skates. Some species may copulate on the sea floor (*T. obesus, G. cirratum, S. retifer*), in mid water (the larger rays) or at the sea surface (*N. brevirostris, C. carcharias* and *N. cepedianus*).

At the conclusion of insemination in *Ginglymostoma cirratum*, thrusting stops, both are suddenly still and then the male very slowly withdraws his clasper. A viscous mucus mass, apparently containing some sperm, stretches out between clasper and female cloaca, and then breaks and retracts. The female remains motionless, still cupping her pelvic fins. The pair then becomes active again and either breaks apart rapidly to swim away separately (catsharks, *G. cirratum*), or the male may release his pectoral grasp and remain motionless on the sea floor as if in recovery (*G. cirratum*). Male *G. cirratum* have been observed to leave the mating area and mate again the next day. Females may rest in the area of mating, may seek shallow water immediately and rest for hours or may mate with another male within seconds.

5.7 CONCLUSIONS

An understanding of the ultimate and proximate causes of specific elasmobranch behaviors can only be developed from an integration of techniques from many disciplines such as physiology, biochemistry, molecular genetics and population ecology. The limited studies of life history strategies of elasmobranchs (Hoenig and Gruber 1990; Pratt and Casey 1990) reveal that these animals typically have slow growth, mature relatively late in life and have low fecundity, making them particularly vulnerable to fishing pressure.

To effectively manage elasmobranch stocks, it is essential to determine size and age at maturity, fecundity and reproductive cycles for all common and/or exploited elasmobranch species. Studies must also include discerning the habitat requirements for both mating and pupping grounds as well as focusing on social dynamics and mating behavior. An example of their importance is found in our new understanding of the male group 'helping' behavior. If such group behaviors prove to be common among elasmobranchs and if, as we suspect, they lead to greater reproductive success, then suggestions to disproportionately harvest males (Conrath and Musick 2002) based on the discovery of female sperm storage, may not be appropriate for species that utilize and surely profit from multiple-male group interactions.

The extent to which a population is able to recover from a directed fishery or from indirect bycatch, has a profound impact on the rate at which animals can be removed from the breeding stock. Such removal should not jeopardize the natural ability of the population to recover. Most studies of reproduction concentrate on one or two aspects of life history rather than on a comprehensive examination of all factors that ultimately influence reproductive success. Notwithstanding the challenging nature of such studies, investigations into reproductive behavior and its required habitat use must be conducted in concert with studies of movement patterns, segregation, distribution, genetics, anatomy, age and growth, food habits, anthropogenic influences and research on captive animals. Only then will the larger picture showing the full scope and complexities of elasmobranch reproduction come into focus.

5.8 ACKNOWLEDGEMENTS

Grateful appreciation is extended to W. Landrum and P. and J. Taylor of the National Park Service, to Frank Murru, Ray Davis and the veterinarians and research staff from SeaWorld Adventure Parks, Orlando, Florida. Capt. T. Taylor provided support on his expedition vessel Tiburon. We thank B. Causey, J. Halas, S. Baumgartner of NOAA's Florida Keys National Marine Sanctuary Program, the Newfound Harbor Marine Institute at Seacamp and the United States Coast Guard. N. Kohler of NOAA/NMFS Narragansett Lab provided tags and some equipment and we thank T. Pratt and C. Carrier for many hours of shark observations and field assistance. N. Whitney improved the manuscript and provided figures and observations from our unpublished manuscript and S. Waterman supplied video footage of *Triaenodon obesus* that is the basis of Whitney's analysis for Figs 5.7 and 5.8. Table 5.1, Appendix 5.1, and text from 'The Emerging Behavior Template' are reproduced from the Environmental Biology of Fishes with permission from Kluwer Academic Publishers. This project was supported in part by Mote Marine Laboratories Center for Shark Research, Summerland Key and Sarasota, FL. and funds to HLP came from NOAA's Highly Migratory Species Management Division and to JCC from the Hewlett-Mellon Faculty Development funds and also from the A. Merton Chickering and W.W. Diehl Endowed Professorships of Albion College.

5.9 LITERATURE CITED

Agassiz, L. 1871. On the method of copulation among selachians. Proceedings of the Bostonian Society of Natural History 14: 339-341.

Bolau, H. 1881. Uber die Paarung and Fortpflanzung der Scllium-Arten. Zeitschrif für Wissenschlaftliche Zoologie 35: 321-325.

Bres, M. 1993. The behaviour of sharks. Reviews in Fish Biology and Fisheries. 3: 133-159.

Callard, I. P., Klosterman, L. and Callard, G. 1988. Reproductive Physiology. Pp. 277-317. In T. J. Shuttleworth (ed.), *Physiology of Elasmobranch Fishes*. Springer-Verlag, Berlin.

Carrier, J. C. and Pratt, H. L., Jr. 1998. Habitat management and closure of a nurse shark breeding and nursery ground. Fisheries Research 39: 209-213.

Carrier, J. C., Pratt, H. L., Jr. and Martin, L. K. 1994. Group reproductive behaviors in free-living nurse sharks, *Ginglymostoma cirratum*. Copeia 1994(3): 646-656.

Carrier, J. C., Pratt Jr, H. L. and Castro, J. I. 2004. Elasmobranch Reproduction. Pp. 269-286. In J. C. Carrier, J. A. Musick and M. R. Heithaus (eds), *Biology of Sharks and Their Relatives*. CRC Press, LLC. Boca Raton.

Casey, J. G. and Kohler, N. E. 1992. Tagging studies on the shortfin mako shark (*Isurus oxyrinchus*) in the western North Atlantic. Australian Journal of Marine and Freshwater Research 43(1): 45-60.

Castro, J. I. 1993. The biology of the finetooth shark, *Carcharhinus isodon*. Environmental Biology of Fishes 36(3): 219-232.

Castro, J. I. 1996. Biology of the blacktip shark, *Carcharhinus limbatus*, off the southeastern United States. Bulletin of Marine Science 59: 508-522.

Castro, J. I., Bubucis, P. M. and Overstrom, N. A. 1988. The reproductive biology of the chain dogfish, *Scyliorhinus retifer*. Copeia 1988: 740-746.

Clark, E. 1963. The maintenance of sharks in captivity, with a report on their instrumental conditioning. Pp. 115-149. In P. W. Gilbert (ed.), *Sharks and Survival*. D.C. Heath and Co., Boston.

Clark, E. 1975. The strangest sea. National Geographic Magazine 148: 338-343.

Compagno, L. J. V. 1984. *Sharks of the World*. FAO Fisheries Synopsis, no. 125, vol. 4, part 1 (Hexanchiformes to Lamniformes): Pp. 1-249, part 2 (Carcharhiniformes): Pp. 251-655 .

Connor, R. C., Heithaus, M. R. and Barre, L. M. 2001. Complex social structure, alliance stability and mating access in a bottlenose dolphin 'super-alliance'. Proceedings of the Royal Society Biological Sciences Series B 268(1464): 263-267.

Conrath, C. L. and Musick, J. A. 2002. Reproductive biology of the smooth dogfish, *Mustelus canis*, in the northwest Atlantic Ocean. Environmental Biology of Fishes 64(4): 367-377.

Darwin, C. 1859. *On the origin of species*. J. Murray, London, 502 pp.

DeLoach, N. 1999. *Reef Fish Behavior: Florida, Caribbean, Bahamas*. New World Publications, Inc., Jacksonville, FL 359 pp.

Dempster, R. P. and Herald, E. S. 1961. Notes on the hornshark, *Heterodontus francisci*, with observations on mating activities. Occasional Papers of the Californian Academy of Sciences 33: 1-7.

Demski, L. S. 1990. Elasmobranch reproductive biology: implications for captive breeding. Journal of Aquariculture and Aquatic Sciences 5: 84-95.

Demski, L. S. 1990. Neuroendocrine mechanism controlling the sexual development and behavior of sharks and rays. Journal of Aquariculture and Aquatic Sciences 5: 53-67.

Dobson, S. and Dodd, J. M. 1977. The role of temperature and photoperiod in the response of the testis of the dogfish, *Scyliorhinus canicula*, L. to partial hypophysectomy (ventral lobectomy). General and Comparative Endocrinology 32: 114-115.

Dodd, J. M. 1983. Reproduction in cartilaginous fishes (Chondrichthyes). Pp. 31-87. In W. S. Hoar, D. J. Randall and E. M. Donaldson (eds), *Fish Physiology*. Academic Press, New York.

Dral, A. J. 1980. Reproduction en aquarium du requin de fond tropical, *Chiloscyllium griseum* Mull. et Henle (Orectolobides). Revue Française d'Aquariologie et Herpetologie 7: 99-104.

Dugatkin, L. A. 1997. *Cooperation among animals: An evolutionary perspective.* Oxford University Press, London, 221 pp.

Dugger, A. J. 1987. Mating stingrays. Sea Frontiers 33: 352-354.

Eberhard, W. G. 1985. *Sexual selection and male genitalia.* Harvard University Press, Cambridge, 244 pp.

Eberhard, W. G. 1998. Female roles in sperm competition. Pp. 91-116. In T. R. Birkhead and A. P. Moller (eds), *Sperm competition and sexual selection.* Academic Press, San Diego.

Ebert, D. A. 1996. Biology of the sevengill shark *Notorynchus cepedianus* (Peron, 1807) in the temperate coastal waters of southern Africa. South African Journal of Marine Science-Suid-Afrikaanse Tydskrif Vir Seewetenskap 17: 93-103.

Economakis, A. E. and Lobel, P. S. 1998. Aggregation behavior of the grey reef shark, *Carcharhinus amblyrhychos*, at Johnston Atoll, Central Pacific Ocean. Environmental Biology of Fishes 51(2): 129-139.

Feder, H. M., Turner, C. H. and Limbaugh, C. 1974. Observations on fishes associated with kelp beds in Southern California. California Department of Fish and Game Bulletin 160: 1-144.

Feldheim, K. A., Gruber, S. H. and Ashley, M. V. 2001. Multiple paternity of a lemon shark litter (Chondrichthyes: Carcharhinidae). Copeia 2001(3): 781-786.

Fowler, H. W. 1906. Some cold-blooded vertebrates of the Florida Keys. Proceedings of the Academy of Sciences of Philadelphia 58: 77-113.

Francis, M. P. 1996. Observations on a pregnant white shark with a review of reproductive biology. Pp. 157-172. In A. P. Klimley and D. G. Ainley (eds), *Great White Sharks: the Biology of Carcharodon carcharias.* Academic Press, San Diego.

Gilbert, P. W. 1981. Patterns of shark reproduction. Oceanus 24: 30-39.

Gilbert, P. W. and Heath, G. W. 1972. The clasper-siphon sac mechanism in *Squalus acanthias* and *Mustelus canis.* Comparative Biochemistry and Physiology 42A: 97-119.

Gilmore, R. G. 1993. Reproductive biology of lamnoid sharks. Environmental Biology of Fishes 38: 95-114.

Gilmore, R. G., Dodrill, J. W. and Linley, P. A. 1983. Reproduction and embryonic development of the sand tiger shark, *Odontaspis taurus* (Rafinesque). United States Fishery Bulletin 81: 201-225.

Gordon, I. 1993. Pre-copulatory behavior of captive sand tiger sharks, *Carcharius taurus.* Environmental Biology of Fishes 38: 159-164.

Gudger, E. W. 1912. Summary of work done on the fishes of Tortugas. Carnegie Institute of Washington Yearbook. 11: 148-150.

Hamilton, W. D. 1971. Geometry for the selfish herd. Journal of Theoretical Biologie 31: 295-311.

Hamlett, W. C. 1999. Male Reproductive System. Pp. 444-471. In W. C. Hamlett (ed.), *Sharks, Skates, and Rays: The Biology of Elasmobranch Fishes.* Johns Hopkins Press, Baltimore.

Hamlett, W. C. and Koob, T. J. 1999. Female Reproductive System. Pp. 398-444. In W. C. Hamlett (ed.), *Sharks, Skates, and Rays: The Biology of Elasmobranch Fishes.* Johns Hopkins Press, Baltimore.

Hamlett, W. C., Musick, J. A., Hysell, C. K. and Sever, D. M. 2002. Uterine epithelial-sperm interaction, endometrial cycle and sperm storage in the terminal zone of the oviducal gland in the placental smoothhound, *Mustelus canis*. Journal of Experimental Zoology 292(2): 129-144.

Hardy, A. 1959. *The open sea: its natural history*. Houghton-Mifflin Co., Boston, 322 pp.

Harvey-Clark, C. J., Strobo, W. T., Helle, E. and Mattson, M. 1999. Putative mating behavior in basking sharks off the Nova Scotia coast. Copeia 1999: 780-782.

Henningsen, A. D. 2000. Notes on reproduction in the southern stingray, *Dasyatis americana* (Chondrichthyes : Dasyatidae), in a captive environment. Copeia 2000(3): 826-828.

Hoenig, J. M. and Gruber, S. J. 1990. Life history patterns in the elasmobranchs: implications for fisheries management. Pp. 1-16. In H. L. Pratt Jr, S. H. Gruber and T. Taniuchi (eds), *Elasmobranchs as Living Resources: Advances in Biology, Ecology, Systematics, and Status of the Fisheries*. NOAA Technical Reports NMFS 90. U.S. Department of Commerce.

Johnson, R. H. and Nelson, D. R. 1978. Copulation and possible olfaction-mediated pair formation in two species of carcharhinid sharks. Copeia 1978: 539-542.

Jones, S. N. and Jones, R. C. 1982. The structure of the male genital system of the Port Jackson shark, *Heterodontus potusjacksoni* with particular reference to genital ducts. Australian Journal of Zoology 30: 523-541.

Kajiura, S. M., Sebastian, A. P. and Tricas, T. C. 2000. Dermal bite wounds as indicators of reproductive seasonality and behaviour in the Atlantic stingray, *Dasyatis sabina*. Environmental Biology of Fishes 58(1): 23-31.

Klimley, A. P. 1980. Observations of courtship and copulation in the nurse shark, *Ginglymostoma cirratum*. Copeia 1980: 878-882.

Klimley, A. P. 1985. Schooling in the large predator, *Sphryna lewini*, a species with low risk of predation: a non-egalitarian state. Zeitschrift für Tierpsychology 70: 297-319.

Klimley, A. P. 1987. The determinants of sexual segregation in the scalloped hammerhead shark, *Sphyrna lewini*. Environmental Biology of Fishes 18(1): 27-40.

Krebs, J. R. and Davies, N. B. 1993. *An introduction to behavioural ecology*. Blackwell Science, Ltd., Cambridge, 420 pp.

Leigh-Sharpe, W. H. 1920. The comparative morphology of the secondary sexual characteristics of elasmobranch fishes. Memoir 1. Journal of Morphology 34: 245-265.

Libby, E. L. and Gilbert, P. W. 1960. Reproduction in the clear nosed skate, *Raja eglanteria*. Anatomical Record 138: 365.

Luer, C. A. and Gilbert, P. W. 1985. Mating behavior, egg deposition, incubation period and hatching in the clearnosed skate, *Raja eglanteria*. Environmental Biology of Fishes 13: 161-171.

Maruska, K. P., Cowie, E. G. and Tricas, T. C. 1996. Periodic gonadal activity and protracted mating in elasmobranch fishes. Journal of Experimental Zoology 276(3): 219-232.

Matthews, L. H. 1950. Reproduction in the basking shark, *Cetorhinus maximus* (Gunner). Philosophical Transactions of the Royal Society of London Series B-Biological Sciences 234: 247-316.

McKibben, J. R. and Nelson, D. R. 1986. Patterns of movement and grouping of grey reef sharks, *Carcharhinus amblyrhynchos*, at Enewetak, Marshall Islands. Bulletin of Marine Science 38: 89-110.

Metten, H. 1941. Studies on the reproduction of the dogfish. Philosophical Transactions of the Royal Society of London B Biological Sciences 230: 217-238.

Michael, S. 1993. *Reef sharks and rays of the world*. Sea Challengers, Monterey, 107 pp.

Musick, J. A. 1999. *Life in the Slow Lane: Ecology and Conservation of Long-Lived Marine Animals*. American Fisheries Society Symposium 23, Bethesda, MD 23, 265 pp.

Myrberg, A. A. and Gruber, S. H. 1974. The behavior of the bonnethead shark, *Sphyrna tiburo*. Copeia 1974: 358-374.

Nordell, S. E. 1994. Observations of the mating behavior and dentition of the round stingray, *Urolophus halleri*. Environmental Biology of Fishes 39(3): 219-229.

Northcutt, R. G. 1977. Elasmobranch central nervous system organization and its possible evolutionary significance. American Zoologist 17(2): 411-429.

Northcutt, R. G. 1978. Brain organization in the cartilaginous fishes. Pp. 117-194. In E. S. Hodgson and R. F. Mathewson (eds), *Sensory Biology of Sharks, Skates, and Rays*. Office of Naval Research, Arlington.

Ohta, Y., Okamura, K., McKinney, E. C., Bartl, S., Hashimoto, K. and Flajnik, M. F. 2000. Primitive synteny of vertebrate major histocompatibility complex class I and class II genes. Proceedings of the National Academy of Sciences, USA. 97(9): 4712-4717.

Olsen, A. M. 1954. The biology, migration, and growth rate of the school shark, *Galeorhinus australis* (Macleay) (Carcharhinidae) in southeastern Australian waters. Australian Journal of Marine and Freshwater Research 5: 353-410.

Packer, C. 1977. Reciprocal altruism in *Pappio anubis*. Nature 265: 441-443.

Prasad, R. R. 1944. The structure, phylogenetic significance, and function of the nidamental glands of some elasmobranchs of the Madras coast. Proceedings of the National Science Institute of India, Part B, Biological Sciences. 11: 282-302.

Prasad, R. R. 1945. Further observations on the structure and function of the nidamental glands of a few elasmobranchs of the Madras coast. Proceedings of the Indian Academy of Sciences, Section B. 22: 368-373.

Pratt, H. L. 1979. Reproduction in the blue shark, *Prionace glauca*. Fishery Bulletin 77(2): 445-470.

Pratt, H. L., Jr. 1993. The storage of spermatozoa in the oviducal glands of Western North Atlantic sharks. Environmental Biology of Fishes 38: 139-149.

Pratt, H. L., Jr. and Carrier, J. C. 1995. Wild mating of the nurse shark. National Geographic Magazine 187: 44-53.

Pratt, H. L., Jr. and Carrier, J. C. 2001. A review of elasmobranch reproductive behavior with a case study on the nurse shark, *Ginglymostoma cirratum*. Environmental Biology of Fishes 60(1-3): 157-188.

Pratt, H. L., Jr. and Casey, J. 1990. Shark reproductive strategies as a limiting factor in directed fisheries, with a review of Holden's method of estimating growth parameters. Pp. 97-111. In H. L. Pratt Jr, S. H. Gruber and T. Taniuchi (eds), *Elasmobranchs as Living Resources: Advances in the Biology, Ecology, Systematics, and the Status of the Fisheries*. NOAA Technical Report NMFS 90. U.S. Department of Commerce.

Pratt, H. L., Jr., Casey, J. and Conklin, R. E. 1982. Observations of large white sharks, *Carcharodon carcharias*, off Long Island. U.S. Fishery Bulletin 80: 153-156.

Pratt, H. L. and Tanaka, S. 1994. Sperm storage in male elasmobranchs: A description and survey. Journal of Morphology 219(3): 297-308.

Rasmussen, L. E. L. and Gruber, S. H. 1990. Serum levels of circulating steroid hormones in free-ranging carcharhinoid sharks. Pp. 143-157. In H. L. Pratt Jr, S. H. Gruber and T. Taniuchi (eds), *Elasmobranchs as Living Resources: Advances in the Biology, Ecology, Systematics, and the Status of the Fisheries*. NOAA Technical Report NMFS 90. U.S. Department of Commerce.

Rasmussen, L. E. L., Hess, D. L. and Gruber, S. H. 1992. Serum steroid hormones during reproduction in elasmobranchs. Pp. 19-42. In W. C. Hamlett (ed.), *Reproductive Biology of South American Vertebrates*. Springer-Verlag, New York.

Reed, J. K. and Gilmore, R. G. 1981. Inshore occurence and nuptial behavior of the roundtail stingray, *Dasyatis centroura* (Dasyatidae), on the continental shelf, east central Florida. Northeast Gulf Sci. 5: 59-62.

Rouse, N. 1985. Nurse sharks' mating ballet. Sea Frontiers: 54-57.

Saville, K. J., Lindley, A. M., Maries, E. G., Carrier, J. C. and Pratt, H. L., Jr. 2002. Multiple paternity in the nurse shark, *Ginglymostoma cirratum*. Environmental Biology of Fishes 63(3): 347-351.

Schensky, F. 1914. Tier-und Pflanzleben der Norsee, Leipzig: Werner Klinkhardt (cited by Matthews, 1950). .

Sims, D. W., Southall, E. J., Quayle, V. A. and Fox, A. M. 2000. Annual social behaviour of basking sharks associated with coastal front areas. Proceedings of the Royal Society Biological Sciences Series B 267(1455): 1897-1904.

Springer, S. 1960. Natural history of the sandbar shark, *Eulamia milberti*. U.S. Fishery Bulletin 61: 1-38.

Springer, S. 1967. Social organization of shark populations. Pp. 149-174. In P. W. Gilbert, R. F. Mathewson and D. P. Rall (eds), *Sharks, Skates, and Rays*. Johns Hopkins Press, Baltimore.

Stevens, J. D. 1974. The occurrence and significance of tooth cuts on the blue shark, *Prionace glauca* (L.) from British waters. Journal of the Marine Biological Association of the United Kingdom 54: 373-378.

Sumpter, J. P. and Dodd, J. M. 1979. The annual reproductive cycle of the female lesser spotted dogfish, *Scyliorhinus canicula* L., and its endocrine control. Journal of Fish Biology 15: 687-695.

Taborski, M. 1984. Broodcare helpers in the cichlid fish *Lamprolongus brichardi*: their costs and benefits. Animal Behaviour 32: 1236-1252.

Tanaka, S., Shiobara, Y., Hioki, S., Abe, H., Nishi, G., Yano, K. and Suzuki, K. 1990. The reproductive biology of the frilled shark, *Chlamydoselachus anguineus*, from Suruga Bay, Japan. Japanese Journal of Ichthyology 37(3): 273-291.

Taniuchi, T. and Shimizu, M. 1993. Dental sexual dimorphism and food habits in the stingray *Dasyatis akajei* from Tokyo Bay, Japan. Nippon Suisan Gakkaishi 59(1): 53-60.

Taylor, L. R. 1993. *Sharks of Hawai'i: their cultural and biological significance*. University of Hawaii Press, Honolulu, 126 pp.

Tricas, T. C. 1980. Courtship and mating-related behaviors in myliobatid rays. Copeia 1980: 553-556.

Tricas, T. C. and Lefeuvre, E. M. 1985. Mating in the reef white-tip shark *Triaenodon obesus*. Marine Biology 84(3): 233-237.

Tricas, T. C., Maruska, K. P. and Rasmussen, L. E. L. 2000. Annual cycles of steroid hormone production, gonad development, and reproductive behavior in the Atlantic stingray. General and Comparative Endocrinology 118(2): 209-225.

Tricas, T. C., Michael, S. W. and Sisneros, J. A. 1995. Electrosensory optimization to conspecific phasic signals for mating. Neuroscience Letters 202(1-2): 129-132.

Trivers, R. L. 1971. The evolution of reciprocal altruism. Quarterly Review of Biology 46: 35-57.

Uchida, S., Toda, M. and Kamei, Y. 1990. Reproduction of elasmobranchs in captivity. Pp. 211-237. In H. L. Pratt Jr, S. H. Gruber and T. Taniuchi (eds), *Elasmobranchs as Living Resources: Advances in the Biology, Ecology, Systematics, and Status of the Fisheries*. NOAA Tech. Rep. NMFS 90. U.S. Department of Commerce.

West, J. G. and Carter, S. 1990. Observations on the growth and development of the epaulette shark, *Hemiscyllium ocellatum* (Bonnaterre) in captivity. Journal of Aquariculture and Aquatic Sciences 5: 111-117.

Wourms, J. P. 1977. Reproduction and development in chondrichthyan fishes. American Zoologist 17: 379-410.
Wourms, J. P. 1981. Viviparity: The maternal-fetal relationship in fishes. American Zoologist 21: 473-515.
Wright, D. E. and Demski, L. S. 1993. Gonadotropin-releasing hormone (Gnrh) pathways and reproductive control in elasmobranchs. Environmental Biology of Fishes 38(1-3): 209-218.
Yano, K., Sato, F. and Takahashi, T. 1999. Observation of the mating behavior of the manta ray, *Manta birostris*, at the Ogasawara Islands. Ichthyological Research 46: 289-296.

APPENDIX 5.1

A glossary of elasmobranch reproductive behaviors.
From Pratt and Carrier (2001) used with permission.

The following is a list of species-typical patterns of reproductive behavior noted by researchers. Following Myrberg and Gruber (Myrberg and Gruber 1974) we have tried to include records of postures and patterns of movement that were sufficiently stereotyped in form and orientation to allow observers to consistently note their occurrence.

Behaviors of Females

Accept - accepting females, permit the approach and precopulatory behaviors of sexually active males. They may engage in 'parallel swimming', and exhibit 'submissive postures', body 'arching', 'cupping' and 'flaring' pelvic fins and will usually permit copulation'(Klimley 1980, Gordon 1993, Carrier *et al.* 1994). See 'back arching' and 'pectoral fin undulations'.

Arch - female sharks that 'avoid' males may first attempt to escape a 'pectoral grasp' by 'arching' away from the male often twisting the body and cloaca away from the male and, at times, out of the water (Carrier *et al.* 1994). Accepting females may also arch toward the male (Carrier *et al.* 1994).

Avoid - females retreating from sexually active pursuing males may 'avoid' an approach or copulation. If engaged, the 'avoiding' female will struggle to escape (Carrier *et al.* 1994).

Back arching and *pectoral fin undulations* - female skate precopulatory behavior to attract the attention of males (Luer and Gilbert 1985).

Cup - a female folds her pelvic fin margins to form a cup shape (Gordon 1993).

Flare - a female curves and spreads her pelvic fins exposing the cloaca (Gordon 1993).

Refuge - a female retreating behavior to limit male reproductive access (Pratt and Carrier 2001).

Lay on back - the female is motionless and rigid with back to the substrate and pectoral fins outstretched ('lying on back', (Klimley 1980)).

Shield - the female swims very close to the substrate 'shielding' to prevent males from approaching her cloaca (Gordon 1993).

Submissive behavior - the female swims slowly with the head lowered about 15 degrees exposing the pelvic region just before copulation (Gordon 1993).

Behaviors of Males

Bite - see 'courtship bite', similar to the 'grasp' (Carrier *et al.* 1994).

Block - a single male positions his body in front of the heads of the mating pair of sharks and thus presumably aids the 'pectoral grasp' and copulation by keeping the pair from moving forward (Carrier *et al.* 1994).

Bob and *sway* - male rays swim in a sinuous path and 'bob' vertically or 'sway' horizontally in pursuit of a female (Tricas 1980).

Carry - after a successful 'pectoral grasp' in shallow water (< 0.5 m) the male will attempt to carry the 'accepting' female to water deep enough to permit copulation (Pratt and Carrier 2001).

Clasper flexion - the erection of claspers. It may be the movement of individual claspers dorsally before a possible insertion attempt by a male bat ray (Tricas 1980) or backward and forward by a 'patrolling' *C. taurus* (Gordon 1993). Myrberg and Gruber (1974) noted that *Sphyrna tiburo* could pivot its claspers separately, alternately, and sometimes rapidly while swimming alone. They also noted 'clasper flexion with thrust'; a side roll with clasper extension and body arching in *S. tiburo.* See 'splaying'.

Competition - in a pre-copulatory reproductive context, a scramble or contest to see which male will achieve the grasp, positioning and alignment to allow copulation.

Courtship bite - a non-feeding bite usually made by a male shark on a female's body to facilitate mating, often leaving mating scars (Stevens 1974, Pratt 1979). The bite may act as a releaser for female mating behaviors and may stimulate ovulation and other physiological changes (Maruska *et al.* 1996). Biting in some species may also be an expression of reproductive hierarchy (Gordon 1993).

Crossing claspers - a male overlapping both claspers until they point toward opposite side (Dral 1980, Gordon 1993).

Follow - often a 'precoupling' or precopulatory behavior (Carrier *et al.* 1994) in which the male closely follows the female, usually within one body length changing direction frequently as needed. It may be brief or prolonged (Carrier *et al.* 1994). 'Following' is similar to the 'close follow' of Johnson and Nelson (1978), 'parallel swimming' (Klimley 1980), 'chasing' (Yano *et al.* 1999). Following is not always of a reproductive nature. Myrberg and Gruber (1974) observed that the largest *S. tiburo*, frequently follow large members of the opposite sex within 1 m without reproductive intent. Sharks of the same sex will also closely follow each during the mating season (Pratt and Carrier 2001).

Gouge - male spotted eagle rays, *Aetobatus narinari*, dive upon and 'gouge' the female's back with their lower toothplate and bite the caudal margins of the female's pectoral fins (Tricas 1980), called 'nibbling' by Uchida *et al.* (1990).

Grasp - a bite used for holding rather than feeding or defense. It may be directed to the female pectoral fin or body (Carrier *et al.* 1994).

Group behavior - multiple males compete to grasp a female pectoral fin (Tricas 1980, Uchida *et al.* 1990, Carrier *et al.* 1994) or possibly cooperate to aid other male's attempts to copulate by 'blocking' the movement of the mating sharks (Carrier *et al.* 1994).

Insertion and copulation - if a male is successful in 'positioning and alignment' clasper insertion and copulation will usually follow (Carrier *et al.* 1994).

Nosing - a response to 'cupping'. A male comes from behind and beneath a female and places his snout just below the cloaca (Gordon 1993). Similar to 'nudging' (Klimley 1980).

Nudge - a male moves the female from the perpendicular to the parallel position by placing his head under and in contact with the female (Klimley 1980).

Pectoral bite - a type of 'courtship bite', the male bites and holds a female's pectoral fin. A common male maneuver to hold and position a female while mating (Springer 1960, Klimley 1980, Carrier *et al.* 1994). It may lead to the 'grasp' (Carrier *et al.* 1994) and is similar to pectoral nipping (Yano *et al.* 1999).

Pivot and roll - an avoidance behavior originated by the female to roll over the back of a male and escape his grasp (Klimley 1980).

Positioning and alignment - a prelude to 'insertion and copulation' (Carrier *et al.* 1994) in which the male successfully aligns the female's body for copulation. The resultant body positions are similar to 'nudging' (Klimley 1980).

Post-copulation - the male removes the clasper from the cloaca, but may briefly maintain his oral hold of the female pectoral (Yano *et al.* 1999). Carrier *et al.* (1994) noted that the pair sometimes remains together quiescently, more typically, one or both rapidly depart in different directions.

Precoupling - may include 'following' behavior (Carrier *et al.* 1994) and 'parallel swimming' (Klimley 1980).

Lay on back (male) - after the 'grasp', the male rolls over on his back, then both sharks remain motionless, side-by-side on the substrate. At this time in Klimley's observations, the right clasper, the one closest to the female, is inserted into the female's cloaca ('lying on back (male)' (Klimley 1980, Uchida *et al.* 1990).

Separating - the male releases the pectoral fin of the female, setting her free (Yano *et al.* 1999).

Snapping - a male aggressive behavior. A quick physical bite directed toward other resident species, then retraction from them (Gordon 1993).

Splayed claspers - a male contorts and opens his claspers laterally up to ninety degrees to the body axis (Gordon 1993).

Stalking - an aggressive display. Males circle and closely pass other fish and divers in a captive environment (Gordon 1993).

Tailing - one male follows another male so closely that the lead shark's tail movement is restricted (Gordon 1993).

Tail tuck - a male tucks his ventral post-pelvic surface behind the female's first dorsal fin to guide clasper alignment.

Torso thrust (with clasper flexion) - an exaggerated swimming behavior in *Sphyrna lewini*, perhaps to communicate with schooling females and possibly to fill the siphon sacs with seawater prior to copulation (Klimley 1985).

Wrap - males of small, limber shark species may wrap their body around the female to facilitate copulation (Bolau 1881, Hardy 1959, Dral 1980, Uchida *et al.* 1990).

Behaviors of Males and Females

Copulation - the male typically inserts one clasper into the cloaca of the female and may transfer sperm to her uteri (Carrier *et al.* 1994, Yano *et al.* 1999).

Lay parallel on substrate - sharks lay with bodies abreast, less than two pectoral widths apart, prior to a 'grasp' ('lying parallel on substrate' (Klimley 1980)).

Mating event - a 'precopulatory' encounter usually starting with a pectoral 'grasp', which may or may not lead to and include 'copulation' (Carrier *et al.* 1994).

Mating scars - wounds left by a 'courtship bite'. Lesions may be fresh and open, with or without bleeding, or healed marks typified by disrupted denticle patterns that may persist for some time. Scars are usually present on the pectoral fins and often on the trunk around the pelvic fins of females. Scars may occur anywhere on body but are rare around the head of most species (Stevens 1974, Pratt 1979). Similar to 'scarring' (Gordon 1993). See 'courtship bite'.

Parallel swimming - male and female swim together less than two pectoral fin widths apart (Klimley 1980). Occurs in 'precoupling' (Carrier *et al.* 1994). See 'following'.

Patrolling - relatively straight line swimming. The most common mode of swimming for *S. tiburo* (Myrberg and Gruber 1974). Patrolling may or may not have reproductive consequences and is included here because it is used by several authors in a reproductive context (Springer 1967, Gordon 1993). Contrast with 'following'.

Stall - an individual shark stops all forward movement and hovers above the bottom. This may only be a behavior of those few sharks like *C. taurus,* whose body densities are very near that of sea water which permits them to hover without sinking (Gordon 1993).

CHAPTER 6

The Testis and Spermatogenesis

Kelly Bonner Engel and Gloria Vincz Callard

6.1 INTRODUCTION

Due to their phylogenetic position and many interesting reproductive strategies, sharks, skates and rays (elasmobranchs) offer an exceptional opportunity to obtain insights into the origin, evolution and basic functioning of vertebrate reproductive and endocrine systems. In addition, elasmobranchs have a unique testicular organization that makes them ideal animal models for investigating germ cell-Sertoli cell interactions, the role of hormones, and effects of environmental toxicants stage-by-stage during the spermatogenic progression.

Elasmobranchs have remained virtually unchanged since their emergence from jawless fishes over 400 million years ago and, in many ways, they are considered ancestral in their design compared to extant vertebrate species (Helfman *et al.* 1997). Yet, remarkably, they have evolved diverse, highly complex mechanisms for assuring reproductive success. In contrast to many bony fish species, elasmobranchs are considered K-strategists, exhibiting slow growth, high maternal investment, and production of only a few, often well-developed, offspring. Added problems of seasonality and a migratory life style have made the acquisition of successful mating and reproduction strategies crucial to their survival across evolutionary time. Examples of such anatomic and functional adaptations include internal fertilization (all species), viviparity (some sharks and all rays), oviparity (some sharks and all skates), and elaborate mechanisms of excessive yolk production for intrauterine cannibalism (some sharks) (see Chapter 8 for references and details). In addition, sperm storage in males and/or females has evolved in many different elasmobranch species (Pratt and Tanaka 1994). The ultimate success of all such strategies, nonetheless, is predicated on the production of sufficient numbers of functionally competent spermatozoa.

Department of Biology, Boston University, Boston, Massachusetts 02215, United States

Comparison of widely separated vertebrate taxa reveals that the fundamental molecular, cellular and physiological processes required for spermatogenesis are already in place in elasmobranchs (Roosen-Runge 1977).

I. <u>GERM CELLS</u>

Elasmobranchs and Mammals

A. Stages of development are primordial germ cells, gonocytes, spermatogonia, spermatocytes, spermatids, spermatozoa.

B. Major events are stem cell renewal, commitment to a differentiative pathway, mitotic proliferation Meiosis, spermiogenesis, spermiation.

C. Cytological changes during development are similar, with the exception of spermiogenesis and the final form of the mature spermatozoon, which has species-specific characteristics

D. Germ cells develop as isogenetic clones: all cells of a clone are derived from a single spermatogonium, remain connected by intercellular bridges, and are synchronized in development

Elasmobranchs	Mammals
F-1. A single germ cell clone, together with a second clonal population of Sertoli cells, forms a spherical unit (spermatocyst) bounded by an acellular basal lamina	F-2 A germ cell clone is flattened with irregular borders and no boundary membrane but is associated with a cohort of Sertoli cells.

II. <u>SERTOLI CELLS</u>

Elasmobranchs and Mammals

A. Somatic cells termed follicle, cyst, nurse or Sertoli cells have an intimate anatomic arrangement with germ cells and mediate between the soma and germ cell clones in all respect.

B. Structural and functional changes are coordinated with germ cell development.

Elasmobranchs	Mammals
C-1. Sertoli cells undergo cycles of proliferation, development, and generation in the adult testis.	C-2. Sertoli cell proliferation is generally limited to early development.
D-1. A given Sertoli cell is associated with a single germ cell generation and displays stage-dependent characteristics	D-2. A sertoli cell of adult testis simultaneously nurtures 4 or 5 germ cell generations and displays base to lumen differentiation in relation to stage
E-1. Sertoli cells are the primary steroidogenic elements of the testis	E-2. Sertoli cells possess a limited number of steroidogenic activities but are not believed to synthesize steroids <u>de novo</u>.

III. <u>SECONDARY GERMINAL COMPARTMENT</u>

Elasmobranchs	Mammals
A-1. None present; spermatocysts are attached to the terminae of collecting ducts and embedded directly in testicular storma.	A-2. Seminiferous tubules are contiguous with the intratesticular collecting ducts and encompass germ cell clones and associated Sertoli elements

IV. <u>LEYDIG CELLS</u>

Elasmobranchs	Mammals
A-1. Definitive Leydig cells are absent through all spermatogenic stages but small, undifferentiated Leydig-like precursor elements are seen.	A-1. Leydig cells, located in the interstices of tubules are the primary steroid secretory elements of the testis and vary structurally and functionally in relation to stage

Fig. 6.1 Comparison of testicular organization and spermatogenesis in sharks and mammals: a summary of conserved vs. variable features. From Callard, Betka and Jorgensen 1994. Pp. 27-54. In A. Bartke (ed.), *Function of Somatic Cells in the Testis*, Springer-Verlag, New York, Fig. 2.1.

To appreciate how novel information from studies in elasmobranchs can support and extend concepts derived from mammals and, conversely, how information from mammals can be used to devise intervention strategies for endangered elasmobranch populations, it is useful at the outset to summarize the main conserved and variable features of spermatogenesis in vertebrates (Fig. 6.1).

Conserved features include (i) the major events of spermatogenesis; (ii) the progression of germ cell stages and cytological features; (iii) the synchronous development of differentiating germ cells as syncitially connected clones; (iv) the organization of succeeding germ cell generations in a strict temporal and spatial order; (v) a central role for somatic elements termed Sertoli cells (*supporting, nutrient, companion, sustentacular, nurse* or *cyst* cells) and (vi) a steroid-rich environment (Roosen-Runge 1977; Pilsworth and Setchell 1981; Setchell and Pilsworth 1989; Callard 1991a,b; Callard *et al.* 1994). Species-specific differences are relatively superficial by comparison: e.g., the final form of the mature spermatozoon (see Chapter 7 of this volume), the number of spermatogonial divisions, the timing between stages, the primary cellular source of high intratesticular steroid levels (Sertoli vs. Leydig cells), and the organization of different cell types and developmental stages within the testis. In this chapter, we review the production and development of mature male germ cells (spermatogenesis) in adult elasmobranchs, and describe related testicular functions and hormonal control and toxicological aspects.

6.2 OVERVIEW OF THE MALE REPRODUCTIVE SYSTEM

The testes and reproductive tract of male elasmobranchs resemble those of mammals and other vertebrates in their basic organization and embryonic origins and are typified by the blue shark (*Prionace glauca*) (Pratt 1979), basking shark (*Cetorhinus maximus*) (Matthews 1950), and spiny dogfish shark, *Squalus acanthias*, (Gilbert 1973). The adult male reproductive tract of the Atlantic stingray (*Dasyatis sabina*) is diagrammed in Fig. 6.2 (Maruska *et al.* 1996). Paired testes have two main functions, spermatogenesis and hormonogenesis, which are functionally interdependent (see below). Mature spermatozoa, the end-products of spermatogenesis, exit the testis via collecting efferent ducts, which are embedded in the epigonal organ that caps the mature pole of the testis. Sperm then travel via the epididymis and vas deferens to the seminal vesicles. Sperm storage occurs in the terminal ampullae of the epididymis for an indeterminate period, as seen in *Cetorhinus maximus, Scyliorhinus stellaris, Torpedo marmorata,* and *T. torpedo* (Pratt and Tanaka 1994). Upon ejaculation, the alkaline gland contributes a fluid to the semen that is high in electrolytes and is thought to maintain sperm motility (Hug *et al.* 2000). The clasper gland secretes a viscous fluid that also adds to the semen but a definitive function is yet unknown. Additionally, Sertoli cell remnants have been found in semen along with high levels of 11-deoxycorticosterone (DOC), 17-hydroxyDOC, and a

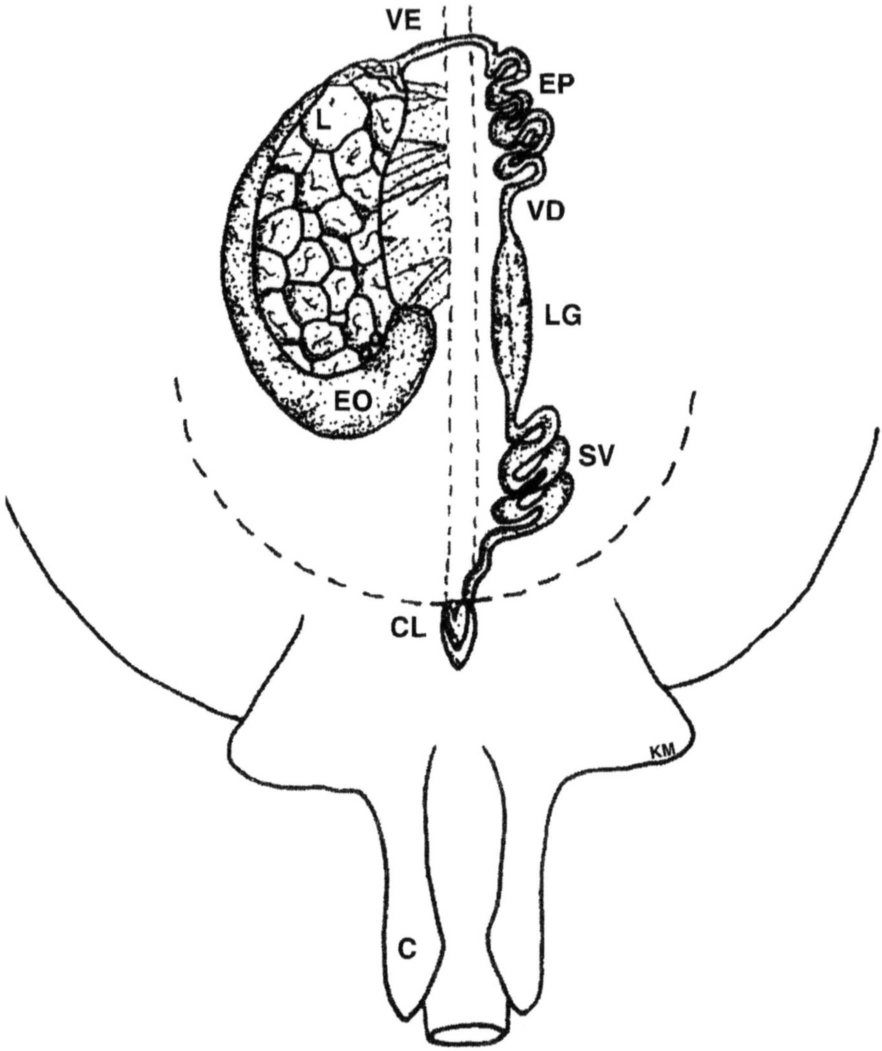

Fig. 6.2 Reproductive anatomy of the mature Atlantic stingray (*Dasyatis sabina*), ventral view. Testicular lobes (L) of males are in close association with the epigonal organ (EO). Mature sperm leave the testis via the vas efferens (VE) and pass to accessory reproductive structures composed of the highly coiled epipidymus (EP), vas deferens (VD), leydig gland (LG) and seminal vesicles (SV). Sperm is then transferred to the female during copulation via the clasper organs (C). From Maruska, K. P., Cowie E. G., and T. C. Tricas 1996. Journal of Experimental Zoology 276: 219-232, Fig 1A.

variety of other hydroxylated progestins (Simpson *et al.* 1964; Pudney and Callard 1986; Barry *et al.* 1993). The functional significance, if any, is not known, although hydroxylated progestins, acting on the outer cell membrane, are involved in final oocyte maturation in teleosts and may play a similar role in spermiation in male elasmobranchs. Another possibility is that high progestin levels in semen somehow promote the acrosome reaction, as in mammals (Kobori *et al.* 2000). The semen is delivered into the female tract via the male intromittent organ, termed a clasper. Mating is seasonal in many species, and is discussed in further detail in Section 6.11.

6.3 THE MALE HYPOTHALAMIC-PITUITARY-GONADAL (HPG) AXIS

It is well established that the initiation of sexual maturation and the maintenance and seasonality of spermatogenesis and hormonogenesis in the vertebrate testis, are dependent on the proper functioning of the hypothalamic-pituitary-gonadal HPG axis (Johnson and Everitt 2000). In turn, hormonal secretions from the testis support spermatogenesis per se and the growth and development of the reproductive tract, the development and maintenance of secondary sex characters, and sex behavior. Additionally, testicular hormones exert feedback control at the level of the brain and pituitary. Fig. 6.3 depicts a prototypical vertebrate HPG axis, and the

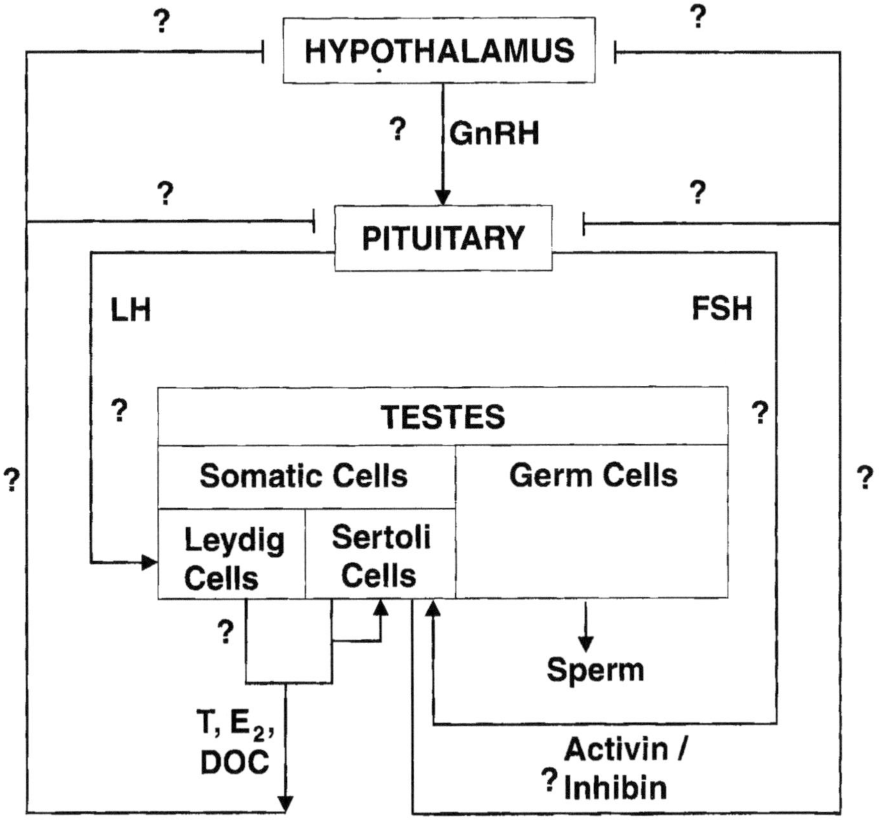

Fig 6.3 Depiction of a typical vertebrate hypothalamic-pituitary-gonadal (HPG) axis. GnRH, gonadotropin releasing hormone; LH, lutenizing hormone; FSH, follicle-stimulating hormone; T, testesterone; E_2, estradiol. Question marks delineate vertebrate pathways that have not yet been characterized in elasmobranchs. Original.

components and pathways that have not yet been definitively identified in elasmobranchs.

Several lines of evidence indicate that a functional pituitary-testicular axis exists in elasmobranchs. Selective ablation of portions of the pituitary gland (Sumpter *et al.* 1978; Dodd and Sumpter 1984), and results of heterologous immunocytochemical (Mellinger and Dubois 1973) and radioimmunoassay studies (Scanes *et al.* 1972) point to the ventral lobe as the source of gonadotropins. In marked contrast to the pituitary of other vertebrates, this lobe is anatomically separated from the remainder of the pituitary gland and quite far removed from the hypothalamus. Bioassay of the ventral lobe using testosterone production by dispersed testicular cells of the turtle demonstrated directly the presence of gonadotropic activity, but intermediate/median lobe extracts were 10- to 100-fold less active (Lance and Callard 1978). Ventral lobe extract administered *in vivo* to female dogfish (*Squalus acanthias*) increased plasma progesterone and testosterone sevenfold and plasma estradiol-17β fourfold (Tsang and Callard 1988). Ventral lobe extract administered to male dogfish *in vivo* significantly increased plasma androgen and neurointermediate lobe extract produced a less pronounced rise, whereas mammalian gonadotropins were ineffective (Sumpter *et al.* 1978). Paradoxically, hypophysectomy did not significantly reduce circulating androgen levels (Dobson and Dodd, 1977a), nor was ventral lobe extract effective in stimulating steroid production by cultured spermatocysts although cAMP and 3-isobutyl-1-methyl-xanthine (MIX) significantly increased testosterone and progesterone ouput (Cuevas and Callard 1989, 1992a).

With respect to control of spermatogenesis, removal of the ventral lobe in *Scyliorhinus canicula* caused regression of the testis with the appearance of a localized zone of degeneration between the last generation of spermatogonia and the first appearance of spermatocytes (Dobson and Dodd 1977b, c), suggesting that this transition may be gonadotropin-sensitive. On the other hand, replacement with human chorionic gonadotropin (hCG: LH-like) or pregnant mares serum goadotropin (PMS: FSH-like) did not prevent the effects of hypophysectomy, confirming that putative gonadotrophin receptors in elasmobranchs are unable to recognize heterologous mammalian hormones. In the same studies, ventral lobectomy reduced [^{3}H]thymidine incorporation into whole testis. Although no attempt was made to determine whether this effect was stage-specific, it was seasonal. Only fish collected in summer (April to September) displayed a zone of degeneration after ventral lobectomy (Dobson and Dodd 1977b, c). In a further study, Dobson and Dodd showed that temperatures resembling those in nature in summer (10-15°C) are critical for division of spermatogonia and for demonstrating effects of hypophysectomy (Dobson and Dodd 1977a). In *Squalus*, a zone of degeneration appears in the testis annually in early spring (Simpson and Wardle 1967). This observation, together with the aforementioned experimental studies, can be interpreted as an indication that gonadotropin levels fluctuate seasonally as part of the normal breeding cycle. Nonetheless, it is important to note that animals surviving as long as two years after ventral lobectomy, although having testes markedly reduced in weight, have all spermatogenic stages present (Dobson and Dodd 1977b). One explanation is that spermatogenesis is not absolutely dependent on pituitary hormones but is rendered more 'efficient' by their presence, a concept supported by evidence in mammals as well (Vernon *et al.* 1975).

Multiple forms of gonadotropin releasing hormone (GnRH) have been identified and partially characterized in elasmobranchs (Lovejoy *et al.* 1992; D'Antonio *et al.* 1995). The presence of vertebrate-type gonadotropins was demonstrated recently by the cloning of a glycoprotein α subunit and two β subunits from the ventral lobe of the pituitary gland of *Scyliorhinus canicula* (Querat *et al.* 2001). The β1 subunit had characteristics like FSH β subunits, and the β2 subunit had similarities to that of LH β. Thus, evidence points to the ventral lobe as the source of gonadotropins. As indicated in Fig. 6.3, many elements of the HPG axis remain to be identified in elasmobranchs. Additional information describing the HPG axis in elasmobranchs is addressed in Chapter 9 and has been reviewed previously (Callard 1991a; D'Antonio *et al.* 1995; Querat *et al.* 2001).

6.4 THE TESTIS

6.4.1 Gross Anatomy

Paired testes are approximately equivalent in size and comprise 1-5% of body weight in adults. In the skates and rays (Batoidea) the testes are dorsoventrally flattened, with the right testis often slightly smaller than the left, probably due to the arrangement of other internal organs (Maruska *et al.* 1996). The testes are located anteriorly in the body cavity and are suspended from the dorsal body wall by double mesorchia. From the dorsal aspect and opposite the epigonal organ in the spiny dogfish and other sharks a pregerminal fold appears as a lucent streak running along the length of the testis and serves as a convenient external landmark indicating the beginning of the spermatogenic sequence and orientation of the spermatogenic wave. The position of the germinal zone is not readily visualized in skates and rays (see 6.7.1). Each testis is embedded in an irregular shaped epigonal organ, which is lymphomyeloid in nature and a major source of leukocytes in the general circulation, elasmobranchs having no bone marrow or lymph nodes (Mattisson and Fange 1982; Fange and Pulsford 1983; Callard 1991a). The epigonal organ in the spiny dogfish shark is upstream in the testicular vascular pathway and the source of a factor, termed epigonal growth-inhibitory factor (EGIF), that inhibits DNA synthesis of premeiotic spermatocysts (Piferrer and Callard 1995). Effects of EGIF are dose- and time-dependent, and completely reversible. The epigonal organ probably also contributes immune cells to the intratesticular population. Although interactions between the immune system and the gonads have been extensively researched in mammals, the functional significance of the association between the testis and epigonal organ of elasmobranchs is largely unknown.

6.4.2 Cell Types

The testis has two main cell types, germ cells and somatic cells, which differ in their embryonic origin, subsequent cell lineage, and primary roles within the testis: respectively, spermatogenesis and hormonogenesis. From the earliest stages of testicular development, however, the activities of germ cells and somatic cells are coordinated and interdependent.

6.4.2.1 Germ cells

Germ cells derive from a self-renewing population of stem cells that originate in the embryonic yolk sac and subsequently migrate to the gonadal ridge, where they take up permanent residence in the testis (Chieffi 1949). Elegant studies in zebrafish using mRNA markers (vasa) show that germ plasm destined to form the definitive germ cell lineage is recognizable in the fertilized egg before the earliest cleavages, and can be visualized in four or more subsequent cleavages as definitive germ cells are formed and proliferate in the larval stage of development (Yoon *et al.* 1997). In the mature elasmobranch testis, the stem cell population is located in a distinct germinal zone or ridge (GZ). Stem cells are few in number but are the source of successive waves of dividing and differentiating germ cells required for spermatogenesis. Mechanisms that determine renewal versus commitment to the differentiative pathway are entirely unknown in elasmobranchs.

What is noteworthy about male vertebrate germ cell development, as compared to female germ cells, is that they develop, not as individual cells, but as isogenetic clones, That is, daughter cells formed by successive divisions of a single primary spermatogonium remain connected, structurally and functionally, by intercellular bridges. This is due to incomplete cytokinesis. The clonal syncitium of germ cells is synchronized in development and, in elasmobranchs, continues in an all-or-none fashion to the final mature spermatid stage. Sperm bundles then dissociate into individual sperm or small groups in the lumen of the collecting ducts (Mellinger 1965). Only in certain species do the sperm bundles reassociate into spermatophores.

6.4.2.2 Sertoli cells

Sertoli cells are the most prominent somatic element of the elasmobranch testis, and share a common origin with cells destined to form the intratesticular collecting duct system (Callard *et al.* 1994). From the earliest stem cell stage and all subsequent stages of development, Sertoli cells are intimately associated with the germ cells within an anatomically distinct "spermatocyst" (see 6.5). Also gap junctions are seen between adjacent Sertoli cells in the basal regions where the plasmalemmas of the two cells appear to be fused, and seem to be reinforced by microfilaments and by parallel arrays of tubulues of the smooth reticulum (Moyne and Collenot 1982). In mammals, Sertoli cells have a crucial role in physically supporting and sequestering the germinal elements, in controlling their microenvironment through secreted products and formation of the blood-testis barrier, and as both a source and target of molecules involved in the regulation of spermatogenesis (Fawcett 1975; Griswold *et al.* 1988; Ritzen *et al.* 1989; Callard 1991a). Available information suggests that elasmobranch Sertoli cells perform similar functions, but several important characteristics distinguish them from those of mammals. First, Sertoli cells of the mature elasmobranch testis undergo cycles of proliferation, differentiation and degeneration (Pudney and Callard 1986). Mammalian Sertoli cells, by contrast, generally cease dividing early in development and (at least in continuous breeders) become permanent elements of the seminiferous epithelium. Second, at any one point in time, elasmobranch Sertoli cells are associated, and developmentally synchronized, with a single germ cell clone. By contrast, a single mammalian Sertoli cell is associated simultaneously with 4-5 different stages of germ cell development. This is because new germ cell clones proliferate and advance into development before completion of preceding generations. A third, important distinction is the capacity of elasmobranch Sertoli cells for steroid production (Callard *et al.* 1978). Whereas the Sertoli cell of mammals is qualitatively and quantitatively limited in its ability to synthesize steroids, Sertoli cells are the primary, and possibly exclusive, source of intratesticular androgen, estrogen and progestins in the spiny dogfish during spermatogenesis. This is marked by development of a prominent Golgi, an increase in rough endoplasmic reticulum, and the accumulation of secretory products in its cisternae (Collenot and Damas 1975, 1980). In addition, markers of steroid synthesis, including a well-developed agranular reticulum, and mitochondria with tubulovesicular cristae and numerous lipid droplets, were observed to increase dramatically during spermatogenesis in Sertoli cells of *S. acanthias* (Pudney and Callard 1984a).

6.4.2.3 Leydig cells

The presence or absence of Leydig cells in elasmobranchs is a matter of debate. Electron microscopic studies in this laboratory indicate that true Leydig cells of the type seen in the interstitium of the testis of adult vertebrates are not present in the spiny dogfish *Squalus acanthias;* however, undifferentiated Leydig-like cells resembling those in fetal mammalian testis are observed (Pudney and Callard 1984b). These cells are small, few in number and mesenchymal in appearance, and have a sparse agranular reticulum, tubulovesicular mitochondria, and lipid droplets, organelles that are typical of steroid producing cells and much more highly developed in Sertoli cells of the same species. These authors postulated, therefore, that the Leydig-like cells of dogfish sharks are the evolutionary forerunners of true vertebrate Leydig cells (Callard 1991a), a conclusion that is consistent with the absence of seminferous tubules (and an interstitium) at this phyletic level (see below). On the other hand, Marina and coworkers (2002), also using electron microscopy, reported Leydig cells in the mature *Torpedo marmorata*. These were located in the interstices between spermatocysts, and were steroidogenic in appearance, with the prescence of smooth endoplasmic reticulum, many lipid droplets, glycogen, and mitochondria with tubular cristae, during stages prior to spermiation. After spermiation, the cellular characteristics demarking steroid secretion were reduced. This study also found Sertoli cells to have steroidogenic organelles at other stages of spermatogenesis (Marina *et al.* 2002). Earlier studies are contradictory as to the presence or absence of Leydig cells, even within the same species (Stephan 1902; Matthews 1950; Chieffi *et al.* 1961; Della *et al.* 1961; Collenot 1970; Callard 1991a), but the techniques used in these studies (light microscopy and cytochemistry) could be misleading as to the location and identity of reactive cells. Whether or not the cells at issue are Leydig-like or true Leydig cells, it is clear that their contribution to intratesticular steroid production is not quantitatively important when compared to Sertoli cells. It cannot be ruled out, however, that these cells are the source of small amounts of functionally important steroid products within the testis per se or contribute to the circulating steroid milieu (Callard 1991b).

6.4.2.4 Other somatic cell types

In addition to somatic elements comprising the blood and lymphatic vasculature of the testis, there is a transient population of immune cells derived mainly from the epigonal organ. Also, cells of the intratesticular collecting duct system within the testis are interspersed among the germinal elements, and may be the forerunners of mammalian peritubular myoid and Leydig cells.

6.5 SPERMATOCYSTS: THE PRIMARY GERMINAL UNITS

In all vertebrates, the syncytial germ cell clone, and its cohort of associated Sertoli cells, comprises the primary spermatogenic unit. In elasmobranchs, and other anamniotes, the germ cell-Sertoli cell unit is an anatomically discrete, closed sphere, bounded by an acellular basal lamina, termed a spermatocyst ("follicle", "ampulla", or "lobule") (Callard 1991b). In elasmobranchs, spermatocysts are embedded directly in the testicular matrix, attached to the termini of a system of intratesticular collecting ducts, which are not patent until the end of spermatogenesis when they empty into the efferent ducts (Callard *et al.* 1994). All other vertebrates have evolved a secondary germinal compartment, the seminiferous tubule, in which the boundaries of individual germinal clones and their associated Sertoli cells are more or less recognizable. In frogs, for example, definitive spermatocysts are found within tubular structures in early spermatogenesis, but release their contents into the tubular lumen toward the end of development. In mammals and birds, identification of a germ cell clone is difficult (serial electron microscopy) because they appear to be "in open communion" with each other. Also, clones in succeeding generations form a stratified epithelium within the tubule and share their relationships with the same subset of Sertoli cells.

6.6 THE SPERMATOGENIC PROGRESSION

Spermatogenesis comprises the life history of a male germ cell through a series of stages (gonocyte→, spermatogonium→, spermatocyte→, spermatid→, spermatozoon) and a defined sequence of complex cellular events: commitment to spermatogenesis proper, mitotic proliferation; apoptosis; meiosis (the central event of spermatogenesis); and spermiogenesis (metamorphosis of an immotile round cell to a free-swimming spermatozoon) (Roosen-Runge 1977). The major cellular events and the cytological changes displayed by individual germ cells at each developmental stage are very similar throughout the Vertebrata, as first noted in 1895/96 by Moore of stage-synchronized cells. In elasmobranchs, each germinal clone and its associated Sertoli cells form an anatomically distinct spermatocyst.

In the following sections, we describe in detail the stage-by-stage progression of spermatocysts through development. Also See Fig 6.4.

6.6.1 Spermatogonial Stage (Premeiotic) Spermatocysts

The spermatogenic process begins in the GZ with the association of a single spermatogonium and a single somatic element (pre-Sertoli cell) to form a primitive spermatocyst. Through succeeding stages, both spermatogonia and Sertoli cells undergo a series of coordinated mitotic divisions, which are accompanied by stepwise growth and differentiation of the two cell types. Thus, the ratio and number of cells of each type in a given spermatocyst are predictable at each stage of spermatogenesis for each elasmobranch species (Callard, 1991b). The number of spermatogonial divisions before entry into meiosis begins is a fixed, species-specific characteristic and ranges from 4 to 14 in different vertebrate species (Roosen-Runge 1977). The number of spermatogonial divisions is 13 in *Squalus acanthias*. During the first nine spermatogonial divisions in *Squalus* (Callard, 1991a), Sertoli cell divisions are tightly coordinated with those of spermatogonia, as indicated by actual counts of each of the two cell types and their ratio per spermatocyst, which is constant during these early stages (1 : 1) (Stanley 1966). Thereafter, spermatogonia undergo three more mitoses, whereas Sertoli cells cease dividing. This results in a germ cell/Sertoli cell ratio of 16 : 1 in spermatocysts entering meiosis, and the total number of germ cells and Sertoli cells at the end of spermatogenesis is, respectively, 32,000 and 500 per spermatocyst. In newly formed spermatocysts, Sertoli cells and primary spermatogonia line the interior of the basal lamina in no apparent order but, as spermatogenesis progresses the two cell types segregate into concentric layers, with Sertoli cells in the adluminal position and spermatogonia located peripherally (Callard 1991a). Still later in development (mid – late spermatogonial stages), Sertoli cell nuclei migrate peripherally to assume a position adjacent to the basement membrane of the spermatocyst, which is maintained through all subsequent meiotic and postmeiotic stages. Sertoli cell migration from the lumen towards the periphery marks the transition from primary to secondary spermatogonia (Callard 1991a, b).

6.6.3 Spermatocyte Stage (Meiotic) Spermatocysts

At the onset of meiosis, the spermatocysts increase sharply in diameter due to the increase in size of the germ cells and a marked increase in cytoplasmic

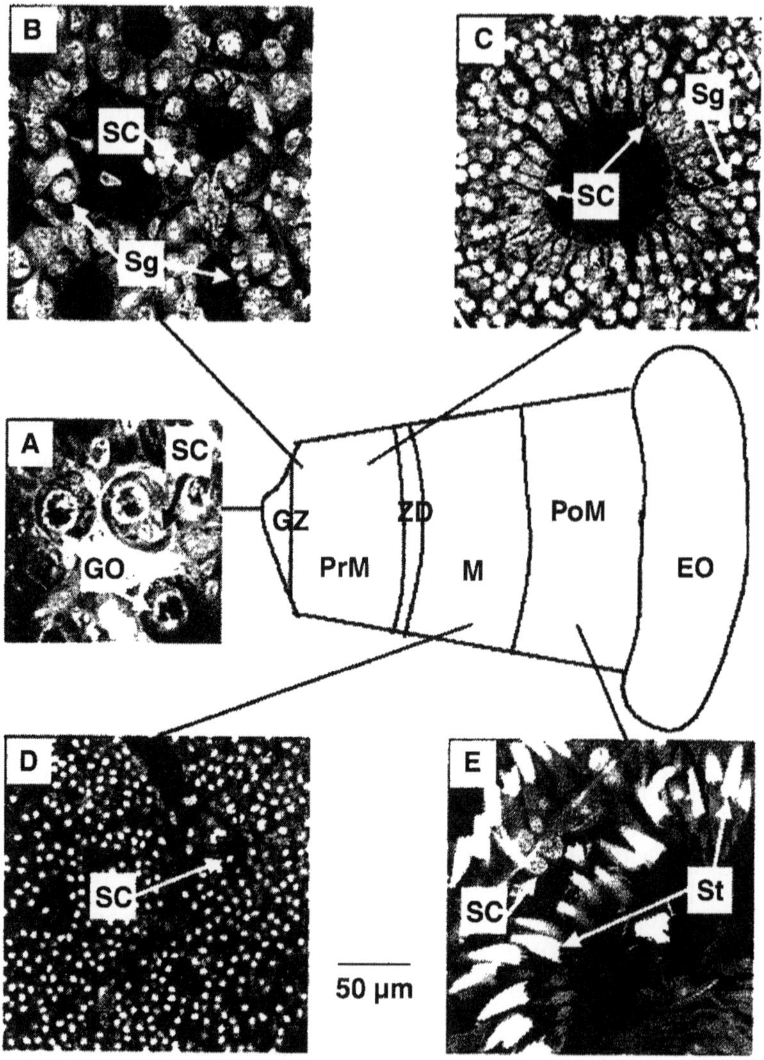

Fig. 6.4 Organization of spermatogenesis and stages of spermatocyst development in the dogfish shark testis. The diagrammatic cross section shows the simple diametric progression of spermatogenesis and demarcation of staged: GZ, PrM, M, PoM. ZD, with a band of apoptotic cysts reflecting the preceding winter's period of spermatogenic inactivity, appears between PrM and M zones in May/June. Fluorescence confocal images show spermatocysts representative of staged tissues after staining with acridine orange and their approximate position in the developmental progression: **A.** GZ with a nest of primitive cysts containing gonocytes (Go). **B.** PrM zone with cyst containing immature spermatogonia (Sg). **C.** PrM zone with cyst containing mature Sg. **D.** M zone with cyst containing spermatocytes (Sc). **E.** segment of PoM cyst containing bundles of mature spermatids (St), with heads pointed peripherally and tails projecting into the lumen. Sertoli cells (SC) nuclei are not readily evident in **D** but are approximately equivalent in number to germ cells in **A** and **B**, located adluminally at the stage of maturation on **C**, and are adjacent to the basement membrane in **E**. From Betka and Callard 1999. Biology of Reproduction 60:14-22, Fig.1.

volume of Sertoli cells (Stanley 1966).The primary spermatocytes are characterized by a large nucleus with distinct, elongate synaptic chromosomes. Two meiotic divisions occur in short succession without an interphase DNA synthesis. The brief interval between the first and second meiotic divisions is termed the secondary spermatocyte stage. The secondary spermatocyte stage, marking completion of meiosis I, is short, as indicated by relatively few cysts seen in this stage compared to those stages immediately preceding and following. By the end of the spermatocyte stage, Sertoli cells are located along the periphery of the cyst (Callard 1991a, b).

6.6.4 Spermatid Stage (Postmeiotic) Spermatocysts

Spermatids mark the completion of meiosis. Early in the spermatid stage, or slightly earlier, fluid-filled spaces develop along the germ cells and the cyst as a whole enlarges somewhat (Stanley 1966). Some preparations show that each Sertoli cell has formed a single large pocket containing a group of spermatids. It is at this stage that the relationship between a single Sertoli cell and its synchronously developing complement of germ cells first becomes obvious. With further development, the spermatids elongate and gradually form loose bundles, oriented with heads facing the basal lamina and tails projecting towards the lumen.

This transformation into elongated, flagellated spermatozoa, occurs in the haploid state and, in mammals, is accompanied by dramatic morphological and biochemical changes, including expression of several testis-specific nucleoprotein and other genes (Sakai *et al.* 1978; Hecht 1986; Griswold *et al.* 1988). Owing to the dramatic cytological changes evident even under the light microscope, spermiogenesis has captured the attention of numerous investigators, and a detailed description is available for several vertebrates including elasmobranchs (Mattei 1970; Stanley 1971a, b; Zirkin 1975; Russell 1984; Sprando and Russell 1988a, b; see also Chapter 7 of this volume). Although this stage exhibits the most variability when different species are compared, the following events are common to all: (1) transformation of the nucleus from a sphere to an elongated form; (2) formation of an acrosomal cap from the Golgi complex; (3) redistribution of loss of cytoplasmic organelles when the "cytoplasmic lobe" is pinched off at spermiation; and (4) differentiation of a flagellum (Russell 1984). Specialized features include the size and shape of the nucleus, presence or absence of a flagellular undulating membrane, shape of the acrosomal cap, and length of the flagellum. By convention, the term spermatozoon is reserved for the mature gamete after it has been released from the germinal epithelium; therefore, haploid cells undergoing spermiogenesis are designated as early (round) or late (elongated, maturation phase) spermatids (Callard 1991b). During spermiogenesis in the spiny dogfish, there is a dramatic increase in abundance of agranular reticulum in adjacent Sertoli cells, which fills Sertoli cell cytoplasm as a mass of tubulues (Pudney and Callard 1984a). Concomitant with an increase in smooth reticulum is a proliferation and development of other organelles associated with steroidogenesis.

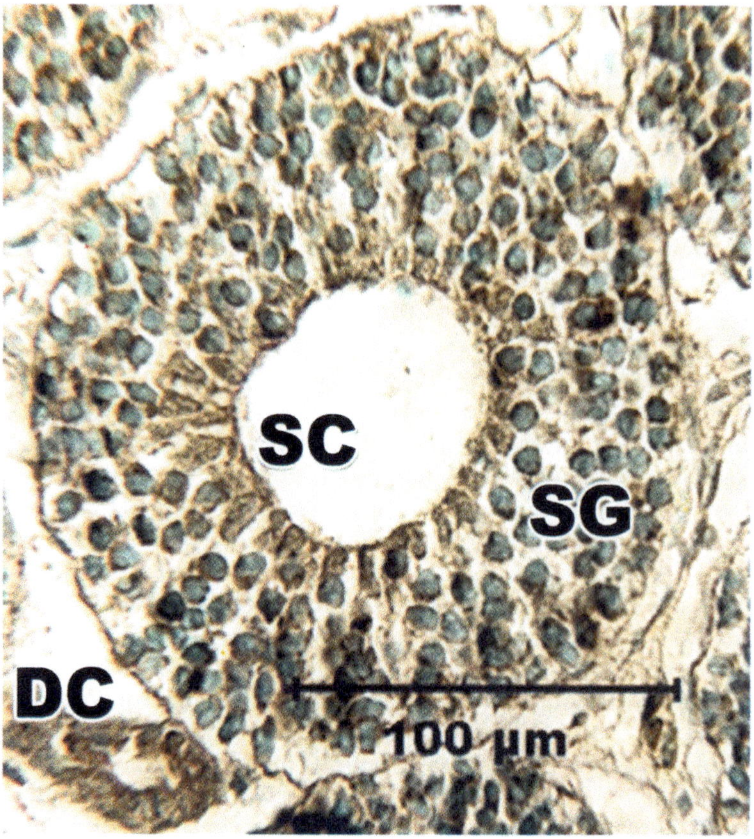

Fig. 6.5 Immunocytochemical localization of androgen receptors in the spiny dogfish, *Squalus acanthias*, testis. A cross-section of a mature spermatogonial stage spermatocyst. Methods as in Gelinas (1997). Germ cell nuclei (SG, spermatogonia) are counterstained with methyl green. Brown staining, demarking location of androgen receptors stains the Sertoli cell nuclei (SC), located adluminal at this stage, as well as ductule cells (DC) (Engel, unpublished data). Original.

6.6.5 Degenerate Spermatocysts

Among the mature spermatogonial stage cysts, depending on time of year, many are seen to have germ cells in various stages of degeneration. Degeneration of germ cells in a subset of spermatocysts entering meiosis has been identified as the apoptotic form of programmed cell death (a) by visualizing condensed chromatin after vital staining with acridine orange; and (b) by terminal deoxynucleotidyl transferase (TdT)-mediated dUTP nick end-labeling of DNA "ladders" (Callard *et al.* 1998). Degeneration affected germ cells but not Sertoli cells, and was all-or-none within an individual cyst. Thus, some cysts had just a few degenerate germ cells, whereas other were entirely filled with degenerate cells, and still others had Sertoli cells only containing remnants of germ cell corpses.

6.7 SPATIAL ORGANIZATION OF SPERMATOGENESIS WITHIN THE TESTIS

The progression of germ cells through development is not random. In all vertebrates, successive stages of germ cell maturation are arranged in a strict spatial order, reflecting the strict temporal order of the developmental progression. In mammals, the spatial progression of stages has two dimensions: (1) lengthwise along each tubule (spermatogenic wave) and (2) at a given point, centripetally from base to lumen within the seminiferous epithelium. Together, the progression through these two dimension in rats, mice and other mammals is termed "the cycle of the seminiferous epithelium" (Fig. 6.6). As a manifestation of the temporal coordination of developing germ cells in elasmobranchs, the spermatocysts are arranged in consecutive order in space, but only in one dimension: i.e., radiating outward from the GZ. The spatial pattern formed by spermatocysts in successive stages of development can be categorized into three distinct subtypes, as first defined by Pratt (1988): (1) compound; (2) radial and (3) linear (Fig 6.6). In all three subtypes, the production of new spermatocysts occurs in one or more germinal zone (GZ). As spermatocysts progress through development, they do not actually migrate but are displaced by successively less mature elements. What differs in the three different spatial subtypes, however, is the number, and location (relative to the testis as a whole) of germinal zone(s). This, in turn, sets the intratesticular pattern formed by the waves of maturing spermatocysts. In sharks (Superorder Selachaii), representatives of both linear and radial forms exist, whereas in rays and skates (Superorder Batoidea) only the compound form exists.

6.7.1 Compound (Rajid, e.g. Skate)

In the testis of batoids, multiple GZ's are located on the dorsal surface of the testis. Spermatocysts derived from each GZ are arranged in columns that radiate away from the center and also traverse the diameter of the testis. Each GZ and its spermatocyst progeny form a lobe-like structure, irregular in shape. Collectively, the different lobes comprise the testis. The diameter of each lobe and the degree of separation between lobes differs in different species.

6.7.2 Radial (Lamniformes, e.g. Mako, White Sharks)

In some species, the testis is subdivided into lobes, each separated from the other by connective tissue as described above. However, in contrast to the compound testicular type, GZ's are not limited to the dorsal surface of the testis. Instead, they are located at various depths within the body of the testis. Each GZ is located in the center of a lobe. As spermatocysts mature, they migrate away from GZ in a radial pattern, with the most mature spermatocysts located at the periphery.

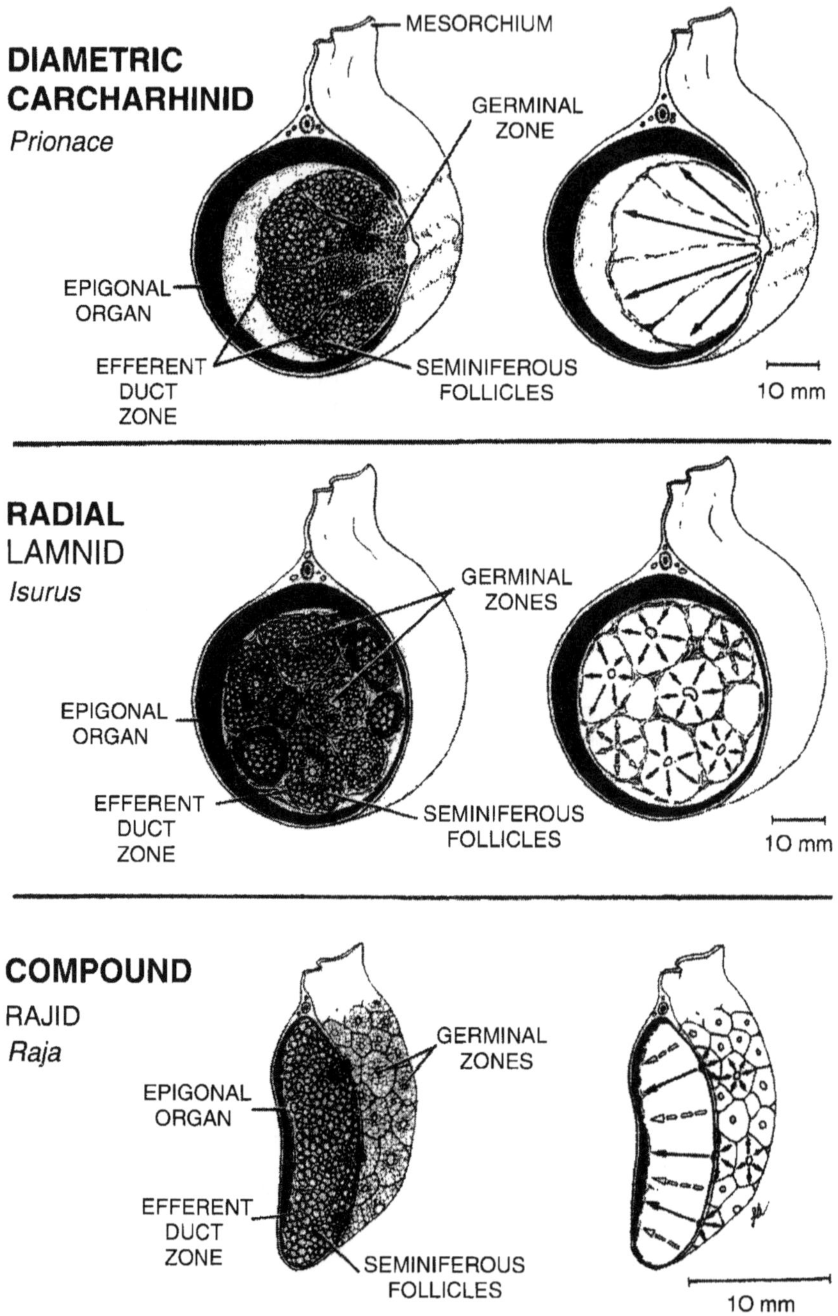

Fig. 6.6 The three forms of testicular organization (the spatial pattern in which spermatocysts migrate as they mature) that is present in mature elasmobranchs, cross-section, anterior view. Left side: size of follicles exaggerated to show development. Right side: arrows indicate paths of seminiferous follicle development. The arrows indicate paths of spermatocyst development. From Pratt 1988. Copeia 3: 719-29, Fig. 2.

6.7.3 Linear (Squalomorphs, Galeomorph, Carcharhinid, e.g. Dogfish, Horn & Sandbar Sharks, Respectively)

In some species, a single GZ is visualized as a visible pregerminal fold that runs down the length of the testis opposite the epigonal organ. This pregerminal fold appears as a lucent streak and serves as a convenient external marker of the orientation of the spermatogenic wave. As spermatocysts mature they are displaced away from GZ in a coordinated and linear fashion with respect to neighboring spermatocysts, and the maturation sequence traverses the entire diameter of the testis. In *Squalus acanthias* and other elasmobranchs with this type of testicular ogranization, parallel "columns" of maturing spermatocysts radiate outward from GZ. All spermatocysts, at a given distance from the germinal zone are in the same stage of development, resulting in concentric "rows" of cysts identical in appearance and thus imparting a visible "zonation" in testicular cross-sections.

6.8 STAGE-BY-STAGE ANALYSIS OF SPERMATOGENESIS

Elasmobranchs exhibiting the linear form of testicular organization are advantageous animal models for stage-by-stage analysis of the spermatogenic progression. Although development is a continuum, several zones of maturation are readily distinguishable on the basis of opacity, color, and position relative to the germinal zone and epigonal tissue even without the aid of a dissecting microscope. In the spiny dogfish, these are: GZ, PrM (premeiotic = spermatogonial stage cysts), M (meiotic = spermatocyte stage cysts); and PoM (postmeiotic = spermatid stage cysts). Also, ZD the zone of degeneration appears at the transition of spermatogonial to spermatocyte stages at the end of the winter period of spermatogenic inactivity (Simpson and Wardle 1967). The position of this degenerative zone relative to the germinal region marks the advance of newly formed cysts from the beginning of the current season's proliferative activity and has been used to estimate the duration of a complete spermatogenic sequence (9 months) (Simpson and Wardle 1967). This simple arrangement of cysts in successive stages of maturation has been used to study morphological, biochemical, and molecular correlates of the different spermatogenetic stages. Experimental strategies using elasmobranchs to elucidate factors and mechanisms that regulate spermatogenesis are summarized in Fig 6.7.

6.9 STEROIDOGENESIS

In all vertebrates, spermatogenesis proceeds in a steroid-rich environment. Steroid biosynthetic pathways and the final steroid products are highly conserved within the vertebrate phylum (Bourne 1991; Selcer and Leavitt 1991), and some of the cytochrome P450 enzymes that regulate key steps in steroid biosynthesis have been cloned, sequenced and characterized in elasmobranchs: e.g., cholesterol side-chain cleavage enzymes (Nunez and

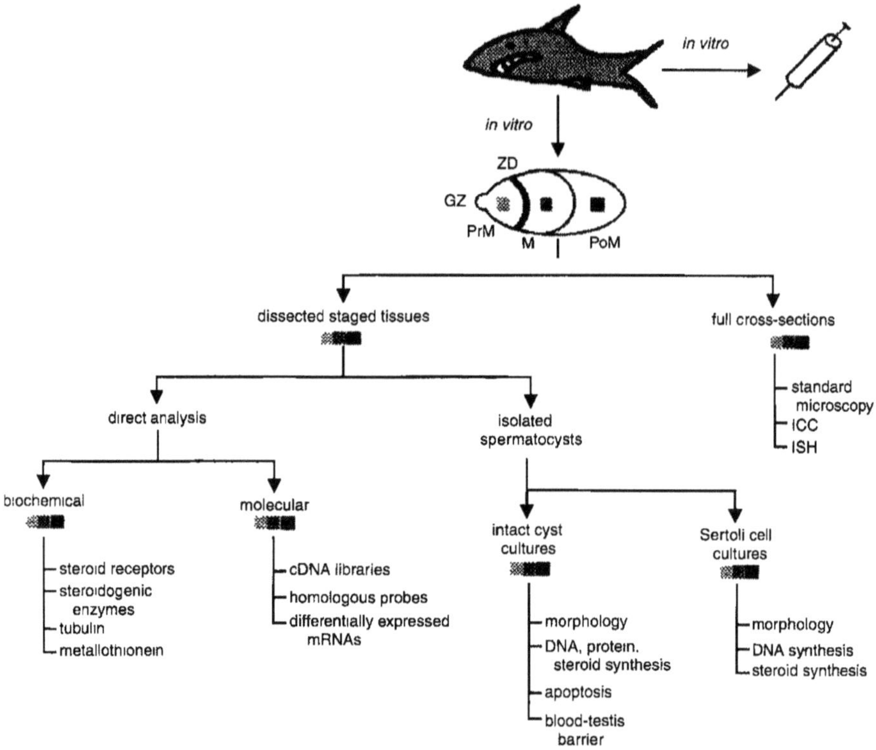

Fig. 6.7 Strategies for investigating spermatogenesis using the shark testis model. GZ, germinal zone; PrM, pre-meiotic zone; M, meiotic zone; PoM, postmeiotic zone; ZD, zone of degeneration. For reference to specific applications, see Callard, 1991a; Piferrer and Callard, 1995). From Callard, G. V., McClusky L. M., and M. Betka 1998. In: Le Gal and Halvorson (eds). *New Developments in Marine Biotechnology*, Plenum Press. New York, Fig. 1.

Trant 1997); 17α-hydroxylase (Trant 1995); and aromatase (Ijiri *et al.* 2000)(Wang, Sikora and Callard unpubl. data) as well as in other nonmammalian species (McPhaul *et al.* 1988; Takahashi *et al.* 1993; Miyashita *et al.* 2000). The primary cellular site of steroid synthesis in elasombranchs is the Sertoli cell, with possible contributions by the Leydig cell (see above).

6.9.1 Steroidogenic Enzymes

Radiolabeled tracer analysis of homogenates and subtractions, together with biochemical and radioimmunoassay analysis of organic extracts and secreted products, has definitively established that steroidogenic pathways leading from cholesterol to biologically active C-21, C-19 and C-18 steroids are identical in the testis of sharks and mammals. However, some notable differences in the quantity of certain metabolites are seen (Callard 1991a,b). For example, 11-deoxycorticosterone (DOC), 17α- hydroxyDOC (Reichstein's Substance S), and a variety of other progestins hydroxylated at C-17, C-20 or C-21 are relatively abundant products of shark testis and are also found in high amounts in semen (Simpson and Wardle 1967; Barry *et al.* 1993). The functional significance, if any, is not known, although hydroxylated progestins acting on the outer cell membrane, are involved in final oocyte maturation in teleost fish and may play a role in spermiation and/or the acrosome reaction in males. Also, conjugating pathways leading to steroid glucuronides and sulfates are exceedingly active in shark testis, such that a 1-h, 1-pass perfusion of the testis in situ with [^{3}H]androgen or [^{3}H]estrogen results in >90% recovered from testicular cytosol as polar metabolites (Cuevas *et al.* 1992). This mechanism of steroid inactivation may have utility in regulating ligand binding in organs like the testis where steroids are synthesized in close proximity to receptors.

Using radiolabeled tracer methodology and microsomal subtractions derived from staged testicular tissues, it was observed that 3βHSD (3β-hydroxysteroid dehydrogenase) the enzyme catalyzing conversion of 5 precursors to the 4, 3-ketone A ring of hormonally active steroids, increases progressively during germ cell maturation (5-fold, PoM > M > PrM) (Cuevas *et al.* 1993). Maturation-related increases in microsomal 3βHSD are also obtained when microsomes from isolated spermatocysts or Sertoli cells are assayed (Dubois *et al.* 1989; Cuevas *et al.* 1993). Moreover, comparison of specific enzyme activities in whole testis versus spermatocysts or Sertoli cells indicates that virtually 100% of testicular 3βHSD can be accounted for by that associated with Sertoli cells. Thus, biochemical results correspond exactly to the development of smooth endoplasmic reticulum seen in Sertoli cells by electron microscopy and support the view that this cell type is the exclusive, or primary, site of steroid biosynthesis. Other enzymes that display maturation-related increases in activity, although with different fold changes, are 17α-hydroxylase/C-17,20-lyase, the key enzyme in androgen biosynthesis (2- to 12-fold); 21-hydroxylase, an enzyme that utilizes progesterone to form DOC (3- to 20-fold); and androgen/estrogen sulfotransferase (2-fold) (Callard *et al.* 1985; Cuevas *et al.* 1992; Barry *et al.* 1993). In addition, the cDNA encoding cholesterol side chain cleavage cytochrome was cloned in the southern stingray (*Dasyatis americana*) although intratesticular localization has not yet been addressed (Nunez and Trant 1997).

A distinctly different stage-related distribution pattern is seen with aromatase, the rate-limiting enzyme in androgen to estrogen conversion, which is 2- to 7-fold higher where germ cells are undergoing meiosis than in less mature or more mature regions (Callard *et al.* 1985). Although aromatase cDNA has been cloned in the stingray and the spiny dogfish (Wang and Callard, unpublished data), testicular staged- analysis has not yet been reported (Ijiri *et al.* 2000). By contrast, 5α-reductase, which regulates testosterone (T) to 5α-dihydrotestosterone (DHT) transformation, differs from all other patterns (PrM > PoM ™ M), although in contrast to the prostate and other androgen targets in mammals, the necessity of this transformation for spermatogenesis has not been established (Cuevas *et al.* 1993).

It is important to note here that the timing of maximal androgen production and low aromatization in shark testis (mature spermatid stage) agrees with observations in rats showing that stage VIII of the seminiferous cycle, a time approximately coincident with the release of the most advanced germ cell generation, is associated with Leydig cells significantly larger than in other stages, highest tubular androgen levels, and the presence of an aromatase inhibitor (Parvinen 1982). Taken together, these data support the view that biologically active versus inactive steroid ligands are developmentally regulated and result in quantitatively and qualitatively unique steroidal microenvironments at each germ cell stage.

6.9.2 Steroid Production

An obvious question is whether steroidogenic enzyme activities, as measured by radiolabelled tracer analysis, predict steroid output from endogenous substrates. To address this question the production of immunoreactive progesterone, testosterone, DOC and 20β-hydroxy-S was measured in isolated, staged spermatocyst cultures (Cuevas and Callard 1989; Barry and Callard 1990; Cuevas and Callard 1992a; Barry *et al.* 1993). Addition of 25-hydroxycholesterol, a soluble form of cholesterol, was essential to maximize steroid output. Under these conditions, all products accumulated in media for up to six days in culture, and output increased by addition of cylic AMP and/or phosphodiesterase inhibitor (MIX, 1-isobutyl-methyl-xanthine) (Cuevas and Callard 1992a; Sourdaine and Garnier 1993). Although traces of immunoreactive estradiol are measurable when testosterone is added as a substrate, pregnenolone, dehyroepiandrosterone, and androstenedione are undetectable (Callard 1991a). Similar *in vitro* studies of staged spermatocyst cultures also found testosterone and progestin levels highest in more mature spermatogenic stages (Sourdaine *et al.* 1990; Sourdaine and Garnier 1993).

6.10 INTRATESTICULAR STEROID RECEPTORS

The strongest indication that steroid regulation of gene expression has been widely conserved as a mechanism of control in vertebrate testis is the presence in several nonmammalian species of steroid receptors that resemble those of mammals in their important physiochemical characteristics, despite minor species differences in ligand specificity and chromatin binding characteristics (Callard *et al.* 1985; Callard and Mak 1985; Ruh *et al.* 1986; Cuevas and Callard 1992b). Furthermore, studies using staged shark tissues have been useful in providing clues to the steroid-sensitive control points during spermatogenesis. Steroid receptors have been cloned in elasmobranch species: estrogen receptor-β (Sikora and Callard, unpubl. data); androgen receptor (Engel and Callard, unpubl. data).

6.10.1 Estrogen Receptors

Estrogen receptors (ER), as measured by binding activity or mRNA, are highest in regions with stem cells and spermatogonia and are virtually nondetectable in regions with mature germ cells (Callard *et al.* 1985; Sikora and Callard, unpublished data). Additionally, the highest percentage of occupied ERs and the highest number of ER-specific chromatin binding sites are present in PrM stages (Callard *et al.* 1985; Ruh *et al.* 1986). Administration of 17β-estradiol to *Squalus acanthias* inhibited DNA synthesis in a dose dependent manner in the premeiotic region (Betka and Callard 1998). Because the vascular flow in the dogfish is directed from more mature to less mature stages, and aromatase has been localized to meiotic regions, estrogen may be involved in a negative feedback loop as part of a growth control mechanism (Betka and Callard 1998). Mammalian studies support this hypothesis. In a recent study, *in vitro* administration of estradiol to fetal rat testicular cultures resulted in a decrease in mitotic proliferation and an increase in apoptosis in a concentration dependent manner (Lassurguere *et al.* 2003). Nonetheless, mammalian data suggests that ER may mediate regulation of spermatids as well. For example, estrogen administration resulted in a reduction in spermatids and sperm in the rat (Newbold 1998), and knockout mice for ER are infertile, producing sperm that are unable to fertilize eggs, with the deficiency isolated to somatic testicular cells (Mahato *et al.* 2001). These *in vivo* studies cannot rule out estrogen effect on the HPG axis, however.

6.10.2 Androgen Receptors

Androgen receptor (AR) binding activity and mRNA are highest in regions with stem cells and spermatogonia (premeiotic stages) in spiny dogfish (Cuevas and Callard 1992). Virtually all testicular ARs are occupied, as determined by exchange assays (Cuevas and Callard 1992). The patterned distributions of the androgen receptor in shark testis are consistent with available data in rodents. AR binding activity is 50% higher in cell nuclei from stages IX-XII and XIII-I when compared to other stages (Isomaa *et al.* 1985). This has been interpreted as an effect of androgen on meiotic divisions in stage XIV, but does not exclude an influence on earlier germ cell generations that are present in the same segments. Most profoundly affected by androgen deprivation in rodents are early meiotic stages of development, specifically pachytene spermatocytes. In androgen-resistant (AR-defective) and androgen receptor knockout mice, spermatogenesis is arrested in the first meiotic divisions (Lyon *et al.* 1975; Yeh *et al.* 2002), buttressing the idea that androgen has actions early in spermatogenesis, although the possibility of effects at later stages cannot be ruled out. Attempts to pinpoint androgen-sensitive steps *in vivo* using hypophysectomy, gonadotropin, or steroid replacement to manipulate testicular steroid levels have revealed that hormones do not influence the duration of the cycle, but somehow increase the efficiency of the process (Vernon *et al.* 1975).

Using an AR antipeptide antibody and a procedure previously used to localize AR in goldfish brain (Gelinas and Callard 1997), immunoreactive staining was observed in somatic cells but not germ cells in the dogfish testis (Fig. 6.5). Specifically, AR immunoreactivity was localized in adluminal Sertoli cell nuclei in spermatogonial stage cysts, and also labeled ductule cells but not peritubular cells in the premeiotic region, supporting mammalian research as well as binding studies in the shark that suggest the importance in premeiotic stages as an androgen target (Cuevas and Callard 1992b; Zhou *et al.* 2002; Yeh *et al.* 2002). The distribution of ER and AR support the hypothesis that endogenous androgen and estrogen may cooperate in regulating gene expression prior to meiosis and are consistent with observations showing that hypophysectomy most profoundly alters the transition from spermatogonia to spermatocytes in the shark (Simpson and Wardle 1967; Dobson and Dodd 1977b, c).

6.10.3 Progesterone Receptors

Progesterone receptor (PR) binding activity was identified in shark testis (Cuevas and Callard 1992). PR clearly differed from AR in their zonal distribution (PoM » M = PrM) and steroid binding characteristics, but was unable to distinguish between progesterone (P) and DOC as ligands.

6.10.4 Non-receptor Steroid Binding Protein

In addition to conventional steroid receptors that are detected in both nuclear and cytosolic extracts and that adhere to DNA-cellulose affinity columns, a nonreceptor steroid binding protein has been characterized in cytosolic subfractions of shark testis (Mak and Callard 1987). In its physicochemical properties (high affinity, broad specificity, apparent molecular weight, isoelectric point, and dimeric structure), it resembles androgen binding protein (ABP) of mammalian testis and is presumed to be the shark counterpart. When its intratesticular distribution is examined, ABP concentration increases progressively through spermatogenic development, corresponding exactly to the hypertrophy and differentiation of Sertoli cells and maximal androgen biosynthetic potential. These data are consistent with the protein secretory appearance of Sertoli cells at mature stages and the known Sertoli cell origin and androgen dependence of ABP in mammals. Interestingly, dissected rat tubular segments secrete ABP maximally at stage VIII, just prior to release of the most advanced cohort of germ cells (Parvinen 1982). Although secretion of ABP by shark Sertoli cells has not been definitively established, this protein may serve as a steroid reservoir in the testis or during long-term sperm storage in the excurrent ducts and may thus account for high endogenous steroid levels in shark semen (Simpson *et al.* 1964).

6.10.4 Intratesticular Vasculature

Analysis of the enzyme and receptor distribution patterns described above reveals a paradox. With the exception of PRs, which are maximally concentrated in testicular regions where P-synthesizing potential is maximal, maximal androgen and estrogen biosynthesis (PoM and M stages, respectively) is spatially (temporally) separated from regions in which ARs and ERs are concentrated (PrM stages). To investigate whether steroids synthesized in more advanced stages of development have access to cognate receptors in less mature stages, the testis was perfused in situ with Evans blue dye via the genital artery. It was observed that blood enters the testis via the capsule immediately beneath the epigonal organ. The pattern of flow was PoM → M → PrM (Callard *et al.* 1989; Cuevas *et al.* 1992). Data from sharks are consistent with the idea that steroids are part of an intratesticular signaling system that operates within (autocrine) and between (paracrine) stages of development. Is this between-stage communication pathway in shark testis unique, or is there a comparable communication pathway in mammals? The different stages comprising the spermatogenic wave in the rat tubule are separated by only a few millimeters and, hence, can readily communicate by simple diffusion of chemical messengers. However, capillaries are arranged longitudinally, and blood entering the testis proper via the region of the efferent ducts flows parallel to the spermatogenic wave in decreasing order (mature→ immature) along the length of the tubule (Setchell 1978).

6.11 SEASONALITY

Although intervals between succeeding spermatogenic steps are invariant in mature males of continuous breeding mammalian species like the rat, mouse, and man, a period of spermatogenic inactivity is a normal part of the testicular cycle in seasonal breeders. One problem in assessing seasonality in male elasmobranchs is that the majority of reports are based on observations such as ovulatory cycles, and presence of young in females, parameters that may or may not be extrapolated to males (Wourms 1977). Also, seasonality of spermatogenesis takes multiple forms in elasmobranchs: In some species (e.g., bonnethead shark), there is a loss of all but immature germ cell stages and a dramatic decline in testicular weight, (Parsons and Grier 1992; Maruska *et al.* 1996) as occurs in seasonal mammals. In other species (spiny dogfish shark), all stages are present throughout the year. There is no obvious change in testicular weight, but there is a seasonal "arrest" of the developmental progression (Simpson and Wardle 1967). Of the studies that have addressed males, even fewer are comprehensive that document reproductive activity, steroid hormone levels, and spermatogenesis at all stages (Parsons and Grier 1992; Maruska *et al.* 1996; Manire and Rasmussen 1997; Heupel *et al.* 1999; Tricas *et al.* 2000).

Steroid levels in the general circulation have often been measured in an attempt to determine whether the HPG axis and testicular functions are cyclic, but changes are often subtle or not clearly related to time of year or not reproducible in succeeding years. Also, because the primary secretory cell type of the mammalian testis (Leydig cells) is poorly developed, few in number or absent in elasmobranchs, it is necessary to demonstrate that peripheral steroid levels accurately reflect testicular secretion and are not merely a byproduct of Sertoli cell efflux.

6.12 TOXICOLOGY

The list of chemicals known or suspected to be spermatotoxicants continues to expand, and include heavy metals, dioxins, polychlorinated biphenyls, polycyclic aromatic hydrocarbons, as well as many other established endocrine disrupting chemicals (Akingbemi and Hardy 2001). Although research in this area using elasmobranchs is sparse, the favorable organization of the testis of the spiny dogfish facilitates identification of stage-specific toxicants. In this species, cadmium, a known mammalian spermatoxicant, was studied. *In vivo* administration of cadmium (5 mg/kg BW) had no effect on cell proliferation, DNA or protein synthesis (Callard *et al.* 1998). However, using a cadmium tracer it was discovered that cadmium accumulates preferentially in premeiotic regions. Also, metallothionein-like binding activity was stage-dependent (PrM = PoM > GZ = M) and was induced by Cd in all but meiotic stages of spermatogenesis (Betka and Callard, 1999). A metallothionein-like protein was partially characterized in the liver of the spotted dogfish shark (*Scyliorhinus*), and trace amounts were detected in the testis (Hidalgo *et al.* 1985; Hidalgo and Flos 1986a; Hidalgo and Flos 1986b; Planas *et al.* 1991). Although many elasmobranchs inhabit inshore waters that are heavily polluted, to our knowledge they have rarely been used to study spermatotoxicity and, in some studies of testicular physiology, the polluted habitat has not been considered.

6.13 CONCLUSIONS AND FUTURE PERSPECTIVES

From the foregoing, it should be apparent that elasmobranch species have considerable potential for studying spermatogenesis, mechanisms of regulation and toxicant action stage-by-stage. Obviously, drawbacks exist concerning the use of an unconventional animal model (e.g. seasonal availability, a dearth of background information, difficulties in laboratory maintenance and experimental manipulation); however, the potential for gain of new information of general relevance to male reproductive biology, coupled with the growing field of research on reproduction in elasmobranchs makes continued investigation in the field warranted and necessary.

6.14 ACKNOWLEGMENTS

Original studies in this report were carried out in part at the Mount Desert Island Biological Laboratory and supported by a center grant to MDIBL (NIEHS ES03828), research grants to GVC (NIEHS P42 ES07381; and EPA STAR R825434), and a predoctoral traineeship to KBE (2T32 HD073897).

6.15 LITERATURE CITED

Akingbemi, B. T. and Hardy, M. P. 2001. Oestrogenic and antiandrogenic chemicals in the environment: effects on male reproductive health. Annals of Medicine 33(6): 391-403.

Barry, T. and Callard, G. 1990. Identification and stage-related synthesis of 11-deoxycorticosterone (DOC) by the dogfish (*Squalus acanthias*) testis. Mount Desert Island Biological Lab Bulletin 29: 131-132.

Barry, T., Thomas, P. and Callard, G. 1993. Stage-related production of 21-hydroxylated progestins by the dogfish (*Squalus acanthias*) testis. Journal of Experimental Zoology 265(5): 522-32.

Betka, M. and Callard, G. V. 1998. Negative feedback control of the spermatogenic progression by testicular oestrogen synthesis: Insights from the shark testis model. Acta pathologica, Microbiologica et Immunologica Scandinavica 106(1): 252-7; discussion 257-8.

Betka, M. and Callard, G. V. 1999. Stage-dependent accumulation of cadmium and induction of metallothionein-like binding activity in the testis of the Dogfish shark, *Squalus acanthias*. Biology of Reproduction 60(1): 14-22.

Bourne, A. 1991. Androgens. Pp. 115-147. In P. Pang and M. Schreibman (eds), *Vertebrate endocrinology: fundamentals and biomedical implications*. Acadmeic Press, New York.

Callard, G. 1991a. Reproduction in male elasmobranch fishes. Pp. 104-154. In R. Kinne, E. Kinne-Saffran and K. Beyenbach (eds), *Oogenesis, Spermatogenesis and Reproduction*. Karger, Basel.

Callard, G. 1991b. Spermatogenesis. Pp. 303-41. In P. Pang and M. Schreibman (eds), *Vertebrate endocrinology: fundamentals and biochemical implications*. Academic Press, New York.

Callard, G., Betka, M. and Jorgensen, J. 1994. Stage-related function of Sertoli cells: lessons from lower vertebrates. Pp. 24-54. In A. Bartke (ed.), *Function of Somatic Cells in the Testis*. Springer Verlag, New York.

Callard, G., Mak, P., DuBois, W. and Cuevas, M. 1989. Regulation of spermatogenesis: the shark testis model. Journal of Experimental Zoology(suppl 2): 353-64.

Callard, G., McClusky, L. and Betka, M. 1998. Apoptosis as a normal mechanism of growth control and target of toxicant actions during spermatogenesis: insights using the shark testis model. Pp. 125-128. In H. H. LeGal Y (ed.), *New Developments in Marine Biotechnology*. Plenum.

Callard, G. V. and Mak, P. 1985. Exclusive nuclear location of estrogen receptors in *Squalus* testis. Proceedings of the National Academy of Sciences U S A 82(5): 1336-40.

Callard, I. P., Callard, G. V., Lance, V., Bolaffi, J. L. and Rosset, J. S. 1978. Testicular regulation in non-mammalian vertebrates. Biology of Reproduction 18: 16-43.

Chieffi, G., Corte, F. D. and Botte, V. 1961. Osservazioni sul tessuto interstiziale del testicolo dei Selaci. Bolletino di Zoolologia 28: 211-217.

Collenot, G. 1970. Apparition el evolution de l'activite endocrine du testicle de *Scyliorhinus canicula* L. (Elasmobranche). Ann. Embryol. Morphogen 2: 461-477.

Collenot, G. and Damas, D. 1975. Mise in evidence de la nature proteique de corps enigmatiques presents dans le testicule de *Scyliorhinus canicula* (elasmobranche). Cahiers de Biologie Marine 16: 39-46.

Collenot, G. and Damas, D. 1980. Etude ultrastructurale de la cellule de Sertoli au cours de la spermiogenese chez *Scyliorhiuns canicula*. Cahiers de Biologie Marine 21: 209-19.

Cuevas, M. and Callard, G. 1989. In vitro steroid secretion by Sertoli/germ cell units (spermatocysts) derived from dogfish (*Squalus acanthias*) testis. Mount Desert Island Biological Lab Bulletin 28: 30-31.

Cuevas, M., Collins, L. and Callard, G. 1993. Stage-related changes in steroid converting enzyme activities in *Squalus* testis: Synthesis of biologically active metabolites via 3(β)-hydroxysteroid dehydrogenase/isomerase (3βHSD/isomerase) and 5(α) -reductase. Steroids 58: 87-94.

Cuevas, M., Miller, W. and Callard, G. 1992. Sulfoconjugation of steroids and the vascular pathway of communication in dogfish testis. Journal of Experimental Zoology 264(2): 119-29.

Cuevas, M. E. and Callard, G. 1992. Androgen and progesterone receptors in shark (*Squalus*) testis: characteristics and stage-related distribution. Endocrinology 130(4): 2173-82.

D'Antonio, M., Vallarino, M., Lovejoy, D., Vandesande, F., King, J., Pierantoni, R. and Peter, R. 1995. Nature and Distribution of Gonadotropin-Releasing Hormone (GnRH) in the Brain, and GnRH and GnRH Binding Activity in Serum of the Spotted Dogfish *Scyliorhinus canicula*. General and Comparative Endocrinology 98: 35-49.

Della, F. C., Botte, V. and Chieffi, G. 1961. Ricerca istochemica dell' attivita della steroide 3(Beta)-olo-deidrogenase nel testiclo do *Torpedo marmorata* Risso e di *Scyliorhiunus stellaris* (L). Atti dell Societa Peloritana di Scienze Fisiche Matematiche e Naturali 7: 393-397.

Dobson, S. and Dodd, J. 1977a. The roles of temperature and photoperiod in the response of the testis of the dogfish, *Scyliorhinus canicula*, to partial hypophysectomy (ventral lobectomy). General Comparative Endocrinology 32: 114-115.

Dobson, S. and Dodd, M. 1977b. Endocrine control of the testis in the dogfish *Scyliorhinus canicula* L., I. Effects of partial hypophysectomy on gravimetric, hormonal and biochemical aspects of testis function. General and Comparative Endocrinology 32: 41-52.

Dobson, S. and Dodd, M. 1977c. Endocrine control of the testis of the dogfish *Scyliorhinus canicula* L., II. Histological and ultrastructural changes in the testis after partial hypophysectomy (ventral lobectomy). General and Comparative Endocrinology 32: 53-71.

Dodd, J. and Sumpter, J. 1984. Fishes. Pp. 21-126. In G. E. Lamming (ed.), *Marshall's Physiology of Reproduction*. Churchill Livingstone, New York.

Dubois, W., Mak, P. and Callard, G. 1989. Sertoli cell functions during spermatogenesis: the shark testis model. Fish Physiology and Biochemistry 7: 221-7.

Fange, R. and Pulsford, A. 1983. Structural studies on lymphomyeloid tissues of the dogfish, *Scyliorhinus canicula* L. Cell Tissue Research 230: 337-351.

Fawcett, D. W. 1975. Ultrastructure and fuction of the Sertoli cell. Pp. 21-56. In Hamilton and Greep (eds), *Handbook of Physiology, sect. 7. Endocrinology, vol. 5: Male Reproductive System*. American Physiological Society, Washington.

Gelinas, D. and Callard, G. 1997. Immunolocalizationof aromatase- and androgen receptor-positive neurons in the goldfish brain. General and Comparative Endocrinology 106(2): 155-68.

Gilbert, S. G. 1973. *Pictoral Anatomy of the Dogfish*. University of Washington Press, Seattle, 59 pp.

Griswold, M., Morales, C. and Sylvester, S. 1988. Molecular biology of the Sertoli cell. Oxford Review of Reproductive Biology 10: 124-161.

Hecht, N. 1986. Regulation of gene expression during mammalian spermatogenesis. Pp. 151-193. In J. Rossant and R. Pedersen (eds), *Experimental Approaches to Mammalian Embryonic Development*. Cambridge University Press, New York.

Helfman, G., Collette, B. and Facey, D. 1997. *The Diversity of Fishes*. Blackwell Science, Malden, pp. 151-178.

Heupel, M., Whittier, J. and Bennett, M. 1999. Plasma steroid hormone profiles and reproductive biology of the Epaulette shark, *Hemiscyllium ocellatum*. Journal of Experimental Zoology 284: 586-594.

Hidalgo, J. and Flos, R. 1986a. Dogfish metallothionein-I. Purification and characterization and comparison with rat metallothionein. Comparative Biochemistry and Physiology 83(C): 99-103.

Hidalgo, J. and Flos, R. 1986b. Dogfish metallothionein - II. Electrophoretic studies and comparison with rat metallothionein. Comparative Biochemistry and Physiology 83(C): 105-109.

Hidalgo, J., Tort, L. and Flos, R. 1985. Cd-, Zn-, Cu-binding protein in the elasmobranch *Scyliorhinus canicula*. Comparative Biochemistry and Physiology 81(C): 159-65.

Hug, M., Gangyopadhyay, N. N. and Frizzell, R. 2000. Regulated Ion Transport by the Alkaline Gland of the Little Skate (*Raja erinacea*). Mount Desert Island Biological Lab Bulletin: 48-50.

Ijiri, S., Berard, C. and Trant, J. 2000. Characterization of gonadal and extra-gonadal forms of the cDNA encoding the Atlantic stinray (*Dasyatis sabina*) cytochrome P450 aromatase (CYP19). Molecular and Cellular Endocrinology 164(1-2): 169-81.

Isomaa, V., Parvinen, M., Janne, O. and Bardin, C. 1985. Nuclear androgen receptors in different stages of the seminiferous epithelial cycle and the interstitial tissue of rat testis. Endocrinology 116: 132-7.

Johnson, M. and Everitt, B. 2000. *Essential Reproduction* (Fifth Edition (ed.)). Blackwell Sciences, Malden, MA, Pp. 66, 103-104.

Kobori, H., Miyazaki, S. and Kuwabara, Y. 2000. Characterization of intracellular Ca(2+) increase in response to progesterone and cyclic nucleotides in mouse spermatozoa. Biology of Reproduction 63(1): 113-20.

Lance, V. and Callard, I. 1978. Gonadotropic activity in pituitary extracts from an elasmobranch (*Squalus acanthias*). Journal of Endocrinology 78: 149-150.

Lassurguere, J., Livera, G., Habert, P. and Jegou, B. 2003. Time- and dose-related effects of estradiol and diethylstilbestrol on the morphology and function of the fetal rat testis in culture. Toxicological Sciences 73(1): 160-9.

Lovejoy, D.A., Stell, W.K. and Sherwood, N.M. 1992. Partial characterization of four forms of immunoreactive gonadotropin-releasing hormone in the brain and terminal nerve of the spiny dogfish (Elasmobranchii; *Squalus acanthias).* Regulatory Peptides 37(1): 39-48.

Lyon, M., Glenister, P. and Lamoreux, M. 1975. Normal spermatozoa from androgen resistant germ cells of chimeric mice and the role of androgen in spermatogenesis. Nature 258: 620-2.

Mahato, D., Goulding, E., Korach, K. and Eddy, E. 2001. Estrogen receptor alpha is required by the supporting somatic cells for spermatogenesis. Molecular and Cellular Endocrinology 178: 57-63.

Mak, P. and Callard, G. 1987. A novel steroid binding protein in the testis of the dogfish *Squalus acanthias*. General and Comparative Endocrinology 68: 104-12.

Manire, C. and Rasmussen, E. 1997. Serum concentrations of steroid hormones in the mature male bonnethead shark, *Sphyrna tiburo*. General and Comparative Endocrinology 107: 414-420.

Marina, P., Annamaria, L., Barbara, D., Loredana, R., Piero, A. and Francesco, A. 2002. Fine structure of Leydig and Sertoli cells in the testis of immature and mature Spotted Ray *Torpedo marmorata*. Molecular Reproduction and Development 63: 192-201.

Maruska, K., Cowie, E. and Tricas, T. 1996. Periodic gonadal activity and protracted mating in elasmobranch fishes. Journal of Experimental Zoology 276: 219-232.

Mattei, X. 1970. Spermiogénèse comparée des poissons. Pp. 57-69. In B. Baccetti (ed.), *Comparative Spermatology*. Academic Press, New York.

Matthews, H. L. 1950. Reproduction in the basking shark, *Cetorhinus maximus*. Philosophical Transactions of the Royal Society, London (Biology) 234: 247-316.

Mattisson, A. and Fange, R. 1982. The cellular structure of the Leydig organ in the shark, *Etmopterus spinax* (L). Biological Bulletin 162: 182-194.

McPhaul, M., Noble, J., Simpson, E., Mendelson, C. and Wilson, J. 1988. The expression of a functional cDNA encoding the chicken cytochrome P450arom (aromatase) that catalyzes the formation of estrogen from androgen. Journal of Biological Chemistry 263(31): 16358-63.

Mellinger, J. 1965. Stades de la spermatogenesechez *Scyliorhinus caniculus*: description, données histochimiques, variations normales et éxperimentales. Zeitschrift für Zellforschung 67: 653-673.

Mellinger, J. and Dubois, M. 1973. Confirmation, par l'immunofluorescence, de la fonction corticotrope du lobe rostral et de la fonction gonadotrope du lobe ventral de l'hypophyse d'un poisson cartilagineux, la torpille marbrée (*Torpedo marmorata*). Comptes Rendues de l'Académie des Sciences 276: 1879-1881.

Miyashita, K., Shimizu, N., Osanai, S. and Miyata, S. 2000. Sequence analysis and expression of the P450 aromatase and estrogen receptor genes in the *Xenopus* ovary. Journal of Steroid Biochemistry and Molecular Biology 75(2-3): 101-7.

Moore, J. E. S. 1895. On the structural changes in the reproductive cells during the spermatogenesis of elsmobranchs. Journal of Microscopy 38: 275-313.

Moyne, G. and Collenot, G. 1982. Unusual nucleolar fine structure in the Sertoli cells of the dogfish *Scyliorhinus canicula*. Biology of the Cell 44: 239-48.

Newbold, R. 1998. Influence of estrogenic agents in mammalian male reproductive tract development. Pp. 531-51. In K. Korach (ed.), *Reproductive and Developmental Toxicology*. Marcel Dekker Inc., New York.

Nunez, S. and Trant, J. 1997. Isolation of the putative cDNA encoding cholesterol side chain cleavage cytochrome P450 (CYP11A) of the southern stingray (*Dasyatis americana*). Gene 187: 123-129.

Parsons, G. and Grier, H. 1992. Seasonal changes in shark testicular structure and spermatogenesis. Journal of Experimental Zoology 261: 173-184.

Parvinen, M. 1982. Regulation of the seminiferous epithelium. Endocrine Reviews 3: 404-17.

Piferrer, F. C. and Callard, G. V. 1995. Inhibition of deoxyribonucleic acid synthesis during premeiotic stages of spermatogenesis by a factor from testis-associated lymphomyeloid tissue in the dogfish shark (*Squalus acanthias*). Biology of Reproduction 53(2): 390-8.

Pilsworth, L. and Setchell, B. 1981. Spermatogenic and endocrine functions of the testis of invertebrate and vertebrate animals. Pp. 9-38. In H. Burger and D. D. Kretser (eds), *The testis*. Raven Press, New York.

Planas, J., Tort, L., Torres, P. and Flos, R. 1991. Cadmium induction of metallothioneins in several dogfish organs. Revista Espanola de Fisiologia 47: 75-80.

Pratt, H. and Tanaka, S. 1994. Sperm Storage in Male Elasmobranchs: A Description and Survey. Journal of Morphology 219: 297-308.

Pratt, H. L. 1979. Reproduction in the blue shark, *Prionace glauca*. Fishery Bulletin 77: 445-470.

Pudney, J. and Callard, G. 1984a. Development of agranular reticulum in Sertoli cells of the testis of the dogfish *Squalus acanthias* during spermatogenesis. Anatomical Record 209(3): 311-21.

Pudney, J. and Callard, G. 1984b. Identification of Leydig-like cells in the testis of the dogfish *Squalus acanthias*. Anatomical Record 209(3): 323-30.

Pudney, J. and Callard, G. 1986. Sertoli cell cytoplasts in the semen of the spiny dogfish (*Squalus acanthias*). Tissue and Cell 18: 375-82.

Querat, B., Tonnerre-Doncarli, C., Genies, F. and Salmon, C. 2001. Duality of Gonadotropin in Gnathosomes. General and Comparative Endocrinology 124: 308-314.

Ritzen, E. M., Hansson, V. and French, F. S. 1989. The Sertoli cell. Pp. 269-302. In B. deKrester (ed.), *The Testis*. Raven Press, New York.

Roosen-Runge, E. 1977. *The Process of Spermatogenesis in Animals*. Cambridge University Press 1-214 pp.

Ruh, M. F., Singh, R. K., Mak, P. and Callard, G. V. 1986. Tissue and species specificity of unmasked nuclear acceptor sites for the estrogen receptor of *Squalus* testes. Endocrinology 118(2): 811-8.

Russell, L. 1984. Spermiation-the sperm reease process: Ultrastructural observations and unresolved problems. Pp. 46-65. In J. V. Blerkom and P. Motta (eds), *Electron Microscopy in Biology and Medicine*. Plenum Press, New York.

Sabatier, A. 1896. De la spermatogénèse chez les poissons selaciens. Travaux Institut Zoologie, Université Montpellier et Stat Marit. Cette, NS Memoires. No 4: 1-191.

Sakai, M., Fujii-Kuriyama, Y. and Muramatsu, M. 1978. Number and frequency of protamine genes in rainbow trout testis. Biochemistry 17: 5510-5515.

Scanes, C., Follett, B. and Goos, J. 1972. Goandotropic activity in the pituitary gland of the dogfish (*Scyliorhinus canicula*). Journal of Endocrinology 54: 343-344.

Selcer, K. and Leavitt, W. 1991. Estrogens and progestins. Pp. 67-114. In P. Pang and M. Schreibman (ed.), *Vertebrate endocrinology: Fundamentals and biochemical implcations*. Academic Press, New York.

Setchell, B. 1978. The mammalian testis. Cornell University Press, Ithica, NY., pp. 61–5.

Setchell, B. and Pilsworth, L. 1989. The functions of the testes of vertebrate and invertebrate animals. Pp. 1-66. In H. Burger and D. D. Kretser (eds), *The testis*. Raven Press, New York.

Simpson, T. and Wardle, C. 1967. A seasonal cycle in the testis of the spurdog, *Squalus acanthias*, and the sites of 3(Beta)-hydroxysteroid dehydrogenase activity. Journal of the Marine Biological Association of the UK 47: 699-808.

Simpson, T., Wright, R. and Hunt, S. 1964. Steroid biosynthesis on the testis of the dogfish (*Squalus acanthias*). Journal of Endocrinology 31: 29-38.

Sourdaine, P. and Garnier, D. 1993. Stage-dependent modulation of Sertoli cell steroid production in dogfish (*Scyliorhinus canicula*). Journal fo Reproduction and Fertility 97(1): 133-42.

Sourdaine, P., Garnier, D. and Jegou, B. 1990. The adult dogfish (*Scyliorhinus canicula*) testis: a model to study stage-dependent changes in steroid levels during spermatogenesis. Journal of Endocrinology 127(3): 451-60.

Sprando, R. L. and Russell, L. D. 1988. Spermiogenesis in the bluegill (*Lepomis machrochirus*): A study of cytoplasmic events including cell volume changes and cytoplasmic elimination. Journal of Morphology 198: 165-177.

Sprando, R. L. and Russell, L. D. 1988. Spermiogenesis in the red-ear turtle (*Pseudymys scripta*) and the domestic fowl (*Gallus domesticus*): A study of cytoplasmic elimination. Journal of Morphology 198: 95-118.

Stanley, H. P. 1971a. Fine Structure of spermiogenesis in the elasmobranch fish *Squalus suckleyi*. I. Acrosome formation, nuclear elongation and differentiation of the midpiece axis. Journal of Ultrastructure Research 36: 86-102.

Stanley, H. P. 1971b. Fine structure of spermiogenesis in the elasmobranch fish *Squalus suckleyi*. II. Late stages of differentiation and structure of the mature spermatozoon. Journal of Ultrastructure Research 36: 103-118.

Stephan, M. P. 1902. L'evolution de la cellule de Sertoli des selaciens apres la sperm genese. Comptes Rendues de la Societé de Biologie, Paris 54: 775-776.

Sumpter, J., Jenkins, N. and Dodd, J. 1978. Gonadotropic hormone in the pituitary gland of the dogfish (*Sycliorhinus canicula*): distribution and physiological significance. General Comparative Endocrinology 36: 275-285.

Takahashi, M., Tanaka, M., Sakai, N., Adachi, S., Miller, W. and Nagahama, Y. 1993. Rainbow trout ovarian cholestrol side-chain cleavage cytochrome P450 (P450-scc). cDNA cloning and mRNA expression during oogenesis. FEBS Letter 319(1-2): 45-48.

Trant, J. 1995. Isolation and characterization of the cDNA encoding the spiny dogfish shark (*Squalus acanthias*) form of cytochrome P450c17. Journal of Experimental Zoology 272(1): 25-33.

Tricas, T., Maruska, K. and Rasmussen, L. 2000. Annual cycles of steroid hormone production, gonad development, and reproductive behavior in the Atlantic stingray. General and Comparative Endocrinology 118: 209-225.

Tsang, P. and Callard, I. 1988. In vivo Steroidogenic Effects of Homologous Pituitary Ventral Lobe Extract in the Spiny Dogfish, *Squalus acanthias*. General and Comparative Endocrinology 70: 164-168.

Vernon, R., Go, V. and Fritz, I. 1975. Hormonal requirement of the different cycles of seminferous epithelium during reintiation of spermatogenesis in long-term hypophysectomized rats. Journal of Reproductive Fertility 42: 77-90.

Wourms, J. P. 1977. Reproduction and development in chondrichthyan fishes. American Zoology 17: 379-410.

Yeh, S., Tsai, M., Xu, Q., Mu, X., Lardy, H., Huang, K., Lin, H., Yeh, S., Altuwaijri, S., Zhou, X., Xing, L., Boyce, B., Hung, M., Zhang, S., Gan, L., Chang, C. and Hung, M. 2002. Generation and characterization of androgen receptor knockout (ARKO) mice: an in vivo model for the study of androgen functions in selective tissues. Proceedings of the National Academy of Sciences USA 99(21): 13498-503.

Yoon, C., Kawakami, K. and Hopkins, N. 1997. Zebrafish vasa homolog RNA is localized to the cleavage planes 2- and 4-cell stage embryos and is expressed in the primordial germ cells. Development 124(16): 3157-65.

Zhou, Q., Nie, R., Prins, G., Saunders, P., Katzenellenbogen, B. and Hess, R. 2002. Localization of androgen and estrogen receptors in adult male mouse reproductive tract. Journal of Andrology 23(6): 870-81.

Zirkin, B. R. 1975. The ultrastructure of nuclear differentiation during spermiogenesis in the salmon. Journal of Ultrastructure Research 50: 174-184.

CHAPTER 7

Chondrichthyan Spermatozoa and Phylogeny

Barrie G. M. Jamieson

7.1 INTRODUCTION

7.1.1 General

Two subclasses are here recognized: the Holocephali and the Elasmobranchii. Examples of the three chief types of organization in the Chondrichthyes: Order Chimaeriformes (rat fish and chimaeras), and superorders Selachimorpha (sharks) and Batidoidimorpha (rays) are shown in Fig. 7.1. Modern forms have internal fertilization, the male with pelvic claspers, and oviparity or viviparity; yolk sac viviparity is considered the plesiomorph reproductive state (Chapter **3** of this volume). Where the eggs are desposited externally, the embryo is encapsulated in a leatherlike case in which gestation is the longest in vertebrates, taking up to two (Nelson 1984) or even (Tanaka *et al.* 1990) 3.5 years. Monophyly of chondrichthyans (see Chapter **1** of this volume) is corroborated by, *inter alia*, the presence of coracobranchial muscles of hypobranchial origin (Wiley 1979) and is fully confirmed by sperm ultrastructure.

Accounts of chondrichthyan sperm ultrastructure are often fragmentary. They are not, therefore, separately reviewed for individual species. A combined account for Holocephali and Elasmobranchii is given below. It is chiefly drawn from accounts of the well-described sperm of two species, the holocephalan *Hydrolagus colliei* (Stanley 1983) and the elasmobranch *Squalus suckleyi* (Stanley 1971) and a study, in this chapter, of *Chiloscyllium punctatum* (Brown-banded Catshark), *Dasyatis fluviorum* (Estuary Stingray) and *D. kuhlii* (Blue-spotted Stingray), augmented from an analysis of six characters in 35 species by Tanaka *et al.* (1995). It is preceded by a brief account of spermiogenesis (see also Chapters **4** and **6** of this volume).

School of Integrative Biology, University of Queensland, Brisbane 4072, Australia

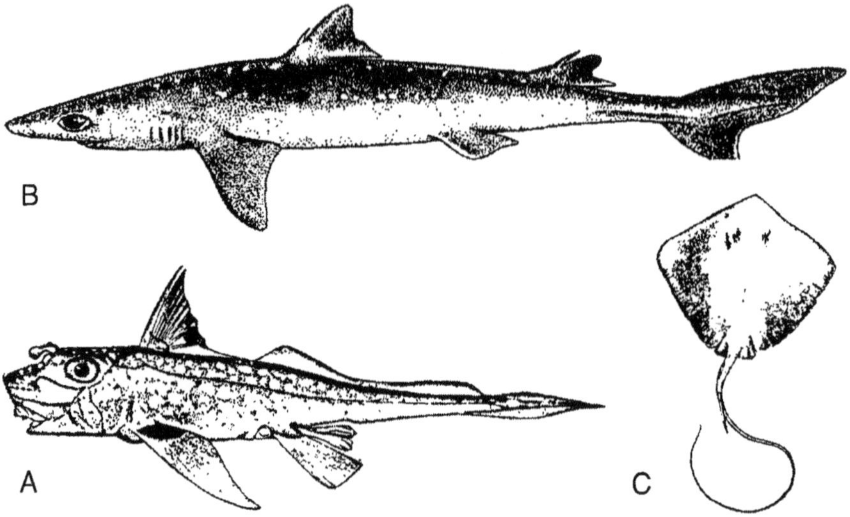

Fig. 7.1 Three chondrichthyans. *Fish Evolution and Systematics: Evidence from Spermatozoa.* Cambridge University Press, Cambridge. 319 pp., Fig. 7.2. **A.** *Hydrolagus* (=*Chimaera*) *colliei.* After Jordan, D.S. 1907. *Fishes.* Henry Holt, New York, NY, Fig. 158. **B.** *Squalus acanthias.* After Jordan, D.S. 1907. *Fishes.* Henry Holt, New York, NY, Fig. 144. **C.** *Dasyatis.* After Romer, A.S. and Parsons, T.S. 1977. *The Vertebrate Body.* W. B. Saunders Company, Philadelphia, Fig. 24D. After Garman.

7.1.2 Subclass Holocephali

Holocephali have separate anal and urinogenital openings, and there is no cloaca; males have a clasping organ on the head, in addition to pelvic claspers. There is a single extant order, the Chimaeriformes (Nelson 1984). Relationship of Holocephali with the Elasmobranchii has been questioned but Nelson (1984) considers that the two groups form a monophyletic entity, the Chondrichthyes.

We will see that sperm ultrastructure confirms the unique relationship of the two subclasses and indicates that putative synapomorphies such as the two axonemal elements and interpolation of the mitochondria between the nucleus and basal body were present in the common ancestor (placoderm). The mutuality of sperm structure and internal fertilization in the holocephalan-elasmobranch assemblage further suggests that this ancestor was internally fertilizing as has been proposed (Nelson 1984) for the extinct holocephalan-like placoderm order Ptyctodontiformes in which the male had claspers (see Fig. 7.15).

7.1.3 Subclass Elasmobranchii

Nelson (1984) recognizes five extant orders: the Hexanchiformes, Lamniformes, Squaliformes, comprising the sharks (Superorder Selachimorpha or Pleurotremata) and the Rajiformes, containing the rays and sawfish (Superorder Batidoidimorpha or Hypotremata) (Nelson 1984). Northcutt (1989) regards the sharks as a paraphyletic group one branch of which contains the batoids. For a detailed analysis of relationships, with references, see Chapter **1** of this volume.

7.2 ULTRASTRUCTURE AND DEVELOPMENT OF CHONDRICHTHYAN SPERM

7.2.1 Species Examined

The chondrichthyan species that have been examined for sperm morphology are listed in Table 7.1. The list does not include all examples of studies which were restricted to optical microscopy or low magnification scanning electron microscopy (SEM) but is intended to be comprehensive for transmission electron microscopy (TEM).

7.2.2 Spermiogenesis

The salient features of spermiogenesis in Chondrichthyes are here briefly summarized, and facilitate interpretation of the structure of the mature spermatozoon. The chief stages have been described, for the guitarfish *Rhinobatos cemiculus*, by Mattei (1970) (Fig. 7.2) and are essentially in agreement with those described for the shark (spiny dogfish) *Squalus suckleyi* by Stanley (1964, 1971a, b) and the stingray *Himantura signifer* by Chatchavalvanich *et al.* (2004) (see also Hamlett 1999, and further references in Chapter **4** of this volume). Tanaka *et al.* (1978) described spermatogenesis in the squalen shark, *Centrophorus atromarginatus*. Hara and Tanaka (1986) also gave a brief abstract on spermatogenesis in the shark *Chlamydoselachus anguineusus, Centroscymus owstoni* and *Prionace glauca*; two species of rays, *Dasyatis kuhlii* and *D. garouensis*; and a species of chimaera, *Chiamaera phantasma*. Gusse and Chevaillier (1978) made an ultrastructural and chemical study of chromatin during spermiogenesis in *Scyliorhinus caniculus*. Aspects of spermiogenesis in *Hydrolagus colliei* were described by Stanely *et al.* (1984) and Stanley and Lambert (1990).

Stanley (1962) and Hara and Tanaka (1986) report 64 spermatids per bundle. The young spermatid is a spherical cell, in *Rhinobatos cemiculus* (Mattie 1970), *Squalus suckleyi* (Stanley 1971a) and *Himantura signifer* (Chatchavalvanich *et al.* 2004) (Fig. 7.3A), 9 μm in diameter in each case. Two or more spermatids have been shown to be interconnected by cytoplasmic bridges (Stanley 1971a; Chatchavalvanich *et al.* 2004). (Fig. 7.3A). Each spermatid contains a large nucleus, a diplosome (two centrioles) near the plasma membrane, a Golgi apparatus in the vicinity of the centrioles, a fibrous rod (not reported for *S. suckleyi* or *H. signifer*), and widely dispersed mitochondria in the cytoplasm (Figs. 7.2A, 7.3A). Subsequently, the Golgi apparatus moves to the anterior pole of the nucleus and secretes an acrosome vesicle which attaches to the nucleus. At the opposite pole of the spermatid the centrioles, which are mutually at right angles, remain close to the plasma

Table 7.1 Chondrichthyes examined for spermatozoal morphology

Class, Order and Family	*Species*	*Author*
Holocephali Chimaeriformes		
Chimaeridae	*Chimaera phantasma*	Hara and Tanaka 1986[2]; Tanaka *et al.* 1995[1]
	Hydrolagus colliei	Stanley 1965[2], 1970[2], 1983[2]; Stanley and Lambert 1990[2]
	Hydrolagus mitsukurii	Tanaka *et al.* 1995[1]
	Neoharriotta pinnata	Mattei 1988[2]
Callorhychidae	*Callorhynchus milii*	Hamlett et al. 2002a[2]; Reardon *et al.* 2002[1]
Rhinochimaeridae	*Rhinochimaera pacifica*	Tanaka *et al.* 1995[1]
Elasmobranchii		
Heterodontiformes		
Heterodontidae	*Heterodontus japonicus*	Tanaka *et al.* 1995[1]
	Heterodontus portusjacksoni	Jones *et al.* 1984[2]
Orectolobiformes		
Orectolobidae	*Orectolobus japonicus*	Tanaka *et al.* 1995[1]
Hemiscyllidae	*Chiloscyllium punctatum*	Present study[2]
Lamniformes		
Odontaspididae	*Carcharias taurus*	Tanaka *et al.* 1995[1]
Alopiidae	*Alopias pelagicus*	Tanaka *et al.* 1995[1]
Lamnidae	*Isurus oxyrinchus*	Tanaka *et al.* 1995[1]
Carcharhiniformes		
Carcharhinidae	*Carcharhinus plumbeus*	Tanaka *et al.* 1995[1]
	Galeocerdo cuvier	Tanaka *et al.* 1995[1]
	Galeorhinus galeus	Van der Horst and McClusky 1986[1]
	Prionace glauca	Hara and Tanaka 1986[2]; Tanaka *et al.* 1995[1]
Scyliorhinidae	*Cephaloscyllium umbratile*	Tanaka *et al.* 1995[1]
	Galeus eastmani	Tanaka *et al.* 1995[1]
	Galeus nipponensis	Tanaka *et al.* 1995[1]
	Scyliorhinus caniculus	Stanley 1964[3], 1966; Gusse and Chevaillier 1978[2]
	Scyliorhinus sp	Stanley 1971[2]
	Cephaloscyllium umbratile	Tanaka *et al.* 1995[1]
Sphyrnidae	*Sphyrna lewini*	Parsons and Grier 1992[3]; Tanaka *et al.* 1995[1]
	Galeorhinus galeus	McClusky 2003[1]
Triakidae	*Hemitriakis japanica*	Tanaka *et al.* 1995[1]
	Mustelus canis	Hamlett *et al.* 2002b[2]; Tanaka *et al.* 1995[1]
	Mustelus griseus	Tanaka *et al.* 1995[1]
	Mustelus manazo	Tanaka *et al.* 1995[1]
Hexanchiformes		
Chlamydoselachidae	*Chlamydoselachus anguineus*	Hara and Tanaka 1986[2]; Tanaka *et al.* 1995[1]

Table 7.1 *Contd.* ...

Table 7.1 *Contd.* ...

Class, Order and Family	*Species*	*Author*
Squaliformes		
Squalidae	*Centrophorus atromarginatus*	Tanaka *et al.* 1978[2], (spermatogenesis)
	Centroscymnus owstoni	Hara and Tanaka 1986[2]; Tanaka *et al.* 1995[1]
	Dalatias licha	Tanaka *et al.* 1995[1]
	Deania calcea	Tanaka *et al.* 1995[1]
	Deania historicosa	Tanaka *et al.* 1995[1]
	Etmopterus brachyurus	Tanaka *et al.* 1995[1]
	Etmopterus molleri	Tanaka *et al.* 1995[1]
	Etmopterus pusillus	Tanaka *et al.* 1995[1]
	Squalus acanthias	Pudney and Callard 1984[2]
	Squalus brevirostris	Tanaka *et al.* 1995[1]
	Squalus japonicus	Tanaka *et al.* 1995[1]
	Squalus suckleyi	Stanley 1964[2], 1965[2], 1970[2], 1971a,b[2]
Squatiniformes		
Squatinidae	*Squatina japonica*	Tanaka *et al.* 1995[1]
Rhinobatiformes		
Rhinobatidae	*Rhinobatos cemiculus*	Boisson *et al.* 1968; Mattei 1970[2]
	Rhinobatos schlegelii	Tanaka *et al.* 1995[1]
Torpediniformes		
Torpedinidae	*Torpedo marmorata*	Stanley 1964[3], 1966[2]
	Torpedo nobiliana	Mattei 1999[2]
	Torpedo tokionis	Tanaka *et al.* 1995[1]
Rajiformes		
Dasyatidae	*Dasyatis fluviorum*	Present study[2]
	Dasyatis kuhlii	Hara and Tanaka 1986[2]; Present study[2]
	Dasyatis garouensis	Hara and Tanaka 1986[2]
	Dasyatis margarita	Mattei 1991[2]
	Himantura signifer	Chatchavalvanich 2004[2] (chiefly spermiogenesis)
Rajidae	*Raja clavata*	Nicander 1968[2]
	Raja eglanteria	Hamlett *et al.*[1] 1999; Tanaka *et al.* 1995[1]
	Raja miraletus	Mattei 1991[2]
	Raja sp.	Stanley 1983[2]
Rhinobatidae	*Rhinobatos cemiculus*	Boisson *et al.* 1968[2]; Mattei 1970[2]
Myliobatiformes		
Urolophidae	*Urolophus aurantiacus*	Tanaka *et al.* 1995[1]
Myliobatididae	*Myliobatis tobijei*	Tanaka *et al.* 1995[1]

[1]Light and/or scanning electron microscopy
[2]Transmission electron microscopy
[3]Light microscopy only

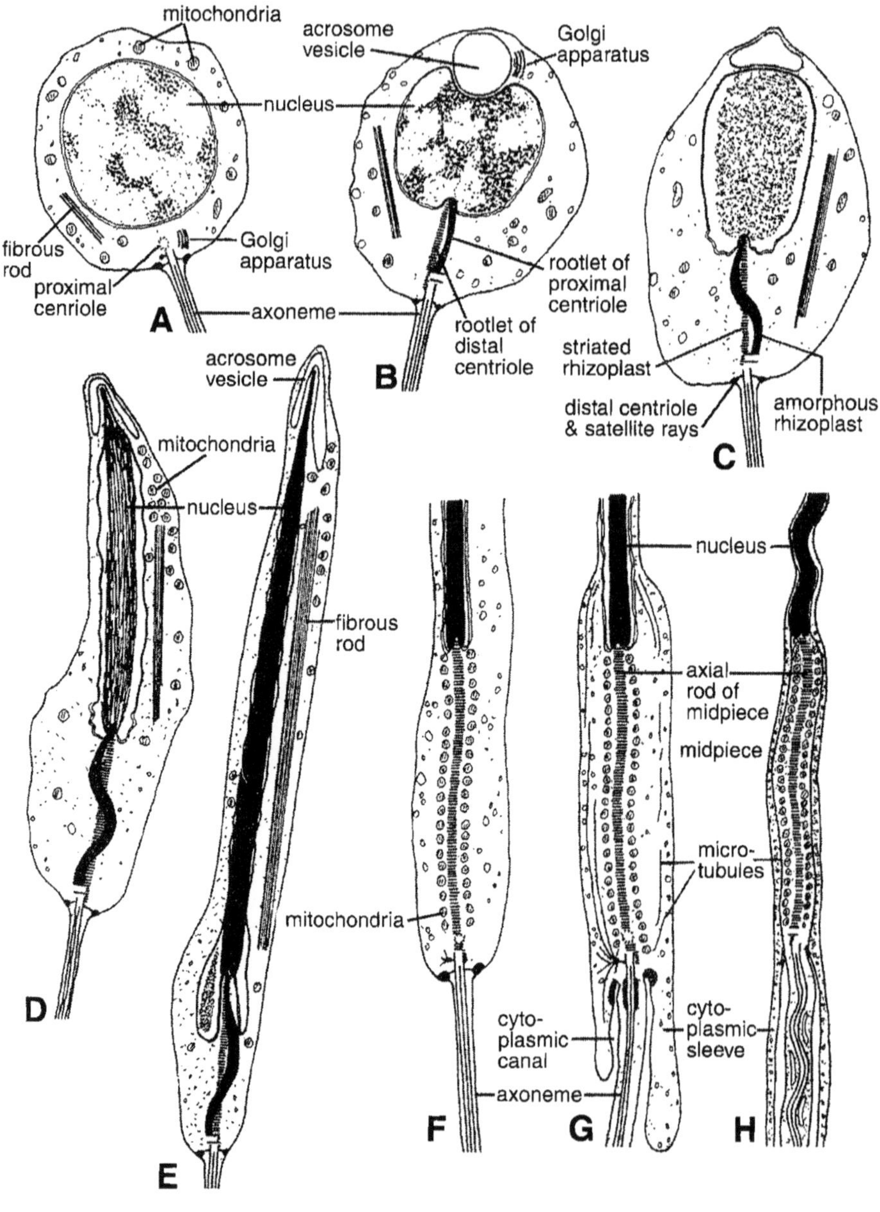

Fig. 7.2 Stages of spermiogenesis in a chondrichthyan, *Rhinobatos cemiculus* (Rhinobatidae). Adapted from Mattei, X. 1970. Spermiogenèse comparée des poissons. pp. 57-69. In B. Baccetti (ed.), *Comparative Spermatology*. Academic Press, New York, NY., Fig. 1.

membrane (Mattei 1970). Stanley (1971a) stated that "the nucleus apparently rotates about 180° taking the acrosome to the pole opposite the flagellum" but this rotation has not been invoked by other authors.

In spermatids of non-chondrichthyans the centrioles usually become displaced towards the nucleus and lodge in the basal nuclear fossa. In chondrichthyans, however, two processes arise from the centrioles, extend progressively in the direction of the nucleus and insert their anterior extremity into the nuclear fossa (Fig. 7.2B). These processes have been termed the amorphous and structured (here termed striated) rhizoplasts [flagellar rootlets] (Boisson *et al.* 1968; Mattei 1970). Stanley (1971a) described and illustrated the striated bundle arising from the distal centriole and shows the amorphous bundle arising from a dense body shortly anterior to the proximal centriole but does not actually claim origin from the latter centriole. Homology of the striated structure with a ciliary rootlet is well argued by Stanley (1971a), chiefly on the basis of the similarity of the striation pattern and length of the repeating period, together with their similar morphological association with the centriole. He refers in this connection to the suggestion of Grobben (1899) that spermatozoa are freed ciliated epithelial cells. A somewhat conflicting account of formation of the midpiece rod is given, for *Centrophorus atromarginatus*, by Tanaka *et al.* (1978). Nevertheless, they show in a micrograph (their Fig. 8) clear transverse striation of the definitive rod. Their transient, unstriated 'X organ' appears to be the amorphous rhizoplast of the other authors. Striated and non-striated longitudinal components of the developing midpiece rod are reported for *Himan tura signifer* (Chatchavalvanich *et al.* 2004) (Fig. 7.4D). Disappearance of the striation at the final stage of development, follows fusion of the two elements (Stanley 1971b). Lengthening of the rod might make this structure more compact and stable (Stanley 1971b; Chatchavalvanich *et al.* 2004). However, some accounts (see 7.2.3) suggest that the striation persists at maturity whereas the amorphous component certainly disappears.

The nucleus elongates concomitantly with the rhizoplasts (and in *Rhinobatos cemiculus*, at least, the fibrous rod). The acrosome caps the anterior extremity of the nucleus as the chromatin filaments individualize. In *Himantura signifer* and *Squalus suckleyi* it has been shown that small vesicles fuse with a larger vesicle to form the acrosome vesicle (Fig. 7.3C). A major finding, demonstrated for both species, has been that the nuclear membrane becomes modified anteriorly *prior to* adhesion of the acrosome vesicle at the site of modification (Fig. 7.3B). Chatchavalvanich *et al.* (2004) point out that there is no evidence of nuclear modification before acrosomal adherence in spermiogenesis of teleost fish (in which the acrosomal vesicle is usually absent), mammals, birds, tuatara or lizards and in both studies it is suggested that this change of the nuclear envelope to specify the site of the acrosomal adherence and the differentiation poles of the spermatid is unique to chondrichthyans. At the site of adhesion of the acrosome, the envelope forms a fibrous layer owing to deposition of electron-dense fibrous material and the fibrous nuclear sheath extends to both sides of acrosomal-nuclear attachment area (Fig. 7.3D), as noted by Stanley (1971a) for *S. suckleyi* the sheath spreads over the entire surface of the nucleus during nuclear elongation but completely disintegrates by maturity. The middle of this attachment region

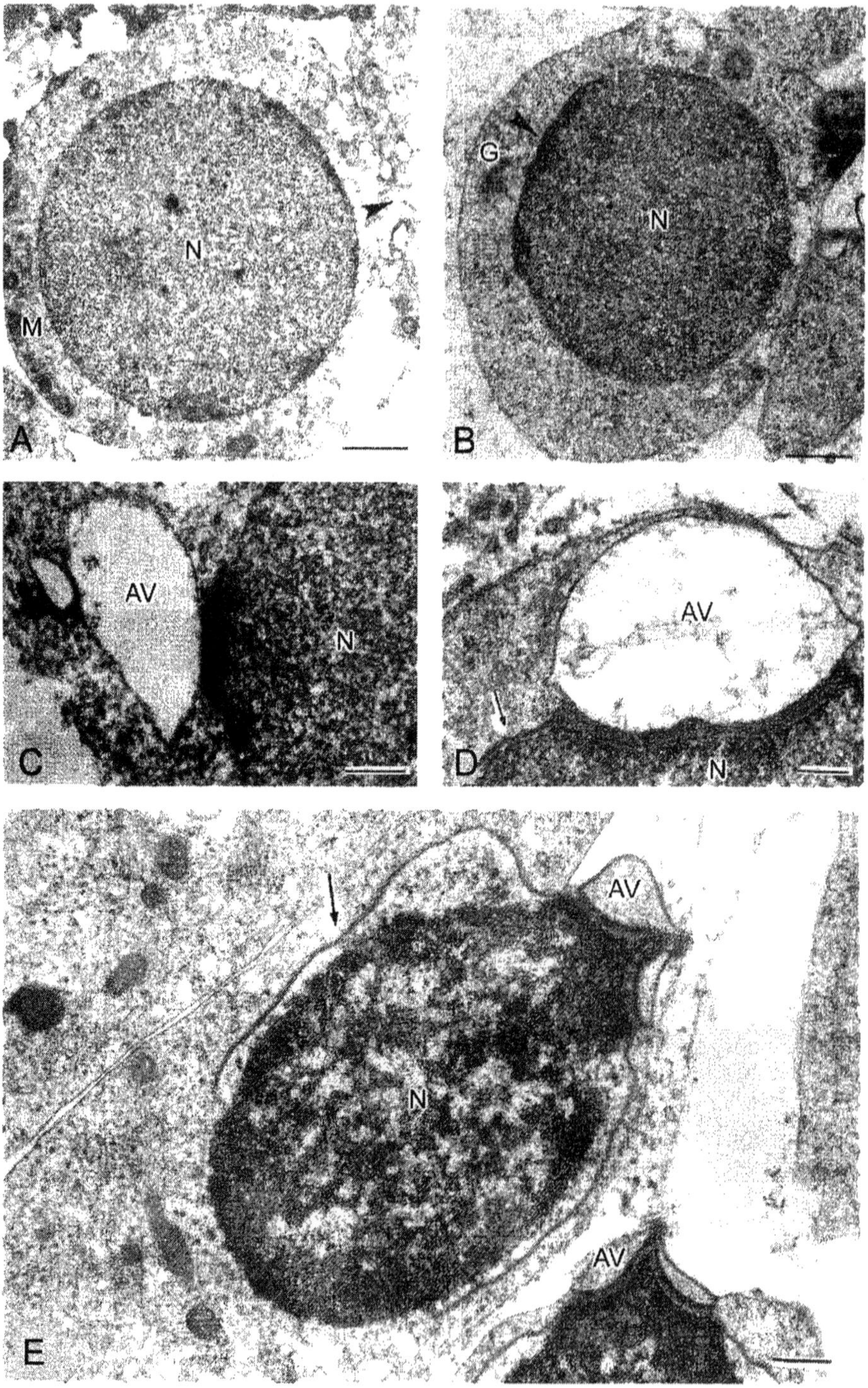

Fig 7.3 *Contd. ...*

grows as a small protrusion into the acrosomal vesicle. Inside the acrosomal vesicle, itself, there is scattered granular material in the early stage which subsequently forms a dense accumulation.

In the following stage, the acrosome begins to spread anteroposteriorly and its protuberant tip contains denser granular material (Figs. 7.3E, 7.4B). The nucleus of the spermatid at this stage has changed from a round to an oval shape. Coarse chromatin granules are arranged in irregular aggregates. The fibrous nuclear sheath is conspicuous and now covers nearly all of the nucleus except for the posterior end (Fig. 7.3E); in *Squalus suckleyi* it shows this extent when the nucleus has reached a length of about 11 μm. During the next stage of development the nucleus undergoes elongation (Fig. 7.4A). The anterior end of nucleus becomes conical or cylindrical. The acrosome acquires a posterior indentation, which fits the projecting anterior end of the nucleus (Fig. 7.4C). Nuclear chromatin exhibits anastomosing configurations of long, slightly twisted nonhomogeneous chromatin fibers (Figs. 7.4D, E). For *Squalus suckleyi*, Stanley (1971a) observes that intranuclear fibrils, initially orientated at random, become aligned longitudinally, then join to form anastomosing longitudinal sheets, which in turn become helically orientated, coomencing posteriorly, before final condensation. In *Squalus suckleyi*, accompanying nuclear spiralization are dissolution of the fibrous sheath, segregation of parachromatin material, and sequential changes in position of a band of nuclear pores (Stanley 1971b).

Chatchavalvanich *et al.* (2004) observe that the irregularly distributed chromatin, which become interconnected as long and slightly twisted fibers appears to be the characteristic of very elongated and often spiral spermatids (Stanley, 1971a). They note that the later spiralization of the nucleus in *Himantura* sperm is earlier foreshadowed by the helical nuclear fibers. They therefore speculate that the nucleoprotein fibers might have specific binding sites for the cross-linkage to make this spiral configuration. Subsequent

---

Fig, 7.3 Spermiogenesis in *Himantura signifer* (Dasyatidae). **A.** Electron micrograph of an early spermatid showing a spherical nucleus with finely granular, rather evenly dispersed chromatin. There is no modification of the nuclear envelope. The arrowhead marks a cytoplasmic bridge between two spermatids. M, mitochondria. Bar = 1 mm. **B.** Section through a spermatid at a slightly later stage than A, showing modified region of the nuclear envelope (arrowhead). A Golgi apparatus is present in the cytoplasm near the nuclear membrane modification. Bar = 1 mm. **C.** The acrosomal vesicle of a spherical spermatid is shown at the time of its first adherence to the modified area of nuclear envelope. Small vesicles (arrowhead) are in the process of fusing with the larger acrosomal vesicle. Bar = 0.5 mm. **D.** Acrosomal-nuclear adhesion site, illustrating the occurrence of fibrous material on either side of the acrosomal membrane. A small projection occurs at the midpoint of the acrosomal-nuclear adhesion area. The fibrous nuclear sheath (arrow) extends over an area of the nucleus peripheral to the acrosomal attachment site. Bar = 0.25 mm. **E.** Oval spermatid in which the acrosome is shown at one end of the nucleus. The fibrous nuclear sheath (arrow) covers nearly all of the nucleus except for the posterior end. Abbreviations: AV, acrosome vesicle; G, Golgi apparatus. Bar = 0.5 mm. After Chatchavalvanich, K., Thongpan, A. and Nakai, M. 2004. Journal of Marine and Freshwater Research (In press), Figs. 1-5.

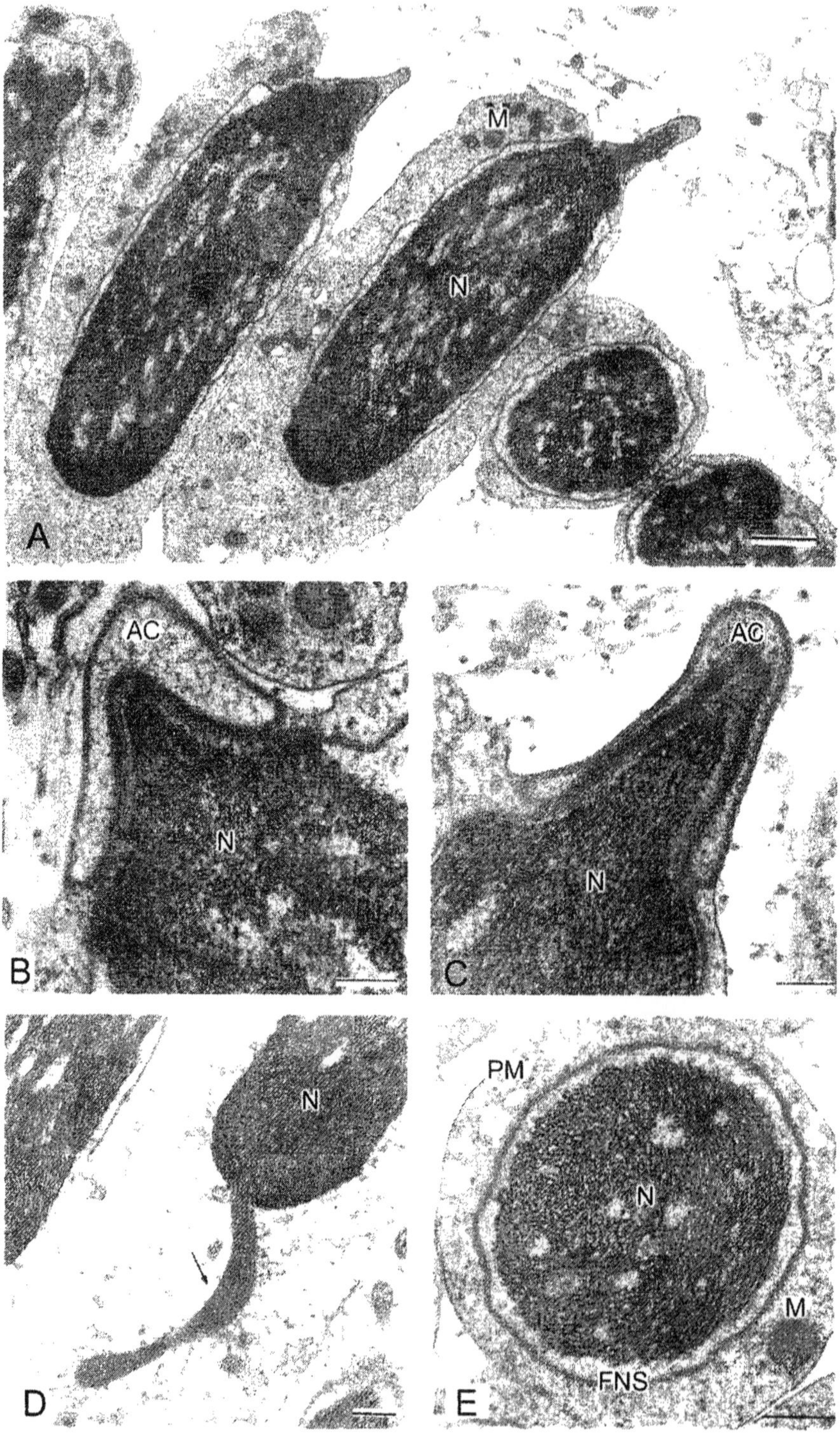

Fig 7.4 *Contd. ...*

condensation of these fibers should then result in the twisting of the sperm head that confirms the close relationship of molecular morphology to the structural formation of the sperm head (Stanley, 1971a). This, they consider, suggests that spiral chromatin fibers might be the cause of spiralization since the fibrous nuclear sheath develops in synchrony with changes in nuclear shape from a round to elongate and becomes disintegrated at the final stage of spermatid development. They observe that the advancement and subsequent regression of the fibrous nuclear sheath in *S. suckleyi* has been linked (Stanley 1971b) with nuclear envelope alterations while dissolution of the fibrous nuclear sheath occurs as the number of microtubules increases. They consider it likely that fibrous nuclear sheath formation is related to nuclear elongation and the alteration of the spermatid shape.

The mitochondria assemble in the anterior region of the spermatid (Fig. 7.2C, D). At the end of nuclear elongation as the contents of the nucleus become condensed, the fibrous rod, seen in *Rhinobatos cemiculus*, reaches its fullest development (Fig. 7.2E). The fibrous rod is then said to disappear (Mattei 1970) as does the amorphous rhizoplast; the mitochondria migrate towards the distal region and group around the striated rhizoplast throughout its length (Figs. 7.2F-H, 7.5A). For *Squalus suckleyi*, Stanley (1971b) states that the spherical mitochondria which assemble around the axial midpiece rod are derived by fragmentation of the elongated mitochondria of the earlier spermatid. The fate of the mitochondria has been examined in detail in *Hydrolagus colliei*. They follow one of three fates: some are retained as functional spermatozoal mitochondria clustered in the midpiece; some are enclosed in large vacuoles and are phagocytosed by Sertoli cells and the remaining mitochondria are segregated to the remnant [sloughed off] cytoplasm as mitochondrial derivatives where they retain their activity as indicated by rhodamine 123 fluorescence studies (Stanley and Lambert 1990).

Some cytoplasmic microtubules associated with a satellite of the distal centriole and develop in the direction of the nucleus to constitute a manchette which comes to surround the mitochondria of the midpiece (Fig. 7.2G). This manchette persists in the mature spermatozoon (Fig. 7.2H), [presumably as the fibrous midpeice sheath] whereas the manchette is transient in sperm of

Fig. 7.4 Spermiogenesis in *Himantura signifer* (Dasyatidae), continued. **A.** In elongated spermatids, the conical shaped acrosome covers the extreme anterior end of the nucleus. Many mitochondria are dispersed throughout the cytoplasm. Bar = 1 mm. **B.** In oval spermatids, a posterior indentation of the acrosome fits over the pointed anterior tip of the nucleus. The acrosome appears as an umbrella-like structure. Bar = 0.25 mm. **C.** Conical acrosome of an elongated spermatid, showing the acrosome covering the anterior end of the elongated nucleus. The chromatin granules are seen as long fibers. Bar = 0.25 mm. **D.** Longitudinal section through elongated spermatid, reveals the axial midpiece rod (arrow) composed of striated and nonstriated portions. These two filamentous bundles are twisted around each other. Bar = 0.5 mm. **E.** A transverse section of the nucleus of elongated spermatid, showing anastomosing patterns of chromatin fibers. Abbreviations: AC, acrosome; FNS, fibrous nuclear sheath; M, mitochondria; PM, plasma membrane. Bar = 0.5 mm. After Chatchavalvanich, K., Thongpan, A. and Nakai, M. 2004. Journal of Marine and Freshwater Research (In press), Figs. 6-10.

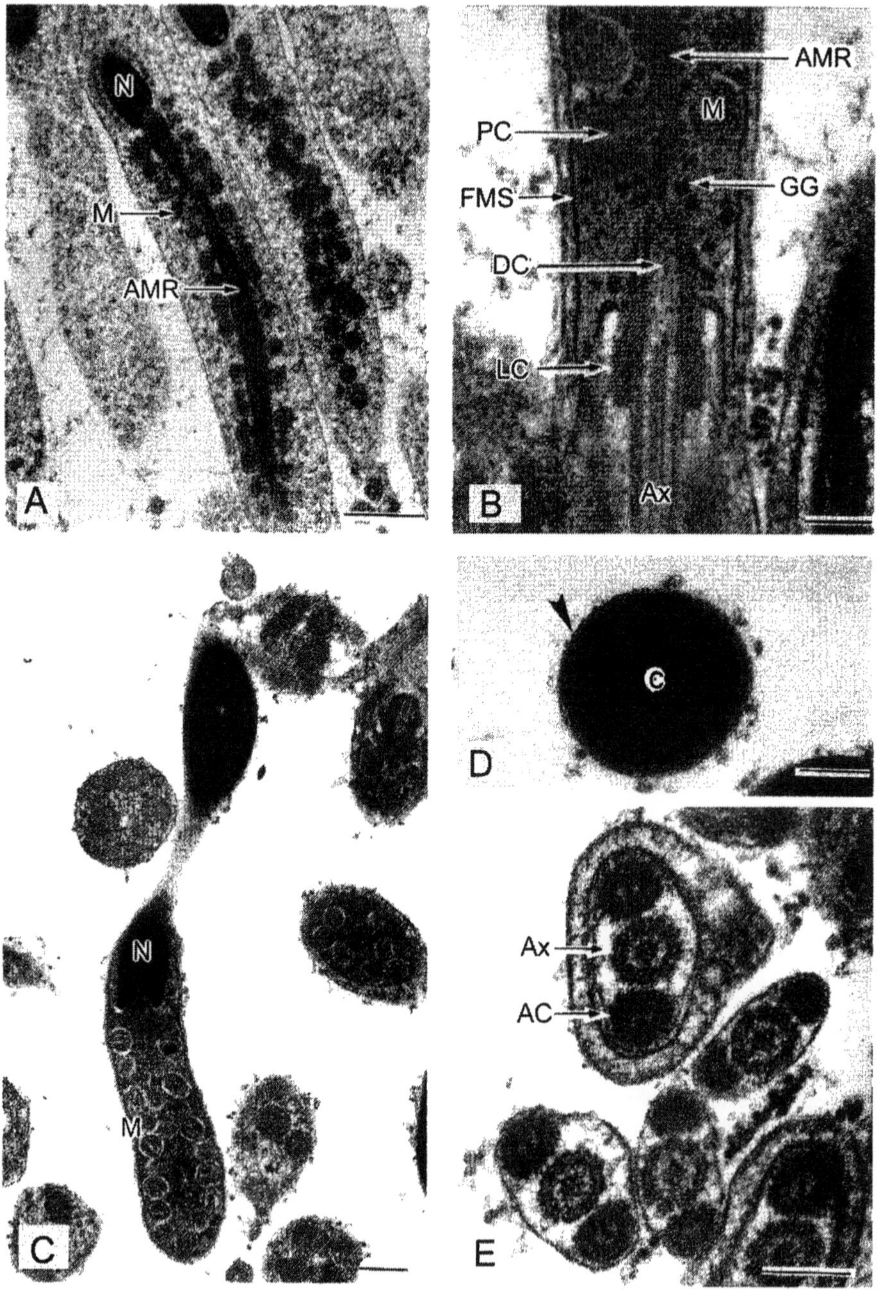

Fig 7.5 *Contd. ...*

most of those animals in which it exists (Mattei 1970). Stanley (1971a) described a complex of structures in association with the flagellar base: an array of nine linear densities extends from the sides of the basal body around the base of the flgaellum; peripheral to the struts lies the annulus; ill-defined spikes of granular material extend radially on its outer surface; and a striated centriolar satellite body extends [uni]laterally from the side of the basal body. Tanaka *et al.* (1978), for *Centrophorus atromarginatus*, also describe development of, and illustrate, an annulus as an electron-dense ring which surrounds the distal centriole and is formed when the axial midpiece rod is completed.

7.2.3 Ultrastructural Characteristics of the Spermatozoa

The ultrastructure of the spermatozoa of a holocephalan, a shark and a ray is illustrated in Fig. 7.6. The 35 species examined, by light and SEM, in Tanaka *et al.* (1995) were described for only six characters pertaining to spermatozoa (Table 7.2): type of sperm aggregate; number of gyres; total length in µm; head length in µm (and ratio to total sperm length); midpiece length in µm (and ratio); flagellum length in µm (and ratio). Chatchavalvanich *et al.* (2004) also described ultrastructural features of the mature spermatid which clearly apply to the spermatozoon. The data of these workers are incorporated in the following review which is drawn from Jamieson (1991) and the three species in the present study (Figs 7.7-7.11).

The total length of the sperm of the 35 species examined by Tanaka *et al.* (1995) ranged from 93 µm in *Galeocerdo cuvier* to 224 µm in *Squatina japonica*. The head in most species was helical, but in the chlamydoselachid *Chlamydoselachus anguineus*, the squalids *Dalatias licha* and three species of the genus *Etmopterus*, and the squatinid *Squatina japonica*, the sperm in the ampulla did not display a clear helical form in the head; the tip of the head was bent like a gaff. The number of gyres of the sperm with the helical form ranged from 3 to 24 (Table 7.2).

Acrosome. The length of the acrosome is 3.5 µm in *Hydrolagus colliei* and 4.7 µm in *Squalus suckleyi*.

In the sharks *Squalus suckleyi* (Stanley 1971b) and *Chiloscyllium puncatum* (present study) (Fig. 7.7A, B), the stingrays *Dasyatis fluviorum* and *D. kuhlii*

Fig. 7.5 Spermiogenesis in *Himantura signifer* (Dasyatidae), continued. **A.** At the junction of the head and midpiece of a late spermatid, there is an indentation of the posterior end of the nucleus into which the axial midpiece rod inserts. Many mitochondria surround the midpiece axis. Bar = 1 mm. **B.** Midpiece-tail junction showing origin of longitudinal columns at the sides of the distal centriole (basal body)-axoneme junction. Bar = 0.25 mm. **C.** Twisted head and midpiece of a spermatozoon in the lumen of spermatocyst. Bar = 0.5 mm. **D.** Transvers sections through the nucleus of a spermatozoon showing that the parachromatin (arrowhead) forms a sheath around the chromatin. Bar = 0.25 mm. **E.** Transverse section through the tail of spermatozoon, showing the 9+2 axoneme and two longitudinal columns. Bar = 0.25 mm. Abbeviations: AMR, axial midpiece rod; Ax, axoneme; DC, distal centriole; FMS, fibrous midpiece sheath; GG, glycogen granules; LC, longitudinal accessory axonemal columns; M, mitochondria; PC, proximal centriole,. After Chatchaval-vanich, K., Thongpan, A. and Nakai, M. 2004. Journal of Marine and Freshwater Research (In press), Figs. 11-15.

Table 7.2 Summary of the measurement and condition of the sperm in 35 chondrichthyan fishes. After Tanaka *et al.* (1995). Memoires du Museum National d'Histoire Naturelle 166: 313-320, Table 2.

Species	*Type of sperm aggregate*	*Number of gyres in µm*	*Total length in µm*	*Head length in µm (ratio)*	*Midpiece length in µm (ratio)*	*Flagellum length (ratio)*
Heterodontus japonicus	1	6	156	41 (26)	14 (9)	102 (65)
Orectolobus japonicus	3	5	136	49 (36)	14 (10)	73 (54)
Carcharias taurus	5	17	110	37 (34)	13 (12)	60 (54)
Alopias pelagicus	4	12	100	37 (37)	9 (9)	54 (54)
Isurus oxyrinchus	4	7	134	60 (45)	12 (9)	62 (46)
Cephaloscyllium umbratile	2	24	186	48 (26)	21 (12)	116 (62)
Galeus eastmani	2	13	197	45 (23)	19 (10)	133 (67)
Galeus nipponensis	2	14	197	44 (22)	19 (10)	134 (68)
Hemitriakis japanica	2	23	120	46 (38)	14 (12)	60 (50)
Mustelus canis	2	11	98	41(42)	7 (7)	50 (51)
Mustelus griseus	2	10	114	47 (41)	9 (8)	58 (51)
Mustelus manazo	2	12	122	38 (31)	10 (8)	74 (61)
Carcharhinus plumbeus	2	7	101	33 (33)	9 (9)	59 (58)
Galeocerdo cuvier	2	17	93	35 (38)	9 (10)	49 (52)
Prionace glauca	3	6	116	34 (30)	9 (7)	73 (63)
Sphyrna lewini	3	12	109	41 (38)	8 (7)	60 (55)
Chlamydoselachus anguineus	1	0	175	46 (26)	19 (11)	110 (63)
Centroscymnus owstoni	1	23	210	93 (44)	20 (10)	97 (46)
Dalatias licha	1	0	130	50 (38)	12 (10)	68 (52)
Deania calcea	1	3	141	56 (40)	13 (9)	72 (51)
Deania historicosa	1	3	162	55 (34)	15 (9)	92 (57)
Etmopterus brachyurus	1	0	193	63 (33)	16 (8)	114 (59)
Etmopterus molleri	1	0	178	58 (33)	16 (8)	104 (59)
Etmopterus pusillus	1	0	172	49 (28)	17 (10)	106 (62)
Squalus brevirostris	1	10	166	36 (22)	11 (7)	119 (71)
Squalus japonicus	1	10	153	35 (23)	9 (6)	109 (71)
Squatina japonica	1	0	224	64 (29)	16 (7)	143 (64)
Rhinobatos schlegelii	1	6	131	46 (35)	9 (7)	76 (58)
Torpedo tokionis	2	5	130	51 (39)	10 (8)	69 (53)
Raja eglanteria	2	6	130	39 (30)	17 (13)	74 (57)
Urolophus aurantiacus	1	7	129	55 (44)	11 (8)	63 (48)
Myliobatis tobijei	1	5	96	35 (36)	10 (11)	51 (53)
Chimaera phantasma	1	3	137	11 (8)	11 (8)	115 (84)
Hydrolagus mitsukurii	1	3	164	11 (7)	18 (11)	135 (82)
Rhinochimaera pacifica	1	3	112	20 (18)	6 (5)	86 (77)

Type of sperm aggregate: 1, Solitary or sperm clumps, 2, Single-layer spermatozeugmata. 3, Compound spermatozeugmata, 4, Spermatophores (rice-grain type), 5, Spermatophores (atypical rod). Values in parentheses indicate a ratio to total length.

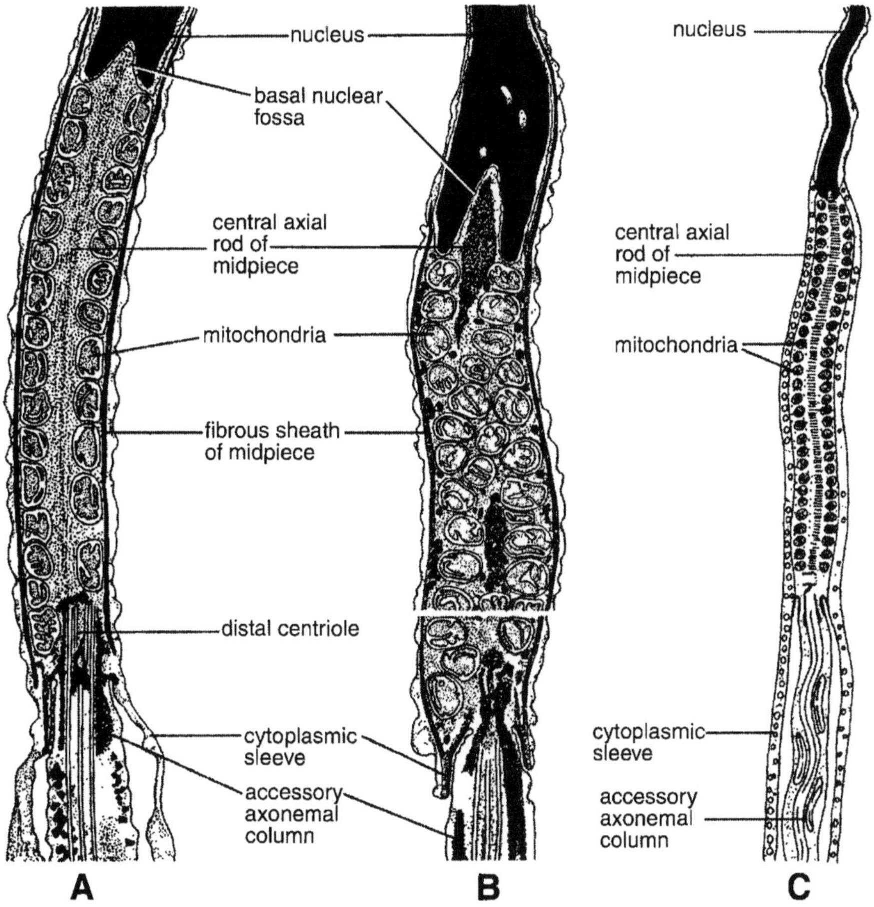

Fig. 7.6 Sperm of three chondrichthyans. From Jamieson 1991. *Fish Evolution and Systematics: Evidence from Spermatozoa.* Cambridge University Press, Cambridge. 319 pp., Fig. 7.5 . **A.** *Hydrolagus colliei.* Longitudinal section through the midpiece with the posterior end of the nucleus at the top and the basal body of the flagellum at the bottom. After a micrograph by Stanley, H.P. 1983. Journal of Ultrastructure Research 83: 184-194, Fig. 11. **B.** *Squalus suckleyi.* Longitudinal section through the nucleus-midpiece junction of a spermatozoon from the ampulla ductus deferentis. The striation pattern on the axial midpiece rod is no longer visible. Mitochondria interspersed with glycogen granules are tightly compressed about the midpiece axis, restricted by the closely apposed fibrous midpiece sheath. After a micrograph by Stanley, H.P. 1971. Journal of Ultrastructure Research 36: 103-118, Fig. 18. **C.** *Rhinobatus cemiculus.* Longitudinal section of spermatozoon from posterior region of nucleus to anterior region of midpiece. The outer, vacuolated sleeve around the midpiece and flagellum is now known to be a transient structure, seen in all three chondrichthean groups. Redrawn from Mattei 1970 After Boisson *et al.*, 1968b. Pp 59-69. In B. Baccetti (ed.), *Comparative Spermatology.* Academic Press, New York, NY, Fig. 1h.

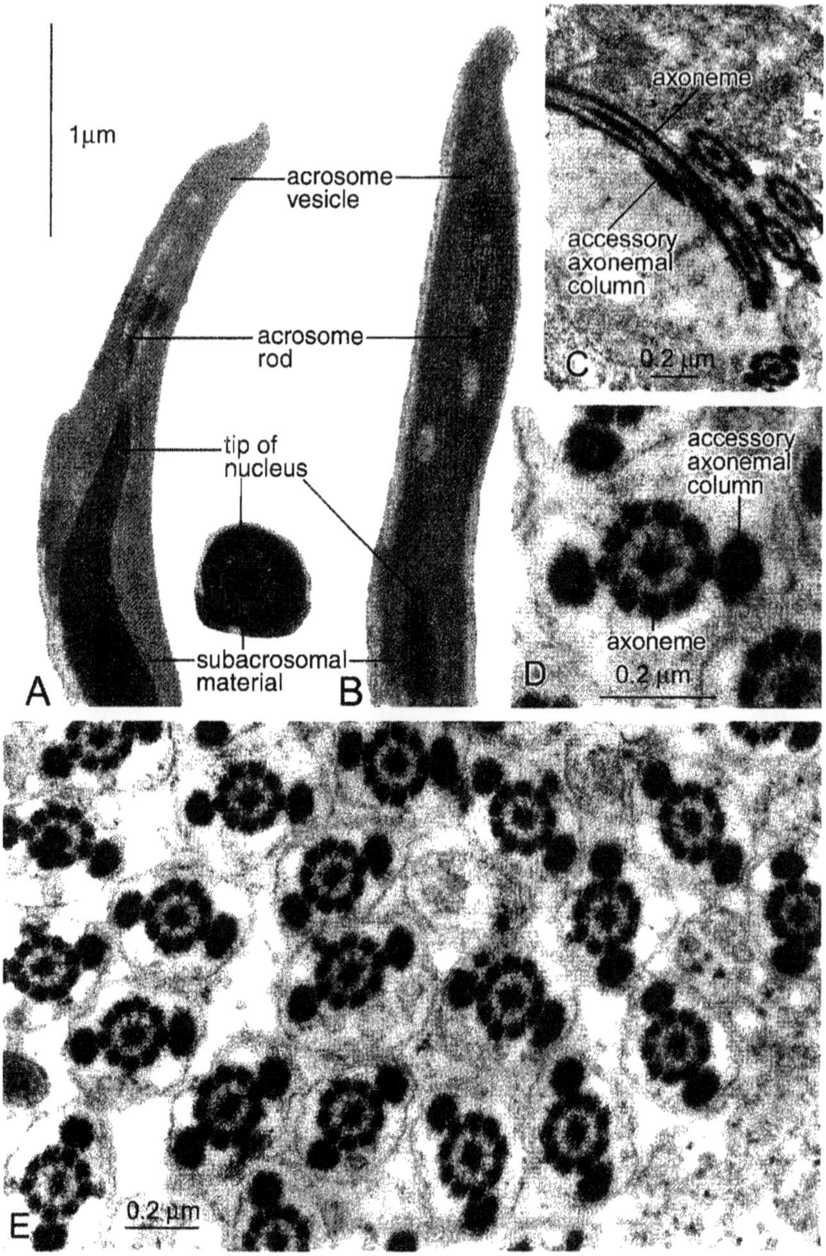

Fig. 7.7 *Chiloscyllium punctatum* (Hemiscyllidae). Transmission electron micrographs (TEM) of testicular spermatozoa. **A, B.** Longitudinal section (LS) of part of the acrosome and tip of nucleus, with accompanying transverse section (TS), showing acrosome rod in a narrow subacrosomal space. **C.** Oblique longitudinal and transverse sections of the axoneme with accessory longituindal columns. **D, E.** TS of same, showing the median flattening, or kidney-shaped outline, characteristic of sharks, of the accessory longitudinal columns. Original.

(present study) (Fig. 7.8H) and the holocephalan *Hydrolagus colliei* (Stanley 1983), the acrosome has a deep posterior indentation which fits over the pointed tip of the nucleus. This contains a subacrosomal rod (putative perforatorium) in *Squalus, Chiloscyllium* and *Dasyatis*. In *H. colliei* it is eccentric and contains a heterogeneous assemblage of medium electron density in

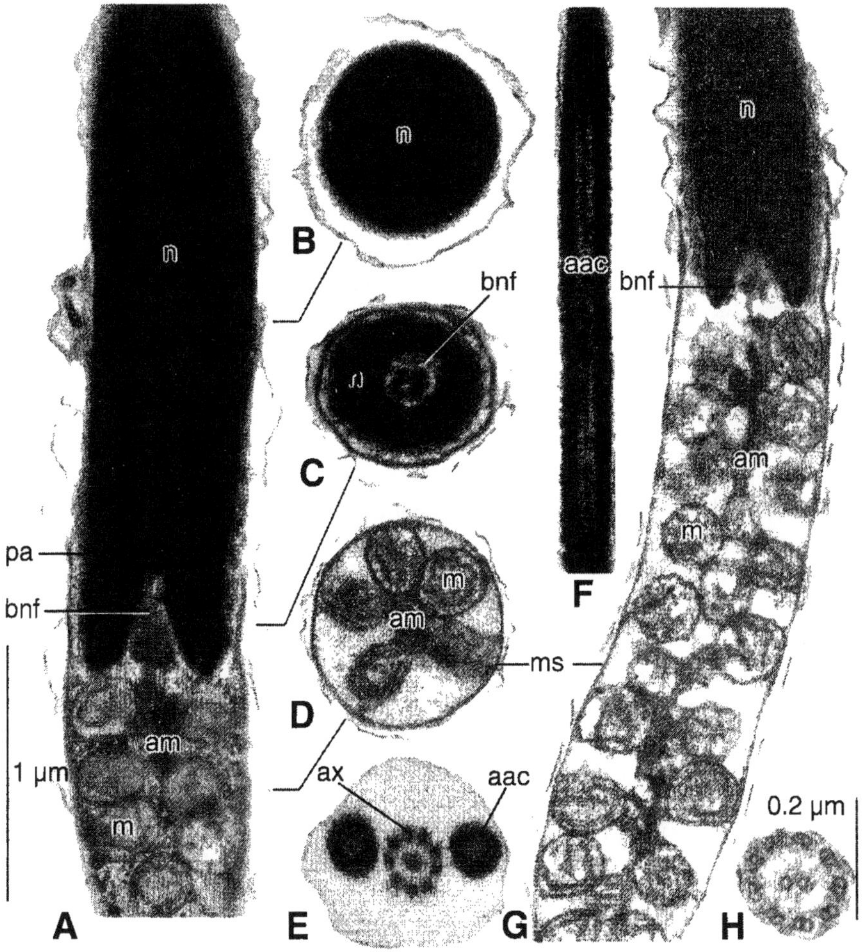

Fig. 7.8 *Dasyatis fluviorum* (Dasyatidae). TEM of testicular spermatozoa. **A and G**. LS base of nucleus and anterior region of midpiece. **B-D.** Corresponding transverse sections. **B.** Nucleus. **C.** Nucleus through the basal fossa. **D.** Midpiece, showing mitochondria around axial midpiece rod. **E.** TS. Through axoneme and accessory columns, showing approximately circular TS of the latter in Rajiformes. **F.** LS straight portion of an acccessory axonemal column. **H.** TS axoneme posterior to axonemal columns. Original. Abbreviations: aac, accessory axonemal column. am, axial rod of midpiece ax, axoneme. bnf, basal nuclear fossa. m, mitochondrion of midpiece. ms, fibrous sheath of midpiece. n, nucleus. pa, parachromatin. Original.

which the denser material consists of several longitudinally orientated strands also tentatively identified as a perforatorium; in addition, a 'ring of dense material' occupies the region between the acrosome and the nuclear tip.

The acrosome in *Squalus suckleyi* is helical and in transverse section shows a unilateral shelflike expansion also with a spiral course, features suggested in at least *Chiloscyllium punctatum* (Fig. 7.7A, B). The posterior end of the acrosome is slanted in these species and *D. kuhlii* (Fig. 7.8H), as in *Hydrolagus colliei*. The external surface of the plasma membrane over the acrosomal region in *S. suckleyi* is covered by a series of low ridges orientated at about 12° to the long axis of the spermatozoon (Stanley 1971b).

In *Hydrolagus colliei* the anterior half of the acrosome is bent at about 30° from the sperm axis and this anterior region bears parallel arrays of extracellular fibrous material (Stanley 1983).

As shown for *Chiloscyllium punctatum* (Fig. 7.7A, B) and *Dasyatis kuhlii* (Fig. 7.11I, H), the acrosome rod occupies a narrow subacrosomal space in which it is surrounded by subacrosomal material. This material extends around the pointed tip of the nucleus where it is weakly demarcated from the parachromatin sheath. The greater posterior extent of the acrosome vesicle on one side of the nucleus compared with the other is clearly seen in these species.

Nucleus. In *Raja, Scyliorhinus* and *Squalus*, as in *Hydrolagus* (see Jamieson 1991), *Himantura signifer* (Chatchavalvanich *et al.* 2004) (Figs. 7.5C, D), and the three species here examined, the dense chromatin of the nucleus is enveloped by the less dense parachromatin sheath, a differentiation which is presumably general in chondrichthyans. The recorded head length (apparently including the acrosome) in elasmobranchs is more than 30 μm, while that in holocephalans is less than 25 μm, both very long compared with the sperm of most animals. The longest chondrychthyan sperm head known attains 93 μm in the squalid *Centroscymnus owstoni*. The standard deviation of the head length in each species is 1.03 to 2.91 μm. The proportion of the head to the total sperm length ranges from 7 to 45 % (Table 7.2) (Tanaka *et al.* 1995). In *Squalus suckleyi* the nucleus has a length of 37 μm. The length of the nucleus is not distinguished from that of the head (nucleus and acrosome) by Tanaka *et al.* (1995).

The nucleus in sperm of sharks and rays has the form of an attenuated cone, narrowly pointed at the anterior end. Posteriorly it is rounded and has a depression, here termed the basal nuclear fossa, which accommodates the tip of the axial midpiece rod. It has highly condensed, electron dense chromatin and, with the known exceptions above, is helical. At maturity the nuclear envelope is closely applied to the nucleus but parachromatin material, gray in appearance, forms an apparently continuous sheath over the outer surface of the chromatin (*Squalus suckleyi*, Stanley 1971b; *Chiloscyllium punctatum*, Fig. 7.7A, B; *Dasyatis fluviorum*, Fig. 7.8A-C; *D. kuhlii*, Fig. 7.10A, 7.11A, C, H). Nuclear pores, believed to be redeveloped in place of those present in the earlier spermatid, are present in the posteriormost part of the nuclear envelope (Stanley 1971b).

In *Hydrolagus colliei* the nucleus is shorter than that of the *Squalus* sperm although still long, at 18 µm. It again forms a loose helix, here of three to four gyres. The moderately dense parachromatin is clearly differentiated from the dense chromatin. Anteriorly, in transverse section of the nucleus, the parachromatin forms a crescent, only partially surrounding the chromatin but posteriorly it forms a thin ring around the whole circumference (Stanley 1983).

A very rare phenomenon for the animal kingdom has been described for *Hydrolagus colliei* in which approximately one tenth of the diploid chromatin content accumulates as a mass on one side of the metaphase plate in the primary spermatocytes and is later eliminated from the spermatid (Stanley *et al.* 1984).

Midpiece. The midpiece seems always to consist of many approximately isodiametric mitochondria. It is much shorter than the head in elasmobranchs (see Table 7.2). This is confirmed for *Himantura signifer* where the lengths of the head and midpiece are respectively 45 µm and 15 µm. In contrast, all examined species of Holocephali (*Chimaera phantasma; Hydrolagus colliei; Hydrolagus mitsukurii; Neoharriotta pinnata; Rhinochimaera pacifica*) have a long midpiece compared to the head. The chondrichthyan midpiece differs from the usual vertebrate condition in interpolation of the mitochondria between the nucleus and the basal body of the axoneme around a rhizoplast-derived rod, rather than distribution around the proximal region of the axoneme.

The width of the midpiece is slightly greater than the head. The midpiece length ranges from 6 µm (in *Rhinochimaera pacifica*) to 21 µm (in *Cephaloscyllium umbratile*). Intraspecific differences are small; the standard deviation of the midpiece length being 0.37 to 1.71 µm (Tanaka *et al.* 1995; see also Jamieson 1991). There are approximately 70 mitochondria in the midpiece of *Hydrolagus colliei* in which they are pressed together to form small polyhedral units with mostly concentric cristae. The midpiece has a central axial rod which fits into an indentation at the posterior end of the nucleus whereas its posterior end inserts on the basal body (distal centriole) of the flagellum. In *Dasyatis fluviorum* the mitochondria have concentric and linear cristae (Fig. 7.8A, D, G) whereas in *D. kuhlii* (Fig. 7.10A, B, 7.11A, B, C) only transverse cristae have been recognized with certainty; whether this is a constant difference between the two species is unknown.

In *Squalus suckleyi* the central rod, like the entire midpiece, is helical in register with the nucleus. Its dual origin, as a transient amorphous rhizoplast and a persistent cross striated rhizoplast, is also seen in other chondrichthyans, (see Spermiogenesis, 7.2.2). Presence in the axial core of the midpiece in *Hydrolagus* of an inner core and outer shell may also relate to the dual origin of the structure (Stanley 1983). For *Raja clavata* the midpiece rod is accurately described by Nicander (1968) as "a pile of dense discs" (see his Fig. 17), here considered to represent the striated component which, alone, persists.

The midpiece examined in *Dasyatis fluviorum* (Fig. 7.8A, D, G) and *D. kuhlii* (Figs. 7.10A, B, 7.11A-C) conforms in general structure to the above but the axial rod, though electron dense, is less well defined and might better be described as a rachis of dense material. Its apparently unstriated nature probably represents a final stage in modification, as observed by Stanley (1971a) for *Squalus suckleyi*.

According to (Stanley 1971b), for *Squalus suckleyi*, the mitochondria are formed by fragmentation of elongate mitochondria but in *Rhinobatos cemiculus* they are discrete throughout spermatogenesis (Boissin *et al.* 1968b; Mattei 1970).

In *Squalus suckleyi* the mitochondria are interspersed with clusters of 200 Å putative glycogen granules, this region giving a PAS positive reaction which is mostly absent after diastase treatment. Putative glycogen granules are visible in the midpieces of *Dasyatis fluviorum* and *D. kuhlii*. The midpiece is surrounded by a fibrous sheath which overlaps the posterior end of the nucleus by 2 or 3 µm. Posteriorly the fibrous sheath attaches to the granular layer of the transient cytoplasmic sleeve (Stanley 1971b). A fibrous midpiece sheath, consisting of spirally orientated filaments also occurs in *Hydrolagus colliei* (see Stanley 1983), is reported for *Rhinobatos cemiculus* by Boissin *et al.* (1968b) and for *Centrophorus atromarginatus* by Tanaka *et al.* (1978) and is here confirmed for *D. fluviorum* and *D. kuhlii*. In *Rhinobatos*, at least, it consists of microtubules extending anteriorly from a satellite of the distal centriole.

Cytoplasmic sleeve. The cytoplasm of the midpiece is reflected as a sleeve around the proximal portion of the tail. The sleeve is about 6.5 µm long in *Squalus suckleyi* and 10 to 16 µm long in *Hydrolagus colliei*.

The contents of the sleeve in *Hydrolagus colliei* consist of smooth membranes, double membrane vesicles, coated invaginated vesicles, and masses of dense material up to 0.2 µm in diameter (Stanley 1983; Stanley and Lambert 1990). As noted above, Rhodamine 123-fluorescence confirms that the vesicles are modified but still functional mitochondria (Stanley and Lambert 1990).

In *Hydrolagus colliei*, on the side of the flagellum nearest doublets 4 through 8, broad, shelflike connections extend, in a series, from the outer surface of the plasma membrane at the flagellar base to the medial surface of the plasma membrane of the remnant sleeve. In some longitudinal sections the series of 9 or 10 transverse bands appears ladder-like but resemblance to a septate junction noted by Stanley (1983) must be regarded as superficial.

Mature sperm in the ampulla ductus deferentis have mostly lost this cytoplasmic remnant; the sleeve is pinched off at the site of junction of the midpiece sheath fibers and the granular layer (*Squalus suckleyi*, Stanley 1971b) and eventually slips off the terminal portion of the sperm tail (*Hydrolagus colliei*, Stanley 1983). Even detached sleeves in the lower male reproductive tract (ampulla) show rhodamine 123 fluorescence (Stanley and Lambert 1990).

The cytoplasmic sleeve is present in *Dasyatis fluviorum* (Fig. 7.9), *D. kuhlii* (Figs. 7.10B, 7.11B, E, F) and *Himantura signifer* (Fig. 7.5B) but its transient nature is demonstrated by its detachment (Fig. 7.11B) and its absence from the mature midpiece (Figs. 7.8A, D, G, 7.10A). However, its basal reflexed portion surrounding a short cytoplasmic canal persists. The contents of the sleeves are consistent with *Hydrolagus colliei*: large vesicles with double membranes, looped double membranes presumably derived from these, and small coated vesicles are present. Some of the large vesicles have shelf-like cristae consistent with their derivation from mitochondria. They may exhibit dense masses in their interiors (Figs. 7.9. 7.11F), as demonstrated by Stanley and Lambert (1990) for *H. colliei*.

Centrioles. The distal centriole (basal body) is in the longitudinal axis of the sperm in *Squalus suckleyi* and *Dasyatis kuhlii* (Fig. 7.10B) or is at 15° to this in *Hydrolagus colliei*. It has the usual nine triplets. In *Squalus*, as in *Hydrolagus colliei* and *Rhinobatos cemiculus* (Fig. 7.2F), from the basal body arise satellite rays which appear to attach to the plasma membrane at the point of its reflection from the flagellum to the inner surface of the cytoplasmic sleeve. Their persistence to maturity is questionable but it is possible that they are contribute to the annulus (see Spermiogenesis, 7.2.2).

In *Squalus suckleyi*, just posterior to the reflection, the basal body is attached to the plasma membrane by nine Y-links (independent of the satellite rays). Anteriorly the 9 + 2 axoneme is surrounded by a dense ring, here considered to be distinct from the annulus, which subdivides to form the two longitudinal accessory columns. A proximal centriole lies anteriorly and almost perpendicularly to the distal centriole. From a micrograph it is seen

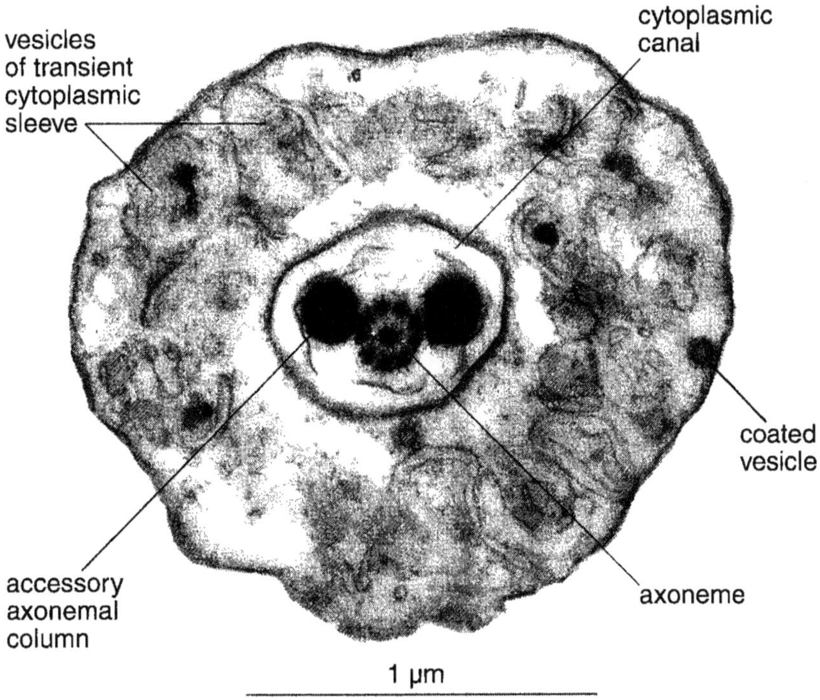

Fig. 7.9 *Dasyatis fluviorum*. TEM of testicular spermatozoa through transient cytoplasmic sleeve, surrounding the axoneme and accessory axonemal columns. The sleeve contains small coated vesicles, large vesicles with double membranes (putative modified mitochondria), and separate membranes. Original.

that the two centrioles intrude slightly into the posterior region of the mitochondrial sheath, the proximal centriole being entirely within it (Stanley 1971b). In *Rhinobatos* a proximal centriole is present in the spermatid (Mattei 1970) and persists in the mature spermatozoon (Fig. 7.2), as apparently in *Dasyatis* (present study, Fig. 7.10). Both centrioles, of which the proximal is transverse to the sperm axis, are clearly shown for the spermatozoon of *Himantura signifer* (Chatchavalvanich *et al.* 2004) (Fig. 7.5B).

No proximal centriole has been observed in the Holocephali (*Chimaera phantasma, Hydrolagus colliei* and *Neoharriotta pinnata*). In *H. colliei* dense material adheres to the periphery of the basal body and almost fills its interior. The proximal ends of the two central singlets of the axoneme are embedded in a plate of dense material [basal plate] at the basal body-flagellar junction.

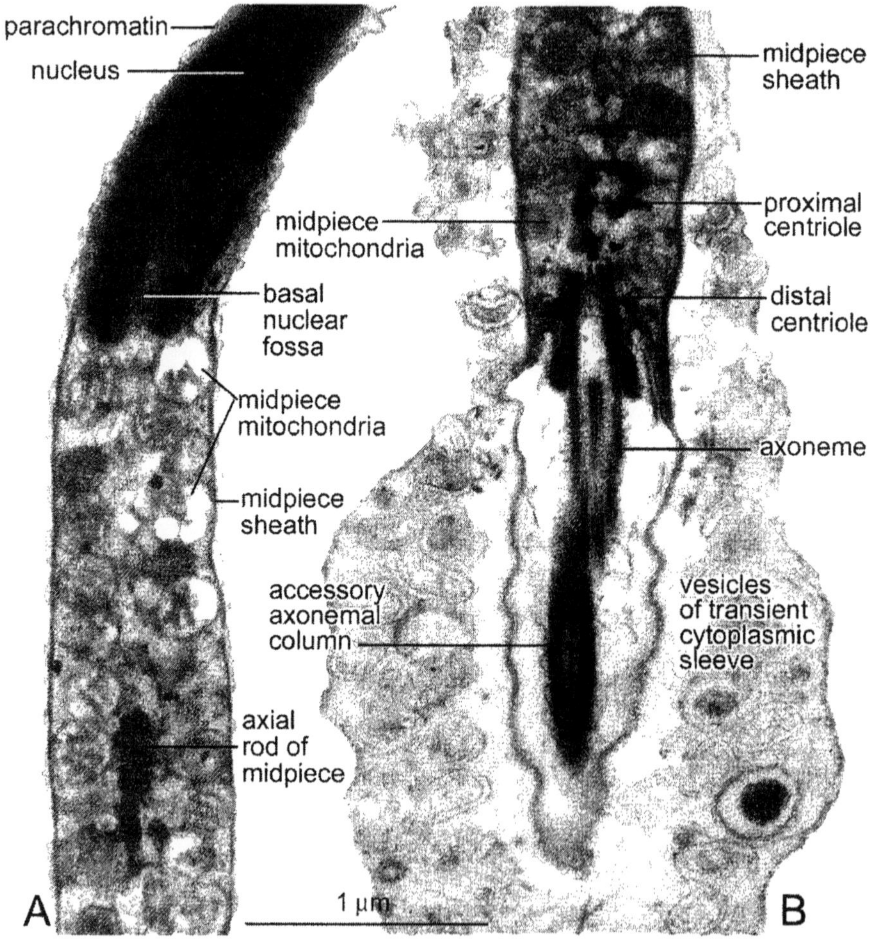

Fig. 7.10 *Dasyatis kuhlii.* TEM of testicular spermatozoa. **A.** LS base of nucleus and anterior region of midpiece. Note that the cytoplasmic sleeve has been lost. **B.** LS base of midpiece centriolar complex and anterior region of axonemal, showing intact cytoplasmic sleeve around nucleus and midpiece (contrast 7.11B). Original.

A ring of fibrous material attached to the anterior end of the basal body extends posteriad as a truncated cone to line a reflected portion of the plasma membrane which is continuous with the inner remnant sleeve membrane. This is considered by Stanley (1983) to be probably homologous with the annulus of mammalian sperm. Nucleus, midpiece and tail elements are helical in accordance with the observed rotation of the sperm about its long axis which allows anteriorward or posteriorward locomotion. Lateral undulation is minimal (Stanley 1964, 1971b).

Flagellum. The axoneme lies along the central axis of the spermatozoon in *Squalus* but is abaxial in *Hydrolagus* (Stanley 1965). The flagellum length ranges from 49 μm in the carcharhinid *Galeocerdo cuvier* to 143 μm in the squatinid *Squatina japonica*, with a standard deviation of 1.42 to 3.45 μm. In the lamnid *Isurus oxyrinchus*, the squalid *Centroscymnus owstoni* and the urolophid *Urolophus aurantiacus*, the flagellum is almost the same length as the head. The proportion of the flagellum in Chimaeriformes is more than 75% (Tanaka *et al.* 1995).

In *Squalus suckleyi* (Stanley 1971b), as in *Chiloscyllium punctatus* (Fig. 7.7C), the tail, originating from the posterior end of the midpiece, contains the usual 9 + 2 axoneme but this is accompanied by two longitudinal columns (accessory axonemal columns) external to doublets 3 and 8, as also noted by Chatchavalvanich *et al.* (2004) for *Himantura signifer* (Fig. 7.5E). The two columns (discussed below) thus lie in a line passing approximately through the two central singlets. The axoneme as a unit is straight but the axonemal doublets and the two columns are helical, the columns describing a double helix. However, in at least *Dasyatis fluviorum* (Fig. 7.8F) long sections of an accessory column may be straight. In *Hydrolagus* the entire flagellum is helical (Stanley 1983).

Axonemal columns. The spermatozoa of the sharks *Squalus suckleyi* (Stanley 1971b) (Fig. 7.13), *Chlamydoselachus anguineus*, *Centroscymnus owstoni*, *Prionace glauca* (Hara and Tanaka 1986) and *Chiloscyllium punctatum* (present study) (Fig. 7.7C-E) have two longitudinal columns in the tail, oval in cross section with a flattened interior surface, sometimes appearing kidney-shaped owing to the fact that the median (adaxonemal) wall is often thinner than the lateral wall and may be indented. However, the flattening in *S. suckleyi* is more pronounced than that in *C. punctatum* in which the transverse sections of the columns differ little from those of rays, though narrower. In rays, exemplified by *Dasyatis kuhlii* (Fig. 7.11F, G), *D. garouensis* and *D. fluviorum* (Figs. 7.8E, 7.9), the two columns are rounded in cross section or if, in individual flagella, they are slightly flattened medially, the flattening is less than in *S. suckleyi* and the median wall is not usually thinner. In *Himantura signifer* the columns are described as round in cross section (Chatchavalvanich *et al.* 2004) but again may individually be flattened or concave on the adaxonemal surface (Fig. 7.5E). In holocephalan sperm the rods are rounded in cross section.

In *Hydrolagus colliei* (Stanley 1983) (Fig. 7.12), and *Neoharriotta pinnata* (Mattei 1988) the column close to doublet 3 is similar in diameter and length

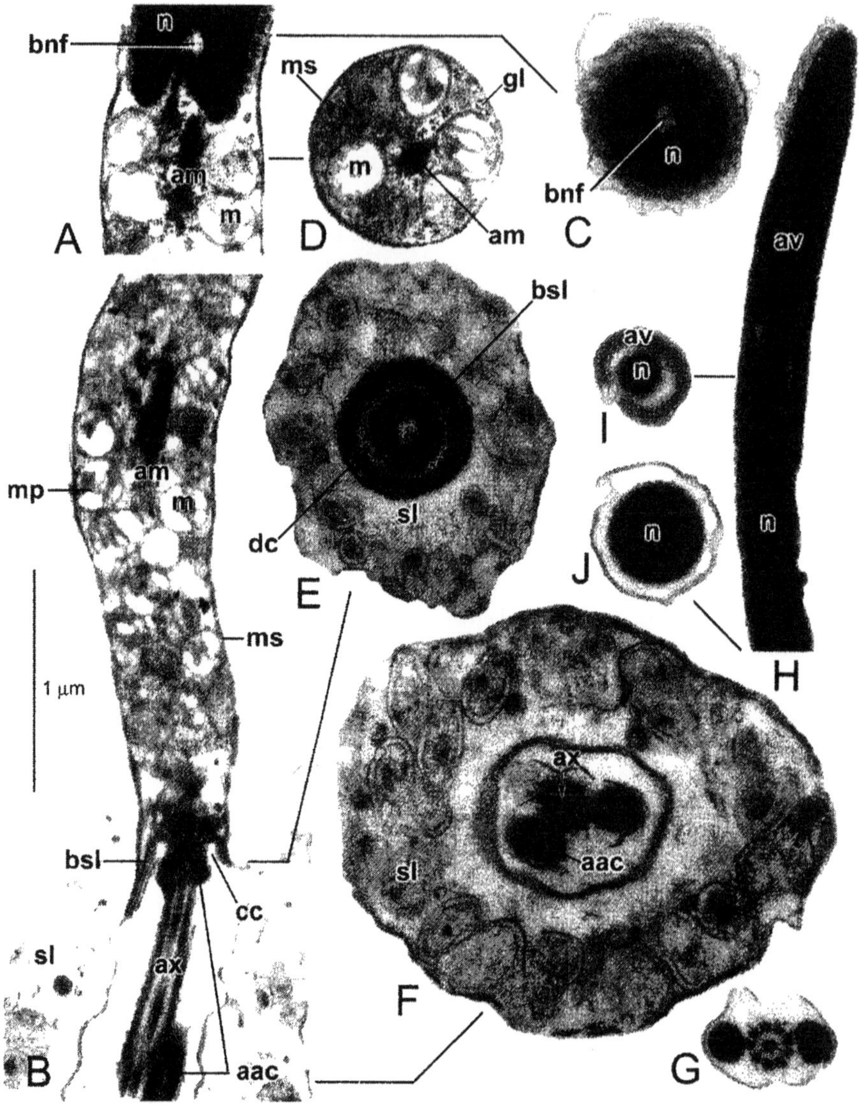

Fig. 7.11 *Dasyatis kuhlii*. TEM of testicular spermatozoa, continued. **A.** LS base of nucleus and anterior region of midpiece. **B.** LS base of midpiece centriolar complex and anterior region of axonemal, showing shedding of cytoplasmic sleeve. **C.** TS nucleus through basal nuclear fossa. **D-F.** corresponding transvers sections. **D.** Midpiece, **E.** Centriolar region with investing cytoplasmic sleeve. **F.** Transient cytoplasmic sleeve, surrounding the axoneme and accessory axonemal columns. **G.** Axoneme and accessory columns posterior to cytoplasmic sleeve. **H.** LS Tip of nucleus and part of overlying acrosome, showing oblique base of latter. All to same scale. Abbreviations: aac, accessory axonemal column. am, axial rod of midpiece. av, acrosome vesicle. ax, axoneme. bnf, basal nuclear fossa. bs, persistent base of cytoplasmic sleeve. cc, cytoplasmic canal. dc, distal centriole. gl, putative glycogen. m, mitochondrion of midpiece. ms, fibrous sheath covering midpiece. n, nucleus, sl, cytoplasmic sleeve.

to those of elasmobranchs; the other, close to doublet 8, is smaller in diameter than its partner and very short while in *Chimaera phantasma* there is a single rod (Hara and Tanaka 1986), one having apparently been lost. In *Hydrolagus colliei*, at least, glycogen is apposed to the axoneme throughout its length. The phylogenetic significance of these arrangements, within the Chondrichthyes, was considered by Jamieson (1991) (Fig. 7.15) (see also Mattei 1991) but the difference between sharks and rays is now, as shown above, less clear than suggested from previous studies. Similarly located accessory axonemal elements occur in the Anura, Actinistia and Dipnoi (see Jamieson 1999; Scheltinga and Jamieson 2003).

The accessory axonemal columns have been described in some detail for *Hydrolagus colliei* by Stanley (1983) (Fig. 7.6A, 7.12). The smaller column, at doublet 8, is approximately circular in cross section, 0.09 μm in diameter and about 1.5 μm long. The larger column, at doublet 3, is about 0.22 x 0.27 μm (erroneously given as 100 times larger) in diameter anteriorly and slightly ovoid in cross section. Both columns are similar in structure. A cylinder of dense material surrounds a less dense core, the latter consisting of a ribbon attached to the inner surface of the cylinder by its medial edge and containing material of moderate electron density on either side. The outer surface of the cylinder has a thin layer adherent to but distinct from it. The larger column appears attached, at least intermittently, to doublet 3 by a short bridge. This column and the axoneme form a double helix with a short pitch: a 180° turn is made in every 4 μm of the flagellar length. It extends to within 2.0 μm of the flagellar tip, gradually tapering to a terminal diameter about equal to that of the smaller column. Glycogen granules are present along the entire

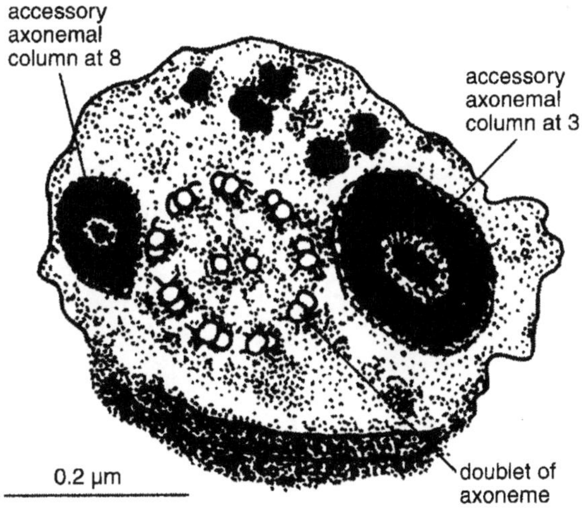

Fig. 7.12 *Hydrolagus colliei* (Chimaeridae). TS of the axoneme, showing the two accessory longitudinal columns. The column at doublet 8 is reduced and posteriorly disappears. From Jamieson 1991. *Fish Evolution and Systematics: Evidence from Spermatozoa*. Cambridge University Press, Cambridge. 319 p., After a micrograph by Stanley, H. P. 1983. Journal of Ultrastructure Research 83: 184-194, Fig. 17.

length of the tail, excepting the terminal 1 μm, lying in the cytoplasm at right angles to the line through the central singlets.

In *Squalus suckleyi* (Stanley 1971b) the two columns are similar to each other (Figs 7.13, 7.15) and extend nearly to the posterior tip of the flagellum from their origin from the ring which surrounds the axoneme anterior to the commencement of the doublets. Each column is composed of a cylinder (elliptical but axially flattened) of dense material with lateral ridges of less dense material. Each cylinder has a lighter staining center in which a longitudinal membrane extends from the side nearest the axoneme into the interior; attached to each side of this membrane are rows of granules, about 100 Å in diameter (Stanley 1971b), an arrangement similar to that described above for *Hydrolagus*.

The earliest step in development of the longitudinal columns involves the appearance of a low ridge extending radially from the centrifugal side of the A microtubule of each of doublets 3 and 8. A similar process occurs in Amphibia exemplified by the urodele *Triturus vulgaris* (Stanley 1970).

7.2.4 Seasonal Changes in Testicular Structure

Seasonal changes in testicular structure and spermatogenesis in several shark species have been reviewed by Parsons and Grier (1999) with particular reference to the bonnethead shark, *Sphyrna tiburo*. They separate shark species into four categories based on the relationship between testicular gonadosomatic index (GSI) and mating season: 1) species that have a defined

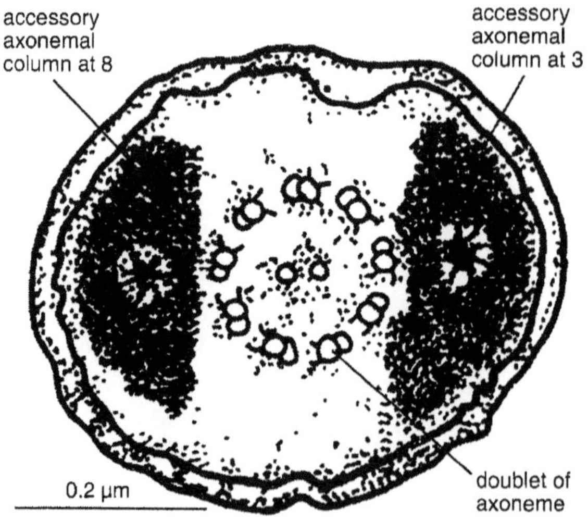

Fig. 7.13 *Squalus suckleyi* (Squalidae). Transverse section of the flagellum through the two accessory axonemal columns. From Jamieson 1991. *Fish Evolution and Systematics: Evidence from Spermatozoa*. Cambridge University Press, Cambridge. 319 p. After a micrograph by Stanley, H. P. 1971. Journal of Ultrastructure Research 36, 103-118, Fig. 7f.

seasonal cycle in GSI and a defined mating season: *Rhizoprionodon terraenovae, Sphyrna tiburo, Mustelus manazo, M. griseus, Carcharinus limbatus, C. brevisrostris, C. acronotus, Negaprion brevirostris, Squalus acanthias;* 2) species that have a defined seasonal cycle in GSI and that mate year-round: no species known; 3) species that show no seasonal change in GSI but that have a defined mating season: *Prionacea glauca;* and 4) species that show no seasonal change in GSI but that mate year-round: *Scyliorhinus caniculus, Carcharinus leucas, Alopias vulpinus, A. superciliosus, Galeorhinus australis.* The testis of *S. tiburo* differs from the few species that have been examined in that 1) not all spermatogenic stages are present at all times of the year, 2) an annual zone of degeneration that progresses through the testis is not present and 3) there appears to be complete testicular regression and recrudescence.

7.3 SPERMATOCYSTS, SPERMATOZEUGMATA AND SPERMATOPHORES

In Chondrichthyes, the process of spermatogenesis produces a spermatocyst composed of Sertoli cells and their cohort of associated spermatozoa linearly arrayed and laterally embedded in the apical end of the Sertoli cell (Hamlett *et al.* 1999, 2002a, b; Reardon *et al.* 2002; Chapter 6 of this volume); 64 spermatids are embedded in a single Sertoli cell (Stanley 1962). The extratesticular ducts consist of paired epididymis, ductus deferens, isthmus, and seminal vesicles. In transit through the ducts, spermatozoa undergo modification by secretions of the extratesticular ducts and associated glands, i.e., the Leydig gland (see Chapter 4 of this volume). The anterior portion of the mesonephros is specialized as the Leydig gland that connects to both the epididymis and ductus deferens and elaborates seminal fluid and matrix that contribute to the spermatophore or spermatozeugmata, depending on the species. The Leydig gland has a simple columnar epithelium with secretory and ciliated cells. Secretory cells have periodic acid-Schiff positive (PAS+) apical secretory granules. In the holocephalan elephant fish, *Callorhynchus milii*, Sertoli cell fragments and spermatozoa enter the first major extratesticular duct, the epididymis. In the epididymis, spermatozoa soon begin to laterally associate so that they are aligned head-to-head. The epididymis is a highly convoluted tubule with a narrow lumen and an epithelium consisting of few ciliated and relatively more secretory cells. Secretory activities of the Leydig gland and the epididymis contribute to the nascent spermatophores, which begin as gel-like aggregations of secretory product in which sperm are embedded. Fully formed spermatophores occur in the ductus. The simple columnar epithelium has both ciliated and secretory cells. The spermatophore is regionalized into a PAS+ and Alcian-blue-positive (AB+) cortex and a distinctively PAS+, and less AB+ medulla. Laterally aligned sperm occupy the medulla and are surrounded by a clear zone separate from the spermatophore matrix. The seminal vesicles are characterized by spiral partitions of the epithelium that project into the lumen, much like a spiral staircase. Each partition is staggered with respect to adjacent partitions and the aperture is eccentric. The generally nonsecretory epithelium of the seminal vesicle is simple columnar with both microvillar and ciliated cells (Hamlett *et al.* 2002).

In *Hydrolagus colliei*, sperm are clustered into spermatozeugmata in the ampulla ductus deferentis. All of the sperm in a spermatozeugma, each approximately 143 μm long, are orientated in the same direction and they are adherent chiefly by the, albeit transient, cytoplasmic sleeves at the anterior end of the flagella (Stanley 1983).

Tanaka *et al.* (1995) recognize five types of sperm aggregate: 1, solitary or sperm clumps, 2, single-layer spermatozeugmata, 3, compound spermatozeugmata, 4, spermatophores of rice-grain type, 5, spermatophores of atypical rod type. The occurrence of these in 35 species of Chondrichthyes is listed in Table 7.2 (see also Pratt and Tanaka 1994).

7.4 MOTILITY

In Holocephali and Elasmobranchs, as exemplified by *Hydrolagus* and *Squalus*, the swimming sperm rotate about their long axis with little lateral bending of their tails. Despite the differences in their tail structure, little difference in their motility patterns is observed. The spiral conformation of the head and tail structures is clearly related to the rotational movement of the sperm and the fact that progress is restricted in fluids of low viscocity is consistent with the hypothesis that the propellar or screw principle is involved. The rapid rotation of the corkscrew-shaped sperm head presumably contributes to the progressive movements of sperm in essentially straight lines through fluids of higher viscocity [as in the female tract] (Stanley 1983). Tanaka *et al.* (1978) state that the midpiece rod in *Centrophorus atromarginatus* is capable of expanding, contracting and bending and, together with the flagellum, contributes largely to sperm movement but Stanley (1983) was unable to corroborate this in any of the material (*Squalus, Scyliorhinus, Raja* and *Hydrolagus*) that he had studied.

7.5 PHYLOGENETIC CONSIDERATIONS

7.5.1 Phenetic, UPGMA Clustering

Tanaka *et al.* (1995) made a phenetic (UPGMA) analysis of the 35 species on the basis of the six characters listed in Table 7.2. As this analysis was not cladistic and potentially allowed groupings on symplesiomorphies it is not reviewed in detail here. However, the coefficient of similarity was high, and this is here considered to indicate low homoplasy and considerable phylogenetic utility. Therefore the principal groupings obtained will be discussed.

Four chief groups were obtained.

Group I. This group included four orders, five families; Scyliorhinidae of the Carcharhiniformes, Heterodontiformes, two families of

Chimaeriformes, and Squaliformes. Only the Scyliorhinidae was distant from the other families of Carcharhiniformes. The sperm of three species of Scyliorhinidae were much longer than those of the other species of Carcharhiniformes. *Cephaloscyllium umbratile*, in which the sperm had 24 gyres, differed from the other species in the group. The proportion of the head and flagellum in *Squalus* spp. was smaller and larger than that in the other genera of Squaliformes, respectively.

Group II. This group consisted of five orders, with five families; Squatiniformes, Squaliformes, Hexanchiformes, Urolophidae of Myliobatiformes and Rhinobatiformes. Most of the species of Squalidae belonged to the group. Urolophidae was separated from Myliobatididae of the same order. The group was divided into two subgroups reflecting the number of gyres. The sperm of *Centroscymnus owstoni*, alone in the Squalidae, had a large number of gyres.

Group III. This group contained only one species, *Raja eglanteria*. The proportion of the midpiece in this species was large compared to the other species, while the sperm had a small number of gyres.

Group IV. This group included five orders and nine families; Torpediniformes, Orectolobiformes, three families of Lamniformes, three families of Carcharhiniformes, and the Myliobatididae in the Myliobatiformes. All species of Carcharhiniformes except Scyliorhinidae belonged to this group. The group was divided into two subgroups on the basis of total length of the spermatozoon. Only *Isurus oxyrinchus* was separated from the other species of Lamniformes. In the Triakidae, *Hemitriakis japanica* was separated from the other species. The sperm of *H. japanica* had a larger number of gyres than other species of the same family.

The families of Group I, except two species of Squalidae, and only one family, Rajidae, of Group III are oviparous. The sperm aggregates were found to be of the same type within the order. This may be related to the similarity of reproductive modes within the order.

Conclusion. The external features of the sperm in chondrichthyan fishes were recognized by Tanaka *et al.* (1995) to be species specific. They showed a similarity within the genus and/or family. The four groups based on the cluster analysis also suggested a similarity of sperm morphology within the order. Compagno (1977) and Shirai (1992) divided elasmobranchs into four and two groups, respectively. Both authors recognized the four orders Heterodontiformes, Orectolobiformes, Lamniformes and Carcharhiniformes, as one group. The study of Tanaka *et al.* (1995) placed three of these orders in one group but excluded the Heterodontiformes from it. The remaining three groups of Compagno: 1. Hexanchiformes, Squaliformes and Pristiophoriformes, 2. Batoids (Skates and Rays), and 3. Squatiniformes, are equal to the group 2 of Shirai. Compagno (1977) considered that the three groups are independently derived, while Shirai (1992) regarded them as one of the two groups derived from a basal group. Group II of Tanaka *et al.* (1995) includes five orders and is close to Shirai's grouping. Compagno (1988) divided Carcharhiniformes into two suborders; Scyliorhinoidei and

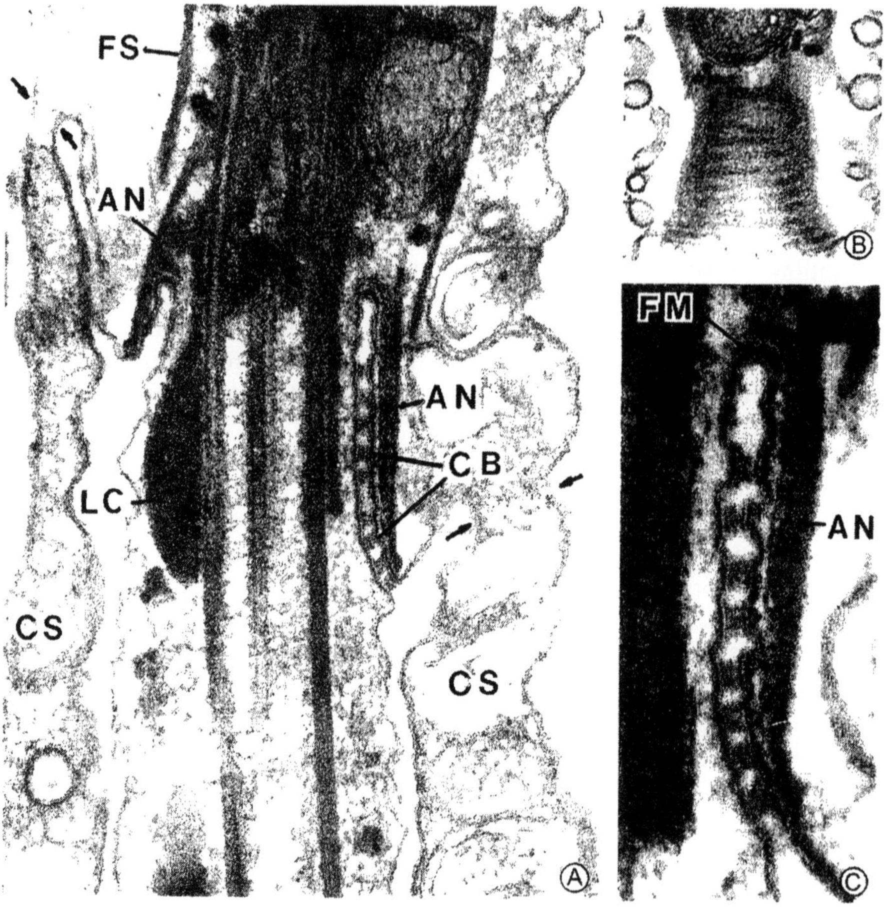

Fig. 7.14 *Hydrolagus colliei* (Chimaeridae). **A.** Section through the basal flagellum region with attached cytoplasmic sleeve. Detachment areas of the remnant cytoplasm are indicated by arrows. The posterior portion of the fibrous midpiece sheath is shown overlapping the annulus. The anterior-most section of the large longitudinal column is indicated. Cross bridges extend between the flagellar plasma membrane and a posteriorly reflected continuation of the same membrane. X 87,100. **B.** Tangential section through cross bridges, showing the continuous, shelf-like form of the connections. X 46,300. **C.** High-magnification micrograph through membrane cross bridges showing some internal structure reminiscent of a desmosome. The flagellar plasma membrane is reflected back upon itself at arrow. X 164,600. Abbreviations: An, annulus. CB, cross bridges. CS, cytoplasmic sleeve. FM, flagellar membrane. FS, fibrous sheath of midpiece. LC, longitudinal column. From Stanley, H. P. 1983. Journal of Ultrastructure Research 83: 184-194, Figs. 26-28.

Carcharhinoidei. The Scyliorhinidae of Group **I** belongs to the former, and the three families of Group **IV** belong to the latter.

The families of Group **I**, except two species of Squalidae, and only one family, Rajidae, of Group **III** are oviparous (Compagno 1990). The formation of sperm aggregates has been demonstrated in various species of elasmobranchs (Pratt and Tanaka 1994). In the study of Tanaka *et al.* (1995), it was found to be of the same type within the order. This may be related to the similarity of reproductive modes within the order (Compagno 1990). Thus, the grouping of chondrichthyan fishes based on the external features of the sperm was considered by Tanaka *et al.* (1995) to reflect the systematics and phylogeny derived from consideration of the external, skeletal and muscular systems of the body and the female reproductive modes. However, inclusion of Holocephali within one group (Group I) of sharks (scyliorhinid Carcharhiniformes, Heterodontiformes and Squaliformes.) requires comment here. It has heuristic value in possibly suggesting paraphyly of extant sharks and origin of Holocephali within this group but more probably indicates, if we accept the strong evidence for a sister group relationship of Holocephali and Elasmobranchii, that the spermatozoon of the common ancestor was more similar in structure to that of Group I sharks than to that of other chondrichthyans.

7.5.2 Spermatozoal Synapomorphies of Chondrichthyes (Holocephali and Elasmobranchii)

The Holocephali and Elasmobranchii are unified by a suite of spermatozoal characters (Fig. 7.15): 1) moderately elongate conical apical acrosome; 2) long, usually helical nucleus; 3) long midpiece, composed of many subspherical mitochondria; 4) fibrous axial midpiece core or rod of rhizoplast origin; 5) basal body situated behind the midpiece mitochondria (a very rare condition, seen also, with no phylogenetic construction, in Onychophora and Euclitellata, see Jamieson 1991); 6) location of two longitudinal columns in the axoneme opposite doublets 3 and 8; and 7) sloughing of much of the cytoplasm by the formation of a remnant sleeve which eventually detaches from the spermatozoon over a ring-shaped area at the posterior end of the midpiece (Stanley 1964, 1965, 1971a, b, 1983; Tanaka *et al* 1978; Stanley and Lambert 1990; Jamieson 1991; Mattei 1991; present study). All of these appear to be apomorphies relative to the ect-aquasperm of invertebrates and lower chordates. However, only 4 and 5 (fibrous axial midpiece core or rod of rhizoplast origin; basal body situated behind the midpiece mitochondria), which are mutually correlated, are certainly synapomorphies for the Chondrichthyes. It is uncertain whether 6 (longitudinal columns) is a synapomorphy of Chondrichthyes homoplastic with Amphibia and lower Sarcopterygii (Actinistia and Dipnoi) or a symplesiomorphy retained from a common ancestor of these groups.

7.5.3 Distinctions between Holocephalan and Elasmobranch Sperm

Holocephali show several distinct modifications of the common structural plan outlined above when compared with elasmobranchs (Stanley 1983; Mattei 1991). It is, however, uncertain whether these differences are constant for the Holocephali. The acrosome in *Hydrolagus colliei* is oval in cross section and bent in its long axis, instead of being straight with a spiral ridge as in

Squalus suckleyi. The midpiece rod differs in being differentiated into a core and outer shell of more distinctly staining densities but this difference, like others, requires confirmation from more detailed investigation of chondrichthyan sperm. As observed by Mattei (1991) the proximal centriole, known for *Squalus*, has not been observed in *Chimaera phantasma*, *Hydrolagus colliei* and *Neoharriota pinnata*. This may be a real difference between holocephalans and elasmobranchs but it has yet to be demonstrated for mature sperm of *Chiloscyllium*. Only in Holocephali (*Hydrolagus*) is the cytoplasmic sleeve joined, in its proximal region, to the flagellum by ladder-like cross connections (Fig. 7.14). The cross bridges remain with the sperm after sloughing of the sleeve. The longitudinal column at doublet 8 is slender and very short in *H. colliei* (Stanley 1965, 1983) and *Neoharriotta pinnata* (Mattei 1988) and possibly absent in *C. phantasma* (Hara and Tanaka 1986) (Fig. 7.15), rendering the flagellum asymmetrical. The flagellar cylinder as a whole is spiral (Stanley 1983) whereas in elasmobranchs it appears that only the

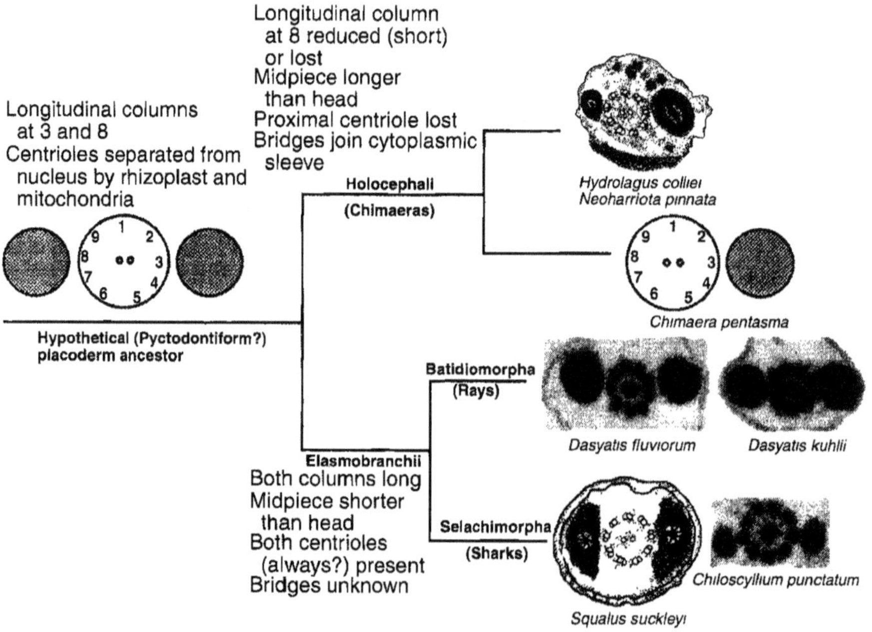

Fig. 7.15 Phylogram of known arrangements of longitudinal axonemal columns in Chondrichthyes. It is inferred that a circular cross section is plesiomorphic for the columns and that the column at doublet 8 in Holocephali is undergoing reduction or loss. In selachimorphs, the accessory axonemal columns in *Chiloscyllium* are smaller and do not show the pronounced median flattening seen in those of *Squalus*; they differ little from those of the ray *Dasyatis* but still show some of the selachimorph median flattening. Unlike those of *Squalus* they are smaller than those of *Dayatis*. Modified for newly examined taxa (*Chiloscyllium and Dasyatis*) from Jamieson 1991. *Fish Evolution and Systematics: Evidence from Spermatozoa*. Cambridge University Press, Cambridge. 319 p., Fig. 7.6, based on data of Stanley 1971b, Hara and Tanaka 1986, Mattei 1988, and present study.

microtubules of the axoneme are spiral; in both groups the accessory columns are spiral, though, in *Chiloscyllium* at least, they may be straight for part of their course (present study). *Hydrolagus* has glycogen distributed along more than 100 μm of the tail while elasmobranchs show a small number of glycogen granules, among the mitochondria of the midpiece (Stanley 1983).

7.5.4 Distinctive Characters of Elasmobranch Sperm

Despite considerable investigation of elsasmobranch sperm ultrastructure, it is still not possible to recognize spermatozoal synapomorphies for the Elasmobranchii with certainty, partly because the direction of character change has not been subjected to parsimony analysis and because data are patchy or lacking for many chondrichthyans. Characters which appear to distinguish Elasmobranchii from Holocephali (Fig. 7.15) are as follows:

(1) The equal or near equal development of the longitudinal axonemal columns, a presumed plesiomorphy relative to the apomorphic reduction of the column at 8 in Holocephali. (2) The midpiece is shorter than the head whereas it is significantly longer in Holcephali. Which of the alternative states is apomorphic is uncertain. (3) Both centrioles are present in most, possibly all, elasmobranchs whereas the proximal centriole is apomorphically lost in Holocephali. (4) The desmosome-like bridge connecting the remnant of the cytopalsmic sleeve to the flagellum in *Hydrolagus* is unknown in elasmobranchs. Presence is clearly apomorphic but it has yet to be established whether the bridge is always absent in elasmobranchs.

No spermatozoal features are known with certainty to separate rays from sharks. This is not surprising if we accept placement of batoids in a squalomorph + squatinomorph clade on the basis of brain anatomy by Northcutt (1989).

It was previously suggested (Jamieson 1991) that sharks differed from other chondrichthyans in median flattening of the accessory axonemal columns, as in *Squalus suckleyi* (Fig. 7.13), but we have seen that in *Chiloscyllium punctatum* (Fig. 7.7C-E) flattening is scarcely appreciable. Further investigation of sperm ultrastructure, for instance that of the midpiece rod which shows variable structure, may yet reveal firm differences between major taxonomic groups. It has, nevertheless, been established by Tanaka *et al.* (1995) that sperm morphology, at least in metric characters, is species specific in chondrichthyans.

7.6 ACKNOWLEDGEMENTS

Micrographs not previously published were taken by Lina Daddow, Julius Miller and the author. The School of Integrative Biology, University of Queensland, is thanked for financial support of the research. Kannika Chatchavalvanich, Amara Thongpan and Masaaki Nakai are thanked for generously making their unpublished account and micrographs available.

7.7 LITERATURE CITED

Boisson, C., Mattei, X. and Mattei, C. 1968. La spermiogenèse de *Rhinobatus cemiculus* Geof. St-Hilaire [Sélacien Rhinobatidae]. Étude au microscope électronique. Institut Fondamental d'Afrique Noire. Bulletin Série A. (Sciences Naturelles) 30: 659-673.

Chatchavalvanich, K., Thongpan, A. and Nakai, M. 2004. Ultrastructure of spermiogenesis in freshwater stingray, *Himantura signifer* Compagno and Roberts, 1982. Journal of Marine and Freshwater Research (In press).

Compagno, L. J. V. 1977. Phyletic relationships of living sharks and rays. American Zoologist 17: 303-322.

Compagno, L. J. V. 1988. *Sharks of the Order Carcharhiniformes.* Princeton University Press, New Jersey 486 pp.

Compagno, L. J. V., 1990. Alternative life-history styles of cartilaginous fishes in time and space. Environmental Biology of Fishes 28: 33-75.

Gusse, M. and Chevaillier, P. H. 1978. Etude ultrastructurale et chimique de la chromatine au cours de la spermiogenèse de la roussette *Scyliorhinus caniculus* (L). Cytobiologie 16: 421-443.

Hamlett, W. C. (1999). Male reproductive system. Pp. 444-470. In W. C. Hamlett (ed.), *Sharks, skates and rays : Biology of Elasmobranch Fishes.* The Johns Hopkins University Press: Baltimore.

Hamlett, W. C., Hysell, M. K., Rozycki, T., Brunette, N., Tumilty, K., Henderson, A. and Dunne, J. 1999. Sperm aggregation and spermatozeugmata formation in the male genital ducts in the clearnose skate, *Raja eglanteria.* Société Francaise d'Icthyologie, Paris, 1999: 281-291.

Hamlett, W. C., Reardon, M., Clark, J. and Walker, T. I. 2002a. Ultrastructure of sperm storage and male genital ducts in a male holocephalan, the elephant fish, *Callorhynchus milii.* Journal of Experimental Zoology. 292(2): 111-128.

Hamlett, W. C., Musick, J. A., Hysell, K. and Sever, D. M. 2002b. Uterine epithelial-sperm interaction, endometrial cycle and sperm storage in the terminal zone of the oviducal gland in the placental smoothhound, *Mustelus canis.* Journal of Experimental Zoology 292: 129-144.

Hara, M. and Tanaka, S. 1986. Fine structure of spermatogenesis and mature spermatozoa in elasmobranch and chimaera fishes: A systematic consideration. Development Growth and Differentiation 28 (suppl): 114.

Jamieson, B. G. M. 1991. *Fish Evolution and Systematics: Evidence from Spermatozoa.* Cambridge University Press, Cambridge. 319 pp.

Jones, R. C., Jones, W. and Djakiew, D. 1984. Luminal composition and maturation of spermatozoa in the male genital ducts of the Port Jackson shark, *Heterodontus portusjacksoni.* Journal of Experimental Zoology 230: 417-426.

Mattei, X. 1970. Spermiogenèse comparée des poissons. Pp. 57-69. In B. Baccetti (ed.), *Comparative Spermatology.* Academic Press, New York, NY.

Mattei, X. 1988. The flagellar apparatus of spermatozoa in fish. Ultrastructure and evolution. Biology of the Cell 63: 151-158.

Mattei, X. 1999. Spermatozoon ultrastructure and its systematic implications in fishes. Canadian Journal of Zoology 69: 3038-3055.

Mc Clusky, L. M. 2003. A scanning electron microscope study of germ cell maturation in the reproductive tract of the male soupfin shark (*Galeorhinus galeus*). Acta Zoologica (Stockholm) 84: 69-76.

Nelson, J. S. 1984. *Fishes of the World.* 2nd edition, John Wiley and Sons, New York, NY, 319 pp.

Nicander, L. 1968. Gametogenesis and the ultrastructure of germ cells in vertebrates. Proceedings of the VIth International Congress of Animal Reproduction and Artificial Insemination, Paris, Volume 1: 89-107.

Northcutt, R. G. 1989. Brain variation and phylogenetic trends in elasmobranch fishes. Journal of Experimental Zoology Supplement 283: 83-100.

Parsons, G. R. and Grier, H. J. 1999. Seasonal changes in shark testicular structure and spermatogenesis. The Journal of Experimental Zoology 261: 173-184.

Pratt, H. L., Jr., and Tanaka, S., 1994. Sperm storage in male elasmobranchs: A description and survey. Journal of Morphology 219: 297-308.

Pudney, J. and Callard, G. V. 1984. Development of agranular reticulum in Sertoli cells of the testis of the dogfish *Squalus acanthias* during spermatogenesis. Anatomical Record 209: 311-321.

Reardon, M. B., Walker, T. I. and Hamlett, W. C. 2002. Microanatomy of spermatophore formation and male genital ducts in the holocephalan, *Callorhynchus milii*. Marine and Freshwater Research 53: 591-600.

Scheltinga, D. M. and Jamieson, B. G. M. (2003). Spermatogenesis and the mature spermatozoon: form, function and phylogenetic implications. Pp. 119-251. In B. G. M. Jamieson (ed.), *Reproductive Biology and Phylogeny of Anura*. Science Publishers, Inc., Enfield, New Hampshire, USA.

Shirai, S 1992. *Squalean Phylogeny: A New Framework of "Squaloid" Sharks and Related Taxa*. Hokkaido University Press, Sapporo 151 pp.

Stanley, H. P. 1962. Morphological relationships between Sertoli cells and germinal cells in the testes of chondrichthyan fishes. American Zoologist 2(4): 561.

Stanley, H. P. 1964. Fine structure and development of the spermatozoan midpiece in the elasmobranch fish *Squalus suckleyi*. Journal of Cell Biology 23: 88A.

Stanley, H. P. 1965. Fine structure of the tail flagella in the spermatozoa of two chondrichthyan fishes, *Squalus suckleyi* and *Hydrolagus colliei*. Anatomical Record 151: 419.

Stanley, H. P. 1966. The structure and development of the seminiferous follicle in *Scyliorhinus caniculus* and *Torpedo marmorata* (Elasmobranchii). Zeitschrift für Zellforschung und Mikroskopische Anatomie 75: 453-468.

Stanley, H. P. 1970. Differential development of homologous structures accessory to the axoneme in sperm of several vertebrate types. Abstracts of the Tenth Annual Meeting American Society of Cell Biology: 201a.

Stanley, H. P. 1971a. Fine structure of spermiogenesis in the elasmobranch fish *Squalus suckleyi*. I. Acrosome formation, nuclear elongation and differentiation of the midpiece axis. Journal of Ultrastructure Research 36: 86-102.

Stanley, H. P. 1971b. Fine structure of spermiogenesis in the elasmobranch fish *Squalus suckleyi* II. Late stages of differentiation and structure of the mature spermatozoon. Journal of Ultrastructure Research 36: 103-118.

Stanley, H. P. 1983. The fine structure of spermatozoa of *Hydrolagus colliei* (Chondrichthyes, Holocephali). Journal of Ultrastructure Research 83: 184-194.

Stanley, H. P. and Lambert, C. C. 1990. Differential fate of mitochondria during spermiogenesis in the ratfish *Hydrolagus*. Tissue and Cell 22(4): 471-476.

Stanley, H. P., Kasinsky, H. E. and Bols, N. C. 1984. Meitoic chromatin diminution in a vertebrate, the holocephalan fish *Hydrolagus colliei* (Chondichthyes, Holocephali). Tissue and Cell 16(2): 203-215.

Tanaka, S., Hara, M. and Mizue, K. 1978. Studies on sharks-XIII. Electron microscopic study on spermatogenesis of the squalen shark *Centrophorus atromarginatus*. Japanese Journal of Icthyology 25(3): 173-180.

Tanaka, S., Kurokawa, H. and Hara, M. 1995. Comparative morphology of the sperm in chondrichthyan fishes. Pp. 313-320. In B. G. M. Jamieson, J. Ausio and J.-L. Justine (eds), *Advances in Spermatozoal Phylogeny and Taxonomy*. Mémoires du Muséum National d'Histoire Naturelle, Paris, 166.

Tanaka, S., Shiobara, Y., Hioki, S., Abe, H., Nishi, G., Yano, K. and Suzuki, K. 1990. The reproductive biology of the frilled shark, *Chlamydoselachus anguineus*, from Suruga Bay, Japan. Japanese Journal of Ichthyology 37(3): 273-291.

Van der Horst, G. and McClusky, L. 1986. Scanning electron microscopy of the testicular ampullae and spermatocysts of the soupfin shark, *Galeorhinus galeus*. Proceedings of the Electron Microscopy Society of South Africa 16: 71-72.

Wiley, E. O. 1979. Ventral gill arch muscles and the interrelationships of gnathostomes, with a new classification of the Vertebrata. Zoological Journal of the Linnean Society 67: 149-179.

CHAPTER 8

The Elasmobranch Ovary

Bram V. Lutton, Joseph St.George, C. R. Murrin, L. A. Fileti and Ian P. Callard

8.1 INTRODUCTION

Extant elasmobranchs (subclass Elasmobranchii; superorders Selachii and Batoidei) are considered to have evolved from placoderm ancestors, about 350-450 million years ago, during the Devonian period. However, it should be noted that the batoids (skates, rays, guitarfish and sawfish) diverged from their shark relatives nearly 180 million years ago, during the Jurassic period. Along with dorso-ventral flattening of the body and other morphological dissimilarities, the batoids also exhibit reproductive differences from the sharks. Therefore, we will describe the reproductive characteristics of these two subgroups separately.

In the diverse spectrum of elasmobranch reproductive modes there are morphological and physiological variations, and no paradigm is common among reproductive modes even within a single group. That is, ovulatory cycles do not correlate directly with reproductive cycles. For all species that have been adequately studied to date, it is clear that morphology of the ovary and secretion of steroids are restricted to certain temporal periods of the reproductive cycle. As in other vertebrates, the elasmobranch ovary produces oocytes through oogenesis, acquires and accumulates nutrients through vitellogenesis, and synthesizes and secretes hormones for various pleiotropic functions throughout the follicular and luteal cycles. However, much less is known of these processes in elasmobranchs than in higher vertebrates. Nonetheless, various modes of reproduction, including oviparity and diverse versions of viviparity have been well-defined for the elasmobranchs. These reproductive strategies have been classified traditionally as oviparous, ovoviviparous, and viviparous and are summarized for species known to date in Table 8.1 (Koob and Callard 1991).

Department of Biology, Boston University, Boston, Massachusetts 02215, USA

Table 8.1 Reproductive strategies in elasmobranchs of the North Carolina shelf

Oviparous

Scyliorhinidae
- *Apristurus laurisonni* -flathead catshark
- *Scyliorhinus meadi*
- *S. retifer - chain dogfish*

Rajidae
- *Bathyraja richardsoni*
- *Breviraja atripinna*
- *B. plutopinna*
- *B. spinosa*
- *Dactylobatus armatus*
- *Leucoraja erinacea* - little skate
- *Raja eglanteria* - clearnose skate
- *R. floridana*
- *R. garmani* - rosette skate
- *R. laevis - barndoor skate*
- *R. ocellata* - winter skate
- *R. radiata* - thorny skate
- *R. senta* - smooth skate

Viviparous

Internal incubators with or without uterine villi (includes yolk sac viviparity and incipient/minimal histotrophy)

Hexanchidae
- *Hexanchus griseus* - sixgill shark

Orectolobidae
- *Ginglyomostoma cirratum* - nurse shark

Rhincodontidae
- *Rhincodon typus* - whale shark

Squalidae
- *Dalatius licha* - kitefin shark
- *Deania profundorum*
- *Etmopterus bullisi*
- *E. gracilispinis*
- *E. hillianus - black belly shark*
- *Somniosus microcephalus* - greenland shark
- *Squalus acanthias* - spiny dogfish

Squatinidae
- *Squatina dumerili* - Atlantic angel shark

Pristidae
- *Pristis pectinata* - sawfish

Rhinobatidae
- *Rhinobatus lentiginosus* - Atlantic guitarfish

Internal incubators with trophonemata (lipid histotrophy)

Torpedinidae
- *Benthobalis marcidae*
- *Narcine brasiliensis* - lesser electric ray
- *Torpedo nobiliana* - Atlantic torpedo

Dasyatidae
- *Dasyatis americana* - southern stingray
- *D. centroura - roughtail* stingray

D. sabina - Atlantic stingray
D. sayi - *bluntnose* stingray
Gymnura altavela - spiny butterfly ray
G. micrura - smooth butterfly ray
Urolopkus jamaicensis - yellow stingray

Myliobatidae

Aetobatus narinari - spotted eagle ray
Myliobatis freminvillei - bullnose ray
M. goodei - southern eagle ray
Rhinoptera bonasus - cownose ray

Mobulidae

Manta birostris - Atlantic manta
Mobula hypostoma - devil ray
M. mobular - devil ray

Internal incubators with oophagy and intrauterine cannibalism

Odontaspidae

Odontaspis taurus - sand tiger

Alopiidae

Alopias superciliosus - bigeye thresher
A. vulpinus - thrasher shark

Lamnidae

Carcharodon carcharias - white shark
Cetorhinus maximus - basking shark
Isurus oxyrinchus - *shortfin mako*
I. paucus - longfin mako
Lamna nasus - porbeagle

Internal incubators with yolk sac placenta

Carcharhinidae

Carcharhinus acronotus - blacknose shark
C. altimus - bignose shark
C. brevipinna - spinner shark
C..falciformis - silky shark
C. isodon - finetootb shark
C. leucas - bull shark
C. limbatus - blacktip shark
C. longimanus - oceanic whitetip shark
C. obscurus - dusky sbark
C. plumbeus - sandbar shark
C. signatus - night shark
Galeocerdo cuvier- tiger shark[1]
Mustelus canis - smooth dogfish
Negaprion brevirostris - lemon shark
Prionace glauca - blue shark
Rhizoprionodon terranovae - Atlantic sharpnose shark[2]

Sphyrnidae

Sphyrna lewini - scalloped hammerhead
S. mokarran - great hammerhead
S. tiburo – bonnethead[2]
S. zygaena - smooth hammerhead

---

[1]This is the only carcharhinid internal incubator that does not develop a yolk sac placenta
[2] With appendiculae.

8.2 DEVELOPMENTAL ASPECTS

In 1876, Balfour published a developmental series of *Scyliorhinus canicula*, which was the first of its kind for an elasmobranch species. During the last 100 years, a number of scientists have contributed studies of elasmobranch development. Most recently, in 1993 Ballard *et al.* revised and described normal developmental stages for *S. canicula*. In *S. canicula*, eggs are laid in pairs, with numbers of pregnant females varying from 10% in August and September to 40% in June and July. Fertilized eggs are encased in a tough, flexible egg case, formed by the oviducal gland (see Chapters 9 and 10 of this volume), which begin to develop during ovulation. The signals leading to proper timing of egg case formation are as yet unknown, although the development of the oviducal gland can be correlated with plasma estradiol levels in the little skate, *Leucoraja (=Raja) erinacea*. Inside the egg case, the fertilized egg is surrounded by a jelly like carbohydrate substance, akin to avian egg white, secreted by the club and papillary zones of the oviducal gland, which protects and supports the developing embryo (Koob and Strauss 1998). A large yolk is present and is the only source of nutrition for the embryo. During early development, the egg cases remain sealed off from the environment; later, presumably due to an increased need for oxygen or waste removal, the egg case opens to the external environment, allowing seawater to flush the compartment. Ballard *et al.* (1993) divide embryonic development in *S. canicula* into 34 stages from fertilization to hatching. Early embryonic development (Stages 1-3) takes place in the oviduct, while most growth and differentiation of the embryos occurs post oviposition. Sex differentiation occurs during the period of organogenesis, between stages 19 and 24. During the pre-hatch period, egg yolk moves from the yolk sac into the fetal yolk stomach, a reserve of nutrients, which is visible through the abdomen for 7-10 days post hatch.

Few studies of sex differentiation in elasmobranchs exist. Chieffi's comparative work on *Torpedo ocellata*, *T. marmorata*, and *Scyliorhinus canicula* are the most complete, and his comparisons with other published accounts of sex differentiation (Picon 1962) show parallels. The undifferentiated gonad of elasmobranchs derives from two separate embryonic tissues, the cortex from the peritoneum, and the medulla from the interrenal blastema (Chieffi 1951, 1952a, 1955). In a typically vertebrate manner, the germ cells migrate into the cortex via the dorsal mesentery after differentiation from early primitive endoderm, segregating very early, possibly even before embryo formation (Beard 1902). As in amniotes, differentiation of the male precedes that of females. Sex differentiation of the genital ridge in males occurs as the germ cells migrate from the cortex to the medulla (Stage 19). The cortex then disappears, and the newly populated medulla is characterized by sex cord formation. At this time, Mullerian duct regression occurs in males. In females, the germ cells remain in the cortex until the embryo has increased in size (Stage 20). The cells then migrate into the medulla, where they remain stable. Formation of primary follicles does not occur until just prior to oviposition in *Torpedo* spp. (Stage 34).

Sex determination appears to be genetic, but can be influenced by steroid hormone exposure early in development in some species (Chieffi 1952b, 1953a, 1953b, 1954; Wourms 1977). Hormone injection into the yolk sac of *Scyliorhinus canicula* prior to sex differentiation alters the path of development and sex differentiation. Thus, a high dose testosterone (T) (0.01-1.0 mg) yields all female gonads, but increased the percentage of embryos with claspers. A low dose testosterone (0.001-0.005 mg) has no effect, yielding equal numbers of male and female gonads, and appropriate differentiation of claspers in males. Estradiol (E_2) (0.1 mg) injections yields 100% female gonads, 75% of which had claspers. Injection of progesterone (P_4) (0.02 mg) also yields all female gonads, but had no effect on clasper development. Thus, all three steroid hormones appears to prevent the differentiation of the testis, although clasper development is sensitive to both T and E_2, but not P_4. It is possible that clasper development is dependent on the expression of aromatase in the normal male. Subsequent elongation of the claspers at sexual maturity appears to be correlated with androgen levels (Collenot 1969). This suggests that while the steroid hormones inhibit the formation of male gonads, they also impact the formation of claspers.

8.3 THE ADULT OVARY

8.3.1 Gross Ovarian Morphology and the Epigonal Organ

Studies of elasmobranch ovaries demonstrate that they are either paired, as in most skates (Koob *et al.* 1986), in which both ovaries are functional, or single, as in others such as the cat sharks (Dodd 1972; Castro *et al.* 1988), where one ovary predominates and the second is rudimentary. In viviparous sharks, the right ovary usually produces eggs for fertilization, while in some shark species both ovaries are functional. Whether the right or left ovary is more fully developed and produces oocytes appears to be species specific. Species studied to date in which the left ovary is rudimentary or absent include *Pristophorus* (Daniel 1928), *Galeus* (Daniel 1928), *Mustelus* (Daniel 1928), *Sphyrna* (Schlernitzauer and Gilbert 1966), *Carcharhinus* (Jensen 1976), and *Prionace* (Pratt 1979). Sharks with two functional ovaries include *Notorynchus masculatus* (Daniel 1928), *Carcharhinus anguineus* (Gudger 1940), *Pristis cuspidatus*, and *Rynchobatus djiddensis* (Setna and Sarangdhar 1948), *Squalus acanthias* (Hisaw and Albert 1947) and *Squalus brevirostris* (Kudo 1956). The Pacific angel shark, *Squatina californica*, is an exception. In this species, the ovaries resemble the situation in most stingrays. Unlike other sharks, the contralateral ovary may contain developing follicles. However, these do not appear to reach ovulatory size, and they are thought to function in the production steroids.

The gonads of elasmobranchs are directly associated with, and invested in an autonomous lymphomyeloid tissue, the epigonal organ. Within the subclass Elasmobranchii there exist at least two different types of epigonal-ovarian associations as established by Pratt (1988). These include internal and external ovaries, and are characterized with respect to the placement of the epigonal organ. Unfortunately, the specific relationship between these two tissues has been investigated very little among these diverse vertebrate taxa and only recently have studies begun to shed light on the potential functions underlying the direct morphological relationship. In females of the family Lamnidae, the germinal epithelium is encapsulated within the epigonal organ. In other families, ovaries have been described as lying on the distal surface of the epigonal organ, or suspended directly from the mesovarium and a short hilum in species where the epigonal organ appears to be reduced. Pratt's informative findings conclude that an investigation of other elasmobranch species may lead to findings of taxonomically significant variations on these themes. It appears to the present authors that the apparent taxonomic differences in epigonal-ovarian associations may be a reflection of the state of oogenesis at the time of autopsy of a particular specimen. It cannot be said with any clarity that there are elasmobranch species without an epigonal-ovarian association. As Yoffey and Courtice (1970) pointed out, lymphomyeloid organs are dynamic structures that change size and cellular composition in relation to endocrine factors. That is, as the ovary develops and vitellogenesis results in enlarged oocytes, the epigonal tissue becomes relatively less obvious.

8.3.2 Folliculogenesis and the Follicular Epithelium

Ovarian histology has been examined for almost 20 species, including representative oviparous and viviparous species of Squaliformes, Rajiformes, and the Holocephali (Giacomini 1896; Wallace 1903; Champy and Gley 1923; Samuel 1943, 1946; Babel 1967, Dodd 1983). These studies show that follicles of various sizes, atretic follicles and corpora lutea are held together by a network of connective tissue stroma. Also, during folliculogenesis and oogenesis the granulosa differentiates into one of two types. In many species (*Squalus acanthias, Chiloscyllium griseum, Heptanchus cinereus, Mustelus laevis, Scyliorhinus canicula, Scyliorhinus stellaris, Scymnus lichea, Spinax niger*) granulosa cells remain cuboidal or become highly columnar and have thin cytoplasmic extensions that cross the widening acellular zona pellucida between the oocyte and the follicle wall. In other species (*Chimaera monstrosa, Myliobatus bovina, Raja spp., Scyllium, Torpedo marmorata, Trygon violacea*) some small cuboidal cells differentiate into much larger cells, with large vesicular nuclei, while other cells remain cuboidal or become columnar throughout follicular development.

Several ultrastructural studies have further elaborated on the processes of folliculogenesis (Andreucetti *et al.* 1999; Hamlett *et al.* 1999; Prisco *et al.* 2002a; Davenport 2003), oogenesis (Prisco *et al.* 2001), and vitellogenesis (Prisco *et al.* 2002b). Ultrastructural studies of the ovary of the yellow spotted stingray, *Urobatis (=Urolophus) jamaicensis*, illustrate a direct association of the epigonal organ with dome-shaped cuboidal epithelial cells, which cover the ovary and constitute the germinal epithelium (Fig. 8.1) (Hamlett *et al.*1999). In both *U. jamaicensis*

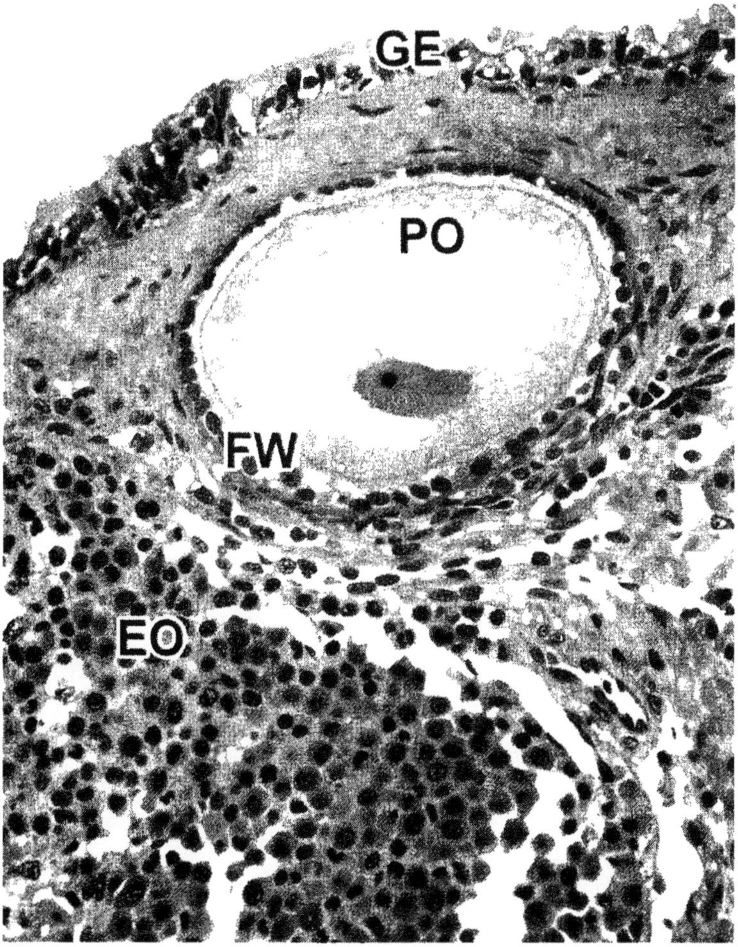

Fig. 8.1 Cross section through the ovary of *Urobatis jamaicensis*. The primary oocyte (PO) lies under the germinal epithelium (GE) and tunica albuginea (TA), and is surrounded by a layer of squamos cells of the follicle wall (FW). The direct association of ovarian components and epigonal organ (EO) can also be seen. X 600. Courtesy of W. Hamlett.

and the spotted ray, *Torpedo marmorata* (Prisco *et al.* 2002), the epithelium becomes multilaminar as vitellogenesis proceeds. Beneath the epithelium lies the tunica albuginea, a thin layer of connective tissue. The outer layer of the ovarian stroma consists of many primordial follicles just below the tunica albuginea. The primordial follicles consist of a primary oocyte, surrounded by a layer of large and small squamous follicle cells. Vascularization is first noted just peripheral to the follicular epithelium within a fibrous layer and a thecal layer as vitellogenesis proceeds. Also, in *U. jamaicensis,* an inward follicular folding, including vascular elements, was noted once ova reach a particular size. This is believed to increase surface area for the transport of yolk precursors to the oocytes (Babel 1967; Hamlett *et al.*1999). All of the species studied to date also have a zona pellucida separating the oocytes from the follicle wall, as well as a basal lamina between granulosa and thecal layers. Interestingly, the zona pellucida of the two shark species whose follicular ultrastructure has been studied to date, *Mustelus canis* and *Rhizoprionodon terraenovae,* is substantially thicker than that of the batoids (Davenport 2003). It also exhibits intricate, long filamentous projections throughout folliculogenesis (Fig. 8.2).

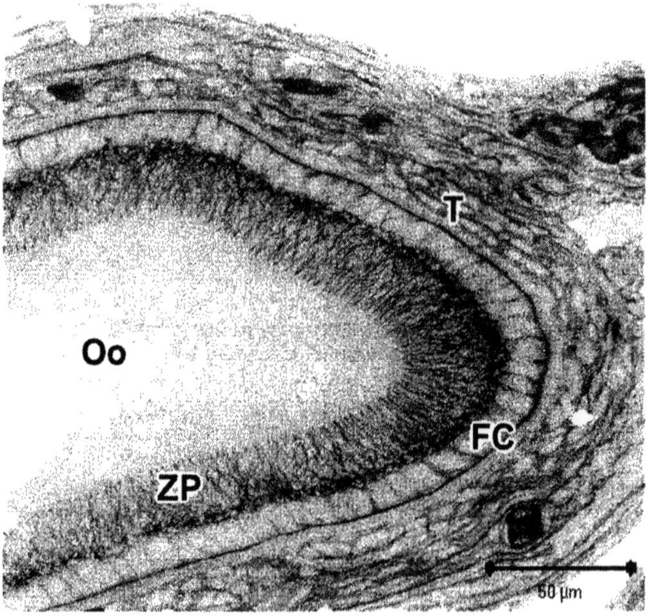

Fig. 8.2 Actin staining (Texas-Red Phalloidin) of 1 mm oocyte from *Mustelus canis.* Follicle cells (FC) are cuboidal, with follicle cell processes spanning the 40 µm-wide zona pellucida (ZP). Oo, Oocyte; T, theca. Courtesy of I. Davenport 2004.

There are some interesting species-specific differences noted regarding the cell types of the follicle wall. Both the theca and granulosa have been shown to change significantly during folliculogenesis. For instance, in *Torpedo marmorata*, the follicle wall is initially a single layer of squamous cells and a network of collagen fibers separated by a basal lamina. This early collagenous thecal layer of the follicle then develops into a theca interna, comprised of flattened fibroblastic cells, and a cuboidal theca externa (Fig. 8.3). This is also the case for *Scyliorhinus canicula* (Dodd and Dodd 1980; Dodd 1983) and *Squalus acanthias* (Lance and Callard 1969), while in *Rhinobatus granulatus* (Samuel 1943) and *Scoliodon sorrakowah* (Guraya 1978) the thecal layers appear to be reversed. In *S. sorrakowah*, the cells of the theca interna are described as polyhedral in shape, containing lipid droplets and large vesicular nuclei. The theca externa is composed of fibroblastic cells embedded in collagen fibers and blood vessels. Contrary to this, a vascularized theca with fibroblasts embedded in a collagenous network is also seen in *Urobatis*

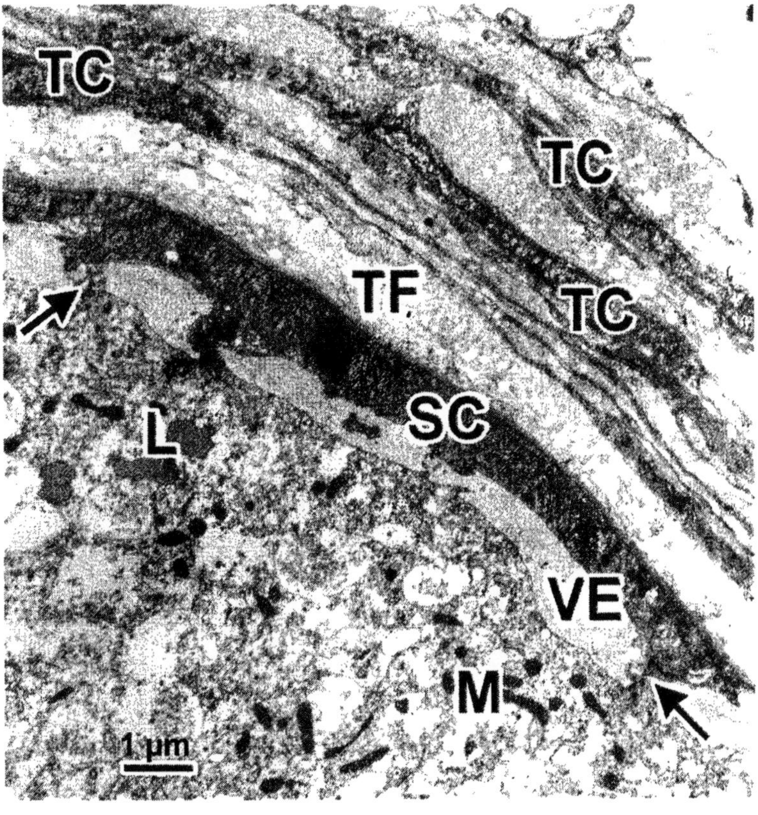

Fig. 8.3 *Torpedo marmorata*. Cross section through follicle wall layers from a 120 mm follicle. Fibroblastic theca cells (TC) and collagenous theca fibers (TF) lie outside a small follicle cell (SC) with protrusions (arrows) through the vitelline envelope (VE) and into the oocyte (Oo). Prisco *et al.* 2002. Molecular Reproduction and Development 61: 78-86, Figs 6 and 7.

jamaicensis, but there is no differentiation between theca interna and externa (Hamlett *et al.* 1999).

The granulosa cells of the follicle wall also demonstrate substantial species-specific variations. That is, Squaliform and Carchariniform elasmobranchs retain a single columnar cell type, whereas Rajidae, Dasyatidae and Chimaeridae develop a granulosa that is heterogeneous. For example, in *Urobatis jamaicensis* (Hamlett 1999) and *U. halleri* (Babel 1967), columnar cells alternate with large, round cells containing lipid-like substances comprising the granulosa layer (Fig. 8.4), but as folliculogenesis proceeds these cells decrease in size and disappear prior to ovulation (Hamlett *et al.* 1999). A different situation exists in *Torpedo marmorata* (Prisco *et al.* 2002), in which the same composition of follicle wall cells is seen as in *Raja asterias* (Andreucetti *et al.*1999). The situation in the latter two species illustrates a system similar to that of squamate reptiles. In vertebrate species that develop large, yolky eggs, a direct

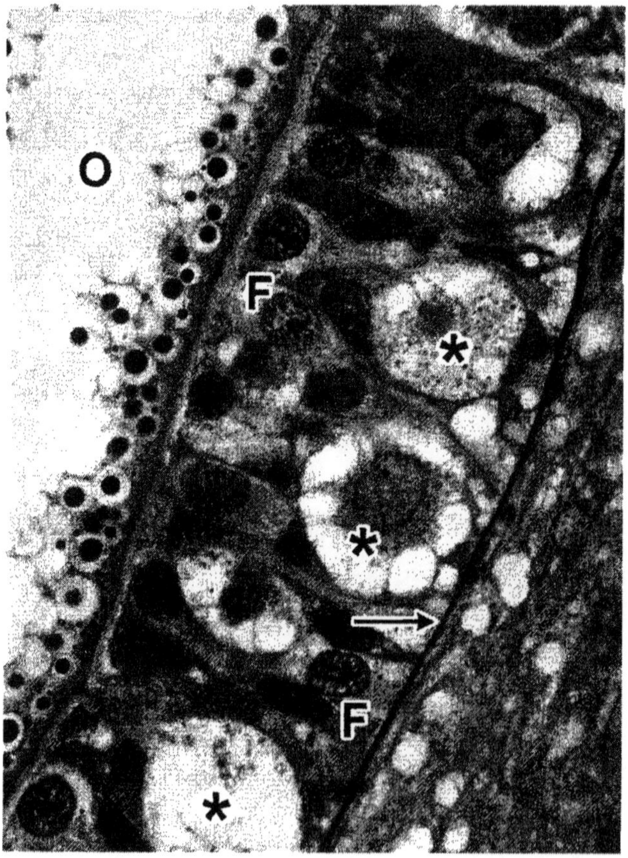

Fig. 8.4 Heterogeneous granulosa cells of the follicle wall in *Urobatis jamaicensis.* Large, lipid-containing cells (asterisks) alternate with columnar follicle cells (F) beneath the basal lamina (arrow). X 600. Courtesy of W. Hamlett.

relationship between structure and function of the follicular epithelia has been noted (Gabaeva 1980). Similarities of pseudostratified follicular epithelia in sharks, skates, birds and reptiles represent evolutionary parallelism, while large cells of anisomorphous follicular epithelia in these species, and trophocytes of the polytrophic ovarioles of insects, are analogous structures representing histological convergence. In *T. marmorata* and *R. asterias*, small cells differentiate, via intermediate or large cells, into pyriform cells after the formation of intercellular bridges (Andreucetti *et al.* 1978; Filosa *et al.* 1979) (Fig. 8.5). In this example, intermediate and pyriform cells are present during pre-vitellogenesis and

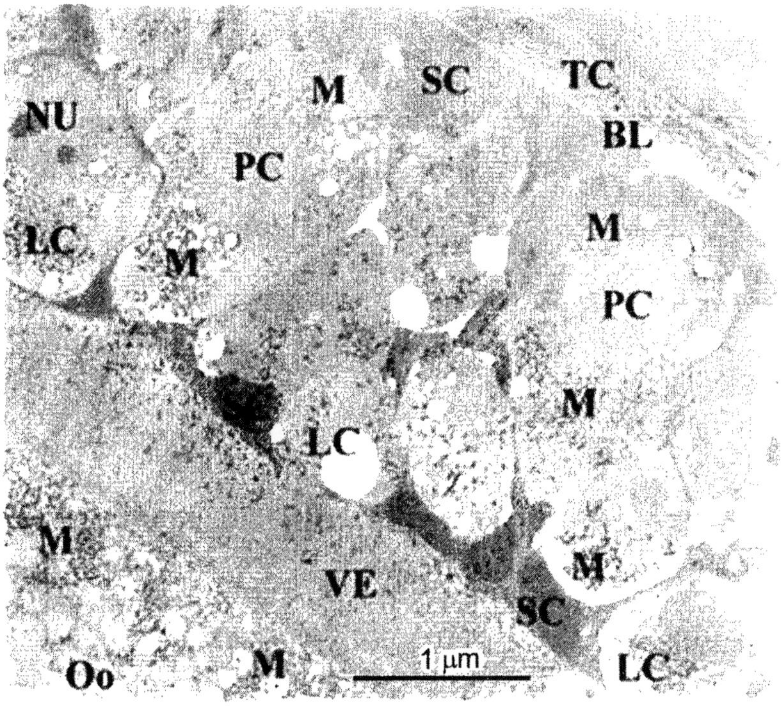

Fig. 8.5 Transmission electron micrograph through a 1,400 mm follicle, illustrating small (SC), intermediate or large (LC), and pyriform-like (PC) cells. The PC basal region contains the nucleus with a prominent nucleolus (NU). Many mitochondria (M) are present within the follicle cells and the oocyte (Oo). BL, basal lamina; VE, vitelline envelope; TC, theca cell. Andreuccetti *et al.* 1999. The Anatomical Record 255: 180-187, Fig. 8.

vitellogenesis, but by the end of oocyte growth only a few layers of small columnar cells remain in the follicle wall. In the shark species studied to date, only one cell type comprises the granulosa layer of the follicle wall. These cells begin as squamous cells (Fig. 8.6) and then differentiate into cuboidal (Fig. 8.7) at approximately 0.5 mm in diameter, followed by an

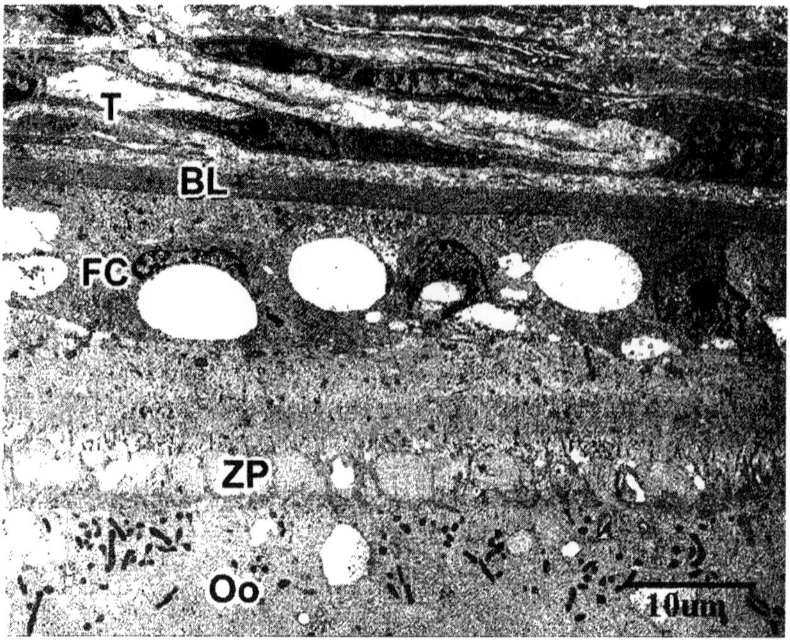

Fig. 8.6 Transmission electron micrograph of squamos follicle cells (FC) from 400 μm oocyte of *Mustelus canis*. T, theca; BL, basal lamina; ZP, zona pellucida; Oo, oocyte. Courtesy of I. Davenport 2004.

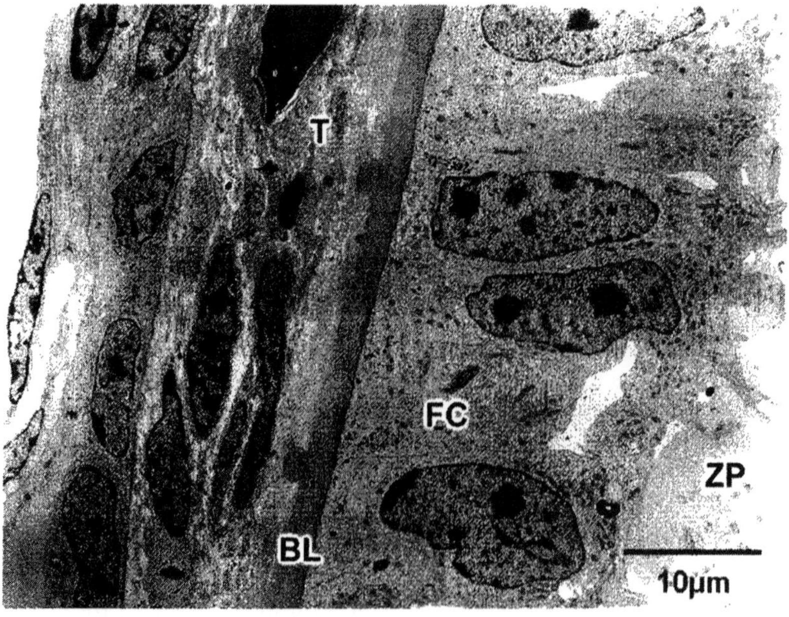

Fig. 8.7 Transmission electron micrograph of cuboidal follicle cells (FC) from 2 mm oocyte of *Mustelus canis*. T, theca; BL, basal lamina; ZP, zona pellucida. Courtesy of I. Davenport.

elongation into columnar cells (Fig. 8.8) at approximately 2.0 mm in diameter (Davenport 2003).

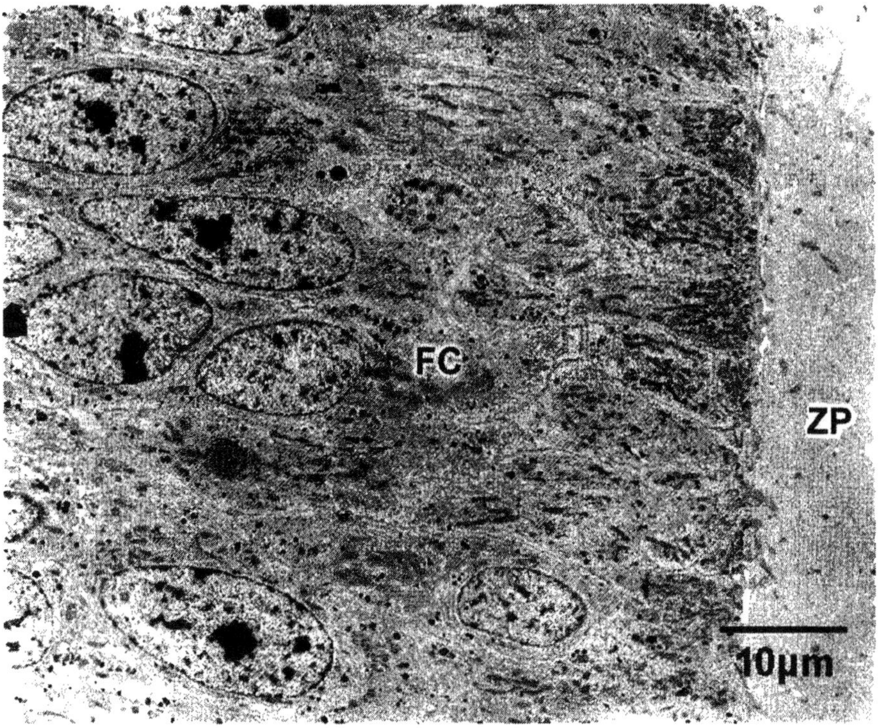

Fig. 8.8 Transmission electron micrograph of columnar follicle cells (FC) from 10mm oocyte of Mustelus canis. T, theca; BL, basal lamina; ZP, zona pellucida. Courtesy of I. Davenport.

8.3.3 The Oocyte and Ovulation

Similarly to other vertebrates, the oocytes of elasmobranchs have elaborate surface microvilli that extend toward the granulosa and form a conspicuous narrow band in histological sections (zona radiata), surrounded by an acellular matrix (zona pellucida). The ooplasm of yolk-filled oocytes has been shown to contain several proteins, including livetin, lipovitellin, and phosvitin localized in yolk platelets. Other proteins and fats, as well as about 6% urea by wet weight are also constituents of the yolk (Faure-Fremiet 1933; Needham 1950; Fuji 1960). Using electron microscopy, Hamlett *et al.* (1999) described the presence of transosomes, or follicle cell extensions that indent and pinch off inside the oocyte membrane (Fig. 8.9), and in the same year Andreuccetti and colleagues discovered intercellular bridges between granulosa cells and the oocyte in *Raja asterias* (Fig. 8.10). A similar transport system was reported for *Torpedo marmorata* (Prisco *et al.*

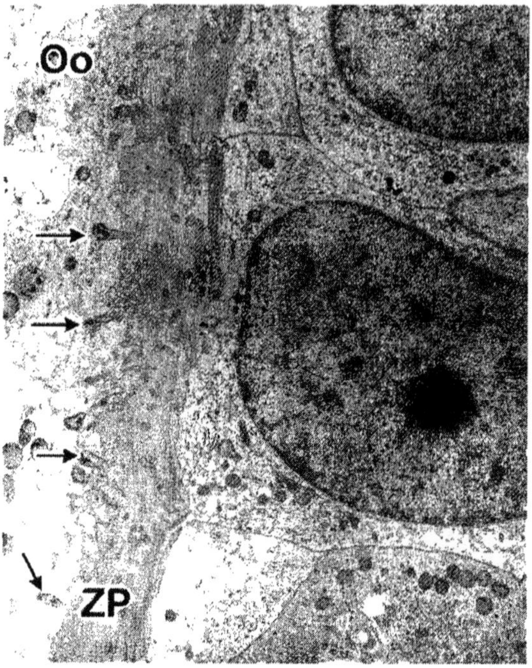

Fig. 8.9 Granulosa cells of *Urobatis jamaicensis* extending transosomes (arrows) through the zona pellucida (ZP) and into the oocyte (Oo). X 4,000. Courtesy of W. Hamlett.

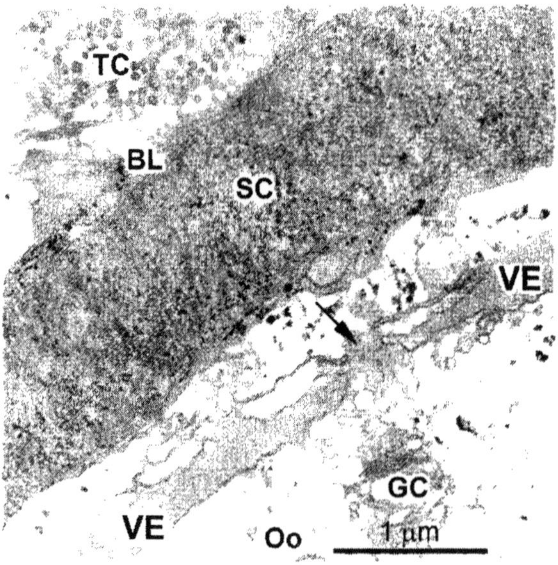

Fig. 8.10 Transmission electron micrograph of an intercellular bridge (arrow) between a small follicle cell (SC) and the oocyte (Oo) in a 150 µm follicle from *Torpedo marmorata*. BL, basal lamina; GC, golgi complex; TC, theca cell; VE, vitelline envelope. Andreuccetti *et al.* 1999. The Anatomical Record 255: 180-187, Fig. 5.

2002), where first small cells, and then intermediate and pyriform cells demonstrate connections with the oocytes (Fig. 8.11). The only information on selachians, other than the batoids, to date also demonstrate similar protrusions from the granulosa cells into the oocytes throughout follicular development (Fig. 8.12) (Davenport 2003). It is clear that as the follicles grow and vitellogenesis proceeds, granulosa cells differentiate and lipid-rich inclusions begin to appear and enlarge. Projections of the follicles into the oocytes have been seen in all ultrastructural studies thus far. And by the time of ovulation these cells have decreased significantly in size in all species, while the oocytes become filled with yolk. It was previously considered that the inclusions within the granulosa cells could be the source of precursors for steroid synthesis and/or they may indicate that these cells synthesize yolk precursor granules and supply them to the oocytes (Tsang and Callard 1983). Though the liver is generally considered the source of yolk precursor proteins, there are some reports that the granulosa cells of elasmobranchs may be capable of vitellogenesis based on ultrastructural observations. The organization of the granulosa seems to exclude transcytosis during the time of vitellogenesis in *T. marmorata*, and morphological organization of metabolically active granulosa cells suggests

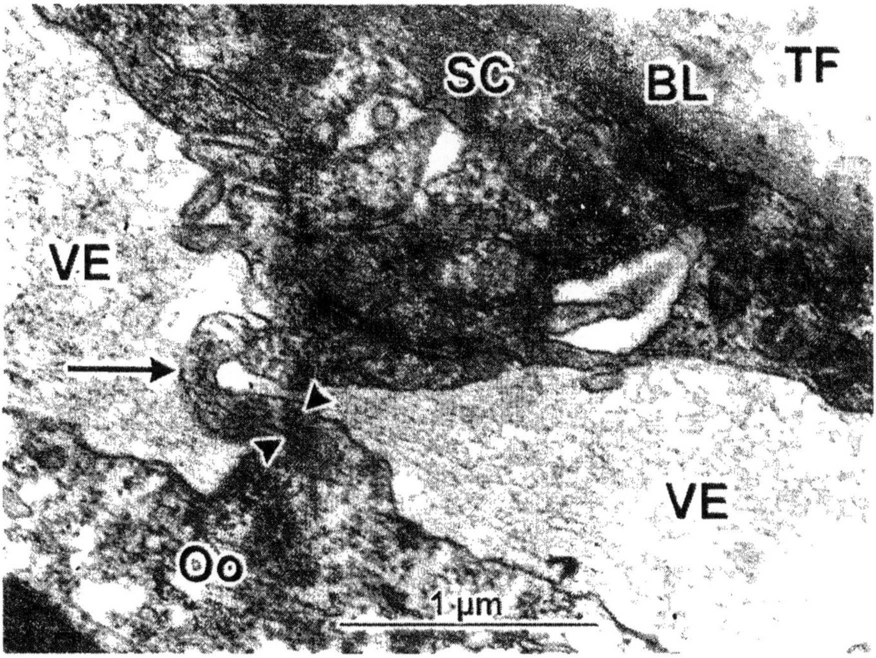

Fig. 8.11 Transmission electron micrograph of the vitelline envelope (VE) from a 120 µm follicle of *Torpedo marmorata* detailing an intercellular bridge (arrow), connecting a small follicle cell (SC) with the oocyte (Oo), and containing bundles of filaments (arrowheads). BL, basal lamina. Prisco *et al.* 2002. Molecular Reproduction and Development 61: 78-86, Fig. 8.

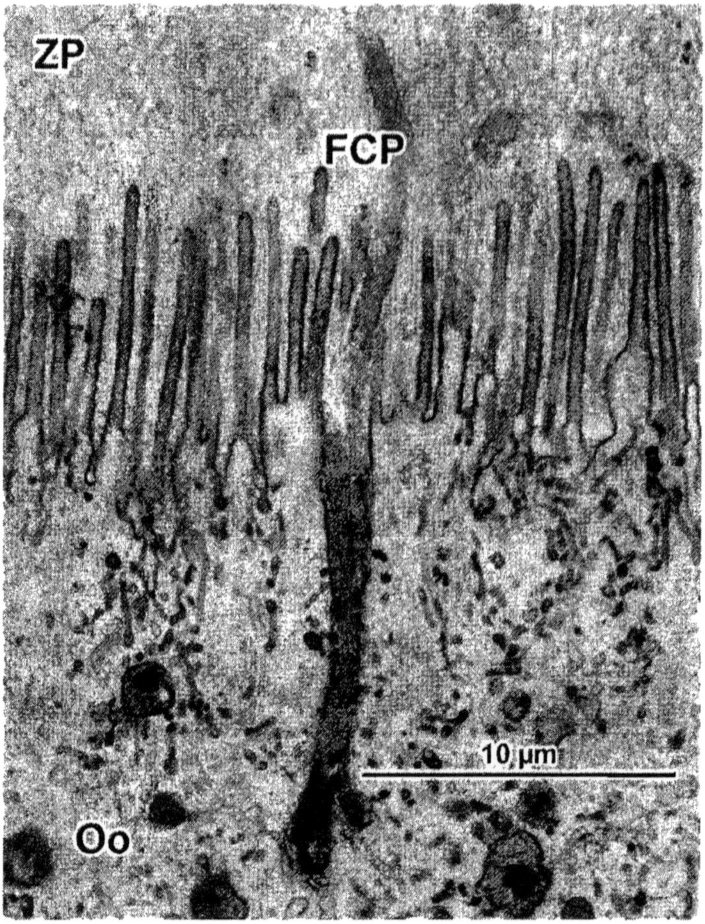

Fig. 8.12 Follicle cell processes (FCP) of 3 mm oocyte from *Mustelus canis*. ZP, zona pellucida; Oo, oocyte. Courtesy of I. Davenport.

that they may be engaged in vitellogenin synthesis (Prisco *et al.* 2002). This supports the hypothesis that oocyte growth in Chondrichthyes is not entirely supported by the liver (Maruska *et al.* 1996).

The number and size of ova vary greatly between species. As the follicles become more mature a stroma develops in the peritoneum through which oocytes pass on their way to the ostium. This process has been carefully studied in some species where, for instance, ripe follicles burst through the outer germinal epithelium, discharging ova into the peritoneal cavity. At this point there can be variability in oocyte transport depending upon what appears to be reproductive strategy of a particular species. For instance, while the ova pass, via ciliary motion along the peritoneum and mesovarium, from the anterior end of the ovary to the ostium in *Scyliorhinus canicula* (Metten 1939), the strategy reported for the basking shark, *Cetorhinus maximus*, is different. In the latter of the two species, the germinal epithelium is invaginated, and smaller ova, without the large quantities of yolk necessary for oviparous and aplacental viviparous species, pass through a tubule network into the peritoneal cavity en route to the oviduct (Matthews 1950). The oocyte then passes to the oviducal (nidamental, or shell) gland, where it is fertilized and, in oviparous species, encapsulated within an egg case.

8.3.4 Pre- and Post-ovulatory Follicles: Corpora Atretica and Corpora Lutea

The degeneration of follicles prior to ovulation is commonly seen in most elasmobranchs. It can be difficult to distinguish pre-ovulatory follicles, or corpora atretica (CA), from post-ovulatory follicles, or corpora lutea (CL) in some species, as they share a number of similar characteristics. In *Squalus acanthias* (Tsang and Callard 1987), it was noted that CA are globular in shape, grainy in texture, and whitish-yellow due to lipid-rich granulosa-derived cells and degenerating oocyte yolk. CL, on the other hand, appear as greenish-yellow to yellow globular masses, and can often be identified, if ovulation has recently taken place, beneath small holes in the outer surface of the ovary. CA have a rich vascular supply to the central lumen, surrounded by a lipid-filled, infolded and degenerating granulosa layer enclosed within the thecal layer. Phagocytosis of yolk by the small cells of the granulosa layer can make it particularly difficult to distinguish CA from CL. Nonetheless, the course of atresia in elasmobranchs is generally divided into four separate stages (TeWinkel 1972; Chieffi *et al.* 1992) and dissolution of the oocytes by phagocytosis makes distinguishing stage 1 and stage 2 possible. That is, granulosa villi are long and granular by stage 2 due to ingestion of oocyte components. In stage 3, the granulosa epithelium is transformed into an active glandular structure accompanied by the development of smooth endoplasmic reticulum in *Torpedo marmorata* (Chieffi *et al.* 1992). In *Mustelus canis* it was shown that theca cells fill with granules and begin to increase in size by stage 3, followed by encroachment of the granulosa and dispersal into small groups of cells, which finally disappear in stage 4 (TeWinkel 1972). Sclerosis and pigmentary degeneration of the atretic follicle characterizes the fourth stage in *Torpedo marmorata* (Chieffi *et al.* 1992). Atresia can occur with any size follicle, depending on mechanisms yet to be elucidated.

In the case of post-ovulatory follicles, once ovulation has occurred and the oocyte moves toward the ostium, the follicle wall immediately collapses to form the CL. Like the CA, the CL also consists of lipid-filled cells derived from the granulosa and is similar in oviparous and viviparous species (Champy and Gley 1923; Samuel 1943, 1946; Hisaw and Hisaw 1959; Chieffi 1961, 1967; Lance and Callard 1969). However, thecal cells do not contribute to the CL other than acting as structural elements and supporting a vascular network to the granulosa-filled central portion of the structure, as with the CA (Chieffi *et al.* 1992). The thick, acellular basal lamina also seems to persist, delimiting the lobes of the granulosa. In *Squalus acanthias*, ultrastructural investigation demonstrates an abundance of lipids, vesicles, vacuoles, smooth endoplasmic reticulum, and mitochondria. Also, small, membrane-bound and electron dense granules are often localized beneath the cell surface. In oviparous species, the size and weight of CL decrease gradually after ovulation, though more rapidly than in viviparous species. The longevity of CL in the latter species can range from a few months to two years, persisting throughout gestation. As the CL degenerate a number of characteristics have been noted The overall size is reduced, cells die, lipid content decreases within individual cells, cellular lobes become increasingly disorganized, and there is an accumulation of degenerating luteal cells and blood cells in the central lumen (Wallace 1903; Samuel 1943; Hisaw and Albert 1947; Lance and Callard 1969).

8.4 CONTROL OF OVARIAN CYCLICITY

8.4.1 Hypothalamic and Pituitary Relationships

Periodicity in reproductive physiology is shown to some degree in all vertebrates and there is evidence that light and temperature are the external stimuli most commonly involved (Johnson and Everitt 2000). In addition, it is possible that individual animals respond to changing population density by a compensatory acceleration of reproductive maturation. A physiological shift towards earlier maturation is due to a heightened activity of the hypothalamic-pituitary-gonadal (HPG) axis. For instance, pulses of GnRH and gonadotropins are associated with ovulatory cycles in mammals. It has been suggested that a pulse generator in the medial basal region of the hypothalamus controls mammalian episodic gonadotropin and GnRH secretion (Karsch 1984). Initially, the pulse generator becomes active at puberty, however changes to GnRH pulses can occur in adult animals in response to acute external stimuli from visual, olfactory and tactile cues. Any change in the onset of reproduction in response to an environmental variable commences in a sensory organ, and sensory neurons relay the perceived change in the environment to subservient physiological systems. Modifications of reproductive responses are integrated at the level of the hypothalamus and eventually cascade to the gonads to regulate oogenesis and steroidogenesis (Karsch 1984).

In the hypothalamus, the neuropeptide GnRH is the primary hormone involved in neural control of reproduction. GnRH has been found in the brains of all chondrichthyans examined to date (Sherwood and Lovejoy 1993) and four forms of GnRH-like molecules have been identified in the spiny dogfish, *Squalus acanthias* (Sherwood and Sower 1985, Lovejoy *et al.* 1992a). GnRH immunoreactivity has been found in all major divisions of the brain of *S. acanthias*. In the diencephalon, most of the GnRH immunoreactive fibers were found in the hypothalamus. These fibers were also found to be intertwined within the fibers of the preoptico-hypophyseal tract, dorso-caudal to the optic chiasm and extending towards the median eminence (Lovejoy *et al.* 1992b). One of the most highly developed of elasmobranch senses is the olfactory system. The greatest density of GnRH fibers in *S. acanthias* is found in the telencephelon, which appears to be associated with additional clusters at the point of entry of the terminal nerve in the dorso-medial area of the rostral telencephalon (Lovejoy *et al.* 1992b). In *S. acanthias* the terminal nerve enters the rostral telencephalon in the medial dorsal hemisphere near the olfactory tract (Smeets *et al.* 1983). In addition, one branch of GnRH immunoreactive fibers extend through the medial pallium following the medial olfactory tract while a second less defined branch is associated with the lateral olfactory tract (Lovejoy *et al.* 1992b). The locations of GnRH in the brain of *S. acanthias* indicate that both the visual and olfactory sensory units could be involved in regulation of reproduction, including onset of reproduction.

Studies of elasmobranch pituitary responsiveness indicates that the ventral lobe is the main if not only source for the synthesis of gonadotropins and responsible for subsequent follicular development. Ventral lobectomy causes follicular atresia and anovulation, but removal of the rostral or median lobe had no detectable effect in elasmobranchs (Dodd *et al.* 1960). Similar results were seen in *Squalus acanthias* (Hisaw and Abramowitz 1939). Immunocytochemical studies using the smooth dogfish, *Scyliorhinus canicula*, have located an LH-like gonadotropin in the ventral lobe of the pituitary (Scanes *et al.* 1972) and Lance and Callard (1978) showed that ventral lobe extract stimulates androgen synthesis in an isolated testicular cell system. Later studies indicated that ventral lobe extracts from *S. acanthias* stimulate progesterone, testosterone and estradiol production in mid-pregnancy, but not early-pregnancy (Tsang and Callard 1988).

Although the pituitary-ovarian axis is essential to ovarian function, no neural or vascular connection between the hypothalamus and the ventral lobe of the pituitary has been described in elasmobranchs (Dodd and Sumpter 1984). Figure 8.13 shows a diagrammatic representation of this lack of a direct connection. Nonetheless, GnRH has been identified in the hypothalamus and plasma (Sherwood and Lovejoy 1993). This indicates that although there is no direct connection, that releasing hormones from the hypothalamus could have an effect on the pituitary or ovary via the general circulation. Injections of GnRH in elasmobranchs have also been shown to induce an increase in plasma steroids (Jenkins and Dodd 1980). Furthermore, cycles of GnRH levels have been found to correlate with reproductive activity (Powell *et al.* 1986). Although GnRH may have direct action on gonadal steroidogenesis (Callard *et al.* 1993), further clarification is needed to determine the role of the hypothalamus in reproductive activity. Nonetheless, evidence for the role of the pituitary and the existence of seasonality in breeding indicates that these pathways are important in controlling reproductive activity in *Squalus acanthias.*

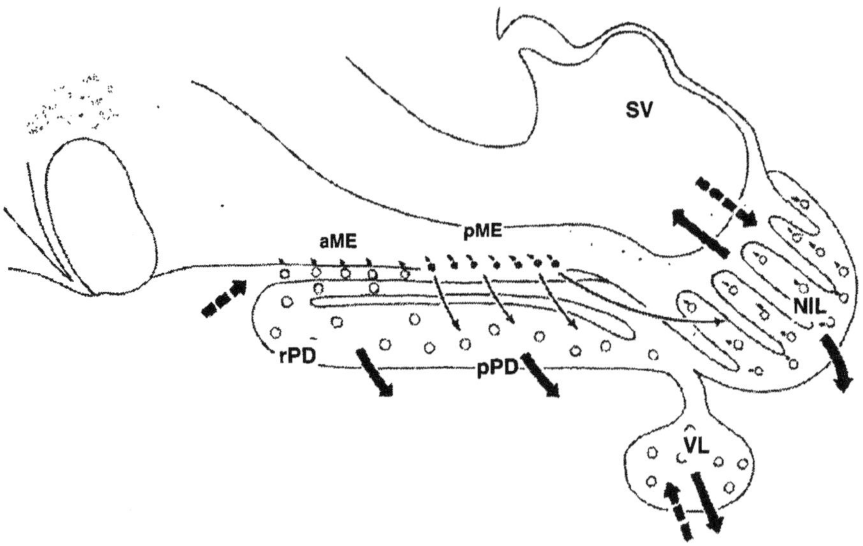

Fig. 8.13 Selachian pituitary, diagrammatic sagittal section showing no direct neural or vascular connection between the hypothalamus and ventral lobe of the pituitary. SV, saccus vasculosus; AME anterior median eminence; PME, posterior median eminence; RPD, rostral pars distalis; PPD, proximal pars distalis; NIL, neuro-intermediate lobe; VL, ventral lobe. Shaded circles, neurosecretory cells of the PON; dotted lines, neurosecretory axons; filled circles, primary portal capillaries; empty circles, secondary portal capillaries and other intrahypophysial capillaries; interrupted arrow, arteries; solid arrow, veins; thin arrows, portal veins. From Jasinski, A. 1969. General and Comparative Endocrinology. Suppl. 2, 510-521, Fig. 3.

8.4.2 Ovarian Cycles

In the oviparous species, the general trend is for production of egg cases to occur throughout the year with a greater proportion of oviposition taking place during particular seasonal periods. In viviparous species, the reproductive cycles are generally characterized by entire populations undergoing nearly synchronous mating, gestation, and parturition, suggesting that ovulation may be coordinated by environmental signals. The steroid profiles for these various reproductive strategies are generally established by thoroughly studied examples of species in Fig. 8.14 and Fig. 8.15.

8.4.2.1 The skates and rays

In contrast to shark species, the skates demonstrate only oviparous modes of reproduction. Also, pregnancy only lasts several months in most rays compared to viviparous sharks (e.g., *Dasyatis sp*: Ranzi 1934; Snelson *et al.* 1989; *Pteromylaeus bovina*: Ranzi 1934; *Urobatis halleri*: Babel, 1967). However, with great variability in habitat and evolutionary diversification, exceptions to each of these trends exist. For instance, unusual cycles for rays include that of *Dasyatis centoura* (Struhsaker 1969), *Rhinoptera bonasus*

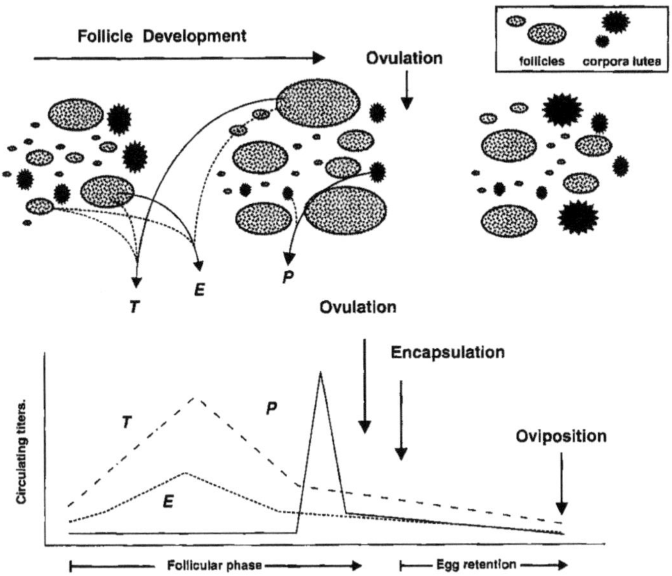

Fig. 8.14 Steroid profile, and tissue source, of circulating estradiol (E), testosterone (T), and progesterone (P) during the ovulatory cycle in Leucoraja erinacea. The principal source of each steroid is indicated by the solid line in the upper panel; additional sources are indicated with dashed lines. Koob and Callard 1999. Journal of Experimental Zoology 284: 557-574, Fig.4.

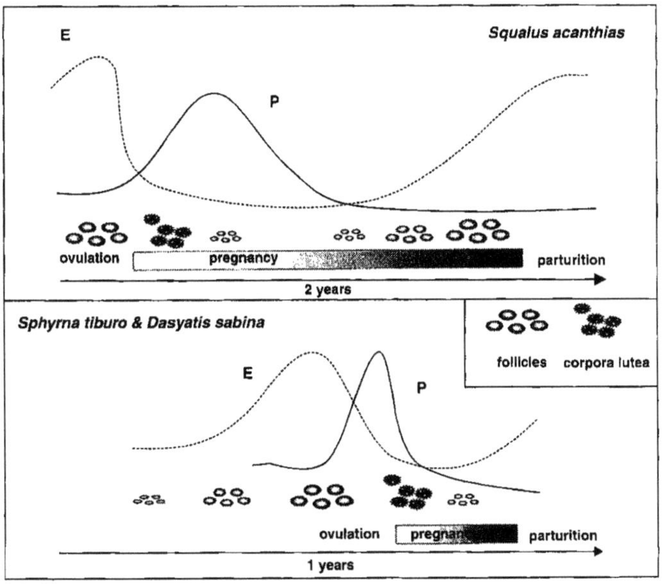

Fig. 8.15 Patterns of circulating estradiol (E), and progesterone (P) titers during the reproductive cycles of three viviparous species of elasmobranchs. Koob and Callard 1999. Journal of Experimental Zoology 284: 557-574, Fig.5.

(Smith and Merriner 1986), and *Myliobatus californicus* (Martin and Caillet 1988), in which gestation lasts almost one year.

Of the batoids, the most thoroughly studied species is the oviparous skate, *Leucoraja erinacea*, an intermittent ovulator in which follicular development continues throughout an extended ovulatory period. These animals produce pairs of eggs every few days and ovulation is thought to occur when egg capsules are one-quarter to one-third formed. Egg capsule formation requires approximately twelve hours and the capsules remain in the uterus for several days prior to oviposition. Several additional days pass before the subsequent ovulation, encapsulation and oviposition. Increasing serum levels of estradiol (E_2) and testosterone (T) in the circulation correlate with ovarian recrudescence and follicular development in preparation for egg laying. The pre-ovulatory production of E_2 and T is a common endocrine mechanism among batoid elasmobranchs, including those in which embryos do not rely entirely on yolk as a source of nutrients. For instance, while Fasano and coworkers (1992) demonstrated high E_2 and T from follicles of *Torpedo marmorata*, and plasma E_2 increased dramatically before ovulation, T was undetectable in plasma. This species is known to have a one-year gestation period followed by two non-pregnant years (Mellinger 1974; Capape 1979), exhibiting one pattern of punctuated reproductive cycles in elasmobranchs.

In a study of *Raja eglanteria*, heightened titers of E_2 and T were also seen between October and November, coinciding with enlargement of follicles, as well as in January and February, when mating activity peaks and egg laying begins (*Rasmussen et al.* 1999). Dihydrotestosterone (DHT) was also at elevated levels during the latter of the two periods, at the time of oviposition. Furthermore, in different populations of *Dasyatis sabina*, increasing levels of E_2 and T were demonstrated during two separate periods. In this species, viviparity is characterized by production of histiotroph by specialized uterine trophonemata following depletion of yolk reserves. The reproductive timing of estuarine *D. sabina* from the east of central Florida (Tricas *et al.* 2000) and Georgia (Schwartz and Dahlberg 1978), and of freshwater *D. sabina* from Lake Monroe, Florida (Snelson *et al.* 1997) is all in close agreement. Estuarine and freshwater populations exhibit increased levels of E_2, first associated with synchronous ovulations between March and April, and then a second either related to the transition from yolk-dependent to histotroph-dependent embryonic nutrition mid-June to mid-July (Snelson *et al.* 1997), or to parturition between July and August (Tricas *et al.* 2000). Increases in E_2 were preceded by heightened T and DHT levels. However, androgens from the two populations showed distinct differences. While freshwater stingrays showed no elevation in T associated with a major reproductive event, increases of T in March and May did correlate with increases of DHT in April and June, respectively. The former study was coincident with the time of ovulation and the latter with a second recruitment of follicles, the function of which remains undetermined. However, this set of follicles may serve as a sink from which developing embryos can obtain nutrients during initial developmental stages. This would be similar to the case of intermittently ovulated eggs in oophagus species where extra eggs are made available early in pregnancy and serve as a nutritional source for developing embryos (Gilmore 1993). The changing levels of ovarian steroids throughout reproductive cycles is likely to be a reflection of enzymatic conversion (T $\rightarrow$ DHT $\rightarrow$ E_2) illustrating specific functional progressions.

In *Leucoraja erinacea*, 10% of the population is gravid year round, while approximately 40% of the population is gravid in June and July, and in October and November (Bigelow and Schroeder 1953). Essentially, the luteal and follicular phases of the cycle overlap. In every species studied to date, corpora lutea (CL) differentiate from follicles after ovulation. However, less is known about the cyclic dynamics of this steroidogenic tissue, and its functional life differs between species with distinct reproductive cycles. For example, although they are very closely related morphologically, *Raja eglanteria* and *Leucoraja erinacea* are separated by significant climatic differences and do not share the same reproductive cycle. *R. eglanteria* is a seasonal breeder, reproductively active for only a portion of the annual cycle, exhibiting only one period of gravidity per year over the entire population. Follicles become vitellogenic between October and mid-December, maximum mating occurring in January, and oviposition occurring between January and mid-July. Furthermore, *in vitro* studies have shown P_4 production occurs from the CL of *L. erinacea* (Fileti and Callard 1988), and it seems that functional P_4 originates from CL of previous ovulatory cycles. In contrast, Rasmussen and Luer (1999) have demonstrated elevations in P_4 after each oviposition of a pair of eggs every four days from *R. eglanteria*, potentially occurring prior to subsequent ovulations, suggesting that P_4 could be the primary maturation- inducing steroid in this species.

Unlike P_4 production by oviparous species, viviparous species demonstrate elevated P_4 titers during the peri-ovulatory period through the initial stages of gestation. Some of these studies provide evidence that P_4 has at least a cooperative role in ovulation as well. For instance, in *Dasyatis sabina*, P_4 peaks between early March and early April, which coincides with the surge of E_2 at the time of final oocyte maturation and ovulation (Tricas *et al.* 2000). Like the elevated titers of P_4 during oviposition in *Leucoraja erinacea* and *R. eglanteria*, however, the estuarine population of *D. sabina* demonstrated a second rise in P_4 between July and August, suggesting that it may play a role in parturition in this species as well. Furthermore, while postovulatory follicles appear to be responsible for P_4 production in *L. erinacea*, investigations of *Torpedo marmorata* indicate that preovulatory atretic follicles are responsible for this function (Fasano *et al.* 1992). The most likely conclusion is that there is a similar relationship between ovarian steroids, as seen in studies of E_2

and P_4 during parturition and stimulation of gonadotropin release from the pituitary in rodents. That is, for certain steroid functions it is necessary for the cognate receptors to be primed by the other steroid and its receptor. One of the questions remaining to be answered in future investigations is, which steroid primes the other for a particular function in these species?

Little information is available regarding ovarian peptide hormones in elasmobranchs. However, the ovaries of *Leucoraja erinacea* have also been shown to contain the hormone relaxin (Bullesbach *et al.* 1987). The B-chain in this species differs from the two reported cases in sharks in that it has 40 amino acids while the shark B-chains possess only 23 amino acids. It shows an overall homology of 42% with *Odontaspis taurus*, 48% with *Squalus acanthias*, 35% with human, and 31% with porcine relaxin. Discussion of the potential role of this hormone in elasmobranch physiology can be found in Chapter 9 (Endocrine Regulation of the Reproductive Tract).

8.4.2.2 The sharks: Oviparous species

The sharks demonstrate similar reproductive cycles to the skates. For instance, in *Scyliorhinus canicula* spawning can be observed at any time of the year in the northwestern Atlantic Ocean. However, based upon a number of parameters, including gonadosomatic index and circulating estradiol (E_2) and testosterone (T) levels, there is a peak breeding season in late winter and spring (Sumpter and Dodd 1979). E_2 and T were shown to rise during ovarian recrudescence, observations which were supported by another study with the same species (Craik 1979) and by steroid synthesis *in vitro* correlating with size of follicles (Dodd 1983). Small vitellogenic follicles produced large amounts of E_2 but only small quantities of T. However, as follicle size increased, the production of T increased while E_2 remained approximately the same. Overall, steroidogenesis increased throughout folliculogenesis in *S. canicula*. It has been suggested that steroid concentrations in females of oviparous elasmobranchs peak more than once during an annual cycle (Koob *et al.* 1986; Callard *et al.* 1993). However, in the epaulette shark, *Hemiscyllium ocellatum*, only one E_2 peak was seen in the months of September through November, coinciding with maximum sizes of ova, ovulation, and the egg-laying period (Heupel *et al.* 1999). This may be considered another difference between batoids and sharks, as it is the case for both oviparous shark species, which have been studied most thoroughly (Sumpter and Dodd 1979; Heupel *et al.* 1999). Similar to the batoids, however, *H. ocellatum* also produces eggs in pairs and appears to ovulate ova into egg capsules after they are at least half formed.

Egg-laying is variable between species as well. For instance, the batoids produce a pair of eggs from 0-2 days, as with *Raja clavata* (Ellis and Shackley 1995), to every 4-5 days in *Leucoraja erinacea* (Koob and Callard 1999) and *Raja eglanteria* (Luer and Gilbert 1985) after the previous pair. The chain dogfish shark, *Scyliorhinus retifer*, on the other hand, requires 14-16 days between egg-laying pairs (Castro *et al.* 1988). Other oviparous sharks also demonstrate a clear seasonal period of egg-laying. For instance, off the waters of southern California, *Apristurus bruneus* and *Parmaturus xaniurus* were reproductively active year-round with a peak midwinter in egg production (Cross 1988). Furthermore, McLaughlin and O'Gower (1971) reported, for *Heterodontus portjacksoni* off the coast of southeastern Australia, that most eggs were laid between August and September.

8.4.2.3 The sharks: Viviparous species

The viviparous elasmobranchs exhibit marked annual reproductive cycles, demonstrating synchronized periods of mating, gestation, and parturition. Most commonly seen among both the placental and aplacental viviparous reproductive strategies is a year-long pregnancy followed immediately by mating and the subsequent pregnancy. In *Mustelus canis*, females ovulate after mating during the first three weeks of June (Dodd 1983). Parturition occurs in May, following an 11-month gestational period, the next ovulation taking place immediately after parturition. The hammerhead sharks (Sphyrnidae) also demonstrate this pattern, with ovulation in September and parturition the following July in the larger species within this group (Chen *et al.* 1988; Castro 1989). Subsequent ovulations occur directly after parturition due to the fact that oocytes mature within the ovary at the same time of embryonic development. In the smaller bonnethead shark, *Sphyrna tiburo*, however, we see one of the shortest gestation periods (4.5-5 months) of all elasmobranchs (Manire *et al.* 1995). After mating in November, sperm is stored until ovulation the following March-April. Placentation begins around mid-gestation, when the embryos are approximately 100 mm in length, and parturition follows in August. E_2 and T levels are high during the mating and preovulatory periods, while P_4 is significantly elevated during the preovulatory-postovulatory period. Dihydrotestosterone is also known to increase significantly during the preovulatory stage. The interesting exception among the viviparous sharks is the Pacific angel shark, *Squantina californica*, due to its primary functional left ovary, and unusual dorso-ventrally flattened morphology. This species, as well as the Atlantic sharpnose shark, *Rhizoprionodon terranovae*, exhibit gestational periods of 10-11 months in duration (Parsons 1981; Natanson and Caillet 1986, respectively). *R. terranovae* is a placental viviparous species, and during most of the year its ovary contains small oocytes (2-5 mm in diameter) and small follicles (greater than 2 mm in diameter). As ovulation approaches, 4-8 follicles rapidly accumulate yolk, while the remainder become atretic. Yolk accumulation occurs predominantly after parturition, and follicle development is slow during pregnancy. This mode of reproduction, in which both oocyte and embryonic growth coincide, allows these species to produce offspring every year.

In other species, such as *Prionace glauca* (Pratt 1979) and *Carcharhinus milberti* (Springer 1960; Wass 1973), females produce offspring every two years, with a year-long gestation and an intervening year of non-pregnancy. In the spiny dogfish, *Squalus acanthias*, with one of the longest gestational periods of any vertebrate, pregnancy lasts nearly two years (Hisaw and Albert 1947). Ovulation and mating occur in the fall of one year and parturition takes place approximately 22 months later. Soon after parturition ovulation and mating are able to occur because a new cohort of oocytes have developed throughout the gestational period. After ovulation, only small follicles (1-5 mm in diameter) are present in the ovary. Approximately a quarter of the way through pregnancy, when the embryos are about 3-5 cm in length, follicles begin to grow and accumulate yolk. Between 6 and 9 months after ovulation the follicles that will grow to ovulatory dimensions have been selected and grow to approximately 15 mm in diameter. Late in pregnancy, oocytes have grown to about 80% ovulatory size. Therefore, most of the follicular growth and yolk acquisition occurs in the last third of gestation. Because of the long gestational period in this species, the gestational follicle development allows advantages with regard to optimizing fecundity by permitting subsequent ovulation immediately after parturition. The ovaries of *S. acanthias* were also shown to possess the peptide hormone, relaxin (Bullesbach *et al.* 1986), after earlier findings by Steinetz *et al.* (1959) that relaxin was bioassayable in the ovaries of *Odantaspis tauras*, the sand tiger shark. After this original finding an authentic relaxin molecule, based upon amino acid sequence, was isolated from ovarian extracts of the same species (Reinig *et al.* 1981). Peptide analysis demonstrated that this molecule was actually closer in homology to porcine insulin than porcine relaxin, and the relaxin from *S. acanthias* showed 75% homology with that of *O. tauras*, and 45% homology with mammalian relaxin molecules. Both of these elasmobranch relaxins have been demonstrated to exhibit small effects on the pubic symphysis of mice.

8.5 STEROID SYNTHESIS

In *Squalus acanthias*, Estradiol (E_2) is secreted by both the granulosa and theca cells, the efficiency of steroidogenesis increasing when exogenous substrate is added (Tsang and Callard 1987a). The granulosa cells appear to be autonomously capable of producing progesterone (P_4), as well as E_2 and testosterone (T) without the theca cells, which is contrary to observations in other vertebrate groups. In the same species, it was definitively shown for the first time, with *in vitro* studies, that elasmobranch corpora lutea were capable of P_4 production (Tsang and Callard 1987b). In *Leucoraja erinacea*, isolated cells from medium (10-15 mm) and large (20-25 mm) follicles are capable of producing both E_2 and T. However, large follicles produced primarily T and medium follicles produced five-fold greater E_2 than small (<10 mm) or large follicles (Fileti and Callard 1990; Callard *et al.* 1993).

Little information on pituitary control of steroidogenesis is available, though it is known that the ventral lobe is the primary, if not the only, source of gonadotropins (Lance and Callard 1978; Tsang and Callard 1988). Recent studies indicate that both follicle stimulating hormone and luteinizing hormone moieties are present in the ventral lobe (French 2001), and it has also been shown that 7 weeks after ventral lobectomy (VLX) of *Leucoraja erinacea* testosterone and estradiol production by small, medium, and large follicles is significantly depressed (Fileti *et al.* 1994). Furthermore, in actively ovulating *L. erinacea*, 9 days after VLX or total hypophysectomy (THYPOX), only 20% of 9 animals had ovulated, compared to 92% of 18 controls.

The effects of chicken II GnRH (ch II GnRH) and ventral lobectomy (VLX) on ovarian steroidogenesis in theca and granulosa cells incubated separately have also been investigated (see Table 8.2). In both cell types, testosterone (T) production increases significantly ($p < .05$) 9 days post VLX. Also, in tissue from intact animals, prior exposure to ch II GnRH has no effect on either theca or granulosa cell T synthesis. In VLX animals, ch II GnRH significantly decreases thecal cell T production ($p < .05$), but is without effect on granulosa cell T production. These observations are in marked contrast to the effect of VLX on theca and granulosa cell estradiol (E_2) production. Granulosa cell E_2 production after GnRH treatment is approximately 12 times greater than production by the theca cells, indicating the granulosa cells are the primary source of E_2. Ch II GnRH treatment, in vivo, has no effect on T production from either cell type from intact Leucoraja erinacea. Nine days after VLX granulosa and theca cell E_2 production decreased markedly compared to that of intact granulosa cells ($p < .001$). Ch II GnRH treatment of VLX animals induced a modest 2-fold increase ($p < .05$) in granulosa cell E_2 production, without effecting theca cell E_2 production.

Table 8.2 *Leucoraja erinacea* steroid production (T, E_2) by theca and granulosa cells *in vitro*: Effect of ventral lobectomy (VLX) and chicken II GnRH.

		Basal *(unstimulated)*		GnRH (*in vivo*)	
		E_2	T	E_2	T
Intact	Theca	224 ± 24	444 ± 64	211 ± 11	424 ± 28
	Granulosa	298 ± 62	423 ± 17	3157 ± 182	409 ± 21
VLX	Theca	109 ± 9	665 ± 26	89 ± 10	287 ± 9
	Granulosa	184 ± 4	1173 ± 51	460± 27	1359 ± 30

Steroid Values Are pg/mg ± SE
Triplicate incubations
Ventral Lobectomy: 9 days in advance of cell isolation
Ch II GnRH: 3 intraveneous injections, days 7-9; 5 ug each.
25 hydroxy-cholesterol substrate, 100 ng/incubation per tube

In summary, chicken II GnRH has no effect on testosterone (T) or estradiol (E_2) production by theca cells from intact animals, but increases granulosa E production. However, after VLX, when E_2 production by granulosa cells is reduced, granulosa cell T production is increased by ch II

GnRH relative to controls. This is presumably because of a failure of aromatization and the build up of aromatase substrate(s). It appears therefore that the removal of one or two gonadotropins by VLX has a marked effect on steroidogenesis, particularly the activity of aromatase.

8.5.1 Second Messengers and the Control of Ovarian Steroidogenesis

The influence of second messengers on steroidogenesis in *Leucoraja erinacea* using fragments (80mg wet wt/incubation well) of isolated follicular wall (FW) from 8 – 23 mm follicles, supplemented with 25-hydroxycholesterol in Eagles basal medium, has recently been investigated. Addition of dibutyryl cyclic AMP (db cAMP) or 3-isobutyl methyl isoxanthine (MIX) at the doses indicated (Figure 8.16) increase both testosterone (T) ($p < .01$) and progesterone (P_4) ($p < .001$) from basal levels. A combination of MIX and db cAMP do not further stimulate steroid synthesis, and no effects are seen on estradiol (E_2) production. Using luteal tissue, db cAMP alone is not stimulatory for either P_4 or T production, but MIX significantly increase P_4 ($p < .001$), and MIX plus db cAMP produce a

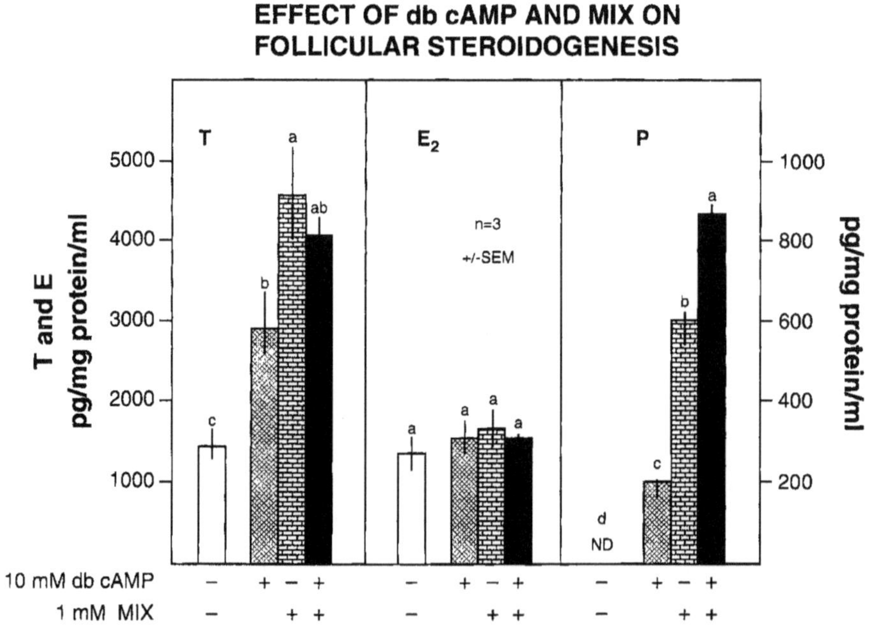

Fig. 8.16 Pooled follicular tissues isolated from intact skates and incubated with 1ug 25-hydroxy cholesterol as substrate with or without dibutyryl 3' 5'-cyclic adenosine monophosphate (db cAMP) +/- the phosphodiesterase inhibitor, 3-isobutyl-methyl xanthine (MIX) for 8 hours at 23°C. Data were analyzed by ANOVA and Duncan's multiple comparison analysis. Values represent means ± SEM of three replicates. Bars with different letters are significantly different ($p < .05$). Unpublished data.

greater stimulation ($p < .001$) than db cAMP alone. No effects were seen on T production (Figure 8.17). These results indicate that the adenyl cyclase system is probably a functional second messenger system in the ovarian tissue of *L. erinacea*. Further, as found in earlier studies of luteal steroid production, P_4 is the primary product (Fileti and Callard unpublished).

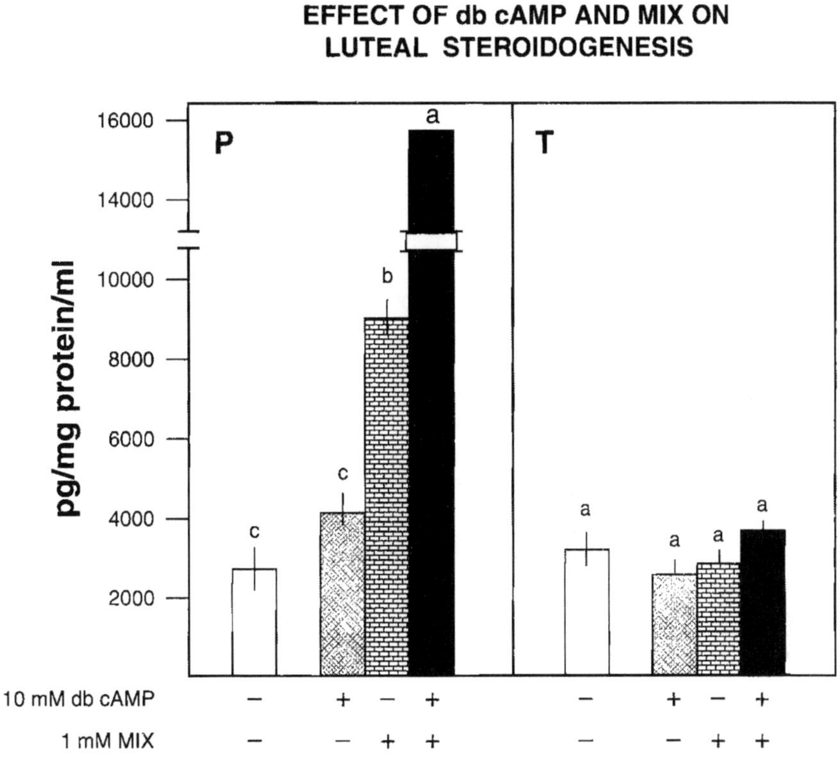

Fig. 8.17 Pooled luteal tissues isolated from intact skates and incubated with 1ug 25-hydroxy cholesterol as substrate with or without dibutyryl 3' 5'-cyclic adenosine monophosphate (db cAMP) +/- the phosphodiesterase inhibitor, 3-isobutyl-methyl xanthine (MIX) for 8 hours at 23°C. Data were analyzed by ANOVA and Duncan's multiple comparison analysis. Values represent means ± SEM of three replicates. Bars with different letters are significantly different ($p < .05$). Unpublished data.

In separate experiments, the role of the phosphoinositide (PI) system in follicular steroidogenesis was investigated (Figure 8.18). 100 nM phorbol dibutyrate (PD) significantly increase testosterone (T) ($p < .01$) and estradiol (E_2) ($p < .01$) production. Only 500 nM PD significantly increase progesterone (P_4) production ($p < .05$). Calcium ionophore A 23187 was also effective in significantly elevating T, E_2 ($p < .01$) and P_4 ($p < .05$) at the 1 ug level. These studies implicate both of these major

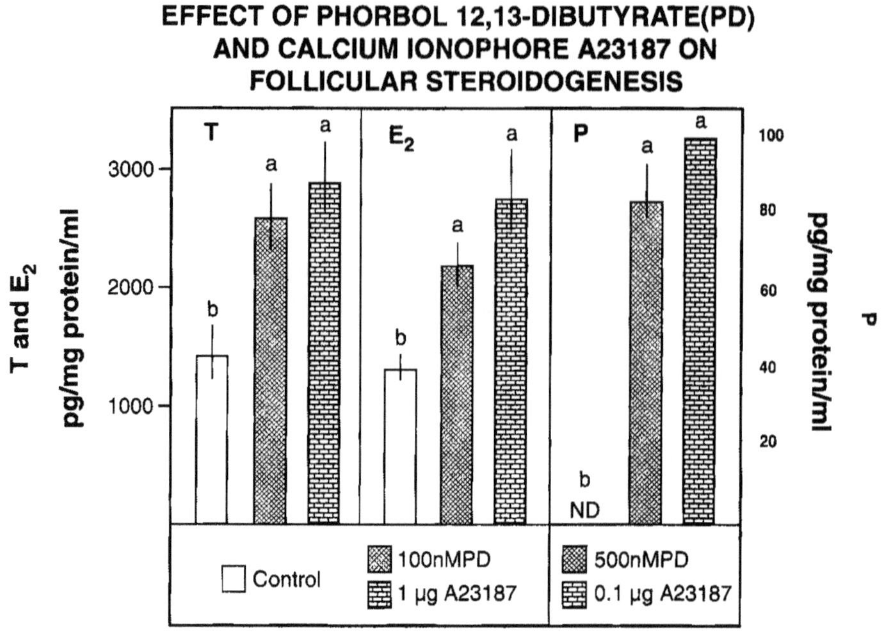

Fig. 8.18 Pooled follicular tissues isolated from intact skates and incubated with 1ug 25-hydroxy cholesterol as substrate with or without the protein kinase C inhibitor, phorbol 12,13-dibutyrate (PD), or calcium ionophore A23187 8 hours at 23°C. Data were analyzed by ANOVA and Duncan's multiple comparison analysis. Values represent means ± SEM of three replicates. Bars with different letters are significantly different ($p < .05$). Unpublished data.

second messenger systems in the control of steroidogenesis in the elasmobranch ovary (Fileti and Callard unpublished).

8.6 REGULATION OF VITELLOGENESIS

High circulating titers of estradiol (E_2) and testosterone (T) during folliculogenesis correspond with elevations in plasma vitellogenin (Vtg) levels, when oocytes are accumulating yolk. Therefore, it is possible that one or both of these steroids regulate nutrients produced by the liver for the oocyte (Craik 1978a, b, c, d,1979; Ho *et al.* 1980; Perez and Callard 1992). Using a homologous antibody to lipovitellin, a cycle of vitellogenin production during the ovulatory cycle of *Leucoraja erinacea* has been described (Perez and Callard 1993). This study showed that Vtg levels were generally high during the cycle in ovulatory females, but consistently low in anovulatory females (Figure 8.19), the nadir of the Vtg level being just after ovulation. In the same study, it was shown that Vtg was not present in normal males, but could be induced by estradiol (E_2), detectable as early as 5 days post-injection. This was not true of progesterone (P_4), which is suggested to be anti-vitellogenic in reptiles

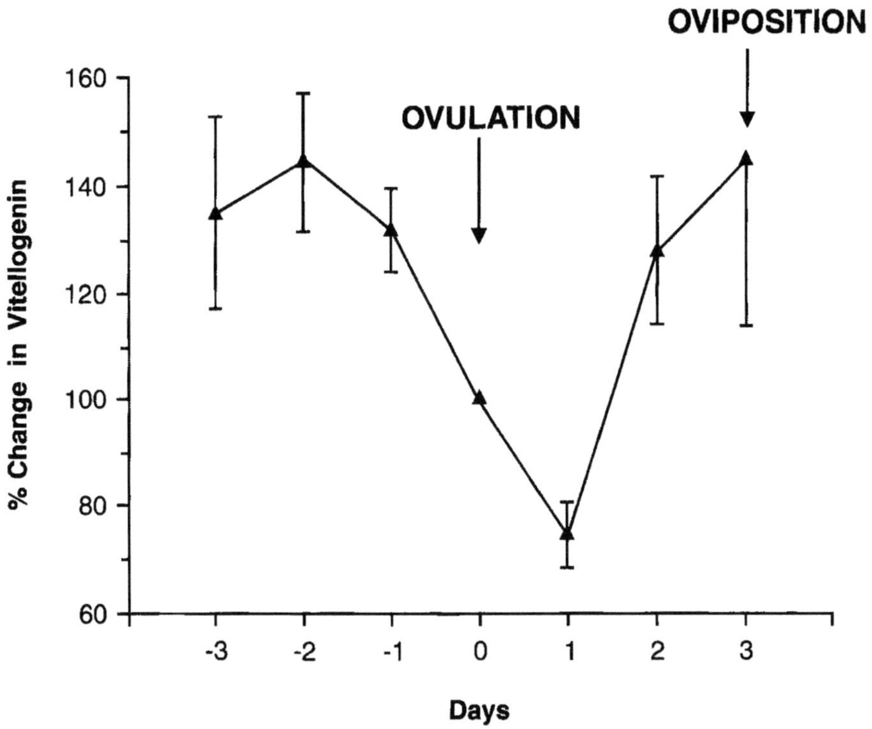

Fig. 8.19 Composite curve showing % change in plasma vitellogenin during the normal ovarian cycle. Values shown are for 5 skates ± SEM. Ovulation = half or fully formed capsules in reproductive tract; oviposition = capsules expelled. Perez and Callard 1993. The Journal of Experimental Zoology 266: 31-39. Fig. 5.

and significantly inhibited the E_2 stimulatory effect on vitellogenesis (Callard *et al.* 1994). In hypophysectomized males, E_2 also stimulated vitellogenesis, an effect that was also inhibited by P_4. These results suggest that P_4 may have a counter-regulatory role in the control of vitellogenesis in elasmobranchs. In an earlier study (Ho *et al.* 1980) it was shown that Vtg could not be induced in *Squalus acanthias* during stage A of gestation, when P_4 levels are high (Tsang and Callard 1987), but could be induced during stage C when P_4 levels are low. Thus, it appears that P_4 can modulate Vtg production in elasmobranchs, an effect that is likely physiologically important in ovarian cycling and the timing of egg production.

8.7 GENERAL EFFECTS OF STEROIDS ON THE REPRODUCTIVE TRACT

The subject of hormone action on the reproductive tract is considered in Chapter 9 of this volume. The following general comments serve only to provide a link between the two related subjects. The functions of estradiol (E_2), progesterone (P_4), and testosterone (T) during gestation and ovarian recrudescence are not fully understood, but enough information is available to indicate that the ovarian hormones regulate reproductive tract function. Relaxin, which has been characterized in elasmobranchs (Bullesbach *et al.* 1987), appears to play a role in connective tissue compliance and is involved in regulating passage of eggs and young from the oviduct, a function that appears to be co-ordinated by E_2, as in mammals. E_2 may also be involved in regulating morphological properties of specialized regions of the reproductive tract, such as the "closing device" at the anterior end of the uterus (Koob *et al.* 1983). Pretreatment with E_2 has been shown to potentiate the effect of relaxin and insulin on the size of the cervix and on fetal retention in *Squalus acanthias* (Koob *et al.* 1984), probably through a regulatory effect on oviductal relaxin receptors. It is also likely that E_2 plays a part in the synthesis of egg capsule precursors (Koob and Callard 1982). Manire *et al.* (1995) suggest that elevations of T and related androgens, such as dihydrotestosterone (DHT), may play a part in direct regulation of quiescence of the myometrium and activation of sperm stored in the oviducal gland during the follicular phase. P_4 is also thought to play a role in ovulation, encapsulation, and oviposition due to timing of heightened titers, and it may also play a role in egg retention (Koob *et al.* 1986). It seems clear that decreasing P_4 titres in the second half of pregnancy allows increasing E_2 levels to upregulate vitellogenin production in the liver.

8.8 ENDOCRINE-IMMUNE SYSTEM INTERACTIONS

The hypothalamic-pituitary-gonadal (HPG) axis provides the major framework for ovarian regulation by positive and negative feedback loops. However, during the past two decades it has become increasingly clear from mammalian studies that ovarian activity, including ovulation and steroidogenesis, are regulated in part by the cells and factors of the immune system (Adashi 1992; Bukulmez and Arici 2000). It is possible that the variation in reproductive strategies demonstrated amongst the various subgroups of elasmobranchs, and even between populations of the same species separated only by habitat, could be the result of more complex regulatory mechanisms than those studied to date. Understanding such relationships is critical for a comprehensive view of the mechanisms governing reproduction, and elasmobranchs offer a unique perspective. These are the only species known in which there is a direct morphological association between the gonads and a an autonomous lymphomyeloid tissue. This immune tissue, the epigonal organ, effectively acts in the same manner as bone marrow of higher vertebrates by the production of leukocytes. However, while the cellular composition of the epigonal organ and the ultrastrucuture of its leukocytes have been thoroughly investigated, little is known about its function. The direct association of the epigonal organ with the gonads in elasmobranchs suggests a functional relationship that may allow for a rapid and specific response of the gonads. In *Leucoraja erinacea*, perfusion analysis and corrosion casting illustrate a vascular pathway from the dorsal aorta to the genital arteries, through the epigonal organ, and into the ovarian follicles (Lutton and Callard 2004). Venous return is into sinuses surrounding the ventro-medial portion of the ovary-epigonal organ complex. This directional flow supports the possibility that the epigonal organ cells may filter antigenic material entering the ovary. A portal system like that of the hypothalamic-pituitary axis, similarly involving hormonal control mechanisms, may exist within this immune-endocrine tissue complex. This would allow communication involving low output of immune cells/factors, enabling reproductive effects without somatic responses or potential diffusion of signals. Light and electron microscopy illustrate a direct morphological association between the cells of the epigonal organ and ovary in several species, including *Urobatis jamaicensis* (Hamlett *et al.* 1999), *Raja asterias* (Andreucetti *et al.* 1999), *Torpedo marmorata* (Prisco *et al.* 2002), and *Leucoraja erinacea* (Lutton and Callard 2004). The presence of leukocytes within the thecal layer has been demonstrated in three of these species (Figs. 8.20, 8.21, 8.22) (Hamlett *et al.* 1999; Prisco *et al.* 2002; Lutton and Callard 2004). Ultrastructural

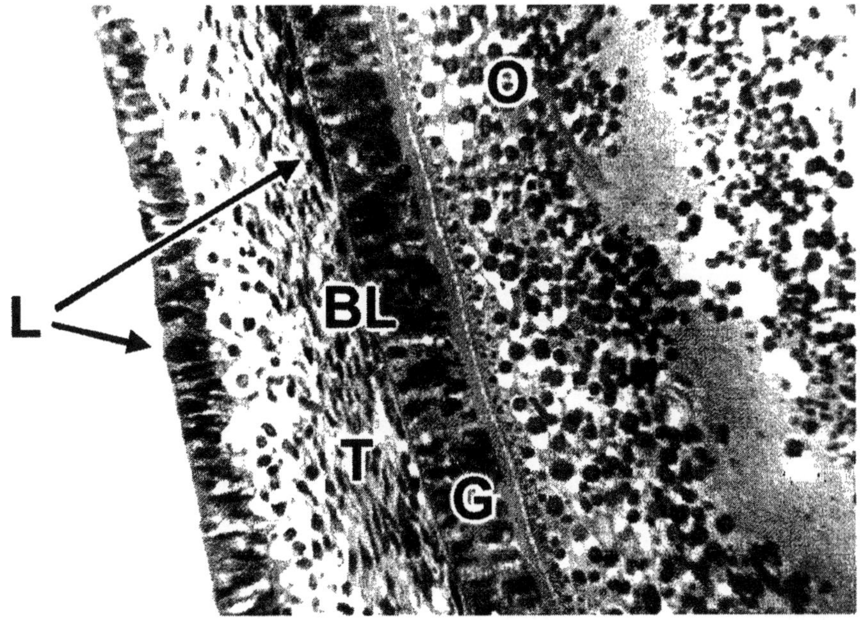

Fig. 8.20 Cross section through the ovary of *Leucoraja erinacea*. Leukocytes (L) within the germinal epithelium and between theca (T) and granulosa (G) layers of the follicle wall. BL, basal lamina; O, oocytes. Lutton and Callard 2004. X 20. Original.

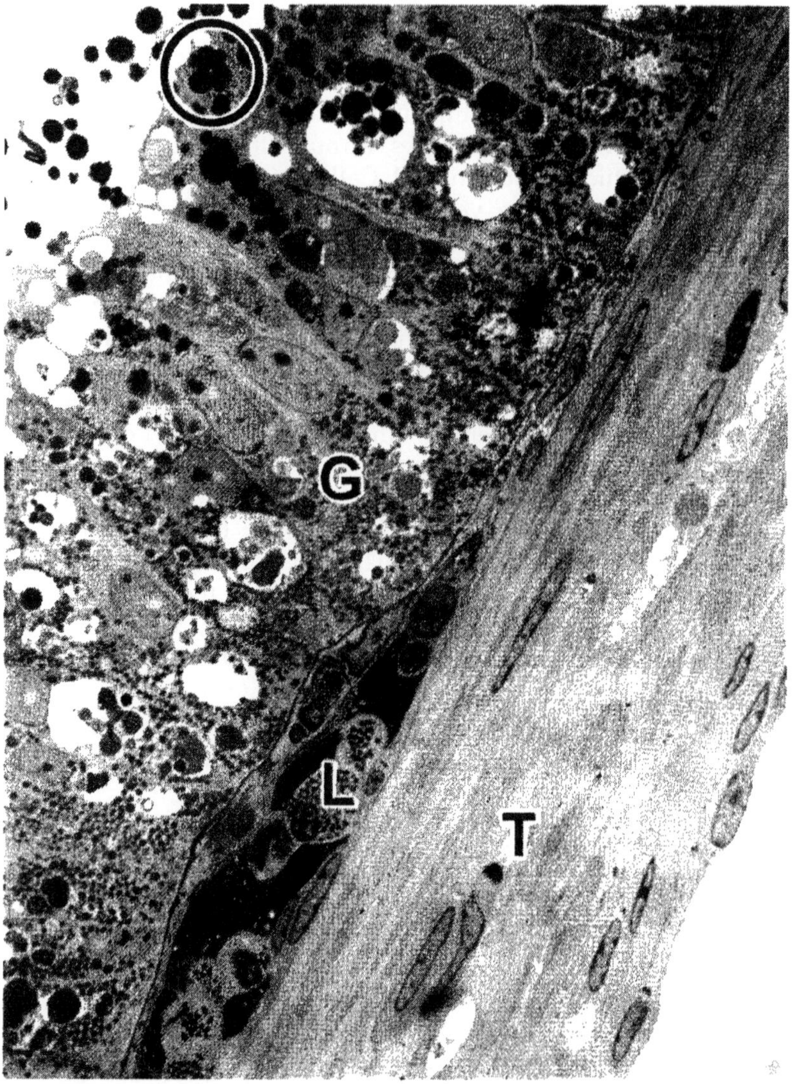

Fig. 8.21 Leukocytes (L) containing numerous granules within the follicle wall of *Urobatis jamaicensis*. A blood vessel between theca cells (T) and granulosa cells (G), containing prominent yolk precursor vesicles (circle), carries the leukocytes. X 600. Courtesy of W. Hamlett.

observations by Fänge and Pulsford (1983) have also illustrated the ability of elasmobranch leukocytes, like those of mammals, to move through fenestrated blood vessels by the process of diapedesis. These studies indicate that cellular contact between immune cells and granulosa/theca cells of the follicle wall is a natural phenomenon, demonstrating the

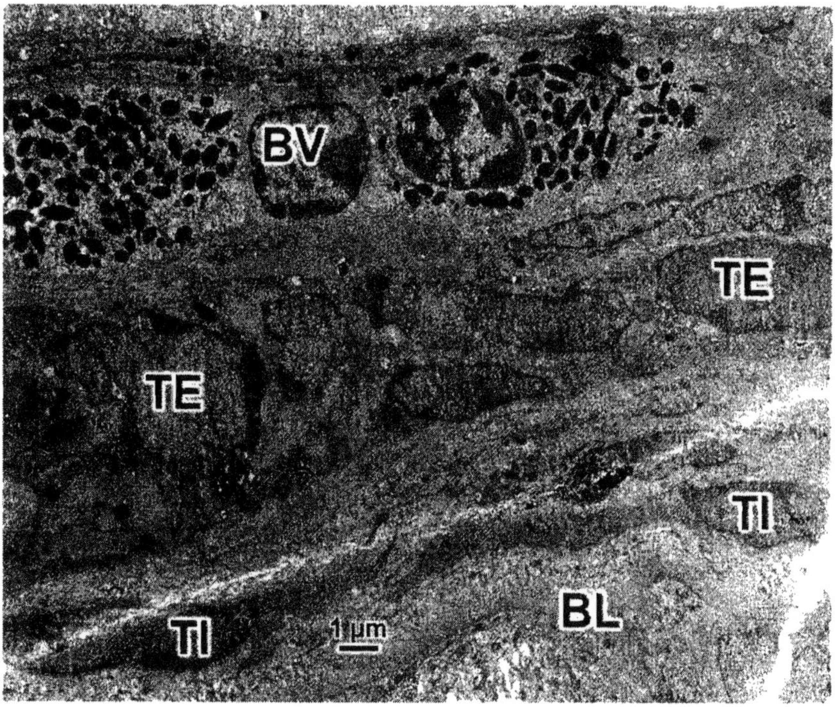

Fig. 8.22 Granulocytic leukocytes within a blood vessel (BV) of a 2500 µm follicle from *Torpedo marmorata*. TI, theca interna; TE, theca externa; BL, basal lamina. Prisco *et al.* 2002. Molecular Reproduction and Development 61: 78-86, Fig. 20.

potential for a functional interaction between the cells of the epigonal organ and steroidogenic cells of the follicle wall. Indeed, production of E_2 and T from the follicle wall cells of *L. erinacea* is modulated by cells and conditioned media of the epigonal organ (Lutton and Callard 2004). It has also been shown that sex steroids (P_4 and T, but not E_2) are capable of inducing apoptosis of epigonal leukocytes to the same extent as corticosteroids (Lutton and Callard 2005). Furthermore, prior investigation with *Squalus acanthias* suggests that an epigonal organ derived substance inhibits spermatogenesis (Piferrer and Callard 1995), shown by inhibition of DNA synthesis in pre-meiotic spermatocysts. Since a significant modulation of E_2 and/or T production by the epigonal organ could potentially induce or prevent ovulation during the reproductive period, this inhibitory effect supports evidence that a secreted factor from the epigonal organ may have the ability to regulate reproductive function *in vivo*. It is also likely that proper functioning of the epigonal organ-ovary complex is maintained by a bi-directional feedback system between these two tissues.

8.9 FISHERY MANAGEMENT: THE EFFECTS OF STRESS ON REPRODUCTION

Although a great deal of attention has been placed on the effects of stress in teleost fish reproduction, there is no information to date for elasmobranchs. At the level of the gonads in teleosts, adrenal corticoids, pro-opiomelanocortin (POMC)-like peptides, and corticotropin-releasing factor (CRF) are all reported to interfere with the stimulatory action of gonadotropins on sex steroid-producing cells (reviewed in Rivier and Rivest 1991). Also, pituitary responsiveness to GnRH is thought to decrease with increasing amounts of circulating corticosteroids. Although neither acute nor prolonged effects of stress on elasmobranch reproduction have been identified, a vast amount of literature illustrates that population stress has been leading to declines in overall numbers of these species.

Recently, concern has been raised about overexploitation of elasmobranch populations. Exploitation of elasmobranchs as bycatch and as targets in directed fisheries has increased in the past few decades (Bonfil 1994; Rose 1996). These species are particularly vulnerable to overfishing due to their K-selected life-history strategy, which is characterized by production of few well developed young throughout reproductive life. This low fecundity is influenced by many factors such as slow growth rate, late age at maturity, long gestation period and small litter size. Due to the close connection between adult stocks and juvenile recruitment, when mature females are fished out of the population, depletion is faster than replacement, leading to a rapid decline in catch rates and possible complete collapse of the fishery (Ripley 1964; Holden 1968). Without new management strategies, sustainable fisheries for elasmobranchs are unlikely (Holden 1974, 1977). Other factors make elasmobranch fisheries difficult to monitor and maintain. Walker *et al.* (2000) noted that few countries have management strategies for these species, and due to their highly migratory nature they are placed outside the responsibility of individual countries and outside the mandate of international bodies (Stevens *et al.* 2000). As a consequence, the reported catch is expected to be only half of the actual global catch (Bonfil 1994).

Unchecked fishing pressure on elasmobranchs has led to a decrease in stock size and changes in population structure of most species. It has been estimated that almost all shark species have declined by more than 50% in the past 8 to 15 years (Baum *et al.* 2003). One population that has received considerable attention is *Squalus acanthias* from the Northwest Atlantic. In this region, adult female *S. acanthias* populations have been severely reduced by overfishing. Since 1987, there has been a five-fold increase in commercial landings of *S. acanthias* with 95% being mature females (Rago 1998). It is estimated that the biomass of mature females has decreased by more than 50% since 1989 and recruitment of juveniles is being negatively impacted (NMFS 2001). Although the life history of this species is well established, little is known of individual responses to severe population reduction. In the 1970's, Holden suggested that flux in stock size would result in a change in fecundity in order to maintain the population (Holden 1977). Since this prediction, changes in size at maturity have been correlated with changes in stock abundance. Increases in *S. acanthias* abundance have been linked with an increase in size at maturity (Silva 1993). In addition, a four year study on the current Northwest Atlantic population has shown that the size at onset of maturity for female *S. acanthias* has declined from 75 cm as seen in the late 1980's to 66 cm and the median size at maturity has declined from 85 cm to 79 cm. These changes have been correlated with the current decreasing trends in population size (Sosebee 2002).

While stock assessment and analysis of population structure continue, it will be important for future investigators to include specific parameters of acute and prolonged stress in these species. The importance of top predators within food webs has been understood for many years, but the reproductive fragility of elasmobranchs has yet to be adequately defined. Gathering information regarding the effects of stress on the reproductive systems of these animals will hopefully become another factor in conservation efforts designed to protect them.

8.10 ACKNOWLEDGEMENTS

The writers are grateful for the support of NSF 8606344, PCM 78-08201, 8104144, 86-06344, and 88-01301; NIH RO1-RR 06633;and NIH ES 07381 to IPC during the conduct of research reported here.

8.11 LITERATURE CITED

Adashi, E. Y. 1990. The potential relevance of cytokines in ovarian physiology: the emerging role of resident ovarian cells of the white blood cell series. Endocrine Reviews 11(3): 454-564.

Andreuccetti, P., Iodice, M., Prisco, M. and Gualtieri, R. 1999. Intercellular bridges between granulosa cells and the oocyte in the elasmobranch, *Raya asterias*. The Anatomical Record 255: 180-187.

Andreuccetti, P., Taddei, C. and Filosa, S. 1978. Intercellular bridges between follicle cells during the differentiation of follicular epithelium in *Lacerta sicula*. Journal of Cell Science 33: 341-350.

Babel, T. S. 1967. Reproduction, life history, and ecology of the round stingray, *Urolophus halleri*. Cooper. California Department of Fish and Game Bulletin 137: 1-104.

Balfour, E. M. 1876. On the development of Elasmobranch Fishes: The general features of the Elasmobranch embryo at successive stages. Journal of Anatomy and Physiology 10: 555-576 and Plates XXIV-XXVI.

Ballard, W. M., Mellinger, J. and Lechenault, H. 1993. A series of normal stages for development of *Scyliorhinus canicula*, the lesser spotted dogfish (Chondrichthyes: Scyliorhynidae). Journal of Experimental Zoology 267: 318-336.

Baum, J. K., Myers, R. A., Kehler, D. G., Worm B., Harley S. J. and Doherty P. A. 2003. Collapse and Conservation of Shark Populations in the Northwest Atlantic. Science 299: 389-392.

Bigelow, H. and Schroeder, W. 1953. Fishes of the Western North Atlantic, Part *Two: Sawfish, Guitarfish, Skates and Rays*. Sears Foundation for Marine Research, New Haven. 502 pp.

Bonfil, R. 1994. Overview of world elasmobranch fisheries. FAO Fisheries Technical Paper, 341., 199 pp.

Bonham, K., Sanford, B., Clegg, W. and Bucher, G. 1949. Biological and Vitamin A Studies of the dogfish landed in the state of Washington (*Squalus suckeyi*). State of Washington Department of Fishery Biology Rep. 49A: 83-114.

Bukulmez, O. and Arici, A. 2000. Leukocytes in ovarian function. Human Reproduction Update 6: 1-15.

Bullesbach, E. E., Gowan, L. K., Schwabe, C., Steinetz, B. G., O'Byrne, E. and Callard, I. P. 1986. Isolation, purification, and the sequence of relaxin from the spiny dogfish (*Squalus acanthias*). European Journal of Biochemistry 161: 335-341.

Bullesbach, E. E., Schwabe, C. and Callard, I. P. 1987. Relaxin from and oviparous species, the skate (*Leucoraja erinacea*). Biochemistry and Biophysics Research Communications 143: 273-280.

Callard, I. P., Giannoukos,G., Charnock-Jones, D. S., Benson, S. and Paolucci, M. 1994. Hormone regulation of vitellogenin genes and the evolution of viviparity, In *Perspectives in Comparative Endocrinology*, National Resource Council of Canada. Pp. 325-332.

Callard, I. P., Filetti, L. A. and Koob, T. J. 1993. Ovarian steroid synthesis and the hormonal control of the elasmobranch reproductive tract. Environmental Biology of Fishes 38: 175-185.

Callard, I. P., Klosterman, L. L., Sorbera, L. A. Fileti, L. A., and Reese, J. C. 1989. Endocrine regulation of reproduction in elasmobranchs: Archetype for terrestrial vertebrates. Journal of Experimental Zoology Supplement 2: 12-22.

Capape, C. 1979. La torpille marbrée, *Torpedo marmorata* Risso, 1810 (Pices, Raji-Formes) des côtes tunisiennes: Nouvelles donnés sur l'écologie et la biologie de la reproduction de l'espèce, avec une comparaison entre les populations mediterranéennes et atlantiques. Annales Des Sciences Naturelles. Zoologica et Biologie Animale 1: 79-97.

Castro, J. I. 1989. The biology of the golden hammerhead, *Sphyrna tudes*, off Trinidad. Environmental Biology Fishes 24: 3-11.

Castro, J. I., Bubucis, P. M. and Overstrom, N. A. 1988. The reproductive biology of the chain dogfish, *Scyliorhinus retifer*. Copei pp. 740-746.

Champy, C. and Gley, P. 1923. Obervations cytologiques sur les ovocytes des poisons et de quelques autres vertébrés. Archives d'Anatomie Microsopique et de Morphologie Experimentale 19: 241-308.

Chen, C.-T., Leu, T.-C. and Joung, S.-J. 1988. Notes on reproduction in the scalloped hammerhead, *Sphyrna lewini*, in northeastern Taiwan waters. Fishery Bulletin 86: 389-393.

Chieffi, B. G., Minucci, S., Di Matteo, L. and Chieffi, G. 1992. Ultrastructural investigation of the corpora atretica of the Electric Ray, *Torpedo marmorata*. General and Comparative Endocrinology 86: 72-80.

Chieffi, G. 1951. Sull'organogenesi della medulla della gonade in *Torpedo ocellata* e in *Scyliorhinus canicula*. Bolletina Zoologica 18: 183-187.

Chieffi, G. 1952a. Sull'organogenesi de3ll'interrenale della medulla della gonade in *Torpedo ocellata* e in *Scyliorhinus canicula*. Pubblicazioni della Stazione Zoologica di Napoli 22: 186-200.

Chieffi, G. 1952b. Azione del testosterone e della diidrofollicolina sul differenziamento sessuale di *Scyliorhinus canicula*. Bolletina Zoologica 19: 117-122.

Chieffi, G. 1953a. Azione del testosterone sul differenziamento sessuale di *Scyliorhinus canicula*. Ricerches Scientificos 23: 111-117.

Chieffi, G. 1954. L'inversione del sesso ottenuta con gli ormoni sessuali ecorticosurrenale in *Scyliorhinus canicula*. Pubblicazioni della Stazione Zoologica di Napoli 25: 477-498.

Chieffi, G. 1955. Nuove osservazioni sull'organogenesi della medulla della gonade nei Vertebrati: Ricerche istochimiche in *Rana escuelenta, Bufo viridis*, e *Scyliorhinus canicula*. Pubblicazioni della Stazione Zoologica di Napoli 27: 62-72.

Chieffi, G. 1961. La luteogenesi nei selaci ovovivipari. Richerche istologische e istochimiche in *Torpedo marmorata* e Torpedo ocellata. Pubblicazioni della Stazione Zoologica di Napoli 32: 145-166.

Chieffi, G. 1967. The reproductive system of the elasmobranchs: development and endocrinological aspects. Pp. 553-580. In Gilbret, Mathewson, and Rall (eds), *Sharks, skates and rays* John Hopkins Press, Baltimore.

Collenot, G. 1969. Ètude biomètrique de la croissance relative des ptèrygopodes chez la roussette *Scyliorhinus canicula* (L.). Cahiers de Biologie Marine 10: 309-323.

Craik, J. C. A. 1978a. Plasma levels of vitellogenin in the elasmobranch, *Scyliorhinus canicula*. L. Comparative Biochemistry and Physiology 60B: 9-18.

Craik, J. C. A. 1978b. Kinetic studies of vitellogenin metabolism in the elasmobranch *Scyliorhinus Canicula*. L. Comparative Biochemistry and Physiology 61A: 355-361.

Craik, J. C. A. 1978c. An annual cycle of vitellogenesis in the elasmobranch, *Scyliorhinus Canicula*. L. Journal of the Marine Biological Station, UK. 58: 719-726.

Craik, J. C. A. 1978d. The effects of oestrogen treatment on certain plasma constituents associated with vitellogenesis in the elasmobranch, *Scyliorhinus canicula*. L. General and Comparative Endocrinology 35: 691-702.

Craik, J. C. A. 1979. Simultaneous measurement of rates of vitellogenin synthesis and plasma levels of oestradiol in an elasmobranch. General and Comparative Endocrinology 38: 264-266.

Cross, J. N. 1988. Aspects of the biology of two Scyliorhinid sharks, *Apristurus brunneus* and *Parmaturus xaniurus*, from the upper continental slope off southern California. Fishery Bulletin 86: 691-702.

Daniel, J. F. 1928. *The elasmobranch fishes*. University of California Press, Berkeley.

Davenport, I. R. 2003. Comparative morphology of oogenesis pertaining to the evolution of extreme egg size in Chondrichthyan fishes. Ph.D. Thesis. Clemson University, Clemson, South Carolina.

Dodd, J. M., Evenett, P. J. and Goddard, C. K. 1960. Reproductive Endocrinology in Cyclostomes and Elasmobranchs. Symposium of the Zoological Society London 1: 77-103.

Dodd, J. M. 1972. Ovarian control in cyclostomes and elasmobranchs. American Zoologist 12: 325-339.

Dodd, M. H. I. and Dodd, J. M. 1980. Ultrastructure of the ovarian follicle in the dogfish *Scyliorhinus canicula*. General and Comparative Endocrinology 40: 330.

Dodd, J. M. 1983. Reproduction in cartilaginous fishes (Chondrichthyes). Pp. 31-95. In Hoar, W. S., Randall, D. J., Donaldson, E. M. (eds), *Fish physiology, vol. IX. Reproduction, part A. Endocrine tissues and hormones*. Academic Press.

Dodd, J. M. and Sumpter, J. P. 1984. *Fishes*. Pp. 1-126. In G. E. Lamming (ed.), *Marshall's Physiology of Reproduction*, Volume 1: *Reproductive Cycles of Vertebrates*. 4th Edition. Churchill Livingstone Inc., New York NY.

Fänge, R. and Pulsford, A. 1983. Structural studies on lymphomyeloid tissues of the dogfish, *Scyliorhinus canicula* L. Cell and Tissue Research 230: 337-351.

Fasano, S., D'Antonio, M., Pierantoni, R. and Chieffi, G. 1992. Plasma and follicular tissue steroid levels in the elasmobranch fish, *Torpedo marmorata*. General and Comparative Endocrinology 85: 327-333.

Faure-Fremiet, E. 1933. Les constituents cytoplasmiques de l'oeuf de raie. Protoplasma 19: 63-77.

Fileti, L. A. and Callard, I. P. 1988. Corpus luteum function and regulation in the little skate, *Leucoraja erinacea*. Bulletin of the Mount Desert Biological Lab 27: 37-39.

Fileti, L. A. and Callard, I. P. 1990. Regulation of *in vitro* ovarian steriodogenesis in the little skate, *Leucoraja erinacea*. Bulletin of the Mount Desert Biological Lab 29: 129-130.

Filosa, S., Taddei, C. and Andreuccetti, P. 1979. The differentiation and proliferation of the follicle cells during oocyte growth in *Lacerta sicula*. Journal of Embryology and Experimental Morphology 54: 5-15.

Fujii, T. 1960. Comparative biochemical studies of the egg-yolk proteins of various animal species. Acta Embryologiae et Morphologiae Experimentalis 3: 260-285.

Gabaeva, N. S. 1980. Morphodynamics of follicular epithelium during elasmobranch oogenesis (the problem of histological paralellisms and convergences in the development of vertebrate follicular epithelia. Arkhiv Anatomii, Gistologii, i Embriologii 78(4): 70-78.

Giacomini, E. 1896. Contributo all'istologia dell'ovario dei Selaci. Rish. Lab. Anat. Norm. Univ. Roma. 5: 221-274.

Gilmore, R. G. 1993. Reproductive biology of lamnoid sharks. Environmental Biology of Fishes 38: 95-114.

Gudger, E. W. 1940. The breeding habits, reproductive organs and external embryonic development of Chlamydoselachus, based on notes and drawings by Bashford Dean. The Bashford Dean Memorial Volumes, Archaic Fishes, Art. VII. American Museum of Natural History.

Guraya, S. S. 1978. Maturation of the follicular wall of nonmammalian vertebrates. Pp. 261-329. In R. E. Jones (ed.), *The vertebrate ovary*. Plenum Press, New York.

Hamlett, W. C., Jezior, M. and Spieler, R. 1999. Ultrastructural analysis of folliculogenesis in the ovary of the yellow spotted stingray *Urolophus jamaicensis*. Annals of Anatomy 181: 159-172.

Hisaw, F. L. and Abramowitz, A. A. 1939. Physiology of Reproduction in the dogfishes, *Mustelus canis* and *Squalus acanthias*. Pp. 22. Reports of the Woods Hole Oceanographic Institute for 1938.

Hisaw, F. L. and Albert, A. 1947. Observations on the reproduction of the spiny dogfish, *Squalus acanthias*. Biological Bulletin, Woods Hole 92: 187-199.

Hisaw, F. L. Jr. and Hisaw, F. L. Corpora lutea of elasmobranch fishes. Anatomical Record 135: 269-277.

Ho, S.-M., Wulczyn, G. and Callard, I. P. 1980. Induction of vitellogenin synthesis in the spiny dogfish *Squalus acanthias*. Bulletin of the Mount Desert Biological Lab 19: 37-38.

Holden, M. J. 1968. The Rational Exploitation of the Scottish-Norwegian stock of spurdogs (*Squalus acanthias* L.). Fishery Investigations, London, Ser II 25(8): 1-27.

Holden, M. J. 1974. Problems in the Rational Exploitation of Elasmobranch Populations and Some Suggested Solutions. Pp. 177-138. In F. R. Harden-Jones (ed.), *Sea Fisheries Research*. Logos Press, London.

Holden, M. J. 1977. Elasmobranchs. Pp. 187-215. In Gulland J. A. (ed.) *Fish Population Dynamics*. John Wiley and Sons, New York, NY.

Huepel, M. R., Whittier, J. M., and Bennett, M. B. 1999. Plasma steroid hormone profiles and reproductive biology of the Epaulette Shark, *Hemiscyllium ocellatum*. Journal of Experimental Zoology 284: 586-594.

Jasinski, A. 1969. Vascularisation of the hypophyseal region in lower vertebrates (cyclostomes and fishes). General and Comparative Endocrinology. Supplement 2: 510-521.

Jenkins, N. and Dodd, J. M. 1980. Effects of Synthetic Mammalian Gonadotropin Releasing Hormone and Dogfish Hypothalamic Extracts on Levels of Androgens and Estradiol in the Circulation of the Dogfish (*Scyliorhinus canicula* L.) Journal of Endocrinology 86: 171-177.

Jensen, N. H. 1976. Reproduction of the bull shark, Carcharhinus leucas, in the Lake Nicaragua-Rio San Juan system, in Thorson, Investigations of the ichthyofauna of Nicaraguan lakes. University of Nebraska, Lincoln.

Johnson, M. H. and Everitt, B. J. (ed.) 2000. *Essential Reproduction*. Second Edition, Blackwell Science Ltd., London. 285 pp.

Karsch, F. J. The hypothalamus and anterior pituitary gland. Pp. 1-20. In C. R. Austin and R. V. Short (eds), *Reproduction in mammals: 3, Hormonal Control and Reproduction*. Cambridge University Press, London.

Koob, T. J. and Callard, I. P. 1982. Relaxin: Speculations on its physiological importance in some non-mammalian species. Annals of the New York Academy of Science 380: 163-173.

Koob, T. J. and Callard, I. P. 1999. Reproductive endocrinology of female elasmobranchs: lessons from the little skate (*Leucoraja erinacea*) and spiny dogfish (*Squalus acanthias*). Journal of Experimental Zoology 284: 557-574.

Koob, T. J. and Callard, I. P. 1991. Reproduction in female elasmobranches. Pp. 155-209. In R. K. H. Kinne (ed.), *Oogenesis, spermatogenesis and reproduction*. Basel, Switzerland: Karger AG.

Koob, T. J., Laffan, J. J. and Callard, I. P. 1984. Effects of relaxin and insulin on reproductive tract size and early fetal loss in *Squalus acanthias*. Biology of Reproduction 31: 231-238.

Koob, T. J., Laffan, J. L., Elger, B. and Callacrd, I. P. 1986. Effects of estradiol on the vershlussvorrichtung of *Squalus acanthias*. Bulletin of the Mount Desert Biological Lab 23: 67-68.

Koob, T. J. and Strauss, J.W. 1998. On the role of egg jelly in *Leucoraja erinacea*. Bulletin of the Mount Desert Island Biological Lab 37: 117-119.

Koob, T. J., Tsang, P. and Callard, I. P. 1986. Plasma estradiol, testosterone, and progesterone levels during ovulatory cycle of the skate (*Leucoraja erinacea*). Biology of Reproduction 35: 267-275.

Kudo, S. 1956. On *Squalus brevirostris* in Hyuga-Nada. Report. Nankai Regional Fisheries Research Laboratory 3: 66-72.

Lance, V. and Callard, I. P. 1969. A histochemical study of ovarian function in the ovoviviparous eleasmobranch, *Squalus acanthias*. General and Comparative Endocrinology 13: 255-267.

Lance, V. and Callard, I. P. 1978. Gonadotrophic activity in pituitary extracts from a elasmobranch (*Squalus acanthias* L.). Journal of Endocrinology 78: 149-150.

Lovejoy, D. A., Stell, W. K. and Sherwood, N. M. 1992a. Partial characterization of four forms of immunoreactive gonadotropin-releasing hormone in the brain and terminal nerve of the spiny dogfish (Elasmobranchii; *Squalus acanthias*). Regulatory Peptides 37: 39-48.

Lovejoy, D. A., Ashmead, B. J., Coe, I. R. and Sherwood, N. M. 1992b. Presence of gonadotropin-releasing hormone immunoreactivity in dogfish and skate brains. Journal of Experimental Zoology 263: 272-283.

Luer, C. A. and Gilbert, P. W. 1985. Mating behavior, egg deposition, incubation period, and hatching in the clearnose skate, *Raja eglanteria*. Environmental Biology of Fishes 13: 161-171.

Lutton, B. and Callard, I. P. 2004. Inhibition of steroidogenesis by a unique lymphomyeloid tissue in the little skate, *Leucoraja erinacea*. Abstract: Proceedings of the Society for the Study of Reproduction.

Lutton, B. and Callard, I.P. 2005. The elasmobranch epigonal organ-ovary complex (EOC): regulation of apoptosis by sex hormones. Abstract: Proceedings of the International Conference of Comparative Endocrinology.

McLaughlin, R. H. and O' Gower, A. K. 1971. Life history and underwater studies of heterodontid sharks. Ecological Monographs 41: 271-289.

Manire, C. A., Rasmussen, L. E. L., Hess, D. L. and Hueter, R. E. 1995. Serum steroid hormones and the reproductive cycle of the female bonnethead shark, *Sphyrna tiburo*. General and comparative Endocrinology 97: 366-376.

Martin, L. K. and Cailliet, G. M. 1988. Aspects of reproduction of the bat ray, *Myliobatis californica*, in central California. Copeia pp. 754-762.

Matthews, L. H. 1950. Reproduction in the basking shark, *Cetorhinus maximus* (Gunner). Philosophical Transitions. Royal Society of London 234: 247-316.

Mellinger, J. 1974. Criossance et reproduction de la torpille (*Torpedo marmorata*) III. L'appaeil genital femelle. Bulletin Biologique de la France et de Belgique 108: 107-150.

Metten, H. 1939. Studies on the reproduction of the dogfish. Philosophical Transactions of the Royal Society of London: 230: 217-238.

Natanson, L. J. and Cailliet, G. M. 1986. Reproduction and development of the Pacific angel shark, *Squantina californica*, off Santa Barbara, California. Copei. 987-994 pp.

Needham, J. 1988. *Biochemistry and morphogenesis*. University Press, Cambridge.

NMFS 2001. 2001-2002 Spiny Dogfish Specifications: Draft Environmental Assessment. Regulatory Impact Review, Initial Regulatory Flexibility Analysis, EFH Assessment. Department of Commerce, NOAA, NMFS, MAMFC & NEFMC, Silver Spring MD: 1-90.

Parsons, G. R. 1981.The reproductive biology of the Atlantic sharpnose shark Rhizoprionodon terraenovae (Richardson). Fishery Bulletin 81: 61-73.

Perez, L. E. and Callard, I. P. 1989. Evidence for progesterone inhibition of vitellogenesis in the skate. American Zoologist 27: 357A.

Perez, L. E. and Callard, I. P. 1993. Regulation of hepatic vitellogenin synthesis in the little skate (*Leucoraja erinacea*): use of a homologous enzyme-linked immunosorbent assay. The Journal of Experimental Zoology 266: 31-39.

Picon, R. 1962. Recherches sur la diffèrentiation sexualle de l'embryon de *Leptocharias smithii* (mullet et henle) elasmobranche. Archives d'Anatomie microscopique et de Morphologie experimentale 51: 541-576.

Piferrer, F. C., and Callard, G. V., 1995. Inhibition of deoxyribonucleic acid synthesis during premeiotic stages of spermatogenesis by a factor from testis-associated lymphomyeloid tissue in the Dogfish Shark (*Squalus acanthias*). Biology of Reproduction 53: 390-398.

Powell, R. C., Millar, R. P. and King, J. A. 1986. Diverse molecular forms of gonadotropin-releasing hormone in an elasmobranch and a teleost fish. General and Comparative Endocrinology 63: 77-85.

Pratt, H. L. 1979. Reproduction in the blue shark, *Prionace glauca*. Fishery Bulletin 77: 445-470.

Pratt, H. L.1988. Elasmobranch gonad structure: A description and survey. Copeia 3: 719-729.

Prisco, M., Loredana, R. and Piero A. 2002. Ultrastructural studies on the developing follicles of the spotted ray *Torpedo marmorata*. Molecular Reproduction and Development 61: 78-86.

Prisco, M., Romano, M., Ricchiaro, L., Limatola, E. and Andreuccetti, P. 2002. An ultrastructural study on the vitellogenesis in the spotted ray, *Torpedo marmorata*. General and Comparative Endocrinology 128: 171-179.

Prisco, M., Ricchiari, L. and Andreuccetti, P. 2001. An ultrastuctural study of germ cells during ovarian differentiation in *Torpedo marmorata*. The Anatomical Record 263: 239-247.

Rago, P. J., Sosebee, K. A., Brodziak, J. K. T., Murawski, S. A. and Anderson E. D. 1998. Implications of recent increases in catches on the dynamics of Northwest Atlantic spiny dogfish (Squalus acanthias). Fisheries Research 39: 165-181.

Ranzi, S. 1934. Le basi fisio morphologiche dello sviluppo embryionale dei Selaci, parts 2, 3. Pubblicazioni. Stazione Zoologica di Napoli 13: 331-437

Rasmussen, L. E. L., Hess, D. L., and Luer, C. A. 1999. Alterations in serum steroid concentration in the clearnose skate, *Raja eglanteria*: correlations with season and reproductive status. Journal of Experimental Zoology 284: 575-585.

Reinig, J. W., Daniel, L. N., Schwabe, C., Gowan, L. K., Steinetz, B. G and O'Byrne, E. 1981. Isolation and characterization of relaxin from the sand tiger shark (*Odontaspis taurus*). Endicrinology 109: 537-543.

Ripley, W. E. 1964. The Soupfin Shark and The Fishery. California Division of Fish and Game, Fishery Bulletin 64: 7-37.

Rivier, C., and Rivest, S. 1991. Effect of stress on the activity of the hypothalamic-pituitary-gonadal axis: Peripheral and central mechanisms. Biology of Reproduction 45: 523-532.

Rose D. A. 1996. An overview of world trade in sharks and other cartilaginous fishes. Traffic International, Cambridge, UK. 106 pp.

Samuel, M. 1943. Studies on the corpus luteum in *Rhinobatus granulatus* Cuvier. Proceedings of the Indian Academy of Science 18B: 133-157.

Samuel, M. 1946. The corpus luteum in *Chiloscyllium griseum* (Muller and Henle). Proceedings of the Indian Academy of Science 22B: 113-123.

Scammon, R. E. 1911. Normal plates of development of *Squalus acanthias*. Normentafeln zur Entwicklungsgeschichts der Wirbeltiere F. Keibel, ed. G. Fischer, Jena 12: 1-140.

Scanes C. G., Dobson S., Follett B. K. and Dodd J. M. 1972. Gonadotrophic activity in the pituitary gland of the dogfish (*Scyliorhinus canicula*). Journal of Endocrinology 54: 343-344.

Schlernitzauer, D. A. and Gilbert, P. W. 1966. Placentation and associated aspects of gestation in the bonnethead shark, *Shyrna tiburo*. Journal of Morphology 120: 219-232.

Schwartz, F. J. and Dahlberg, M. D. 1978. Biology and ecology of the Atlantic stingray, *Dasyatis sabina* (Pisces: Dasyatidae), in North Carolina and Georgia. Northeast Gulf Science 2: 1-23.

Setna, S. B. and Sarangdhar, P. N. 1948. Observations on the development of *Chiloscyllium griseum* M. & H., *Pristus cuspidatus* L and *Rhynchobatus dijiddensis* (Forsk). Records. Indian Museum 46: 1-24.

Sherwood, N. M. and Sower, S. A. 1985. A new family member for gonadotropin-releasing hormone. Neuropeptides 6: 205-214.

Sherwood, N. M. and Lovejoy, D. A. 1993. Gonadotropin-releasing hormone in cartilaginous fishes: structure, location, and transport. Environmental Biology of Fishes 38: 197-208.

Silva, H. M. 1993. Population Dynamics of Spiny Dogfish, *Squalus acanthias*, in the NW Atlantic. Amherst, MA: University of Massachusetts Dissertation.

Smeets, W. J. A. J., Nieuwenhuys, R. and Roberts, B. L. 1983. *The Central Nervous System of Cartilaginous Fishes. Structure and Functional Correlations*. Springer-Verlag, New York. 1-266 pp.

Smith, J. W. and Merriner, J. V. 1986. Observations on the reproductive biology of the cownose ray, *Rhinoptera bonasus* in Chesapeake Bay. Fishery Bulletin 84: 871-877.

Sosebee, K. A. 2002. Are Density-dependent Effects on Elasmobranch Maturity Possible? Scientific Council Meeting – September 2002. NAFO SCR Doc. 02/120.

Snelson, F. F. Jr., Rasmussen, L. E. L., Johnson M. R., and Hess, D. L. 1997. Serum concentrations of steroid hormones during reproduction in the Atlantic stingray, *Dasyatis Sabina*. General and Comparative Endocrinology 108: 67-79.

Snelson, F. F. Jr., Williams-Hooper, S. E., and Schmid, T. H. 1989. Biology of the bluntnose stingray, *Dasyatis sayi*, in Florida coastal lagoons. Bulletin Marine Science 45: 15-25.

Springer, S. 1960. Natural history of the sandbar shark, *Eulamia milberti*. Fishery Bulletin 61: 1-38.

Steinetz, B. G., Beach, V. H. and Kroc, R. L. 1959. The physiology of relaxin in laboratory animals. Pp. 389-428. In C. W. Lloyd (ed.), *Recent progress in the endocrinology of reproduction*. Academic Press, New York.

Stevens, J. D., Bonfil, R., Dulvy, N. K., Walker, P. A. 2000. The effect of fishing on sharks, rays, and chimeras (chondrichthyans), and the implications for marine ecosystems. ICES Journal of Marine Science 57: 476-494.

Sumpter, J. P. and Dodd, J. M. 1979. The annual reproductive cycle of the female lesser spotted dogfish, *Scyliorhinus canicula* L., and its endocrine control. Journal of Fish Biology 15: 687-695.

TeWinkle, L. E. 1972. Histological and histochemical studies of post-ovulatory and pre-ovulatory atretic follicles in *Mustelus canis*. Journal of Morphology 136: 433-458.

Tricas, T. C., Maruska, K. P. and Rasmussen, L. E. L. 2000. Annual cycles of steroid hormone production, gonad development, and reproductive behavior in the Atlantic stingray. General and Comparative Endocrinology 118: 209-225.

Tsang, P. and Callard, I. P. 1982. Steroid production by isolated skate ovarian follicle cells. Bulletin of the Mount Desert Biological Lab 22: 96-97.

Tsang, P. and Callard, I. P. 1983. In vitro steroid production by ovarian granulosa cells of *Squalus acanthias*. Bulletin of the Mount Desert Biological Lab 23: 78-79.

Tsang, P. and Callard, I. P. 1987a. Morphological and endocrine correlates of the reproductive cycle of the aplacental viviparous dogfish, *Squalus acanthias*. General and Comparative Endocrinology 66: 182-189.

Tsang, P. and Callard, I. P. 1987b. Luteal progesterone production and regulation in viviparous dogfish, *Squalus acanthias*. Journal of Experimental Zoology 241: 377-382.

Tsang, P. C. and Callard, I. P. 1988. *In Vivo* Steroidogenic Effects of Homologous Pituitary Ventral Lobe Extract in the Spiny Dogfish, *Squalus acanthias*. General and Comparative Endocrinology, 70: 164-168.

Wallace, W. 1903. Observations on ovarian ova and follicles in certain teleostean and elasmobranch fishes. Quarterly Journal of Microscopical Science 47: 161-213.

Wass, R. C. 1973. Size, growth, and reproduction of the sandbar shark, *Carcharinus milberti*, in Hawaii. Pacific Science 27: 305-318.

Wourms, J. P. 1977. Reproduction and development in Chondrichthyan fishes. American Zoologist 17: 379-410.

Yoffey, J. M. and Courtice, F. C. 1970 *Lymphatics, Lymph and the Lymphomyeloid Complex*. Academic Press, New York. 942 pp.

CHAPTER 9

Endocrine Control of the Female Reproductive Tract

Ian P. Callard[1], Joseph St. George[1] and Thomas J. Koob[2]

9.1 THE CHONDRICHTHYAN REPRODUCTIVE TRACT (FIG. 9.1)

The vertebrate reproductive tract is a complex structure derived from the Mullerian duct during sexual differentiation in most vertebrates, although lacking in the Agnatha and Teleosts. Its functions are the conveyance of gonadal products to the exterior, the provision of additional nutrient and protective coverings for the egg, and to provide a protected environment for development in the case of viviparous species. In elasmobranchs, well-developed Mullerian ducts arise early in ontogeny from the pronephros and its duct, an origin somewhat different than that of tetrapods in which the funnel is believed to be of pronephric origin and the oviduct of Wolffian derivation. Nevertheless, there is considerable structural and functional similarity and a suite of primary regulatory hormones appears to have been established in the elasmobranches and to have been retained throughout the vertebrates. With regard to the mammalian uterus, Heller (1966) suggested that it is the target of more hormones than any other single organ. This is likely also true for the chondrichthyan reproductive tract.

In the present chapter we review evidence supporting the conclusion that ovarian derived hormones regulate morphological and physiological events associated with specific functions of the female reproductive tract. Before discussing this evidence, however, we first briefly examine reproductive tract events and processes that are undoubtedly under endocrine control. While there is a relative abundance of gross and microscopic morphological observations on the reproductive tract during reproductive cycles, particularly in viviparous species, little is known about precise biochemical and physiological mechanisms that have

[1]Department of Biology, Boston University, Boston MA 02215,
[2]Skeletal Biology, Shriner's Hospitals for Children, Tampa, FL 33612

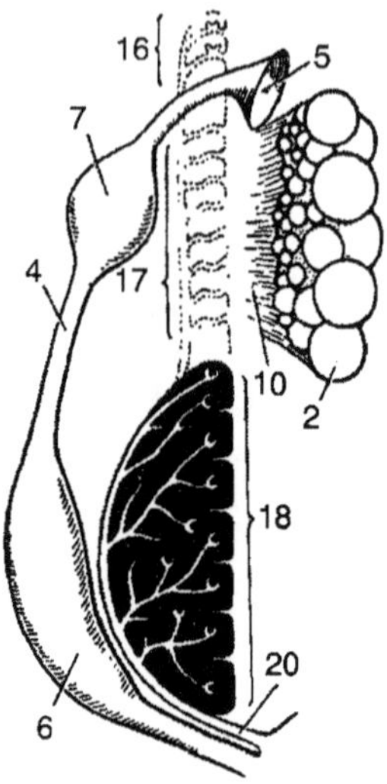

Fig. 9.1 The Müllerian duct and its derivatives in elasmobranchs.
Key: 2 = ovary, 4 = oviduct, 5 = funnel/isthmus, 6 = uterus, 7 = oviducal gland, 10 = mesovarium, 15 = rectal remnant, 16 = pronephros, 17 = opisthonephros, gonadal region, 18 = functional opisthonephros, 20 = ureter.
Reproduced from Hoar, W. S. and Randall, D. J. (1969) Reproduction. In *Fish Physiology* Academic Press and Portman A. (1969). *Einfürung in die vergleichende morphologie die wierbieltiere.* Schwabe and Co. AG. Verlag, Basel.

facilitated evolution of the diverse reproductive modes in elasmobranchs, since data are available for a limited number of species. Moreover, endocrine studies have been have been reported for even fewer species. This chapter describes what little is known about the endocrine control of the female reproductive tract in elasmobranchs.

9.2 DEVELOPMENTAL ASPECTS

The Müllerian ducts develop early in embryonic development from a pronephric origin, possibly programmed genetically (Chieffi 1967). The Wolffian ducts have already begun their regression when the Mullerian ducts first begin their differentiation (Picon 1962). The anterior region of each duct is characterized by a funnel like structure, called the ostium tubae, which is derived from a pronephric nephrostome (Wourms 1977). Frequently, the two funnel structures form a single central ostium. The Mullerian duct forms as the pronephric duct divides to produce a dorsal and ventral tube (Wourms 1977). The dorsal tube is populated by tubules from the kidney, and opens separately into the cloaca. The ventral tube, which is continuous with the ostium, differentiates as the Mullerian duct. The duct is composed of several distinct regions. The ostium is at the anterior end, opening into the body cavity. A short narrow region follows which merges with the oviducal gland. The oviducal gland, a small kidney shaped organ is ultimately responsible for secreting the egg capsule material as well as the protective egg jelly. In turn, this opens into a long segment of oviduct which is characterized by two regions: the thin walled tube that is directly posterior to the oviducal gland and the more muscular region which connects to the cloaca.

Few studies of an endocrine nature have been made on elasmobranch sexual differentiation and development. Injection of steroid hormones (testosterone and progesterone) through the shell membrane into the yolk sac of *Scyliorhinus canicula* embryos induced both the oviduct and the Wolffian ducts to hypertrophy (Chieffi 1967). After hormone treatment, the Wolffian ducts, which normally atrophy in females, undergo differentiation, change cell structure, and become feminized by a high dose of testosterone (1.0 mg) or progesterone (.02 mg). Injection of estrogen causes hypertrophy of the primordial oviduct in both genetic males and females.

The female elasmobranch tract undergoes further development just prior to final sexual maturation. At this time, there is a period of growth during which the oviducal gland further differentiates into three distinct, functional zones (See below). Further physiological and biochemical changes occur in the oviducal gland after sexual maturation, which allow secretion and assembly of egg case precursors and maintenance of live sperm. The uterine region in both oviparous and viviparous species similarly differentiates coincident with sexual maturation in preparation for endocrine regulated events specific to individual reproductive modes.

9.3 ADULT MORPHOLOGY AND FUNCTION

The reproductive tract has evolved distinct regional functions in support of the divergent reproductive modes in female elasmobranchs. Common to all species is its role as a conduit for transporting the future progeny from the ovary to the external environment. Along the passage, eggs are fertilized, encapsulated, retained, and delivered. For oviparous species, encapsulation is the dominant process, though transitory egg retention is necessary for the egg capsule to achieve the necessary physicochemical properties before oviposition. For viviparous species, encapsulation, while still maintained in most species, takes on a subordinate role to intrauterine mechanisms adapted for embryonic development to term, including morphogenetic differentiation of the uterine wall, regulation of the intrauterine milieu, and provision of nutrients beyond those supplied in the ova. Distinct forms of viviparity operate in sharks and rays and these are described in detail in other chapter of this volume (Chapters **8**, **10**, **13**, **14**, and **15**). For the purposes of the present chapter, we focus on the following since precise endocrine data are limited to species listed: oviparous species (little skate, *Leucoraja erinacea*; clearnose skate, *Raja eglanteria*; small-spotted catshark, *Scyliorhinus canicula*), lecithotrophic viviparous sharks (aplacental yolk sac viviparity, spiny dogfish; *Squalus acanthias*), and matrotrophic species (placental viviparity, bonnet head shark, *Sphyrna tiburo*; aplacental viviparity, Atlantic stingray, *Dasyatis sabina*).

9.3.1 The Fimbria and Ostium

There is no better description of the fimbria, ostium and early migration of the ovulated oocyte than that of Metten (1939) for the dogfish, *Scyliorhinus canicula*. New researchers in the area of elasmobranch reproductive physiology should begin with this classic paper. Metten (1939) describes the ciliated tracts leading from the single ovary, between the hepatic lobes, and into the hepatic sinus to the fimbrial septa. These septa together form the ostium which then leads into the left and right oviducts. The abdominal ciliated tracts are confined to the adult female, and are absent in the male and immature female, suggesting hormonal control, probably by ovarian steroids, as in mammals (Boling and Blandau 1971). Ciliary activity is not influenced by the presence of eggs in the oviduct (defined as 'pregnancy' in this oviparous species). In a series of detailed descriptions, Metten describes the path of migrating ova through the coelomic spaces to the ostium. According to Metten's direct observations in living dogfish, passage of the ovum through the ostium takes about 30 minutes and does not involve muscular activity; the fimbrial septa which form the ostium 'become wrapped round the ovum, thereby bringing more of it into contact with their cilia and accelerating engulfment'. Due to an asymmetry between the fimbria, passage into the right oviduct is favored, the egg reaching the oviducal gland in about two hours. Metten (1939) observed that these times are probably shorter during normal ovulation and egg transport. Although no direct evidence is available, it seems likely that the activity of the cilia is regulated from ovarian derived factors secreted in the peri-ovulatory period.

9.3.2 The Oviducal Gland (Fig. 9.2)

Nearly all elasmobranchs encapsulate eggs in materials synthesized, secreted, and assembled by the oviducal gland (Hobson 1930; Metten 1939; Hamlett *et al.* 1998). The shell glands in all oviparous species, as well as some lecithotrophic viviparous species like the whale shark and nurse shark, produce substantial egg capsules that, because of the rapid and repeated production of capsules, require considerable morphological and metabolic investments. In matrotrophic viviparous species, oviducal glands manufacture thin egg envelopes that persist either throughout gestation, as in placental sharks such as the bonnet head shark, *Sphyrna tiburo*, or function only during the initial stages of pregnancy, as in the lecithotropic spiny dogfish, *Squalus acanthias*, and some rays. The oviducal gland is almost certainly under endocrine control, first to induce and maintain synthesis of the structural proteins and enzymes, and, second, to cause secretion and assembly of the capsule precursors around the descending oocytes. What factor induces egg capsule secretion and assembly is entirely unknown and unexplored, but it is likely to involve

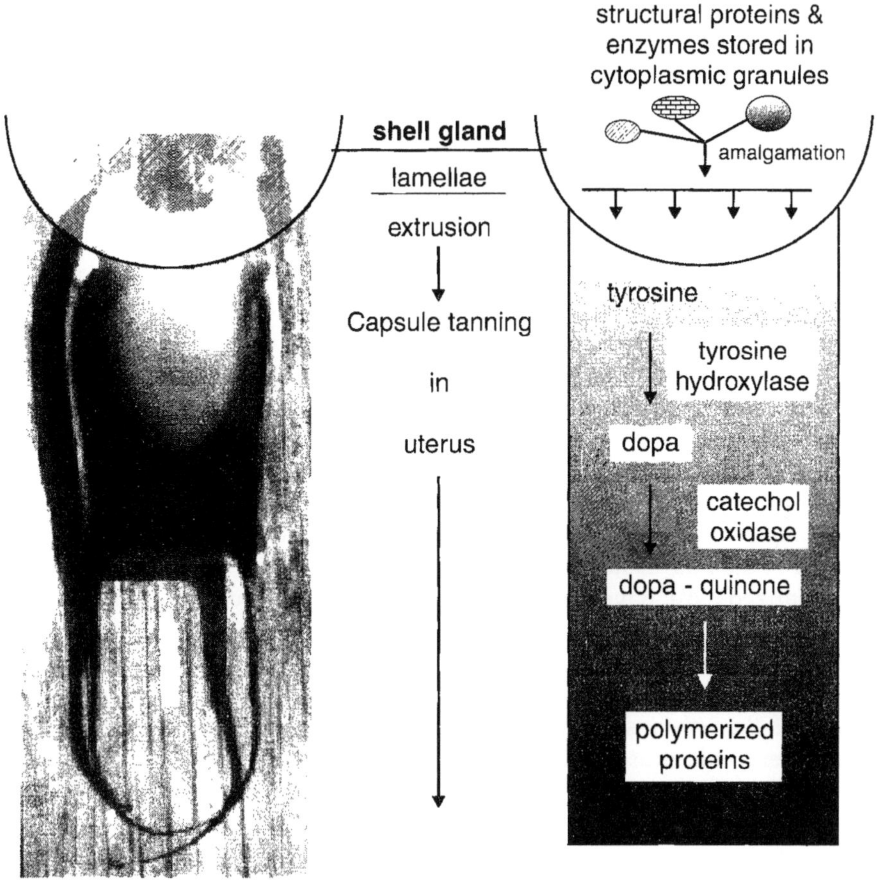

Fig. 9.2 Illustration of capsule formation in the little skate *Leucoraja erinacea*. The seven tyrosine-rich structural proteins and two enzymes are stored in cytoplasmic granules in the epithelial cells bordering the gland tubules. Upon endocrine induction, the capsule precursors are secreted from the cells into the tubules where they are mixed to form an amalgam. The amalgam is propelled to lamellae in which the capsule is assembled during extrusion into the shell gland lumen. After exiting the gland lumen, half of the tyrosine residues are converted by a tyrosine hydroxylase activity to peptide bound dopa. Approximately half of the dopa residues are oxidized by a catechol oxidase to dopa quinone, which then polymerizes the capsule proteins. The capsule at left shows that assembly of the capsule takes place in the shell gland and capsule tanning occurs in the uterus. Original.

endocrine signals from the ovary, although it does not require ova since egg capsule formation can occur without passage of the egg through the gland. The synchronous formation of capsule pairs in oviparous species begins prior to ovulation and is approximately half completed when the ova enter the capsule (Hobson 1930). Moreover, oviposited capsules that lack ova are occasionally produced by oviparous sharks and skates. These capsules, while smaller, are otherwise structurally complete, including the egg jelly that occupies the capsule lumen.

In addition to egg capsule production, the oviducal gland is thought to be the principal site for sperm storage and fertilization (Pratt 1993). Sperm were always found in the oviducal gland of adult female dogfish, whether actively ovulating or not, but never in immature females (Metten 1939). Extended storage of sperm is advantageous in that it de-couples mating from ovulatory events. This is particularly important in oviparous species that ovulate, encapsulate, and oviposit eggs over several months. The same is true for populations of viviparous species, such as *Sphyrna tiburo*, in which mating is temporally separated from ovulation and pregnancy. Whether endocrine factors are involved in sperm storage and activation is uncertain but based on information from other vertebrates and the suite of hormones available and functional in various processes in elasmobranchs (see below) it is likely that hormones are involved.

9.3.3 Isthmus

The region of the reproductive tract immediately caudal to the oviducal gland in the little skate and spiny dogfish is morphologically and functionally specialized. We have previously referred to the region connecting the oviducal gland with the specialized uterus as the isthmus since its function differs from that of the region which houses the egg capsules and embryos. In *Leucoraja erinacea*, immediately before ovulation and encapsulation, the wall of the isthmus is predominated by loose, hydrated connective tissue with very little muscle (Koob and Hamlett 1998). Compliance of this region is necessary in order to accommodate passage of the mechanically friable and chemically untanned capsule (Koob *et al.* 1981). Based on morphological examination, the isthmus here is also specialized for biosynthesis and secretion of metabolites and for facilitating vascular exchange. The morphology of this region changes during the ovulatory cycle and preliminary observations suggest that progesterone influences the changes. In the little skate, the isthmus is richest in estrogen receptors (see below). This region of the spiny dogfish reproductive tract must function similarly for capsule formation. It also functions as a valve to close off the major uterine compartment in which the embryos develop to term (Koob and Callard 1991). This region of the reproductive tract has received scant observation since Widakowich (1907) first drew attention to these structures nearly 100 years ago. Gross morphological observations indicate that this region of the reproductive tract likely functions in all female elasmobranchs during ovulatory cycles, egg retention and pregnancy, though the function may differ.

9.3.4 The Uterine Region

The principal function of the uterus is to house the encapsulated oocytes and developing fetuses prior to delivery. In oviparous species, egg retention allows completion of the capsule sclerotization process. Eggs are held *in utero* for only a brief period before oviposition, e.g., 2-7 days in *Leucoraja erinacea* (Callard and Koob 1993). The activity of the uterus during this time is critical for the development of the physicochemical properties of the capsule, since the tanning process is extinguished if the capsule is oviposited early or untanned capsule material is removed and incubated in sea water. The uterus must also propel the capsule along the tract, eventually moving it to the urogenital sinus for oviposition. Based on the increased relative proportion of smooth muscle over ciliated epithelium, activity of the muscularis must be responsible for movement in the distal region of the uterus (Koob and Hamlett 1998). Since the capsule moves into this region, is retained for several days, and is then moved out, uterine muscular activity must be under humoral control (see below). In viviparous species, the uterus must attain a size sufficient to accommodate the large volume of ovulated eggs. Uterine morphogenesis occurs during sexual maturation and ovulatory cycles coincide with follicle development, suggesting that ovarian derived factors mediate morphogenetic differentiation. During pregnancy in some species, the size and physical properties of the uterus are altered in order to accommodate the increasing volume of the conceptus. In a few viviparous species, particularly placental species, the uterus undergoes further structural adaptations in the formation of uterine compartments that separate individual embryos.

The uterus in viviparous species not only retains and harbors the developing fetuses through pregnancies lasting months, but also differentiates pathways for regulating the intrauterine milieu. During pregnancy in *Squalus acanthias* there is a great increase in surface area of the uterus caused by longitudinal folding of the uterine surface and an increase in blood flow to the uterine region (Jollie and Jollie 1967). In the basking shark, complex villiar modifications which synthesize and secrete embryotroph, grow continuously resulting in a constant supply of food for the embryo. In the spiny dogfish, pregnant females may flush sea-water into the uterus intermittently, a process thought to maintain the electrolyte environment and remove wastes (Kormanik 1993; Chapter 13 of this volume), and possibly renewing the oxygen gradient. However, additional mechanisms are likely to be of importance. In this species, during the 3rd and 4th gestational quarters, the internal lining of the uterine chamber develops an extremely rich supply of vascular loops, 0.8-1.2 mm in length, seen best in the empty contracted uterus (Fig. 9.3). This vascularization is most probably an important adaptation for oxygenation of the uterine fluids surrounding the developing embryos and is probably controlled by endocrine signals (angiogenic factors), such as vascular endothelial growth factor (VEGF) from within the uterine wall, which are under estrogen control (see below) or are derived from the developing embryos. In matrotrophic species, morphological and biochemical mechanisms for providing nutrients operate through much of gestation, but function principally after the yolk reserves have been depleted. Placentae operate in

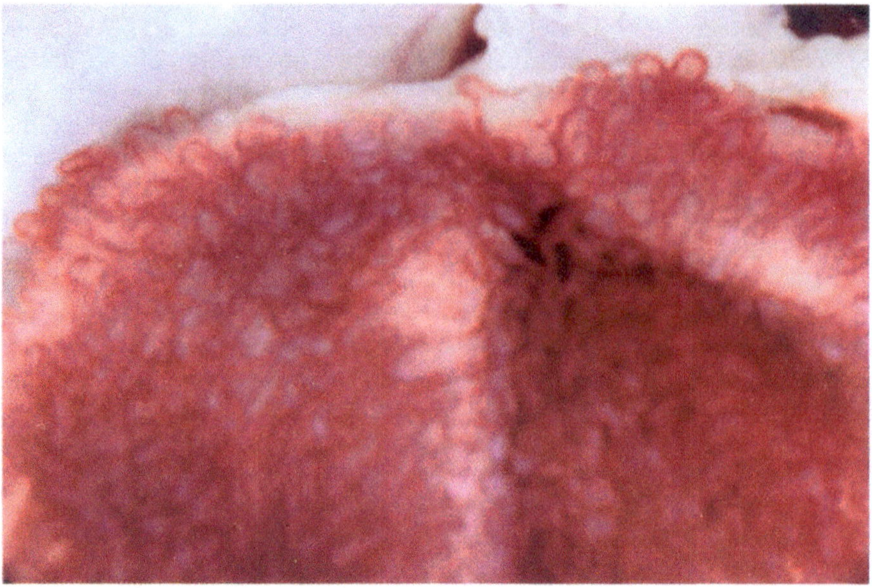

Fig. 9.3 The luminal surface of the uterine cavity of a stage C pregnant *Squalus acanthias* showing the prominent capillary loops that develop as pregnancy progresses. Individual loops are approximately 0.8-1.2 mm in length. Original.

matrotrophic sharks to mediate nutrient transfer. Trophonemata are the immediate source of nutrients in rays. Given the abundant observations on these structural specializations in a variety of viviparous species, it is surprising that nothing is known about the endocrine control of the specific mechanisms involved. It is not known whether ovarian derived endocrine factors regulate uterine vascularization, morphogenesis and activity of placenta, or development and metabolism of trophonemata.

Although the myometrium is perhaps the least interesting tissue from a morphological point of view, its contractile function is critical beyond periodic flushing of the uterine contents as described above for *Squalus acanthias*. The control of oviposition and parturition by neural and endocrine orchestration of contractile responses is undoubted in elasmobranches as in other species (discussed below).

9.3.5 The Uterine Cervix

Oviposition and parturition require alterations in the properties of the uterine cervix joining the uterus with the urogenital sinus. The cervix maintains the isolation of the uterine contents from the urogenital sinus and the external environment. In the little skate, the cervix is inextensible during egg retention, but becomes rapidly compliant at oviposition (Callard and Koob 1993) allowing passage of the 7.0 cm circumference egg. It seems likely that similar structures are hormonally controlled in all viviparous species to prevent ingress of environmental elements and to allow parturition. In the spiny dogfish the cervix prevents premature loss of the fetuses, and then must allow delivery of the term fetuses at parturition (Koob *et al.* 1984). It is likely that the cervix is a structure common to all elasmobranchs and that it functions under endocrine control to maintain the integrity of the uterine contents. The endocrine factors are pivotal for the regulation of its mechanical properties, and therefore the proper timing of oviposition and parturition.

9.4 ENDOCRINE CONTROL OF THE REPRODUCTIVE TRACT 9.4-9.7

We have in the past presented justification based on correlative and experimental studies for postulating functional roles for estradiol and progesterone in regulating reproductive tract activities (Koob and Callard 1991; Callard and Koob 1993; Callard *et al.* 1993). We have previously detailed these correlations (Koob and Callard 1999), so only a brief overview will be given here. For detailed analyses, the reader is directed to these reviews. For the purposes of this discussion, a brief description of the conclusions reached in these studies will be necessary to place the present discussion on endocrine regulation in proper context.

Studies correlating circulating levels of steroids with reproductive events provide insights from which initial hypothesis can be generated on the functional roles of each steroid. Estradiol predominates in the pre-ovulatory period during follicular development and correlates with morphogenesis and growth of the reproductive tract tissues in both oviparous and viviparous species. Circulating estradiol diminishes to low levels soon after ovulation, in both oviparous and viviparous species. In

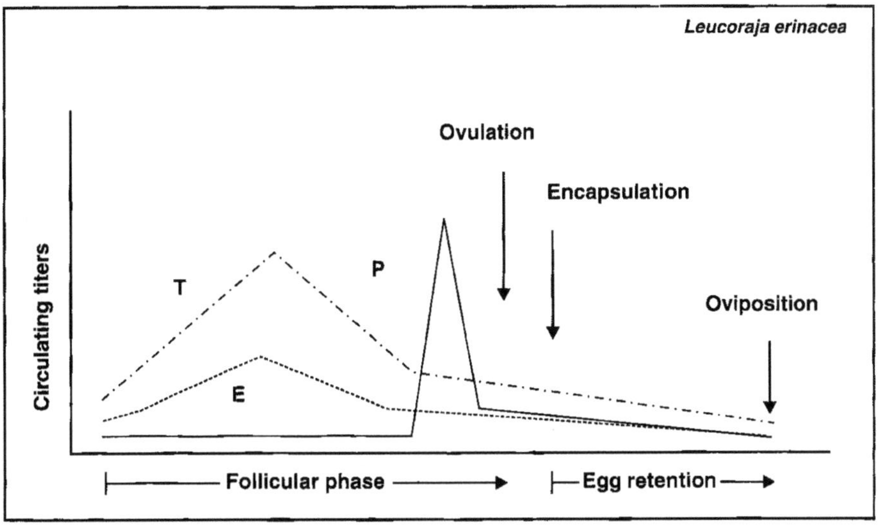

Fig. 9.4 Circulating titers of estradiol (E), testosterone (T) and progesterone (P) during the ovulatory cycle of the little skate, *Leucoraja erinacea*. Modified from Koob, T. J. and Callard, I. P. 1999. Journal of Experimental Zoology 284(5): 557-574, Fig. 4.

oviparous species, estradiol levels remain low during egg retention, and then begin to rise after oviposition during development of the follicles destined for the next ovulation. In viviparous species, estradiol levels are low during the initial stages of pregnancy. Estradiol levels rise coincident with the subsequent follicle development. In continuous breeders such as *Squalus acanthias*, estradiol levels rise during the latter half of pregnancy; whereas, in seasonal breeders such as *Sphyrna tiburo*, estradiol remains low when follicles begin growth and begins to increase during the period of follicle development preceding the next pregnancy.

Testosterone levels are unusually high in elasmobranchs compared to other female vertebrates. Changes in circulating testosterone levels correlate with those of estradiol in a precursor/product relationship. However, the role of circulating testosterone in female elasmobranchs is not known.

Cyclic alterations in progesterone levels do not correlate with estradiol and testosterone. In oviparous species, progesterone levels spike just before ovulation and fall to low levels at ovulation. In viviparous species, *Squalus acanthias*, *Sphyrna tiburo* and *Dasyatis sabina*, progesterone levels rise after ovulation and remain elevated while pregnancy is established, after which they decline to pre-ovulatory levels. In *Squalus acanthias*, the postovulatory corpora lutea are a principal source of progesterone (Tsang and Callard 1987).

9.4.1 The Oviducal Gland

With few exceptions, notably some viviparous rays, oviducal gland functions to a greater or lesser extent in both oviparous and viviparous species. Based on correlative and experimental studies, the oviducal gland is regulated, at least in large part, by estradiol during the follicular phase of each cycle. Oviducal gland growth is coincident with follicle development and elevated estradiol titers in the little skate (Koob *et al.* 1986) and estradiol treatment causes oviducal gland enlargement and capsule protein synthesis in *Scyliorhinus canicula* (Dodd and Goddard 1961). In addition, since elevated estradiol titers are characteristic of the pre-ovulatory phase when oviducal gland growth is required in preparation for capsule formation, as described above for *Raja eglanteria*, *Squalus acanthias*, *Sphyrna tiburo* and *Dasyatis sabina*, it seems reasonable to conclude that estradiol is critically involved in oviducal gland function in these species. However, although this conclusion is warranted, there are many aspects of regulation of oviducal gland function, including differentiation of the complex array of cells required for capsule biosynthesis and assembly (Hamlett *et al.* 1998); induction of gene expression for the structural proteins and enzymes that must be coordinately produced in the proper relative amounts (Koob and Cox 1993); and morphogenesis of the extensive tubules, baffle plates, spinnerets, and lamellae that assemble the capsule precursors (Hamlett *et al.* 1998). At the present time, little evidence exists

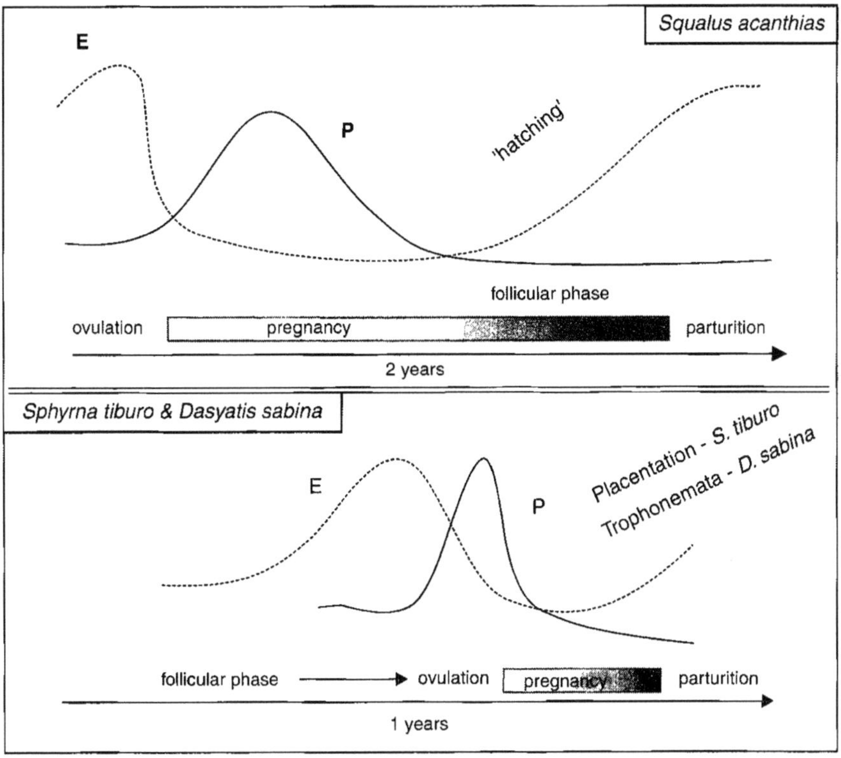

Fig. 9.5 Circulating titers of estradiol (E) and progesterone (P) in viviparous species during the follicular phase and pregnancy in *Squalus acanthias*, *Sphyrna tiburo* and *Dasyatis sabina*. *Squalus acanthias* is a continuous breeder employing a lecithotrophic viviparous reproductive mode. The first half of embryonic development proceeds inside a thin egg capsule while progesterone levels are elevated. The embryos then 'hatch' out of the capsule to complete development free within the uterine lumen when progesterone levels are relatively low. During this second half of pregnancy estradiol levels rise and the ovarian follicles develop so that ovulation occurs soon after parturition. Lowered progesterone levels and the rise in estradiol titers also corresponds with the extensive vascularization of the uterine wall. *Sphyrna tiburo* is a placental viviparous species with an intermittent yearly cycle. Early embryonic development occurs in thin egg capsules and relies on yolk provided in the ovulated egg. Progesterone levels are elevated during this time. Placentae form as the yolk is expended and the remainder of fetal development relies on supply of nutrients through placental exchange. Placentae form after progesterone levels decline. Circulating estradiol and progesterone remain relatively low until the subsequent follicular phase and pregnancy. *Dasyatis sabina* utilizes an intermittent matrotrophic viviparous reproductive mode. In this mode early embryonic development takes place in thin egg envelopes when serum estradiol levels have declined from a maximum prior to ovulation to relatively low concentrations and the pre- to post- ovulatory surge in progesterone titers remains elevated. After the yolk is consumed and circulating progesterone titers have decreased, the fetuses exit the capsule and develop surrounded by a uterine derived 'milk' produced and secreted from trophonemata. Morphogenesis and functional activity of trophonemata occur after progesterone levels decline. Estradiol levels are low through most of pregnancy, but rise near term coincident with resumption of ovarian follicle development. Modified from Koob, T. J. and Callard, I. P. 1999. Journal of Experimental Zoology 284(5): 557-574, Fig. 5.

that allows assignment of a specific regulatory activity to a particular ovarian derived hormone.

As mentioned previously, secretion of the capsule proteins and enzymes from the tubule cells, propulsion of the capsule amalgam through the tubules to the spinnerets, and eventual extrusion from the luminal lamellae are not induced by passage of the egg through the oviducal gland. Capsule formation in the gland must be controlled by endocrine factors. These factors are likely derived from the ovarian follicles since the timing of capsule formation is intimately coupled with ovulation. Control of secretion and assembly of the capsule precursors is regionally regulated since the horns are produced from the lateral aspects of the gland first, then the body of the capsule is produced by the bulk of the gland, followed by addition of horns again from the lateral glandular regions. It should also be recognized that the region of the gland that produces the egg jelly is tightly coordinated with capsule production. The complexity of regionally specific secretion and assembly of the morphologically specialized parts of the capsule argues for an equally complex regulatory system.

As mentioned above, sperm are stored in the oviducal gland, in some cases for months. What factors regulate sperm viability are not known. However, it has been suggested that one function of the high testosterone and related androgens during the follicular phase of ovulatory cycles (see Koob and Callard 1999 for details on circulating levels of testosterone) is direct regulation of the quiescence and activation of stored spermatozoa (Manire *et al.* 1995). Both activities must be under endocrine regulation; however, little evidence is available to support speculation.

9.4.2 The Isthmus

The isthmus in both oviparous and viviparous species would seem to require endocrine control as discussed above, and experimental evidence supports this hypothesis. We have shown that estradiol administration increases the compliance of this region of the reproductive tract in *Squalus acanthias* (Koob *et al.* 1983). In addition, we reported that estradiol alone increases the circumference of the isthmus in *Leucoraja erinacea*, and when administered together with a relaxin analogue, it produces the size necessary for passage of the egg capsule (Callard and Koob 1993).

9.4.3 The Uterine Region

We have speculated based on circulating steroid titers that progesterone is involved in regulating egg retention in the little skate; moreover, experimentally manipulated chronic elevations of progesterone caused early oviposition in this species (Callard and Koob 1993).

While the initial morphogenesis of the reproductive tract correlates with elevated estradiol and testosterone titers, subsequent structural modifications during pregnancy are less well correlated with endocrine factors due to the absence of specific studies. Embryo-related structural specializations in the uterus such as uterine folds, general growth and

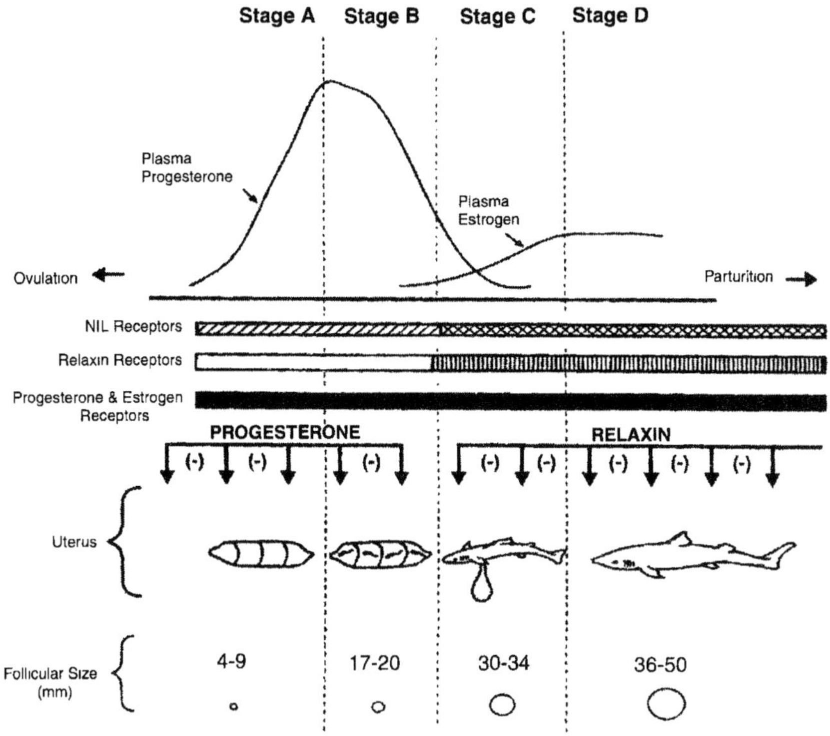

Fig. 9.6 Model of circulating progesterone, estrogen, relaxin, and neurointermedate lobe hormones and their receptors in uterus during the reproductive cycle in *Squalus acanthias.* Presence of relaxin and the NIL hormone receptors is inferred from biological actions of hormones *in vitro* and *in vivo.* Estrogen receptors have previously been identified in elasmobranch (Reese and Callard 1991), and progesterone receptors have also been identified (Paolucci and Callard 1998). However, receptors have not been measured over a complete two-year reproductive cycle. From Sorberra, L. and Callard, IP. 1995. American Journal of physiology (Regulatory Integrative Comparative Physiology 38: R 389- R397.

development, as well as placentation are probably caused by the interaction of embryonic, uterine and endocrine signals, as in the mammal (Johnson and Everett 2000). The most basic elements necessary for endocrine regulation of the reproductive tract, estrogen and progesterone receptors, have been demonstrated (Reese and Callard 1991; Paolucci and Callard 1998) and reproductive tract responses in elasmobranchs are very likely to set the evolutionary pattern for the vertebrates.

One uterine tissue that has received experimental study is the myometrium in the spiny dogfish (Sorbera and Callard 1995). Regulation of uterine contractions is important for two reasons: during early pregnancy, the capsule containing the developing embryos is easily broken by mechanical stresses; later in pregnancy after the embryos have broken out of the capsule, the uterus periodically flushes the lumen with sea water. Endogenous spontaneous myometrial activity in early pregnancy when circulating progesterone levels are elevated is relatively low, suggesting progesterone may modulate contractions. Similarly, the frequency of spontaneous uterine contractions both *in vivo* and *in vitro* was inhibited in late pregnancy by administration of homologous, ovarian-derived, dogfish relaxin, an effect potentiated by estradiol. The proposed model explaining these observations includes progesterone inhibition of uterine contractions in early pregnancy. Attenuated uterine contractions would protect the encapsulated embryos early in pregnancy. Estrogen facilitated relaxin modulation of the frequency of contractions in late pregnancy would allow myometrial generated uterine flushing but at the same time prevent early parturition (Sorbera and Callard 1995).

9.4.5 Functional Roles of Progesterone and Estradiol in Embryo-fetal Nutrition

In all three viviparous species in which circulating steroids have been accurately tracked with reproductive events, progesterone titers are elevated during the initial phase of pregnancy (Tsang and Callard 1987; Manire *et al.* 1995; Snelson *et al.* 1997). As indicated above, elevated progesterone may function to inhibit hepatic vitellogenin production and thereby follicle and oocyte development at an inopportune time for some species (i.e. early pregnancy in the spiny dogfish; see Ho *et al.* 1980). Progesterone inhibits estrogen induced vitellogenin synthesis in the little skate also (Perez and Callard 1993). As progesterone rises in the bonnethead and Atlantic stingray, estradiol levels decline and while progesterone levels are elevated, estradiol and testosterone are low, suggesting counter regulation of the steroidogenic enzymes necessary for testosterone and estradiol synthesis. Typically, the rise in progesterone and fall in estrogen is correlated with ovulation and the onset of luteal function (Tsang and Callard 1987a, b). Progesterone is also likely to regulate uterine metabolic activity, but no evidence exists to suggest which specific activities might be affected by progesterone.

Estradiol titers rise during late pregnancy in *Squalus acanthias, Sphyrna tiburo,* and *Dasyatis sabina.* This elevation is quantitatively most significant in the spiny dogfish and Atlantic stingray, suggesting estradiol might function to regulate events occurring *in utero* at this time. In the Atlantic stingray, the slightly enlarged follicles (Snelson 1997) may be the source of elevated estradiol during late pregnancy and parturition, suggesting a collateral role related to gestation and parturition. Concurrent with elevated estradiol titers is a shift in embryonic nutrient supply from the yolk reserves to the histotroph produced by the trophonemata in the stingray. Estradiol in concert with progesterone probably regulates trophonemata function by controlling the secretion of histotroph, much as these hormones act in mammalian species. In the bonnethead shark, the small elevation in estradiol levels correlates with the formation of the placenta which may be a contributing source of estradiol, and the resulting shift from yolk to placental exchange as the principal source of nutrients.

Thus for both of these species (stingray, bonnethead) estradiol is likely to control availability of nutrients through uterine structures (placenta or trophonemata).

Taken together, these observations suggest that a major role of estradiol in elasmobranch reproductive cycles is to regulate structures providing nutrients to the developing offspring. In short cycle egg laying species (*Leucoraja erinacea*), or the aplacental dogfish, *Scyliorhins canicula*, estradiol regulates nutrient principally via hepatic yolk synthesis and transport to the ovary. In viviparous species with yolk reduction, the role of estradiol in nutrient provisioning is two fold (a) vitellogenin for maintenance of early development and (b) placental growth and development or trophonemata growth and function, for later nutrient transfer (e.g. stingray, *Dasyatis sabina*; bonnethead, *Sphyrna tiburo*).

9.4.4 The Cervix

Estradiol secretion towards term in elasmobranchs may also be important in facilitating parturition or egg laying. Based on experimental studies,

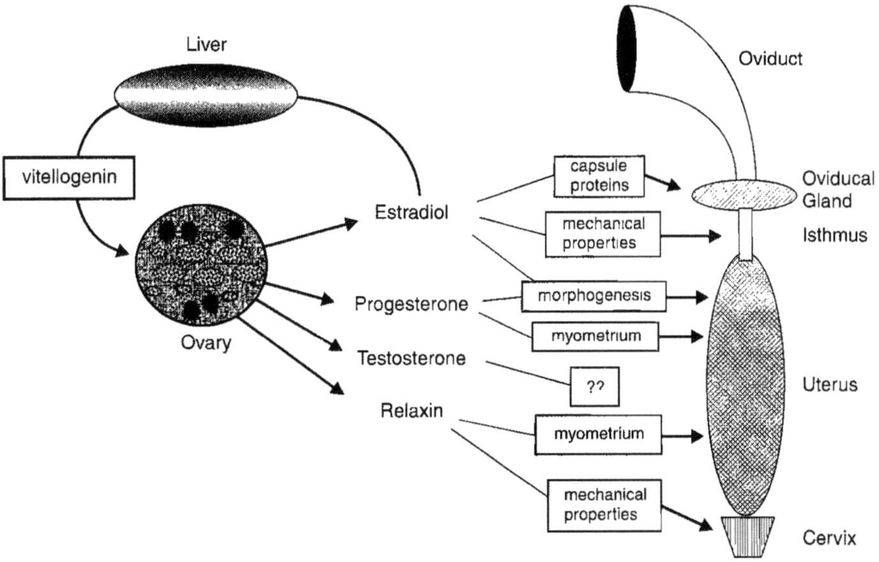

Fig. 9.7 Putative functions of ovarian derived endocrine factors on the female reproductive tract in elasmobranchs. Estradiol is the principal factor that promotes differentiation, morphogenesis and growth of the entire reproductive tract in preparation for ovulatory cycles. It induces biosynthesis of capsule precursors in the oviducal gland, increases compliance of the isthmus and uterus, and prepares the uterus for egg retention and pregnancy. It also facilitates relaxin mediated increases in compliance of the uterine cervix at parturition. Progesterone is involved in regulating egg rentention in both oviparous and viviparous species by inhibiting myometrial activity. Testosterone may be involved in regulating quiescence and activation of spermatozoa in the oviducal gland, otherwise, nothing is known about its function. Relaxin mediates compliance of the cervix at parturition. From Koob, T. J. and Callard, I. P. 1999. Journal of Experimental Zoology 284(5): 557-574, Fig. 7.

ovarian derived relaxin may be involved in regulating cervical compliance in *Leucoraja erinacea*. *In vivo* treatment with the structurally related analogue, insulin, also caused a significant increase in the extensibility of the cervix, and estradiol potentiated the insulin effect (Koob and Callard 1982; Callard and Koob 1993). In the spiny dogfish, insulin had similar effects on rendering the cervix compliant and, in addition, caused early delivery (Koob *et al.* 1984). Thus, regulation of cervical compliance is critically important in maintaining pregnancy in this viviparous shark. The isolation of relaxin from the ovary of the spiny dogfish (Bullesbach *et al.* 1986) and the little skate (Bullesbach *et al.* 1987) emphasizes the importance of this hormone in reproductive tract accommodation at this early stage of vertebrate evolution.

9.5 SUMMARY (FIG. 9.7)

Of all the vertebrate groups, we know least about the role of ovarian derived steroids in regulating events in the female reproductive tract of chondrichthyan fishes. While it is abundantly clear that estradiol, testosterone and progesterone are associated with ovulatory cycles and pregnancy and parturition in elasmobranchs, no specific function can be attributed to individual steroids. This deficiency is due in no small part to the lack of knowledge about the biochemical and physiological mechanisms employed by the diverse array of reproductive modes in the elasmobranchs. Nevertheless, broad conclusions are possible from correlative studies on circulating titers of steroids and reproductive events, and from the few experimental studies performed on *Leucoraja erinacea* and *Squalus acanthias*.

9.6 ACKNOWLEDGEMENTS

Supported by ES 07381 to IP Callard and Shriner's Hospital for Children 8610 to T. J. Koob.

9.7 LITERATURE CITED

Boling, J. L. and Blandau, R. J. 1971. Egg transport through the ampullae of the oviducts of rabbits under various experimental conditions. Biology of Reproduction 4(2): 174-184.

Bullesbach, E. E., Gowan, L. K., Schwabe, C., Steinetz, B. G., O'Byrne, E. and Callard, I.P. 1986. Isolation, purification, and the sequence of relaxin from spiny dogfish (*Squalus acanthias*). European Journal of Biochemistry 161(2): 335-41.

Bullesbach, E. E., Schwabe, C. and Callard I. P. 1987. Relaxin from an oviparous species, the skate (*Raja erinacea*). Biochemistry and Biophysics Research Community 143(1): 273-80.

Callard, I. P., Fileti, L. A. and Koob, T. J. 1993. Ovarian derived steroids and the hormonal control of the elasmobranch reproductive tract. Environmental Biology of Fishes 38: 175-185.

Callard, I. P. and Koob, T. J. 1993. Endocrine regulation of the elasmobranch reproductive tract. Journal of Experimental Zoology 266: 368-377.

Chieffi, G. 1967. The Reproductive System of Elasmobranchs: Developmental and endocrinological aspects. Pp. 553-580. In P. W. Gilbert, R. F. Mathewson and D. P. Rall (eds), *Sharks, Skates and Rays*. The Johns Hopkins Press Baltimore, MD.

Dodd, J. M. and Goddard, C. K. 1961. Some effects of oestradiol benzoate on the reproductive ducts of the female dogfish, *Scyliorhinus canicula*. Proceedings of the Zoological Society of London 137: 325-331.

Hamlett, W. C., Knight, D. P., Koob,, T. J., Jezior, M., Luong, T., Rozycki, T., Brunette, N. and Hysell, M. K. 1998. Survey of oviducal structure and function in elasmobranchs. Journal of Experimental Zoology 282: 399-420.

Heller, H. 1966. Introduction. In V. R. Pickles and R. J. Fitzpatrick (eds), *Endogeneous substances affecting the myometrium*. Memoirs of the Society for Endocrinology 14: 145-146.

Ho, S.M., Wulczyn, G. and Callard, I. P. 1980. Induction of vitellogenin synthesis in the spiny dogfish, *Squalus acanthias*. Bulletin of the Mount Desert Island Biological Laboratory 19: 37-38.

Hobson, A.D. 1930. A note on the formation of the egg case of the skate. Journal of the Marine Biology Association of the United Kingdom 16: 577-581.

Johnson, M. H. and Everitt, B. J. 2000. *Essential Reproduction*. Blackwell Science, Ltd. London, UK.

Jollie, W. P. and Jollie, L. G. 1967. Electron microscopic observations on accomodations to pregnancy in the uterus of the spiny dogfish, *Squalus acanthias*. Journal of Ultrastructural Research 20: 161-178.

Koob, T. J. and Callard, I. P. 1982. Relaxin: Speculations on its physiological importance in some nonmammalian species. Annals New York Academy of Sciences 380: 163-173.

Koob, T. J. and Callard, I. P. 1991. Reproduction in Female Elasmobranchs. Pp. 155-209. In R. K. H. Kinne (ed.), *Oogenesis, Spermatgenesis, and Reproduction*. Volume X. Comparative Physiology, Basel, Karger.

Koob, T. J. and Callard, I. P. 1999. Reproductive endocrinology of female elasmobranchs: Lessons from the little skate (*Raja erinacea*) and spiny dogfish (*Squalus acanthias*). Journal of Experimental Zoology 284(5): 557-574.

Koob, T. J. and Cox, D. L. 1993. Stabilization and sclerotization of *Raja erinacea* egg capsule proteins. Environmental Biology of Fishes 38: 151-157.

Koob, T. J. and Hamlett, W. C. 1998. Microscopic structure of the gravid uterus in the little skate, *Raja erinacea*. Journal of Experimental Zoology 282: 421-437.

Koob, T. J., Laffan, J. J. and Callard, I. P. 1984. Effects of relaxin and insulin on reproductive tract size and early fetal loss in *Squalus acanthias*. Biology of Reproduction 31(2): 231-238.

Koob, T. J., Laffan, J. J., Elger, B. and Callard, I. P. 1983. The effects of estradiol on the Verschlussvorrichtung in *Squalus acanthias*. Bulletin of Mount Desert Island Biological Laboratory 23: 67-68.

Koob, T. J., Laffan, J. L. and Callard, I. P. 1981. Egg-oviduct size relationships in *Raja erinacea*. Bulletin of the Mount Desert Island Biological Laboratory 21: 46-48.

Koob, T. J., Tsang, P. and Callard, I. P. 1986. Plasma estradiol, testosterone, and progesterone levels during the ovulatory cycle of the skate (*Raja erinacea*). Biology of Reproduction 35(2): 267-275.

Kormanik, G. A. 1993. Ionic and osmotic environment of developing elasmobranch embryos. Environmental Biology of Fishes 38: 233-240.

Manire, C. A., Rasmussen L. E. L., Hess D. L. and Heuter R. E. 1995. Serum steroid hormones and reproductive cycle of female bonnethead shark, *Sphyrna tiburo*. General and Comparative Endocrinology 97: 366-376.

Metten, H. 1939. Studies on the reproduction of the dogfish. Philosophical Transactions of the Royal Society of London 230B: 217-238.

Paolucci, M. and Callard I. P. 1998. Characterization of progesterone-binding moieties in the little skate *Raja erinacea*. General and Comparative Endocrinology 109(1): 106-118.

Perez, L. E. and Callard, I. P. 1993. Regulation of hepatic vitellogenin synthesis in the little skate (*Raja erinacea*): Use of a homologous enzyme linked immunosorbent assay. Journal of Experimental Zoology 266: 31-39.

Picon, R. 1962. Recherches sur la differenciation sexualle de l'embryon de *Leptocharias smithii* (Muller et Henle) Elasmobranche. Archives d'Anatomie Microscopique et de Morphologie Experimentale 51(4): 541-576.

Pratt, H. L. 1993. The storage of spermatazoa in the oviducal glands of western North Atlantic sharks. Environmental Biology of Fishes 38: 139-149.

Reese, J. C. and Callard, I. P. 1991. Characterization of a specific estrogen receptor in the oviduct of the little skate, *Raja erinacea*. General and Comparative Endocrinology 84: 170-181.

Snelson, F. F. Jr., Rasmussen, L. E. L., Johnson, M. R. and Hess, D. L. 1997. Serum concentrations of steroid hormones during reproduction in the Atlantic stingray, *Dasyatis sabina*. General and Comparative Endocrinology 108(1): 67-79.

Sorbera, L.A. and Callard, I. P. 1995. Myometrium of the spiny dogfish *Squalus acanthias*: peptide and steroid regulation. American Journal of Physiology 269(2): R389-R397.

Tsang, P. and Callard, I. P. 1987a. Morphological and endocrine correlates of the reproductive cycle of the aplacental viviparous dogfish, *Squalus acanthias*. General and Comparative Endocrinology 66(2): 182-189.

Tsang, P. and Callard, I. P. 1987b. Luteal progesterone production and regulation in the viviparous dogfish, *Squalus acanthias*. Journal of Experimental Zoology 241: 377-382.

Tsang, P. and Callard, I. P. 1992. Regulation of ovarian steroidogenesis *in vitro* in the viviparous shark, *Squalus acanthias*. Journal of Experimental Zoology 261: 97-104.

Widakowich, V. 1907. Uber eine Verschlussvorrichtung im Eileiter von *Squalus acanthias*. Zoologische Anzeiger 31: 636-643.

Wourms, J. P. 1977. Reproduction and development in Chondrichthyan fishes. American Zoologist 17: 379-410.

CHAPTER 10

Oviducal Glands in Chondrichthyans

William C. Hamlett[1], *D. P. Knight*[2], *F. T. V. Pereira*[3], *J. Steele*[1] *and David M. Sever*[4]

10.1 INTRODUCTION

Oviducal glands (OG), also referred to as shell or nidamental glands (Prasad 1945, 1948; Knight *et al.* 1993) are discrete, specialized regions of the anterior portion of the oviduct in cartilaginous fishes. Oviducal glands produce the components of the egg jelly that surrounds the fertilized egg in the early stages of embryogenesis. They also produce the tertiary egg envelopes including the rigid egg capsule of oviparous species, the thin pliable transient egg candle case of yolk sac species and the thin pleated egg envelope in most placental sharks (Hamlett *et al.* 1998a, 1999; Hamlett and Koob 1999). OG have also been implicated in sperm storage (Metten 1939; Prasad 1945; Pratt 1993; Hamlett *et al.* 2002a, b, 2003). Historically the terms shell, nidamental and oviducal gland have been used imprecisely and interchangeably. The region of the oviduct that produces a tough egg case that is deposited to the exterior in oviparous species is correctly termed the shell gland since this term denotes its function. A shell is defined as a hard, outer covering, hence the designation. The term nidamental gland is best used to refer to the gland that secretes the thin egg coverings of some viviparous species. The term nidamental is derived from *nidus* L. for nest. In many placental species each embryo is surrounded by its own egg covering and the embryo and its coverings develop in its own uterine compartment, hence the nest. Neither of the

[1]Department Anatomy and Cell Biology, Indiana University School of Medicine, Notre Dame, IN 46556, USA
[2]Department of Zoology, Oxford University, South Parks Road, Oxford OX1 3PS, UK
[3]Paulista State University – UNESP, Rua Bahia, 332 Centro, Dracena – SP, CEP 17.900-000, Brazil
[4]Department of Biological Sciences, Southeastern Louisiana University, Hammond, LA 70402, USA

previous terms can be correctly applied to the gland when referring to the region of the oviduct in some rays where no egg covering is produced. To establish a consistent terminology, we have chosen not to use the terms shell or nidamental but to use only the term OG to refer to any of the aforementioned glands since they are all derived from the oviduct.

Early descriptions of OG in elasmobranchs were given by Borcea (1904, 1906), Widakowich (1906), Filhol and Garrault (1938), Nalini (1940), Prasad (1945, 1948). The vast majority of work elucidating the structure and function of elasmobranch OG has centered on the oviparous dogfish, *Scyliorhinus canicula*. Metten (1939) described the structure of the gland and several authors considered its histochemical characteristics (Threadgold, 1957; Rusaouën 1976). Krishnan (1959) added information on the histochemistry of the OG in *Chiloscyllium griseus*. The formation and nature of the components of the egg capsule of *S. canicula* has been well studied (Rusaouën *et al.* 1976; Rusaouën-Innocent 1985,1990; Hunt, 1985; Feng and Knight, 1992, 1994a, b; Knight and Feng, 1992, 1994a, b; Knight *et al.*, 1993;Hepworth *et al.*, 1994; Knight *et al.* 1996). Thomason *et al.* (1994) investigated the antifouling properties of the egg case and Koob and Cox (1993) reported on tanning of the egg capsule in the little skate, *Leucoraja erinacea*. Pratt (1993) discussed sperm storage in the OG of some elasmobranchs. OG structure and function were surveyed in a variety of elasmobranchs (Hamlett *et al.* 1998a) and the fundamental zones discussed (Hamlett *et al.* 1999). A comprehensive treatment of reproductive biology in female elasmobranchs, including a consideration of the OG, was recently published (Hamlett and Koob 1999).

Older literature variously referred to the OG zones as albumen, mucous and shell secreting (Metten 1939; Prasad 1945) and in *Scyliorhinus canicula* terminology was based on histochemistry and zonation was classed A zone, B zone, B1 zone, B2 zone, etc. (Rusaouën 1976; Knight *et al.* 1996). Recently, terminology relating to the zonation in elasmobranch OG has been reconsidered (Hamlett *et al.* 1998a, 1999). These authors noted morphological features common in species of various reproductive modes. They described zones from anterior to posterior as club, papillary, baffle and terminal. The terminology is based on light microscopy (LM) of frontally sectioned glands. Club and papillary refer to the profile of the surface layer when viewed via LM. Baffle refers to the baffle plates that form the lips of the extrusion dies in the region that produces the various types of tertiary egg envelopes. Terminal refers to its caudad position. Terminology based on morphology allows reliable comparisons to be made across reproductive modes, whereas the histochemical compositions of OG cells vary during gestation, reflecting the synthetic and secretory activity of the cells, hence are not reliable. Older descriptions generally refer to the club zone as the albumen zone (despite the lack of evidence for albumen secretion in this region), the papillary zone as the mucous zone and the baffle zone as oblique plates (Metten 1939) and tufts (Prasad 1945). Previous descriptions of the zonation in *S. canicula* have reported five to nine zones depending on the author and the means of analysis, ie. histochemistry, morphology via scanning electron microscopy, etc. (Threadgold, 1957; Rusaouën 1976).

Species that produce an egg case, candle case or egg envelope all share general design characteristics of the OG. The club and papillary zones produce the various types of egg jelly that initially surround the egg. The baffle zone produces the tertiary egg envelope and has a highly conserved structure (Fig. 10.1A, B). Secretory cells in simple tubular glands produce nascent egg envelope material that moves to the secretory duct by ciliary movement and secretion pressure exerted within the tubule from new secretion. From the secretory duct egg envelope material is passed to the spinneret which is composed of paired baffle plates that help to manipulate and orient the material as it passes to transverse grooves that extend the full width of the OG. Both *Urobatis halleri* (Babel 1967) and *U. jamaicensis* lack a discrete functional baffle zone and consequently does not produce an egg envelope. Members of the family *Narcinidae,* the numbfishes, are reported to have no OG at all (Prasad 1945).

10.2 CLUB ZONE

The nature of the secretions of the club zone has been examined by histochemical means in *Scyliorhinus canicula*. Threadgold (1957) noted periodic acid-Schiff positive (PAS+) staining as well as positive staining with sulfotoluidine blue. Stains for proteins were weak or absent. He concluded that the secretory product was a carbohydrate, possibly a neutral mucopolysaccharide and not albumen as it was believed to be by Filhol and Garrault (1938). Rusaouën (1976) noted that secretory materials of the club zone of *S. canicula* stained alcian blue positive (AB+) at acid pH, indicating strong polyanions. Toluidine blue staining yielded a strong metachromatic reaction suggesting that the products might be sulfated polysaccharides. She concluded that secretions of the club zone were neutral mucopolysaccharides and acidic polysaccharides not bound to protein.

10.3 PAPILLARY ZONE

Staining properties suggested to Nalini (1940) that the product of the papillary zone in *Chiloscyllium griseum* was a type of mucin, the purported function was to separate the fluid component of the club zone from the capsule materials and to serve as a lubricant between the fluid and the forming capsule during encapsulation. In *S. canicula,* Threadgold (1957) noted metachromatic staining and concluded that the products contained carbohydrate. Rusaouën (1976), however, indicated that the zone was a strongly sulfated mucopolysaccharide. Feng and Knight (1992) observed AB+ staining and suggested that the material was a sulfated glycosaminoglycan. In *Iago omanensis* (Hamlett *et al.* 2002b) the papillary zone was PAS+, indicating glycoprotein or any mucous substance containing

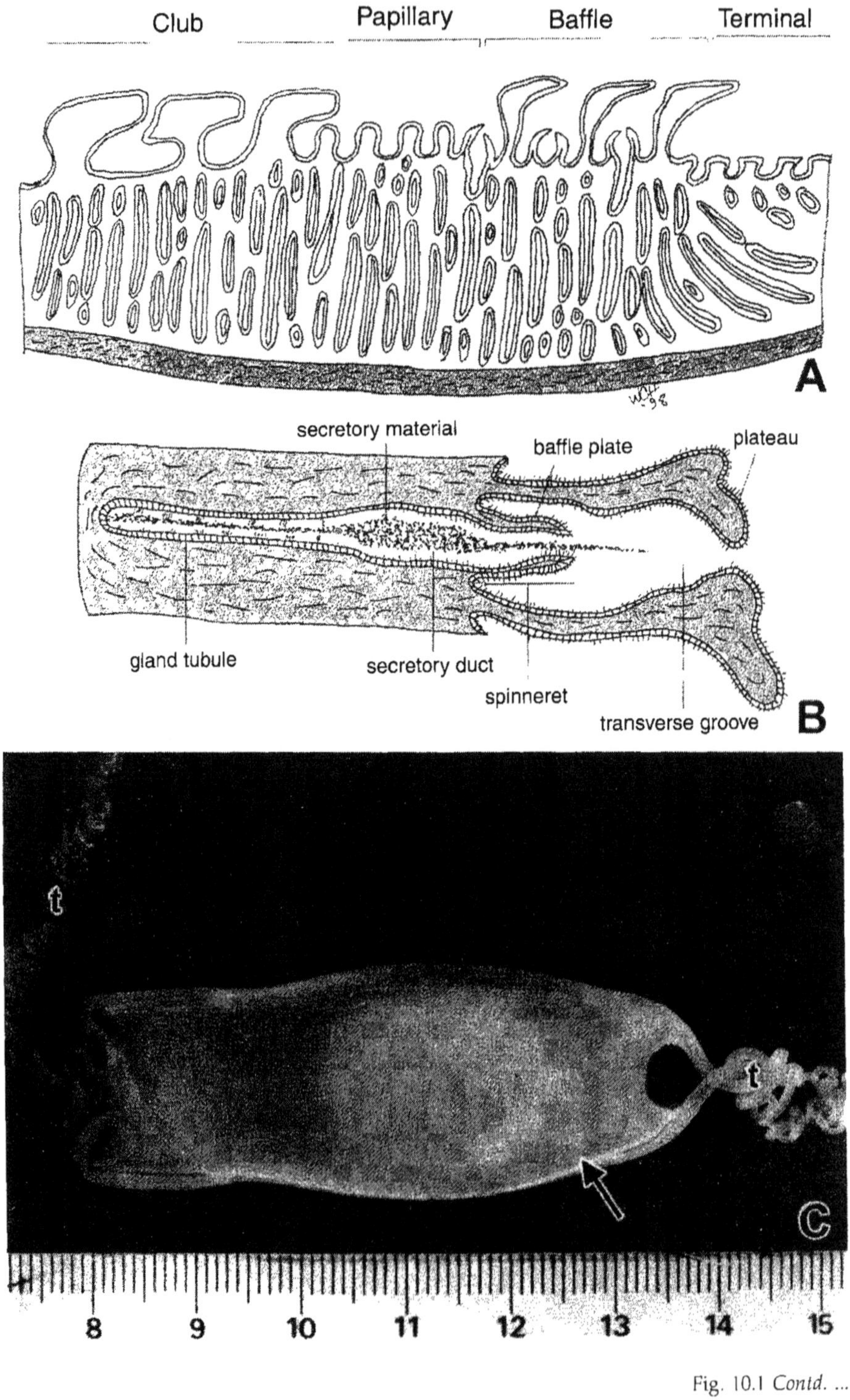

Fig. 10.1 *Contd. ...*

neutral sugars. The caudal-most papillary lamella was AB+ at pH 2.5 indicating the presence of sulfated and unsulfated acid glycosaminoglycans and sialoglycoproteins.

Even though the exact chemical composition of both the club and papillary zones is unknown, it is clear that this region produces the more-or-less fluid jelly compartment surrounding eggs contained within the tertiary egg envelopes of all elasmobranchs that encapsulate their eggs. *Urobatis jamaicensis* does not have a club or papillary zone and no egg jelly is produced (Hamlett unpublished).

Koob and Straus (1998) concluded that egg jelly in the little skate, *Leucoraja erinacea*, functions as a structural device to hydrodynamically support the egg and developing embryo. Diversity in the nature of the jelly investment also exists in species with differing reproductive modes. In some sharks and rays, the egg and jelly fill the entire envelope; the developing embryos break out of the envelope to complete development free in the uterus. In other viviparous species the jelly occupies only a small proportion of the envelope. In placental sharks the modest jelly component surrounds the fertilized egg only during the early stages of gestation.

10.4 BAFFLE ZONE

The baffle zone produces tertiary egg envelopes. Simple tubular gland secretory cells produce a liquid crystalline material (Knight *et al.* 1996). It emerges from the secretory duct and is manipulated by the divergent die and baffle plates of the spinneret that function at extrusion. The liquid crystal material is assembled into a highly ordered fibrillar component of each subsequent layer within transverse grooves of the gland. Although the morphology of the baffle zone is similar in all species thus far examined, that produce a tertiary egg envelope, diversity exists in the nature of the egg covering, hence diversity exists in the composition and function of the egg coverings formed (Hamlett *et al* 1998a; Hamlett and Koob 1999). *Urobatis jamaicensis* lacks baffle plates and cells of its highly modified OG is non-secretory, hence no egg envelope is formed (Hamlett unpublished).

10.5 TERMINAL ZONE

The last zone of the OG is the terminal zone which is the site of formation of surface hairs that adorn the exterior of the capsule in

Fig. 10.1 A. Composite line diagram of a "generic" oviducal gland showing club, papillary, baffle and terminal zones. **B.** Baffle zone unit consisting of secretory gland tubule, secretory duct, spinneret with baffle plates and transverse groove between plateau projections. Figs. 10.1A-B from Hamlett, W. C., Knight, D. P., Koob, T., Jezior, M., Luong, Rozycki, T., Brunette, N. and Hysell, M. 1998a. Journal of Experimental Zoology 282: 399-420, Fig. 1. **C.** Eggcase of *Scyliorhinus canicula*. t = tendrils, arrow = marginal rib. Original.

species that have hairs and sperm storage. A feature of some oviparous OG is the secretion of hair filaments from the terminal zone. In the holocephalan *Callorhynchus milii* (Smith *et al.* 2004) and the batoid *Raja eglanteria* (Hamlett *et al.* 1999), the terminal zone differentiates into a region of hair production and a region of sperm storage. The hair region has mucoid tubules near the lumen and the base of the same tubules contains secretory cells that resemble baffle zone secretory cells. Initially, the basal portion of the tubules produce a secretion that extrudes up the tubule by secretion pressure from below. As the emerging hair filament passes cells of the luminal mucoid region, it is coated with secretions. There are similar observations of two cell types in the same tubule in the holocephalan *Hydrolagus colliei* (Prasad 1945; Stanley 1963). The dorsal surface of the egg case in *C. milli* has abundant hairs that aid adhesion of sand to the egg case. The ventral surface of the egg case is smooth with lateral flanges that are concave. This formation results in a suction cup arrangement that helps to attach the egg case to the muddy ocean floor. Hence the egg case is attached to the mud and the sticky dorsal surface with hairs is rapidly coated with sand thereby providing camouflage. Similar arrangements occur in skates where the egg case is smooth but elongate. Spiral tendrils extend from the lateral margin and ends of the egg case. They are sticky and serve the same purpose of attaching sand for camouflage. Additionally the tendrils coil with sea grass to attach and support the egg case.

10.6 DIVERSITY OF OVIDUCAL GLAND MORPHOLOGY AND TERTIARY EGG ENVELOPE

In oviparous species including *Scyliorhinus canicula* (Fig. 10.1C), *Raja eglanteria* (Fig. 10.2C) and the holocephalan *Callorhynchus milii* (Fig. 10.3D) a tough firm egg case is formed. Oviducal glands in oviparous species are the largest in chondrichthyes. In *S. canalicula* the egg case has tendrils to attach the capsule to a substrate, frequently sea plants and marginal ribs for stability. *Raja eglanteria* lacks tendrils. *Callorhynchus milii* has a convex surface covered with sticky hairs that attach sand thereby camouflaging the case. The opposite side is smooth and acts as a suction cup to keep the capsule planted in the sand.

Yolk sac viviparous species have thin transient candle cases. In the spiny dogfish, *Squalus acanthias*, (Figs 10.2A, B) and common sawshark, *Pristiophorus cirratus* (Fig. 10.2D) the OG is short and barrel shaped (Stevens 2002). The candle case encloses multiple eggs in *S. acanthias* (Fig. 10.2B) but each candle contains a single egg in *P. cirratus* (Fig. 10.2D). In both cases a small volume of egg jelly initially coats the egg inside the candle. In both species the candle dissolves and the embryo completes development free in the uterus.

Species that have uterine compartments have a thin egg envelope that is initially pleated but unfolds as the embryo grows (Fig. 10.3B). The

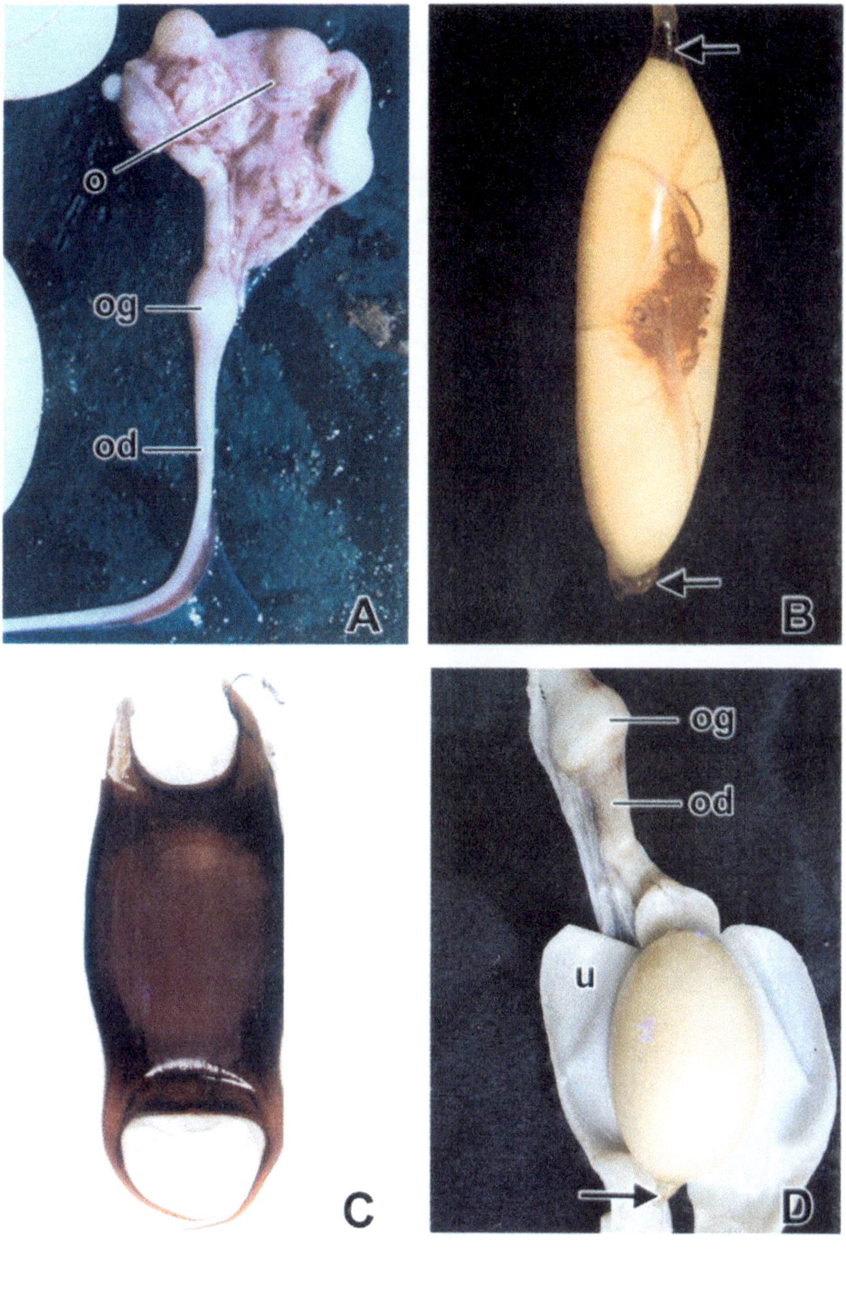

Fig. 10.2 A. Oviducal gland of *Squalus acanthias*. o = ovary, og = oviducal gland, od = oviduct. Original. **B**. Candle case of *S. acanthias* with two embryos. Arrows = egg jelly inside candle. **C.** Egg case of *Raja eglanteria*. **D.** Reproductive tract of *Pristiophorus cirratus* containg candle case with one egg. og = oviducal gland, od = oviduct, u = uterus, arrow = egg jelly inside candle. Original.

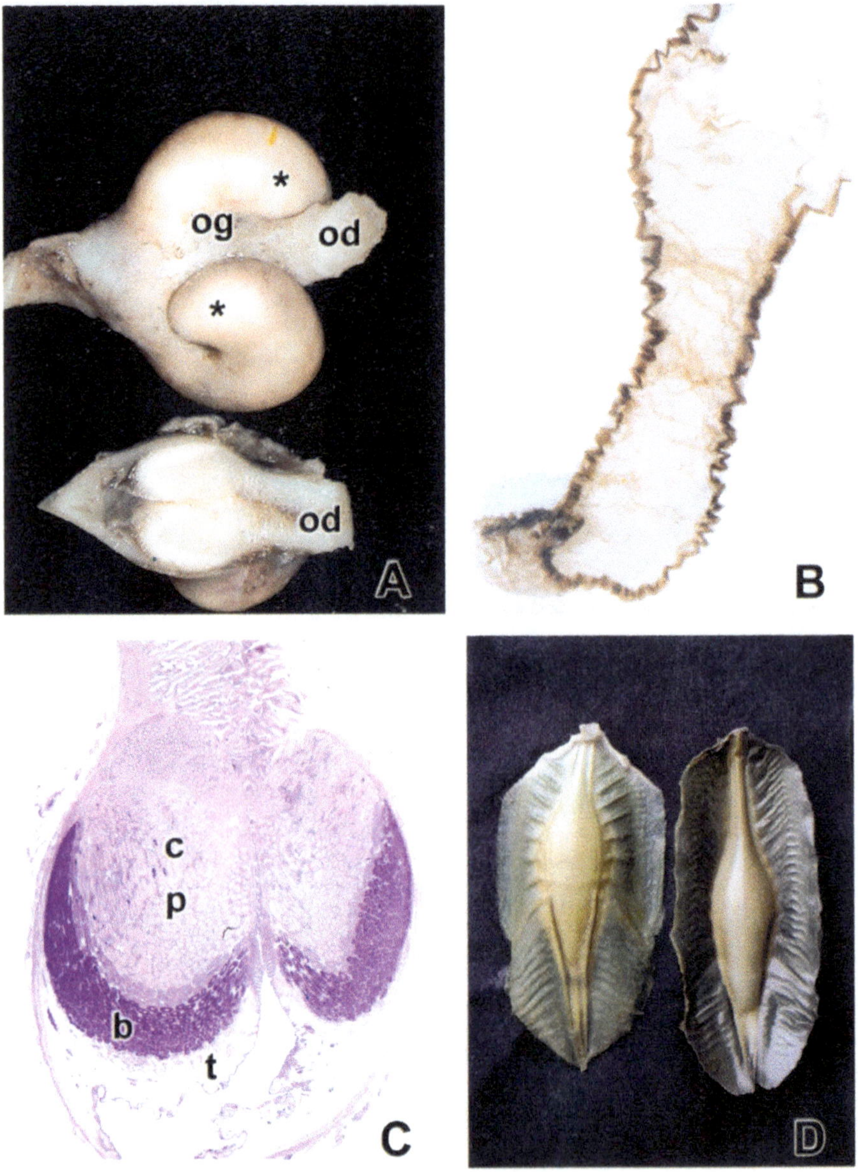

Fig. 10.3 A. Oviducal glands of *Mustelus canis*. Upper oviducal gland (og) has coiled lateral extensions (asterisks). The lower gland has been sectioned. od = oviduct. **B.** Partly unfolded pleated egg envelope of *M. canis*. **C.** LM of longitudinally sectioned OG of *M. canis* similar to the bottom gland in Fig. 10.3A. c = club zone, p = papillary zone, b = baffle zone, t = terminal zone. **D.** Egg case of *Callorhynchus milii* showing central oval portion that houses the embryo and lateral flanges that anchor the capsule to the sandy bottom. Original.

placental smooth dogfish shark *Mustelus canis* and gummy shark *M. antarcticus* that has yolk sac viviparity(Storrie 2004) both have OG that are virtually identical at the gross level (Fig. 10.3A) and microscopically (Fig. 10.3C). To produce a wide egg envelope, the gland does not extend laterally but coils like a ram's horn. This allows for the production of a large, pleated envelope while restricting the overall size of the gland. Microscopically the club, papillary, baffle and terminal zones are evident (Fig. 10.3C). In both *M. canis* and *M. antarcticus* the terminal zone sweeps laterally and anteriorly such that sperm storage tubules are located below baffle zone tubules.

10.7 HOLOCEPHALON OVIDUCAL GLANDS

Observation of the OG of holocephalans is limited, but there are early reports of sperm in the upper oviduct of a *Chimaera* (Dean 1895, 1906) and a description of the morphology of the egg capsules (Dean 1912). An early description of the OG in *Hydrolagus collei* includes an anterior 'albumen zone' that secretes into cranial transverse bands and a few mucous lamellae between the 'albumen' and shell zones (Prasad 1948). The shell zone was divided into a cranial region with baffle plates and a caudal shell zone with a few mucous tubules. Results for *Callorhynchus milii* verify and extend these observations, but current terminology is adopted on the basis of morphological characteristics of the zones (Hamlett *et al.* 1998a) rather than the purported nature of the secretions. As there are presently no biochemical studies of the composition of the secretions, the terminology of albumen and mucous is not warranted and the types of secretions may not be consistent across the reproductive modes. Subtle variations in the type of secretions undoubtedly occur in the production of different types of egg coverings, whereas the basic morphological zonation persists. The club zone is synonymous with the earlier described 'albumen' zone, the papillary zone with the earlier middle mucous zone, the baffle zone with the earlier cranial shell zone and the terminal zone with the earlier caudal shell zone.

In *Callorhynchus milii*, the club and papillary zones elaborate egg jelly that coats the fertilized egg, initially filling the lumen of the egg capsule. It has been suggested that its function must be critical to the developing embryo for at least the early stages of its development (Hamlett *et al.* 1998a). The egg jelly in *Leucoraja erinacea* functions as a structural device that hydrodynamically supports the egg and developing embryo, and the jelly from various regions have differing carbohydrate compositions. The egg jelly supports the embryo during the fragile period of embryogenesis and while it has external gill filaments. The jelly is progressively liquefied by secretory activity of eclosion glands located on the rostrum of the embryo (Hamlett unpublished). There is currently no evidence that the jelly layers are nutritive for the embryo (Koob and Straus 1998).

Secretions produced by the club and papillary zones have been examined histochemically, adequately for only one species, the oviparous catshark *Scyliorhinus canicula*. Secretions of the club zone were initially reported to be as PAS+ carbohydrate (Threadgold 1957), but the secretions were subsequently reported as both PAS+, AB+ polysaccharide (Rusaouen 1976) while the club zone secretions in *Callorhynchus milii* stained both PAS+, AB+ (Smith *et al.* 2004).

Secretory material of the papillary zone in *Callorhynchus milii* is AB+ with more intense staining in the caudal-most papillary tubules. The exact chemical composition of papillary secretions has yet to be determined, but one suggestion is that the material is a type of mucin, functioning to separate and lubricate the region between the egg and the forthcoming capsule (Nalini 1940). Other suggestions include: the secretion of the caudal tubules functions to bind the perimeter of egg jelly to the egg capsule (Knight *et al.* 1996; Hamlett *et al.* 1998a). In *Scyliorhinus canicula* papillary secretions contain carbohydrate and stained metachromatically (Threadgold 1957). Feng and Knight (1992) identified the material as a sulfated glycosaminoglycan as it stained AB+. While we cannot assume that the material secreted in *C. milii* is the same as that of oviparous elasmobranchs, its function of surrounding the egg and filling the egg capsule is thought to be similar.

10.8 STRUCTURE AND FUNCTION OF THE EGG CASE OF *SCYLIORHINUS CANICULA*

The selachian egg case serves mainly as a shock absorber to protect the embryo. The wall of the case is a highly cross-linked collagenous composite material with a complex, multi-lamellar, highly hierarchical construction. In *S. canicula* much of the thickness of the case wall is formed from about 20 lamellae. Each lamella is built from a single layer of flattened collagenous ribbons. These overlap within lamella like tiles on a roof. Ribbons are held together by small quantities of amorphous matrix. The ribbons are continuous throughout the length of the egg case. Each ribbon is formed from numerous transversely banded collagen fibrils arranged in a remarkably regular way that varies somewhat in different regions of the egg case. Each collagen fibril is practically crystalline being constructed from a tetragonal array of collagen molecules (Knight and Hunt 1974) regularly kinked approximately midway between their ends and with the kinked segments running at approximately 20° to the long axis of the fibril. Bands of low protein density within the axial periodic structure of the fibril may allow the egg case wall to act as a semi-permeable membrane, highly permeable to oxygen, carbon dioxide and nitrogenous waste (Knight and Feng 1994b; Knight, Feng *et al.* 1996). The collagen molecules of the egg case are held together by extensive covalent cross-linking (Luong *et al.* 1997), which contributes to the material's mechanical and thermal stability, its extreme insolubility and resistance to enzymatic degradation. The detailed structure of the egg case collagen fibril and its relation to similar fibrils of type VI collagen has been analyzed by Knupp and co-workers

(Knupp *et al.* 1996; Knupp *et al.* 1998; Knupp and Squire 1998, 2001; Knupp *et al.* 2002a; Knupp *et al.* 2002b).

10.8.1 Properties of the Egg Case of *Scyliorhinus canicula*

The mechanical properties of wet longitudinal strips of the *Scyliorhinus canicula* egg case wall have been investigated (Hepworth *et al.* 1994). The strips were remarkably strong (tensile strength 11.9 ± 0.7 MPa, n = 8) and extensible (strain at fracture 0.39 ± 0.03, n = 8). The combination of high tensile strength and extensibility gives them a high toughness (6×10^4 J m^{-2} or 6 kJ kg^{-1} for a specimen measuring 4 x 1 x 10 mm). This toughness is in the range of that of steels. Three mechanisms of energy storage and dissipation probably help to account for the material's toughness. Firstly an increase in axial periodicity in moderately strained material demonstrated by low angle X-ray diffraction and TEM suggests that the kinks in the collagen molecules straighten out progressively as the material is strained and spring back when released. Recovery is substantially complete up to a global strain of up to 24%. This suggests an entropic mechanism of energy storage. A less marked kinking in mammalian tendon collagen molecules may have a similar function. Secondly as the material is progressively strained, the composite construction allows different layers of fibrils to rotate and fail successively depending on their initial orientation. Energy is thought to be dissipated by deformation of the matrix between the collagen fibrils as they rotate and by the formation of numerous cracks as matrix and fibrils fail. Thirdly, filler particles are present in high concentrations in the outer layers of the case (Knight and Vollrath 2001). Ultrastructural evidence (Knight and Feng 1994) suggests these are bound into the meshwork of collagen fibrils and stretch when the egg case is strained probably helping to store and/or dissipate energy.

10.8.2 Extrusion of the Egg Case in *Scyliorhinus canicula*

The selachian egg case is secreted as a continuous extrusion from the OG (Knight *et al.* 1996a). The pressure required for extrusion must be fairly small as the OG is only surrounded by a relatively thin connective tissue sheath and is minimally reinforced internally with connective tissue. The extrusion pressure is thought to be derived from the secretion and swelling of a jelly-like material into the lumen of the OG and from the transport of the large ovum into it (Knight *et al.* 1996a). Although the rate of protein secretion is extremely high (an estimated 140 mg of dry collagen per gland per day) (Knight *et al.* 1996a; Knight and Vollrath 2001) the linear extrusion rate appears to be quite small. As the egg case wall (60 mm long) and tendrils have a combined length of approximately 1 m and as the fastest observed rate of egg case production is approximately one egg case per gland every 60 hours, the average extrusion rate (0.3 mm min^{-1}) is very slow (Knight and Vollrath 2001). This calculation, however, assumes that there is no pause for rest and that the case wall is extruded at the same rate as the tendrils. Slow secretion of the case wall is further suggested by the

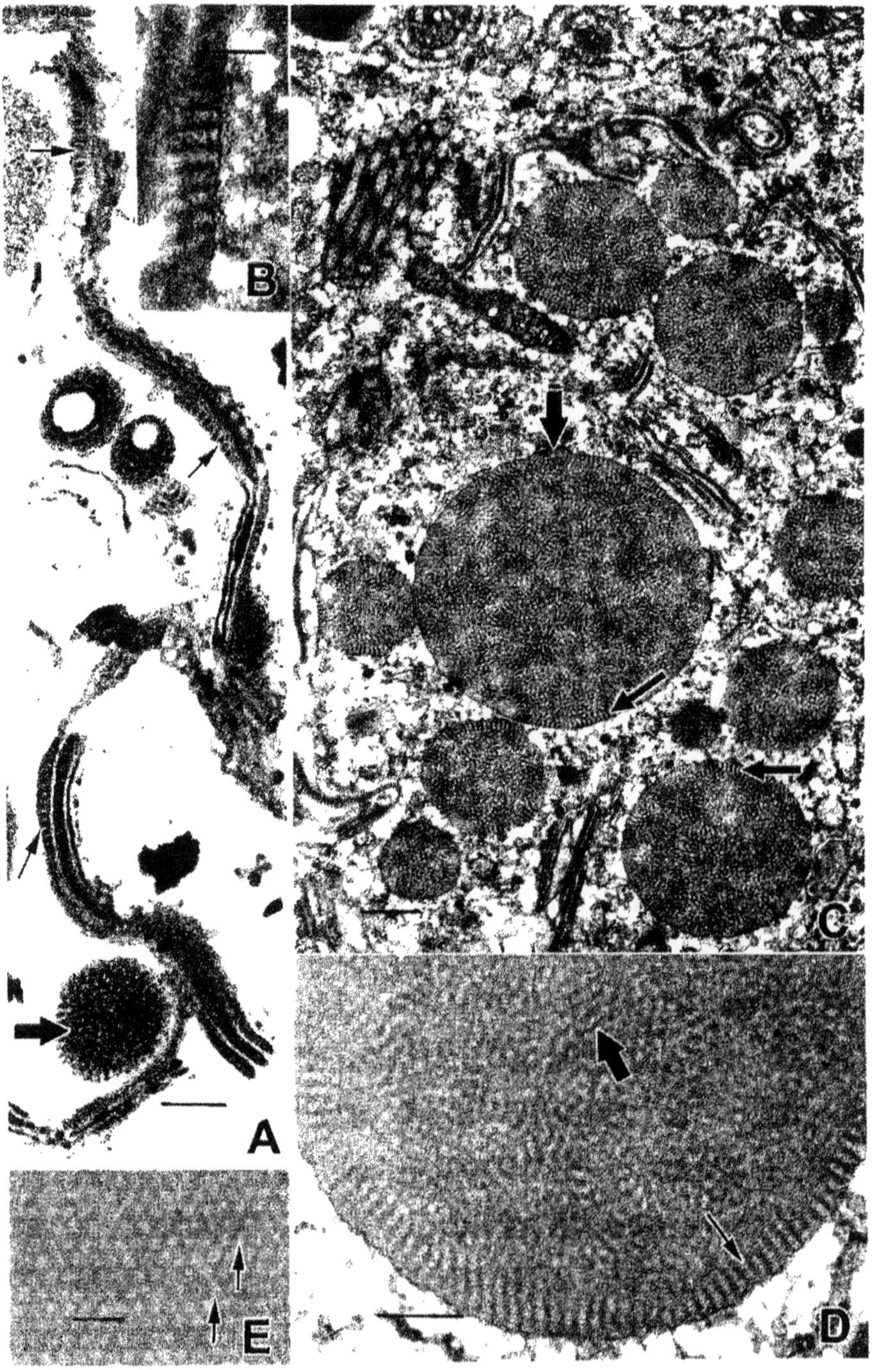

Fig. 10.4 *Contd. ...*

observation that it is not difficult to catch mature fish on their breeding grounds in the act of secreting this structure.

The bulk of the thickness of the OG is constructed from numerous narrow (100 μm diameter) and very long (up to 18 mm) tubular glands. The epithelium of each tubular gland is constructed from ciliated cells and gland cells. The gland cells synthesize the collagen and store it, together with accessory proteins, in secretory vesicles measuring about 1 μm in diameter (Knight *et al.* 1993).

The ultrastructure of storage and secretion has been studied by transmission electron microscopy (Knight *et al.* 1993). Within the rough endoplasmic reticulum cisternae the collagen appears isotropic but becomes assembled into a smetic A or laminar phase in the Golgi cisternae. This phase persists in early secretory granules, where it is found in conjunction with a micellar phase. As these granules mature, the collagen appears to pass through a cholesteric mesophase before adopting a hexagonal columnar arrangement. On merocrine secretion the granule contents revert rapidly to the smectic A-lamellar and micellar phases. As it passes along the OG gland tubules, the collagen is present as a lamellar phase before assembling within the transverse grooves into the final fibrils that constitute the egg capsule. These ultrastructural changes can be divided into distinct phases: (I) collagen assembly within the *trans* Golgi (Figs 10.4A, B), (II) formation of storage granules at the *trans* face of the Golgi (Figs 10.4C), (III) maturation of storage granules (Figs. 10.4D, E; 10.5A-F; 10.6A), (IV) merocrine secretion of the mature granules and their coalescence to produce a strand of secreted material within the lumen of the gland (Figs 10.6B, C), (V) transport of the strand to the transverse grooves and initiation of fibrillogenesis (Figs. 10.7A-D) and

Fig. 10.4 A. Ultrathin section of material stained for acid phosphatase. Banded material (phase II; small arrows) with a D periodicity of approximately 34 nm is revealed by a negative staining effect in some regions of the trans Golgi cisternae. Tangentially oriented material with a similar periodicity is seen in the periphery of a developing granule (broad arrow) close to the trans face. Scale bar 250 nm. **B.** Material similar to that seen in Fig. 10.4A is demonstrated in a trans Golgi cistern any a positive staining effect. Uranyl acetate and lead citrate. Scale bar 100 nm. **C.** A large number of developing granules are seen close to the trans face of the Golgi. Tangentially oriented banded phase II material (small arrows) is seen in the periphery of these granules. The bulk of the granule appears to contain phase I material together with some narrow, short and irregular oriented segments of phase II material. The amorphous material at the periphery of the granule may represent transversely sectioned phase II material. A coated pit (broad arrow) is seen in the limiting membrane of the granule. Uranyl acetate and lead citrate. Scale bar 500 nm. **D.** Part of a developing storage granule. Much of the material (phase I) in the developing storage granule appears to be constructed from rather evenly spaced granules approximately 15 nm in diameter. Tangentially oriented phase II material (small arrow) is seen at the periphery of the granule. A small segment of this material (broad arrow) is seen in the center of the granule. Uranyl acetate and lead citrate. Scale bar 200 nm. **E.** Portion of a hexagonal columnar phase (phase V) seen in a mature granule. A fine dot can be seen at the center of the clear area in some of the hexagons (small arrow). Uranyl acatete and lead citrate. Scale bar 100 nm.

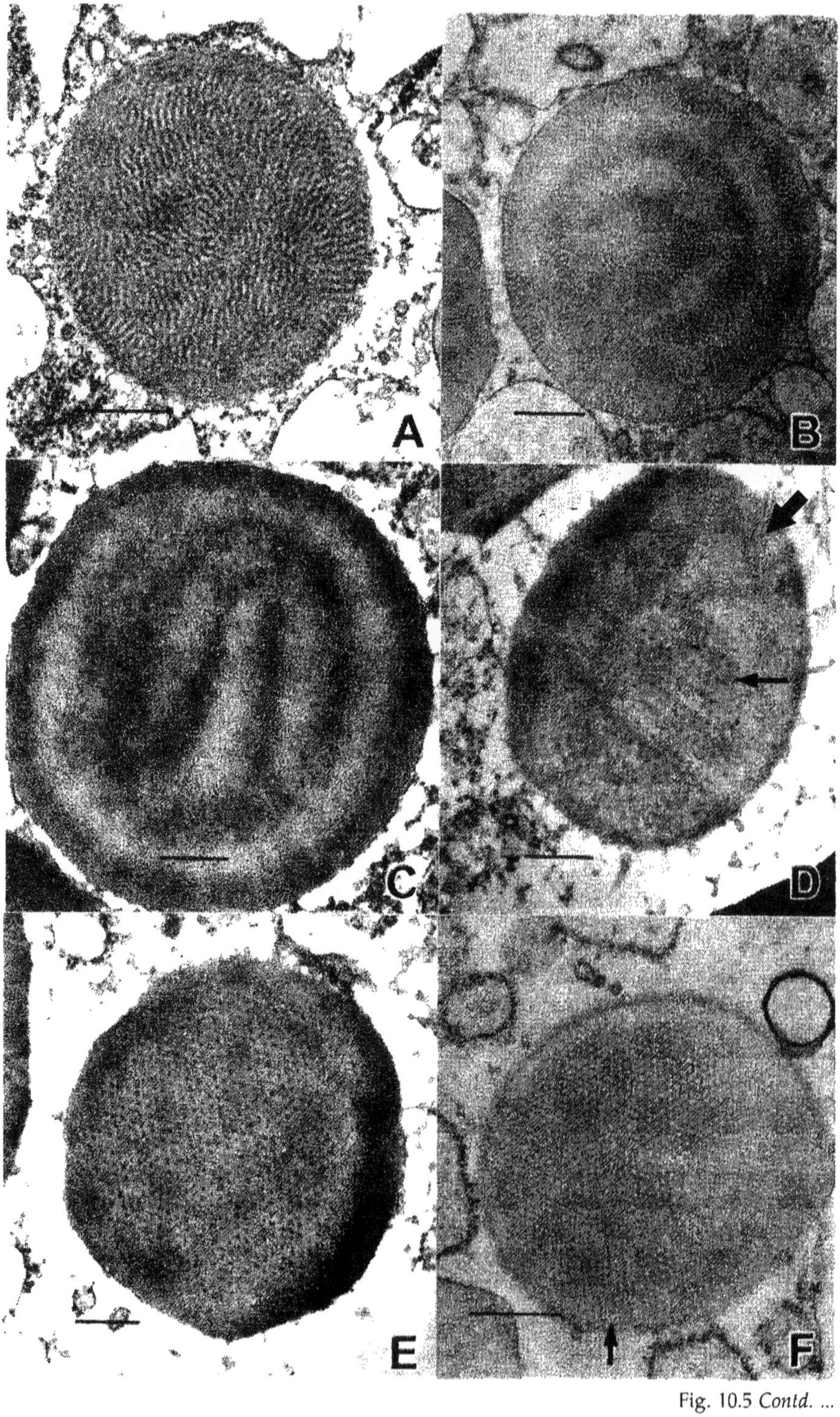

Fig. 10.5 *Contd. ...*

(VI) formation of the white capsule and its transformation into the final clear capsule (Figs 10.7E-F). A summary diagram depicting ultrastructural changes occurring during collagen assembly and secretion is shown in Fig. 10.8) (taken from Knight *et al.* 1993).

Each tubular gland is connected to its own extrusion die (Fig. 10.1A-B) taken from (Knight *et al.* 1997). The latter are arranged in a precisely defined way, overlapping like tiles in a single row at the base of each transverse groove. The lumen of the extrusion die has a circular cross-section (approximately 10μm in diameter) at its input face (connected to a tubular gland) but rapidly expands in one plane so the output face is an elongated slit measuring approximately 10 x 150 to 425 μm (Knight *et al.* 1996c; Knight *et al.* 1997), the actual size depending on location in the OG. The rapid divergence of the die can be described by a hyperbolic function (Knight and Vollrath 2001). The external lips of each die are formed by two flattened roughly semi-circular plate-like structures that protrude into the base of the transverse groove. Each of these plates is formed from two epithelial sheets arranged back to back. This epithelium, like that of the transverse groove contains numerous ciliated cells and some gland cells. The latter are thought to secrete the adhesive that holds individual ribbons together (see below).

The extrusion apparatus is thought to function in the following way (Knight and Vollrath 2001). The dope contains two major components: the collagen as a highly concentrated lamellar liquid crystalline phase and,

Fig. 10.5 A. A granule which appears to be almost entirely packed with phase II material. This is thought to represent a transitional stage between the developing granule (Fig. 10.4C and D) and the mature granule with the double twisted cholesteric mesophase arrangement (Fig. 10.5C.; phase III). Uranyl acatete and lead citrate. Scale bar 0.5 μm. **B.** A granule similar to that seen in Fig. 10.5A which appears to be largely filled with phase II material. A concentric pattern of rather evenly spaced light and dark bands is seen similar to that in Fig. 10.5.C but with the banding seen in phase II material. The pale bands appear to be practically amorphous at this magnification and are thought to represent transversely and obliquely sectioned ribbons of phase II material. The granule is thought to represent a transition stage between the developing and the mature granule. Uranyl acetate and lead citrate. Scale bar 0.5 μm. **C.** A mature granule largely filled with phase III material but with small quantities of what appears to be phase IV material. The pattern of dark and light bands is similar to that seen in Fig. 10.5B. but the constituent material is not D periodic and is seen to be composed of fine protofibrils. Uranyl acetate, phosphotungstic acid and lead citrate. Scale bar 250 μm. **D.** A mature granule containing both phase III (broad arrow) and columnar hexagonal phase IV material (narrow arrow). The hexagonal columnar material is found within a lattice defect in phase III material. Uranyl acetate, phosphotungstic acid and lead citrate. Scale bar 250 μm. **E.** A mature granule which appears to be largely filled with phase IV material but with a small quantity of phase III material visible at the periphery of the granule on the right hand side. The bulk of the contents appear as two hexagonal crystallites with a lattice dislocation between them. Uranyl acetate, phosphotungstic acid and lead citrate. Scale bar 0.5 μm. **F.** When phosphotungstic acid is omitted the phase IV pseudocell has the same spacing as in material stained with phosphotungstic acid but appears as a dark hexagon with a light center. What appears to be a view down the 001 plane (arrow) and several lattice effects can be seen. Uranyl acetate and lead citrate. Scale bar 0.5 μm.

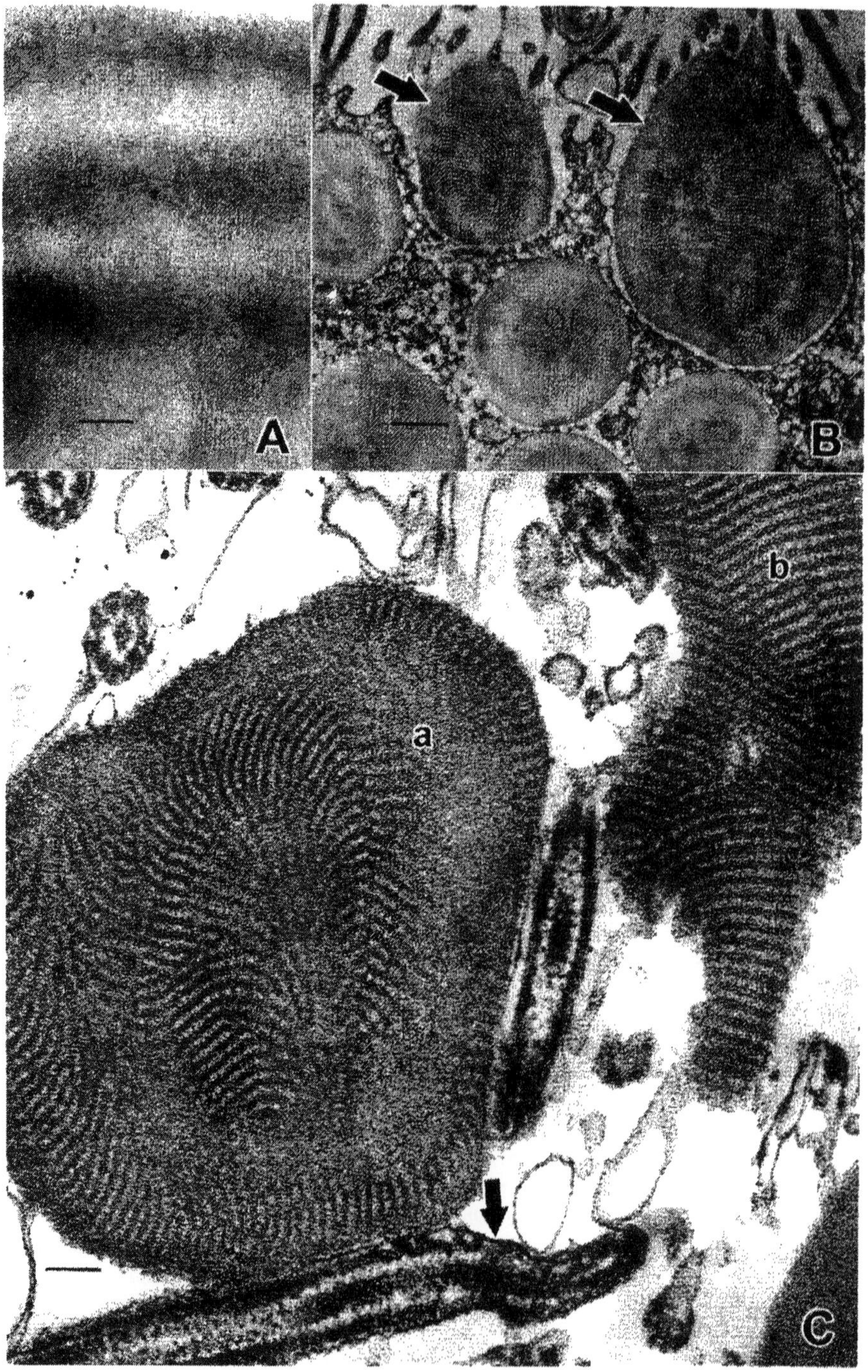

Fig. 10.6 *Contd. ...*

suspended within the collagen solution, droplets of a tyrosine-rich protein that eventually form the filler particles. Ciliary action probably helps to transport the dope to the extrusion dies. The relatively low pH (approximately 6.6) of the lumen of the tubular gland (Feng and Knight 1994b) is probably beneath that of the isoelectric point of the collagen and therefore maintains a positive repulsive charge on the collagen molecules. This opposes weak attractive forces between the collagen molecules, maintaining a liquid crystalline state and preventing premature aggregation and hydrogen bonding of the molecules. On reaching the extrusion die, the rapid divergence of the lumen causes the dope to be subjected to an elongational flow in the hoop (transverse) direction. Here, the liquid crystallinity of short rod-shaped collagen molecules or molecular aggregates causes them to orient rapidly and efficiently in the flow field. This gives rise to a nested arc pattern of molecular orientations (Figs. 10.9A-B). These can be readily demonstrated by polarizing microscopy in aldehyde-fixed dope withdrawn from the transverse grooves and in fully-formed ribbons peeled from the egg case wall. The fluid droplets present in the dope also elongate in the flow field, taking up the same orientation as the surrounding collagen. This can be seen in Nomarski differential interference micrographs of aldehyde-fixed dope withdrawn from the dies (Fig. 10.9A). Thus elongational flow in the hoop direction results in the extrusion of ribbons containing rather precisely defined collagen with similarly orientated elongated droplets of tyrosine-rich protein. As the resulting ribbons pass out from between the curved lips of the extrusion dies they adhere together to give lamellae. Similarly, as the lamellae flow out from the transverse grooves into the main lumen of the gland they adhere together. Adhesion probably results from pulling the ribbons and lamellae together by pumping out water from the transverse grooves and from secretion of an adhesive from numerous gland cells on the plate-like lips of the die and walls of the transverse grooves.

Fig. 10.6 A. Part of a mature granule showing phase III material which has the appearance of a cholesteric mesophase. Uranyl acetate, phosphotungstic acid and lead citrate. Scale bar 100 nm. **B.** Two granules (arrowed) at the apical surface of the cell which appear to have been fixed in the act of merocrine secretion apparently contain only phase II material. The vast majority of granules found in the apex of the cell are mature granules containing a mixture of phase III and IV material. The simplest hypothesis to account for this is that the contents of mature granules show a rapid transition to phase II as soon as the limiting membrane is breached. Uranyl acatete and lead citrate. Scale bar 500 nm. **C.** A section through the lumen of the lower (proximal) part of a tubule shows the coalesced strand of material (B) and the contents of a secreted granule (A) which has not yet coalesced with B. The granule appears to be largely filled with phase II material closely resembling that seen in the grazing section of the coalesced secretion. The transverse order in this material does not appear to be as good as in phase VI. The lumen contains numerous myelin figures. In two instances these appear to adhere to the membrane of a cilium (arrowed). Uranyl acetate and lead citrate. Scale bar 100 nm.

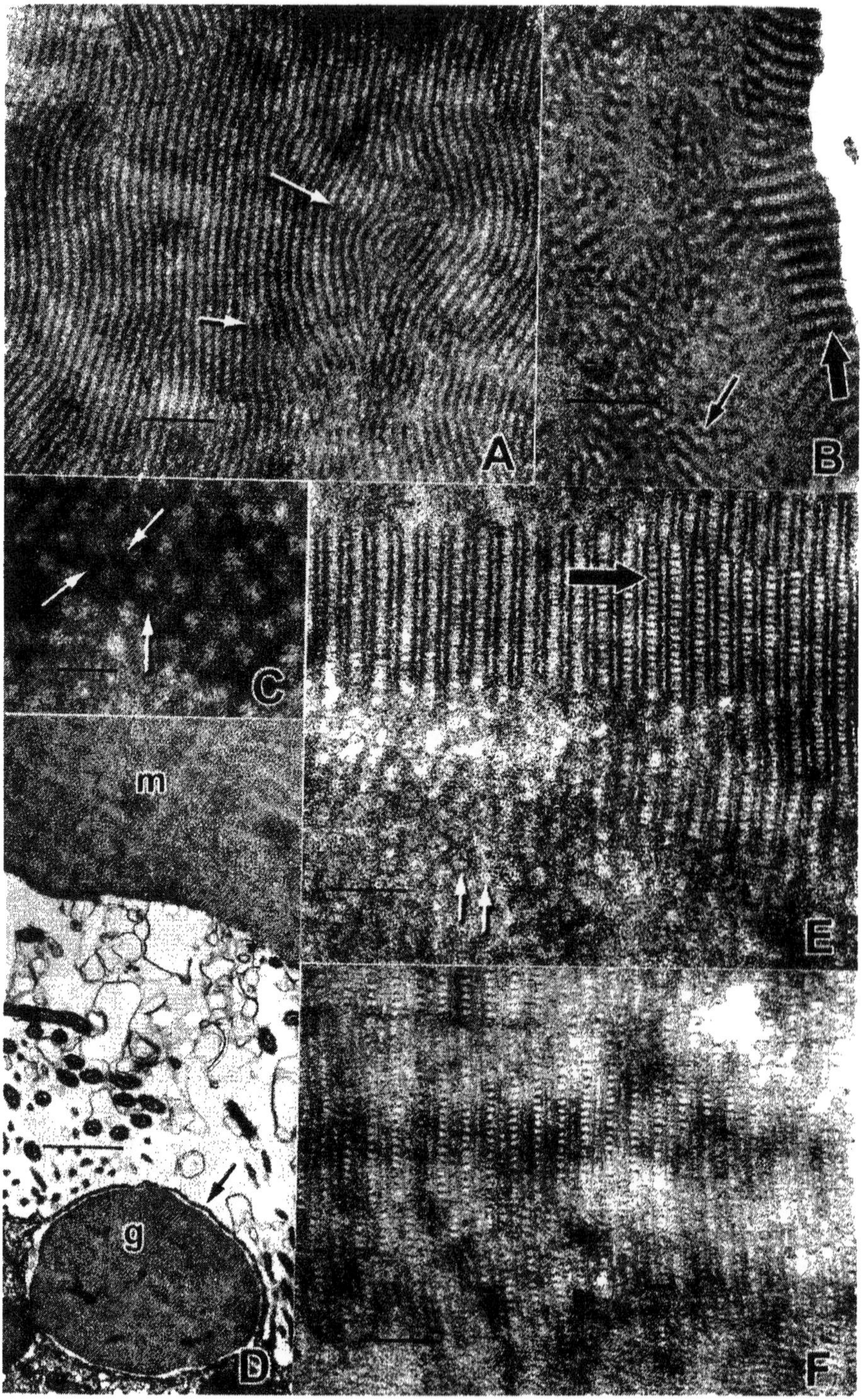

Fig. 10.7 *Contd. ...*

Further processing occurs as the lamellae of extruded collagen leave the transverse grooves and travel into the main lumen of the gland. Here, the pH is thought to increase from about 6.6 to about 8.1 (Feng and Knight 1994b). It is not known how this is achieved but the epithelia may transport hydrogen or bicarbonate ions or the large ovum may produce ammonia. This increase in pH is thought to have two important consequences (Knight *et al.* 1996a): First, it removes repulsive charges and allows the collagen molecules to approach one another closely and hydrogen bond together to form the crystalline fibrils described above. This locks in place the molecular orientations defined by the extrusion dies. Second, an increase in pH brings inactive peroxidase and phenoloxidase previously secreted into the dope, to their pH optima enabling them to form covalent cross-link, thus stabilizing both the collagen and the tyrosine-rich protein. A thin layer of water lubricates the dope as it passes through the die.

10.8.3 What Holds the Ribbons and Lamellae of the Egg Case Together?

The question of what sticks each flattened ribbon from a row of spinnerets into a lamella and each lamella together to form the capsule wall, rib or tendril has received little attention in the literature. A simple observation suggests that there is an additional adhesive over and above the collagenous material that forms the bulk of the egg case. Soaking the egg case for 24 hours in fairly concentrated acetic acid produced only slight swelling of the ribbons and lamellae but enabled them to be easily

Fig. 10.7 A. Grazing section of coalesced material in the lumen as in Fig. 10.6C. showing two different types of disclination (arrowed) Uranyl acetate and lead citrate. Scale bar 200 nm. **B.** Tangentially oriented material (broad arrow) lies at the surface of the coalesced material in the proximal part of the tubular gland while the deeper material shows some phase I material and small irregularly oriented segments of phase II material (narrow arrow). Uranyl acetate and lead citrate. Scale bar 200 nm. **C.** High magnification micrograph of an ultrathin section of columnar phase material stained with uranyl acetate and lead citrate. The dense hexagonal ring appears to be constructed from six subunits (arrows). Scale bar 50 nm. **D.** Proximal part of the tubule. Secreted material (M) which forms the coalesced strand appears to be constructed largely from phase I material and is closely similar to the contents of the granule (G) whose limiting membrane (arrowed) is thought to have recently ruptured. Microvilli, cilia and numerous myelin figures are seen in the lumen. Uranyl acetate and lead citrate. Scale bar 1 μm. **E.** White egg capsule. The material appears to consist of two phases: Fibrils similar to those seen in the final egg case (phase VI: broad black arrow) and micelles (phase V) in between them which appear to consist of a dense ring with a light center in which a dense central dot (tip of white arrows) is sometimes seen. The high degree of transverse order in the fibril can be seen on the right of the black arrow. The dense transverse bands of the fibrils appear to run continuously from fibril to fibril. Uranyl acetate, phosphotungstic acid and lead citrate. Scale bar 100 nm. **F.** Final egg capsule. This appears to be constructed almost entirely from phase VI fibrils. These appear longitudinally sectioned in most of the micrograph but are obliquely sectioned in a narrow strip at the bottom. Material fixed in a solution containing 1% glutaraldehyde and 3% paraformaldehyde and 1% tannic acid. Uranyl acetate, phosphotungstic acid and lead citrate. Scale bar 100 nm.

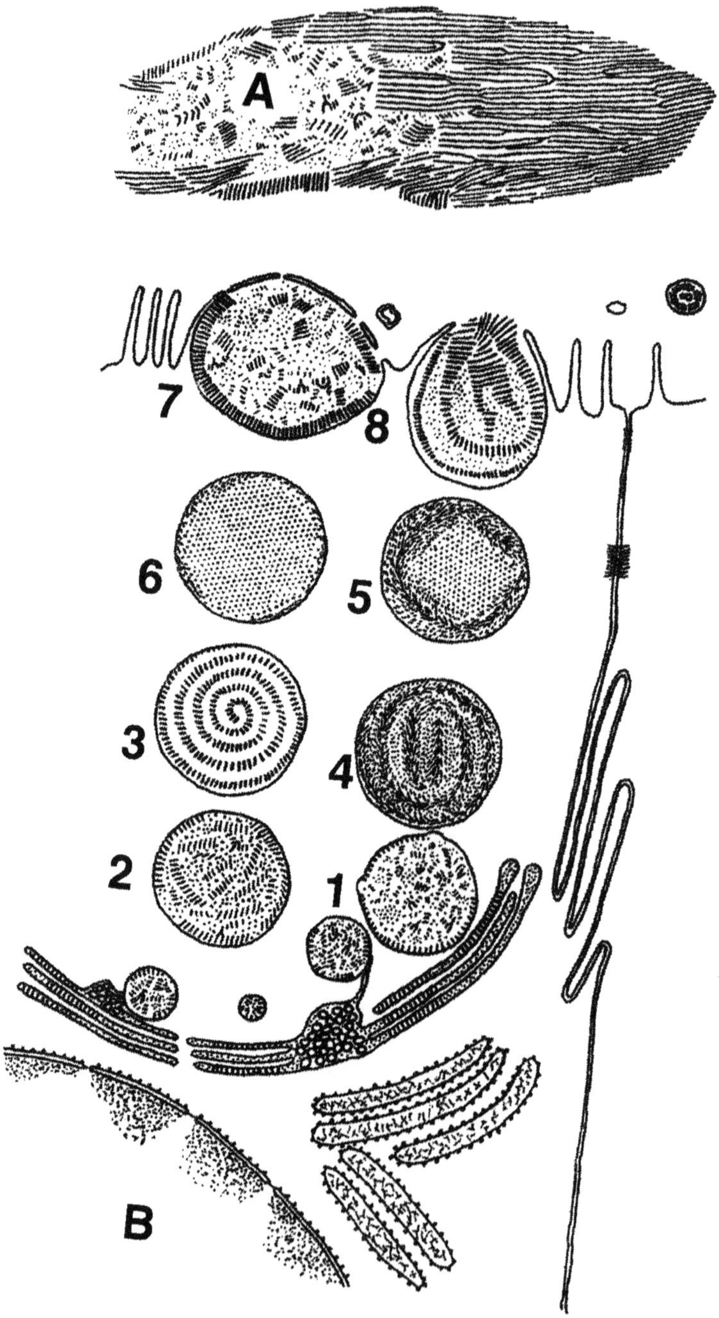

Fig. 10.8 *Contd.* ...

separated from one another (Knight *et al.* 1996) suggesting the extraction of an adhesive secreted in the last part of the secretory pathway, probably from the surface of the baffle plates (to stick the ribbons together) and transverse grooves of the OG (to stick the lamellae together). At least in one location, the middle lamellae of the marginal rib of the egg case, there is good evidence that the baffle plates and transverse grooves do secrete an additional material that coats the ribbons and lamellae. In laser confocal images the surfaces of the individual ribbons are seen to be coated with a thin layer of autofluorescent granules while the lamellae are coated by a slightly thicker layer of the same material (Knight *et al.* 1996b).

In an attempt to locate gland cells which might be responsible for the secretion of an adhesive we have recently examined the epithelium of the baffle plates and transverse grooves in both SEM and TEM. A low power SEM of the surface of the baffle plates and the wall of the transverse groove reveals the presence of a high density of glandular cells almost entirely lacking the numerous cilia seen in the adjacent ciliated cells (Fig. 10.10A). The gland cells are distributed fairly randomly between the columnar ciliated cells and are particularly numerous in the more basal parts of the transverse grooves. A closer view of the anterior edges of the baffle plates shows that the apical surface of these glandular cells is relatively smooth at least in inactive cells except for a thin band of microvilli outlining the junction between the apical and lateral surfaces of the cell. Low bumps with a diameter of about 0.3 μm occur on the apical surface of the gland cells may represent secretory granules which are seen in TEM sections to slightly protrude onto the cell surface in some regions (Fig. 10.10B). A high magnification TEM of the apical part of active gland cells (Fig. 10.10C) reveals numerous secretory vesicles with a diameter of about 0.35 μm containing material with a range of densities perhaps representing secretory material at different packing densities. Where the material within the secretory vesicles is less dense it is seen to be nanofibrillar and resembles the nanofibrillar material lying between the apical surface of the cell and densely staining ribbon of egg case material. In active gland cells some of the microvilli appear to adhere to the ribbon of egg case material, possibly helping to ensure that concentrated adhesive material from the secretory granules is transferred directly to the ribbon.

Fig. 10.8 Diagramatic drawing of a part of a secretory cell in the proximal part of the baffle zone secretory tubule to illustrate the scheme for the sequence of changes that occur during storage and secretion of the (pro)collagen. Nucleus (N). The ER contains the isotropic phase. The Golgi cisternae contain some phase II material: 1. Developing granule containing phases I and II. 2. Developing granule containing a higher proportion of phase II. 3. Granule containing spirally oriented phase II. 4. Double twisted cholesteric phase (IV) and some phase II material. 5. Granule containing both cholesteric (phase III) and phase IV material. 6. Granule containing mainly phase IV material. 7. and 8. Granules undergoing merocrine secretion with consequent reversion to phases I and II. Figs. 10.4A-10.8 from Knight, D. P., Feng, D., Stewart, M. and King, E. 1993. Philosophical Transactions of the Royal Society London B 341: 419436, Figs. 3-16.

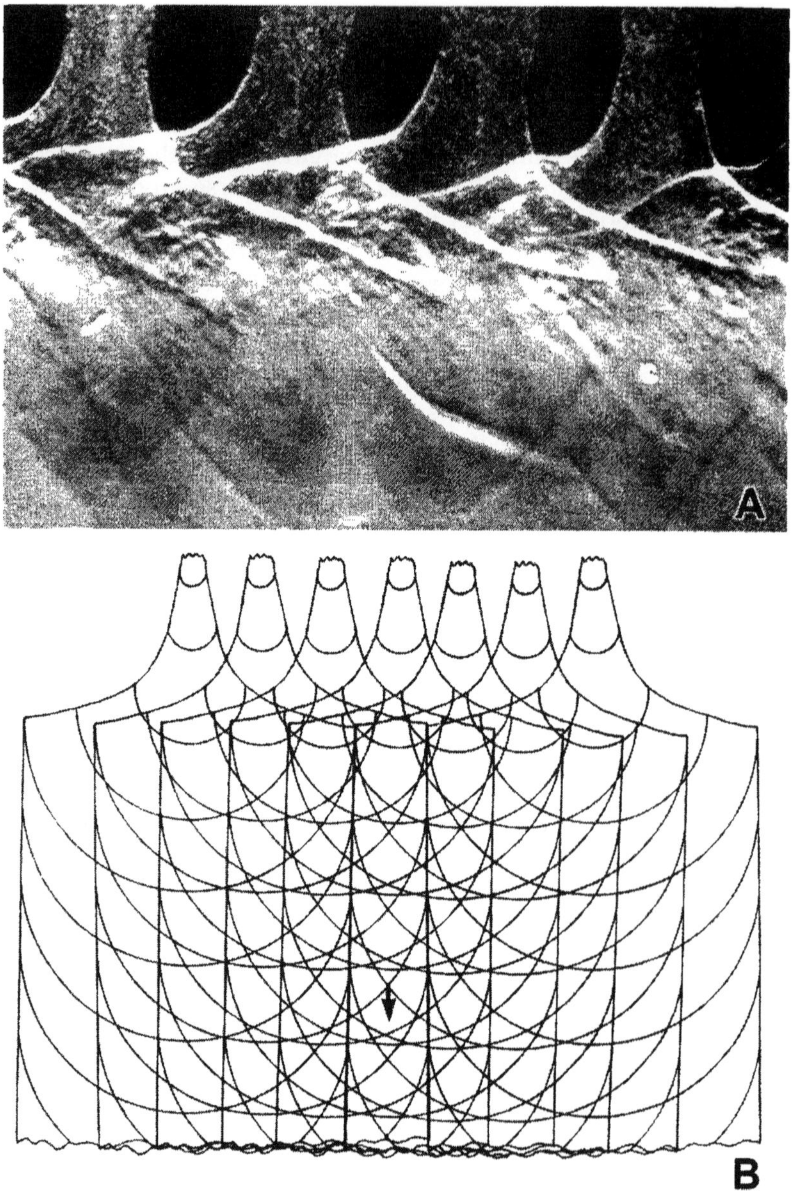

Fig. 10.9 A. Nomarski photomicrograph of a single row of spinneret casts. An overlapping row of ribbons fuses together to form a single lamella. The superposition of the semicircular arced pattern in successive ribbons can be seen in the lamella. The overlapping of 5 ribbons at each point in the lamella gives rise to a plywood-like construction. **B.** Highly simplified diagram showing how a novel plywood-like construction is produced by the superposition of ribbons. The stepped rotation of the fibers in this construction can be seen in the center of the diagram, where 5 ribbons each containing nested arcs of fibers overlap. The width of each ribbon is approximately 425 µm. Figs. 10.9A-B from Knight, D. P., Hu, X. W., Newton, R. H, Cipollone, M., Gathercole, L. J. and Koob, T. 1996. Journal of Biomimetics. 4: 105-120, Figs. 3, 6.

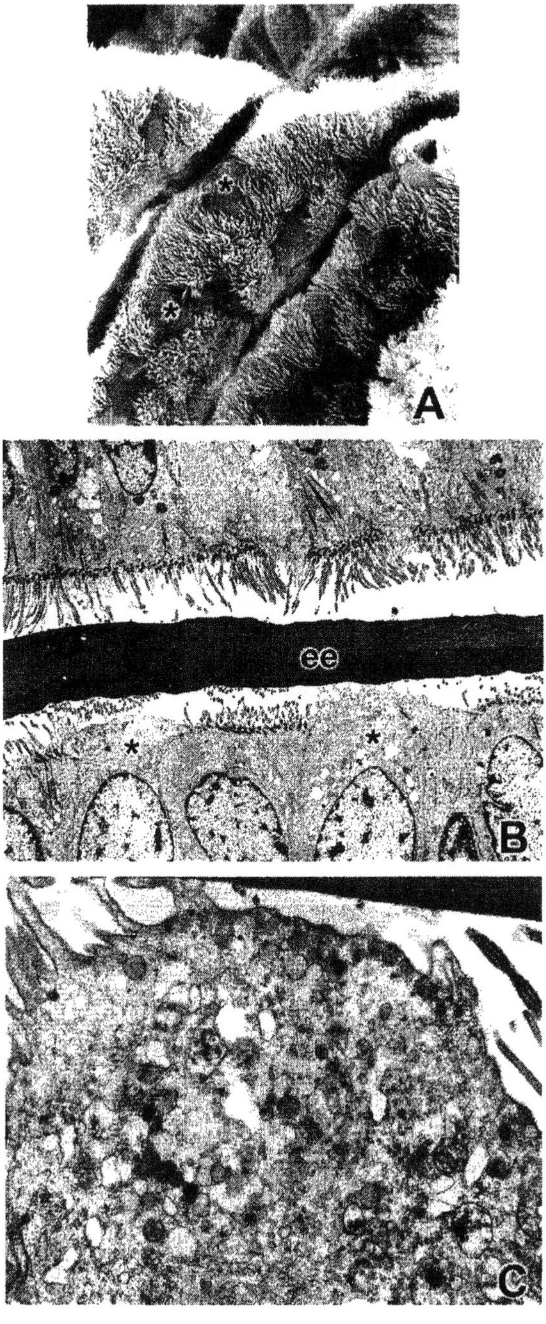

Fig. 10.10 A. SEM of baffle plates of *Scyliorhinus canicula* showing ciliated cells and secretory cells (asterisks). 3,068X. B. TEM of baffle plates of *S. canicula* with ciliated cells, secretory cells (asterisks) and nascent egg envelope. 2,000X. C. TEM of secretory cell with secretory vesicles containing adhesive material in baffle plates of *S. canicula*. 10,000X. Original.

These observations suggest that gland cells on the baffle plates and transverse grooves secrete an adhesive, possibly a glycoprotein soluble in acetic acid, which helps to stick the ribbons and lamellae of the egg case together. Removal of water from the space between the baffle plates and transverse grooves probably helps to bring the ribbons and lamellae together. The water removal may be facilitated by the action of a $Na^{+}K^{+}$ dependent ATPase apparently present in high concentrations between the highly infolded lateral surfaces of the ciliated cells (Knight and Feng unpublished). The epithelium also contains numerous small vesicles which stain vitally with neutral red indicating that a proton pump might also be involved in water transport from the transverse groove (Knight and Feng unpublished). Finally it is likely that the adhesion of the lamellae is assisted by pressing them together once they reach the main lumen of the gland and that the pressure for this is derived mainly from the swelling of the 'jelly' secreted into the lumen of the gland. Thus the sticking together of the ribbons and lamellae is likely to be a fairly complex process in which different parts of the OG play a part. The formation of the layered structure of the egg case shows some resemblance to industrial processes for the formation of glued laminates but is more ingenious in that it is a continuous extrusion process.

10.9 OVIDUCAL GLAND IN *RAJA EGLANTERIA*

The general shape and microscopic organization of the OG of *Raja eglanteria* is fundamentally similar to that of *Scyliorhinus canicula, Leucoraja erinacea* and the holocephalan *Callorhynchus milii.* Examination of the baffle zone in OG actively elaborating egg case material in *R. eglanteria* reveals the extrusion process (Figs. 10.11A, B; 10.12A). Nascent egg case material emerges from a single secretory tubule and passes between baffle plates to emerge in the transverse groove as an individual thread. Adjacent threads adhere to produce broad ribbons. As ribbons widen they form sheets of the egg case (Figs. 10.11A, B). When an actively secreting gland is sectioned longitudinally the extrusion process is revealed. Threads of secretory material pass between baffle plates fuse as ribbons and eventually produce sheets. Figure 10.12A shows how luminal threads (yellow) are transformed into sheets that contribute to the lamellae of the egg case. Examination of a portion of the egg case that has been fractured reveals the lamellar organization of the case and that each lamella is composed of fibrils that generate a plywood-like arrangement (Fig. 10.12B).

10.10 HAIR FORMATION IN THE TERMINAL ZONE

The terminal zone of oviparous species is broad and extensive. Instead of possessing lamellae, simple tubular glands are scattered throughout the terminal zone where they perform two functions, sperm storage and the production of fine hairs that decorate the outside of the egg capsule in

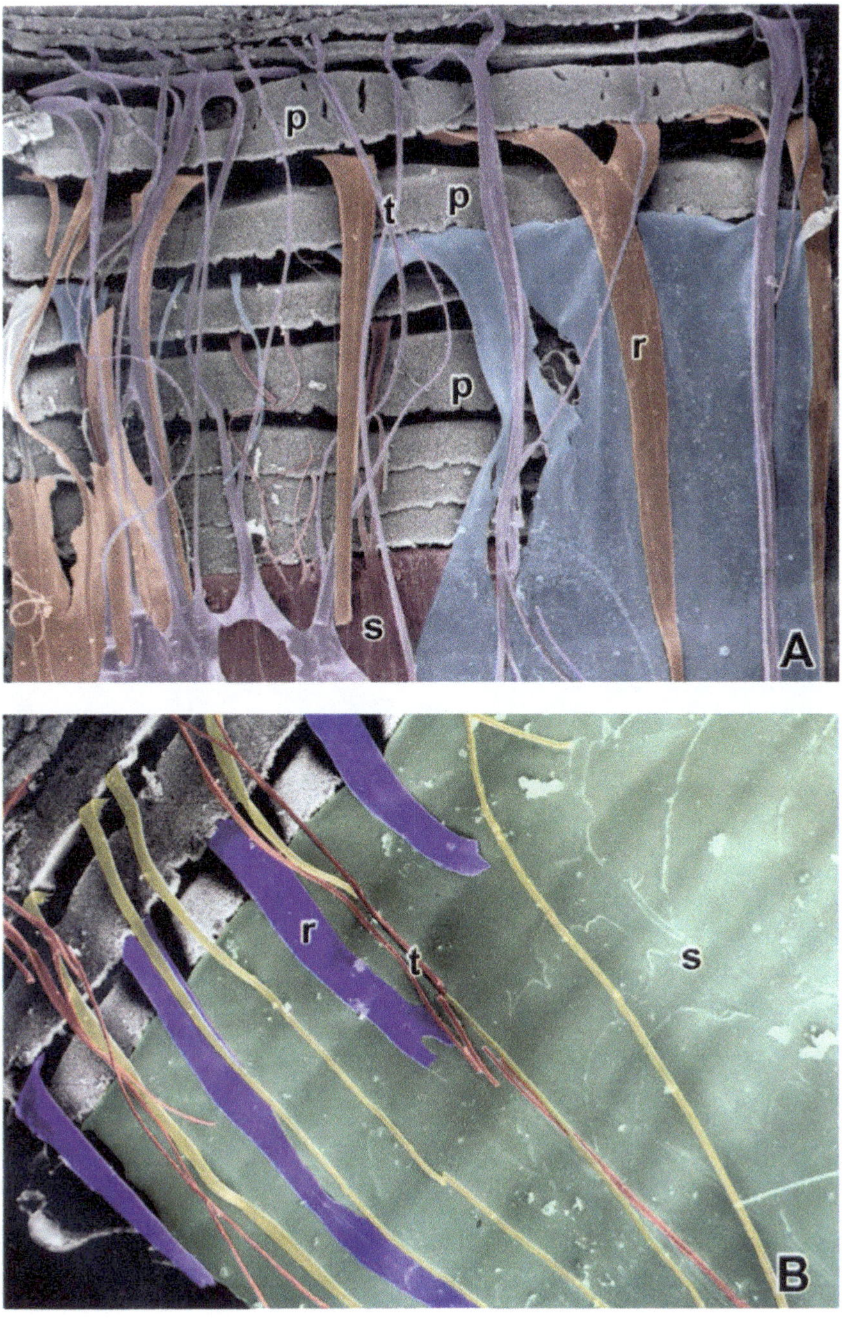

Fig. 10.11 A. SEM of *Raja eglanteria* baffle zone showing egg envelope material emerging from transverse grooves alternating with plateau projections (p). In the more anterior levels egg envelope initially emerges and thin threads (t) that coalesce as flat ribbons (r) that then blend to form laminar sheets (s). 50X. **B.** Baffle zone of *R. eglanteria* showing egg envelope threads (t), ribbons (r) and a completed sheet (s). 75X. Original.

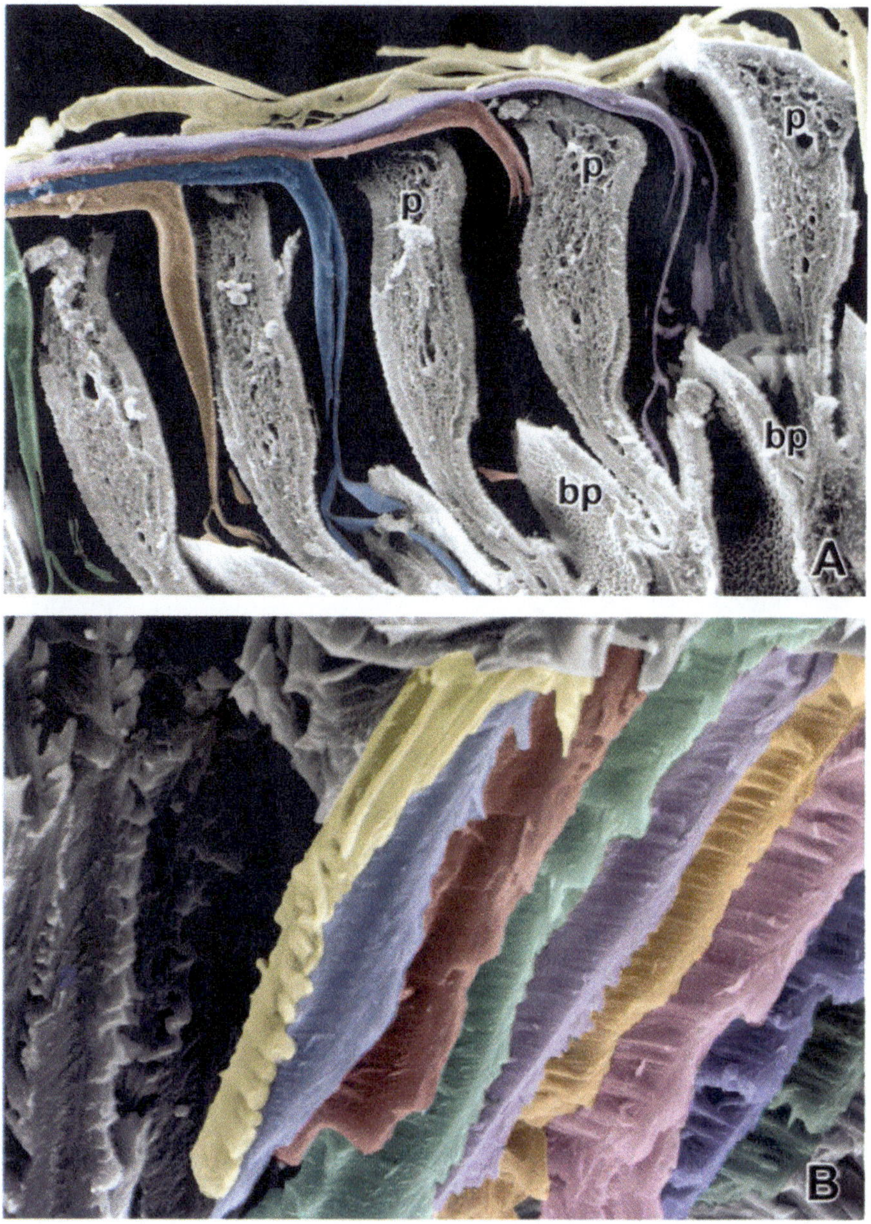

Fig. 10.12 A. SEM of longitudinally sectioned baffle zone of *Raja eglanteria* showing how secretory material from each transverse groove contributed to the laminar organization of the egg case. bp = baffle plates, p = plateau projections. 150X. B. SEM of razor blade section of egg case in *R. eglanteria* to show laminar organization of the egg case and fibrils that constitute the layers. 750X. Original.

species that form hairs. The deepest region of the hair forming tubules in *R. eglanteria* have secretory cells that are virtually identical microscopically to capsule-forming cells of the baffle zone (Fig. 10.13A). Cells close to the opening of the tubules into the main lumen of the OG are mucous. The apparent difference in the secretory product of the baffle zone tubules and hair forming tubules of the terminal zone is that the lamellar architecture of the baffle zone results in a continuous sheet of egg capsule being formed. Secretion of each tubule emerges as a liquid crystal polymer that then blends with secretions of adjacent tubules to form a complete sheet (Knight *et al.* 1996). In the hair forming tubules of the terminal zone the hairs never merge but remain separate. We suggest the surface mucous segment allows the hairs to remain separate as they emerge. The chemical basis of this phenomenon remains unexplored.

10.11 SPERM STORAGE IN THE TERMINAL ZONE

Metten (1939) described sperm storage in *Scyliorhinus canicula* as occurring in the shell-secreting tubules. This is contrary to observations of Hamlett *et al.* (1998a, 1999). Metten examined the OG of the oviparous shark *S. canicula* from animals in various stages of secretion of the tertiary egg envelope. He cited Hobson's (1930) work in the skate and stated that ova were found in the upper oviducts, between the ostium and OG, whilst egg capsules were three-quarters completed. In his own observations in *S. canicula*, Metten reported that in fish with ova in the coelom or upper oviduct, the egg capsule was half secreted or less. He believed that fertilization and egg capsule secretion occur simultaneously and that the shell-secreting tubules provided some nutrient material for the spermatozoa in the capsule substance. He noted that some sperm were incorporated into the egg capsule substance and that it hardened immediately upon leaving the glands. He also claimed that sperm in the bottom of shell-secreting tubules actively secreting egg capsule material were in the process of "turning around" to exit the glands along with the egg capsule material. In studies of the same animal, Knight *et al.* (1996a) examined the structure of the shell secreting tubules of the OG and concluded that fertilization must occur in the upper oviduct or abdominal cavity. They noted sperm in the baffle zone tubules in animals secreting egg capsule but did not notice sperm in the caudad segment of the OG corresponding to the terminal zone.

Metten (1939) did not recognize the terminal zone but pictured a broad caudal region of the OG that he indicated had short mucous glands. This corresponds to the terminal zone. The histological organization of the OG of two other oviparous Chondrichthyes have been studied. In the skate, *Raja eglanteria*, (Hamlett *et al.* 1999) and the elephantfish, *Callorhynchus milii*, (Smith *et al.* 2004) terminal zone tubules are short, broadly dispersed and do not form lamellae. Their surface area is large due to the width of the terminal zone, not, as in viviparous species, to the depth of each gland. In both *Raja* and *Callorhynchus* sperm have been

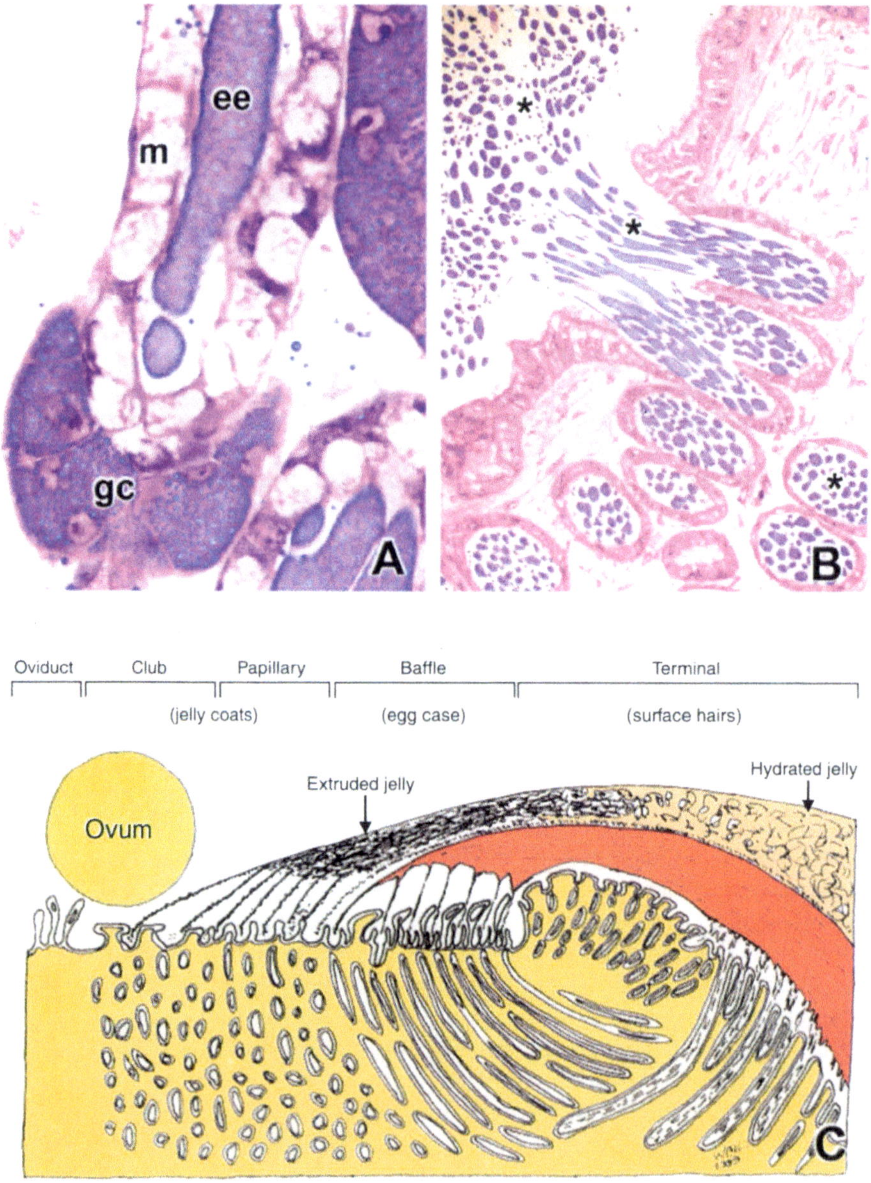

Fig. 10.13 A. LM of hair forming terminal zone tubule in *Raja eglanteria*. The base of the gland has secretory gland cells (gc) virtually identical to baffle zone secretory cells that elaborate egg envelope material (ee) The neck of the tubule has mucous cells (m). 600X. **B.** Luminal aspect of hair forming tubules in terminal zone of *R. eglanteria*. The egg envelope material remains as discrete elements (asterisks) and does not fuse into sheets as in the baffle zone. Presumably the mucous coat contributes to this phenomenon. 200X. **C.** Diagram of OG in *R. eglanteria*, zone sizes not to scale. Original.

observed in the terminal zone and no sperm were seen in the baffle zone. Metten's results show incidental sperm occurrence in baffle zone tubules and sperm being purged from the tubules with secretion of egg capsule. The refilling of gland tubules may be the result of repeated inseminations.

The organization and distribution of terminal zone tubules in *Raja eglanteria*, *Leucoraja erinacea* (Hamlett *et al.* 1998a) and *Scyliorhinus canicula* (Knight *et al.* 1996) are all very similar. The terminal zone tubules in *Callorhynchus milli* strongly resemble the tubules of these oviparous elasmobranchs and perform similar functions.

The organization and distribution of terminal zone tubules in three triakid sharks is in sharp contrast to terminal zone tubules in oviparous Chondrichthyes (Hamlett *et al.* 2003). In *Mustelus canis* (Hamlett *et al.* 2002a), *M. antarcticus* (Storrie 2004) and *Iago omanensis* (Hamlett *et al.* 2002b) terminal zone tubules sweep laterally from the OG lumen and form laterally situated dilated recesses that harbor sperm year round. Recently, Conrath and Musick (2002) studied various aspects of the reproductive biology in *M. canis* and reported observations on sperm storage. They made transverse histological sections of the caudal one-third of the OG from samples collected throughout the year and consistently found sperm in the OG, specifically the terminal zone.

The fate of spermatozoa deposited within the female reproductive tract has been described in the smooth hound, *Mustelus canis* (Hamlett *et al.* 2002a, 2003). Evidence of sperm-uterine association is presented as well as documentation of sperm storage specifically in the terminal zone of the OG. Immediately postpartum placental-uterine attachment sites, now termed uterine or placental scars, begin to remodel to a mucous epithelium for the next gestational cycle. Sperm become embedded in the uterine epithelium adjacent to placental scars. Fertilization is presumed to occur in the anterior oviduct above the OG. Hamlett *et al.* (2002a) suggested uterine embedded-sperm may be in the process of being activated and capacitated as in mammals. This does not explain why it would be necessary to activate sperm before storage in the terminal zone. Alternatively uterine sperm may be phagocytosed by uterine epithelium. The physiological mechanisms that mediate sperm-uterus attachment, release, and storage in the terminal zone of the OG are currently under investigation. Various workers have commented on uterine sperm. Metten (1944) reported uterine digestion of sperm in *Scyliorhinus canicula*. Our observations in *M. canis* do not confirm Metten's (1944) conclusions. Leesa *et al.* (1986) observed sperm in the uterus and OG of *Rhinobatos horkelii* from Brazil after birth and subsequent copulation. The published micrograph of sperm in the OG does not allow determination of what precise zone of the OG was involved but it appears that the sperm are in the gland lumen. Fishelson and Baranes (1998) described the folded endometrium of gravid placental *Iago omanensis* as forming simple tubular glands at the bases of the folds. They saw

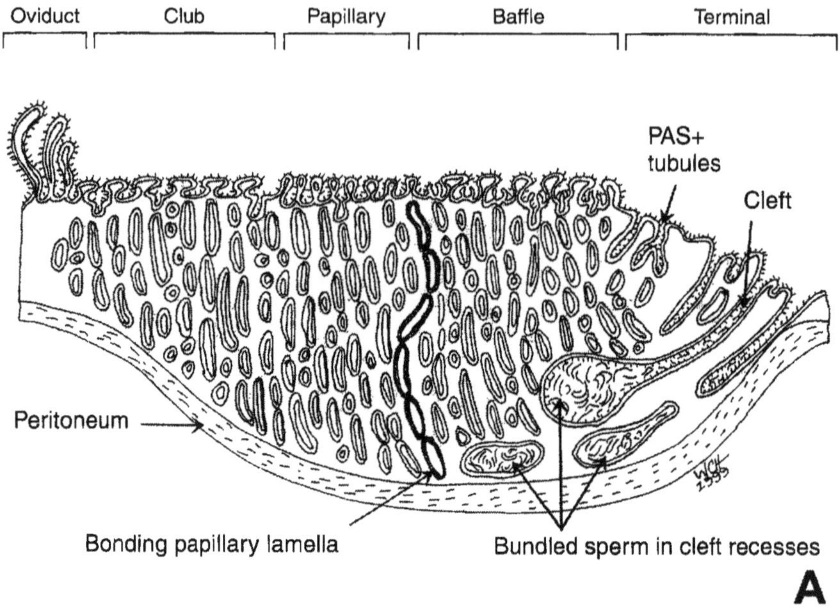

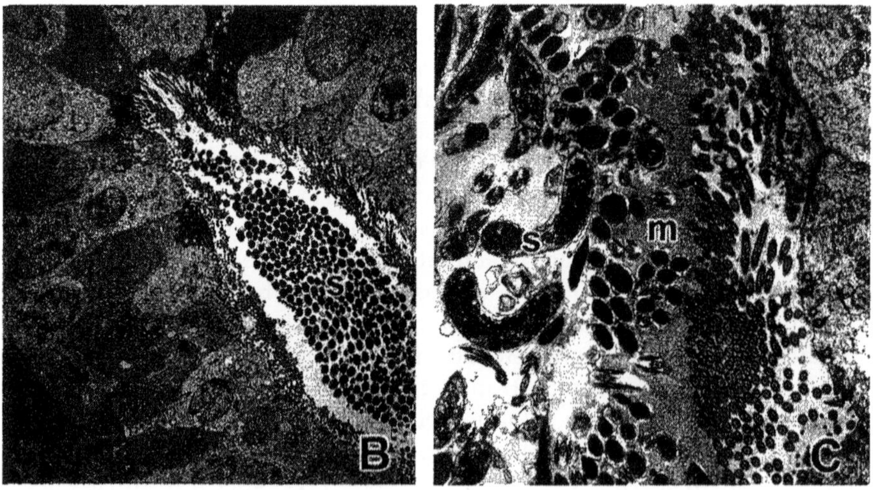

Fig. 10.14 A. Diagram of OG in *Iago omanensis* to show the sperm storage tubules beneath baffle zone tubules. Zone size not to scale. **B.** TEM of sperm storage tubule in *I. omanensis* with luminal sperm (s). 6,000X. **C.** Ciliated epithelium of sperm storage tubules of *Iago omanensis* showing sperm (s) surrounded by luminal matrix (asterisk). 600X. Figs. 10.14A-C from Hamlett, W. C., Fishelson, L., Baranes, A., Hysell, C. K. and Sever, D. M. 2002c. Marine and Freshwater Research 53: 601-613, Figs. 23, 24.

aggregations of sperm in the tubules but did not report sperm embedded in the uterus.

Storrie (2004) reported on sperm in the OG in *Mustelus antarcticus* during different periods of gestation. She demonstrated sperm storage exclusively in the terminal zone, although transient occurrence of sperm was noted in other gland tubules in animals not actively secreting jelly or egg envelope. Noteworthy is the fact that terminal zone sperm were found in both mature (pregnant, non-pregnant and postpartum) and immature (prior to first ovulation) animals throughout the year. Efforts are being directed at elucidating whether terminal zone sperm are being stored for prolonged periods from a single mating event with one or more males or result from multiple matings throughout the year. Feldheim *et al.* (2001) have recently applied genetic analysis using DNA microsatellite loci developed for lemon sharks, *Negaprion brevirostris*, to investigate the possibility of multiple paternity. Their results demonstrated that at least three males sired a single litter. Additionally, using other molecular methods for genetic analysis Saville *et al.* (2002) showed that at least four fathers contributed to a brood of 32 pups in the nurse shark, *Ginglymostoma cirratum*. Histological examination of the OG from neither species was not performed.

Although viviparity is advantageous for nourishment and growth of the offspring, it involves an immunological risk for the sperm containing paternal antigens. The placental smoothhound dogfish, *Mustelus canis* has sperm storage in the terminal zone of the oviducal gland. Sperm storage tubules (SST) in the terminal zone were examined using immunohistochemistry to investigate the tissue expression of the cytokine interleukin IL-1 receptor type I (IL-1R tI) and macrophage migration inhibitory factor (MIF) (Paulseu *et al.* 2003). Immunostaining of the SST were positive for both MIF and IL-1 receptor. The present findings on the expression of the interleukin system and MIF in SST suggest that these cytokines are involved in evolutionarily conserved immunological mechanisms that anticipate the presence of paternal antigens in maternal tissues, specifically the terminal zone.

10.12 CONCLUSIONS

The OG of chondrichthyan fishes is a unique structure that is found in most species where it functions to: (1) produce egg jelly; (2) form the tertiary egg envelope; (3) store sperm; (3) protect sperm from immunological attack by the mother; (4) nourish sperm. There is diversity in the morphology of the tertiary egg envelopes depending on the method of reproduction: oviparous species produce egg cases, yolk sac viviparous species produce transient candle cases and placental species produce an egg envelope. Few species do not encapsulate eggs with jelly and egg envelope, notably *Urobatis halleri* (Babel 1967) and *U. jamaicensis* (Hamlett *et al.* 1998a, 1999). It is rare that the OG is absent as in *Narcine* (Prasad 1945).

10.13 ACKNOWLEDGEMENTS

We wish to thank Brad Johnson and Tamara Fairbanks of the South Bend Medical Foundation for valuable technical assistance with light microscopy and to Joan Clark, University of Melbourne, for technical expertise in electron microscopy. This work was supported by funds provided by Indiana University School of Medicine.

10.14 LITERATURE CITED

Arias, J. L., Fernandez, M. S. Dennis, J. E. and Caplan, A. 1991. The fabrication and collagenous substructure of the eggshell membrane in the isthmus of the hen oviduct. Matrix 11: 13-320.

Arias, J. L., Nakamura, O., Fernandez, M. S., Wu, J. J., Knigge, P., Eyre, D. R. and Caplan, A. I. 1997. Role of type X collagen on experimental mineralization of eggshell membranes. Connective Tissue Research 36: 21-33.

Babel, J. S. 1967. Reproduction, life history, and ecology of the round stingray, *Urolophus halleri* Cooper. California Fish and Game Bulletin 137: 1-104.

Borcea, J. 1904. Sur la glande nidamentaire de l'oviducte des Elasmobranches. Comptes Rendus de l'Académie des Science, Paris 138: 99-101.

Borcea, J. 1906. Recherches sur le système urogénital des elasmobranches. Archives of Experimental and General Zoology. 4: 199-484.

Conrath, C. L. and Musick, J. A. 2002. Reproductive biology of the smooth dogfish, *Mustelus canis*, in the northwest Atlantic Ocean. Environmental Biology of Fishes 64: 367–377.

Dean, B. 1895. *Fishes living and fossil.* Columbia University, N.Y. 300 pp.

Dean, B. 1906. *Chimaeroid fishes and their development.* Washington: Carnegie Institute. 156 pp.

Dean, B. 1912. Orthogenesis in the egg capsules of *Chimaera*. Bulletin of the American Museum of Natural History 31: 35–40.

Feldheim, K. A., Gruber, S. H. and Ashley, M. V. 2001. Multiple paternity of a lemon shark litter (Chondrichthyes; Carcharhinidae). Copiea. 781–786 pp.

Feng, D. and Knight, D. P. 1992. Secretion and stabilization of the layers of the egg capsule of the dogfish *Scyliorhinus canicula*, Tissue and Cell 24: 773-790.

Feng, D. and Knight, D. P. 1994a. Structure and formation of the egg capsule tendrils in the dogfish *Scyliorhinus canicula*. Philosophical Transactions of the Royal Society London B, 343: 285302.

Feng, D. and Knight, D. P. 1994b. The effect of pH on fibrillogenesis of collagen in the egg capsule of the dogfish *Scyliorhinus canicula*. Tissue and Cell 26: 649-659.

Filhol, J., and Garrault, H. 1938. La sécretion de la prokératine et la formation de la capsule ovulaire chez les selaciens. Archives d'Anatomie Microscopie 34: 105-145.

Fishelson, L., and Baranes, A. 1998. Observations on the Oman shark, *Iago omanensis* (Triakidae), with emphasis on the morphological and cytological changes of the oviduct and yolk sac during gestation. Journal of Morphology 236: 151–165.

Hamlett, W. C., Knight, D. P., Koob, T., Jezior, M., Luong, Rozycki, T., Brunette, N. and Hysell, M. 1998a. Survey of oviducal gland structure and function in elasmobranchs. Journal of Experimental Zoology 282: 399-420.

Hamlett, W. C., Hysell, M. K.,Galvin, J. and Spieler, R. 1998b. Reproductive accommodations for gestation in the Atlantic guitarfish, *Rhinobatos lentiginosus*. Journal of the Elisha Mitchell Scientific Society 114 (4): 199-208.

Hamlett, W. C. and Koob, T. 1999. Female reproductive system. Pp. 398-443. In W. C. Hamlett (ed.), *Sharks, Skates and Rays: Biology of Elasmobranch Fishes*, The Johns Hopkins University Press, Baltimore, Maryland.

Hamlett, W. C., Hysell, M., Jezior, M., Rozycki, N. Brunette and Tumilty, K. 1999. Fundamental zonation in elasmobranch oviducal glands. Proceedings of the 5th Indo-Pacific Fish Conference, Nouméa, 3-8 1997. Séret B. and J.-Y. Sire, eds, Paris: Soc. Fr. Ichthyol. and ORSTOM, 1999: 271-280.

Hamlett, W. C., Musick, J. A., Hysell, C. K. and Sever, D. M. 2002a. Uterine epithelial-sperm interaction, endometrial cycle and sperm storage in the terminal zone of the oviducal gland in the placental smoothhound, *Mustelus canis*. Journal of Experimental Zoology 292: 129-144.

Hamlett, W. C., Fishelson, L., Baranes, A., Hysell, C. K. and Sever, D. M. 2002b. Ultrastructural analysis of the oviducal gland and sperm storage in the Oman shark, *Iago omanensis*. Marine and Freshwater Research 53: 601-613.

Hamlett, W. C., Greven, H. and Schindler, J. 2003. Sperm storage in the Class Chondrichthyes and Class Osteichthyes. Pp. 405-413. In A. Legakis, S. Sfenthourakis, R. Polymeni and M. Thessalou-Legaki (eds), *The New Panorama of Animal Evolution*, Proc. 18th Int. Congr. Zoology, Pensoft Publications, Sofia and Moscow.

Hepworth, D. G., Gathercole, L. J., Knight, D. P., Feng, D. and Vincent, J. F. V. 1994. Correlation of ultrastructure and tensile properties of a collagenous composite material, the egg capsule of the dogfish, *Scyliorhinus* spp. A sophisticated collagenous material. Journal of Structural Biology 112: 231-240.

Hobson, A. D. 1930. A note on the formation of the egg-capsule of the skate. Journal of the Marine Biological Association United Kingdom 16: 577-581.

Hunt, S. 1985. The selachian egg case collagen. Pp. 409-434. In A. Bairati and R. Garrone (eds), *Biology of the invertebrate and lower vertebrate collagens, NATO ASI ser. A*, 93.

Jollie, W. P., and L. G. Jollie 1967. Electron microscopic observations on accommodations to pregnancy in the uterus of the spiny dogfish, *Squalus acanthias*. Journal of Ultrastructure Research 20: 161-178.

Knight, D. P. and Feng, D. 1992. Formation of the dogfish egg capsule; a coextruded, multilayer laminate. Journal of Biomimetics 1: 151-175.

Knight, D. P., Feng, D., Stewart, M. and King, E. 1993. Changes in macromolecular organisation in collagen assemblies during secretion in the nidamental gland and formation of the egg capsule wall in the dogfish *Scyliorhinus canicula*. Philosophical Transactions of the Royal Society London B, 341: 419-436.

Knight, D. P and Feng, D. 1994a. Interaction of collagen with hydrophobic protein granules in the egg capsule of the dogfish *Scyliorhinus canicula*. Tissue and Cell. 26: 155-167.

Knight, D. P. and D. Feng, D. 1994b. Some observations on the collagen fibrils of the egg capsule of the dogfish, *Scyliorhinus canicula*. Tissue and Cell. 26: 384-401.

Knight, D. P., Feng, D. and Stewart, M. 1996. Structure and function of the selachian egg case. Biological Reviews. 76: 81-111.

Knight, D. P. Hu, X. W., Gathercole, L. J., Rusaouen-Innocent, M., Ho, M. W., Newton, R. 1996. Molecular orientations in an extruded collagenous composite, the marginal rib of the egg capsule of the dogfish *Scyliorhinus canicula*, A novel lyotropic liquid crystalline arrangement and how it is defined in the spinneret. Philosophical Transactions of the Royal Society London B. 351: 1205-1222.

Knight, D. P., Hu, X. W., Newton, R. H, Cipollone, M., Gathercole, L. J. and Koob, T. 1997. Spinnerets in fish extrude sheet material with complex molecular orientations. Journal of Biomimetics. 4: 105-120.

Knight, D. P. and S. Hunt 1974. Fibril structure of collagen in egg capsule of dogfish. Nature 249: 379-380.

Knight, D. P. and Hunt, S. 1976. Fine structure of the dogfish egg case: A unique collagenous material. Tissue and Cell. 8: 183-193.

Knight, D. P. and Hunt, S. 1986 A kinked molecular model for the collagen-containing fibrils in the egg case of the dogfish, *Scyliorhinus canicula*. Tissue and Cell. 18: 201-208.

Knight, D. P. and F. Vollrath 2001. Comparison of the spinning of Selachian egg case ply sheets and orb web spider dragline filaments. Biomacromolecules 2: 323-334.

Knupp, C., Amin, S. Z., Munro, P. M. G., Luthert, P. J., Squire, J. M. 2002. Collagen VI assemblies in age-related macular degeneration. Journal of Structural Biology 139(3): 181-189.

Knupp, C., Chew, M., Morris, E., Squire, J. 1996. Three-dimensional reconstruction of a collagen IV analogue in the dogfish egg case wall. Journal of Structural Biology 117(3): 209-221.

Knupp, C., M. Chew, Squire, J. 1998. Collagen packing in the dogfish egg case wall. Journal of Structural Biology 122(1-2): 101-110.

Knupp, C., Chong, N. H. V., Munro, P. M. G., Luthert, P. J., Squire, J. M. 2002. Analysis of the collagen VI assemblies associated with Sorsby's fundus dystrophy. Journal of Structural Biology 137(1-2): 31-40.

Knupp, C. and J. Squire 1998. X-ray diffraction analysis of the 3D organization of collagen fibrils in the wall of the dogfish egg case. Proceedings of the Royal Society of London Series B-Biological Sciences 265(1411): 2177-2186.

Knupp, C. and J. M. Squire 2001. A new twist in the collagen story—The type VI segmented supercoil. Embo Journal 20(3): 372-376.

Koob, T. J. and D. L. Cox 1993. Stabilization and sclerotization of *Raja erinacea* egg capsule proteins. Environmental Biology of Fishes. 38: 151-157.

Koob, T. J., and Straus, J. W. 1998. On the role of egg jelly in *Raja erinacea*. Bulletin of the Mt. Desert Island Biological Laboratory 37: 117–119.

Krishnan, G. 1959. Histochemical studies on the nature and formation of egg capsules of the shark *Chiloscyllium griseum*. Biological Bulletin Marine Biological Laboratory, Woods Hole. 117: 298-411.

Leesa, R., Vooren, C. M. and LaHaye, J. 1986. Desenvolvimento e ciclo sexual das femeas, migracoes e fecundidade da viola *Rhinobatos horkelii* (Müller and Henle, 1841) do sul do Brasil. Atlantica, Rio Grande 8: 5-34.

Lo Bianco, S. 1908-1909. Notizie biologiche reguardanti specialmente il periodo di maturità sessuale degli animali del golfo di Napoli. Mitt. Zool., Stat. Napoli 19: 513-761.

Luong, T., M. M Boutillon, R. Garrone, and D. P. Knight 1997a. Partial characterization and sequencing of selachian egg case collagen. *In preparation*.

Luong, T., A. J. Bailey, and D. P. Knight 1997b. Identification of putative cross-links in Selachian egg capsules. FASEB Journal 11:1380.

Metten, H. 1939. Studies on the reproduction of the dogfish. Philosophical Transactions of the Royal Society London B. 230: 217-238.

Metten, H. 1944. The fate of spermatozoa in the female dogfish (*Scyliorhinus canicula*). Quarterly Journal of Microscopical Science 84: 283–295.

Nalini, K. P. 1940. Structure and function of the nidamental gland of *Chiloscyllium griseum* (Mull. and Henle). Proceedings of the Indian Academy of Science 128: 189-214.

Paulesu, L., Cetani, C., Steele, J. and Hamlett, W. C. 2003. Immunoprivileged sites for placentation and sperm storage in a chondrichthyan fish, the smoothhound shark,

Mustelus canis. I Latin-American Symposium on Maternal-Fetal Interaction, Sao Paulo, Brazil.

Prasad, R. R. 1945. The structure, phylogenetic significance, and function of the nidamental glands of some elasmobranchs of the Madras coast. Proceedings of the National Institute of Science India. 11: 282-303.

Prasad, R. R. 1948. Observations on the nidamental glands of *Hydrolagus collei, Raja rhina* and *Platyrhinoidis triseriatus*. Copeia: 1948: 54-57.

Pratt, H. L. 1993. The storage of spermatozoa in the oviducal glands of western North Atlantic sharks. Environmental Biology of Fishes. 38: 139-149.

Rusaouën, M. 1976. The dogfish shell gland, a histochemical study. Journal of Experimental Marine Biology and Ecology 23: 267-283.

Rusaouën, M., Pujol, J. P., Bocquet, J., Veillard, A. and Borel, J. P. 1976. Evidence of collagen in the egg capsule of the dogfish *Scyliorhinus canicula*. Comparative Biochemistry and Physiology B 53: 539-543.

Rusaouën-Innocent, M. 1985. Nidamental gland secreting the dog-fish egg-shell. Pp. 471-476. In A. Bairati and R. Garrone (eds), *Biology of the invertebrate and lower vertebrate collagens*. NATO ASI ser. A 93.

Rusaouën-Innocent, M. 1990. A radiographic study of collagen secretion in the dogfish nidamental gland. Tissue and Cell. 22: 449-462.

Saville, K. J., Lindley, A. M., Maries, E. G., Carrier, J. C., and Pratt, Jr., H. L. 2002. Multiple paternity in the nurse shark, *Ginglymostoma cirratum*. Environmental Biology of Fishes 63: 347–351.

Smith, R. M., Walker, T. I. and Hamlett, W. C. 2004. Microscopic organization of the oviducal gland of the holocephalan elephant fish, *Callorhynchus milii*. Marine and Freshwater Research 55: 155-164.

Stanley, H. P. 1963. Urogenital morphology in the chimaeroid fish *Hydrolagus colliei* (Lay and Bennett). Journal of Morphology 112: 99-125.

Stevens, B. 2002. Uterine and oviducal mechanisms for gestation in the common sawshark, *Pristiophorus cirratus*. Bachelor of Science (Honors) thesis, Department of Zoology, University of Melbourne, Melbourne, Victoria, Australia. 41 pp.

Storrie, M. 2004. Microscopic modifications of the reproductive tissues of the gummy shark (*Mustelus antarcticus*) during maturation and gestation. Ph.D. dissertation, School of Ecology and Environment, Faculty of Science and Technology, Deakin University, Warrnambool, Victoria, Australia. 153 pp.

Thomason, J. C., Davenport, J. and Rogerson, A. 1994. Antifouling performance of the embryo and eggcase of the dogfish *Scyliorhinus canicula*. Journal of the Marine Biological Association United Kingdom 74: 823-836.

Thomason, J. C., S. J. Marrs, and J. Davenport 1996. Antibacterial and antisettlement activity of the dogfish (*Scyliorhinus canicula*) egg case. Journal of the Marine Biological Association of the United Kingdom 12: 577-643.

Threadgold, L. T. 1957. A histochemical study of the shell gland of *Scyliorhinus canicula*. Journal of Histochemistry and Cytochemistry 5: 159-166.

Widakowich, V. 1907. Über Bau und Funktion des Nidamental Organ von *Scyllium canicula*. Zeitschrift für Wissenschaftliche Zoologie. 80: 1-21.

CHAPTER 11

Alkaline Glands and Clasper Glands of Batoids

Eric R. Lacy

11.1 INTRODUCTION

Elasmobranch fish have evolved especially complex anatomical, physiological, biochemical and behavioral specializations as part of their successful reproductive strategy. Phylogenetically, this group of fish appears to be "out of place" because fertilization is internal using unique copulatory organs for the deposition of sperm deep into the female genital tract. This degree of anatomical specialization coupled with internal fertilization is not present, or is rarely approached, in the bony fish, amphibians, reptiles or birds. Given this complexity, it is understandable that little is known about elasmobranch reproduction in general and even less about the mechanisms responsible for delivery of viable sperm into the female reproductive tract. What is known about sperm delivery comes from a relatively small number of species in a few phylogenetic groups. Generalizing the findings from these restricted groups and species to represent larger taxa may, if done judiciously, suggest general patterns of organization and function.

11.2 ALKALINE GLANDS

11.2.1 Background

Batoids lack a defined anatomical structure for the propulsion of sperm from the male reproductive tract into that of the female. This function in sharks is served partly by the siphon sac, an expanded epithelia-lined tube surrounded by a rich tunic of muscle (Chapter 12 of this volume). Sperm propulsion in the most extensively studied group, the mammals,

Marine Biomedicine and Environmental Sciences Center, Medical University of South Carolina, 221 Fort Johnson Road, Charleston, SC 29412 occurs by the contraction of smooth muscle lining the vas (ductus) deferens, ejaculatory ducts, prostate, and the bulbocavernosus muscles. Although some early authors attempted to find homologous structures in elasmobranch fish as in mammals, further study has not revealed any such structures. However, sperm motility in male skates and stingrays may be enhanced by fluid produced in an accessory reproductive "organ," the alkaline gland. It is present only in males and was described in the early anatomical literature as two small, blind-ended sacs that open into the cloaca (Borcea 1906; Daniel 1934). Because these sacs lie adjacent to the most distal and ventral surface of the kidneys and to the terminal portion of the sperm ducts and because they open into the cloaca, they were believed to be urinary bladders or sperm storage sacs. However, there is no anatomical or physiological evidence that the alkaline gland(s) serve as urinary bladders despite adhering to the kidneys. Nor is there evidence that the alkaline gland stores sperm despite its close proximity to the opening of the vas deferens. Unfortunately, these functional misconceptions were repeated for decades in numerous texts and dissecting manuals (e.g., Walker and Hamberger 1982). Subsequently, these paired sacs were termed "Marshall's gland" (Smith 1909; Maren *et al.* 1963) by Smith (1929) based on conversations with his colleague, E. K. Marshall (T. Maren, personal communication). The "gland" actually consists of two anatomically separate sacs, which collectively are now termed "alkaline gland" based on the consistent finding that the pH of its fluid is significantly alkaline in all batoids thus far examined, three species of skates and one species of stingray, the little skate, *Raja erinacea*, barn door skate; *R. stabuliformis*, big skate; *R. ocellata*; and the Atlantic stingray, *Dasyatis sabina* (Maren *et al.* 1963; Grabowski *et al.* 1999).

The effluent tracts (tube[s]) of the alkaline gland sacs empty through the urinary papillae within the cloaca. The close anatomical proximity of the opening of the sac with that of the sperm duct in skates prompted Smith (1929) to speculate that the alkaline pH (8.0-9.2) of the fluid functioned to buffer the acidic pH of the urine, thus protecting the sperm during its exit into the cloaca, presumably laden with urine. However, reports of elasmobranch urine pH are rare and those that are published do not indicate that it is acidic (Choe and Evans 2003).

The elevated pH of the alkaline gland fluid has intrigued investigators because the mucosa generates and maintains a 50-fold gradient of CO_2 and a 100-fold concentration gradient of OH^- ions from plasma to gland lumen. These are some of the steepest alkaline gradients across any epithelium in nature and thus portend the unique if not exaggerated molecular transport processes within and across the glandular mucosa.

11.2.2 Morphology

In skates the alkaline gland is a cream to slightly yellow/brown color (Maren *et al.* 1963). The stingray gland is a port wine color *in situ* and remains that color even after being emptied of its fluid and flushed with physiologic saline (Grabowski *et al.* 1999). Each sac in skates and stingrays is adherent to the most posterior and ventral surface of each respective kidney and thus is covered externally by the peritoneal tissue as part of the coelomic cavity (Fig. 11.1). In the stingray a broad fibrous ligament traverses the ventral surface of the gland, connecting the vertebral column anteriorly to the distal end of the puboischiadic bar.

Each sac of the paired gland narrows in its medial and caudal portion into a single duct that traverses the thick cloacal wall (Figs. 11.2A, B). In the stingray, the duct joins the respective vas deferens (sperm duct) within the cloacal wall, forming a common duct (alkaline gland plus sperm duct), which extends approximately 3 mm, and then opens just lateral to the tip of the urinary papilla (Fig. 11.2B). This anatomical arrangement differs from that in skates in which the alkaline gland does not join the sperm duct but instead opens adjacent to it on the urinary papilla (Maren *et al.* 1963) (Fig. 11.2A). The confluence of both alkaline gland duct and sperm duct to form the common duct in the stingray and the close proximity of their orifices in the skate strongly suggests that the alkaline gland fluid acts on the sperm as it exits the animal's body. In addition, this anatomical arrangement in stingrays explains the observation that the alkaline fluid gland and sperm are extruded simultaneously as a mixture when light pressure is placed on the animal's cloacal region. It is doubtful that the alkaline gland fluid functions primarily to protect sperm by neutralizing the urine, as suggested by Smith (1929), because the fluid meets the sperm prior to its exposure to any residual urine that may be in the cloaca. Anatomically, the alkaline gland appears to be a secondary reproductive organ whose fluid acts directly on the sperm. Further study is needed to test this idea.

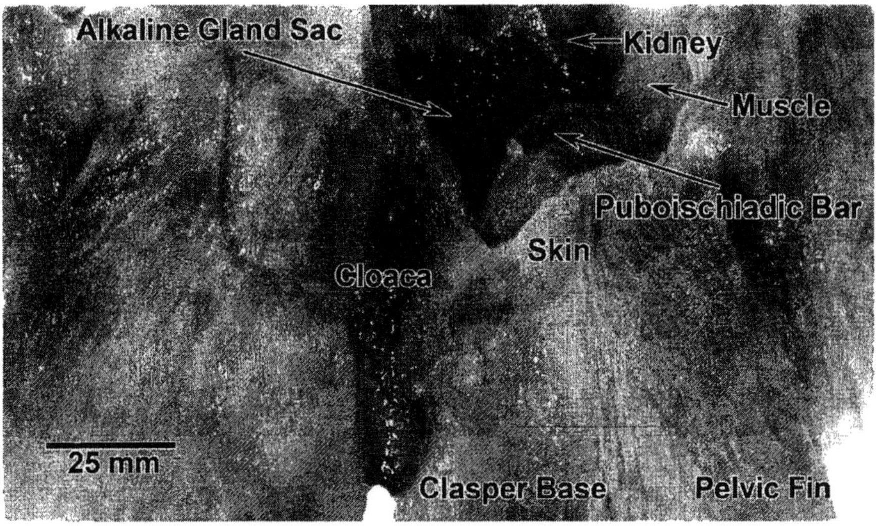

Fig. 11.1 Ventral pelvic region of the Atlantic stingray, *Dasyatis sabina*. The left body wall has been surgically removed to show the animal's left alkaline gland adherent to the posterior portion of the kidney. Original.

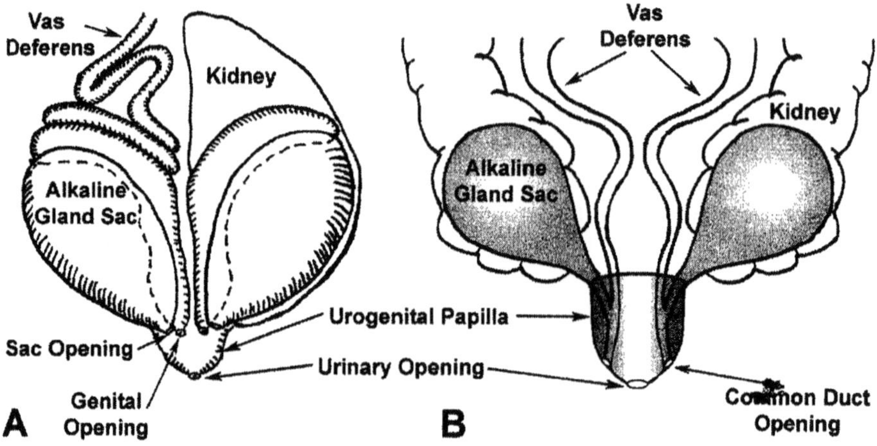

Fig. 11.2 Diagrammatic representation of the alkaline gland showing its relationship to the vas deferens and urinary papilla in **A.** Skates (modified from Maren *et al.* 1963), and **B.** Atlantic stingray (*Dasyatis sabina*). Note the confluence of the alkaline gland duct with the sperm duct (terminus of vas deferens) in the stingray but not the skate. Original.

11.2.3 Microscopic Structure

Light and electron microscopical analyses of the alkaline gland are consistent among skates and stingrays in that the luminal surface is thrown into large folds. In skates the mucosa is formed into villar-like ridges similar to the mammalian small intestine (Maren *et al.* 1963; Hamlett 1999). This differs from stingrays, in which mucosal folds appear as long ridges (Grabowski *et al.* 1999). The gland lumen is lined by simple columnar epithelial cells in both skates and stingrays (Maren *et al.* 1963; Masur 1984; Grabowski *et al.* 1999; Hamlett 1999). In the stingray there are two populations of columnar cells based on size (Fig. 11.3). About 16% of the cells are approximately 15 µm wide whereas approximately 84% of the cells are only about 7 µm wide. They are otherwise morphologically identical, suggesting that these two populations of cells may reflect age differences.

The luminal surface of all columnar epithelial cells is amplified into a brush border (skates) or microplicae (stingray). The lateral cell membranes are relatively straight along the luminal half of each columnar cell but interdigitate extensively with themselves and with the adjacent cells along the basolateral surface (Fig. 11.4). Ruthenium red histochemical staining shows a rich glycocalyx projecting from the luminal plasma membrane (Masur 1984). Epithelial cells of both elasmobranch groups have large basally placed nuclei. Extensive vesicles, both secretory and endocytotic, lie in the apical cytoplasm near the luminal cell surface (Masur 1984; Grabowski *et al.* 1999; Hamlett 1999). These cells have microarchitectural characteristics of both active ion transport (amplified plasma membrane, extensive mitochondria

Fig. 11.3 Scanning electron micrograph showing the luminal surfaces of the alkaline gland columnar epithelial cells which are elaborated into microplicae. Note the differences in cell width denoting the two populations of epithelial cells. From Grabowski 1993. Ph.D. Dissertation, Medical University of South Carolina, Charleston SC, Fig. 3.

with lamellar cristae) and protein secretion (prominent endoplasmic reticulum, Golgi apparatus, and numerous budding secretory vesicles). A distinctive aspect of the stingray epithelium is the presence of large secondary lysosomes (phagosomes) containing stacks of lipid bilayers (myelin figures) and multivesiculated bodies (Figs. 11.4, 11.5), which stain with lipofuscin, indicative of an increased lysomal processing of lipid membranes as observed in other tissues (Reed *et al.* 1965; Harmon 1990).

The glandular epithelium of the stingray consists of a subpopulation of cells (about 5%) which are morphologically distinct from the dominant columnar cells (Fig. 11.4). These cells, termed "basal cells", are generally round, not columnar, and do not contact the alkaline gland luminal surface or the basal lamina. Furthermore, the plasma membranes of these cells do not significantly interdigitate with adjacent columnar cells. Basal cells are easily identified because their cytoplasm generally stains lighter than that of the columnar cells; they have an exceptionally large nucleus/cytoplasm ratio and relatively few organelles. They appear to be undifferentiated cells and thus may be epithelial "stem cells" (Fig. 11.4).

Freeze-facture techniques applied to the stingray alkaline gland epithelium reveal a highly distinctive feature. The "tight junctions," zonulae occludentes, are deep (average 1.4 µm) and consists of about 22 strands (fibrils) (Fig. 11.6). The majority of the fibrils are arranged parallel

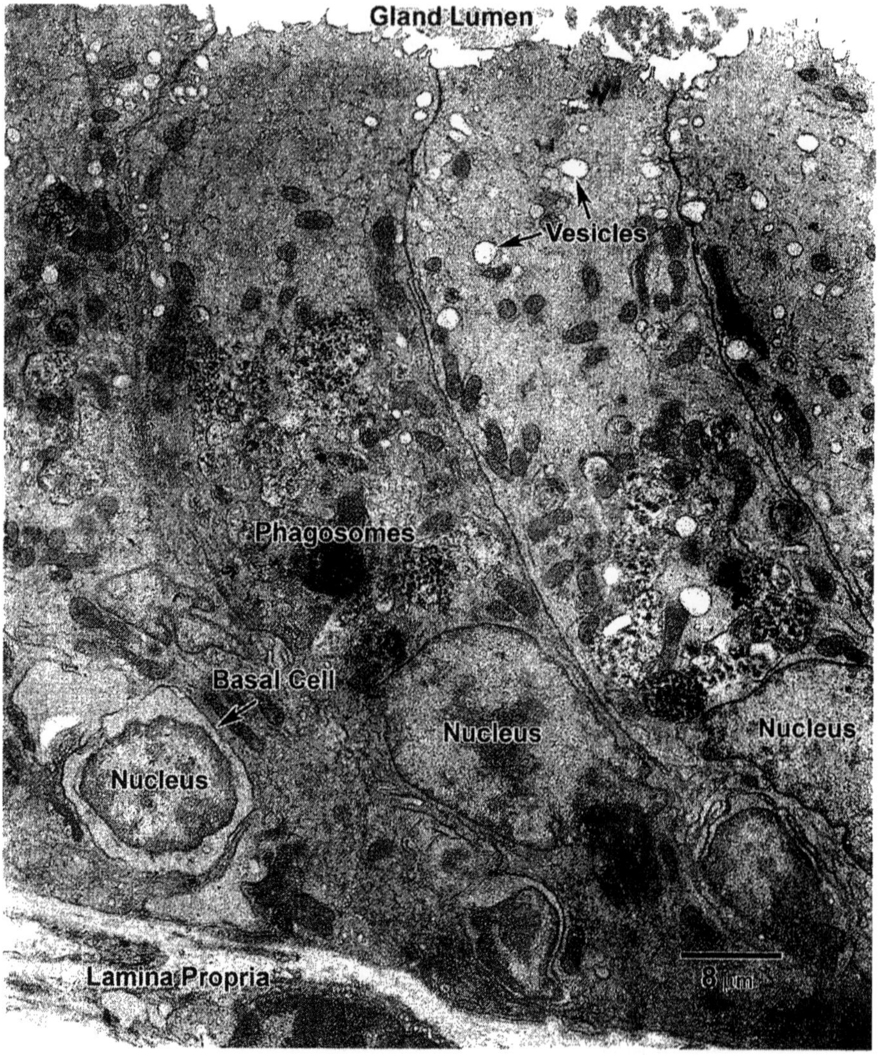

Fig. 11.4 Transmission electron micrograph of the columnar epithelium lining the stingray alkaline gland and extending to the basal lamina. Basal cells are interspersed among the columnar cells. Original.

to the apical plasma membrane but those in the lower part of the junctional complex form a loose anastomosing network (Fig. 11.6) (Grabowski *et al.* 1999). Correlations between tight junctional morphology and one of the defining parameters of epithelial transport, the transepithelial electrical resistance, predict the gland epithelium to be moderately tight to ionic flux (Claude and Goodenough 1973). Several electrophysiological studies in both skates and stingrays (Maren *et al.* 1963; Smith 1981; Grabowski *et al.* 1999) confirm that the electrical resistance is indeed very high across the mucosa. Intramembranous particles (proteins) are abundant particularly in the area

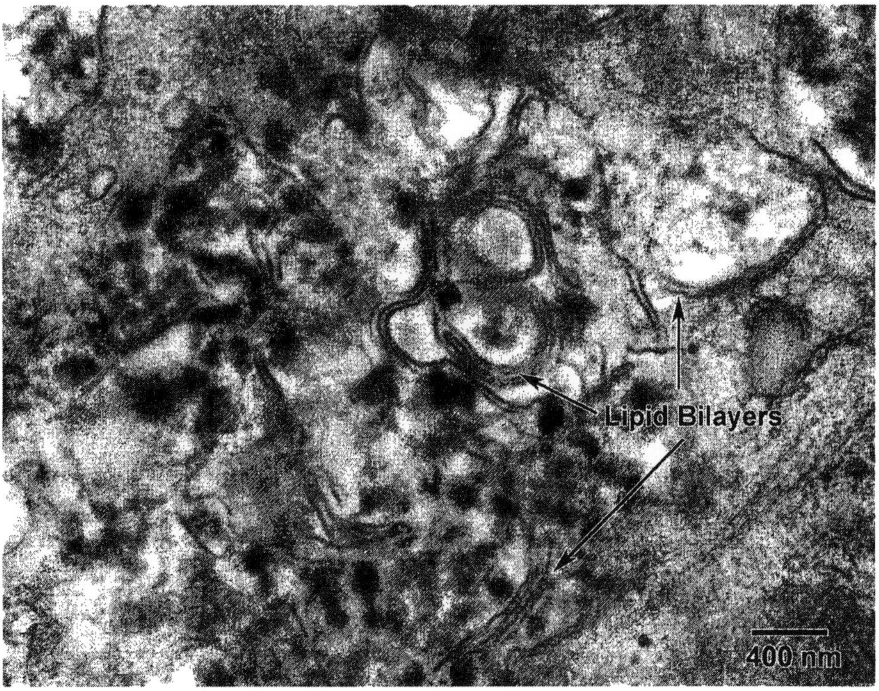

Fig. 11.5 Transmission electron micrograph of phagosome with extensive lipid bilayers and other cell debris. Original.

adjacent to the occluding junctional complex. Gap junctions representing sites of intercellular communication are occasionally present. The apical plasma membrane does not exhibit distinctive intramembranous particles in freeze fracture replicas (Grabowski *et al.* 1999).

The lamina propria of the gland has an abundant network of vessels including capillaries, arterioles and venules. This is further evidence that the epithelium is a secretory tissue responsible for production of the fluid in the sac, *i.e.*, the sac is not purely for storage, the fluid having been produced elsewhere. The rich plexus of nerves observed in the lamina propria of the Atlantic stingray probably functions to contract the sac walls, propelling gland fluid from the sac lumen into the common duct (alkaline gland and sperm duct) where it mixes with sperm, and then into the cloaca. In lightly anesthetized fecund stingrays, slight manipulation of the clasper results in expulsion of sperm mixed with the alkaline gland fluid from the tip of the urinary papilla. There is no evidence for the presence or absence of sphincters in the duct of the alkaline gland that could control the timing and volume of fluid release.

11.2.4 Fluid Composition

Access to the lumen of each sac and the collection of fluid it contains is obtained without surgery by puncturing the lateral wall of the cloaca with

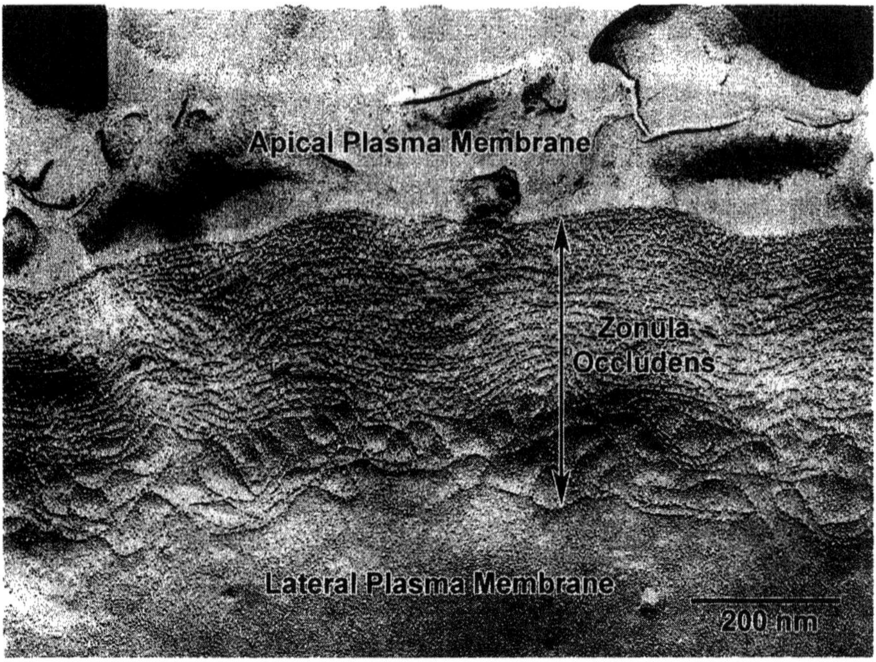

Fig. 11.6 Freeze fracture replica showing the tight junction (zonula occludens) between two columnar epithelial cells in stingray alkaline gland. Note the extensive strands that are closely packed apically and looser basally. From Grabowski *et al.* 1999. Biological Bulletin 197: 82-93, Fig. 6.

a hypodermic needle and syringe. In skates this fluid is light yellow and in stingrays it is deep purple. Microscopical observation of this darkly-colored fluid in stingrays reveals only the remnants of some necrotic sperm and cell organelles (Fig. 11.7). These findings further corroborate the notion that the alkaline gland does not function as a *bona fide* sperm sac. Some residual sperm probably gain access to the alkaline gland lumen through reflux via the common duct (Fig. 11.2B). The conspicuous secondary lysosomes in the lining epithelial cells suggest a robust cellular phagocytosis pathway (lysomal processing of lipid membrane) to remove necrotic sperm from the gland lumen.

The amount of fluid in each gland varies greatly among individuals, but in stingrays captured along the Southeast coast of the United States it is roughly correlated with the season, the greatest amount being present in the late summer and autumn. Adult Atlantic stingrays, *Dasyatis sabina*, measuring 30-35 cm wing width, have from 0.5 ml to as much as 10 ml of fluid per sac. The secretory and turnover rates of fluid are not known except that in skates depleted of fluid through aspiration there is enough fluid in the sacs after 24 hours to do "chemical analyses" (Maren *et al.* 1963).

Analyses of the fluid in each sac, which is most assuredly secreted by the epithelium lining it, have been published for several species of

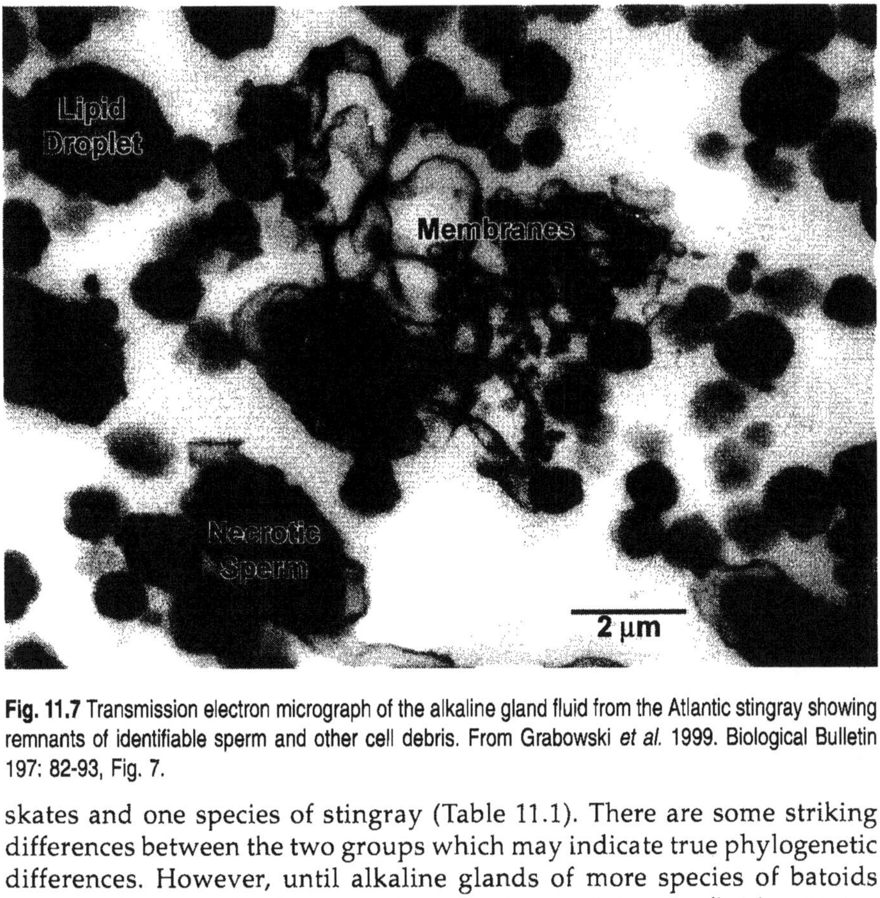

Fig. 11.7 Transmission electron micrograph of the alkaline gland fluid from the Atlantic stingray showing remnants of identifiable sperm and other cell debris. From Grabowski *et al.* 1999. Biological Bulletin 197: 82-93, Fig. 7.

skates and one species of stingray (Table 11.1). There are some striking differences between the two groups which may indicate true phylogenetic differences. However, until alkaline glands of more species of batoids are examined, such ideas remain speculative. Stingray fluid contains significant quantities of protein and urea; skate fluid has no protein and only approximately one-third the concentration of urea as found in the Atlantic stingray (Maren *et al.* 1963; Grabowski *et al.* 1999). Sodium and chloride are the quantatively dominant ions in fluids from all species, but stingrays have only one-half the concentration of these ions (including K^+) compared with skates. Skate fluid contains between 2.5 and 20 mM/L of sulphates and no phosphates. These molecules were not measured in the stingray. In both groups of batoids the fluid osmolality is roughly iso-isomotic with that of the plasma.

11.2.5 Physiology

Because of the large ionic gradient for OH- and CO_2across the alkaline gland, the subcellular mechanisms responsible for these transport processes have been more extensively studied than the function of the fluid itself. There have been two methods used to study the transport of molecules across the epithelium. Both are *in vitro* techniques and rely on electrical measurements to infer ion movement and sequestration.

Table 11.1 Composition of alkaline gland fluid in skates and stingrays

	pH	CO_2	Cl^-	Na^+	K^+	Mg^{++}	Ca^{++}	Cu^{++}	Zn^{++}	Fe^{++}	*Sulph*	*Urea*	*Phos*	*Protein*	*Glu*	*Osm.*
Skates																
R. stabuliforis	8.7	102	303	483	8.0	1.0	1.0	-	-	-	20	72(1)	0	0	-	959
R. ocellata	9.2	210	209	527	7.1	1.0	1.0	-	-	-	2.5	-	0	0	-	-
R. erinacea	9.2	214	241	550	6.6	1.0	1.0	-	-	-	2.5	93	0	0	37.0	961
R. erinacea(2)	9.4	-	-	-	-	-	-	-	-	-	-	-	-	-	-	-
Stingrays																
D. sabina	8.1	-	113	286	3.7	1.7	0.8	0.6	0.9	0.6	-	271	-	5.9	-	875

R. = *Raja*, D. = *Dasyatis*, and Sulph = sulphates, Phos = phosphates, Glu = glucose. Values for all ions, urea, CO_2, are in mM/L. Protein is mg/ml. Osm is mOsm/L. All values in skates are from Maren *et al.* (1963) except (1) (Smith 1929) and (2) (Smith 1981). Values for stingrays are from Grabowski *et al.* (1999). The dash indicates no data available.

In one method, individual epithelial cells are punctured with a microelectrode (Smith 1981, 1985). In the other method, the alkaline gland is mounted between two plexiglass chambers in which the mucosal and serosal sides of the tissue are separated. This chambered technique allows for electrical measurements to be made across the epithelium under conditions where the composition of the fluids bathing the two surfaces can be altered (Ussing and Zerahn 1951). Both the little skate (*Raja erinacea*) and the Atlantic stingray (*Dasyatis sabina*) epithelium secrete bicarbonate and chloride into the fluid. Chloride is the main anion responsible for the short circuit current (Isc) (Table 11.2). The apical plasma membrane is dominated by a large chloride conductance whereas the basolateral membrane contains a potassium channel that is of the barium-sensitive type. There appears to be no transepithelial active transport of Na+ by the batoid alkaline gland. The alkalinization of this unique fluid is dependent upon the presence of chloride ions in the media (Smith 1985; Grabowski *et al.* 1999). The results of these studies cumulatively suggest the presence of bicarbonate exchangers (intramembranous transport proteins) that are dependent on the presence of chloride and/or sodium ions. Furthermore, there is sodium and HCO_3- transport across the basolateral plasma membrane.

The enzyme, carbonic anhydrase, which catalyzes CO_2 and water to H_2CO_3(H^+ and HCO_3^-), is involved in producing the highly alkaline fluid. In skate alkaline glands there is a relationship between the amount of carbonic anhydrase in the tissue and the elevated pH of the fluid (Maren *et al.* 1963). But electrophysiological studies in the stingray alkaline gland fail to show an effect of pharmacological doses of the carbonic anhydrase inhibitor, acetazolamide. Nevertheless, immunohistochemical studies in the stingray show that there is significant and clearly localized carbonic anhydrase activity between the epithelial cells' lateral but not basal plasma membranes (Grabowski *et al.* 1999) (Fig. 11.8). It is not known whether this localization of enzyme activity reflects the intramembranous or the soluble form of carbonic anhydrase. The intercellular localization is consistent with that found in other actively transporting tissues: opercular epithelium of the teleost fish (Lacy 1983a), the elasmobranch

Table 11.2 Electrophysiological parameters in the alkaline gland of skate and stingray

Species (reference)	*PD*	*Isc*	*R*	*Alkalinization*
R. erinacea				
(Maren *et al.* 1963)	16.0	72	-	-
(Smith 1981, 1985)	6.9	52	140	3.0
D. sabina				
(Grabowski *et al.* 1999)	14.5	29.1	732	4.5

Species as described in Table 11.1 PD = potential difference in millivolts. Isc = short circuit current in μAmps/cm^2 of tissue, R = transepithelial resistance in ohms/cm^2 of tissue, Alkalinization in μEq of acid/cm^2 of tissue/hr. Dash indicates no data available.

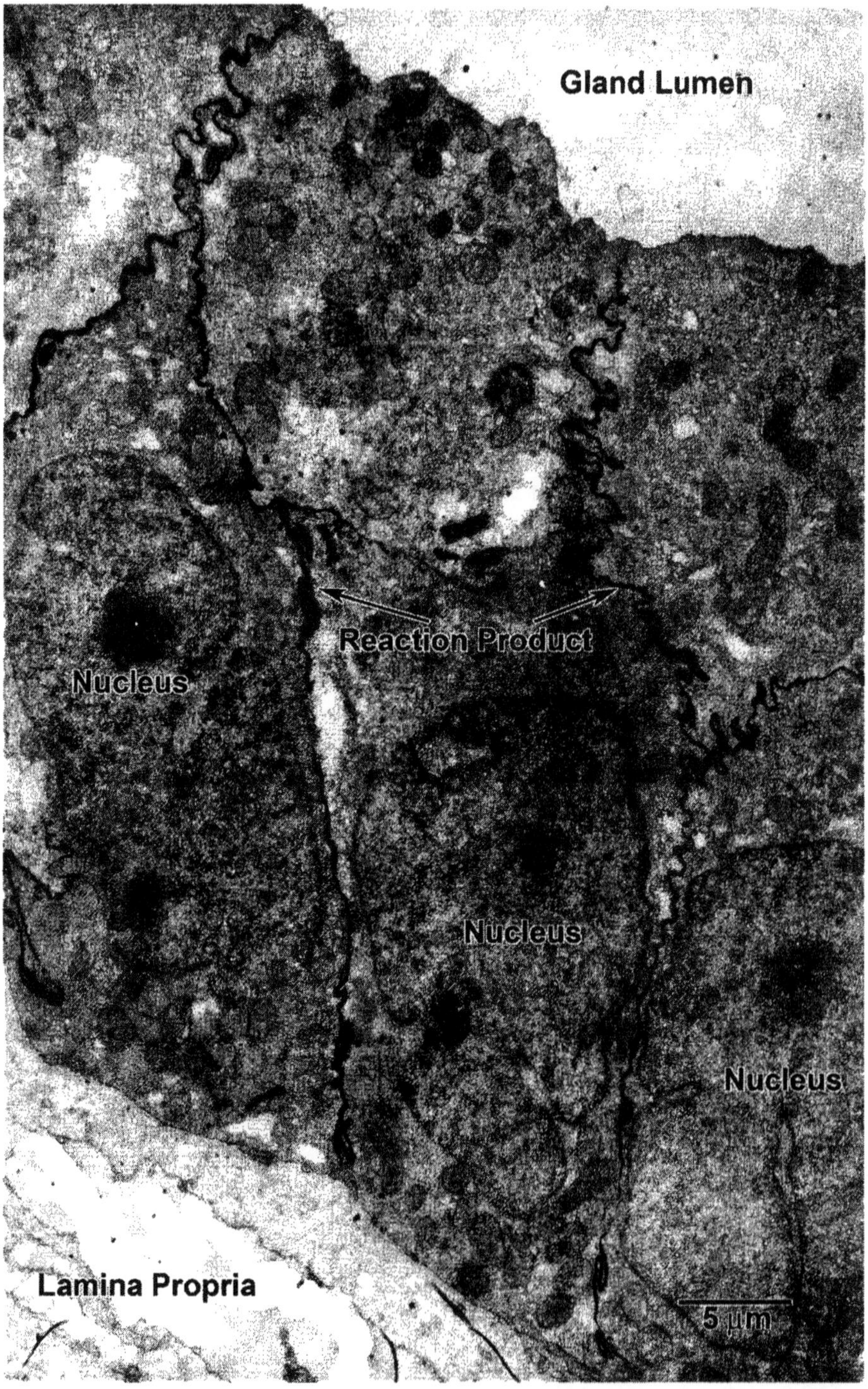

Fig. 11.8 Transmission electron micrograph showing carbonic anhydrase activity demonstrated as dense reaction product between adjacent columnar epithelial cells of the Atlantic stingray. Original.

rectal gland (Lacy 1983b), the mammalian gallbladder, duodenum, and sweat gland (Hansson 1968). Although there are many possibilities for carbonic anhydrase to participate in the alkalinization process of the fluid, these findings are strongly suggestive that the lateral intercellular spaces play a role in the overall ionic transport processes and production of alkaline gland fluid. Furthermore, it would appear that the most significant contribution of this enzyme for transport would be across the baso-lateral cell membranes and not through the extensive tight junctional complexes that unite adjacent cells at their most apical (luminal) position (Fig. 11.6).

11.2.6 Function

Although there is strong circumstantial evidence that alkaline gland fluid plays a role in male batoid reproduction and specifically that it affects sperm, there is little experimental evidence to support this notion. A study by Büllesbach *et al.* (1997) investigated the possibility that the fluid contains a peptide hormone, relaxin, found in reproductive tissues and some secreted fluids of mammals. For example, relaxin in mammalian seminal fluid stimulates sperm motility (Essig *et al.* 1982; Weiss 1989). Furthermore, this hormone, which is produced in female mammals and acts to relax the pelvic ligaments during parturition, was biochemically identified in the ovaries of sharks and skates (Gowan *et al.* 1981; Reinig *et al.* 1981; Büllesbach *et al.* 1986, 1987), suggesting a reproductive function in elasmobranchs. Alkaline gland fluid does contain a unique relaxin-like molecule (Büllesbach *et al.* 1997) termed "raylaxin". It is characteristic of all molecules in the relaxin family by having two chains (A and B) (Fig. 11.9). Raylaxin is the largest molecule yet to be described within the relaxin family. It has 27 amino acids in the A chain and 54 in the B chain (Fig. 11.9). Furthermore, it is the only reported member of the relaxin family

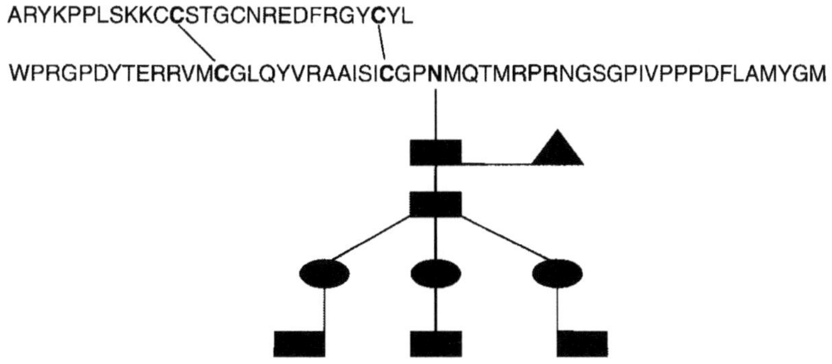

Fig. 11.9 Primary structure and oligosaccharide structure of raylaxin. The upper, A chain is linked to the lower, B chain through disulfide bonds indicated by lines. The N in the B chain shows the site for the branched oligosaccharide binding. Rectangles = N-acetylglucosamine, oblongs = mannose, triangle = fucose. Original.

of molecules that is glycosylated, having an oligosaccharide on the B chain (Fig. 11.9). Raylaxin does not have biological activity in the standard mouse pubic symphysis assay (Büllesbach *et al.* 1997) because the characteristic amino acid sequence defining the receptor binding site for the mammalian relaxin is not present in raylaxin. Raylaxin alone does not produce a quantitative increase in stingray sperm motility. This is different from the initial observation suggesting that native alkaline gland fluid results in a rapid increase in motility and longevity of stingray sperm (Grabowski 1993). These observations suggest that the active factor(s) for fluid-induced increases in sperm motility is not raylaxin but another molecule(s) or that raylaxin acts synergistically with other alkaline gland fluid molecules.

11.3 CLASPER GLANDS

11.3.1 Background

Male skates and stingrays possess a pair of specialized glandular tissue masses, each of which sits within a subdermal sac at the base of the two pelvic fins and the claspers and lateral to the cloaca (Leydig 1852; Petri 1877; Haack 1903; Leigh-Sharpe 1920; Friedman 1935; LaMarca 1964) (Fig. 10A, Table 11.3). Unlike the alkaline glands, which are within the coelom, clasper glands are embedded within the ventral body wall and have no connection with the coelomic cavity. The glandular tissue mass and the epithelial-lined sac within which it is enclosed are collectively called the clasper gland. Each gland is oblong with the long axis, running approximately parallel to the long axis of the cloaca and claspers (Fig. 10A). Each blind-ended sac housing the glandular tissue mass narrows at its most distal region to open directly into the proximal region of the clasper groove, the apopyle, or in a few species it may extend along most of the length of the clasper, terminating at its most distal region, the hypopyle. The glandular tissue mass inside each sac produces a secretory fluid which flows into the respective clasper groove.

11.3.2 Morphology

The glandular cylinder of tissue projects into each sac lumen from the dorsal wall to which it is attached (Fig. 11.10B). The degree to which the glandular tissue is attached to the sac wall varies and thus the extent to which the tissue fills the sac lumen varies among species. In some cases such as the yellow stingray (*Urobatis jamaicensis*), the spotted torpedo (*Torpedo marmoratta*) and the common stingray (*Trygon pastinaca*), it nearly completely fills the sac, leaving the sac lumen as only a "potential space."

Along the apex of the free ventral surface of the tissue mass lies a linear indentation, the central longitudinal groove (Fig. 11.10B). In some batoid species this groove is so deep as to give the erroneous impression from gross observation that the secretory tissue is two separate lobes (Leigh-Sharpe 1920). Also, the glandular tissue may fold over the

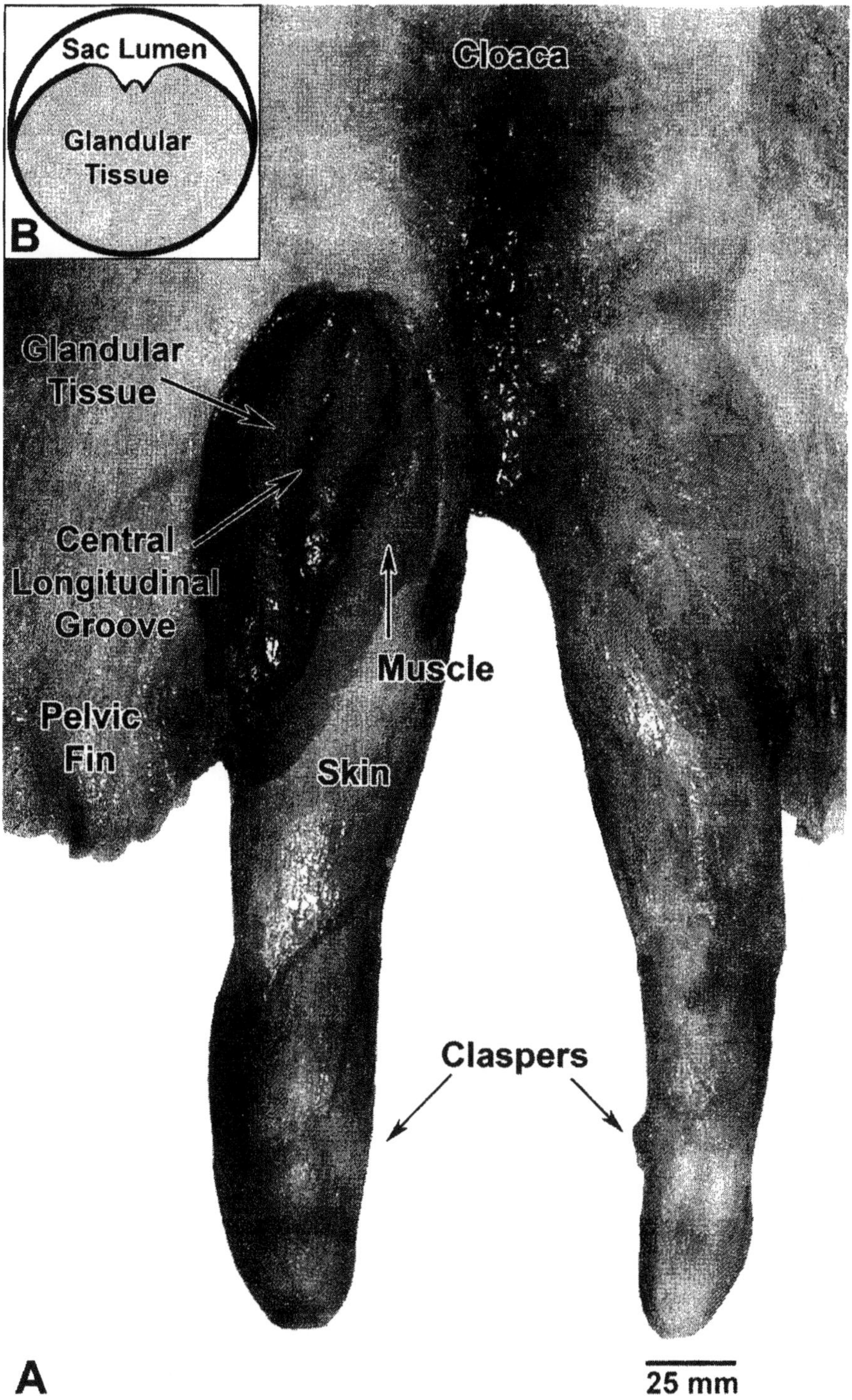

Fig. 11.10 *Contd. ...*

Table 11.3 Batoid species in which clasper glands have been reported

Common Name	*Original Scientific Name*	*Reference*
Yellow Stingray	*Urobatis jamaicensis*	LaMarca 1964
Stingaree	*U. torpedinus*	Leigh-Sharpe 1926c
Cuckoo Ray	*Raia circularis*	Leigh-Sharpe 1926c
Thornback Skate/Ray	*R. clavata*	Leigh-Sharpe 1926c
Blonde Ray	*R. blonda*	Leigh-Sharpe 1926c
Maderia Skate/Ray	*R. madeirensis*	Leigh-Sharpe 1926c
Spotted Ray	*R. maculata*	Leigh-Sharpe 1926c
Owl Ray	*R. microcellata*	Leigh-Sharpe 1926c
Shagreen Ray	*R. fullanica*	Leigh-Sharpe 1926c
Longnose Skate	*R. rhina*	Leigh-Sharpe 1926c
Undulate Skate/Ray	*R. undulata*	Leigh-Sharpe 1926c
Common Skate	*R. batis*	Leigh-Sharpe 1926c
?	*R. nidrosiensis*	Leigh-Sharpe 1926c
California Skate	*R. inornata*	Leigh-Sharpe 1926c
?	*R. lintea*	Leigh-Sharpe 1926c
Starry Ray	*R. radiata*	Leigh-Sharpe 1926c
?	*R. naevus*	Leigh-Sharpe 1926c
?	*R. platana*	Leigh-Sharpe 1926c
?	*R. cyclophora*	Leigh-Sharpe 1926c
Short tailed Ray	*R. brachyura*	Leigh-Sharpe 1926c
?	*R. murrayi*	Leigh-Sharpe 1926c
?	*R. eatoni*	Leigh-Sharpe 1926c
?	*R. marginata*	Leigh-Sharpe 1926c
Winter skate	*R. ocellata*	Leigh-Sharpe 1926c
Bonnet skate, Summer skate	*R. erinacea*	Leigh-Sharpe 1926c
Barndoor skate	*R. laevis*	Leigh-Sharpe 1926c
Lesser Electric Ray	*Narcina brasiliensis*	Leigh-Sharpe 1924a.b
Spotted/Marbled Torpedo	*Torpedo marmorata*	Leigh-Sharpe 1924a,b
Stingray	*Trygon pastinaca*	Leigh-Sharpe 1924a,b
Eagleray	*Myliobatis aquila*	Leigh-Sharpe 1924a,b
Longnosed Skate	*Rhinobatus productus*	Leigh-Sharpe 1924a,b
Devil ray/Ox-horned ray	*Dicerobatis olfersii*	Leigh-Sharpe 1926c
Butterfly ray	*Pteroplatea altavela*	Leigh-Sharpe 1926c
Blind Torpedo ray	*Benthobatis moreshyi*	Leigh-Sharpe 1926c
	Astrape japonica	Leigh-Sharpe 1926c

Fig. 11.10 A. Ventral surface of the male Atlantic stingray with skin and muscle removed to reveal the right clasper gland. The glandular tissue with its central longitudinal groove running the length of the tissue sits in the epithelia-lined sac. The ventral sac wall has been surgically removed to expose the glandular tissue. **B.** Diagram of cross section of clasper gland *in situ* showing attachment of the glandular tissue to the muscular sac wall dorsally and open to the sac lumen ventrally. Original.

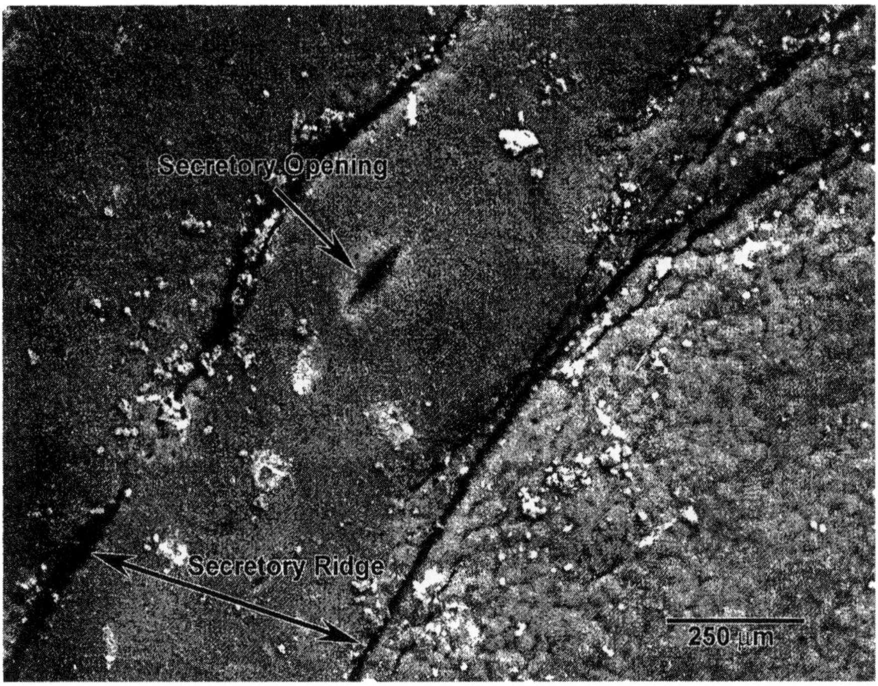

Fig. 11.11 Scanning electron micrograph showing the base of the central longitudinal groove lined by the secretory ridge. Atlantic stingray. Original.

longitudinal groove, giving the false impression of a tube within the gland (Leigh-Sharpe 1922a, b, c). At the base of the groove along its entire length lie the openings for the secretory tubules that constitute the glandular tissue. In some batoid species these openings consist of a series of papillae longitudinally arranged along the base of the groove (Leigh-Sharp 1922a, b, c, 1926a; LaMarca 1964). In the Atlantic stingray the base of the central groove is filled with a small, longitudinal ridge of tissue (secretory ridge) on the apex of which are the slit-like openings for extrusion of secretory fluid (Figs 11.11, 11.12). These openings drain extensive fields of secretory tubules that form the mass of the glandular tissue. In the case where the clasper gland terminates at the apopyle, the papillae (secretory openings) may open directly into the clasper groove (LaMarca 1964).

The clasper gland as well as the tail and clasper receive extensive innervation from spinal nerves, some of which form plexi. Early investigators used electrostimulation of these nerves to determine their distribution and function (Friedman 1935, LaMarca 1964). The exact distribution of spinal nerves to the clasper gland appears to be species-dependent. Spinal nerves numbers 48-59 anastomose and innervate the lateral part of the adductor muscles (compressor) and the "tunic"(sac) of the clasper gland in the yellow stingray (LaMarca, 1964). A separate plexus formed from spinal nerves numbers 60-66 innervates the adductor and

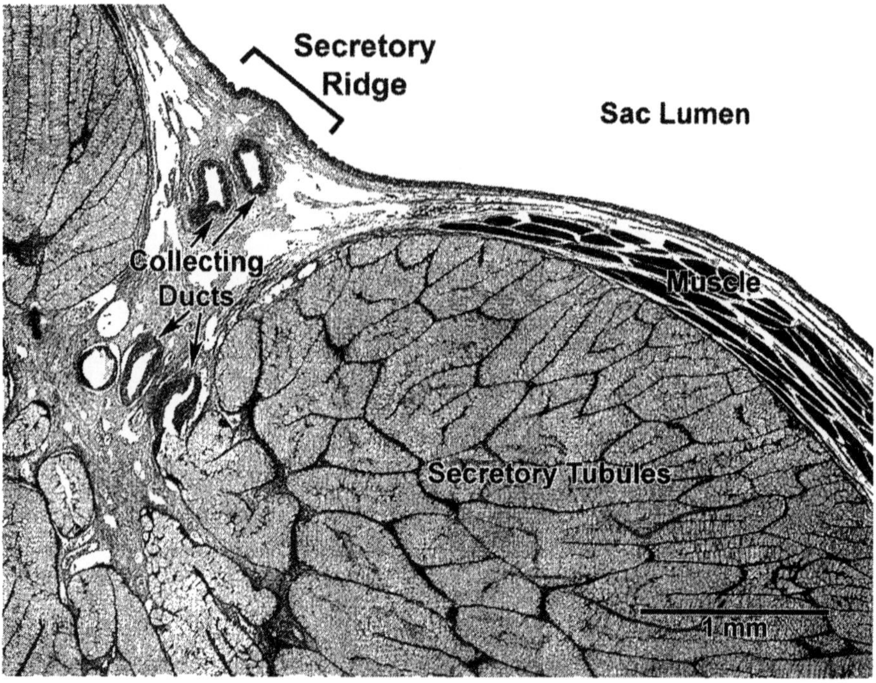

Fig. 11.12 Light micrograph of a cross-section of the central longitudinal groove, secretory ridge, sac lumen and adjacent secretory tubules in the Atlantic stingray. Original.

flexor muscles which form part of the sac's mucosal wall in that species. In *Raja clavata* the sac and its secretory tissue are innervated by a posterior branch of the 52nd spinal nerve (Friedman 1935). In various skate species spinal nerves 48-54 are responsible for clasper erection and gland contraction and secretion (Friedman, 1935, LaMarca 1964). Electro-stimulation *in vivo* of various pelvic muscles in anesthetized stingrays results in a consistent reduction of about 66% of the length of the gland and sac. This contraction results in a subsequent extrusion of secretory fluid from the linear row of papillae or secretory openings along the base of the glandular tissue (Fig. 11.11). Thus fluid secretion appears to be a mechanical event in which the muscles in the surrounding sac squeeze the fluid from tubules in the gland mass, out of the openings (papillae), and then out of the sac and into the clasper groove. Within the clasper groove the gland secretion is mixed with spermatozoa which is transferred from the cloaca to the apopyle by the flexed claspers as part of the copulatory behavior.

11.3.3 Microscopic Structure

Each clasper gland (glandular tissue mass and outer sac) is encased by a tunic of muscle, but it is not as extensive as that of the siphon sac found in sharks. The muscle fibers of the clasper gland run transversely and obliquely around the long axis of the gland. In the Yellow stingray most fibers originate and insert on the outer sac, but a few fibers connect the sac with the underlying cartilaginous skeleton (LaMarca 1964). In the Atlantic stingray the glandular tissue mass is encased in a thick muscular coat except along the central longitudinal groove where the fluid exits (Fig. 11.12).

The epithelial cells lining the lumen of the gland, which consists of those over the secretory tissue mass and those lining the sac, are microscopically similar to those of the apopyle in the yellow stingray, *Urobatis jamaicensis* (LaMarca 1964). In the Atlantic stingray, *Dasyatis sabina*, this epithelium is a pseodostratified columnar epithelium in which nuclei lie in the widest part of the cell and all cells touch the basement membrane (Fig. 11.13). The upper pale cytoplasm is most visible at the light microscope level. The large round nuclei have marginated or clumped chromatin. The epithelium shows prominent terminal bars (tight junctions). There is a sharp transition between collecting duct cells and secretory glandular epithelial cells (Fig. 11.14). In the Atlantic stingray the secretory cells measure about 25 x 100 microns but in other species they are 70-120 micrometers in diameter (Haack 1903, LaMarca 1964).

Each clasper gland papilla (secretory opening) drains a large field of tubules via collecting ducts (Fig. 11.12). The collecting ducts lie within an extensive connective tissue matrix under and adjacent to the secretory ridge at the base of the central longitudinal groove. The epithelial cells lining the secretory ducts are nearly all microscopically similar. They are packed with secretory product consisting of muco-glyco-proteins (LaMarca 1964) (Fig. 11.14). Because approximately 95% of each cell is densely packed with these large secretory granules, the densely staining nuclei are small, flattened and are positioned adjacent to the terminal bars or in some cases adjacent to the luminal plama membrane. In the yellow stingray the nuclei are mostly basally located and contain 8-10 smaller darkly-staining chromatin masses. Many of these nuclei have been identified as pycnotic in the yellow stingray (LaMarca 1964), suggesting that this epithelium has an active cellular turnover rate, possibly including apoptosis. In the Atlantic stingray the flattened nuclei

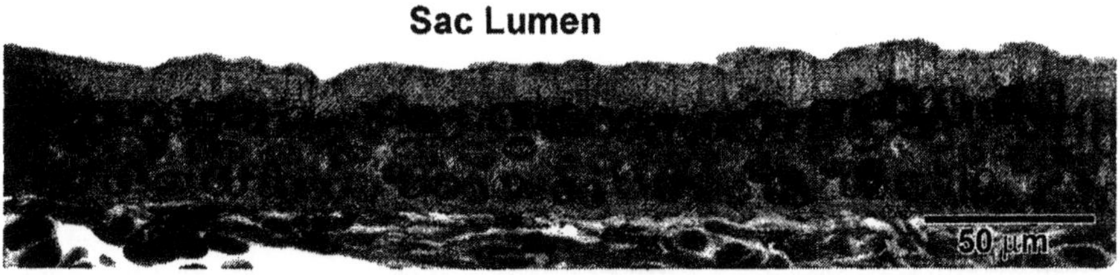

Fig. 11.13 Light micrograph of the pseodostratified columnar epithelium lining the Atlantic stingray clasper gland sac. Original.

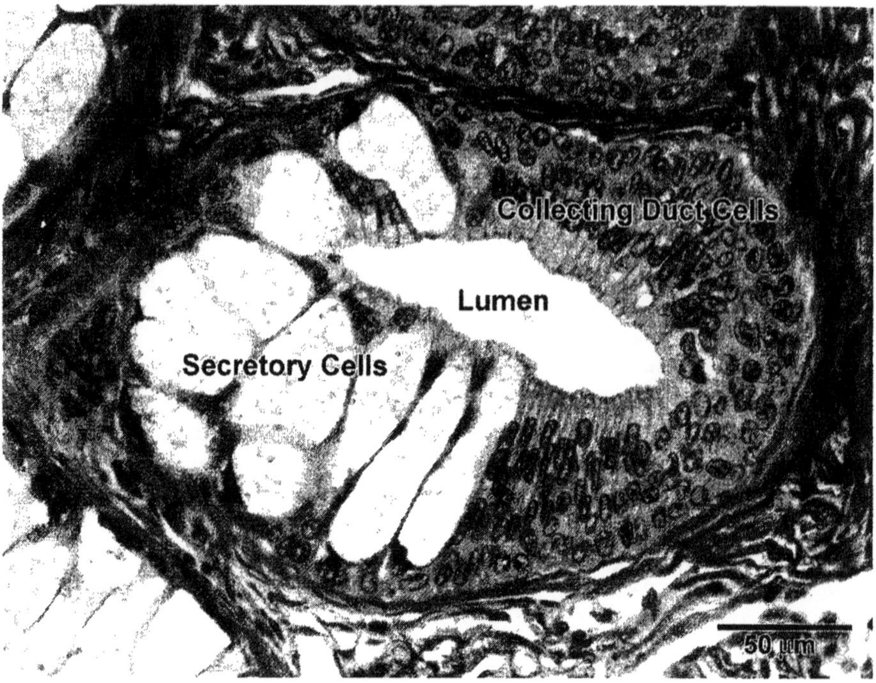

Fig. 11.14 Light micrograph of the junction between the pseudostratified columnar epithelium lining the gland collecting ducts and the simple columnar (secretory) epithelial cells lining the secretory tubules of the Atlantic stingray clasper gland. Original.

and cytoplasm are forced along the apical and lateral cell margins by the extensive and tightly packed secretory granules (Fig. 11.15). Histochemical analyses in the Yellow stingray reveal that approximately 0.5% of the secretory cell population is different from the major cell type lining the tubules. These epithelial cells, termed "end cells" because they are located at the most distal ends of the secretory tubules, contain secretory granules that stain positive for neutral lipids or lipid-containing moieties (LaMarca 1964). This small subset of "end cells" additionally has granules containing phospholipids, probably in the form of lipoprotein (LaMarca, 1964).

11.3.4 Secretory Fluid Composition

Fresh clasper gland secretion exits the secretory tissue mass when the tissue is subjected to physical pressure. The fluid is a highly viscous, white, slightly acidic (pH 6.2 in the yellow stingray) and coagulates when exposed to seawater (LaMarca 1964). It has been likened to albumin by several authors (Leigh-Sharpe 1920; Friedman 1935; LaMarca 1964). The yellow stingray secretes a fluid consisting of muco-, glyco- and lipoprotein (LaMarca 1964). The amino acids tyrosine and tryptophan are present in that fluid.

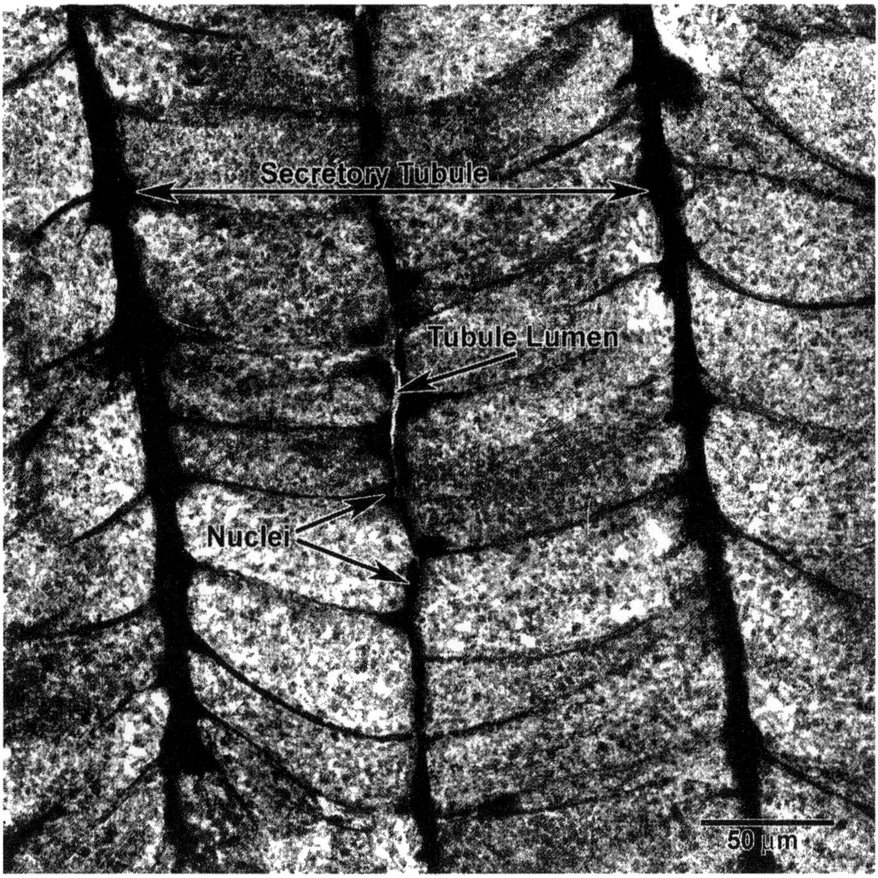

Fig. 11.15 Light micrograph of Atlantic stingray clasper gland secretory tissue. Columnar epithelial cells packed with vesicles line the secretory tubule. Original.

Invertebrate parasites are commonly found on the integument of skates and stingrays, including in the cloaca but none have not been reported in the clasper gland lumen. This may indicate the presence of an unfavorable chemical environment (gland secretion) in the clasper gland sac for these parasites.

11.3.5 Physiology

Unlike the alkaline gland, there have not been any reports on the physiology of the clasper gland secretory tissue. Nonetheless, numerous theories have been presented. Although siphon sacs, present in sharks (see Chapter 12 of this volume), are absent in stingrays and skates, the striking similarity in their size and position, narrowing of the sac into a tubule which opens into the clasper groove and being surrounded by a distinctive band of muscle, led Leigh-Sharpe (1920, 1921, 1922a, b, c, 1924a, b, 1926a, b, c, d) to speculate that skates and stingrays had evolved modified siphon-sacs that included a large secretory tissue mass. By homology then, the clasper glands would function to physically propel sperm into the female during copulation. However, microscopical observations have not revealed as extensive a muscular wall in clasper glands as is found in the siphon sacs. Furthermore, there are no reports of sperm storage or seawater propulsion from the clasper gland. LaMarca (1964) proposed that the clasper gland fluid, which oozes out into the apopyle, seals the clasper groove—acts like a glue—thus forming a tube in the clasper through which the spermatozoa travel protected from the seawater and fluids of the female reproductive tract during copulation. It is also possible but unproven that the fluid may act as a medium for sperm suspension and transport. Leigh-Sharpe (1920, 1921, 1922a, b, c, 1924a, b, 1926a, b, c, d) speculated, based on presumed homology with the hominid prostate gland, that the clasper gland functions would include 1) lubrication and a medium for spermatozoa, 2) control of erection, 3) activation of spermatozoa, 4) provision of nutrients for spermatozoa.

Unfortunately, there is no fossil evidence of clasper glands, which further hinders theories about its function. This, coupled with the absence of reports using modern biochemical and genomics technologies, leaves the function of the clasper gland as enigmatic as ever.

11.4 ACKNOWLEDGEMENTS

The National Science Foundation grant DCB8903369 and IBN 9816747 to Eric R. Lacy, Ms. Marion Hinson for secretarial assistance, Mr. James H. Nicholson for graphics assistance, and Ms. Laurinda Smith, NOAA Biotoxins Program, Hollings Marine Lab, for scanning electron microscopy assistance.

11.5 LITERATURE CITED

Borcea I. 1906. Recherches sur la systeme urogenital des elasmobranches. Archives de zoologie experimentale et generale. 4: 199-484.

Büllesbach, E. E., Schwabe, C., and Lacy, E. R. 1997. Identification of a glycosylated relaxin-like molecule from the male Atlantic stingray, *Dasyatis sabina*. Biochemistry 36: 10735-10741.

Büllesbach, E. E., Gowan, L. K., Schwabe, C., Steinetz, B. G., O'Byrne, E., and Callard, I. P. 1986 Isolation, purification, and the sequence of relaxin from spiny dogfish (*Squalus acanthias*). European Journal of Biochemistry 161: 335-341.

Büllesbach, E. E., Schwabe, C., and Callard, I. P. 1987 Relaxin from an oviparous species, the skate (*Leucoraja erinacea*). Biochemical and Biophysical Research Communications 143: 273-280.

Choe, K. P. and Evans, D. H. 2003. Compensation for hypercapnia by a euryhaline elasmobranch: Effect of salinity and roles of gills and kidneys in fresh water. Journal of Experimental Zoology 297A: 52-63.

Claude, P. and Goodenough, D. A. 1973. Fracture faces of zonulae occludente from "tight" and "leaky" epithelia. Journal of Cell Biology 58: 390-400.

Daniel, F. J. 1934. Urogenital system. Pp. 300-303. In F. J. Daniel (ed.), *The Elasmobranch Fishes*. University of California, Berkeley, CA.

Essig, M., Schoenfeld, C., D-Eletto, R., Amelar, R., Dubin, L., Steinetz, B. G., O'Bryne, M., and Weiss, G. 1982. Relaxin in human seminal plasma. Annals of the New York Academy of Science 380: 224-230.

Friedman, M. H. F. 1935. The function of the claspers and clasper gland in the skate. Journal of the Biological Board of Canada 1: 261-268.

Gilbert, P. W. 1958. The siphon sacs and their secretion in the spiny dogfish *Squalus acanthias*. Anatomical Record 130: 411.

Gilbert, P. W., and G. W. Heath. 1955. The functional anatomy of the claspers and siphon sacs in the spiny dogfish, *Squalus acanthias* and smooth dogfish, *Mustelus canis*. Anatomical Record 121: 433.

Gowan, L. K., Reinig, J. W., Schwabe, C., Bedarkar, S., and Blundell, T. L. 1981. On the primary and tertiary structure of relaxin from the sand tiger shark (*Odontaspis taurus*). FEBS Letters 129: 80-82.

Grabowski, G. G. 1993 Structure and function of the alkaline gland in the male Atlantic stingray, *Dasyatis sabina*. Ph.D. Dissertation, Medical University of South Carolina, Charleston SC.

Grabowski, G. G., Blackburn, J. G., and Lacy, E. R. 1999 Morphology and epithelial ion transport of the alkaline gland in the Atlantic stingray (*Dasyatis sabina*). Biological Bulletin 197: 82-93.

Haack, W. 1903 Uber Mundholhlendrusen bei Petromyzonten. Zeitschrift für wissenschaftliche Zoology 75: 112-146.

Hamlett, W. C. 1999 Male reproductive system. In W. C. Hamlett (ed.), *Sharks, skates and rays: The biology of Elasmobranch fishes*. The Johns Hopkins University Press, Baltimore, MD.

Hansson, H. P. J. 1968. Histochemical demonstration of carbonic anhydrase activity. Histochemie 11: 112-128.

Harmon, D. 1990. Lipofuscin and ceriod formation: The cellular recycling system. Advances in Experimental Medicine and Biology 266: 3-18.

Lacy, E. R. 1983a. Carbonic anhydrase localization in elasmobranch rectal gland. Journal of Experimental Zoology 226: 163-169.

Lacy, E. R. 1983b. Histochemical and biochemical studies of carbonic anhydrase activity in the opercular epithelium of the euryhaline teleost, *Fundulus heteroclitus*. American Journal of Anatomy 166: 19-39.

LaMarca, M. J. 1964. The functional anatomy of the clasper and clasper gland of the yellow stingray, *Urolophus jamaicensis* (Cuvier). Journal of Morphology 114: 303-324.

Leigh-Sharpe, W. H. 1920. The comparative morphology of the secondary sexual characters of elasmobranch fishes, the claspers, clasper siphons, and clasper glands. Mem. I. Journal of Morphology 34: 245-265.

Leigh-Sharpe, W. H. 1921. The comparative morphology of the secondary sexual characters of elasmobranch fishes, the claspers, clasper siphons, and clasper glands. Mem. II. Journal of Morphology 35: 359-380.

Leigh-Sharpe, W. H. 1922a. The comparative morphology of the secondary sexual characters of elasmobranch fishes, the claspers, clasper siphons, and clasper glands. Mem. III. Journal of Morphology 36: 191-198.

Leigh-Sharpe, W. H. 1922b. The comparative morphology of the secondary characters of Holocephali and elasmobranch fishes, the claspers, clasper siphons, and clasper glands. Mem. IV. Journal of Morphology 36: 199-220.

Leigh-Sharpe, W. H. 1922c. The comparative morphology of the secondary characters of Holocephali and elasmobranch fishes, the claspers, clasper siphons, and clasper glands. Mem. V. Journal of Morphology 36: 221-243.

Leigh-Sharpe, W. H. 1924a. The comparative morphology of the secondary sexual characters of elasmobranch fishes, the claspers, clasper siphons, and clasper glands. Mem. VI. Journal of Morphology 39: 553-566.

Leigh-Sharpe, W. H. 1924b. The comparative morphology of the secondary sexual characters of elasmobranch fishes, the claspers, clasper siphons, and clasper glands. Mem. VII. Journal of Morphology 39: 567-577.

Leigh-Sharp, W. H. 1926a. The comparative morphology of the secondary sexual characters of elasmobranch fishes, the claspers, clasper siphons, and clasper glands. Mem. VIII. Journal of Morphology 42: 307-320.

Leigh-Sharpe, W. H. 1926b. The comparative morphology of the secondary sexual characters of elasmobranch fishes, the claspers, clasper siphons, and clasper glands. Mem. IX. Journal of Morphology 42: 321-334.

Leigh-Sharpe, W. H. 1926c. The comparative morphology of the secondary sexual characters of elasmobranch fishes, the claspers, clasper siphons, and clasper glands. Mem. X. Journal of Morphology 42: 335-348.

Leigh-Sharpe, W. H. 1926d. The comparative morphology of the secondary sexual characters of elasmobranch fishes, the claspers, clasper siphons, and clasper glands together with a dissertation on Cowper's glands of human. Mem. XI. Journal of Morphology 42: 349-358.

Leydig, F. 1852 Beitrage zur Microskopischen anatomie und Entwichklungsageschichte der Rochen und Haie. Wihlem Englemann, Leipzig. 125 pp.

Maren, T. H., Rawls, J. A., Burger, J. W., and Myers, A. C. 1963. The alkaline (Marshall's) gland of the skate. Comparative Biochemistry and Physiology 10: 1-16.

Masur, S. K. 1984. Electron microscopy of the alkaline gland epithelium of the little skate, *Raja erinacea*. Bulletin of the Mt. Desert Island Biological Laboratory 24: 96-98.

Petri K. R. 1877 Die Copulationsorgane der Plagiostomen. Inaugural-dissertation vorgeleget der Philosophischen Fakultat der Universitat Leipzig zur Erlangung Der Doktorwurde, Wilhelm Engelmann, Leipzig. 48 pp.

Redeke, H. C. 1898. Onderzoekingen betreffende het urogenitaalsysteem der selachiers en holocephalen. Uitgegeven te helder door. C. de Boer, Jr. 85 pp.

Reed, R., McMillan, G., Hartroft, S., and Porta, E. 1965. Progress of medical science: Pathology and bacteriology. American Journal of Medical Science 250: 116-137.

Reinig, J. W., Daniel, L. N., Schwabe, C., Gowan, L. K., Steinetz, B. G., and O'Bryne, E. M. 1981. Isolation and characterization of relaxin from the sand tiger shark (*Odontaspis taurus*). Endocrinology 109: 537-543.

Smith, H. W. 1929. The composition of the body fluids of elasmobranches. Journal of Biological Chemistry 81: 407-419.

Smith, P. L. 1981. Electrolyte transport by the alkaline gland of the little skate, *Raja erinacea*. Mechanism of luminal alkalinization. Bulletin of the Mt. Desert Island Biological Laboratory 21: 80-83.

Smith, P. L. 1985. Electrolyte transport by the alkaline gland of little skate, *Raja erinacea*. American Journal of Physiology 248: R346-R352.

Ussing, H. H. and Zerahn, H. 1951. Active transport of sodium as the source of electric current in the short-circuited isolated frog skin. Acta Physiologica Scandinavica 23: 110-127.

Walker, W. F. and Homberger, D.G. 1982. *Vertebrate Dissection*, 8th Ed. Harcourt, Brace, Jovanovich College Publishers, NY, USA. 459 pp.

Weiss, G. 1989. Relaxin in the male. Biology of Reproduction 40: 197-200.

CHAPTER 12

Male Genital Ducts and Copulatory Appendages in Chondrichthyans

Carolyn J. P. Jones[1], *Terence I. Walker*[2], *Justin D. Bell*[2,6], *Matt B. Reardon*[2,3], *Carlos E. Ambrosio*[4], *Adriana Almeida*[4] *and William C. Hamlett*[3,5]

12.1 INTRODUCTION

The genital system in Chondrichthyans is specialized for internal fertilization. Paired intromittent appendages termed claspers (Fig. 12.1A, B) are modifications of the pelvic fins. In holocephalans accessory paired prepelvic claspers (Fig. 12.1C, D, 12.2A) and a tenaculum (12.1E, 12.2B) located on the rostrum of the head occur. Female holocephalans also have external specializations associated with copulation. Female Callorhynchids have paired prepelvic abdominal slits (Fig. 12.2C) that receive the prepelvic claspers during mating. In addition, most or all female holocephalans have a sperm pouch or receptacle (Fig. 12.2D) posterior to the cloaca that harbors a sperm mass. Internal organs include paired testes and the extratesticular ducts (Fig. 12.3A, B) *viz.* epididymis, ductus deferens and seminal vesicle. For reviews of the male reproductive system in Chondrichthyes see Callard (1988) and Hamlett (1999).

[1]Academic Unit of Obstetrics and Gynaecology, University of Manchester, St. Mary's Hospital, M130JH, United Kingdom

[2] Primary Industries Research Victoria, PO Box 114, Queenscliff, Victoria, Australia

[3]Department of Zoology, University of Melbourne, Parkville, Victoria 3052, Australia

[4]Dept. De Cirurgia da Faculdade de Med. Vet. E Zoot. Da Universidade de São Paulo, SP, Brazil

[5]Department of Anatomy and Cell Biology, Indiana University School of Medicine, Notre Dame, IN 46530, USA

[6]School of Ecology and Environment, Faculty of Science and Technology, Deakin University, Warnambool, Victoria, Australia

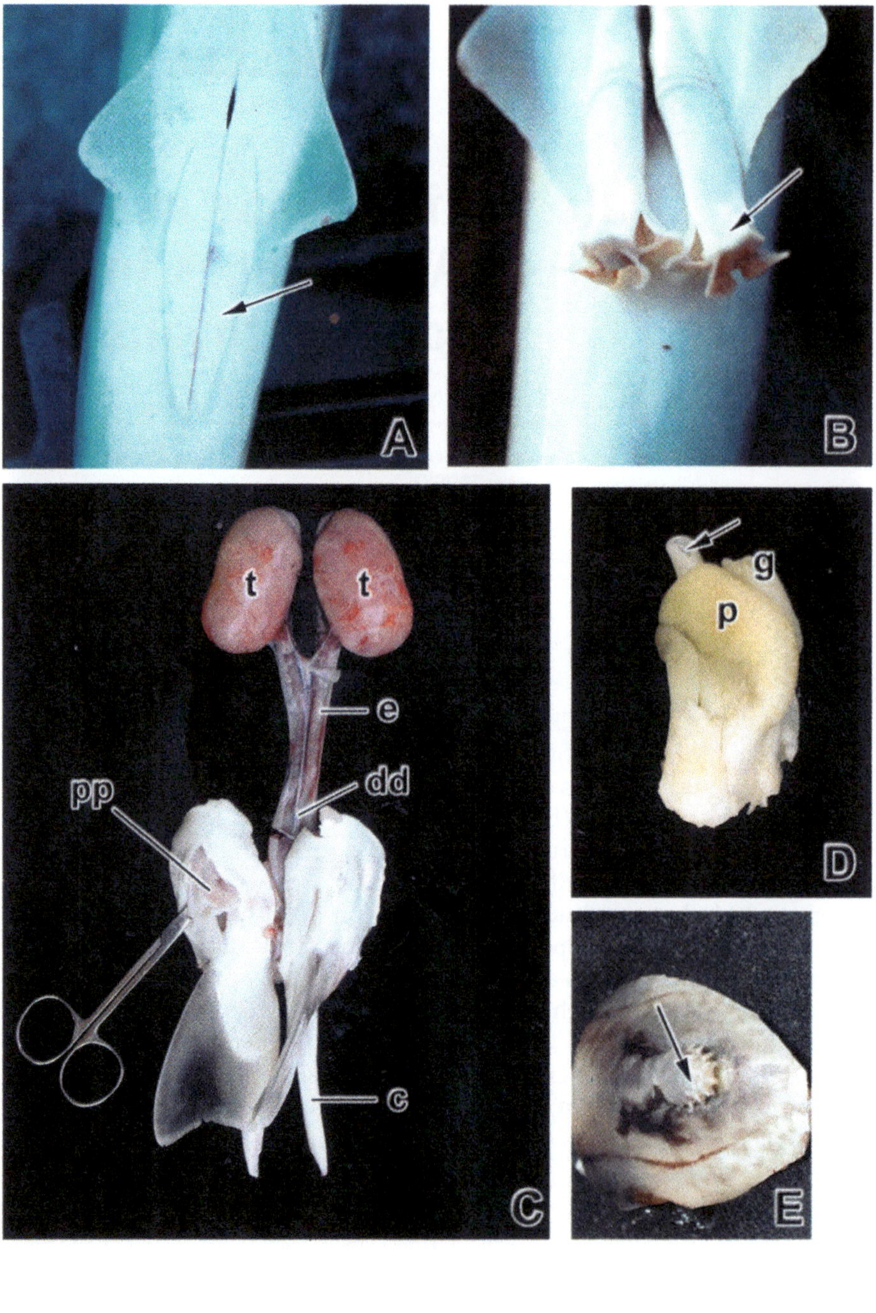

Fig. 12.1 A. Flaccid claspers (arrow) of *Mustelus canis*. **B.** Erect claspers of *M. canis*, note splayed cartilages (arrow). **C.** Male reproductive structures in *Callorhynchus milii*. t = testis, e = epididymis, dd = ductus deferens, pp = prepelvic clasper, c = clasper. **D.** Prepelvic clasper of *C. milii*. Arrow = cartilaginous tube, g = gland, p = placoid scales. **E.** Frontal tenaculum (arrow) of *C. milii*. Original.

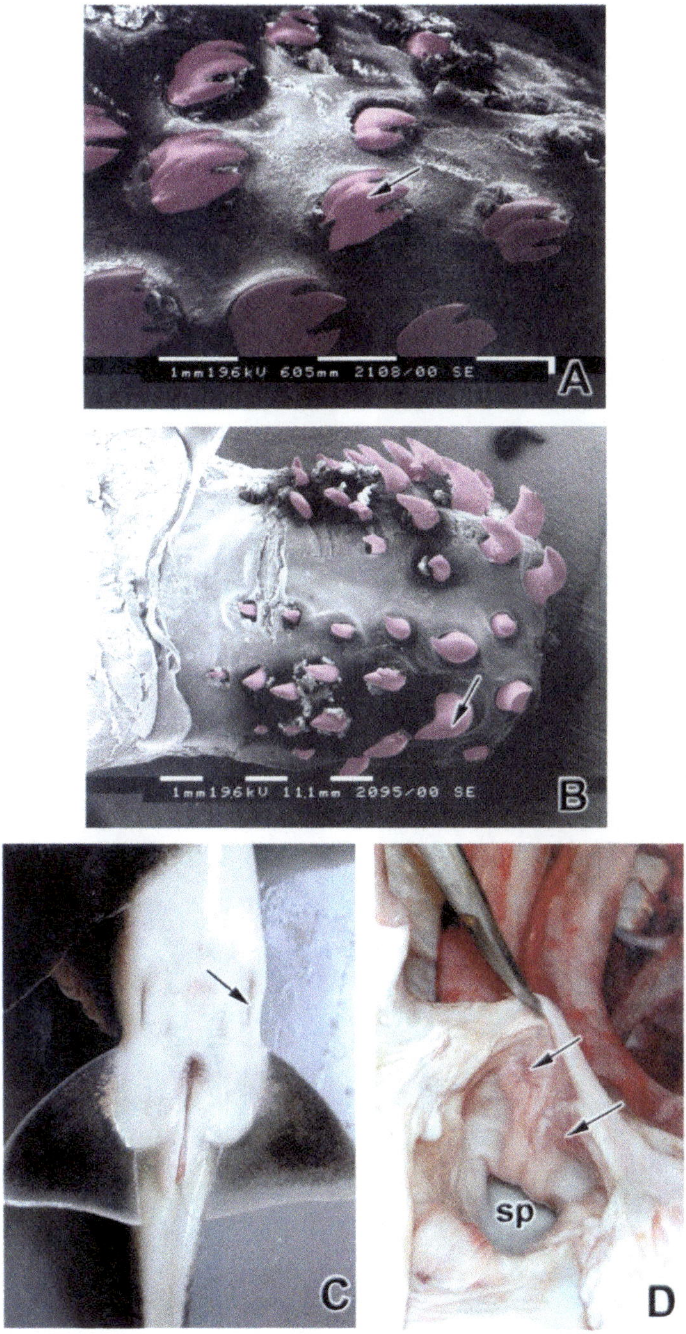

Fig. 12.2 **A.** Scanning electron micrograph of tricuspate placoid scales (arrow) of prepelvic clasper of *Callorhynchus milii*. **B.** Scanning electron micrograph of frontal tenaculum with spines (arrow) of *C. milii*. **C.** Prepelvic abdominal slits (arrow) in female *C. milii*. **D.** Sperm pouch (sp) in female, *C. milii*. Arrows = entrance to uteri. Original.

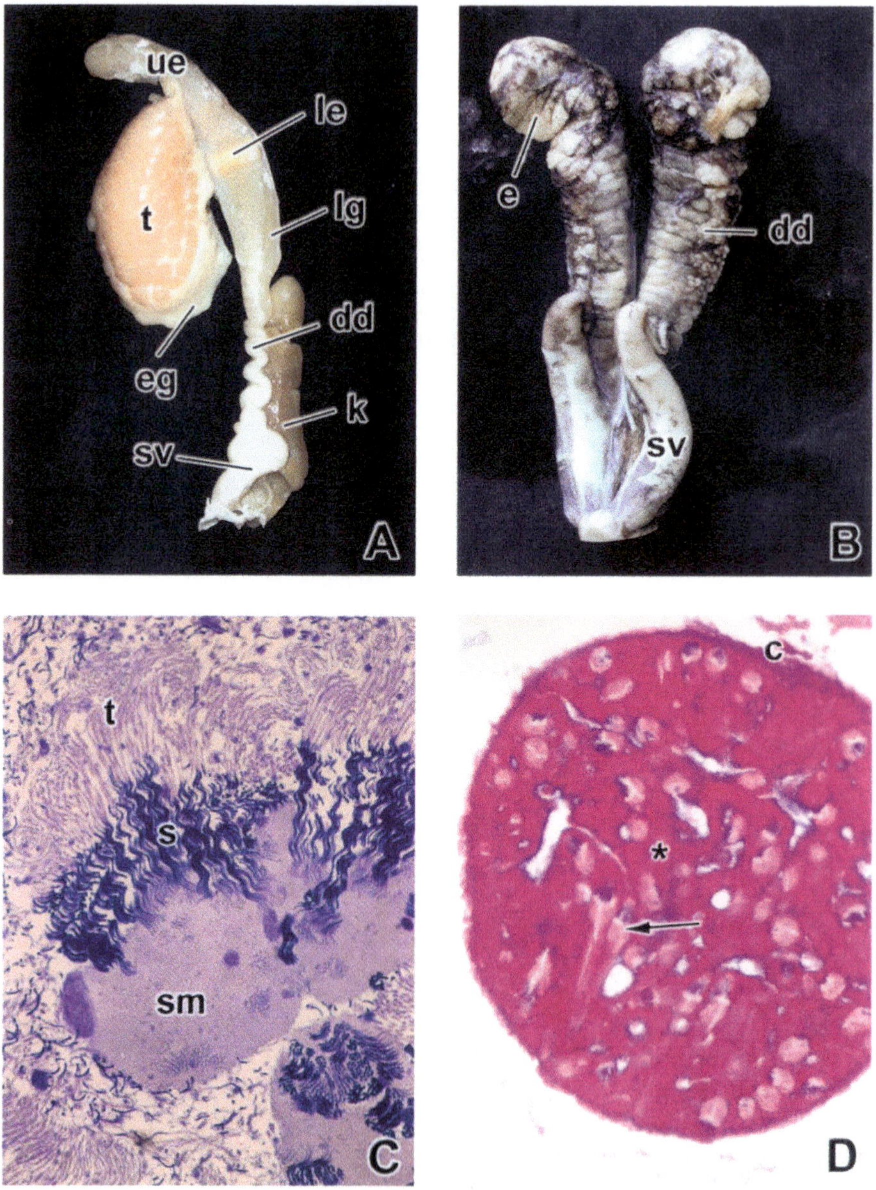

Fig. 12.3 A. Male genital tract of *Raja eglanteria*. t = testis, eg = epigonal gland, ue = upper epididymis, le = lower epididymis, dd = ductus deferens, sv = seminal vesicle, lg = Leydig gland, k = kidney. **B.** Male genital tract of *Callorhynchus milii*. e = epididymis, dd = ductus deferens, sv = seminal vesicle. **C.** Spermatozeugmata of *R. eglanteria*. Laterally aligned sperm heads (s) are embedded in seminal matrix (sm) while their tails (t) project peripherally. 600X. **D.** Spermatophore of *C. milli*. Spermatophores have a medulla that is periodic acid-Schiff positive (asterisk) and an Alcian blue positive cortex (c). Laterally aligned sperm bundles occupy a clear alcove (arrow). x100. Original.

12.2 EXTERNAL GENITALIA

12.2.1 Claspers

Male copulatory appendages of Chondrichthyes, myxopterygia or claspers are paired extensions of the posterior bases of the pelvic fins (Fig. 12.1A). General accounts of clasper structure may be found in Daniel (1928) and Breder and Rosen (1966). Gilbert and Heath (1972) considered their function in *Mustelus canis* and *Squalus acanthias*. Leigh-Sharpe described the external morphology of 87 species (1920, 1921, 1922a, b, c, 1924a, b, 1926a, b, c, d) and Compagno (1988) published a comprehensive study of claspers in the carcharhinids. Although varying in detail (c.f. Compagno 2003) claspers of most species possess a longitudinal groove, presumably through which semen passes to the female, although alternate sperm ducts may ejaculate semen, coating the claspers which are then inserted. The claspers of some species possess a series of cartilages, which once inserted into the female cloaca, splay out to insure that the claspers are not dislodged until transfer of sperm is completed (Fig. 12.1B).

12.3 SIPHON SAC IN SHARKS

In sharks the clasper groove communicates anteriorly with a muscular sac, the siphon sac or siphon, situated ventrally under the skin (Gilbert and Heath 1955). Gilbert and Heath (1972) described the structure and function of the clasper-siphon sac in *Squalus acanthias* and *Mustelus canis* and can be consulted for details. In summary, each siphon sac opens through the apopyle to the clasper. In *S. acanthias* the sacs are small, 12% body length, whereas in *M. canis* they are 30% body length. In mature animals, the sacs contain a sticky fluid secreted by goblet cells of the surface epithelium. The secretion has a pH of 5.8. Mann (1960) demonstrated that siphon sac secretion of mature *S. acanthias* contains a high concentration of serotonin (5-hydroxytryptamine), 437.5 Fg/ml or 6.5% of the dry material. In sexually immature males, serotonin was absent or present only in trace amounts in semen removed directly from the seminal vesicles. Siphons in these immature specimens contained some 200 times less serotonin, 0.017-0.048%, than in sexually mature males. In other vertebrates, serotonin is a powerful stimulator of smooth muscle contraction and is implicated in various vital functions including regulation of blood pressure, pulmonary and renal circulation and cardiac output. Serotonin stimulates isolated rat uterus and when administered to dogs intravenously, elicits strong uterine contractions followed by inhibition (Mann 1960). Serotonin may play a role in copulation and ejaculation in male elasmobranchs. Mann and Prosser (1963) demonstrated that siphon sac secretion of serotonin in *Squalus* uterus *in vitro* caused uterine contractions. They suggested that during copulation serotonin caused uterine contractions that aid sperm transport in the female by stimulating uterine contractions thus influencing sperm transport and fertilization.

Batoids have unique structures to aid in copulation and reproduction. Details of the structure and function of the alkaline and clasper glands can be found in Chapter **11** of this volume.

12.4 COPULATION

Mating behavior in elasmobranchs has not frequently been observed (Gudger 1912; Matthews 1950; Dempster and Herald 1961; Myerberg and Gruber 1974; Johnson and Nelson 1978; Klimley 1980; Luer and Gilbert 1985; Tricas and Le Feuvre 1985; Uchida *et al.* 1990). Gordon described precopulatory behavior in captive sand tiger sharks (Gordon 1993) and Nordell (1994) reported mating in *Urobatis halleri.* A detailed presentation of mating behavior in the nurse shark, *Ginglymostoma cirratum*, may be found in Chapter 5 of this volume.

Clasper innervation and central nervous system control has been studied in *Urobatis halleri* (Liu and Demski 1993). Several large myelinated nerves innervate the clasper muscles and skin. Electrical stimulation at low levels elicits various clasper movements including elevation, medial and lateral extension, rotation and opening. Stimulation of nerve 60 causes the clasper to move laterally and stimulation of nerves 61-62 produces elevation, rotation and opening of the clasper. Retrograde labeling with either horseradish peroxidase or cobalt-lysine confirmed that neurons were located in a well developed motor horn in a discrete segment of the spinal cord. Cobalt labeled sensory cells were ipsilateral in the dorsal root ganglia of nerves 61-63.

Extensive neural pathways have been described in elasmobranchs, including stingrays that are related to immunoreactive (ir) GnRH (Demski 1989; Wright and Demski 1991, 1993). Beaded GnRH-ir fibers occur throughout the spinal cord and in areas where clasper neurons are located. These fibers are thought to arise from a prominent GnRH-ir nucleus in the midbrain (Wright and Demski 1991, 1993).

12.5 REPRODUCTION IN THE HOLOCEPHALI

The elephant fish, *Callorhynchus milii*, is an example of an archetypal holocephalan. *Callorhynchus milii* are oviparous, migrating in large schools during spring or autumn into the bays and inlets of southern Australia and New Zealand to lay approximately 20 egg capsules per female (Bell 2003). The male possesses a cranial tenaculum (Figs 12.1E, 12.2B) and prepelvic claspers (Figs 12.1C, D, 12.2A) and females possess prepelvic slits (Fig. 12.2C) and a sperm receptacle (Fig. 12.2D).

12.5.1 The Male Prepelvic Claspers and Female Prepelvic Slits

The requirement for female immobility during copulation is likely the selective force for the evolution of the prepelvic claspers and tenacula, to supplement the insufficient grasping capabilities of holocephalan dentition. Prepelvic claspers are peculiar structures, their function seemingly apparent when male and female *C. milii* are aligned ventrally. The prepelvic claspers of the male and prepelvic slits of the female complement each other in their morphological positioning on the abdomen. Although copulation in elephant fish has never been observed, prepelvic claspers are probably inserted into the prepelvic slits. Prepelvic slits of the female possess a cartilaginous extension derived from the pelvic fins, which lines the inner portion of the slit and curves to accommodate the prepelvic claspers.

When inspected microscopically, prepelvic claspers possess numerous placoid scale-like (denticle) structures, which are tricuspate, semi-erect, recurved and relatively sharp (Figs 12.1D, 12.2A), providing a rigid surface to grasp and hold the cartilaginous pelvic extensions of the female.

A large gland is present directly beneath the prepelvic claspers with a cartilaginous tube protruding to the external surface (Fig. 12.1D). The gland secretes an unknown substance, perhaps a lubricant or relaxant or both, to ease insertion of the prepelvic claspers or to pacify the female during copulation.

Prepelvic slits of female Callorhynchids are lined with a stratified squamous epithelium and may have evolved to accommodate the placoid scales of the prepelvic claspers. Other Chondrichthyes have evolved sexual dimorphisms to minimize damages incurred during mating. For example, female blue sharks, *Prionace glauca*, have greatly enlarged dermal denticles along their dorsal integument to minimize damage from males biting them during copulation (Pratt 1979). Interestingly, representatives from other holocephalan genera, *Hydrolagus*, *Chimaera* and *Rhinochimaera*, do not possess prepelvic slits and this occurrence seems to be correlated with greatly reduced complexity, size and strength of the male prepelvic claspers.

12.5.2 The Tenacula

Cranial tenacula of male *Callorhynchus milii* have a complex morphology and a complementary crevice in the cranium to house the structure. The tenaculum comprises approximately 38-40 teeth on its anterior portion that vary in length reaching a maximum of 5 mm (Figs 12.1E, 12.2B). Teeth on the ventral surface tend to be the smallest, increasing in size laterally where they reach their maximum lengths and appear to be the most damage prone.

Dermal lacerations commonly occur during mating in many elasmobranch species on one or both pectoral fins as single or multiple jaw outlines and the trailing edge of the fin may be shredded (Jensen *et al.* 2002). Similar damage to the posterior margin of pectoral fins are observed on female *C. milii* and are presumed to be mating scars resulting from the tenacula (Bell 2003).

12.5.3 The Sperm Plug

Although never directly observed, mating probably occurs during the initial stages of the inshore migration due to the presence of a sperm plug in females prior to entering egg laying grounds (Gorman 1963; Reardon 2001; Bell 2003). The sperm plug is an aggregation of spermatophores (Smith *et al.* 2004) placed by the male in a small pouch immediately posterior to the cloaca of the female (Fig. 12.2D).

Early in the egg-laying season, sperm plugs are present in most mature females where their size and mass reach a maximum (Bell 2003). At this stage, sperm plugs are bright green and fill the entire sperm receptacle. The mass of the sperm plug consistently decreases through the egg laying season until the frequency of occurrence begins to decline, reaching zero some three months after they are first observed (Bell 2003).

As the egg laying season progresses, sperm plugs change from bright green to a pale milky green color (Gorman 1963; Bell 2003) as sperm stores are utilized for subsequent ovulations (Bell 2003). Strands of sperm are drawn into each uterus simultaneously (Fig. 12.5A), presumably before each ovulation (Bell 2003), although sperm storage in the oviducal gland (Smith *et al.* 2004) may last several ovulations.

Elasmobranch species possess a cloaca and during copulation one clasper (Hamlett 1999, Hamlett *et al.* 1999) or perhaps both claspers are inserted into the cloaca or one or both uterine sphincters. In Callorhynchids, on the other hand, there are two separate uterine openings and the paired uteri do not fuse (Fig. 12.2D). For both eggs to be fertilized, the male would have to insert both claspers into the uteri or would have to insert a clasper into each uterus individually. Alternatively the presence of the sperm plug appears to provide a single source for sperm to be drawn into each uterus as required. Similarly, the sperm plug may reduce the occurrence of infertility, providing sperm for several ovulations or probably the entire egg-laying season from a single copulation.

Coupled with the sperm storage capabilities of the oviducal gland, the sperm plug may enable sperm from multiple donors to be stored. By increasing the number of sperm donors, females insure genetic diversity of offspring and minimize the possibility of unsuccessful matings with infertile males. Multiple paternity has been demonstrated in elasmobranchs species including the lemon shark, *Negaprion brevirostris*, (Feldheim *et al.* 2001) and the nurse shark, *Ginglymostoma cirratum* (Saville *et al.* 2002). A similar strategy may prove to be active in holocephalans.

12.6 SPERMATOZEUGMATA VS SPERMATOPHORES

The structure taken by ejaculate differs in various species. Spermatozeugmata are characteristic of *Raja eglanteria* (Fig. 12.3C) (Jones and Hamlett 2002). This structure is characterized by a central mass of seminal matrix with sperm heads embedded in it with the sperm tails projecting out from the mass. This is in contrast to spermatophores of *C. milii* in which sperm align head to head to form lateral bundles (Fig. 12.3D) (Reardon *et al.* 2002; Hamlett *et al.* 2002). The bundles are separated from others by seminal matrix that contains unusual vesicular elements.

12.7 GROSS AND MICROSCOPIC STRUCTURE OF THE MALE DUCT SYSTEM

Considerable variation exists regarding terminology of genital ducts (Borcea 1906; Matthews 1950; Botte *et al.* 1963; Stanley 1963). From the testis, spermatozoa and Sertoli cell fragments move through efferent ductules located in the anterior end of the mesorchium, the mesentery that attaches the testis to the body wall.

Most authors refer to the small bore, highly coiled initial segment of the extratesticular ducts as the epididymis. This segment is continuous with the broader diameter, sinuous ductus deferens, also called the vas deferens or Wolffian duct. The ductus deferens is continuous as the very large diameter seminal vesicle also referred to as the ampulla of the ductus deferens. Jones and Jones (1982), Jones *et al.* (1984), Jones and Lin (1993) refer to the derivative of the mesonephric duct as the ductus epididymis rather than the ductus deferens, the term used by most authors.

We have chosen to use the most widely held terminology e.g. epididymis, ductus deferens and seminal vesicle. In *Raja eglanteria*, the epididymis may be further subdivided into an upper division and a lower epididymis that is markedly pigmented yellow (Fig. 12.3A). In *Callorhynchus milii*, no such division of the epididymis is evident (Fig. 12.3B). Both the epididymis and ductus deferens receive a viscous fluid produced by the adjacent Leydig gland. In sexually immature specimens the ductus is a thin straight tube but in sexually mature males it is coiled and covers most of the kidney. The two broad seminal vesicles receive no ducts from the Leydig gland. Seminal vesicles unite to form the single urogenital sinus that ends as the urogenital papilla emptying into the cloaca. The cloaca is common to both sexes and is located between the pelvic fins and receives material from the digestive, urinary and reproductive tracts.

12.7.1 Luminal Contents: Sertoli Cell Bodies, Sertoli Cell Cytoplasts and Leydig Gland Bodies

Three main types of particulate material have been identified in male genital ducts in *Heterodontus portusjacksoni*, namely Sertoli cell bodies, Sertoli cell cytoplasts and Leydig gland bodies. Sertoli cell bodies originate in the supranuclear region of Sertoli cells just prior to spermiation. They have previously been termed problematic bodies (Holstein 1969; Collenot and Damas 1980) and Sertoli bodies (Simpson and Wardle 1967). They are large, oval, membrane bound, eosinophilic bodies. In *Squalus*, they have been recovered from the epididymis and cytochemical techniques reveal that their proteins are rich in lysine, cysteine and tryptophan (Collenot and Damas 1975; Pudney and Callard 1986). Sertoli cell cytoplasts (Pudney and Callard 1986) have also been referred to as "small eosinophilic particles" (Jones and Jones 1982) and "cytoplasmic bodies" (Jones *et al.* 1984). Sertoli cell cytoplasts are remnants of Sertoli cells exclusive of the Sertoli cell bodies (Callard *et al.* 1989). They are membrane bound structures containing mitochondria, ribosomes, lipid and endoplasmic reticulum. Since Sertoli cells may be a source of testicular steroids in elasmobranchs (Pudney and Callard 1986; Callard *et al.* 1989) and Sertoli cell cytoplasts contain organelles appropriate for steroid synthesis, it is possible they are responsible for the steroidogenic activity of shark semen (Simpson *et al.* 1963; Simpson *et al.* 1964a; Simpson *et al.* 1964b). This may provide the mechanism for local control of extragonadal sperm duct activity. Leydig gland bodies are large, eosinophilic, nonmembrane bound secretions of Leydig glands.

12.8 STRUCTURE OF EXTRATESTICULAR DUCTS IN *CALLORHYNCHUS MILII*

Male holocephalans share many of the conserved features of elasmobranchs including an epididymis, ductus deferens and seminal vesicle, and an anterior mesonephros modified into a Leydig gland in mature males (Fig. 12.3B). The structure and function of the male genital ducts have been well documented in the elephant fish, *Callorhynchus milii* and may serve to provide insight into the reproductive strategy of male holocephalans as a whole (Hamlett *et al.* 2002; Reardon *et al.* 2002).

Gross anatomy of the male elephant fish urogenital tract and the presence of spermatophores as a means of sperm storage was first described in 1897, but without recourse to detailed histological analysis of their structure or formation (Parker and Haswell 1897). In both *C. milii* and *C. callorhynchus* spermatogenesis proceeds in a diametric pattern from the germinal zone of the testes via spermatocysts, with mature spermatids being liberated into the epididymis via the efferent ductules (Rossouw 1995; Reardon 2001). The process of spermatogenesis and spermatophore formation is of a seasonal nature, with the maximum output of spermatophores occurring in late summer/early autumn (Reardon 2001).

12.8.1 Leydig Gland

In maturing male elephant fish, the anterior mesonephros is modified into the continuous small diameter tubule, the Leydig gland. The mucosal contour is smooth and the epithelium consists of secretory simple columnar cells interspersed with ciliated pyriform cells (Fig 12.4A). The tubule shows continuous folding and all sections of the tubule contain secretory cells. The extent of the secretory activity is variable along the length, with some cells containing heterochromatin, and others containing euchromatin, indicating a cycling of protein synthesis and secretion, followed by a period of latency and recovery. Secretory cells outnumber ciliated cells and have a basal nucleus. Secretions of the Leydig gland stain strongly with periodic acid-Schiff (PAS+), with homogeneous regions of Alcian blue (AB+) product. Secretory product may fill the tubule lumen. Similar PAS+ and AB+ matrix is seen in the lumen of both the epididymis and ductus deferens, demonstrating the Leydig gland's contribution to the seminal fluid in these regions (Reardon 2001; Hamlett *et al.* 2002).

12.8.2 Epididymis

The epididymis consists of a narrow tubule with smooth unfolded epithelium and a surrounding sheath of vascularized connective tissue. The epithelium includes simple columnar.cells, interspersed with ciliated pyriform cells. Light microscopy reveals distinct apical secretory vesicles within columnar cells, indicating their contribution to the luminal matrix (Fig. 12.4B). Transmission electron microscopy confirms this, demonstrating abundant rough endoplasmic reticulum, with apical secretory vesicles and lipid inclusions corresponding to the region of the Golgi apparatus (Hamlett *et al.* 2002). The lumen of the epididymis contains prominent homogenous vesicular bodies or elements with some of the vesicles having a homogenous lipid-like content (Figs 12.5B, 12.6A-D). The luminal matrix consists of a PAS+ granular material surrounded by AB+ hyaline elements (Fig 12.4C) (Reardon *et al.* 2002).

Sperm are released from spermatocysts singly via the efferent ductules to begin lateral alignment in the epididymis. It is here that nascent spermatophores are first observed, with distinct medullary and cortical regions already apparent. No sperm tails protrude through the cortical region thus distinguishing them from spermatozeugmata. Spermatophores are less than 1 mm in size containing sperm along with matrix and vesicular bodies. It is suggested that the spherical/ovoid structure of the spermatophores presents the lowest energy level, and minimizes the likelihood of sperm loss to the environment during storage and transfer to the female (Reardon 2001).

12.8.3 Ductus Deferens

The genital duct widens posterior to the epididymis to continue as the ductus deferens. Simple columnar and pyriform cells continue to predominate in this region, with microvilli present on all cell types (Fig. 12.4C) (Reardon 2001). Both cell types have basal ovoid nuclei; secretory cells have an array of rough endoplasmic reticulum adjacent to the nucleus and a supranuclear Golgi apparatus. Lipid inclusions and secretory vesicles occupy the apical cytoplasm of these cells indicating a further contribution to the seminal matrix (Hamlett *et al.* 2002).

Sperm are all sequestered in the form of spermatophores in the ductus deferens where mature spermatophores are present (Reardon *et al.* 2002). Transmission electron microscopy shows the spermatophores to consist

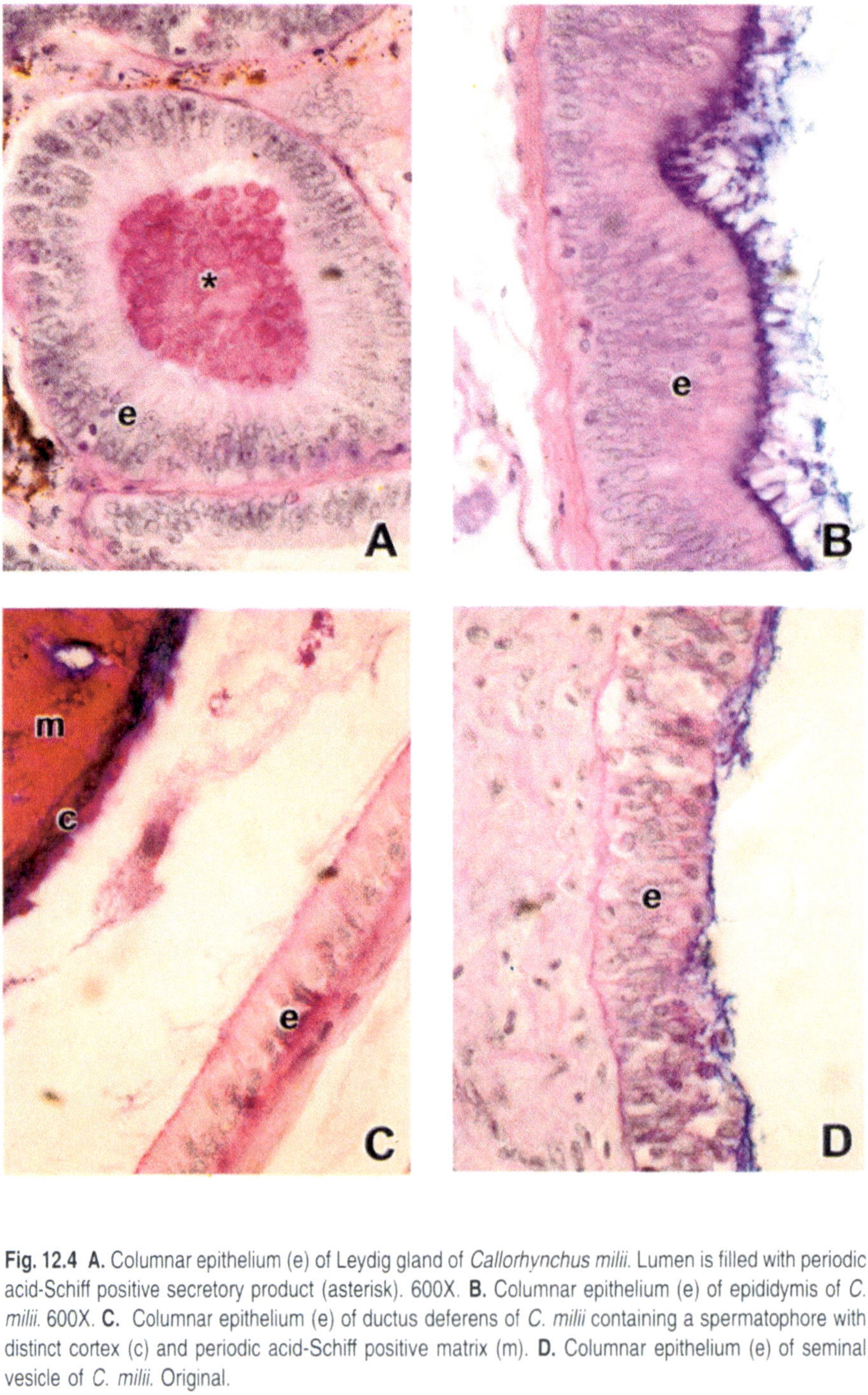

Fig. 12.4 **A.** Columnar epithelium (e) of Leydig gland of *Callorhynchus milii*. Lumen is filled with periodic acid-Schiff positive secretory product (asterisk). 600X. **B.** Columnar epithelium (e) of epididymis of *C. milii*. 600X. **C.** Columnar epithelium (e) of ductus deferens of *C. milii* containing a spermatophore with distinct cortex (c) and periodic acid-Schiff positive matrix (m). **D.** Columnar epithelium (e) of seminal vesicle of *C. milii*. Original.

of regions of tightly bundled spermatozoa immediately surrounded by a homogenous matrix that becomes denser towards the periphery, and is interspersed with vesicular bodies (Hamlett *et al.* 2002). The medullary region of the spermatophores stains PAS+ while the surrounding cortex stains both PAS+ and AB+ (pH 2.5) indicating the presence of acid glycosaminoglycans and 1-2-glycols (Fig. 12.4C) (Reardon *et al.* 2002).

12.8.4 Seminal Vesicle

The structure of the seminal vesicle is best envisioned as a spiral stairway arrangement of transverse septa, and an eccentric aperture continuing throughout the full length. Transmission electron microscopy reveals the epithelium to be composed of simple columnar cells with both ciliated and microvillar cell types, with no evidence of secretion (Hamlett *et al.* 2002). Definitive spermatophores are present, approximately 1 mm or less in size. Spermatophores have PAS-/AB+ cortical region and predominantly PAS+ medulla with some heterogeneity of material in the transitional region. The epithelium rests on a basement membrane and is surrounded by contractile smooth muscle that flushes the stored semen into the claspers during copulation (Fig. 12.4D) (Reardon 2001).

12.9 FATE OF SPERMATOPHORES IN *CALLORHYNCHUS MILII*

The sperm plug initially fills the female sperm receptacle or pouch (Fig. 12.2D). When the entire female is frozen it is possible to remove the sperm mass from the pouch and uteri. In several instances sperm threads extend from the periphery of the sperm plug into both uteri (Fig. 12.5A). Microscopic observation of the core of spermatophores reveals sperm tightly aligned head to head. They are surrounded by a clear zone presumed to be aqueous (Figs 12.5B, 12.6A-D). The PAS+ matrix contains abundant lipid-like vesicular bodies (Figs 12.5B, 12.6A-D). Vesicular bodies break down within the aqueous chamber housing the sperm (Figs 12.6A-D). It is suggested that the vesicular bodies provide sustenance to the sperm. Microscopic examination of the periphery of the sperm plug show that the sperm have lost their close lateral alignment and are released (Figs 12.5C, D). Several spermatophores are empty while some are being solubilized to release sperm. The matrix between spermatophores breaks down to release sperm (Fig. 12.5C).

12.10 ULTRASTRUCTURE OF THE MALE REPRODUCTIVE TRACT IN *RAJA EGLANTERIA*

12.10.1 Leydig Gland

Cells comprising the Leydig gland show many different phenotypes depending on the position of ducts within the gland and their physiological status with respect to secretory activity. In general, cells are

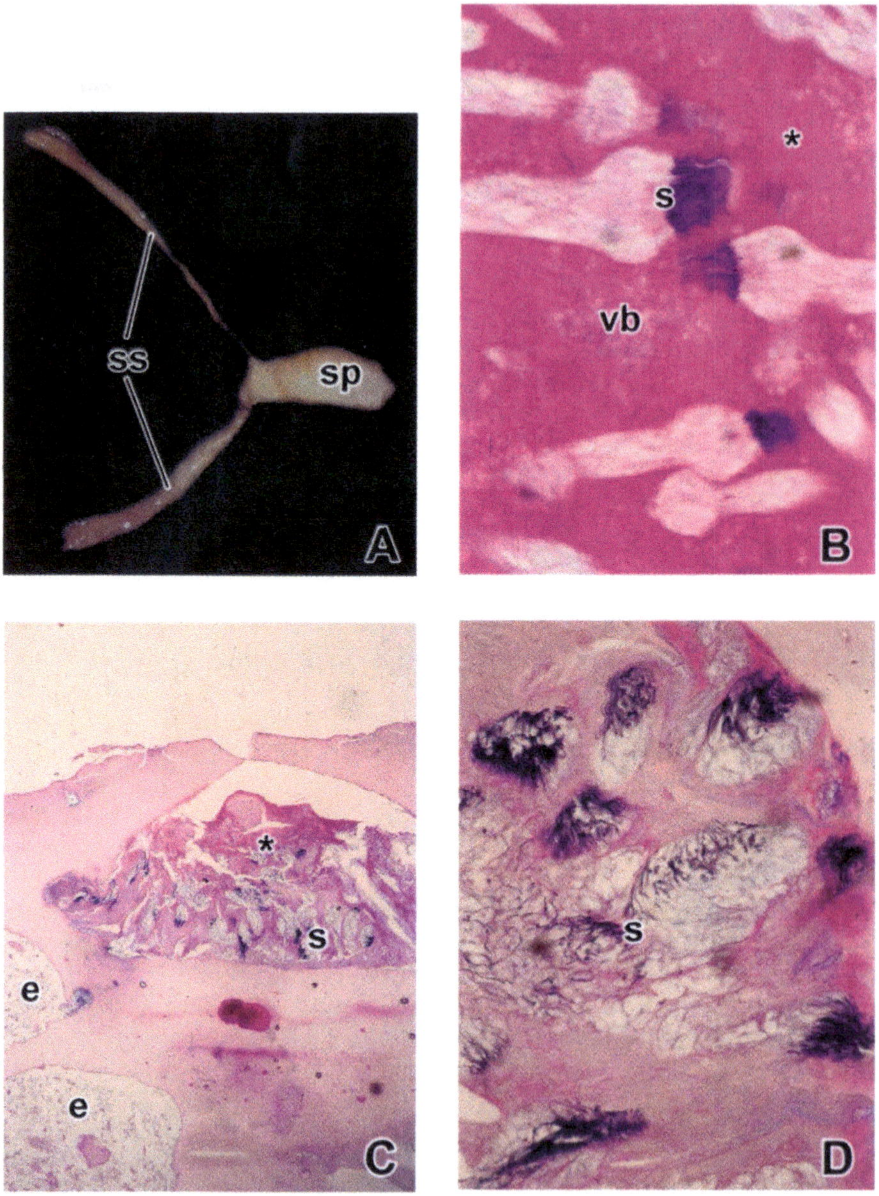

Fig. 12.5 A. Sperm plug (sp) and sperm strands (ss) entering the uteri of *Callorhynchus milii*. This animal was frozen to allow the sperm mass to be removed prior to fixation. **B.** Medullary region of spermatophores of *C. milii* shows vesicular bodies (vb), periodic acid-Schiff positive matrix (asterisk) and laterally aligned sperm bundles (s) occupying an alcove in the matrix. **C.** Periphery of sperm plug in *C. milii* showing two empty areas previously occupied by spermatophores (e). The matrix around one spermatophore is dissolving (asterisk) to release sperm (s). **D.** Light micrograph of cortex of sperm plug in *C. milii* showing dispersal of sperm (s) prior to release from the sperm plug. Original.

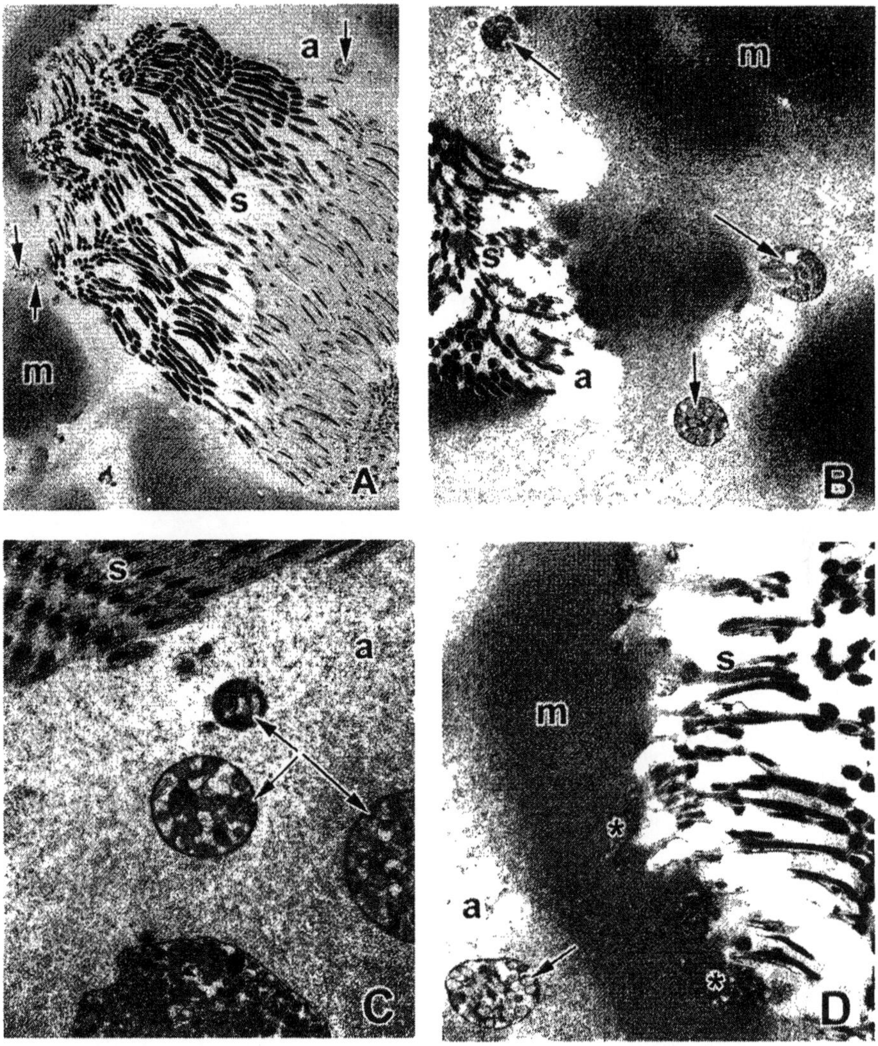

Fig. 12.6 A-D. Transmission electron micrographs of a spermatophore of *Callorhynchus milii*. Laterally aligned sperm bundles (s) occupy a presumably aqueous clear alcove (a) in the more dense seminal matrix (sm). Lipid-like vesicular bodies (arrows) occur in the matrix and some migrate to the alcove where they dissolve (asterisk). This is interpreted as a nutrient source for the sperm. Original.

columnar with either microvillous or ciliated surfaces and a cytoplasm characteristic of a richly secreting cell, with well-developed Golgi bodies, plentiful rough endoplasmic reticulum and varying numbers of secretory droplets (Fig. 12.7A-D). Previous lectin histochemical studies (Jones and Hamlett 2002) revealed two cell populations and this is confirmed at the ultrastructural level in some, but not in all duct profiles. Slender, pyriform cells (Fig. 12.7A) with narrow bases are seen with nuclei apically

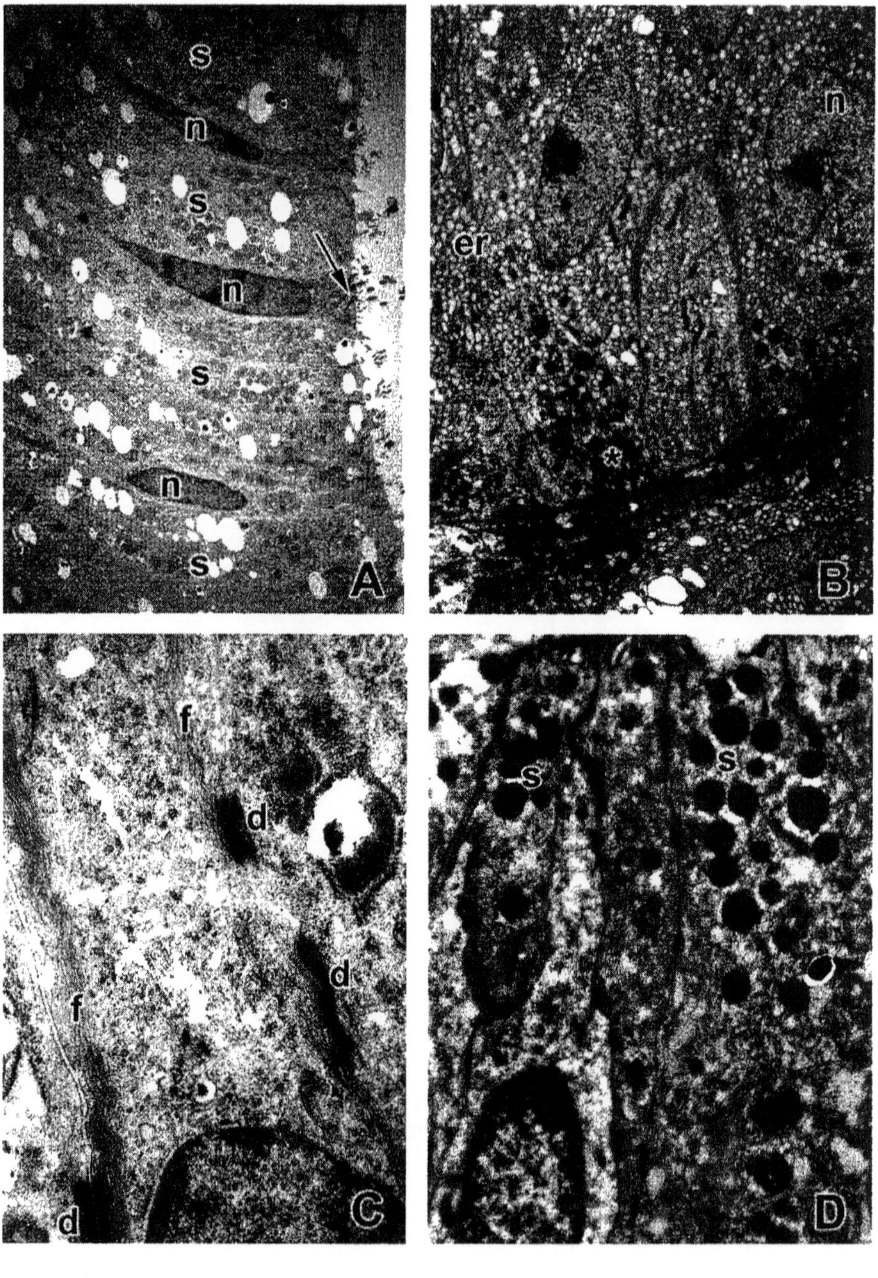

Fig. 12.7 A. Low power overview of Leydig gland epithelium in *Raja eglanteria* showing two cell types, ciliated and secretory (s). Only the nuclei (n) of the ciliated cells are visible here. Note the striated rootlets (arrow) associated with the basal bodies of the cilia. Numerous mitochondria and vacuoles are present but secretory

Fig. 12.7 *Contd.* ...

positioned compared with the basal, cleft nuclei of the main cell type (Fig 12.7B). These cells are ciliated unlike the main cell type which has sparse microvilli or blunt processes, and are distinguished by swathes of cytoplasmic filaments that emanate from the prominent desmosomes (Fig 12.7C) that link them to their neighbors at the upper aspect of the cells. Here, as elsewhere in the genital duct, basal bodies of the cilia are associated with striated rootlets (Fig. 12.7A, arrows). Junctional complexes are present at the apices of the cells while in mid and basal areas, adjoining lateral membranes form interdigitations with each other. There is great variation in the disposition of secretory granules. In the lectin study, prominent basal and some apical granules were found, but these are not common in the material examined in the electron microscope (Fig. 12.7B). Apical secretory droplets are occasionally seen (Fig. 12.7D) but mostly the cytoplasm is filled with parallel cisternae of endoplasmic reticulum interspersed with electron-lucent vacuoles and mitochondria; in some ducts these cisternae form a reticular pattern (Fig. 12.7B). Secretory material can be seen in many of the duct lumina, and the cells rest on a thin basal lamina overlying a dense connective tissue.

Many of the features described here have also been found in *Heterodontus portusjacksoni* (Jones and Jones 1992; Jones and Lin 1993) and *Callorhynchus milii* (Hamlett *et al.* 2002), including the presence of two cell types, one with apical nuclei and cilia and the other with basal nuclei, and well-developed Golgi and endoplasmic reticulum secreting eosinophilic material. In *C. milii*, PAS positive vesicles were observed apically, possibly similar to the glycan rich granules found in *R. eglanteria* (Jones and Hamlett 2002).

12.10.2 Upper Epididymis

This segment is composed of low, cuboidal epithelium with surfaces bearing both microvilli and cilia (Fig. 12.8A). An assortment of irregular electron dense granules with heterogeneous contents are present, together with a population of very large basal granules resembling phagocytic vacuoles. Some of these appear to be contained within basally-situated cells that may be intraepithelial leukocytes (Fig 12.8A), such as have been described in the epididymis of the Port Jackson shark *Heterodontus portusjacksoni* (Jones and Lin 1992, 1993; Jones 1998). Jones and Lin (1992)

Fig. 12.7 *Contd. ...*

droplets are sparse in this part of the gland. 1,900X. **B.** Basal area of a different part of the gland, showing round profiles of rough endoplasmic reticulum (er) and clusters of basal secretory droplets (asterisk). The grooved or cleft nuclei (n) characteristic of this tissue can be seen, and cells have interdigitating lateral plasma membranes. X2,500. **C.** High power of the subnuclear area of a ciliated cell showing the arrays of cytoplasmic filaments (f) emanating from desmosomes (d). X15,000. **D.** Occasionally, apical secretory droplets can be seen in the Leydig gland. Surface microvilli are present at the top of the micrograph. X6,000. Original.

observed large, electron-dense accumulations in these cells at the ultrastructural level, which may be analogous to the large masses seen here. Similarly laden cells are also seen in the connective tissue stroma underlying the epithelium. These epithelial cells are heavily glycosylated (Jones and Hamlett 2002), especially in the cytoplasmic granules; Golgi bodies are not a prominent feature and it is possible that the granules are secondary lysosomes (Fig. 12.8B) and the result of ingestion via the microvillous surface; such activity has also been described in *H. portusjacksoni* (Jones and Lin 1992). Their accumulation in the basal leucocytes may be a mechanism for their egress into the underlying connective tissue. However, the epididymis has been shown to secrete a wide variety of enzymes and proteins in all species studied (Feuchter *et al.* 1987; Dacheux *et al.* 1989; Syntin *et al.* 1996; Cooper *et al.* 1998; Lui *et al.* 2000) and therefore protein biosynthesis must also be active in this tissue.

12.10.3 Lower Epididymis

Cells of the lower epididymis are heavily vacuolated with many different types of structures. As well as electron dense cytoplasmic granules (Fig. 12.8C) there are many electron-lucent vacuoles with particulate and membranous contents (Fig. 12.8D). Around the nuclei there are often clusters of vacuolated lipid droplets (Fig. 12.8C) and in the basal areas, intraepithelial leukocytes with many heterogeneous inclusions. Lipid inclusions have also been described in *Callorhynchus milii* (Hamlett *et al.* 2002) together with secretory vesicles. The surfaces of cells of the lower epididymis have a mixture of cilia and microvilli, often both in the same cell. Apical areas are linked by junctional complexes and well developed desmosomes while the lower parts of cells have occasional interdigitations of their generally smooth lateral membranes. Mitochondria are small and endoplasmic reticulum composed of somewhat short profiles with many free ribosomes. Golgi bodies are not prominent, and the general distribution of the vacuoles, and the increasing electron density in the lower regions of the cell, suggest they are the result of the ingestion of material which is subsequently condensed rather than secretory droplets being extruded from the cell. This function of fluid absorption was also suggested for the epididymis of *Heterodontus portusjacksoni* (Jones *et al.* 1984) as epididymal plasma was found to become more concentrated with transit through the genital duct.

In *Heterodontus*, ultrastructural characteristics of epididymal cells suggest their main function is protein secretion; however, micropuncture studies did not demonstrate an increase in luminal protein concentration. It is interpreted that protein associates with individual spermatozoa in the lumen immediately upon secretion (Jones *et al.* 1984). It is possible that the sperm-associated protein may play a role in sperm aggregation in the tail of the epididymis and into the ductus deferens and seminal

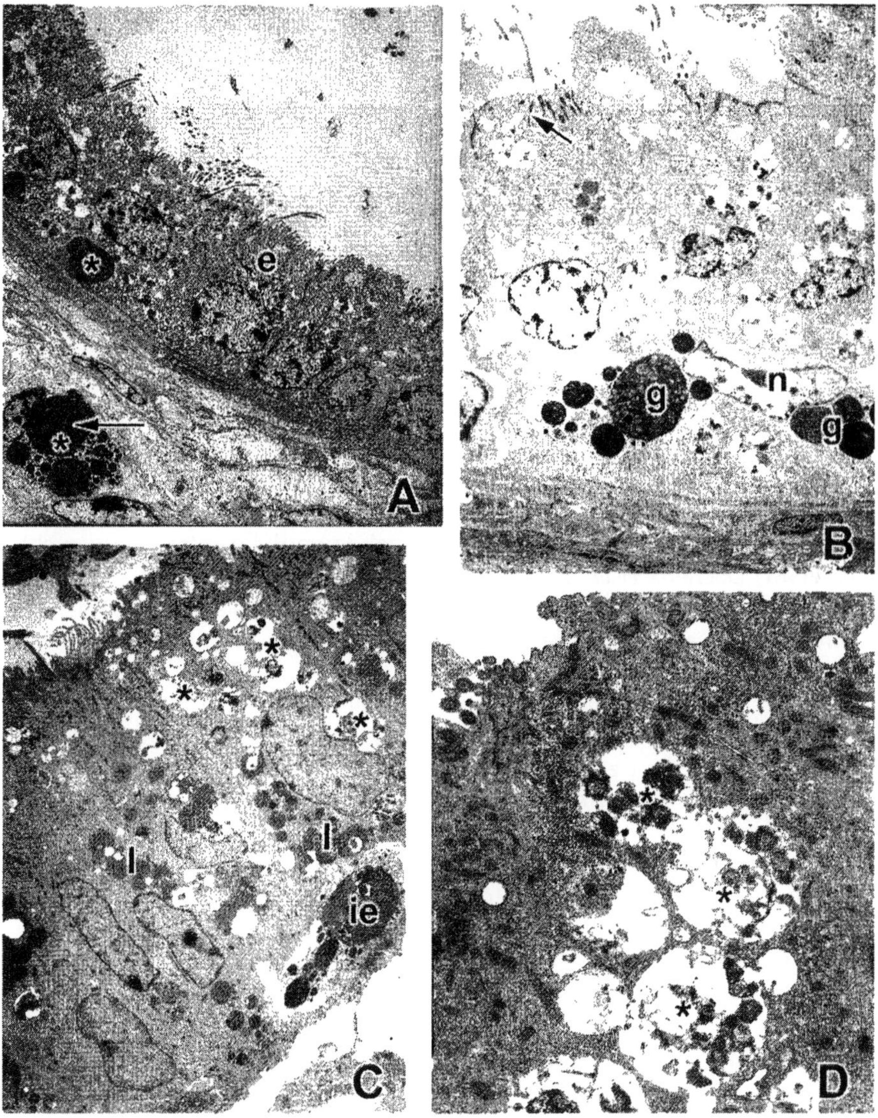

Fig. 12.8 **A.** Upper epididymis in *Raja eglanteria*. Low power overview showing the low, cuboidal epithelium (e) with large heterogeneous basal masses (asterisk). In the connective tissue underlying the epithelium, a granule laden cell can be seen (arrow). X1,650. **B.** In the upper epididymis, granules (g) displace the nucleus (n). Striated rootlets are evident (arrows) and the cell surfaces carry a mixture of cilia and microvilli. Numerous apical vacuoles and granules are present in the cytoplasm. X6,000. **C.** Lower epididymis. Low power overview showing the range of vacuoles present in the cells in *R. eglanteria*, together with an intraepithelial leukocyte (i.e.) with many intracellular inclusions. Vacuoles (asterisks) with heterogeneous contents suspended in an electron-lucent milieu as well as lipid (l) inclusions can be seen. Cilia and microvilli are present on the cell surface. X6,000. **D.** A higher power view of some of the apical vacuoles (asterisks) with ingested contents. X8,900. Original.

vesicle. Net fluid transport was not detected in *Heterodontus*, yet a net resorption of sodium occurred in this region (Jones *et al.* 1984).

12.10.4 Ductus Deferens

The ductus deferens is characterized by columnar cells, some of which have microvilli but others have masses of long cilia which form thick bundles along the surface of the ducts (Fig. 12.9A). Glycosylation studies showed that there are two cell types with respect to the biosynthesis of glycans (Jones and Hamlett 2002), but two morphologically distinct cell populations are not evident at the ultrastructural level. The cells are polarized with basal nuclei and well-developed Golgi bodies in a supranuclear position and sparse, electron-dense secretory droplets. The rough endoplasmic reticulum is extensively developed in these cells (Fig. 12.9B) with widely dilated cisternae, and mitochondria mainly clustered in the basal regions, suggesting these cells are mainly secretory as opposed to absorptive. This is in keeping with studies in mammalian species, which show that the composition of the fluid surrounding sperm changes as they travel down the genital tract (Feuchter *et al.* 1987; Dacheux *et al.* 1989; Syntin *et al.* 1996), and is discussed more fully in the section on *Raja eglanteria* glycosylation. At this stage in the tract, sperm are laterally aligned within the lumina, although the mechanism whereby this alignment is achieved is not resolved.

In *Heterodontus portusjacksoni* ciliated cells seem to be involved in heterophagic digestion as the apical cytoplasm contains numerous vacuoles and dense bodies (Jones and Lin 1992). Sertoli cell bodies are not present, suggesting they either disintegrate or are resorbed in the epididymis. Micropuncture studies reveal that 60% of testicular fluid is resorbed in the ductus deferens and there appears to be no net resorption of sodium from the lumen (Jones *et al.* 1984).

12.10.5 Seminal Vesicle

The epithelium of the seminal vesicle is low and cuboidal, and two cell variants are present in the material examined (Fig. 12.9C), one of which is sparsely distributed, more rounded and electron lucent. Both types have a mixture of microvilli, apical protrusions and cilia on their surfaces, and are linked by junctional complexes and desmosomes laterally, and contain numerous apical mitochondria (Fig. 12.9D) with occasional secretory droplets. There is little evidence of excessive absorptive activity in this area of the duct. Basal areas of the cells are joined by shallow interdigitations of their plasma membranes. It is probable that these morphological variants represent different phases of the cell cycle. Spermatozoa in the seminal vesicle form large aggregates, some of which are embedded in matrix to form spermatozeugmata, in which form they are transferred to the female during copulation.

Based on the micropuncture studies of Jones *et al.* (1984), it is concluded that luminal fluids of efferent ducts and the epididymis when

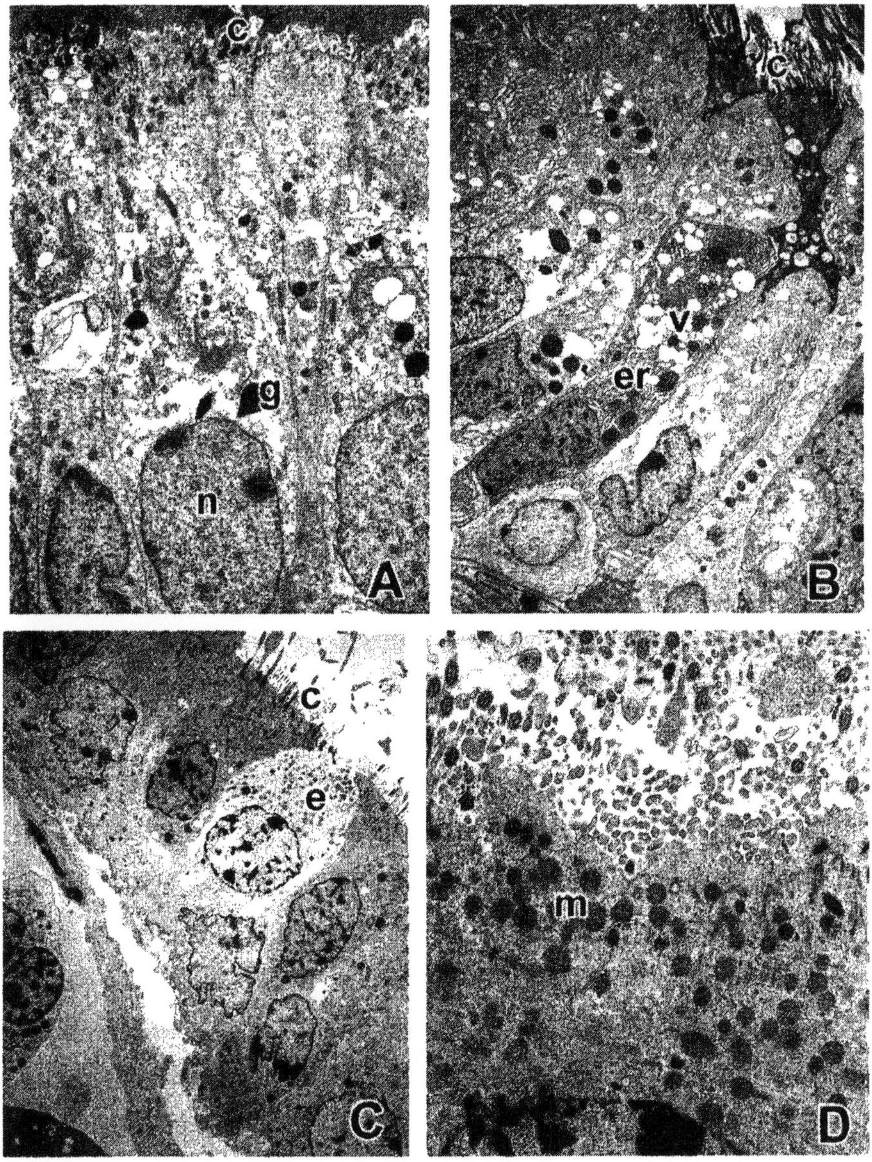

Fig. 12.9 A. In the ductus deferens a mass of cilia (c) overlies the cells which themselves bear microvilli with occasional cilia between them. Prominent Golgi bodies and scattered electron dense granules (g) are present above the basal nuclei (n). X6,000. **B.** Whorls of rough endoplasmic reticulum (er), vacuoles (v) and dense bodies with heterogeneous contents are present in these cells which have variable electron density. Long cilia (c) and some microvilli can be seen on the cell surfaces. X6,000. **C.** In the seminal vesicle a low power view shows an electron-lucent cell (e) situated amidst more electron-dense neighbours. Small, round mitochondria are clearly evident against the pale background, and occasional secretory droplets are present. Cilia (c) and microvilli extend into the lumen. 1,900X. **D.** Small round mitochondria and plentiful free ribosomes indicate a metabolically active cell in the seminal vesicle. Few secretory droplets can be seen, however. X8,900. Original.

compared to blood plasma show higher levels of sodium, potassium, protein and have a greater osmolarity. Spermatocrit samples from the epididymis and ductus deferens reveal that virtually all fluid is resorbed. The per cent sperm motility increases when testicular (9%) samples are compared with epididymal samples (96%) despite any ultrastructural change. These data suggest that the epididymis and Leydig gland secretions play important roles in ion and water transport, protein secretion and maturation of spermatozoa. In *Heterodontus*, only the seminal vesicle has a muscular tunic, all other segments of the duct system convey sperm via ciliary activity (Jones and Lin 1993) and pressure exerted by secretory activity in the more proximal ducts.

12.11 LECTIN HISTOCHEMISTRY OF THE MALE REPRODUCTIVE TRACT IN *RAJA EGLANTERIA*

In the oviparous clearnose skate *Raja eglanteria*, sperm are released from the testis as individual units but during their transit through the male genital ducts they associate laterally into spermatozeugmata and become embedded in a seminal matrix, largely secreted by the Leydig gland but with notable contributions from the epididymis (Pratt and Tanaka 1994). These packets of sperm are stored in the seminal vesicle prior to copulation, during which process they are transferred to the female. We have recently used lectin histochemistry to examine glycosylation of the male reproductive tract in *R. eglanteria* (Jones and Hamlett 2002) in order to identify the various carbohydrates produced by the different regions of the tract, which support and sustain the sperm during their transit. The reader is referred to the above publication for details of the techniques and lectins used. The findings are summarized below with kind permission from Kluwer Academic Publishers.

Each segment of the male genital tract of *Raja eglanteria* has a characteristic pattern of glycan expression, suggesting the secretion of these substances may play an important role in sperm maturation. The fact that spermatozoa from the testis and proximal segments of the male reproductive tract show little movement in physiological media, and that swimming is only evident in sperm that have travelled two-thirds along the tract and beyond (Bedford 1979) implies that, as in the mammalian system, sperm must undergo important physiological and biochemical changes in their transit through the epididymis.

Leydig gland secretions empty directly into the epididymis and ductus deferens and, in *Heterodontus portusjacksoni*, are thought to be the main source of increase in protein concentration of the luminal fluid in the ductus deferens (Jones and Lin 1992, 1993). They may, therefore, play a major role in the maturation of spermatozoa (Jones and Jones 1982, Jones *et al.* 1984, Hamlett *et al.* 2002). The cells contain a well-developed Golgi apparatus and numerous secretory granules (Fig. 12.10A), and share many similarities with those described by Jones and Lin (1993) in

H. portusjacksoni. Prominent clusters of small, basal granules (Figs 12.10A, 12.10B) contain N-glycan of different classes and N-Acetyl glucosamine oligomers, with some sialylation. The Golgi apparatus and supranuclear granules show strong bisected bi/tri-antennary N-glycan expression and N-acetyl lactosamine (Fig. 12.10A) and there are regional differences in the distribution of other glycans, often with somewhat stronger expression near the ductus deferens; this may be connected with the formation of spermatozeugmata. An area of tissue adjoining the upper epididymis also shows a different glycosylation pattern (Fig. 12.10C). These regional variations suggest that different secretory products are associated with different parts of the gland which are also indicated by changes in the appearance of the secretions.

Terminal and α1,2-linked fucosyl residues are sparse in this tissue, though the fucosylated structure GalNAcα1,3(Fucdα1,2)Galß1,3/4-GlcNAcß1- is present; other linkages not detected by the fucose-binding lectins used in our study may also be present. A subpopulation of narrow, ciliated goblet cells express different glycans from the principal cells including N-Acetyl glucosamine oligomers (Fig .12.10B), subterminal N-Acetyl lactosamine and (α2,6-linked sialic acid. Whether these two cell populations represent different cell types or are different phases of a secretory cycle, is not, at present, known. The luminal secretions are richly glycosylated, with glycans similar to those of the Golgi bodies and apical granules.

There is a notable increase in the degree of glycosylation in the upper epididymis compared with the Leydig gland (Fig. 12.10C). Here, basal granules resemble phagocytic vacuoles with heterogeneous contents, consistent with an absorptive function for the epididymal cells, while the vesicular appearance of some of the granules is also consistent with this role. They express abundant N-Acetyl lactosamine residues and N-Acetyl glucosamine oligomers though complex N-glycan (Fig. 12.10D) is sparse. High mannose structures and other glycans are also expressed. Occasional large masses are present in both parts of the epididymis (Figs 12.10D, E), possibly phagosomes or intraepithelial leucocytes as reported in the epididymis of *H. portusjacksoni* (Jones and Lin 1992, 1993; Jones 1998). Large, electron-dense accumulations in these cells have been described at the ultrastructural level (Jones and Lin 1992), which may be analogous to the large masses seen here. Apical granules, some of which have unstained contents, contain more N-Acetyl lactosamine and glucosamine oligomers, sialic acid and subterminal N-Acetyl lactosamine than in the Leydig gland. Non-bisected tri/tetra-antennary complex N-glycan and fucose residues, absent from the granules, are strongly expressed on the apical surface. Numerous apical granules and general glycan expression suggests that, as well as being absorptive, these cells secrete extracellular products, adding to those produced by the Leydig gland.

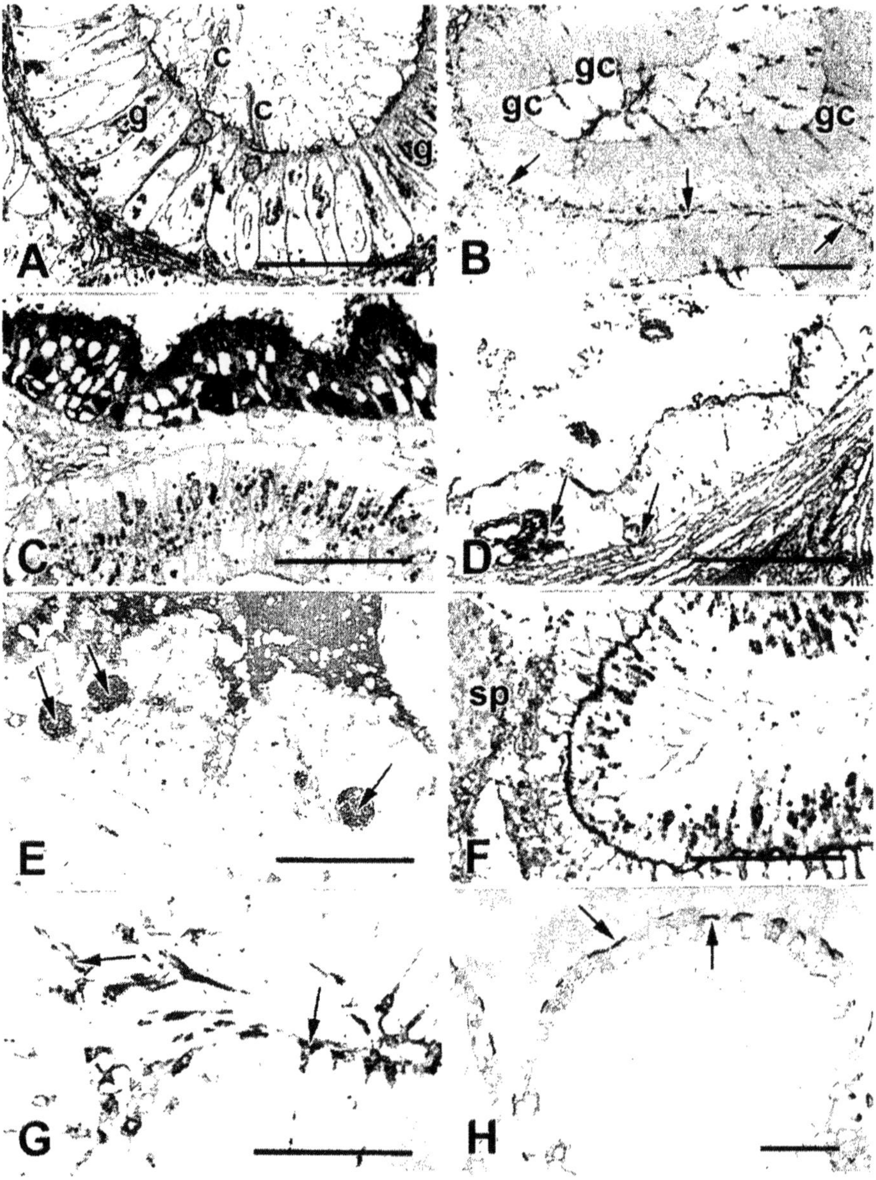

Fig 12.10 A-H. Immunohistochemistry of the genital tract in *Raja eglanteria*. Scale bars: 50 µm. **A.** In the Leydig gland, N-Acetyl lactosamine bound by *Datura stramonium* lectin is expressed by Golgi bodies (g), basal granules and cilia (c); apical granules are not prominent. **B.** The goblet cells (gc) and basal granules (arrows) in the Leydig gland are revealed with *Solanum tuberosum* lectin which binds to N-Acetyl glucosamine oligomers. **C.** Large, heterogeneous granules are present basally in the upper epididymis, strongly stained by *S. tuberosum* lectin which also binds to the ciliated apical surface. Part

Fig. 12.10 *Contd. ...*

Glycosylation of the lower epididymis (Fig. 12.10E) is similar to that of the non-pigmented upper region, in contrast to epididymal secretion in many mammals (Jones *et al.* 1983; Parillo *et al.* 1997) and reptiles (Desantis *et al.* 2002), which changes markedly in different segments of the epididymis (Lee and Damjanov 1984; Dacheux *et al.* 1989; Syntin *et al.* 1996; Cooper 1998) and is associated with many surface changes in the sperm plasma membrane as they travel through the genital duct (Hammerstedt *et al.* 1982; Magargee *et al.* 1988; Dacheux *et al.* 1989; Srivastava and Olson 1991). Some N-glycan expression is stronger but generally staining appears to be slightly decreased. The foamy appearance of the cells indicates an increased absorptive function compared to the upper epididymis, reflecting changes in the composition and consistency of the luminal fluid.

In the mammalian epididymis, glycans become incorporated into the spermatozoon plasma membrane (Arya and Vanha-Perttula 1984; Eddy *et al.* 1985; Burkett *et al.* 1987; Tulsiani *et al.* 1993, Liu *et al.* 2000). They may be secreted from epididymal cells, as in the mouse (Feuchter *et al.* 1987; Saxena *et al.* 2002) and in the human (Tezón *et al.* 1985; Liu *et al.* 2000). Numerous glycosyl transferases have been identified in the epididymal luminal fluid of various mammals (Bernal *et al.* 1980, 1983; Tulsiani *et al.* 1993) as well as glycosidases (Skudlarek *et al.* 1993;Tulsiani *et al.* 1995; Syntin *et al.* 1996) which are secreted from the epididymis and modify the glycans present on sperm. Carbohydrate residues are thus altered as the sperm traverse the male reproductive tract (Hammerstedt *et al.* 1982; Rankin *et al.* 1989; Dacheux *et al.* 1989). In the ram and bull, sialic acid produced by these cells may bind to the sperm surface (Holt 1980; Voglmayr *et al.* 1985), increasing the negative charge on the sperm and masking antigenic groups, thus preventing phagocytosis of sperm by maternal leucocytes. In *Raja eglanteria*, sialyl residues are prominent throughout the cells of the male genital tract, as well as in the matrix components, and could fulfill a similar masking role.

Fig. 12.10 *Contd. ...*

of the Leydig gland seen here contains more granules than in B. **D.** Large basal masses (arrows) containing bi/triantennary bisected complex N-glycan are clearly seen with *Phaseolus vulgaris* erythroagglutinin. **E.** In the lower epididymis, *Sambucus nigra* lectin binds to α2,6-linked sialic acid and shows the basal masses (arrows) and various granules. **F.** N-Acetyl glucosamine oligomers are present on the apical surface and granules in the ductus deferens, shown by staining with *Lycopersicon esculentum* lectin. Masses of spermatozoa (sp) can be seen on the left. **G.** After neuraminidase pretreatment of the ductus deferens, the lectin from *Erythrina cristagalli* binds to subterminal N-acetyl lactosamine in the population of ciliated, thin cells (arrows). **H.** In the seminal vesicle, some apical cell surfaces expressing terminal N-acetyl galactosamine bind *Helix pomatia* lectin strongly (arrows), while the cilia and matrix do not stain. Reproduced from Jones, C. J. P. and Hamlett, W. C. 2002. Glycosylation of the male genital ducts and spermatozeugmata formation in the clearnose skate *Raja eglanteria*. Histochemical Journal 34: 601-615, with kind permission from Kluwer Academic Publishers.

In the ductus deferens (Figs 12.10F, G), there are no basal granules and a weaker cytoplasmic expression of complex N-glycan is found, though surface expression is still strong. Supranuclear granules express many classes of both N- and O-linked glycan (Fig. 12.10A), generally similar to those in the lower epididymis but with a loss of subterminal N-Acetyl lactosamine and terminal and subterminal N-Acetyl galactosamine. The apical surface shows an increase in terminal and subterminal galactose and both show increased (α2,3-linked sialic acid, with a decrease in N-Acetyl galactosamine. The population of narrow goblet cells alone shows strong expression of subterminal N-Acetyl lactosamine (Fig.12.10G), terminal and subterminal N-Acetyl galactosamine and some sialic acid. Whether these cells are constitutively different, or reflect a different phase of a secretory cycle, is not known. Spermatozoa and luminal contents in this part of the genital tract express similar glycans as before, but generally with stronger staining. The initiation of sperm aggregation shows variation between specimens as it is present at different levels of the duct in different animals (*i.e.* the lower epididymis as opposed to the seminal vesicle).

The seminal vesicle contains cells with various glycosylated granules, with loss of tri/tetra-antennate non-bisected structures and few terminal (α-galactosyl residues. Subterminal N-Acetyl galactosamine and lactosamine is expressed, however, whereas sialic acid in various linkages is abundant. Apical cell surfaces show striking variation in expression of N-Acetyl galactosamine (Fig. 12.10H) and lactosamine reflecting differences in the surface plasma membrane or secretory status. Staining of aggregated spermatozoa, their heads embedded in matrix, appear reduced, and the glycans found in the matrix are similarly changed.

As they travel through the male reproductive tract, spermatozoa show subtle changes in their glycan expression. In general, they stain strongly with many lectins, indicating many glycans on their surface, probably adsorbed from the luminal fluid. Terminal and (α1,2-linked fucosyl residues are generally absent, however, in contrast to findings in mouse (Cossu and Boitani 1984), as are bi/tri-antennary non-bisected N-glycans, but otherwise staining is strong in the epididymis and ductus deferens, and in general reflected similar staining patterns to those of the matrix, suggesting a coating effect. In the seminal vesicle, where spermatozoa are aggregated into spermatozeugmata, the staining intensity is diminished. The matrix itself stains with most of the lectins, with aggregation into masses occurring in the lower epididymis.

Different staining properties of tissues in the *Raja* male reproductive tract suggest the composition of the fluid surrounding the spermatozoa changes significantly as they travel along from the testis to the seminal vesicle. This is also found in many mammals (Lee and Damjanov 1984; Arya and Vanha-Perttula 1984, 1986; Feuchter *et al.* 1987; Magargee *et al.* 1988; Parillo *et al.* 1997). Sperm are immotile in the testes of elasmobranch (Jones *et al.* 1984, Minamikawa and Morisawa 1996) and salmonid fish

(Miura *et al.* 1992), with maturation occurring in the epididymis. Studies on teleost and elasmobranch spermatozoa (Rojas and Esponda 2001) show that, as sperm mature, lectin binding sites diminish until only the head is stained. In *Raja*, the physical nature of the seminal fluid also alters, with it becoming more "cheesy" and heterogeneous. The mechanism of sperm alignment in the spermatozeugmata is not clear. One explanation might be the presence of non-junctional adhesion molecules such as cadherins. Preliminary transmission electron microscopy (Hamlett unpublished) of sperm aggregates in various elasmobranchs failed to reveal any type of junctional complex, nor have any been seen between sperm in aggregates in the holocephalan elephant fish *Callorhynchus milii* (Hamlett *et al.* 2002). Non-junctional adhesion is carried out via membrane associated molecules, thus the nature of glycans surrounding the sperm, and consequent glycan-glycan interactions, are probably important factors in its regulation.

In chondrichthyans, spermatozoa undergo modifications during transit through the genital ducts. Initially, individual sperm laterally align in aggregates that associate with luminal matrix, forming spermatozeugmata or spermatophores that support sperm during transfer to the female reproductive tract. The production of a heavily glycosylated matrix together with other luminal secretions of the Leydig glands and contributions from the genital ducts creates an environment that possibly protects and nourishes the sperm. Lectin binding studies have revealed widespread expression of several glycans that no doubt have a critical role in such processes.

12.12 CONCLUSIONS

The male reproductive system of Chondrichthyes shows many common shared traits. In all species internal fertilization has necessitated the evolution of spermatozeugmata or spermatophores as a carrier device for the transfer of sperm from the male to the female. The unique Leydig gland is the source of the majority of seminal matrix. Glycosylation studies (Jones and Hamlett 2002) have revealed different glycan expression associated with each segment of the male genital ducts as well as the matrix of both spermatozeugmata and spermatophores. Copulatory appendages in all chondrichthyes are claspers but in most, or potentially all holocephalans, accessory appendages, namely prepelvic claspers and a cranial tenaculum play a role in copulation.

12.13 ACKNOWLEDGEMENTS

This work was supported by funds provided by Indiana University School of Medicine (WCH) and the Fisheries Research and Development Corporation in Australia (TIW). Gratitude is expressed for the highly skilled technical expertise in electron microscopy by Joan Clark, University of Melbourne.

12.13 LITERATURE CITED

Arya, M. and Vanha-Perttula, T. 1986. Comparison of lectin-staining pattern in testis and epididymis of gerbil, guinea-pig, mouse and nutria. American Journal of Anatomy 175: 449-469.

Arya, M. and Vanha-Perttula, T. 1984. Distribution of lectin binding in rat testis and epididymis. Andrologia 16: 493-508.

Bedford, J. M. 1979. Evolution of the sperm maturation and sperm storage functions of the epididymis. Pp. 7-21. In D. W. Fawcett and J. M. Bedford (eds), *The Spermatozoon*. Urban and Schwarzenberg, Inc., Baltimore-Munich.

Bell, J. D. (2003) Fisheries and Reproductive Biology of the Elephant Fish, *Callorhinchus milii*, in southern Australia. Bachelor of Science honours thesis. School of Ecology and the Environment, Deakin University, Warrnambool, 3280.

Bernal, A., Torres, J., Reyes, A. and Rosado, A. 1980. Presence and regional distribution of sialyl transferase in the epididymis. Biology of Reproduction 23: 290-293.

Bernal, A., Mercado, E., Calzada, L. and Hicks, J. J. 1983. Microsomal and plasma membrane sialyltransferase activity in rat epididymis. Archives of Andrology 11: 33-38.

Borcea, J. 1906. Systeme uro-genital des elasmobranches. Archives Zoology Experimental General 4: 119-484.

Botte, V., Chieffi, G., and Stanley, H. P. 1963. Histological and histochemical observations on the male reproductive tract of *Scyliorhinus stellaris, Torpedo marmorata* and *T. torpedo*. Pubblicazioni della Stazione di Zoologica di Napoli 33: 224-242.

Breder, C. M and Rosen, D. E. 1966. *Modes of Reproduction in Fishes*. Natural History Press, Gardan City, NY.

Burkett, B. N., Schulte, B .A., and Spicer, S. S. 1987. Histochemical evaluation of glycoconjugates in the male reproductive tract with lectin-horseradish peroxidase conjugates 1) Staining of principal cells and spermatozoa in the mouse. American Journal of Anatomy 178: 11-22.

Callard, G. V. 1988. Reproductive physiology, Part B: the female. Pp. 292-317. In T. J. Shuttleworth (ed.), *Physiology of Elasmobranch Fishes*. Springer-Verlag, New York.

Callard, G. V., Mak, P., DuBois, W., and Cuevas, M. 1989. Regulation of spermatogenesis: the shark testis model. The Journal of Experimental Zoology., suppl 2: 353-364.

Collenot, G., and Damas, D. 1975. Mise en evidence de las nature proteique de corps enigmatiques presents dans le testicule de *Scyliorhinus canicula* L. (Elasmobranche). Cahiers de Biologie Marine 16: 39-64.

Collenot, G., and Damas, D. 1980. Etude ultrastructurale de la cellule de Sertoli au cours de la spermiogénèse chez. *Scyliorhinus canicula* L. Cahiers de Biologie Marine. 21: 209-219.

Compagno, L. J. V. 2003. *Sharks of the order Carcharhiniformes*. The Blackburn Press, Caldwell, NJ.

Cooper, T. G. 1998. Interactions between epididymal secretions and spermatozoa. Journal of Reproduction and Fertility, Supplement 53: 119-136.

Cooper, T. G. 1998. Interactions between epididymal secretions and spermatozoa. Journal of Reproduction and Fertility Supplement 53: 119-136.

Cossu, G. and Boitani, C. 1984. Lactosaminoglycans synthesized by mouse male germ cells are fucosylated by an epididymal fucosyl transferase. Developmental Biology 102: 402-408.

Dacheux, F. 1998. Role of epididymal secretory proteins in sperm maturation with particular reference to the boar. Journal of Reproduction and Fertility Supplement 53: 99-107.

Dacheux, J. L., Dacheux, F. and Paquignon, M. 1989. Changes in sperm surface membrane in luminal protein fluid content during epididymal transit in the boar. Biology of Reproduction 40L: 635-651.

Dacheux, J. L., Dacheux, F. and Paquignon, M. 1989. Changes in sperm surface membrane in luminal protein fluid content during epididymal transit in the boar. Biology of Reproduction 40L: 635-651.

Daniel, J. F. 1928. *The Elasmobranch Fishes*. University of California Press, Berleley, CA.

Demski, L. 1989. Pathways for GnRH control of elasmobranch reproductive physiology and behavior. The Journal of Experimental Zoology. Suppl. 2: 4-11.

Dempster, R. P., and Herald, E. S. 1961. Notes on the horn shark, *Heterodontus francisci*, with observations on mating activities. Occasional Papers of the California Academy of Sciences 33: 1-7.

Desantis, S., Labate, M., Labate, G. M. and Cirillo, F. 2002. Evidence of regional differences in the lectin histochemistry along the ductus epididymis of the lizard *Podarcis sicula* Raf. Histochemical Journal 34: 123-130.

Eddy, E. M., Vernon, R. B., Muller, C. H., Hahnel, A.C. and Fenderson, B. A. 1985. Immunodissection of sperm surface modifications during epididymal maturation. American Journal of Anatomy 174: 225-237.

Feldheim, K. A., Gruber, S. H. And Ashley, M. V. 2001. Multiple paternity of a lemon shark litter (Chondrichthyes: Carcharhinidae). Copiea 2001 (3): 781-786.

Feuchter, F. A., Green, M. F. and Tabet, A. J. 1987. maturation antigen of the mouse sperm flagella 2) Origin from holocrine cells of the distal caput epididymis. Anatomical Record 217: 146-152.

Gilbert, P. W., and Heath, G. W. 1955. The functional anatomy of the claspers and siphon sacs in the spiny dogfish (*Squalus acanthias*) and smooth dogfish (*Mustelus canis*). Anatomical Record 121: 433, abstr.

Gilbert, P. W., and Heath, G. W. 1972. The clasper-siphon sac mechanism in *Squalus acanthias* and *Mustelus canis*. Comparative Biochemistry and Physiology 42A: 97-119.

Gordon, I. 1993. Pre-copulatory behavior of captive sandtiger sharks, *Carcharias taurus*. Environmental Biology of Fishes 38: 159-164.

Gorman, T. B. S. 1963. Biological and economic aspects of the elephant fish *Callorhinchus milii* Bory in Pegasus Bay and the Canterbury Bight. New Zealand Marine Department, Fisheries Technical Report, 8, Wellington, New Zealand.

Gudger, E. W. 1912. Summary of work done on the fishes of Tortugas. Carnegie Institution Washington Yearbook 11: 148-150.

Hamlett, W. C. 1999. Male reproductive system. Pp. 444-470. In W. C. Hamlett (ed.), *Sharks, skates, and rays: The biology of Elasmobranch fishes*. The Johns Hopkins University Press, Baltimore, MD.

Hamlett, W. C., Hysell, M. K., Rozycki, T., Brunette, N., Tumilty, K., Henderson, A. and Dunne, J. 1999. Sperm aggregation and spermatozeugmata formation in the male clearnose skate *Raja eglanteria*. Pp. 281-289. In B. Séret and J-Y Sire (eds), *Proceedings of the 5th Indo-Pacific Fish Conference*, Nouméa, 1997. Society of French Ichthyologists, Paris.

Hamlett, W. C., Reardon, M., Clark, J. and Walker, T. I. 2002. Ultrastructure of sperm storage and male genital ducts in a male holocephalan, the elephant fish *Callorhynchus milii*. The Journal of Experimental Zoology 292: 111-128.

Hammerstedt, R. H., Hay, S. R. and Amann, R. P. 1982. Modification of ram sperm membranes during epididymal transit. Biology of Reproduction 27: 745-754.

Holstein, A. F. 1969. Zur fragc der lokalen steuerung der spermatogenese beim dorhai *(Squalus acanthias as* L). Z.eirschrift Zellforsch. 93: 265-281.
Holt, W. V. 1980. Surface-bound sialic acid on ram and bull spermatozoa: Deposition during epididymal transit and stability during washing. Biology of Reproduction 23: 847-857.
Jensen, C. E., Natanson, L. Pratt, H. L., Kohler, N. E., and Campana, S. E. (2002) The reproductive biology of the porbeable shark (*Lamna nasus*) in the Western Atlantic Ocean. Fisheries Bulletin, 100, 727-738.
Jones, C. J. P. and Hamlett, W. C. 2002. Glycosylation of the male genital ducts and spermatozeugmata formation in the clearnose skate *Raja eglanteria*. Histochemical Journal 34: 601-615.
Jones, N. and Jones, R. C. 1982. The structure of the male genital system of the Port Jackson shark *Heterodontus portusjacksoni*, with particular reference to the genital ducts. Australian Journal of Zoology 30: 523-541.
Jones, R. C., Jones, N. and Djakiew, D. 1984. Luminal composition and maturation of spermatozoa in the male genital ducts of the Port Jackson shark *Heterodontus portusjacksoni*. The Journal of Experimental Zoology 230: 417-427.
Jones, R. C. 1998. Evolution of the vertebrate epididymis. Journal of Reproduction and Fertility Supplement 53: 163-181.
Jones, R., von Glos, K. I. and Brown, C. R. 1983. Changes in the protein composition of rat spermatozoa during maturation in the epididymis. Journal of Reproduction and Fertility 67:299-306.
Jones, R. C. 1998. Evolution of the vertebrate epididymis. Journal of Reproduction and Fertility Supplement 53: 163-181.
Jones, R. C. and Lin, M. 1992. Structure and functions of the of the male Port Jackson shark *Heterodontus portusjacksoni*. Australian Journal of Zoology 40: 257-266.
Jones, R. C. and Lin, M. 1993 Structure and functions of the genital ducts of the male Port Jackson shark *Heterodontus portusjacksoni*. Environmental Biology of Fishes 38: 127-138.
Jones, R. C., Jones, N. and Djakiew, D. 1984. Luminal composition and maturation of spermatozoa in the male genital ducts of the Port Jackson shark *Heterodontus portusjacksoni*. Journal of Experimental Zoology 230: 417-427.
Johnson, R. H., and Nelson, D. R. 1978. Copulation and possible olfaction-mediated pair formation in two species of carcharhinid sharks. *Copeia* 1978: 539-542.
Klimley, A. P. 1980. Observations of courtship and copulation in the nurse shark. *Copeia* 1980: 878-882.
Lee, M-C. and Damjanov, I. 1984. Anatomic distribution of lectin-binding sites in mouse testis and epididymis. Differentiation 27: 74-8.
Leigh-Sharpe, W. H. 1920. The comparative morphology of the secondary sexual characters of elasmobranch fishes. Mem. I. Journal of Morphololgy 34: 245-265
Leigh-Sharpe, W. H. 1921. The comparative morphology of the secondary sexual characters of elasmobranch fishes, Mem. III, and V. Journal of Morphology 36: 191-243.
Leigh-Sharpe, W. H. 1922a, b, c. The comparative morphology of the secondary sexual characters of elasmobranch fishes. Mem. III. The comparative morphology of the secondary characters of Holocephali and elasmobranch fishes. Mem. IV. And V. Journal of Morphology 36: 191-243.
Leigh-Sharpe, W. H. 1924a, b. The comparative morphology of the secondary sexual characters of elasmobranch fishes. Mem. VI. And VII. Journal of Morphology 39: 553-577.

Leigh-Sharpe, W. H. 1926a, b, c, d. The comparative morphology of the secondary sexual characters of elasmobranch fishes. Mem. VIII., IX., X. And XI. Journal of Morphology 42: 307-348.

Liu, Q., and Demski, L. S. 1993. Clasper control in the round stingray, *Urolophus halleri*: lower sensorimotor pathways. Environmental Biology of Fishes 38: 219-230.

Liu, H-W., Lin, Y-C., Chao, C-F., Chang, S-Y. and Sun, G-H. 2000. GP83 and GP39, two glycoproteins secreted by human epididymis are conjugated to spermatozoa during maturation. Molecular Human Reproduction 6: 422-428.

Luer, C.A., and Gilbert, P. W. 1985. Mating behavior, egg deposition, incubation period, and hatching in the clearnose skate, *Raja eglanteria*. Environmental Biology of Fishes 13: 161-171.

Magargee, S. F., Kunze, E., Hammerstedt, R. H. 1988. Changes in lectin binding features of ram sperm surfaces associated with epididymal maturation and ejaculation. Biology of Reproduction 38: 667-685.

Mann, T. 1960. Serotonin 5-hydroxytryptamine in the male reproductive tract of the spiny dogfish. Nature 188: 941-942.

Mann, T. and Prosser, C. L. 1963. Uterine response to 5-hydroxytryptamine in the clasper siphon secretion of the spiny dogfish *Squalus acanthias*. Biological Bulletin 125: 384-385

Matthews, H. L. 1950. Reproduction in the basking shark, *Cetorhinus maximus* Philisophical Transactions of the Royal Society of London, Biology. 234: 247-316.

Minamikawa, S. and Morisawa, M. 1996. Acquisition, initiation and maintenance of sperm motility in the shark *Triakis scyllia*. Comparative Biochemistry and Physiology 113A: 387-392.

Miura, T., Yamauchi, K., Takahashi, H. and Nagahama, Y. 1992. The role of hormones in the acquisition of sperm motility in Salmonid fish. The Journal of Experimental Zoology 261: 359-363.

Myerberg, A. A., and Gruber, S. H. 1974. The behavior of the bonnethead shark, *Sphyrna tiburo*. Copeia 1974: 358-374.

Nordell, S. E. 1994. Observations of the mating behavior and dentition of the round stingray, *Urolophus halleri*. Environmental Biology of Fishes 39: 219-229.

Parillo, F., Stradaioli, G., Supplizi, A.V. and Monaci, M. 1997. Detection of glycoconjugates in the ductus epididymis of the prepubertal and adult horse by lectin histochemistry. Histology and Histopathology 12: 691-700.

Parker, T. J. and Haswell, W. A. 1897. *A textbook of zoology* , Macmillan, London.

Pratt, H. L. and Tanaka, S. 1994. Sperm storage in male elasmobranchs: a description and survey. Journal of Morphology 219: 297-308.

Pratt, J. L. Jr. 1979 Reproduction in the blue shark, *Prionace glauca*. US National Marine Fisheries Service Fishery Bulletin 77: 445-470

Rankin, T. L., Holland, M. K. and Orgebin-Crist, M-C. 1989. Lectin-binding characteristics of mouse epididymal fluid and sperm extracts. Gamete Research 24: 439-451.

Pudney, J., and Callard, G. V. 1986. Sertoli cell cytoplasts in the semen of the spiny dogfish (*Squalus acanthias*). Tissue and Cell 18: 375-382.

Reardon, M. (2001) Seasonality and Microanatomy of Spermatophore Formation in a Holocephalan, the Elephant Fish, *Callorhinchus milii*. Bachelor of Science honours Thesis. Department of Zoology, University of Melbourne, Parkville, Melbourne.

Reardon, M. B., Walker, T. I., Hamlett, W. C. 2002. Microanatomy of spermatophore formation and male genital ducts in the holocephalan, *Callorhynchus milii*. Marine and Freshwater Research 53: 591-600.

Rojas, M. V. and Esponda, P. 2001. Plasma membrane glycoproteins during spermatogenesis and in spermatozoa of some fishes. Journal of Submicroscopic Cytology and Pathology 33: 133-140.

Rossouw, G. J. (1995) Spermatogenesis *in Callorhynchus callorhynchus* (Linnaeus) (Pisces: Holocephali), from Chile. Revista Chilena de Historia Natural 68: 101-105.

Saxena, D. K., Oh-Oka, T., Kadomatsu, K., Muramatsu, T. and Toshimori, K. 2002. Behaviour of a sperm surface transmembrane glycoprotein basigin during epididymal maturation and its role in fertilization in mice. Reproduction 123: 435-444.

Saville, K. J., Lindley, A. M., Maries, E. G., Carrier, J. C. and Pratt, Jr., H. L. 2001. Multiple paternity in the nurse shark, *Ginglymostoma cirratum*. Environmental Biology of Fishes 63: 347-351.

Simpson, T. H., and Wardle, C. S. 1967. A seasonal cycle in the testis of the spurdog, *Squalus acanthias*, and the sites of 3B-hydroxysteroid dehydrogenase activity. Journal of the Marine Biological Association of the United Kingdom 47: 699-708.

Simpson, T. H., Wright, R. S., and Hunt, S. V. 1964a. Steroid biosynthesis in the testis of the dogfish *Squalus acanthias*. Journal of Endocrinology 31: 29-38.

Simpson, T. H., Wright, R. S., and Renfrew, J. 1964b. Steroid biosynthesis in the semen of dogfish *Squalus acanthias*. Journal of Endocrinology 31: 11-20.

Simpson, T. H., Wright, R. S., and Gottfried, H. 1963. Steroids in the semen of dogfish *Squalus acanthias*. Journal of Endocrinology 26: 489-498.

Skudlarek, M. D., Tulsiani, D. R. P., Nagdas, S. K. and Orgebin-Crist, M-C. 1993. β-D-Galactosidase of rat spermatozoa: Subcellular distribution, substrate specificity and molecular changes during epididymal maturation. Biology of Reproduction 49: 204-213.

Smith, R. M., Walker, T. I., and Hamlett, W. C. 2004. Microscopic organization of the oviducal gland of the holocephalan elephant fish *Callorhynchus milii*. Marine and Freshwater Research 55: 155-164.

Srivastava, A. and Olson, G. E. 1991. Glycoprotein changes in the rat sperm plasma membrane during maturation in the epididymis. Molecular Reproductive Development 29: 357-364.

Stanley, H. P. 1963. Urogenital morphology in the chimaeroid fish *Hydrolagus colliei* (Lay and Bennett). Journal of Morphology 112: 99-128.

Syntin, P., Dacheux, F., Druart, X., Gatti, J. L., Okamura, N. and Dacheux, J-L. 1996. Characterization and identification of proteins secreted in the various regions of the adult boar epididymis. Biology of Reproduction 55: 956-974.

Syntin, P., Dacheux, F., Druart, X., Gatti, J. L., Okamura, N. and Dacheux, J-L. 1996. Characterization and identification of proteins secreted in the various regions of the adult boar epididymis. Biology of Reproduction 55: 956-974.

Tezón, J. G., Ramella, E., Cameo, M. S., Vazquez, M. H. and Blaquier, J. A. 1985. Immunochemical localization of secretory antigens in the human epididymis and their association with spermatozoa. Biology of Reproduction 32: 591-597.

Tricas, T. C., and Le Feuvre, E. M. 1985. Mating of the white tip shark *Triaenodon obesus*. Marine Biology 84: 233-237.

Tulsiani, D. R. P., Skudlarek, M. D., Holland, M. K. and Orgebin-Crist, M-C.1993. Glycosylation of rat sperm plasma membrane during epididymal maturation. Biology of Reproduction 48: 417-428.

Tulsiani, D. R. P., Skudlarek, M. D., Araki, Y. and Orgebin-Crist, M-C. 1995. Purification and characterization of two forms of ß-D-galactosidase from rat epididymal luminal fluid: evidence for their role in the modification of sperm plasma membrane glycoprotein(s). Biochemical Journal 305: 41-50.

Uchida, S., Toda, M., and Kamei, Y. 1990. Reproduction of elasmobranchs in captivity. Pp. 211-237. In H. L. Pratt, S. H. Gruber and T. Taniuchi (eds), *Elasmobranchs as living resources*, NOAA Technical Report NMFS 90: 211-237.

Voglmayr, J. K., Sawyer, R. F. and Dacheux, J. L. 1985. Glycoproteins: A variable factor in surface transformation of rat spermatozoa during epididymal transit. Biology of Reproduction 33: 165-176.

Wright, D. E., and Demski, L. S. 1991. Gonadotropin hormone-releasing hormone of sharks and rays. Journal of Comparative Neurology 307: 49-56.

Wright, D. E., and Demski, L. S. 1993. Gonadotropin-releasing hormone (GnRH) pathways and reproductive control in elasmobranchs. Environmental Biology of Fishes 38: 209-218.

CHAPTER 13

Chondrichthyan Parity, Lecithotrophy and Matrotrophy

William C. Hamlett[1,4], *Gregg Kormanik*[2], *Megan Storrie*[3], *Bronwyn Stevens*[4] *and Terence I. Walker*[5]

13.1 INTRODUCTION

Viviparity is a reproductive mode, well represented in vertebrates, in which the female retains developing eggs within her reproductive tract and gives birth to living offspring. Oviparity is a mode characterized by the deposition of fertilized eggs that develop and hatch externally. The majority of vertebrate species are oviparous but among the Chondrichthyes and mammals viviparity is widespread. We can further group species according to the manner in which nutrients are provided to embryos. Lecithotrophy is a developmental pattern in which yolk, produced by the maternal liver and sequestered in the yolk sac, provides for embryonic nutrition. Most vertebrate species exhibit lecithotrophic oviparity but many viviparous vertebrates are also lecithotrophic (yolk sac viviparity also called lecithotrophic viviparity). Yolk sac viviparity is the dominant mode of reproduction in Chondrichthyes. Lecithotrophic species generally show a dry weight loss during development on an order of 20%. In contrast to lecithotrophy, matrotrophy is a developmental pattern in which the maternal organism supplements yolk from other sources such as uterine secretions called histotroph (histotrophy), ova (ovatrophy), siblings (adelphotrophy) or placental transfer (placentatrophy). Matrotrophic species generally show a dry weight gain during development and that gain can be minimal to extensive but any weight loss greater than

[1]Department of Anatomy and Cell Biology, Indiana University School of Medicine, Notre Dame, IN 46556, USA
[2]Department of Biology, University of North Carolina Asheville, NC, USA
[3]School of Ecology and Environment, Faculty of Science and Technology, Deakin University, Warrnambool, Victoria, Australia
[4]Department of Zoology, Melbourne University, Melbourne, Victoria, Australia
[5]Primary Industries Research Victoria, P. O. Box 114, Queenscliff, Victoria 3225, Australia

approximately -20 % is considered matrotrophy. This is evidence that the maternal organism is providing supplemental nutrients beyond those accounted for by yolk. There is a gradation from strict lecithotrophic yolk sac viviparity to situations in which the maternal uterus provides a very minimal but positive contribution to the nutrition of embryos although this contribution may not be substantial enough to generate a weight gain. We recognize incipient histotrophy where the histotroph is relatively low in organic content and there is a dry weight loss during development but not a loss of the magnitude seen in yolk sac viviparous species. Some species currently thought to exhibit yolk sac viviparity may be found to exhibit matrotrophic histotrophy or vice versa. Minimal histotrophy is where the uterus produces mucus and transfers water and minerals to the embryo such that there is a dry weight gain during gestation. It must, however, be emphasized that there is a continuum from strict yolk sac viviparity to histotrophy and in the absence of reliable weight data it is impossible to unequivocally place many species in a precise category.

Species that exhibit histotrophy show considerable variation in the quality and quantity of uterine secretions. There is a general trend, however, from mucous to lipid histotrophy that is correlated with increased dry weight gain. There is a fine, indistinct line between lecithotrophic yolk sac viviparity and matrotrophic histotrophy. Structural modifications of the uterus, however, occur in all viviparous species and it is likely that all species, regardless of reproductive mode, have some varying degree of histotrophy. Reliable dry weight determinations between uterine eggs and term fetuses are lacking for most Chondrichthyes. Very few studies (Ranzi 1934; Guallart and Vicent 2001) have appropriately dried samples and subsequently incinerated them so that only water, organic matter and minerals can be accounted for. Few relatively isolated biochemical analyses of uterine secretions exist and ultrastructural analyses of the uterus are fragmentary. These data are critical to defining reproductive modes. Definitive lipid histotrophy with a large dry weight gain is characteristic of the stingrays in which the uterus is festooned with secretory villi termed trophonemata. Uterine secretions or "uterine milk" in stingrays is particularly rich in lipid. Other uterine specializations associated with viviparity are: 1) formation of abundantly vascularized uterine folds, flaps or villi, most of which are non-secretory, 2) uterine compartments, 3) thinning and reduction of epithelial layers separating maternal and fetal vessels thereby reducing gas diffusion distance and 4) presence of dilated intercellular spaces which are involved in water and mineral transfer.

13.2 MATROTROPHY NUTRIENT SOURCES AND UPTAKE SITES

In all matrotrophic Chondrichthyes the embryo is initially nourished by yolk sequestered in the external yolk sac. The yolk sac is a trilaminar structure consisting of an exterior of ectodermal epithelium, a connective tissue core with blood vessels and a lining endodermal epithelium. In both oviparous and viviparous species, early in development yolk sac endoderm absorbs yolk and delivers it to the fetal yolk sac vasculature (Hamlett *et al.* 1987) hence the endoderm mediates yolk digestion and the yolk sac ectoderm is quiescent. Yolk is also transported via the ciliated ductus vitellointestinalis either to an internal yolk sac (IYS) in species that have them or the stomach where digestion occurs with subsequent absorption in the intestine. Mellinger *et al.* (1986) estimate that 90% of yolk is digested in the intestine of the oviparous species *Scyliorhinus canicula*. In placental species after yolk stores are exhausted the yolk sac ectoderm functions to receive maternally derived nutrients and gases via the placental connection. In late development the only remnant of the external yolk sac is an abdominal yolk scar. Many species still have an internal yolk sac with yolk deposits after they are born and use this yolk as an energy reservoir during the first few days after birth while they are beginning to forage on their own.

In matrotrophic Chondrichthyes depending on the species, various nutrient sources supplement yolk (Fig. 13.1). Histotroph, often called uterine milk, is defined as "any nutritive material, including glandular secretions, cells, and cell debris, that is available to the embryo or fetus and is derived directly from maternal tissues other than blood" (Mossman 1987). In addition to uterine secretions from epithelial cells another possible source of nutrients are uterine vessels. Transudate from uterine vessels may gain access to the uterine lumen where it becomes part of the periembryonic fluids. Specialized uterine epithelial cells that are apposed to the distal yolk sac in placental sharks produce nutrients that are transferred to and taken up by the yolk sac placenta (Hamlett and Wourms 1984; Hamlett *et al.* 1985a, b, c; Hamlett 1993; 1993a, b). In egg-eating species, ova occupy the uterine cavity where they may be ingested by the embryo. In *Carcharias taurus*, commonly known as the sandtiger or grey nurse shark, fetuses ingest ova and siblings in a process known as intrauterine cannibalism (See Chapter **14** of this volume).

Depending on the species, various embryonic surfaces function as nutrient uptake sites (Fig. 13.1). Oral ingestion of periembryonic fluids, ova or siblings leads to intestinal absorption of nutrients. In placental sharks the ectoderm of the distal end of the yolk sac placenta is the nutrient exchange site as well as the site of respiratory exchange. Prior to establishment of the placenta, yolk sac may absorb nutrients in periembryonic fluids (See Chapter **15** of this volume). Some placental species possess vascular finger-like filaments that adorn the elongated yolk stalk now called an umbilical cord. These filaments, termed appendiculae, may absorb periembryonic molecules. Early in gestation all chondrichthyan embryos have external branchial or gill filaments. In *Rhizoprionodon terraenovae, in vitro* exposure of external gills to a solution of horseradish peroxidase demonstrated uptake by the gill epithelium of marker the size of protein (Hamlett *et al.* 1985e).

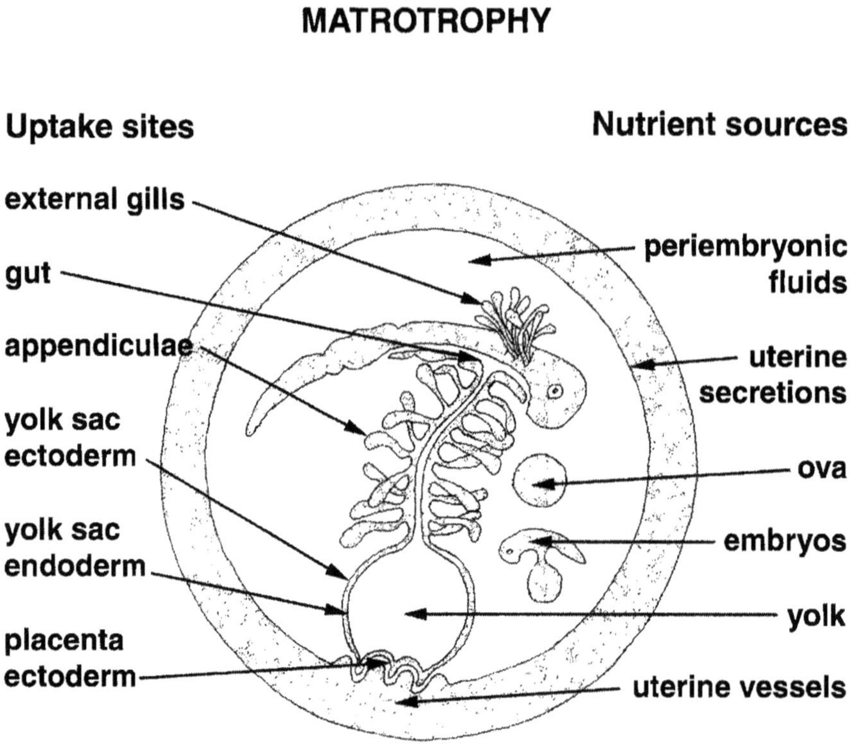

Fig. 13.1 Schematic diagram of nutrient sources and uptake sites in chondrichthyan gestation. Original.

Compagno (1988) recognized the fine line between yolk sac viviparity and incipient histotrophy when he defined "ovoviviparity"as "those species which are livebearing, lack a placenta, and may or may not supplement fetal nutrition with a maternal component." That is to say most species have yolk sac viviparity except oviparous animals, those with a placenta or stingrays. Animals that fit this broad category include a few members of the Family Scyliorhinidae (*Galeus polli*, *G. area*, *Halaelurus lutarius*), and Proscyllidae (*Eridacnis*). About 49% of triakids are ovoviviparous by Compagno's definition including *Triakis*, *Furgaleus*, *Gogolia*, *Galeorhinus*, *Scyliogaleus*, one species of *Hemitriakis* and approximately 33% of the *Mustelus* species (Compagno 1988). Yolk sac viviparity is, therefore, the dominant reproductive mode.

The tiger shark, *Galeocerdo cuvier* is the only non-placental member of the "advanced" carcharinid families (Hemigaleidae, Carcharhinidae and Sphyrnidae) (Compagno 1988). Uteri contain compartments and young are born at large sizes from 50 to 75 cm with each embryo enclosed in a thin tertiary egg envelope. Of note is that the egg envelopes contain a large quantity of periembryonic fluid (Whitley 1940; Sarangdhar 1949).

The very large weight gain from egg to term embryo suggests that there is uterine secretion of nutrients in addition to minerals and water. To our knowledge there are no published reports of uterine histology or chemical analysis of uterine fluids or periembryonic fluid in the tiger shark. Clearly, this is an area that begs investigation.

13.3 ATTEMPTS TO QUANTIFY LECITHOTROPHY AND MATROTROPHY

Ranzi (1932, 1934; summarized in Needham 1942) was the first to attempt to determine the existence of maternal nutrient provision during embryo development and to quantify its importance. Given that water is incorporated into the embryo during development, the comparison of dry weight of eggs and full term embryos provides general data on organic and mineral transfer. General figures given by Capapé (1993) claim that chondrichthyan eggs are 50% and embryos are 75% water. Examination of Ranzi's original data (Table 13.1) clearly shows that such a generalization is not well founded. Water, mineral and organic content vary among reproductive modes and from genus to genus.

Sixty years on, Ranzi's seminal studies stand the test of time and provide a valuable compilation of reliable figures. Many of Ranzi's data were, however, based on only a few ova, eggs and embryos and he realized the probability of error in a small sample size. There is significant variability in the size of ova at ovulation generally correlated with the size of the female, hence special care must be taken to ensure that embryos used for comparison come from ova of similar size. Ranzi dried his samples at 105-120° C and, at that temperature, losses of some of the most volatile organic material are expected. Guallart and Vicent (2001) investigated changes in the composition of ova and new born *Centrophorus granulosus*

Table 13.1 Percentage change from egg to term embryo (Ranzi 1934)

Species	*Weight*	*Water*	*Minerals*	*Organic substance*
Scyliorhinus canicula	108	216	292	-21
Torpedo torpedo	103	300	150	-23
Torpedo marmorata	91	233	186	-31
Dalatias licha	45	100	30	-22
Centrophorus granulosus	17	94	253	-54
Squalus acanthias	78	147	-	-40
Squalus blainvillei	53	88	1250	1
Galeorhinus galeus	118	210	700	11
Mustelus antarcticus	635	1480	1100	110
Mustelus mustelus	1450	2520	173	369
Gymnura micrura	-	-	-	4900
Pteromylaeus bovinus	6150	7200	990	3120
Pteroplatytrygon violacea	5800	10000	950	1680
Prionace glauca	-	-	-	840
Mustelus mustelus	3340	5750	660	1050

and compared their results with that of Ranzi. Ranzi's data suggested a loss of organic matter during development of -50% but this was based on a single ovum and embryo. Guallart and Vicent (2001) dried their samples at 60° C until a constant weight was achieved which took 4–45 days. The samples were then incinerated at successive temperatures of 200° C, 350° C and 550° C. This allowed the following to be determined: water content = wet weight – dry weight; organic content = dry weight – ash weight; inorganic matter = ash weight. The resultant decrease in organic matter was much lower than the data of Ranzi and was similar to that of oviparous species which lead them to conclude that *C. granulosus* is strictly lecithotrophic with no organic matter contributed by the mother. Their results showed the wet weight of ova was significantly correlated with maternal size and embryo weight also increased with maternal size. In their system total weight of the embryo during development was: water +99/+101%; organic matter -18/-25%; inorganic matter +114/170%. Inaccuracies inherent in any given method of evaluation of dry weight, therefore, are related to sample size, correlating ova and embryo sizes and methodology must also factor in ash content to give an accurate figure for organic content.

In a series of papers Capapé and coworkers (Capapé 1993; Capapé *et al*. 1990, 1999, 2000a, b; 2002) reported a chemical balance of development by dividing the dry weight of term embryo with those for uterine ova. Any figure greater than 1.0 was taken to be a sign of matrotrophy. They also concluded that there are "semi-lecithotrophic species" which are neither strictly lecithotrophic nor overtly matrotrophic. This is in line with our position that there is a gradient of reliance on yolk to histotrophy for many species and this is particularly important for species at the cusp between yolk sac viviparity and incipient histotrophy. For many species in this range there are no adequate and reliable weight determinations. Total dry weight figures must be accepted as indicative of general trends only and do not represent true change in organic content.

Another point that has been neglected is the fate of yolk sequestered in the internal yolk sac (IYS). Many species have IYS and young can be born while still retaining substantial yolk in the IYS. Guallart and Vicent (2001) dissected embryos and separated internal vs external yolk to arrive at their organic change of -18 to -25% in *Centrophorus granulosus*. There is no indication Ranzi (1934) used this methodology, hence this could be a source of the disagreement in their respective figures.

Oviparous species have often been used to infer a standard value of organic transfer in viviparous species. Ranzi (1932) used values obtained from *Scyliorhinus canicula*. This diminution was reported as -20.7% and this was the result of energy utilization in converting yolk to tissue and basal metabolism. Mellinger and colleagues (Mellinger *et al*. 1986; Mellinger and Wriez 1989; Lechenault *et al*. 1993) used freeze drying of ova and neonates of *S. canicula* for dry weight determinations in this oviparous species. These data along with unpublished results (Mellinger personal communication) indicate that there is a mean loss of -16.8% in dry weight during development. Other work by Delhaye *et al.* (1992) working with the same species placed the figure at -20.8%. Mellinger's unpublished work was based on a series of 27 pairs of eggs and neonates and he concludes that estimating matrotrophy accurately in viviparous species would require a minimum of 15 fully developed eggs and 15 just born neonates. This is contradictory to figures arrived at by Wourms *et al.* (1988) in which he relied on previous studies to conclude that the loss in organic content from ova to embryo ranged from 25-55% and proposed a value of 35% for reference. Taken together results from oviparous species examined by Mellinger *et al.* (1986) of -16.8% , Delhaye *et al.* (1992) of -20.8% and Ranzi's (1932) figure of -20.7% show the trend. When figures obtained (-18 to -25%) from oviparous species are compared with changes in organic matter of the lecithotrophic viviparous *Centrophorus granulosus* (Guallart and Vicent 2001), we can conclude that a loss of about -20% is appropriate for oviparous and strictly lecithotrophic species. If we must apply one very general figure, then -20% for oviparous species is that figure. Careful methodology to assure accurate weight changes are necessary to establish reliable figures, perhaps then we can purge the literature of "numbers" fraught with error (Wourms *et al.* 1988) based on erroneous extrapolations.

Careful examination of Ranzi's (1934) original data (Table 13.1) clearly demonstrates the importance of determining water and ash content to establish change in organic substance. In the species studied, percent change in water content from egg to term embryo ranged from 17 to 6,150% and minerals from 30 to 1,250%. These results demonstrate that water and mineral content are specific for a given genus and comparisons based only on total dry weight can yield misleading figures. In addition, variation in samples of the same species can result in conflicting figures as in *Mustelus mustelus* below. Henceforth, all determinations of organic substance changes during development should include ash content.

In this chapter we focus on morphological adaptations associated with the various modes of reproduction and use the "numbers" solely for general groupings. It must be emphasized that mere repetition of flawed data perpetuates the problem, therefore, care must be taken when citing figures calculated from extrapolations.

13.4 LECITHOTROPHIC OVIPARITY

13.4.1 Uterus of *Leucoraja (=Raja) erinacea*

The uterus in *Leucoraja erinacea* and other species like *Raja eglanteria* (Fig. 13.2A) is morphologically specialized for well defined functional activities associated with egg capsule formation: (1) regionally distinct structural modifications, both in the uterine wall and the epithelial lining, for active movement of the capsule through the uterine lumen via smooth muscle and cilia, (2) biosynthesis and secretion of materials into the lumen, (3) vascular facilitation for oxygen transfer in order to fuel the oxidation process involved in capsule tanning, and (4) intercellular spaces for removal of water from the uterine lumen. Increased vascularity of the uterus is shared with yolk sac viviparous *Squalus acanthias* (Figs 13.2B-D). The uterus of *L. erinacea* has longitudinal folds punctuated by secretory crypts (Figs 13.3A, B). *Squalus acanthias* has richly vascularied uterine flaps (Figs 13.2B-D). In *Leucoraja*, increased vascularity serves to fuel the relatively short oxidation process associated with capsule tanning and

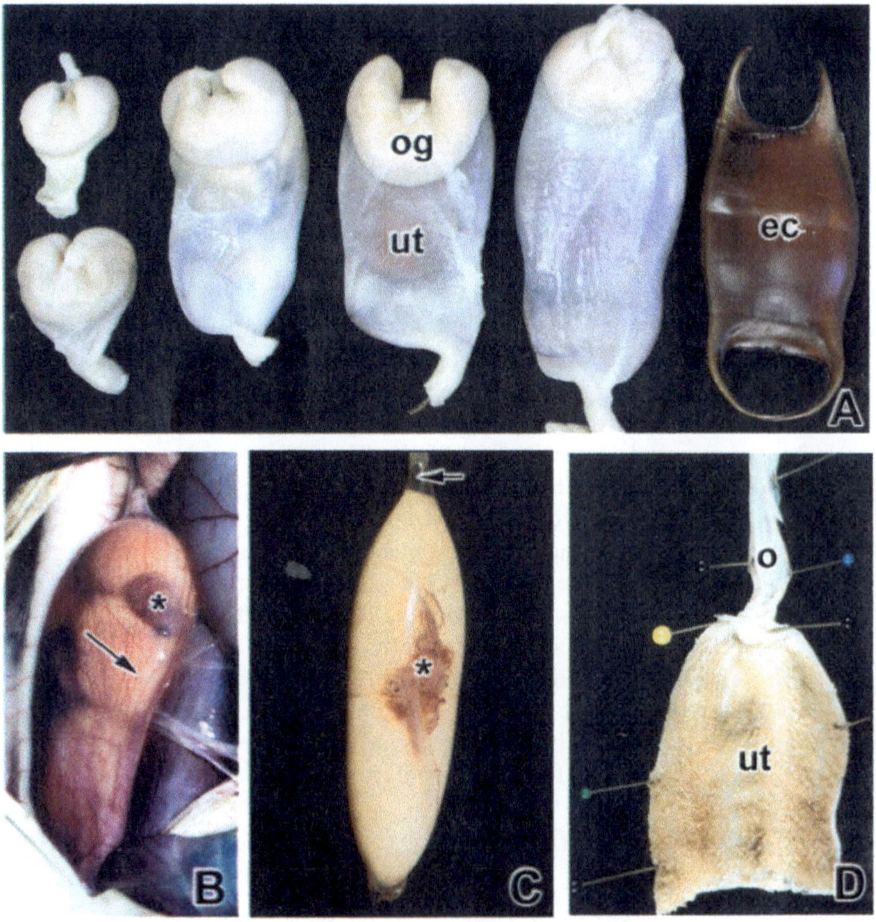

Fig. 13.2 A. Series of samples of *Raja eglanteria* illustrating degree of enlargement of the oviducal gland (og) and uterus (ut) as the egg capsule (ec) is formed. **B**. Gross photograph of *Squalus acanthias* gravid uterus with longitudinal blood vessels (arrow) containing an egg candle with embryos (asterisk). **C**. Egg candle of *Squalus acanthias* containing a thin rim of egg jelly (arrow) and yolk dependent embryos (asterisk). **D**. Longitudinally sectioned anterior oviduct (o) and uterus (ut) of *Squalus acanthias* to show uterine flaps (arrow). Original.

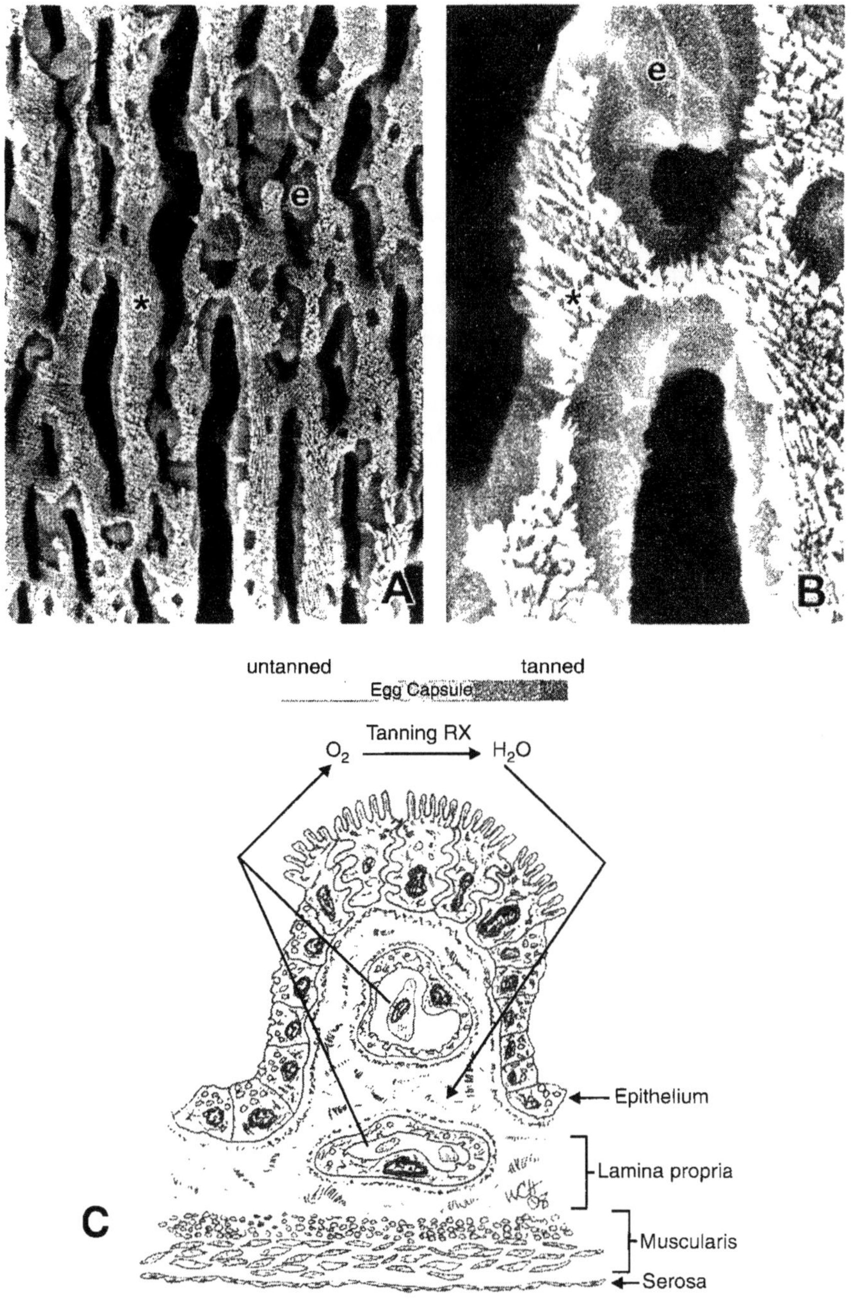

Fig. 13.3 A, B. Scanning electron micrograph of gravid uterus from *Leucoraja erinacea* to show cilia (asterisks), openings of glands and secretory epithelium (e). **C**. Composite line drawing indicating the structural organization of the uterine tissues, source of oxygen to fuel the tanning reaction, and to account for the removal of water created by the reaction. From Koob, T. J. and Hamlett, W. C. 1998. Journal of Experimental Zoology 282: 421-437, Figs 7, 5 and 18.

in *Squalus* vascularity supports respiratory demands of the embryo during a lengthy gestation (Hamlett and Hysell 1998).

Only one other report has appeared in the recent literature on the ultrastructure of the uterus in an oviparous species, *Cephaloscyllium umbratile,* and the data consisted of a single light micrograph and a single transmission electron micrograph (Otake 1990). The uterine epithelium was stratified and did not possess crypts as seen in *L. erinacea* (Koob and Hamlett 1998).

In *L. erinacea* and other skates such as *Raja eglanteria* (Fig. 13.2A) when newly assembled, untanned capsule precursors move out of the oviducal gland lumen and enter the proximal portion of the uterus. The capsule is easily deformed since the quinone tanning process has only just begun. The uterine wall in this region is dominated by a loose, hydrated connective tissue matrix and little muscle. This morphology suggests that the uterine wall in this region is highly extensible under very low loads. Previous biomechanical measurements indicated that this region neighboring the oviducal gland in non-reproductively active, adult little skates does not extend under high loads to a circumference adequate to allow entry of the formed egg capsules, whereas this region in the egg laying females is highly extensible (Koob *et al.* 1981).

The presence of abundantly ciliated, epithelial cells lining the uterine lumen (Figs 13.3A, B, 13.4A, D) suggests that they may participate in moving the capsule. Decline in the relative proportion of ciliated cells and the greater proportion of the uterine wall muscle lower down supports this hypothesis. In the distal region of the uterus, ciliated cells are essentially absent and the muscularis makes up the bulk of the uterine wall. This gradual transition from an epithelium dominated by ciliated cells to one lacking cilia where the uterine wall is predominantly smooth muscle correlates directly with the extent of quinone tanning and sclerotization of the capsule. A likely interpretation of the observed gradual change in the relative proportions of ciliated epithelium and muscularis is that cilia are responsible for moving the newly formed capsule in the upper uterus while the muscularis moves the tanned capsule in the distal regions.

All of the non-ciliated epithelial cells of the uterus in *L. erinacea* contain an extensive Golgi apparatus, abundant secretory granules and mucous secretions (Figs 13.4A, B). Uterine epithelium in this oviparous species synthesizes and secretes material that is likely involved in capsule formation and/or processing in the uterus. The nature and function of the secreted material are presently unknown.

The mechanism for tanning and sclerotizing the capsule matrix involves the catechol oxidase catalyzed oxidation of catechols to quinones and requires molecular oxygen (Koob and Cox 1988, 1990). Since catechol oxidation occurs within the uterus, oxygen must be available within the uterine lumen. While the source of oxygen is unknown, observations on uterine vascularization suggest oxygen is supplied via the circulation

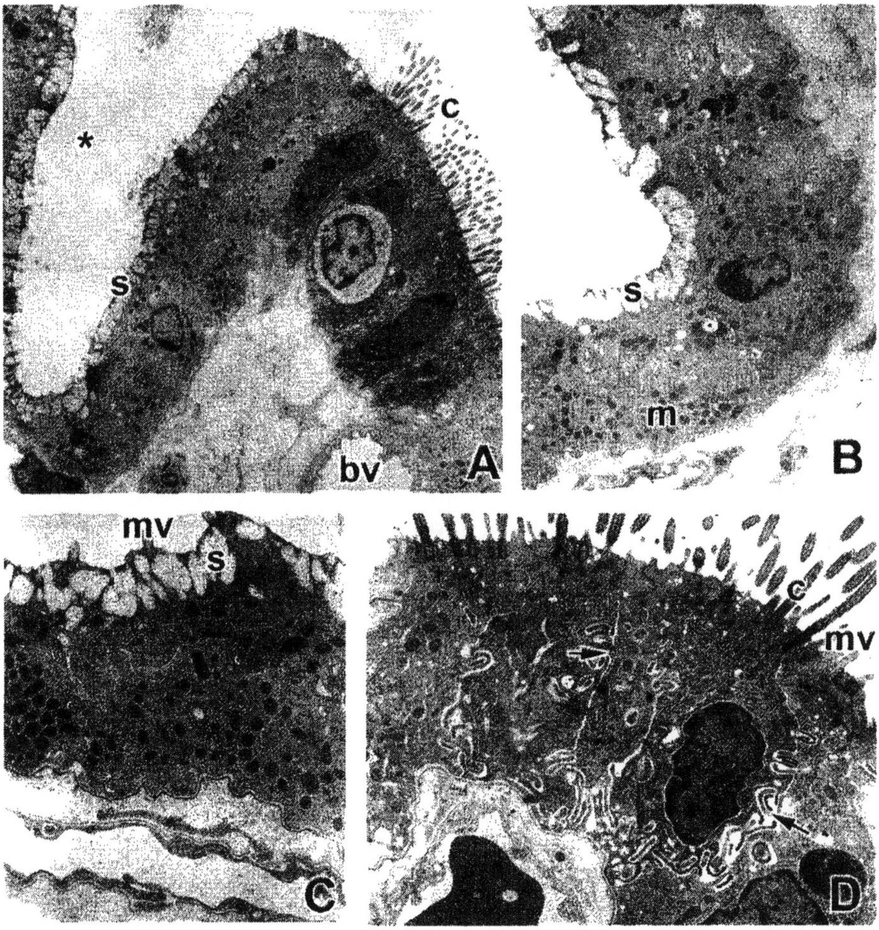

Fig. 13.4 A. Transmission electron micrograph of gravid uterus of *Leucoraja erinacea* showing secretory crypt (asterisk) on left and ciliated cells (c) on the right. S = secretory vesicles, bv = blood vessel. X2,000. **B**. Transmission electron micrograph of gravid uterus of *Leucoraja erinacea* showing basal mitochondria (m) and secretory vesicles (s). X3,500. **C.** Transmission electron micrograph of secretory cell of gravid uterus in *Leucoraja erinacea* showing secretory vesicles (s), microvilli (mv) and basal mitochondria. X8,000. **D.** In gravid uterus of *Leucoraja erinacea*, surface columnar cells have cilia (c), microvilli (mv) and intercellular folds (arrows). X6,000. From Koob, T. J. and Hamlett, W. C. 1998. Journal of Experimental Zoology 282: 421-437, Figs 12, 13, 16 and 17.

(Fig. 13.3C). Each crypt wall is supplied with a substantial vascularization separated from the lumen by a single cell.

A second feature of catechol oxidation is the formation of water from extracted catechol hydroxyl hydrogens and oxygen. The fate of this water is currently unknown, but water does not accumulate in the uterine lumen and the surface of the capsule is remarkably dry (Koob and Hamlett 1998). Intercellular folds (Fig. 13.4D) are common in transporting epithelia where osmotic gradients produce transepithelial water movements by pumping solutes into confined compartments, the lateral cell spaces between lateral folds (Berridge and Oschman 1972). Kaye *et al.* (1966) examined the ultrastructure of the rabbit gall bladder and determined that intercellular spaces distend during transport and then collapse when transport ceases. These observations were extended by Diamond and Bossert (1967, 1968) and Diamond and Tormey (1966a, b) who demonstrated a standing osmotic gradient within intercellular spaces. Thus the standing gradient model has widespread application to transporting epithelia in general, while the main characteristic of transporting cells is extensive intercellular folding, forming long channels. Solute transport from the cell would go into the intercellular spaces and bring about distention of the spaces and exert hydrostatic pressure between the lateral cell membranes. The result would be the distention of the intercellular spaces and creation of hydraulic flow through the spaces and into the connective tissue and associated vascular channels. A similar phenomenon may be occurring in the skate uterus, especially in the upper portion just below the oviducal gland, where these intercellular folds are abundant.

The capacity of the chondrichthyan uterus to develop physiological means in order to satisfy the demands peculiar to each reproductive mode relies on a few, shared basic principles: structural accommodation of the eggs and embryos, supply of oxygen to the uterine lumen and biosynthesis and secretion of structural or nutritional materials. Regardless of the reproductive mode, similar design characteristics are shared by all chondrichthyan uteri (Hamlett and Hysell 1998).

13.5 YOLK SAC VIVIPARITY (LECITHOTROPHIC VIVIPARITY)

Design characteristics of the uterus in yolk sac viviparous species include: 1) increased vascularity, 2) development of longitudinal folds, flaps or villi to increase surface area for respiratory exchange and 3) thinning of surface epithelium to 1 to 2 cells thick. The uterus is specialized for respiratory exchange and osmoregulation, not nutrient provision. Significant work has been done on the function of the spiny dogfish, *Squalus acanthias* uterus and this will be discussed in section 13.5.1.

Ranzi (1934) presented a series of photographs of the gross surface epithelial amplifications and histology of several species with yolk sac viviparity. Generally there is developmental weight loss but all species share a common suite of uterine specializations for respiration, electrolyte balance and osmoregulation. *Squalus acanthias* has longitudinal uterine vessels that can be seen by gross observation (Figs 13.2B, C) and the thin nature of the uterine wall allows the embryos to be seen through the wall. Longitudinal vessels then ramify to supply uterine flaps (Fig. 13.2D). A large vessel courses along the crest of each flap which sprouts capillaries. *Squalus acanthias* has longitudinal uterine folds but the crest contains a large vessel (Ranzi 1934). *Torpedo marmorata* has leaf-like uterine flaps and *Torpedo torpedo* has fine villi with a simple epithelium consisting of one cell layer. In no cases were any substantial secretory cells seen (Ranzi 1934).

13.5.1 Uterus of *Squalus acanthias*

The dogfish, *S. acanthias*, is arguably one of the most studied of all Chondrichthyes, especially with regard to its anatomy and physiology. Embryos are incubated internally in the uteri of the mother for a gestation period that lasts nearly two years (Hisaw and Albert 1947; Woodhead 1976; Nammack, *et al.* 1985). During the first four to six months of gestation, embryos are incubated in a transparent, diaphanous and relatively fragile egg capsule; the whole is termed a candle (Fig. 13.2C). A candle in each of two uteri may hold a few to eight embryos (Jones and Ugland 2001). The egg candle breaks, releasing the embryos and gestation continues to term in the uteri. Nutrients for the developing embryo are provided by yolk from a pendulous external yolk sac. As yolk is digested the sac shrinks and all that is left is a yolk sac scar. The term applied to this mode of nutrition is yolk sac viviparity or lecithotrophic viviparity. Yolk utilization is virtually identical in both oviparous and viviparous species but striking differences are apparent in modifications of the uterus in oviparous *versus* viviparous species (see description of the uterus in *Leucoraja erinacea*). The conditions during uterine incubation hold some important implications and pose many questions especially regarding ionic regulation, acid-base balance and nitrogen metabolism for these developing embryos.

13.5.2 Uterine Structure

The wall of the uteri in immature females and those that are not pregnant is relatively thinner and the uterus is smaller than for gravid females. The diameter of the gravid uterus increases upwards of 20-fold, and the mucosa becomes thicker, with an increase in pronounced longitudinal mucosal folds, or uterine flaps (Fig. 13.2D) a concomitant increase in surface area and reduction in diffusional distances (Jollie and Jollie 1967a; Hamlett and Hysell 1998). The lining also becomes highly vascularized (Jollie and Jollie 1967a; Hamlett and Hysell 1998). At the connective tissue core of the flaps arterial loops are formed, with extensive connections to an underlying network of capillaries to the mucosal epithelium (Jollie and Jollie 1967a). Stratified cuboidal cells with microvilli cover the folded epithelium (Hamlett and Hysell 1998). The lining takes on a very distinctive appearance, with extensive looped blood vessels (Hamlett and Hysell 1998). Uterine blood flow increases via expanded arteries (Fuller *et al.* 1983). The structure of the epithelium, its similarity to other transporting tissues, and the fact that it separates compartments of differing osmotic strength and ionic composition (blood and a seawater-like fluid) suggested to Jollie and Jollie (1967a) a role for the uterine epithelium in osmoregulation and the transport of ions and water. While the uterine epithelium does serve as a barrier between the uterine environment and maternal blood, more direct evidence regarding its specific regulatory role, via physiological, biochemical or molecular identification of specific transporters would be welcome. For example, in order to prevent urea loss across this amplified surface area, especially later during incubation of late-term embryos, a potential role for urea transporters, identified in gill and other tissues (see Wilkie 2002), is suggested, but not yet investigated. Early observations of the extensive uterine vascularization and observations on gestating females also suggested that the uterine lining provides for oxygen delivery and carbon dioxide removal for the developing embryos (Burger 1967; Jollie and Jollie 1967a). Embryos can be readily observed to ventilate their gills while residing *in utero* (Kormanik, personal observations). Burger (1967) also reported that the pregnant female may flush the uteri with seawater, likely to aid the embryos (pups) in respiration, as well as remove any waste products that may accumulate.

Jollie and Jollie (1967b) also examined the ultrastructure of the embryonic yolk sac, which could potentially act as a pseudoplacenta, a conduit for maternal fetal-transfer. These authors described two barriers, the innermost acting as a yolk-blood barrier consisting of a ciliated columnar epithelium, and a respiratory barrier, which thickens and presumably becomes less useful in respiration as the respiratory filaments of the embryo and then the gills become more functional. There is no attachment or specialized structural interaction of the *S. acanthias* yolk sac with the lining of the uterine horns, although close apposition between the yolk sac and uterine wall would serve to reduce diffusional distances. Jollie and Jollie (1967b) concluded that the yolk sac was transiently placental during the early stages of development, serving in respiratory gas and nutritive transport, albeit providing no direct evidence.

13.5.3 Uterine Environment

In early gestation, uterine fluid in which these egg capsules reside resembles a plasma exudate or perhaps transudate with regard to ionic composition (Evans *et al.* 1982). Later, during the last 12 to 15 months of gestation, the fluid changes to one more closely resembling seawater, at least with respect to the major ions (Evans *et al.* 1982; Kormanik and Evans 1986; Kormanik 1992), as a result of increasing exchange with external seawater (Burger 1967). Lombardi *et al.* (1993) found that total protein and glucose of uterine fluids in near term *S. acanthias* were less than 0.01 mg per ml, and total lipids, 0.07 %, indicating minimal maternal-fetal transfer of macromolecules. Taurine (13 nanomol per ml) was the only amino acid detected in uterine fluids (Files and Lombardi 1993). Taurine frequently serves as an intracellular osmolyte in elasmobranchs (see Goldstein and Pearlman 1995) but such a role here would appear to be minimal at this dilution. The uterine fluid is slightly acidic, (pH about 6), and has a total CO_2 content of a few tenths millimole per liter, compared to two millimole per liter for normal seawater. More surprising was the total ammonia (NH_3^+ +NH_4^+) concentration of the fluid, which approached 22 millimole per liter (Kormanik and Evans 1986). This accumulated ammonia suggested to these authors that, if the uteri were flushed as suggested by Burger (1967), it likely was a rather infrequent event.

Ammonia is an inexpensively derived waste form of nitrogen typically utilized by aquatic organisms when it may be diluted by copious volumes of ambient water. Ammonia may exert a toxic effect at relatively low concentrations (Randall and Tsui 2002). Of more interest was the possible toxic nature of the high total ammonia concentration, its origins and a potential role in embryonic nitrogen metabolism. Kormanik (1988) examined the origins of these chemical gradients, and demonstrated that when the uteri of a late-term female are filled with fresh seawater, in the absence of the pups the mother can acidify this uterine seawater environment to a pH of around 6, and reduce the total CO_2 content to the levels typically observed for fresh-caught females in 24 hours. Kormanik and Kremer (1986) also reported that the acidification was inhibited by acetazolamide, implicating carbonic anhydrase, a ubiquitous enzyme responsible for interconversion of CO_2 and bicarbonate, in the process. Low levels of activity for carbonic anhydrase were reported in the uterine tissue (Kormanik and Maren 1988), and later confirmed by Flugel and Lutjen-Drecoll (1991). Total ammonia also accumulates to several millimole per liter during that period, establishing a gradient between maternal blood and uterine seawater. The decreased pH of the uterine environment, while serving to detoxify the ammonia by converting it to the less permeable ionic form, (NH_4^+), also facilitates accumulation of total ammonia by proton trapping (ammonia diffuses in as NH_3, and is then protonated to the less permeable ionic form, NH_4^+). This accumulation of ammonia, a compound usually considered to be relatively toxic to water breathers, in the uteri and of maternal origin suggested a potential role in the nitrogen budget of the developing embryos (see below).

13.5.4 Osmoregulation

During the extended period of incubation, *S. acanthias* embryos experience a changing ionic uterine milieu (Kormanik 1992). Since the egg candle is very thin and permeable, it provides a minimal barrier to solutes (see Kormanik 1992). Evans *et al.* (1982) demonstrated that *S. acanthias* embryos (25-30 mm in length) could regulate major osmolytes during exposure to seawater for several days. At around this length the uterine environment is beginning to take on more seawater-like characteristics, with a decrease in urea concentration and increase in major ion concentrations. Also, the kidneys may begin to function (Price and Daiber 1967) and the anlage of the rectal gland is present (Chan and Phillips

1966). Kormanik *et al.* (1991; unpublished data) demonstrated that mitochondria-rich cells first appear in gill arches of embryos that are 17-20 mm in body length although their potential role in osmoregulation at that developmental stage is unknown. Older embryos (19 cm) have abundant mitochondria-rich cells on the gill filaments and in the rectal gland as well (Kormanik *et al.* 1991). However, the role of mitochondria-rich cells in the gill (= chloride cells) for osmoregulation is debated even for adults (Wilson *et al.* 2002). Nevertheless, embryos (ca. >5 cm in length) can and must independently regulate during the latter portion of the gestation period, when the uterine fluids resemble seawater; embryos likely resemble the adult pattern for osmoregulation during this time.

On a gross level, *S. acanthias* loses a substantial amount of dry organic mass (20-34%) and ash (23-37%) through development, indicating no net transfer of nutrients from the mother to the developing embryo, other than that initially endowed in the egg (Hisaw and Albert 1947).Yet, the possibility of nitrogen transfer from mother to embryo remains intriguing. To be of use to the developing embryo, uterine ammonia provided by the mother must enter the embryo, be modified to some useful form and be retained, or replace nitrogen lost via other pathways. Retention of nitrogen by the embryo may include mechanisms for reducing loss of numerous nitrogenous metabolites, especially via the kidney and gills. Precise values for comparison of changes in total nitrogen during development of embryos are difficult to collect, since it would be difficult to track total N for a specific individual through development. Eggs and neonates also vary in mass. Nevertheless, one may presume that the largest eggs give rise to the largest neonates, and some estimates can be made on that basis. On the one hand, *S. acanthias* embryos for most of the developmental period, do not exhibit a net gain of nitrogen. In fact they appear to lose 16 percent of the endowed nitrogen, at least when comparing that of the egg or earliest of embryos, to later term (stage "C") embryos (Kormanik 1989); total nitrogen in full-term embryos or neonates was not measured. Using the ranges of mass for ova and near term embryos reported by Hisaw and Albert (1947) and the nitrogen content values reported by Kormanik (1989), embryos lose 12 to 28 percent of endowed nitrogen through development (see also Kormanik *et al.* 1992). However, in order to better assess any potential maternal contribution, an appropriate benchmark for comparison is the amount of nitrogen lost by oviparous species, where there is no maternal contribution after oviposition. Observations on paired embryos (to minimize size variation) of the oviparous *Leucoraja erinacea* showed that hatchlings have gained wet mass (132-153%) and lost about 32-40% of their nitrogen during development, while retaining nearly all of their TMAO (83-100%) (Kormanik and Totten 1993). In contrast, *S. acanthias* embryos, a few months short of full development (and birth), have lost less estimated total nitrogen (3-14%), but lost more TMAO (13-22%) while gaining wet mass (124-140% (Kormanik *et al.* 1992). Neither *L. erinacea* nor *S. acanthias* can synthesize TMAO (Goldstein and Pallat 1974). The urea to mass ratio is somewhat higher for *S. acanthias* than for *L. erinacea*, and could be expected to increase during incubation in the high ammonia environment if urea synthesis is stimulated.

Squalus acanthias embryos would appear to lose rather less N than may be expected for an entirely lecithotrophic mode of nutrition. How might this occur? *Squalus acanthias* incubated in the uteri have elevated blood ammonia concentrations, compared to free-swimming embryos and adult sharks; plasma glutamine levels are increased several-fold, which leads to elevation of glutamine in muscle, liver, brain and kidney; glutamine is utilized by liver CPSIII as a source of nitrogen for urea synthesis and ordinarily is at very low levels in elasmobranch plasma (Kormanik and Verity 1995), compared with mammals. The enzyme responsible for glutamine synthesis is glutamine synthetase (GSase), which traps ammonia in the less toxic and less permeable molecule, glutamine. Highest absolute levels of GSase were found in the brain of embryos, followed by the liver, kidney, gill and white muscle. GSase has a low Km for ammonia (15 micromole per liter; Shankar and Anderson 1985) and CPS-III, a low Km for glutamine (ca. 0.16 millimole per liter; Anderson 1980). While the low Km of GSase might suggest that the enzyme would be only minimally affected by an elevated blood ammonia level, plasma glutamine nevertheless increases several-fold under simulated uterine ammonia conditions compared to seawater fish. In contrast to the role of GSase, glutaminase releases ammonia from glutamine, and potentially may result in futile cycling of ammonia. Therefore the ratios of GSase to glutaminase in these embryonic tissues are important and can indicate whether glutamine synthesis or catabolism is favored. GSase: glutaminase ratios for brain (4.4 : 1), liver (83 : 1), kidney (11 : 1), gill (4.4 : 1) and even white muscle (1.1 : 1) suggest that these tissues may be sites for net glutamine production, and is confirmed by the elevated levels of glutamine seen in these tissues in response to ammonia exposure (Kormanik and Verity 1995). In addition, potential sites for ammonia loss (gill and kidney) have GSase/glutaminase ratios that would favor trapping of ammonia and could help account for the relatively low excretion rates for ammonia observed previously in embryos as well as adult elasmobranchs (Kormanik 1989). Infusion of glutamine loads in adult *S. acanthias* demonstrates that the nitrogen load is likely retained (Kormanik *et al.* 1999). On the whole, these data suggest that ammonia from the uterine environment and therefore of maternal origin can enter the late-term embryo, elevate embryonic ammonia, stimulate glutamine and thus urea synthesis and help to replace embryonic stores of urea lost by diffusion. Once synthesized and circulated via the blood, glutamine may also serve as a nitrogen source for other anabolic reactions. While not providing for a substantial or even net gain in nitrogen, maternal ammonia may help offset potential losses of nitrogen by diffusion and other pathways in these developing embryos of lecithotrophic sharks.

13.6 UTERUS OF *PRISTIOPHORUS* SP

In the saw shark *Pristiophorus* sp., eclosion from the tertiary egg envelope occurs well before the external yolk sac is resorbed (Fig. 13.5A) and at approximately midterm the internal yolk sac is roughly the size of the external sac (Fig. 13.5B). The internal yolk sac of no Chondricthyan fish has ever been examined ultrastructurally. The rich vascular supply of the EYS reflects its role in endodermal degradation of yolk and its passage to the embryonic circulation. The sparseness of vascularity in the IYS suggests that yolk is not being digested. Confirmation awaits analysis. When the uterine mucosa is examined grossly, robust red villi are evident (Fig. 13.5C). Examination of uterine villi in *Pristiophorus* was carried out by Stevens (2002) using light microscopy. She evaluated uteri throughout gestation. Uterine villi of an animal near term showed dilated vessels in intimate contact with the squamous surface epithelium (Fig. 13.5D). There were no secretory crypts thus the microanatomy suggests specializations for respiration. Ultrastructural studies (Hamlett and Stevens unpublished) of villi earlier in gestation show a very modest amount of mucous secretion (Figs 13.6A, B). In an animal near term, mucous secretion has all but ceased while dilated intercellular spaces dominate (Figs 13.6C, D). It is concluded that yolk is the primary nutrient throughout development but there is a very modest mucus contribution. Later in development dilated intercellular spaces (Figs 13.6C, D) serve to provide water and minerals to the embryo similar to the system that exists in many epithelial transport models (Berridge and Oschman 1972). No secretory crypts were ever formed unlike *Dasyatis americana* which has definitive lipid histotrophy (Hamlett *et al.* 1996a). In *Pristiophorus* scant mucus production occurs throughout gestation. In addition to histological studies Stevens (2002) made biochemical analyses of uterine fluids and maternal serum. These studies showed that uterine fluid had much lower concentrations of carbohydrate and protein than maternal serum. No significant increases in carbohydrate or protein concentrations in maternal serum or uterine fluid were seen over the periods of gestation studied. This is consistent with incipient histotrophy where there is very slight mucus secretion coupled with water weight gain by the embryo in addition to yolk.

13.6.2 Uterine Specialization in the Gummy Shark, *Mustelus antarcticus*

The gummy shark, *Mustelus antarcticus*, is an example of an animal with minimal histotrophy. Storrie (2004) studied histological and ultrastructural changes in the uterus of females throughout gestation. The gummy shark has several characteristics associated with placental sharks. Foremost is the development of uterine compartments. The uterus does not possess villi as is seen in some of the animals with incipient histotrophy but surface area available to each embryo is provided by individual uterine compartments. Another feature shared with placental

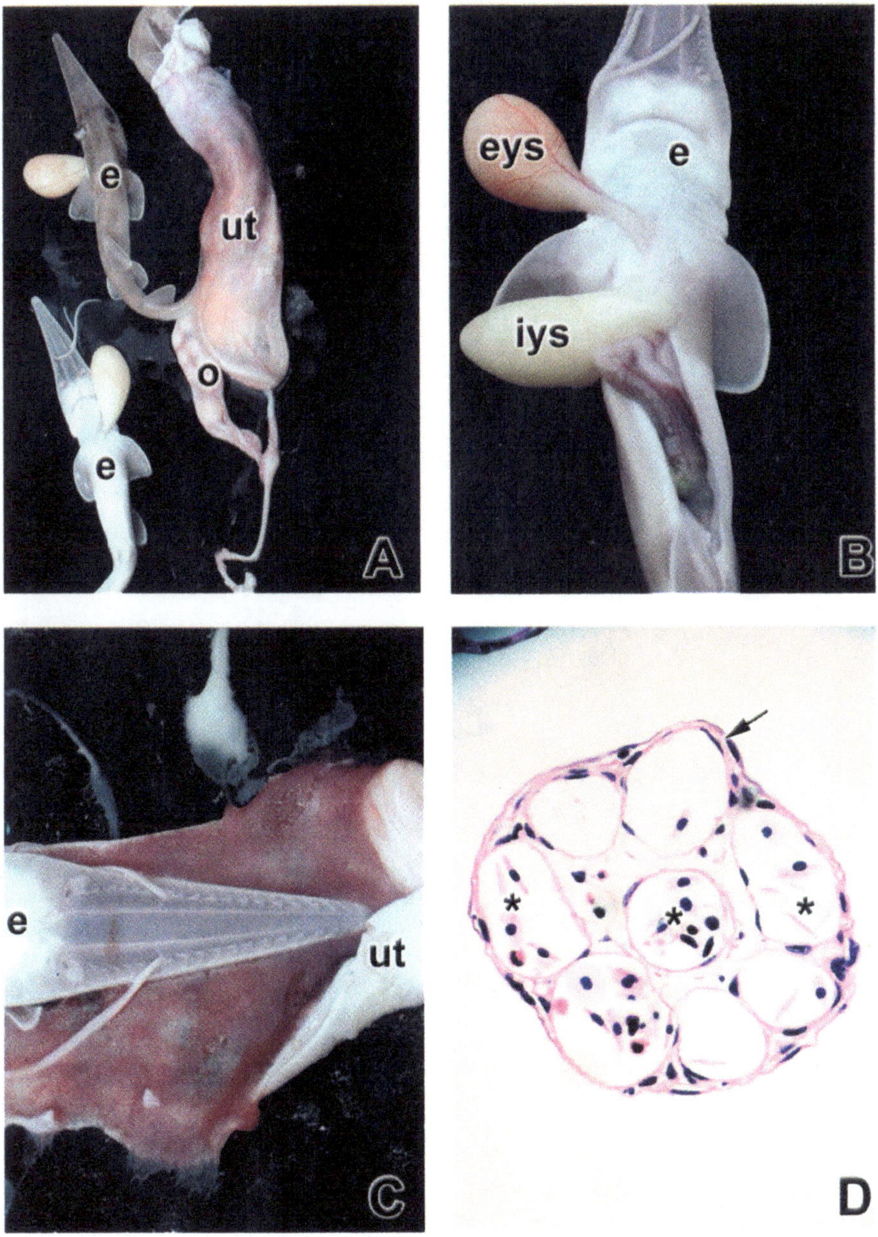

Figs. 13.5 A-C. Gravid uterus (ut) of *Pristiophorus* containing embryos (e) with both external yolk sac (eys), internal yolk sac (iys) and uterine villi. o = ovary. **D**. Light micrograph of transversely sectioned uterine villus of *Pristiophorus*. Large vessels (asterisks) abut the thin squamous epithelium (arrow). X600. Original.

species is a persistent tertiary egg envelope that is present throughout gestation. The oviducal gland is virtually identical both at the gross, light

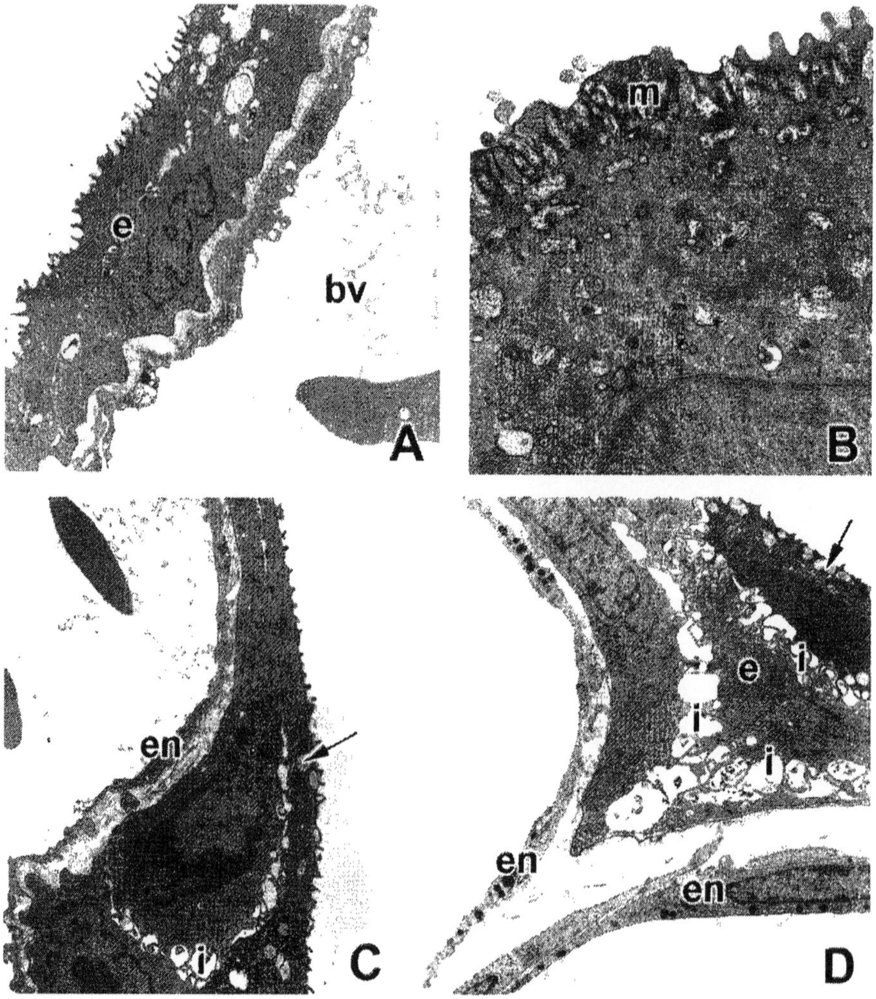

Fig. 13.6 A, B. Transmission electron micrographs show the characteristics of uterine villi of *Pristiophorus* in early gestation which include thin bilayered squamous epithelium (e), modest mucous secretion (m) and close proximity of large blood vessels (bv) to the surface epithelium. Fig 17 = X5,200, Fig 18 = X15,500. C, D. Transmission electron micrographs of uterine villi of *Pristiophorus* containing near term fetuses show reduced mucous vesicles (arrows), close proximity of endothelia (en) to surface epithelium (e) and dilated intercellular spaces (i). Fig. 13.6.C. = X5,200, Fig. 13.6D = X5,200. Original.

microscopic and ultrastructural levels to that in the placental, *M. canis* (Hamlett *et al.* 2002). The average total dry weight of multiple ova samples was 2.267 g. Average embryo total dry weight at approximately mid-gestation was 8.55 g representing an increase in dry mass of 377.21%. Term samples weighed an average of 17.775 g or a 784.19% increase over the egg. In early to mid-gestation the uterine epithelium is stratified and

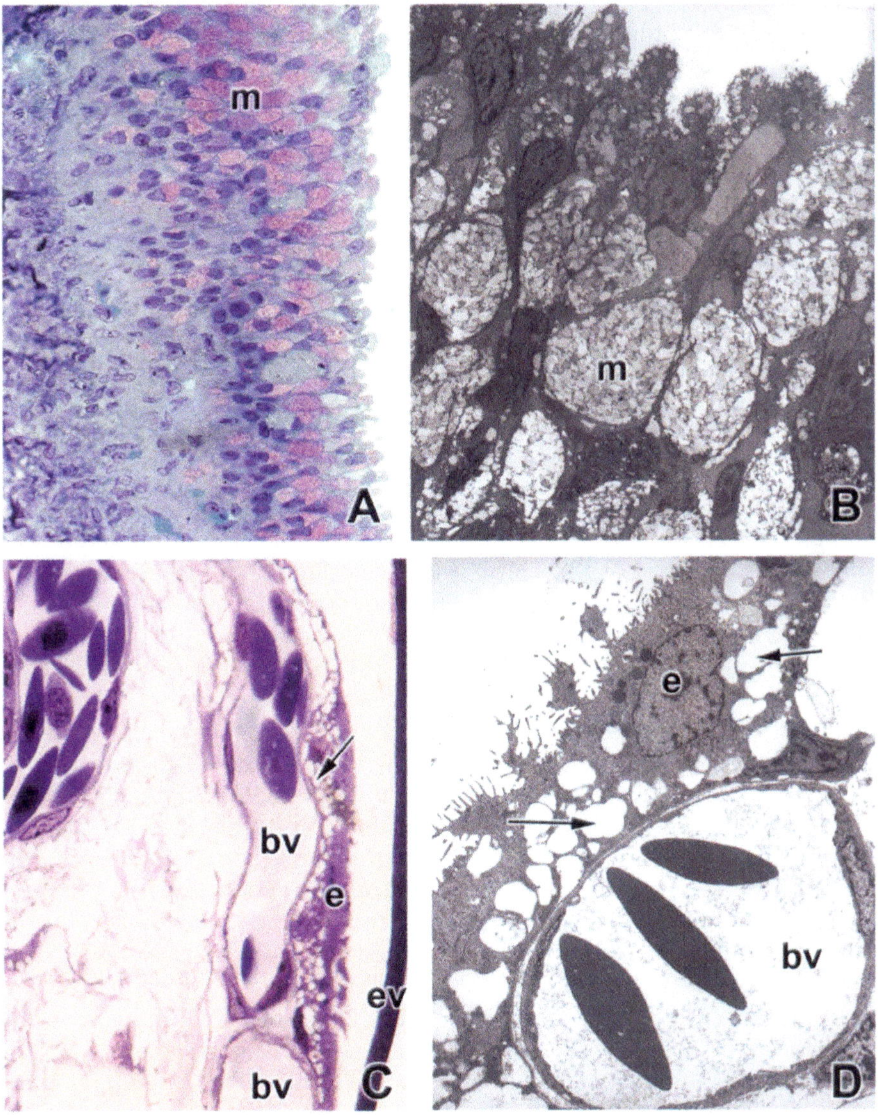

Fig. 13.7 A. Light micrograph of uterus of *Mustelus antarcticus* in early gestation shows toluidine blue metachromatic staining of multiple strata of mucous cells (m). X600. **B**. Transmission electron micrograph of uterus of *Mustelus antarcticus* in early gestation shows robust stratified mucous (m) cells. **C**. Light micrograph of uterine compartment of *Mustelus antarcticus* containing near term fetuses shows persistent egg envelope (ev), close proximity of blood vessels (bv) to squamous surface epithelium (e) with dilated spaces (arrow). X600. Stain toluidine blue. **D**. Transmission electron micrograph of uterine compartment of *Mustelus antarcticus* containing near term fetuses shows close proximity of blood vessels (bv) to squamous surface epithelium (e) with dilated spaces (arrows). X2,200. From Storrie 2004.

is dominated by high mucous secretory activity (Figs 13.7A, B). By near term the uterine mucosa has undergone a striking morphological modification. The epithelium thins to simple squamous and mucous secretion dissipates and disappears. Vessels proliferate, dilate (Figs 13.7C, D) and are in intimate contact with the basal lamina of the epithelium (Fig. 13 7C). A prominent feature is the presence of extensively dilated spaces (Fig. 13.7D). We conclude that abundant mucous secretion augments yolk stores in early and mid-gestation and water and mineral transfer primarily occur in later gestation. This accounts for the substantial dry weight gain over species like *Pristiophorus*.

13.7 DEFINITIVE LIPID HISTOTROPHY

13.7.1 Trophonemata in Stingrays

In stingrays, as in all Chondrichthyes, initial development is yolk dependent. As embryos grow yolk reserves in the EYS are depleted (Fig. 13.8A) and the size of the uterus increases (Fig. 13.8B) to accommodate the intrauterine young (Fig. 13.8C). Vascularized spatulate to villiform uterine villi termed trophonemata (Fig. 13.8C) develop. In most stingrays with definitive lipid historiotrophy, the histotroph (Fig. 13.8D) is elaborated by secretory crypts. In early gestation when embryos still have external gill filaments the vascular supply to the trophonemata is modest (Fig. 19.9A). Secretory crypts alternate with peripherally located vessels covered by a simple cuboidal epithelium (Fig. 13.9B). In later gestation when external filaments are resorbed and respiratory demands of the larger embryos are greater, morphological modifications of the trophonemata become evident. Vascularity increases to a large degree (Fig. 13.9C), peripheral vessels enlarge to sinusoids and the epithelium over the vessels thins to a very attenuated simple squamous epithelium (Figs 13.9C, 13.10D). This effectively reduces the diffusion distance for gas exchange. Snelson and co-workers (Johnson and Snelson 1996; Snelson *et al.* 1988, 1997) presented data on various aspects of the reproductive biology of *Dasyatis sabina* and Hamlett *et al.* (1985d) reported on the ultrastructure of trophonemata in *Rhinoptera bonasus* and on trophonemata, stomach and spiral intestine in *D. americana* (Hamlett *et al.* 1996a, b).

The majority of the secretory cells of the trophonemata are abundantly rich in lipid secretions (Figs 13.9B, D, 13.10B, C) as in *Dasyatis americana*. The term uterolactation (Hamlett *et al.* 1996a) was introduced to refer to production of lipid rich uterine histotroph. The term reinforces the ultrastructural similaries to mammalian lactation. In more evolutionarily basal rays such as *Urobatis jamaicensis* (Fig. 13.10A) (See Chapter **3** of this volume) the uterine milk is considerably more dilute and not as abundant in lipid. The degree of lipid histotrophy is considerably less in *Urobatis* than in *Rhinoptera bonasus* and *D. americana*. Henningsen

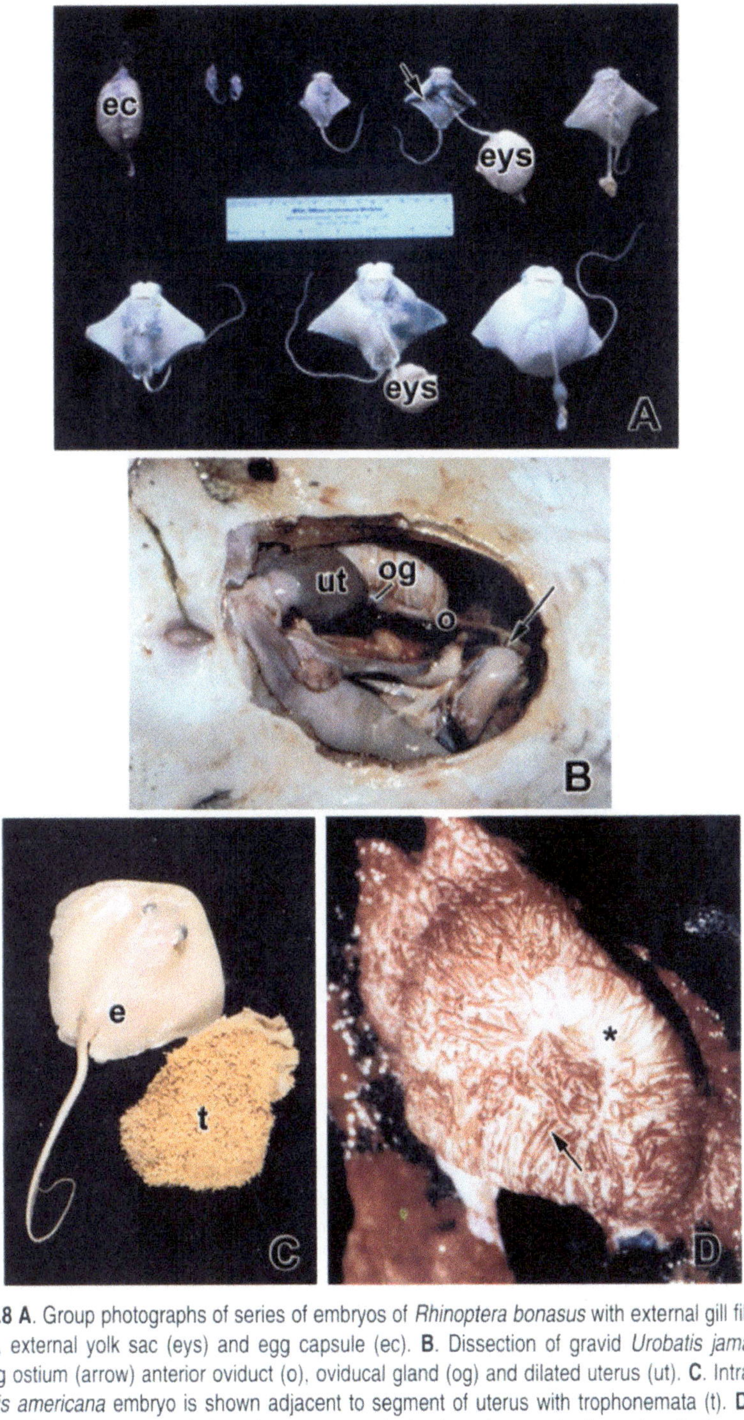

Fig. 13.8 A. Group photographs of series of embryos of *Rhinoptera bonasus* with external gill filaments (arrow), external yolk sac (eys) and egg capsule (ec). **B**. Dissection of gravid *Urobatis jamaicensis* showing ostium (arrow) anterior oviduct (o), oviducal gland (og) and dilated uterus (ut). **C**. Intrauterine *Dasyatis americana* embryo is shown adjacent to segment of uterus with trophonemata (t). **D**. Fresh uterus of *D. americana* depicting elongate, vascularized trophonemata (arrow) actively secreting histotroph (asterisk). Original.

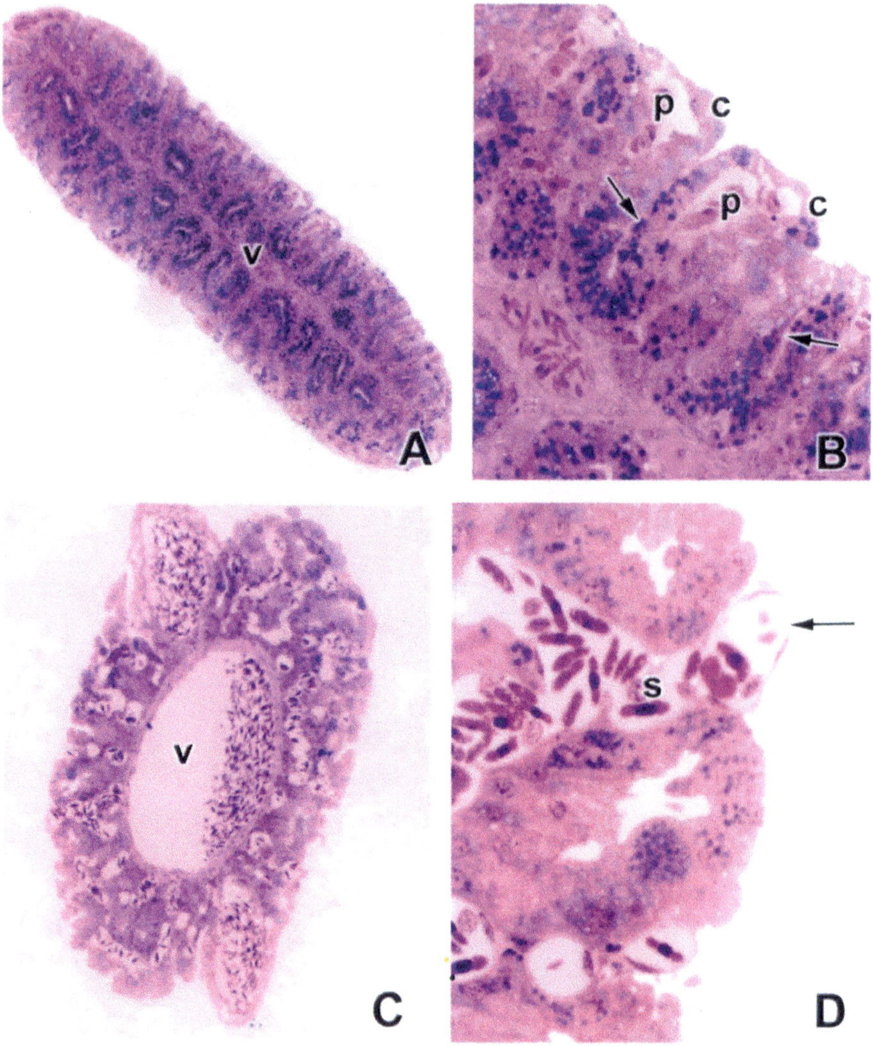

Fig. 13.9 A, B. Light micrographs of early term trophonematum of *Dasyatis americana* showing modest sized central vessel (v), small peripheral vessels (p), secretory crypts (arrows) with lipid (l) and surface cuboidal (c) epithelium atop the peripheral vessels. Fig. 13.9A. = X100, Fig. 13.9B. = X600. **C, D**. Light micrographs of term trophonematum of *D. americana* showing large central vessel (v), similarly dilated peripheral sinusoids (s) and simple squamous epithelium atop the peripheral vessels (arrow). Fig. 13.9C. = X100, Fig. 13.9D. = X600. Original.

(personal communication) measured histotroph protein size and concentration, as well as lipid and fatty acid from three species, *D. americana* (midgestation), *R. bonasus* (late gestation) and *U. jamaicensis* (peripartum). Total lipid in *Dasyatis* was 3 mg/g, 17 in *Rhinoptera* and < 0.5 in *Urobatis*. The sizes of the predominant protein determined via SDS-PAGE were

56.2 kDa, 100.2 kDa and 85.6 kDa for *Dasyatis, Rhinoptera* and *Urobatis* respectively. Predominant proteins in *Dasyatis* and *Rhinoptera* serum were 73.1 kDa and 78.7 kDa, respectively. The protein concentrations of histotroph from *Dasyatis, Rhinoptera* and *Urobatis* were: 20.6 mg/ml, 103.2 mg/ml, and 1.2 mg/ml, respectively. Fractions collected off the gel filtration column yielded a concentration of 39.1 mg/ml and the size of the protein was 55-56 kDa. McKee and Hamlett (unpublished) analyzed the histotroph from *U. jamaicensis* females containing term fetuses. Protein was 9.5 ± 1.7 mg/ml. They did not detect any glucose in the milk.

The predominant lipid class represented in *Dasyatis* histotroph was triglycerides while phospholipids were most abundant in *Rhinoptera* and *Urobatis* histotroph (Henningsen unpublished personal communication). The ratio of saturated to unsaturated fatty acids in *Dasyatis* (1: 1.5) and *Rhinoptera* histotroph (1 : 1.7) were similar and significantly different from histotroph from *Urobatis* (12.5: 1). Differences occurred in protein size and concentration, lipid content and composition, and fatty acid composition in histotroph between species

One of us (WCH) tastes uterine milk of all samples and the milk of *Dasyatis americana* is thick, creamy, sweet and the color and consistency of a vanilla milk shake. *Urobatis jamaicensis* milk is clear, watery and tastes astringent with no sign of sweetness. Production of copious amounts of histotroph can be rapid. When a *D. americana* female was brought on board she aborted her fetuses and when opened, the uterus was bright red all over with no indication of histotroph. The sample was left in the dissecting tray for no more than 5 min, at which time the photograph in Fig. 13.8D was taken showing copious histotroph production.

13.7.2 Uterine specializations in *Dasyatis americana*

The gross gravid uterus of the Southern stingray, *D. americana*, is composed of an inner endometrium or mucosa that forms villous trophonemata, measuring 1.5 cm in length (Figs 13.8C, D). The middle layer is the distensible smooth muscle wall or myometrium which measures 1.0 cm thick when contracted post-partum and fixed in 10% neutral buffered formalin. The outer layer is the serous membrane or perimetrium.

Crypt cells are grouped as secretory acini of up to 50 cells. Light microscopy of trophonemata from uteri containing fertilized eggs reveals prominent lipid crypt cells (Fig. 13.9B). Epithelia cells that overlay vessels are cuboidal while crypt cells are more columnar (Fig. 13.9B. The central vessel is of modest size and the vascular branches are small. The diffusion distance between the endothelium of the vessels and the surface is 3 μm in uteri harboring early term embryos (Hamlett *et al.* 1996a).

In uteri containing term fetuses, the vascular pattern is dramatically changed. The extent of increased vascularity is prominent and vessels at the edge of the trophonemata are large diameter (Fig. 13.9C). Cuboidal epithelium is replaced by squamous epithelium that overlays dilated sinusoids (Fig. 13.9D), thus diminishing the diffusion distance between the fetal vascular bed and the uterine lumen to 0.3 µm near term (Hamlett *et al.* 1996a). Gaseous diffusion can occur across the endothelium of the dilated sinusoid, its basal lamina, a greatly reduced connective tissue zone consisting of sparse collagen fibers, the basal lamina of the surface squamous cell and thin cell process of the surface squamous cell. These squamous areas can thus serve as respiratory foci (Fig. 13.10D).

The squamous nature of epithelial cells that form respiratory foci (Fig. 13.10D) in trophonemata surrounding term fetuses is evident via transmission electron microscopy as they overlay a sinusoid filled with blood cells. The endothelium of the sinusoid is continuous with many marginal folds and has very thin areas where almost all the cytoplasm is excluded and the luminal and basal plasmalemmae nearly approximate each other. The endothelium contains caveolae, mitochondria and a patchy, indistinct basal lamina. A sparse amount of collagen occupies the space between the basal laminae of the endothelium and the surface cells. Pericytes frequently are seen in association with the sinusoids. The barrier between the endothelial lumen and the uterine lumen thus consists of: endothelium, basal lamina, small connective tissue space with pericytes if present, basal lamina of surface cell and surface cell process (Fig. 13.11). Pericytes are stellate cells that encircle capillaries. Their role in the regulation of the diameter of capillaries has not been completely resolved but they are believed to represent reserve cells capable of differentiating into macrophages.

Crypt cells converge on the lumen (Fig. 13.9B) and are joined by contiguous luminal tight junctions. Secretory cells contain an elaborate rough endoplasmic reticulum, Golgi, abundant lipid droplets (Figs 13.10B, C, 13.11) and some dark staining membrane limited proteinaceous secretory granules. Fusion and exocytosis may occur very rapidly. All trophonemata may not be secreting simultaneously, that is, some regions may be actively elaborating secretory product while other areas are quiescent. Similar ultrastructural characteristics of trophonemata from females near term *Rhinoptera bonasus* have been reported (Hamlett *et al.* 1985d).

There is variability in the composition of uterine milk from *Dasyatis* and *Urobatis*, as previously mentioned. When secretory crypt cells from animals near term of both species are compared ultrastructurally, *Urobatis* (Fig. 13.10A) has dramatically fewer lipid droplets than *Dasyatis* (Fig. 13.10B, C). Both species, however, show a roughly 3,000 fold increase in wet weight during development. A study is in progress (Hamlett *et al.*) that examines the biochemical composition of histotroph and trophonemata structure through out gestation. One of the goals is to determine how *Urobatis* achieves roughly the same wet weight gain as *Dasyatis* although possessing a much more dilute histotroph.

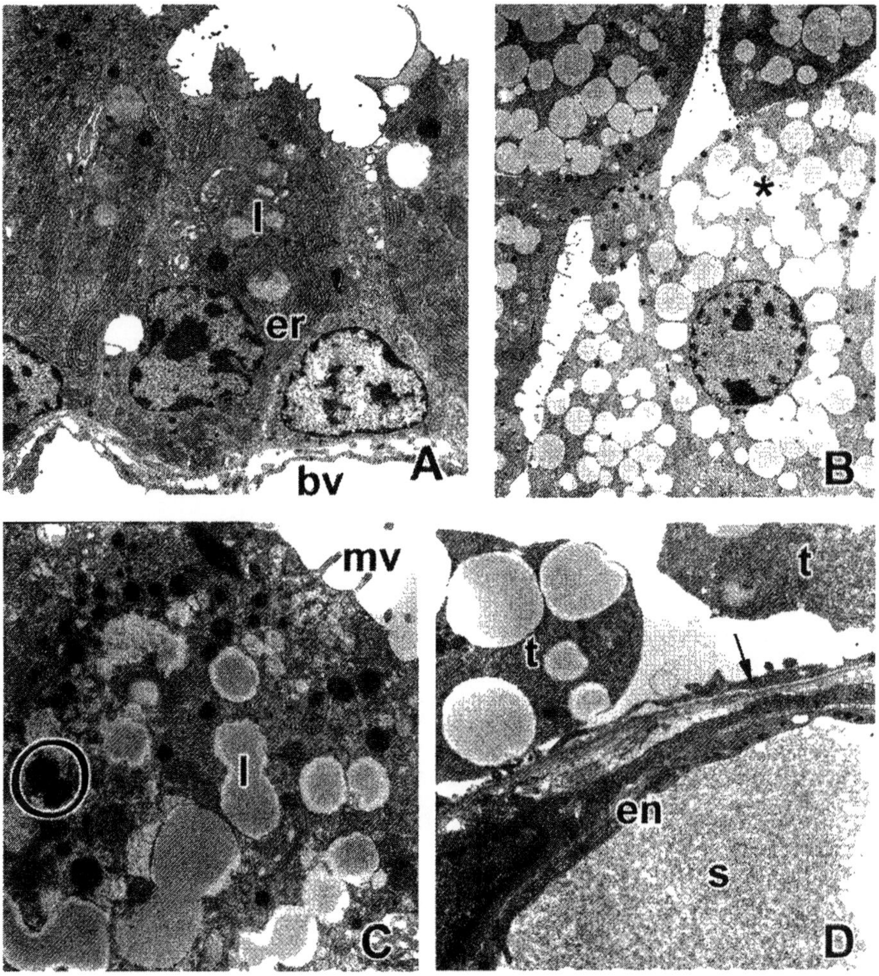

Fig. 13.10 A. Transmission electron micrograph of trophonematum from uterus of *Urobatis jamaicensis* containing near term fetuses. Characteristics include close proximity of blood vessels (bv) to crypt epithelium with abundant rough endoplasmic reticulum (er) and scant lipid (l). X3,000. Original. **B**. Transmission electron micrograph of trophonematum from uterus of *Dasyatis americana* containing near term fetuses is dominated by the abundant lipid content (asterisk). 4,300X. From Hamlett, W. C., Musick, J. A., Eulitt, A. M., Jarrell, R. L. and Kelly, M. A. 1996a. Canadian Journal of Zoology 74: 1417-1430, Fig. 10. **C.** Transmission electron micrograph of trophonematum from uterus of *Dasyatis americana* containing near term fetuses is characterized by microvilli (mv), prominent lipid (l) and miscellaneous secretory vesicles (circle). X8,000. Original. **D**. Transmission electron micrograph of trophonematum from uterus of *Dasyatis americana* containing near term fetuses is characterized by dilated peripheral sinusoids (s) delimited by endothelia (en) and covered by simple squamous epithelium (arrow) Adjacent secretory cells (t). X8,700. From Hamlett, W. C., Musick, J. A., Eulitt, A. M., Jarrell, R. L. and Kelly, M. A. 1996a. Canadian Journal of Zoology 74: 1417-1430, Fig. 14.

A composite line drawing (Fig. 13.11) derived from transmission electron micrographs depicts the situation in uteri of *D. americana* containing fertilized eggs on the left and uteri containing term fetuses on the right. The most prominent features in early gestation are cuboidal epithelium over small diameter vessels. In later gestation, peripheral vasculature increases in size to sinusoids and the epithelium over vessels thin to squamous to function as respiratory foci. The remainder of the surface and crypt cells continue to function in secretion.

Gudger (1912) proposed that early myliobatoid embryos absorbed histotroph through external branchial filaments. Using India ink, Ranzi (1934) showed that early term embryos of *Dasyatis violacea* absorbed histotroph through the yolk sac and external gill filaments. He also demonstrated that later term embryos ingested histotroph orally. The

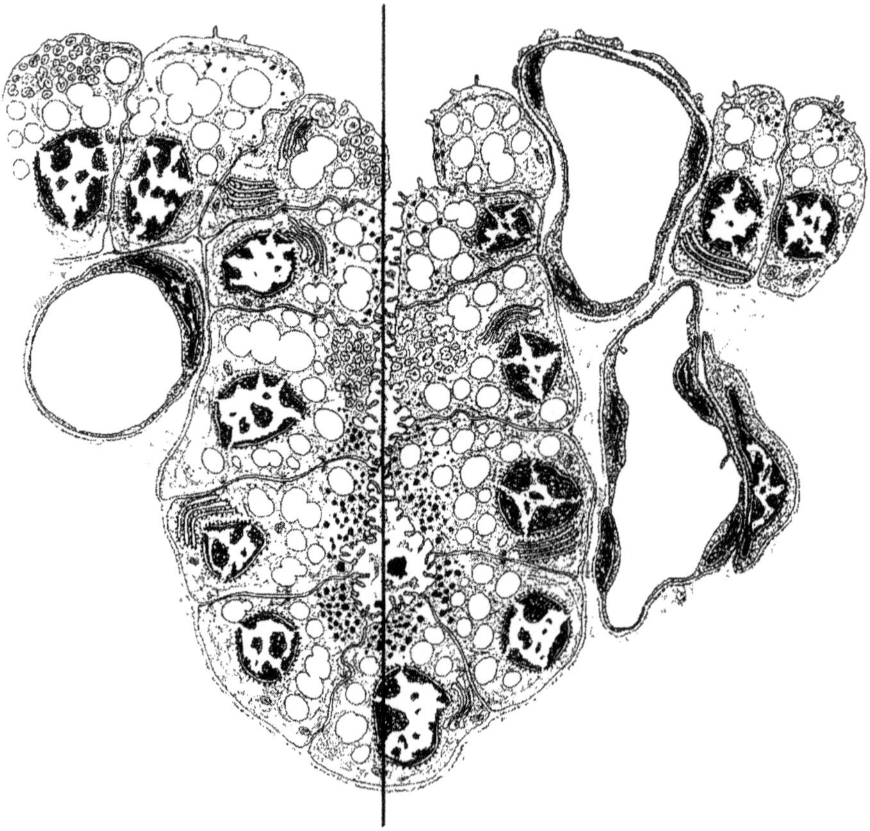

Fig. 13.11 Composite line drawing illustrating a uterus containing fertilized eggs on the left, characterized by cuboidal surface epithelium and capillaries. The right side shows a uterus containing term fetuses, which is characterized by simple squamous epithelium overlying dilated sinusoids. From Hamlett, W. C., Musick, J. A., Eulitt, A. M., Jarrell, R. L. and Kelly, M. A. 1996a. Canadian Journal of Zoology 74: 1417-1430, Fig. 17.

ability of external branchial filaments to endocytose the protein marker, horseradish peroxidase has been demonstrated in the preimplantation stage of the shark, *Rhizoprionodon terraenovae* (Hamlett *et al.* 1985e).

In *D. americana* distinct, direct vascular maternal-fetal connection is lacking but the muscular uterus closely envelope the young and consequently trophonemata are brought into physical contact with the embryo. This is probably the basis for reports of trophonemata entering the spiracles of embryos (Wood-Mason and Alcock 1891; Bearden 1959; Hess 1959). Smith (1980) noticed trophonemata occasionally entering the gill slits or mouth but never the spiracles of *R. bonasus*.

In his description of trophonematal secretion, Ranzi (1934) indicates secretion begins when a leucocyte enters a trophonematal epithelial cell and then disintegrates. Following this, lipid and protein granules are secreted. No evidence of this process was observed in *D. americana* by Hamlett *et al.* (1996a). Ranzi's observations may have been based on fixation artifacts or what he interpreted as a leucocyte within a cell may have been a capillary containing a leucocyte between ridges or an intraepithelial leucocyte.

Hamlett *et al.* (1996a) introduced the term uterolactation to refer to the production of nutrient histotroph by uterine tissues. The process of synthesis and secretion of milk is reminiscent of mammary gland function in other vertebrates. In addition to the nutrient role of trophonemata, the structural remodeling that occurs during gestation provides a mechanism to increase the amount of oxygenated blood reaching the uterus. The thinning of the layers separating maternal blood from the fetus contributes to respiratory efficiency. Viviparous animals acquire oxygen from the maternal vascular system. Oxygen uptake is enhanced by a fetal hemoglobin with higher oxygen affinity than the hemoglobin of the mother. Three mechanisms have been reported to account for the higher oxygen affinity in the fetus (Ingermann 1992). The fetus and adult may have structurally identical hemoglobins but have different types or amounts of erythrocytic organic phosphates which reversibly binds to deoxygenated hemoglobin thereby lowering oxygen affinity (Bonaventura and Bonaventura 1980). Fetal and adult hemoglobins can possess intrinsically different oxygen affinities resulting from structural differences in the hemoglobin molecule. Such intrinsic differences have been found in viviparous animals such as the teleost, *Zoarces viviparous* (Weber and Hartvig 1984). Possessing a higher affinity hemoglobin than the mother would be adaptive to the fetus. Manwell (1958) reported fetal hemoglobin in the skate, *Raja binoculata*, to be slightly higher in affinity than adult hemoglobin. King (1994) reported that fetal hemoglobin had a higher intrinsic oxygen affinity than adults in the oviparous swell shark, *Cephaloscyllium ventriosum*, perhaps as a means of more efficiently extracting oxygen from the relatively stagnant environment within the egg case. A similar situation might be expected in *D. americana* where the aqueous environment surrounding the fetus is a nutrient fluid that must also serve as the medium through which gas exchange is effected. A higher oxygen affinity of fetal hemoglobin would enable the fetus to more efficiently extract oxygen from the maternal vasculature.

13.7.3 Structure of the Stomach and Spiral Intestine in *Dasyatis americana*

In mammals, separate gastric cells produce hydrochloric acid, viz. parietal or oxyntic, and pepsinogen, viz. chief or zymogenic. In many non-mammalian vertebrates, secretion of hydrochloric acid and pepsinogen in the gastric mucosa is performed by a single cell, the oxynticopeptic cell (Ito and Winchester 1967). These cells have an elaborate tubulo-vesicular system (Forte *et al.* 1972).

In *D. americana* (Hamlett *et al.* 1996b) oxynticopeptic cells dominate the crypt of the gastric glands. They have elongate surface microvilli that project into the crypt lumen and round, smooth walled vesicles with homogenous contents subjacent to microvilli. Mitochondria, free ribosomes and Golgi also occur in this region. The apical cytoplasm contains an abundant system of randomly oriented smooth walled tubulovesicles. Lateral cell boundaries interdigitate modestly. The nucleus is ovoid with an occasional indentation. Heterochromatin is present peripherally and as nucleoplasmic clumps. There is a single nucleolus. Oxynticopeptic cells rest on a thin basal lamina which is supported by collagen and connective tissue fibroblasts. Tight junctions seal adjacent cells at the lumen.

Enteroendocrine cells are encountered between oxynticopeptic cells. They are characterized by basal secretion granules whose profile varies from spherical to dumbbell shaped. Rough endoplasmic reticulum and mitochondria are found in proximity to the secretion vesicles. Enteroendocrine cells do not contact the lumen but deliver their secretory products to the basal compartment below the cell.

Grabowski *et al.* (1995) have described oxynticopeptic cells from the stomach of the river ray, *Potamotrygon* sp. Gastric glands were found exclusively in the cardiac and fundic regions, not the pylorus. Lining epithelium was almost exclusively oxynticopeptic cells. A few mucous neck cells and a variety of enteroendocrine cells were also present. In all cases the oxynticopeptic cells were characterized by membrane-bound granules and a tubulovesicular system. Oxynticopeptic cell morphology correlated with the region of the stomach. In the proximal region, rough endoplasmic reticulum and secretory granules predominated whereas in the distal region, tubulovesicles and mitochondria were dominant. These features are characteristic of acid-secreting cells. These findings suggest a gradient of proteolytic enzymes and hydrochloric acid along the proximal-distal axis of the stomach in these freshwater rays.

The spiral or valvular intestine in *D. americana* is the ring type (White 1937, cf. Compagno 1988) which resembles a stack of washers when the intestine is sectioned longitudinally. The anterior six rows of the valve are arranged transversely and are distinctly stout and bushy. The posterior fourteen rows are broad and thin and form a bib-like flap that is much wider than the intestinal lumen. There is more space between rows and the majority of intestinal contents are located between these rows. Transmission electron microscopy of *D. americana* fetal intestine reveals that adjacent epithelial cells are joined by a classic tripartite junctional complex of zonula occludens, zonula adherens and macula adherens. Subjacent to the surface microvilli are an abundant population of smooth walled tubules and vesicles, as well as a heterogeneous assemblage of small, electron dense endocytotic vesicles (Figs 13.12A-D). Lateral cell boundaries are close, fairly straight with little intervening space. Small endocytic vesicles coalesce to form enormous storage vesicles whose contents are generally electron dense with spherical light staining regions and their contents are indented peripherally. Smaller vesicles bud off and are transferred to the lysosomal system.

Babel (1967) reported a concentrated mass of fecal material present in the spiral intestine of the fetal *Urobatis halleri*, and that this mass gradually dissolved following parturition. A similar situation probably exists in all rays. This, along with the enormous storage vesicles in the spiral intestine epithelium of *D. americana*, provides a mechanism to insure that fecal material is not produced that would foul the uterine environment.

The presence of an extensive system of apical tubules and invaginations and enormous supranuclear vacuoles in the ray intestine is similar to the jejunum and ileum in neonatal calves (Staley *et al.* 1972), the small intestine in the suckling rat (Shervey 1966, Cornell and Padykula 1969) and the duodenum in the newborn mouse (Hugon 1971). In certain newborn mammals, including the rat, lamb, goat, dog, cat and pig, the small intestine is transiently capable of absorbing intact proteins (Brambell 1958). This is the major method of transmitting passive immunity from the mother to offspring. In newborn animals, maternal antibodies present in milk are absorbed by the distal small intestine. Lesser amounts of maternal antibodies are received transplacentally (Brambell 1958). In the postnatal rat, this capability is gradually lost as the rate of antibody absorption falls from days 1 to 18 and is completely lost at 21 days (Halliday 1955). In man, nearly all passive immunity is imparted transplacentally but antibody absorption can occur in breast-fed infants (Leissring *et al.* 1962).

In the fetal *D. americana*, both the stomach and intestine function after yolk sac contents have been depleted. In fetuses the intestine is stained dark green by bile. These organs digest and absorb uterine secretions to provide for growth and development of the embryo and fetus. The similarity of the endocytotic vesicles in the intestine to that of suckling mammals suggests that uterine milk may be a source of immunoglobulins.

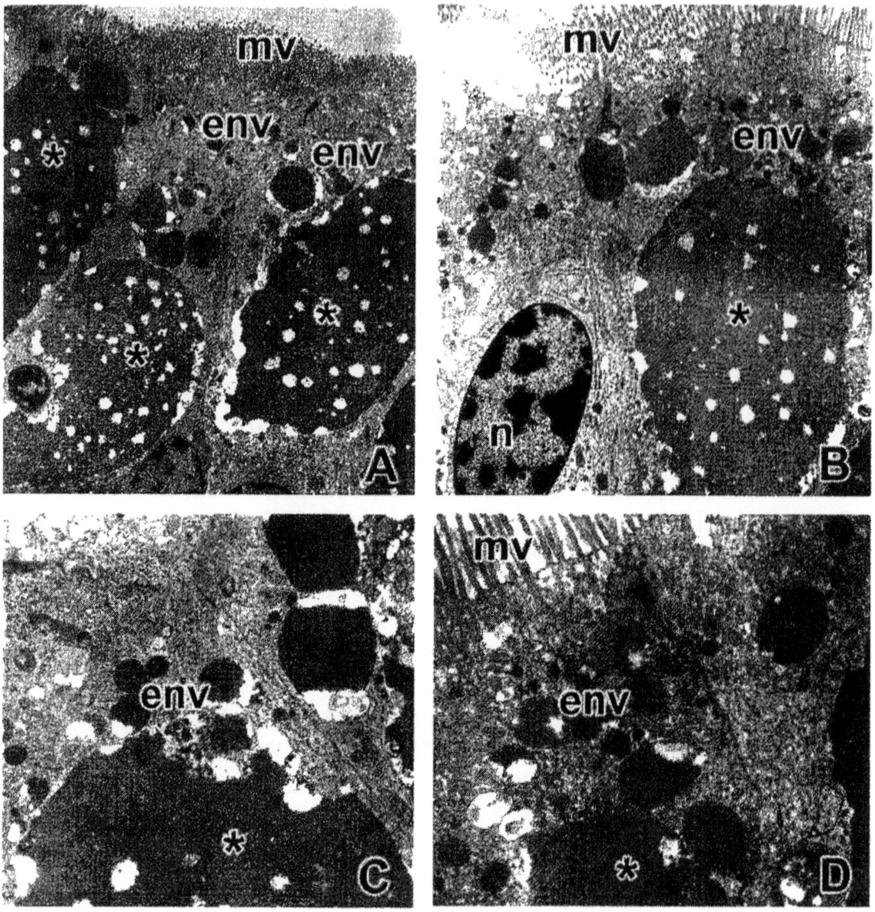

Fig. 13.12 A-D. Transmission electron micrographs of columnar absorptive cells of spiral intestine from embryo of *Dasyatis americana* with microvilli (mv), small apical endocytotic vesicles (env), enormous supranuclear storage vesicles and nuclei (n). Fig. 13.12A. = 2,000X, Fig. 13.12B. = X3,500, Fig. 13.12C. = X6,000, Fig. 13.12C. = X10,000. Original.

13.8 MODIFICATION OF PARITY THEMES IN *Gollum attenuatus* AND *Squatina*

In this chapter we discuss lecithotrophy and matrotrophy and developmental mechanisms involved in them. Placentatrophy is discussed in Chapter **15** and oophagy and intrauterine cannibalism in Chapter **14** of this volume. A brief description of reproduction in the Pseudotrikidae and a comment on *Squatina* species will be given here.

The family Pseudotriakidae includes *Gollum attenuatus* and *Pseudotriakis microdon*. Yano (1993) presented a thorough description of reproductive biology in the slender smoothhound, *G. attenuatus* and *P. microdon* (Yano 1992). In *Gollum* only the right ovary is functional and the uteri contain villi whose length increase during gestation as a transparent liquid accumulates in the uteri of gravid females. Only a single egg capsule occurs in each uterus. It contains a single embryo and 30-80 ova that are 4-8 mm in diameter. The undeveloped ova mostly disintegrate and provide exogenous yolk for the developing embryo. Thus the embryo utilizes yolk in its own external yolk sac and also accumulates ingested yolk in its external yolk sac, not its stomach as in Lamnids. Following eclosion from the egg capsule the embryo develops by using yolk stores both of endogenous (yolk sac) and exogenous (mixed yolk from ova within the same capsule) origin. *Pseudotriakis* is also an egg eater and stores consumed ova in its external yolk sac (Yano 1992). The role of uterine fluid is unresolved but some degree of histotrophy is suspected. A fruitful study would be biochemical analysis of uterine fluid throughout gestation as well as an ultrastructural analysis of the embryonic gut.

Sunye and Vooren (1997) presented observations of what they term as "cloacal gestation" in *Squatina guggenheim* and *S. occulta*. According to them, embryos occupy the uteri for the first 4 months of their 11 month gestation. Initially there is a single egg candle in each uterus containing several ova. Embryos break out of the candle to complete gestation free of their candle. At mid-gestation the embryos drop into a single heart-shaped "uterine-cloacal" chamber where they complete development. The authors describe the uterine portion as highly vascularized but the cloacal component as poorly vascularized and flaccid. They state that the uterine-cloacal chamber communicates with the external environment through the cloacal vent. They conclude that the mother spends much of the time lying still on sandy or muddy bottoms but rise periodically to facilitate flushing of the uterine-cloacal chamber with seawater. Further, the authors state that cloacal gestation does not require the production of histotroph but does necessitate mechanisms to flush the uterine-cloacal chamber. The observations by Sunye and Vooren (1997) in *S. guggenheim* and *S. occulta* are not in agreement with the accepted reproductive mode of yolk sac viviparity in other *Squatina* species (Nathanson and Cailliet 1986). Clearly the system in *S. guggenheim* and *S. occulta* requires verification. Histological study of the uterus and cloaca throughout gestation would provide further insight into this mode of development.

13.9 CONCLUSIONS

The majority of Chondrichthyes have yolk sac viviparity where endogenous yolk provides the majority of nutrients but some degree of histotrophy is probably expressed in all viviparous species. The line between strictly lecithotrophic yolk sac viviparity and incipient histotrophy is still unresolved and in a state of flux, especially for species that show little weight loss or gain. Critical analysis requires separate dry weight determinations of water, ash and organic content. A general trend from mucous to lipid histotrophy occurs with the latter only present in the Myliobatiforms. Morphological modifications of the uterus include: increased vascularity, thinning of the number of cell layers separating maternal and fetal blood to reduce diffusion distance, development of spaces for provision of water and minerals to the embryo and increased uterine secretory activity.

13.10 ACKNOWLEDGEMENTS

This work was supported by funds provided by Indiana University School of Medicine (WCH) and the Fisheries Research and Development Corporation in Australia (TIW). We wish to express gratitude to Joan Clark, University of Melbourne, for technical expertise in electron microscopy.

13.11 LITERATURE CITED

Anderson, P. M. 1980. Glutamine and N-acetylglutamate-dependent carbamoyl phosphate synthetase in elasmobranchs. Science 208: 291-293.

Babel, J. S. 1967. Reproduction, life history, and ecology of the round stingray, *Urolophus halleri* Cooper. California Fish and Game Bulletin 137: 1-104.

Bearden, C. 1959. A life history study of the eagle ray, *Myliobatis freminvilli* Leseur 1824, in Delaware Bay. M. S. Thesis, University of Delaware, Dover, Delaware.

Berridge, M. J., and J. L. Oschman (1972) *Transporting epithelia.*, Academic Press, New York, N. Y.

Bonaventura, C. and Bonaventura, J. 1980. Anionic control of function in vertebrate hemoglobins. American Zoologist 20: 131-138.

Brambell, F. W. R. 1958. The passive immunity of the young mammal. Biological Review 33: 488-531.

Burger, J. W. 1967. Problems in the electrolyte economy of the spiny dogfish, *Squalus acanthias*. Pp. 177-185. In P. W. Gilbert, R. F. Matheson and D. Rall (eds), *Sharks, Skates and Rays*. The Johns Hopkins University Press, Baltimore, MD.

Compagno, L. J. V. 1988. *Sharks of the Order Carcharhiniformes*. Princeton University Press, Princeton, N.J., 486 pp.

Cornell, R., and Padykula, H. A. 1969. A cytological study of intestinal absorption in the suckling rat. American Journal of Anatomy 125: 291-316.

Capapé, C., 1993. New data on the reproductive biology of the thorny stingray, *Dasyatis centroura* (Mitchill, 1815), from off the Tunisian coasts. Environmental Biology of Fishes, 38: 73-80.

Capapé, C., Quignard J. P. and Mellinger, J. 1990. Reproduction and development of two angel sharks, *Squatina squatina* and *S. oculata* (Pisces: Squatinidæ), off Tunisian coasts: semi-delayed vitellogenesis, lack of egg-capsules and lecithotrophy. Journal of Fish Biology 37: 347-356.

Capapé, C., Seck, A. A. and Diatta, Y. 2000a. Reproductive biology of the common torpedo, *Torpedo torpedo* (Linnaeus, 1758) from the coast of Senegal. Miscellaneous Zoology 23 (1): 9-21.

Capapé, C., Seck, A. A., Diatta, Y. and Diop, M. 2000b. Observations on the reproductive biology of *Torpedo (Tetronarce) mackayana* from off the coast of Senegal (Eastern Tropical Atlantic). Cybium 25 (1): 95-99.

Capapé C., Seck, A. A., Gueye-Ndiaye A., Diatta, Y. and Diop, M. 2002. Reproductive biology of the smooth angelshark, *Squatina oculata* (Elasmobranchii: Squatinidae), from the coast of Senegal (eastern tropical Atlantic). Journal of the Marine Biological Association of the United Kingdom. 82: 635-640.

Capapé, C., Seck, A. A. and Quignard, J. P. 1999. Aspects of the reproductive biology of the angular rough shark, *Oxynotus centrina* (Oxynotidae). Cybium, 23 (3): 259-271.

Chan, D. K. O. and Phillips, J. G. 1966. The embryology of the rectal gland of the spiny dogfish *Squalus acanthias*. L. Journal of Anatomy 100: 899-903.

Delhaye, E., Lechenault, H., Wriez, F. Leray, C., Haye, B. and Mellinger, J. 1992. Localisation, composition et utilisation del lipids vitellins chez *Scyliorhinus canicula* (L). Bulletin of the Zoological Society of France 117: 149-156.

Diamond, J. M. and W. H. Bossert, W. H. 1967. Standing-gradient osmotic flow. A mechanism for coupling of water and solute transport in epithelia. Journal of General Physiology 50: 2061-2083.

Diamond, J. M. and Bossert, W. H. 1968. Functional consequences of ultrastructural geometry in "backwards" fluid-transporting epithelia. Journal of Cell Biology 37: 694-702.

Diamond, J. M. and Tormey, J. 1966a. Role of long extracellular channels in fluid transport across epithelia. Nature (London) 210: 817-820.

Diamond, J. M. and Tormey, J. 1966b. Studies on the structural basis of water transport across epithelial membranes. Federation Proceedings of the Federation of American Society of Experimental Biology 25: 1458-1463.

Evans, D. H., Oikari, A., Kormanik, G. A. and Mansberger, L. 1982. Osmoregulation by the prenatal spiny dogfish, *Squalus acanthias*. Journal of Experimental Biology 101: 295-305.

Fange, R., Lundblad, G., Lind, J., and Slettengren, K. 1979. Chitinolytic enzymes in the digestive system of marine fishes. Marine Biology 53: 317-321.

Files, T. and Lombardi, J. 1993. Free amino acids in the uterine fluids of four species of viviparous sharks (*Squalus acanthias, Carcharhinus plumbeus, Mustelus canis and Rhizoprionodon terraenovae*). Comparative Biochemistry and Physiology 104B, 3: 583-588.

Flugel, C. and Lutjen-Drecoll, E. 1991. Distribution of carbonic anhydrase in the uterus of late-term pregnant spiny dogfish (*Squalus acanthias*). Journal of Experimental Biology 158: 531-537.

Forte, J. G., Forte, T. M., and Ray, T. K. 1972. Membranes of the oxyntic cell: their structure, composition and genesis. Pp. 37-49. In G. Sachs (ed.), *Gastric Secretion*. Academic Press, New York.

Fuller, E. O., Greindling, K. K. and Kent, B. 1983. Mechanisms and biochemistry of arterial segments from the pregnant dogfish. Bulletin Mount Desert Island Biological Laboratory 23: 19-22.

Goldstein, L. and Pallat, P. J. 1974. Trimethylamine oxide excretion rates in elasmobranchs. American Journal of Physiology 227: 1268-1272.

Goldstein, L. and Perlman, F. 1995. Nitrogen metabolism, excretion, osmoregulation and cell volume regulation in elasmobranchs. Pp. 91-104. In P. J. Walsh and P. Wright (eds), *Nitrogen Metabolism and Excretion*. CRC Press, Inc., Boca Raton, FL.

Grabowski, G. M., Luciano, L., Lacy, E. R., and Reale E. 1995. Morphologic variations of oxynticopeptic cells in the stomach of the river ray, *Potamotrygon* sp. Journal of Aquariculture and Aquatic Science 7: 38-44.

Guallart J. and Vicent, J. J. 2001. Changes in composition during embryo development of the gulper shark, *Centrophorus granulosus* (Elasmobranchii,

Centrophoridae): an assessment of maternal-embryonic nutritional relationships. Environmental Biology of Fishes 61: 135-150.

Gudger, E. W. 1912. Natural history of some Beaufort, N.C. fishes, 1910-1911. No. I. Elasmobranchii–with special reference to uterogestation. Proceedings of the Biological Society of Washington 25: 141-156.

Halliday, R. 1955. The absorption of antibodies from immune sera by the gut of the young rat. Procceedings of the Royal Society of London B, 143: 408-413.

Hamlett, W. C. 1993. Ontogeny of the umbilical cord and placenta in the Atlantic sharpnose shark, *Rhizoprionodon terraenovae*. Environmental Biology of Fishes 38: 253-267.

Hamlett, W. C. and Wourms, J. P. 1984. Ultrastructure of the preimplantation shark yolk sac placenta. Tissue and Cell 16: 613-625.

Hamlett, W. C., Wourms, J. P., and Hudson, J. S. 1985a. Ultrastructure of the full term shark yolk sac placenta. I. Morphology and cellular transport at the fetal attachment site. Journal of Ultrastructure Research. 91: 192-206.

Hamlett, W. C., Wourms, J. P., and Hudson, J. S. 1985b. Ultrastructure of the full term shark yolk sac placenta. II. The smooth, proximal segment. Journal of Ultrastructure Research. 91: 207-220.

Hamlett, W. C., Wourms, J. P., and Hudson, J. S. 1985c. Ultrastructure of the full term shark yolk sac placenta. III. The maternal attachment site. Journal of Ultrastructure Research. Res. 91: 221-231.

Hamlett, W. C., Wourms, J. P. and Smith, J. W.. 1985d. Stingray placental analogues: structure of trophonemata in *Rhinoptera bonasus*. Journal of Submicroscopic Cytology. 17: 541-550.

Hamlett, W. C., Allen, D. J., Stribling, M. D., Schwartz, F. J. and DiDio, L. J. A. 1985e. Permeability of embryonic shark external gill filaments. Electron microscopic observations using horseradish peroxidase as a macromolecular tracer. Journal of Submicroscopic Cytology. 17: 31-40.

Hamlett, W. C., Schwartz, F. J. and DiDio, L. J. A. 1987. Subcellular organization of the yolk syncytial-endoderm complex in the preimplantation yolk Sac of the shark, *Rhizoprionodon terraenovae*. Cell and Tissue Research 247: 275-285.

Hamlett, W. C., Eulitt, A. M. Jarrell, R. L. and Kelly, M. A. 1993a. Uterogestation and placentation in elasmobranchs. Journal of Experimental Zoology. 266: 347-367.

Hamlett, W.C., Miglino, M.A. and DiDio, L.J.A. 1993b. Subcellular organization of the placenta in the Atlantic sharpnose shark, *Rhizoprionodon terraenovae*. Journal of Submicroscopic Cytology and Pathology. 25: 535-545.

Hamlett, W. C., Musick, J. A., Eulitt, A. M., Jarrell, R. L. and Kelly, M. A. 1996a. Ultrastructure of uterine trophonemata, accommodation for uterolactation and gas exchange in the Southern stingray, *Dasyatis americana*. Canadian Journal of Zoology 74: 1417-1430.

Hamlett, W. C., Musick, J. A., Eulitt, A. M., Jarrell, R. L. and Kelly, M. A. 1996b. Ultrastructure of fetal alimentary organs: Stomach and spiral intestine in the Southern stingray, *Dasyatis americana*. Canadian Journal of Zoology 74: 1431-1443.

Hamlett,W. C. and Hysell, M. K. 1998. Uterine specializations in elasmobranchs. Journal of Experimental Zoology 282: 438-459.

Hamlett, W. C., Musick, J. A., Hysell, C. K. and Sever, D. M. 2002. Uterine epithelial-sperm interaction, endometrial cycle and sperm storage in the terminal zone of the oviducal gland in the placental smoothhound, *Mustelus canis*. Journal of Experimental Zoology 292: 129-144.

Hess, P. W. 1959. The biology of the stingrays, *Dasyatis centroura* Mitchill 1815 and *Dasyatis sayi* Lesuer 1817 in Delaware Bay. M.S. thesis, University of Delaware, Dover, Delaware.

Hisaw, F. L. and Albert, A. 1947. Observations on the reproduction of the spiny dogfish, *Squalus acanthias*. Biological Bulletin Marine Biological Laboratory Woods Hole 92: 187-199.

Hugon, J. S. 1971. Absorption of horse-radish peroxidase by the mucosal cells of the duodenum of mouse. II. The newborn mouse. Histochemie 26: 19-27.

Ingermann, R. L. 1992. Maternal-fetal oxygen transfer in lower vertebrates. American Zoologist. 32: 322-330.

Ito, S., and Winchester, S. 1967. The fine structure of the gastric mucosa in the bat. Journal of Cell Biology. 16: 541-577.

Jollie W. P. and Jollie, L. G. 1967a. Electron microscopic observations on accommodations to pregnancy in the uterus of the spiny dogfish, *Squalus acanthias*. Journal of Ultrastructure Research 20: 161-178.

Jollie W. P. and Jollie, L. G. 1967b. Electron microscopic observations on the yolk sac of the spiny dogfish, *Squalus acanthias*. Journal of Ultrastructure Research 18: 102-126.

Jones, T. S. and Ugland, K. I. 2001. Reproduction of female spiny dogfish in the Oslo fjord. Fishery Bulletin Seattle 99(4): 685-690.

Johnson, M. R. and Snelson, F. F. Jr. 1996. Reproductive life history of the Atlantic stingray, *Dasyatis sabina* (Pisces, Dasyatidae) in the freshwater St. Johns River, Florida. Marine Science 59: 74-88.

Kaye, G. I. , Wheeler, H. O.,Whitlock, R. T. and Lane, N. 1966. Fluid transport in the rabbit gallbladder. A combined physiological and electron microscopic study. Journal of Cell Biology 30: 237-268.

King, L. A. 1994. Adult and fetal hemoglobins in the oviparous swell shark, *Cephaloscyllium ventriosum*. Comparative Biochemistry and Physiology. 109B: 237-243.

Koob, T. J. and Cox, D. L. 1988. Egg capsule catechol oxidase from the little skate *Raja erinacea* Mitchill, 1825. Biological Bulletin, Woods Hole 175: 202-211.

Koob, T. J. and Cox, D. L. 1990. Introduction and oxidation of catechols during the formation of the skate (*Raja erinacea*) egg capsule. Journal of the Marine Biological Association of the United Kingdom 70: 395-411.

Koob, T. J., Laffan, J. L. and Callard, I. P. 1981. Egg-oviduct size relationships in *Raja erinacea*. Bulletin Mount Desert Island Biological Laboratory 21: 46-48.

Koob, T. J and Hamlett, W. C. 1998. Microscopic structure of the gravid uterus in the little skate, *Raja erinacea*. The Journal of Experimental Zoology 282: 421-437.

Kormanik, G. A. 1988. Time course of the establishment of uterine seawater conditions in late-term pregnant spiny dogfish (*Squalus acanthias*). Journal of Experimental Biology 137: 443-456.

Kormanik, G. A. 1989. Nitrogen budget in developing embryos of the spiny dogfish *Squalus acanthias*. Journal of Experimental Biology 144: 583-587.

Kormanik, G. 1992. Ion and osmoregulation in prenatal elasmobranchs: evolutionary implications. American Zoologist 32: 294-302.

Kormanik, G. A., Billings, C., MacKichan, C. and Leach, T. 1999. Glutamine loading and nitrogen excretion in *Squalus acanthias*. Bulletin Mount Desert Island Biological Laboratory 38: 77-78.

Kormanik, G. A. and Evans, D. H. 1986. The acid-base status of prenatal pups of the dogfish, *Squalus acanthias*, in the uterine environment. Journal of Experimental Biology 125: 173-179.

Kormanik, G. A. and Kremer, W. 1986. Effect of acetazolamide on the establishment of uterine sea water conditions in pregnant dogfish (*Squalus acanthias*). Bulletin Mount Desert Island Biological Laboratory 26: 142-144.

Kormanik, G. A., Lofton, A. and O'Leary-Liu, N. 1992. Nitrogen budget in developing elasmobranch embryos. Bulletin Mount Desert Island Biological Laboratory 31: 44-46.

Kormanik, G. A., Lofton, A. J. and Vibbard, D. E. 1991. The ontogeny of mitochondria-rich cells in embryos of the spiny dogfish, *Squalus acanthias*. Bulletin Mount Desert Island Biological Laboratory 30: 4-7.

Kormanik, G. A. and Maren, T. H. 1988. Carbonic anhydrase in the uterine sea water acidification process in *Squalus acanthias:* Localization and the effect of various inhibitors. Bulletin Mount Desert Island Biological Laboratory 27: 25-27.

Kormanik, G. A. and Totten, R. K. 1993. Total nitrogen, trimethylamine n-oxide and urea in developing embryos of the little skate, *Leucoraja erinacea*. Bulletin Mount Desert Island Biological Laboratory 32: 9-10.

Kormanik, G. A. and Verity, R. 1995. Effect of elevated plasma ammonia on tissue levels of glutamine and glutamate in embryos of the dogfish (*Squalus acanthias*). Bulletin Mount Desert Island Biological Laboratory 34: 92-93.

Leissring, J. C., Anderson, J. W., and Smith, D. W. 1962. Uptake of antibodies by the intestine of the new-born infant. American Journal of Disorders of the Child 103: 160-165.

Lechenault, H., Wriez, F. and Mellinger, J. 1993. Yolk utilization in *Scyliorhinus canicula*, an oviparous dogfish. Environmental Biology of Fishes 38: 241-252.

Lombardi, J., Jones, K. B., Garrity, C. A. and Files, T. 1993. Chemical composition of uterine fluid in four species of viviparous sharks (*Squalus acanthias, Carcharhinus plumbeus, Mustelus canis and Rhizoprionodon terraenovae*). Comparative Biochemistry and Physiology 105A, 1: 91-102.

Manwell, C. 1958. Ontogeny of hemoglobin in the skate *Raja binoculata*. Science 128: 419-420.

Mellinger, J. Wriez, F. and Alluchon-Gérard, M.-J. 1986. Developmental biology of the oviparous shark, *Scyliorhinus canicula*. Pp. 310-332. In T. Uyeno, R. Arai, T. Taniuchi and K. Matsuura (eds), *Indo-Pacific Fish Biology*. Ichthyological Society of Japan, Tokyo.

Mellinger, J. and Wriez, F. 1989. Biologie et physiologie compares du développement de deux Sélachiens ovipares, les rousettes *Scyliorhinus canicula* et *Scyliorhinus stellaris*. Evolution de la matière sèche, de l'eau et des ions (Cl^-, NA+, K+) dans le vitellus de *S. canicula* au cours du développement. Bulletin de la Société Zoologique de France. 114: 51-62.

Mossman, H. W. 1987. *Vertebrate fetal membranes: Comparative ontogeny and morphology, evolution, phylogenetic significance, basic functions, research opportunities*. Rutgers University Press, Brunswick, NJ. 383 pp.

Nammack, M. F., Musick, J. A. and Covocoresses, J. A. 1985. Life history of spiny dogfish off the northeastern United States. Transactions American Fisheries Society 114: 367-376.

Nathanson, L. J. and Cailliet, G. M. 1986. Reproduction and development of the Pacific angel shark, *Squatina californica*, off Santa Barbara, California. Copeia 1986: 987-994.

Needham, J. 1942. *Biochemistry and Morphogenesis*. Cambridge University Press, Cambridge.

Otake, T. (1990) Classification of reproductive modes in sharks with comments on female reproductive tissues and structures. Pp. 111-130. In L. Pratt, S. H. Gruber, and T. Taniuchi, (eds), *Elasmobranchs as living resources: advances in the biology, ecology, systematics, and the status of the fisheries*. NOAA Tech. Rpt.90.

Price, Jr. K. S. and Daiber, F. C. 1967. Osmotic environments during fetal development of dogfish, *Mustelus canis* (Mitchell) and *Squalus acanthias* Linnaeus, and some comparisons with skates and rays. Physiological Zoology 40: 248-260.

Randall, D. J. and Tsui, T. K. 2002. Ammonia toxicity in fish. Marine Pollution Bulletin 45(1-12): 17-23.

Ranzi, S. 1932. Le basi fisio-morfologiche dello sviluppo embrionale dei Selaci. Parte I. Pubblication Stazione Zoologica di Napoli 13: 209-290.

Ranzi, S. 1934. Le basi fisio-morfologiche dello sviluppo embrionale dei Selaci. Parti II e III. Pubblication Stazione Zoologica di Napoli 13: 331-437.

Sarangdhar, P. N. 1949. On the breeding of the tiger shark (*Galeocerdo tigrinsus* Muller & Henle. The Journal of the Bombay Natural History Society 46: 192-193.

Shankar, R. A. and Anderson, P. M. 1985. Purification and properties of glutamine synthetase from liver of *Squalus acanthias*. Archives of Biochemistry and Biophysics 239: 248-259.

Shervey, P. 1966. Observations on the development and histochemistry of the intestinal inclusion bodies of the suckling rat. Anatomical Record. 154: 422-435.

Smith, J. W. 1980. The life history of the cownose ray, *Rhinoptera bonasus* (Mitchill 1815), in lower Chesapeake Bay, with notes on management of the species. MS thesis, College of William and Mary, Williamsburg, Virginia.

Snelson, F. F. Jr, Rasmussen, L. E. L. and Hess, D. L. 1997. Serum concentrations of steroid hormones during reproduction in the Atlantic stingray, *Dasyatis sabina*. General and Comparative Endocrinology 108: 67-79.

Snelson, F. F. Jr, Williams-Hooper, T. H. and Schmid, T. H. 1988. Reproduction and ecology of the Atlantic stingray, *Dasyatis sabina*, in Florida coastal lagoons. Copiea 1988 (3): 729-739.

Staley, T. E., Corley, L. D., Bush, L. J. and Jones, E. W. 1972. The ultrastructure of neonatal calf intestine and absorption of heterologous proteins. Anatomical Record. 172: 559-580.

Stevens, B. 2002. Uterine and oviducal mechanisms for gestation in the common sawshark, *Pristiophorus cirratus*. Bachelor of Science (Honors) thesis, Department of Zoology, University of Melbourne, Melbourne, Victoria, Australia, 41 pp.

Storrie, M. 2004. Microscopic modifications of the reproductive tissues of the gummy shark (*Mustelus antarcticus*) during maturation and gestation. Ph.D. dissertation, School of Ecology and Environment, Faculty of Science and Technology, Deakin University, Warrnambool, Victoria, Australia, 153 pp.

Sunye P. S. and Vooren, C. M. 1997. On cloacal gestation in angel sharks from southern Brazil. Journal of Fish Biology 50: 86-94.

Weber, R. E., and Hartvig, M. 1984. Specific fetal hemoglobin underlies the fetal-maternal shift in blood oxygen affinity in a viviparous teleost. Molecular Physiology 6: 27-32.

White, E. G. 1937. Interrelationships of the elasmobranchs with a key to the Order Galea. Bulletin of the American Museum of Natural History. 74: 25-138.

Whitley, G. P. 1940. The fishes of Australia. Part 1: the sharks, rays, devil-fish, and other primitive fishes of Australia and New Zealand. Royal Zoological Society of New South Wales, Sydney, Australia, 280 pp.

Wilkie, M. P. 2002. Ammonia excretion and urea handling by fish gills: present understanding and future research challenges. Journal Experimental Zoology 293(3): 284-301.

Wilson, J. M., Morgan, J. D., Vogle, A. W. and Randall, D. J. 2002. Branchial mitochondrial-rich cells in the dogfish *Squalus acanthias*. Comparative Biochemistry and Physiology 132A(2): 365-374.

Woodhead, A. D. 1976. Reproductive ecology of the spiny dogfish, *Squalus acanthias*. Bulletin Mount Desert Island Biological Laboratory 16: 103-106.

Wood-Mason, J. and Alcock, A. 1891. On the uterine villiform papillae of *Pteroplatea micrura*, and their relation to the embryo. Proceedings of the Royal Society., London. 49: 359-367.

Wourms, J. P., Grove, B. D. and Lombardi, J. 1988. The maternal-embryonic relationship in viviparous fishes. Pp. 1-134. In W. S. Hoar and D. J. Randall (eds), *Fish Physiology* Volume XI Part B. Academic Press, Inc., San Diego, CA.

Yano, K. 1992. Comments on the reproductive mode of the false cat shark *Pseudotrakis microdon*. Copiea 1992: 460-468.

Yano, K. 1993. Reproductive biology of the slender smoothhound, *Gollum attenuatus*, collected from New Zealand waters. Environmental Biology of Fishes 38: 59-71.

CHAPTER 14

Oophagy, Intrauterine Cannibalism and Reproductive Strategy in Lamnoid Sharks

R. Grant Gilmore, Jr.[1] *Oliver Putz*[2] *and Jon W. Dodrill*[3]

14.1 INTRODUCTION

Lamnoid sharks, order lamniformes, include seven families, 10 genera and 16 species whose morphological similarities, genetic characters and reproductive strategy form synapomorphies, revealing a monophyletic ordinal group whose evolutionary history spans over 200 million years. Their reproductive strategy, most notably oophagy, is a homologous character uniting this phyletic group, yet sets them apart from most other elasmobranchs. Oophagy as an embryonic nutritive strategy has had a long history of success within this group, contrasting remarkably with the more apomorphic carchariniform strategy of embryonic placentation. Both carcharhiniform and lamniform reproductive strategies produce large precocious neonates requiring considerable maternal energy expenditures.

The major objective of this chapter is to: (1) review the reproductive biology of lamnoid sharks; (2) review the functional significance of oophagy and embryophagy within the order; (3) review present reproductive patterns relative to species and familial phylogenetic relationships within lamniformes.

The earliest fossil records reveal that the order lamniformes evolved during the Mesozoic, with *Odontaspis, Carcharias, Mitsukurina* and *Pseudocarcharias* being the oldest forms, based on genetic evidence as well

[1] Senior Aquatic Scientist, Dynamac Inc., 5920 1st St. SW, Vero Beach, Florida, 32968, USA
[2] Section for Integrative Biology, University of Texas at Austin, Patterson Building, 24th and Speedway, Austin, Texas 78712 USA
[3] Division of Marine Fisheries, Florida Fish and Wildlife Conservation Commission, 6205 Meridian St. Tallahassee, Florida, 32399-1600, USA

as comparative dental and cranial morphology (Kent 1994; Long and Waggoner 1996; Martin 1996). *Megachasma* and *Cetorhinus* appear to be ancient genera as well (Long and Waggoner 1996; Martin and Naylor 1997; Morrissey *et al.* 1997). These genera still survive today allowing comparative study of their life history strategies, their reproductive biology. Although the entire order needs further study, *Carcharias* is presently the only lamnoid genus whose reproductive strategy is characterized by embryonic cannibalism, or adelphophagy, as well as oophagy.

The chapter will describe lamnoid embryology across phyletic groups within lamniformes, define oophagy and embryonic cannibalism, completing the presentation with a discussion of the evolutionary significance of these reproductive strategies within lamnoids, and within elasmobranchs in general.

14.2 REPRODUCTION IN LAMNOID SHARKS

Although there are several variations in reproductive strategies within lamniformes, the reproductive systems described to date reveal oophagy as a major source of embryonic nutrition throughout the order (Gilmore 1993; Uchida *et al.* 1996; Mollet *et al.* 2000). Embryophagy has only been documented in *Carcharias taurus*. The number of neonates at birth vary from species to species, but the group as a whole is characterized as having low fecundity rates. The maximum number of lamnoid embryos occur in *Isurus oxyrinchus*, 18 documented at parturition (Branstetter 1981; Stevens 1983; potentially larger litters in *I. oxyrinchus* according to Mollet *et al.* 2000) and in *Carcharodon carcharias*, with 14 embryos (Francis 1996), however most other lamnids produce fewer than 10 embryos. *Carcharias taurus* and *Alopias superciliosus*, *A. vulpinus* and *A. pelagicus* all typically give birth to two embryos, one from each uterus though two per uterus have been recorded for *Alopias* spp.(Moreno and Morón 1992; Gilmore 1993) as does *Lamna* spp. (Gilmore 1993). Embryophagy limits *Carcharias taurus* to a single surviving embryo per uterus.

Although considerable progress has been made over the past 20 years reproductive observations are lacking for most lamnoid shark species. Nothing is known of reproduction in *Mitsukurina owstoni* and *Odontaspis ferox* and *O. noronhai*, little of *Megachasma pelagios* and *Cetorhinus maximus* in spite of the latter species having been fished for over a century. Much of the information presented here will be revised as more pregnant specimens of the rarer species are captured. The *Carcharias taurus* information presented is derived from the examination of 47 pregnant female *C. taurus* and their 105 embryos captured from September 1970 to 1 December 1987, between 29° 20' N and 27° 20' N off the central east coast of Florida at depths less than 30-40 m. An additional 146 *C. taurus* captures were recorded from east and northeastern Florida from the St. Mary's River estuary (30° 20' N) south to Salerno Florida (27° 20' N) from 10 April 1946 to December 1977. Only adult female *C. taurus* were captured and all females captured between May and January were pregnant. In contrast, other lamnoids examined from the same study region, 1977-1993, revealed low pregnancy rates. The authors examined two pregnant *Isurus oxyrhinchus*, a pregnant *I. paucus*, and two pregnant *A. supercilosus*. and specimens, letters/literature furnished by our colleagues with interests and experience in lamnoid reproductive biology. The *Isurus* spp., *Carcharodon carcharias* and *Alopias superciliosus* were captured in the same study region, from shallow nearshore waters at depths less than 4 m to depths over 700 m, and revealed pregnancy rates (< 6%) with a very heterogenous age distribution from neonates to exceptionally large adults. Of 26 *Isurus oxyrinchus* and 27 *Isurus paucus* examined by the authors from this region, only two *Isurus. oxyrinchus* and one *I. paucus* were pregnant. Only two *A. superciliosus* of 16 examined were pregnant, no *A. vulpinus* of four examined were pregnant. Of 14 *C. carcharias* examined from this region, none were pregnant and only two were mature females.

The relative abundance of *Carcharias taurus* specimens allows a more detailed analysis of the reproductive biology of this species providing a reproductive model to compare with other lamnoid species. The only other lamnoid which has enough reproductive information available to make some rudimentary comparisons with *C. taurus* is *Isurus oxyrinchus*. *C. taurus* and *I. oxyrinchus* reproductive models contrast significantly presenting an interesting divergence in lamnoid reproductive strategies.

The time scale for *Carcharias taurus* reproductive activity presented below is based on the seasonal synchrony and predictability of reproductive condition for east Florida populations over a 50 yr observation period. Mating has been documented with sperm presence in the oviduct (8-12 May specimens) and sperm presence in early egg capsules lacking ova. Although variations in mating times certainly occur, possibly by as much as 2-3 weeks, May 10th was chosen as the mean date of insemination as a point of reference. All other dates are referenced to this date as day one in the gestation cycle in east Florida *C. taurus*. Therefore all plots and data tables presented below refer to the "days after first insemination" (DAFI, 10 May) as a scale on which to measure approximate changes in reproductive activities for *C. taurus*. This time scale aids in quantifying the *C. taurus* reproductive model which can be used to compare with other lamnoid species.

14.2.1 Functional Reproductive Anatomy

14.2.1.1 Ovary

Only the enlarged right ovary is active in all lamnoids described to date. Ovarian morphology has not been described for *Mitsukorina owstoni* or *Odontaspis* spp. The left ovary is vestigial. The lamnoid ovary is distinct from other elasmobranchs as it's morphology optimizes ova production and high ovulation rates. Pratt (1988) describes the lamnoid ovary type as containing a series of internal cavities increasing the 3-dimensional surface area, associated with an ellipsoidal interior lumen surrounding the oocyte mass (Fig. 14.1). This morphology gives more surface area to generate oocytes than the relatively planar ovary found in other elasmobranch orders. A pocket, or efferent ovarian pore opens to the abdominal cavity and allows ovulated ova to escape to the ostium. This pocket, situated anterior and dorsally on the right side of the ovary, enlarges as the ovary matures and reaches maximum size when ovulation peaks.

Lamnoid ovarian fecundity rates exceed that recorded for other elasmobranchs with over 6 million ova estimated for *Cetorhinus maximus* ovaries (Matthews 1950). This is undoubtedly associated with the lamnoid oophagous embryonic nutritional strategy. In *C. taurus* the ovary reaches a maximum size of 36.0 - 45.5 cm in length and weight of 8.5 kg, containing over 22,000 ova, 1.3 to 10.0 mm in diameter (Figs 14.1, 14.2) (Gilmore *et al.* 1983). In *C. taurus* this typically occurs 60-70 DAFI. Minimum ovarian size is reached 200 DAI, simultaneous with the minimum number of egg capsules in the uterus (Fig. 14.2).

Differential ova size in lamnoids has potential significant influence on embryo growth rates and parturition size. The 10-12 mm *Carcharodon taurus* ovum is eight times the mass of a 5-6 mm ovum recorded for *Cetorhinus maximus, Lamna* spp., *Isurus* spp. and *Alopias* spp. (Fig. 14.2).

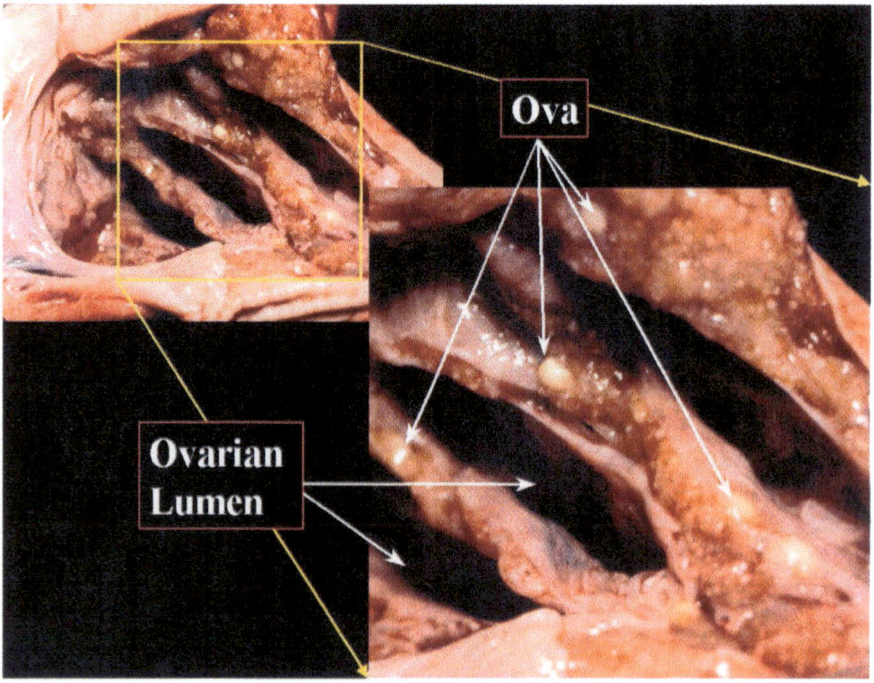

Fig. 14.1 Chambered ovary of *Isurus oxyrinchus*, typical of lamnoid sharks, from pregnant female, 1 April 1985, interior lumen of ovary with ova near ovulation on right. Original.

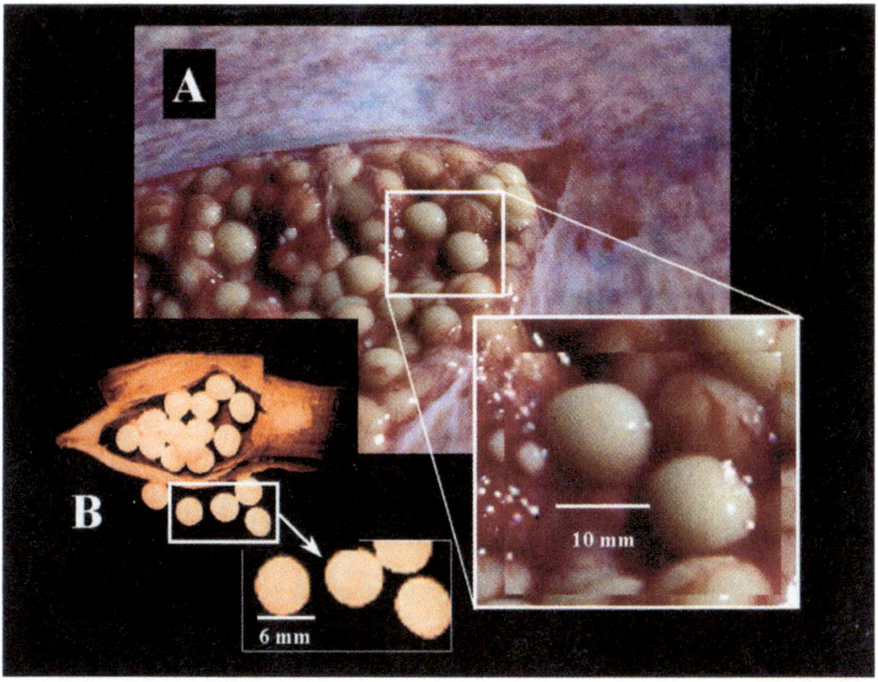

Fig. 14.2 Lamnoid ova comparison: **A.** *Carcharias taurus* ovary with 10 mm diameter ova; **B.** *Isurus oxyrinchus* egg capsule with 6 mm diameter ova. Original.

Hamlett (1983) determined the ova consumption for each *Carcharius taurus* embryo using the caloric content (940 calories/ovum) and an conversion rate of 70% as described by Needham (1942) for oviparous embryos. Hamlett (1983) estimates that embryos of *C. taurus* may ingest 17,000 ova, averaging 10 mm in diameter during gestation. Using our data on reduction in ova number (Fig. 14.3), a consumption rate of 19,157 ova containing about 18,000 kcalories is calculated. We also predict that a mean of 430 ova were not ovulated remaining in the ovary throughout the late gestation phase. This converts to a range of 1,133-1,277 capsules containing a mean of 15 ova each (max of 23/capsule) during the oophagous period of gestation. Oophagy was observed to last around 100 days, ending 200 DAFI. This converts to a mean capsule consumption rate of 11-13 capsules per day by a ravenous 335-1000 mm embryo. Therefore, ovarian ovulation rates are maximum during the 100 day oophagous period in embryonic development, from late August to late November in east Florida *C. taurus* populations. A maximum of 81 capsules were found in the uterus 59 DAFI demonstrating that during the period when embryos are encapsulated or in the cannibalistic developmental stage, the capsules are stored in the uterus for at least 100 days, at which time oophagy begins. Ovulation rates during this early gestation period is considerably lower 1-2 capsules per day than in the oophagous period.

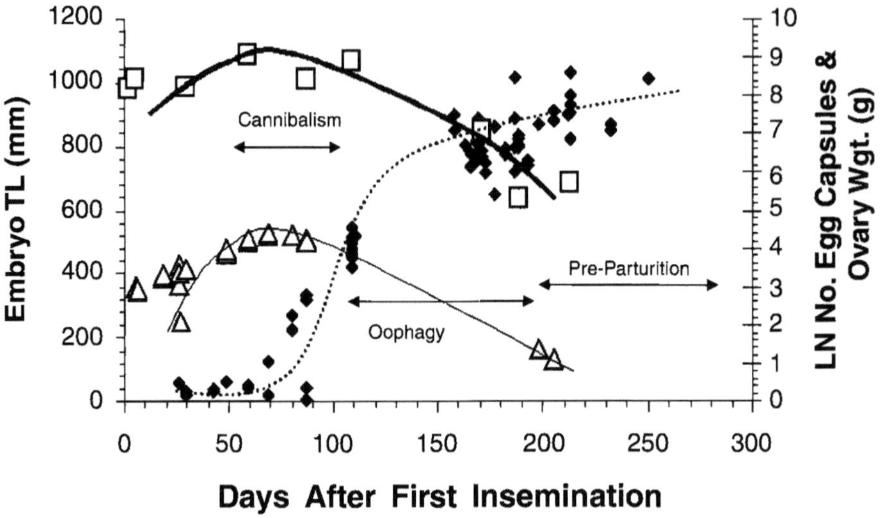

Fig. 14.3 Plot of LN number of *C. taurus* egg capsules (triangles; 84 uteri, 47 specimens), and LN ovarian weight (squares; 8 specimens) examined throughout the year, 1970-1987, including embryo TL data (solid diamonds; 105 specimens). Embryonic nutritional transition periods occur at day 50 when cannibalism begins, day 100 when oophagy begins and day 200 when the pre-parturition phase begins. Original.

There is little information on ovarian fecundity rates and dynamics of ovarian activity in other lamnoid species. The maximum ovarian weight presented for *C. maximus* by Mathews (1950) was 12.2 kg with an estimated six million ova having diameters ranging from 0.5 to 5.0 mm. Therefore, *Cetorhinus maximus* may produce 273 times more ova than *Carcharias taurus*. As Mathews' *C. maximus* specimen was not in gestation, the ovary may not have reached maximum size. However, the maximum ova diameters observed by Mathews could represent the largest diameters attained for *C. maximus* as there was evidence of recent copulation, and therefore ovulation was proximate. Based on ova diameters and ova mass differences, *C. maximus* would have to ovulate eight times more ova to get the same embryonic biomass conversion rates calculated for *C. taurus*. Since the basking shark ovary contains significantly more ova, an increased ovulation rate in *C. maximus* is likely. The smaller ovaries and ova (5-6 mm maximum diameter) of the isuriid and alopiid sharks and term embryos with 50-60% the precaudal length of *C. taurus* term embryos, indicates that embryonic growth rates in isurid and alopiid sharks is slower than in *C. taurus*.

14.2.1.2 Oviducts/Oviducal gland

The function of the paired oviducts in lamnoid sharks differs significantly from other elasmobranchs primarily based on the role of the oviduct in maintaining cannibalistic and oophagous embryos.

The paired oviducts of *Carcharias taurus* may be divided into four basic sections: (1) ostium and anterior tube; (2) oviducal gland; (3) isthmus; (4) uterus. The ostium and anterior tube are lined with ciliated columnar epithelial cells which transport the ova captured at the ostium to the oviducal gland. The anterior tube is 310 mm long, 9 mm in diameter in a 254 cm *C. taurus*. Since sperm is not stored in the oviducal gland of *C. taurus* and active sperm have been isolated from the anterior tube and ostium in recently mated sand tigers (8 and 12 May 1986 specimens), it is most likely that ova fertilization takes place in the anterior oviduct before encapsulation takes place in the oviducal gland. The oviducal gland encapsulates the ova. and secretes mucus, ova-albumin, and the major elasmobranch egg case component, collagen (Wourms 1977; Hamlett 1999).

In spite of extensive histological examination of recently mated and pregnant *Carcharias taurus, Alopias superciliosus, Isurus paucus* and *I. oxyrinchus,* (Pratt 1993, Pratt and Tanaka 1994) there has been no evidence of sperm storage in the oviducal gland of lamnoid sharks. We have stained and sectioned 12 oviducal glands from pregnant *C. taurus, I. oxyrinchus, I. paucus* and *A. supercilosis*, but have found no evidence of sperm storage in these glands. Sperm storage within the tissues of the oviducal gland has been documented in other elasmobranchs (Pratt 1993; Pratt and Tanaka 1994), but not in lamnoids. Active sperm have been observed throughout the oviduct from the cloaca to the ostium in *C. taurus* (Gilmore 1993). Sperm have been documented within the first egg capsules formed by the oviducal gland containing ovalbumin but lacking ova (ibid.). *C. taurus* and *I. oxyrinchus* sperm have remained viable under refrigerated conditions, 15° C, for up 90 days, then activated in seawater at room temperature, 20° C. This implies that mating activity in *C. taurus* does not require sperm storage, possibly not in *Isurus* spp. and *Lamna* spp. as well. The small compact cylindrical spermatophore of *I. oxyrinchus* may allow for prolonged survival of sperm within the oviduct of this species, or at least delayed activation during mating. A survey of sperm concentration throughout the oviduct in a recently mated *C. taurus* revealed the highest concentration of sperm posterior the the oviducal gland in the isthmus, and sperm were present in the uterus, oviducal gland and anterior oviduct to the ostium.

Sperm storage is not necessary in lamnoid sharks if mates are predictably available. If mating for a population is synchronous and at a predictable location and time, mates (sperm) will be available. All evidence collected on *Carcharias taurus* populations off east Florida over several decades indicates synchronous and predictable mating activity at specific locations (Gilmore *et al.* 1983; Gilmore 1993). It is also possible that migrating pelagic lamnoids such as *Isurus oxyrinchus* maintain predictable mating periods yet remain mobile with mixed sex aggregations spread over a wider foraging range in the water column. Fecundity, ovulation rates and egg capsule formation differ significantly between lamnoids and other elasmobranchs whose ova are not used for embryonic consumption. Ephemeral ovulation is common in other elasmobranchs, but does not occur in lamnoid sharks. The *C. taurus* ovary remains active, ovulating, for nearly 200 days, requiring prolonged oviducal gland activity for egg capsule formation. It is unlikely that sperm storage would function in such an active oviducal gland and detailed histological examination does not show sperm storage in *C. taurus, I. oxyrinchus, I. paucus* and *Alopias superciliosus* examined by the authors and others (Pratt 1993). The oviducal gland of *I. oxyrinchus* may remain active for a longer period than *C. taurus* as gestation is estimated to last 15-18 months in the former species (Mollett *et al.* 2000).

Just as ovulation rates change considerably during gestation relative to embryonic activity and nutritional needs, egg capsule morphologies change in capsule contents. The greatest capsule heterogeneity occurs during early gestation. Early capsules may lack ova containing only ovalbumin and, if recently mated, contain sperm (Gilmore *et al.* 1983). The first capsule to contain ova, typically only contains a single ovum. Multiple ova occur in increasing numbers in successive capsules. Early historical observations indicated that although multiple blastodiscs may be seen on several ova in a single egg capsule (Type 1 blastodisc capsule of Gilmore *et al.* 1983), only one embryo developed per capsule (Gilmore *et al.* 1983). However, additional observations reveal that up to three embryos may develop within a single capsule in *Carcharias taurus*. A pregnant *C. taurus* captured 19 June 1985 contained capsules with two and three embryos, 22-31 mm TL, in each of two capsules, respectively. Typically, the first to develop is encapsulated alone. The same 19 June specimen contained an embryo encapsulated with an unfertilized ovum. The unfertilized ovum may provide an additional food source for the single encapsulated embryo. Since embryonic dentition develops while the embryos are encapsulated it is possible for these encapsulated embryos to consume ova larger than their gape. Capsules containing embryos are produced during the first 70 days of gestation. This correlates with long term observations of males syntopic with female *C. taurus* at mating sites south of Cape Hatteras North Carolina (1985-1993) and off Cape Romano South Carolina (May 1993). Males were observed with females for at least 60 days, May to June, possibly longer when prenuptial periods (February to April) are included. Capsules containing only unfertilized ova are produced throughout gestation (Type II & III capsules of Gilmore *et al.* 1983), with peak in abundance of up to 23 ova per capsule during the oophagous period, 100-200 DAFI.

Although fewer specimens of *Alopias* spp., *Isurus* spp., *Pseudocarcharias kamohari, Carcharodon carcharias* and *Lamna nasus* have been examined, capsule formation and variation in capsule morphology resembles that found in *Carcharias taurus* (Gubanov 1972, Fujita 1981, Gruber and Compagno 1981, Gilmore 1983, Gilmore *et al* 1983; Uchida *et al.* 1996; Francis 1996). Capsules produced early in the ovulation cycle do not contain ova, whereas capsules produced during the major oophagous interval in embryonic development contain the largest number of ova.

14.2.1.3 Uterus

The complex uterine wall performs a secretory function producing fluid for both embryonic nutrition, lubrication and oxygenation of embryonic tissues.

Egg capsules leave the oviducal gland and proceed down the elastic narrow isthmus, 250-350 mm in length, 20-34 mm in diameter in pregnant adult *Carcharias taurus*, connecting the gland with the expanded uterus.

Vascularization and folding of the inner epithelial lining of the uterus increases at the isthmus. This has been observed in *Carcharias taurus*, *Isurus paucus*, *I. oxyrinchus*, *Alopias superciliosus*, *A. vulpinus* and *A. pelagicas* as well as in *Carcharodon carcharias* and *Megachasma pelagios* (Gruber *et al.* 1983, Gilmore *et al.* 1983; Moreno and Morén 1992; Gilmore 1993; Tanaka and Yano 1997; Castro *et al.* 1997; Hamlett 1999). This large surface area and proximity of maternal blood supplies apparently insures oxygenation of the uterine fluid and may enhance fluid secretion (Fig. 14.4). Oxygenation of uterine fluid, which is undoubtedly a major source of respiratory oxygen, is necessary as lamnoids produce very large embryos without placentation. Since lamnoids require oxygenation of uterine fluid and lubrication of uterine membranes containing active

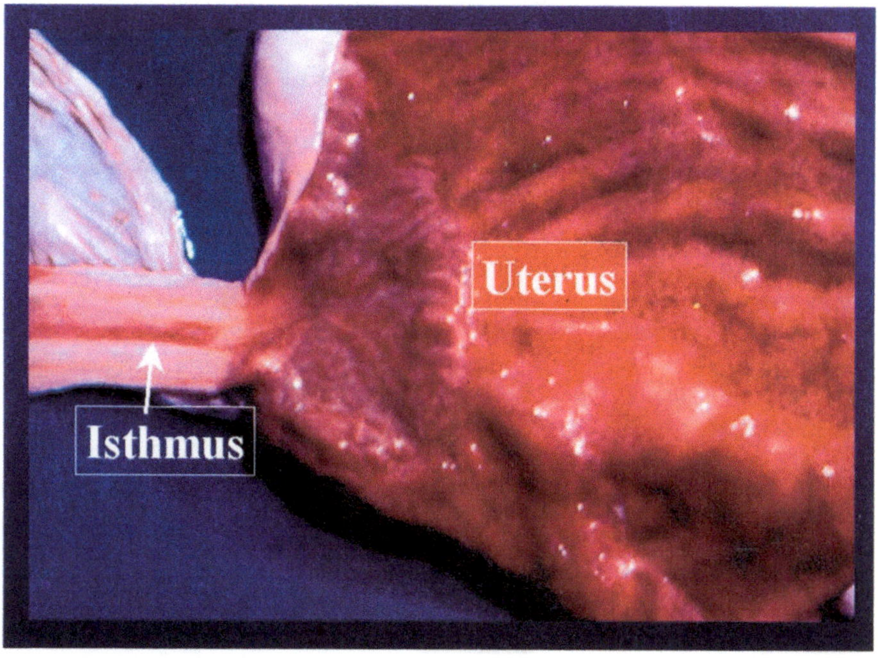

Fig. 14.4 Vascularized uterus at isthmus. This uterus contained a 750 mm TL *Carcharias taurus* embryo from female captured 10 October 1985. Original.

ravenous embryos hunting and attacking egg capsules, sometimes other embryos, all without a direct maternal - fetal tissue connection, the uterine wall must play a function unique to lamnoids. This is particularly true due to the large size of lamnoid embryos relative to the adult female baring them (each of ten embryos may be 10% of the female length in a pregnant *C. carcharias* (Uchida *et al.* 1996; Francis 1996).

In *Carcharias taurus* the size of the uterus, the volume of the uterine fluid, and the length of the isthmus increases during early gestation (June to July). *C. taurus* uterine fluid also increases in relative cloudiness through gestation and contains numerous ruptured egg capsules and yolk fragments. Uterine fluid may have antibiotic properties as egg capsules are stored in the uterus for 70-100 days during early gestation and dead embryos have been found mummified within the uterus during late gestation. Uterine fluid stored in refrigerators for several months has not deteriorated or obviously decomposed indicating potential antibiotic properties.

14.2.2 Embryology

At present lamnoid embryonic development can only be described in detail based on the *Carcharias taurus* model as specimens taken throughout gestation are lacking for the other species. *C. taurus* embryonic development can be classified based on six distinct nourishment stages beginning with (1) nutrition from endocoelomic yolk supplies occupying the pharyngeal, pericardial and coelomic cavities in the 13 mm embryo, proceeding to (2) nutrition from encapsulated fluid and extra-embryonic yolk materials in the 18-57 mm embryos, then to (3) nutrition from embryonic yolk sac, uterine fluid consumption in the 60 to 100 mm embryo, (4) nutrition from embryophagy in the 100 to 335 mm embryo,and (5) oophagy from 335 to 1000+ mm, then finally (6) the pre-parturition metabolic stage, after 200 DAI, in which the 900-1100 mm embryos digest consumed yolk and places most nutritive reserves into the liver and other storage tissues. All lamnoid sharks studied to date have revealed oophagous embryos, *Isurus oxyrinchus, I. paucus, L. nasus, Carcharodon carcharias, Alopias superciliosus, A. vulpinus, A. pelagicus* and *Pseudocarcharias kamohari.* The embryos of *Cetorhinus maximus, Megachasma pelagios, Odontaspis* spp. and *Mitsukorina owstoni* have yet to be described.

14.2.2.1 Stage I-II. Pre-hatch encapsulated embryos, 13-57 mm TL, 0- 43 DAFI

The encapsulated developmental period is characterized by a four fold increase in length and increase in biomass from 0.4 to 0.7-0.8 g. During this period the embryo goes through substantial changes in morphology first resembling an obese rounded amphibian embryo with endometrial yolk (Fig. 14.5A) and ending with a filamentous, dentigerous, ravenous predator that shows sexual dimorphism. Ironically the external yolk sac does not reduce significantly in size during this developmental period indicating energy supplies external to the embryo are being consumed.

The 13 mm embryo is the earliest developmental stage embryo described and was sectioned to examine its internal structure and determine nutritive sources (Gilmore *et al.* 1983) (Fig. 14.5A). Histological sections showed an incomplete connection between internal yolk supplies and an external yolk sac verifying that yolk sac yolk is not utilized in this early developmental stage. A membrane at the junction of the yolk stalk and the yolk sac isolates the yolk-sac yolk from the yolk stalk and coelomic yolk supplies. The coelomic cavity, cardiac stomach, valvular intestine, and pericardial cavity all contained yolk which is most likely the primary source of nutrition. The 13 mm embryo is nearly round in cross-section with a maximum horizontal diameter of 9 mm due to the large internal yolk supply. This diameter was greater than that of the attached yolk sac (6.0 mm). The gill arches and the toothless mouth cavity were open. The 13 mm embryo resembles an amphibian embryo after gastrulation and formation of primary organ rudiments. The 13 mm *Carcharias taurus* embryo does not resemble the early embryos described for other elasmobranchs [e.g., *Heterodontus japonicus* (Smith 1942); *Chlamydoselachus anguineus* (Gudger 1940): *Mustelus canis* (TeWinkel 1950, 1963). The unique morphology of this early embryonic stage in C. *taurus* could also be a lamnoid character. Endocoelomic yolk supplies could give the early

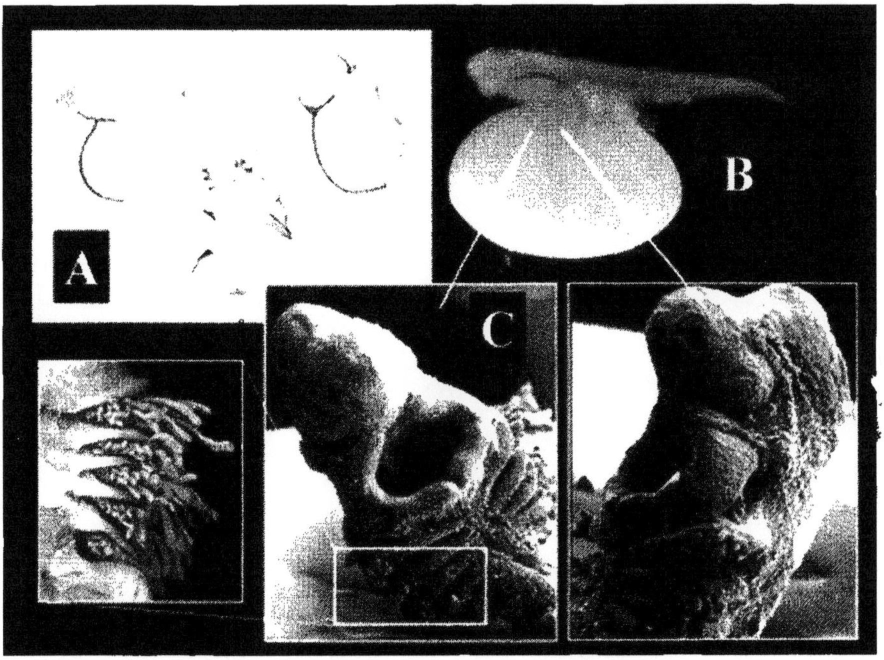

Fig. 14.5 A. Views of 13 *mm C. taurus* embryo, **B.** 22 mm embryo from 18 June 1985, sand tiger; **C,** Electron micrograph of same embryo showing head, gills and toothless jaw detail. Original.

embryo a more proximate easily mobilized energy source which would be another advantage in increasing the rate of embryonic development in *C. taurus*. By 18.5 mm the embryo has less internal yolk and greater differentiation of external features. A spiracle was present as were first and second dorsal, caudal, anal, and pelvic fin buds in addition to the pectoral fin buds which had developed earlier. The external yolk sac is still over 6.0 mm in diameter indicating that it is not being used for nutrition.

At 22-31 mm the *Carcharias taurus* embryo enters the second encapsulation stage which is dependent on external nutritive sources either within the capsule or moving through the capsule membrane from the uterus (Fig. 5B, C). This embryo has extensive well developed gill filaments emerging from the gill arches (Figs 14.5, 14.6). Embryos have been found encapsulated with single unfertilized ova and other embryos (Fig. 14.7). The external yolk sac has not reduced in size, but there is no evidence of internal yolk either, therefore indicating that nutrition must be coming from other sources in the egg capsule or across the egg capsule membrane. As the embryo approaches 49-57 mm TL functional embryonic teeth are obvious with 6-7 sharp conical teeth on either side of the upper and lower jaws (Fig. 14.8). Functional dentition can allow the embryo to escape through the tough collagen membrane surrounding the capsule.

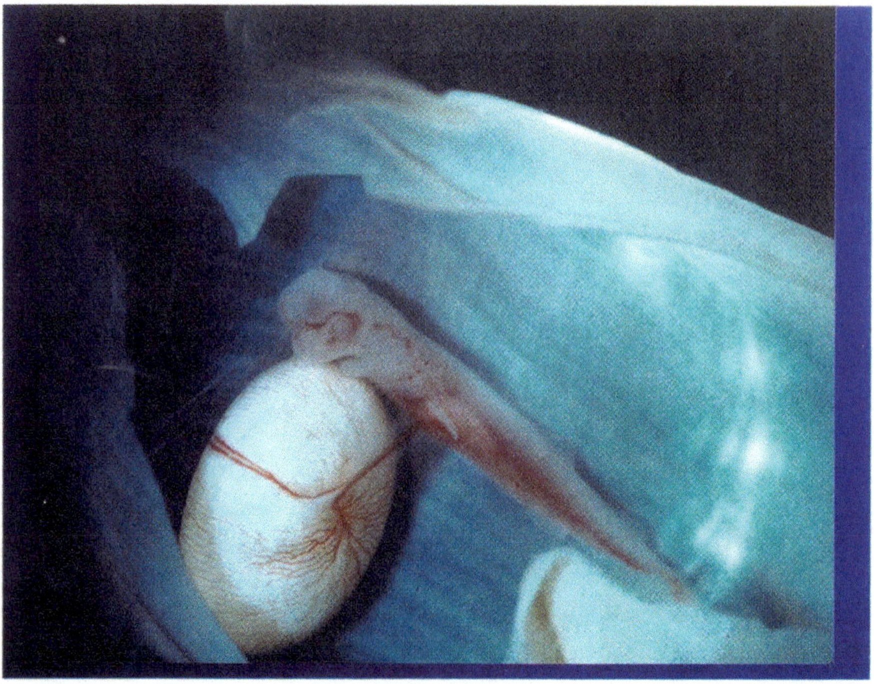

Fig. 14.6 A 33 mm TL *Carcharias taurus* embryo within an egg capsule, showing highly vascular gills and orbits as well as abdominal cavity. Original.

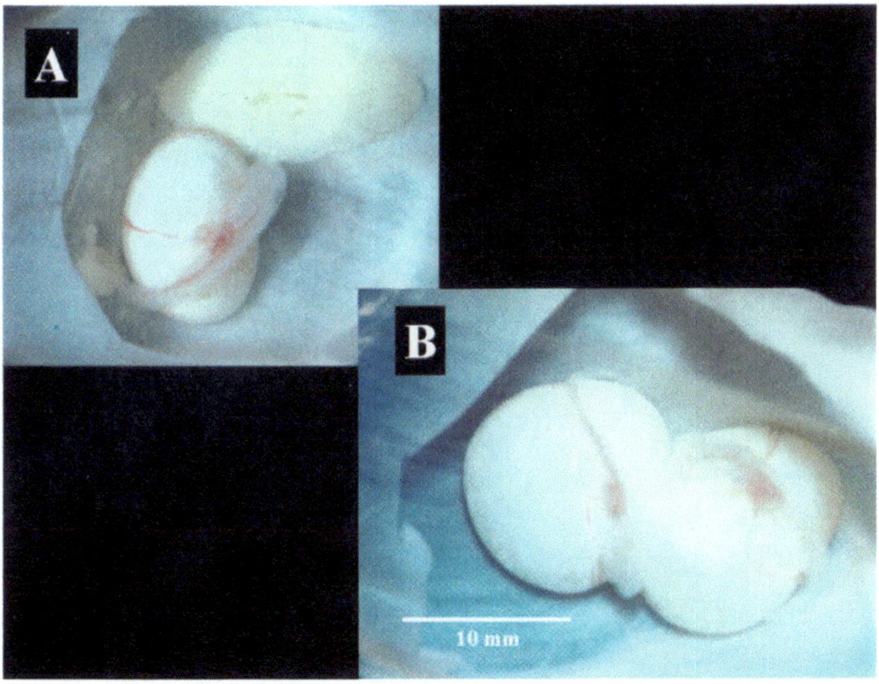

Fig. 14.7 A. 20 mm *Carcharias taurus* embryo encapsulated with a single unfertilized ovum. **B.** two 21-22 mm TL *C. taurus* embryos within the same capsule. Original.

At this pre-hatch stage well developed external filaments extend from the gill arches, and also from the spiracle and orbit indicating a potential function in oxygenation and nutrition for neurological and somatic development. Retinal tissue typically has a high metabolic rate thus may require its own attached respiratory filament. The external yolk sac does not change in diameter throughout the encapsulation phase as it averages between 6-7 mm throughout this phase. Rudimentary claspers are evident on the inner margin of the pelvic fins, indicating secondary sex characteristics have developed prior to hatching. By 60 mm TL the *C. taurus* embryo is free swimming within the uterus (embryonic swimming within the uterus was documented with an endoscope placed in the uterus of a pregnant *C. taurus* by John McKenney of Jack McKenney Productions, Los Angeles, CA).

14.2.2.2 Stage III. Immediate post - hatch free swimming embryos, 60 - 100 mm TL, 43-55 DAFI

Only two embryos, 62 and 63 mm TL have been examined from this brief development stage. Both specimens had external gill filaments, though not as extensively developed as found in embryos in the final encapsulation stages. Also present were yolk sacs 5.5 - 6.0 mm in diameter indicating that yolk-sac yolk supplies were not yet being consumed. However, twelve days later, the 100 mm embryos which are cannibalistic

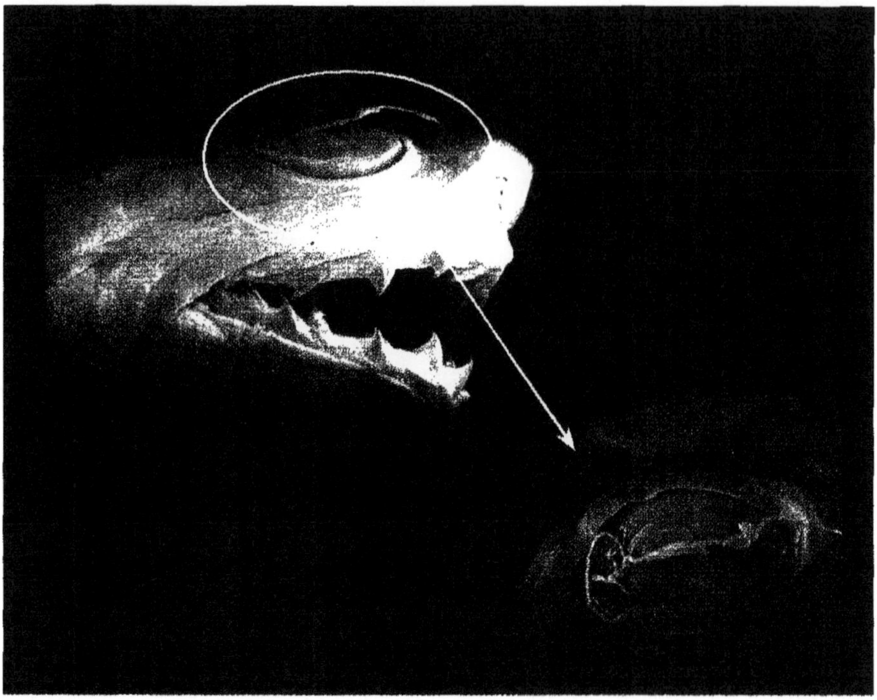

Fig. 14.8 Electron micrograph of *Carcharias taurus* embryo, 55 mm TL from egg capsule, adult captured 10-11 July 1981. Note well developed dentition and orbital filament. Original.

and do not have an external yolk sac and lack external gill filaments indicating that external yolk-sac yolk is totally consumed during this post-hatch, pre-cannibalistic period. The presence of gill filaments in this early post hatch period may allow the recently hatched embryo to utilize uterine fluid nutritive supplies prior to consumption of external yolk sac supplies. Since this was apparently the function of the extensive filaments just prior to hatching, it is realistic to assume gill filaments function in nutrition as well as tissue oxygenation after hatching. The resorption of both gill filaments and yolk sac and the commencement of cannibalism mark the end of this developmental stage.

14.2.2.3 Stage IV. Intra-uterine cannibalism, 100 - 335 mm TL, 55 - 100 DAFI.

After reaching 100 mm and exhausting external yolk sac nutrition sources the well armed *Carcharias taurus* embryo begins to attack siblings within the uterus (Fig. 14.9). Evidence obtained from the examination of three *C. taurus* litters containing multiple embryos indicates selective capture of siblings by the dominate older embryo. All consumed or killed siblings were <51 mm TL indicating that they were encapsulated when attacked. Egg capsule pieces were observed in the cannibal's stomach along with consumed siblings (Fig. 14.8). This is further verified by the observation that lacerated and punctured dead embryos were found encapsulated. Empty capsules had been lacerated and torn open during repeated attacks by the larger embryo. It is of interest that within the same uterus in which sibling consumption had occurred up to 66 ovoid egg capsules containing no embryos, only unfertilized ova were not lacerated or damaged in any way indicating that they were not attacked and that egg capsules containing developing embryos were selectively attacked. The maximum number of embryos found within another embryo at the time of capture was four, 9- 36 mm TL. However, in late periods of the cannibalistic stage, as embryos approach 335 mm TL, yolk is found within the stomach indicating that some oophagy is occurring as well as cannibalism.

The 100 mm TL *Carcharias taurus* embryo has well developed fins with large gill slits without external filaments. Both upper and lower labial furrows are well developed. The yolk sac is absent, however, an attachment scar remains. Erect teeth more slender than in previous embryos are present in multiple rows indicating that teeth are now being used for opening capsules, grasping and swallowing embryos, and may be lost during this process hence multiple rows are necessary. Teeth lack lateral secondary basal cusps (basal denticles) typical of adult *C. taurus*, but observed in embryonic *Carcharodon carcharias* (Uchida *et al.* 1996, Francis 1996).

Cannibalizing embryos do not have a greatly extended abdomen (Fig. 14.8). However, their growth rate is exponential during this period with the greatest length and weight increase per day during gestation (Fig. 14.3). The average daily change in biomass is 14.89%, total length 5.22 % during the 45 day cannibalistic period. This compares with only a 5.57 % daily biomass change for 335 - 1000 mm TL oophagous *Carcharias taurus* embryos, or 1.33 % TL change daily. Between 87 and 100 DAFI mating and fertilization of ova ceases and the remaining embryo begins to consume egg capsules.

Cannibalism has not been observed in any of the other lamnoid species examined to date though some species may only give birth to a single embryo per uterus (ex. *Alopias* spp.). Figure 14.10 shows the flared gill slits and enlarged yolk stomach of a 322 mm TL *Isurus oxyrinchus* embryo, obviously oophagous, but not embryophagous.

14.2.2.4 Stage V. Oophagous period. 335 - 1000 mm TL, 100 - 200 DAFI

Oophagy has been observed in *Isurus* spp., *Lamna* spp., *Alopias* spp, *Pseudocarcharias kamohari*, *Carcharodon carcharias* and *Carcharias taurus* (Fig. 14.11). Ovarian morphology suggests oophagy occurs in *Cetorhinus maximus* and is likely in *Megachasma pelagicus*. Oophagy is undoubtedly a shared character for all lamnoid sharks, representing a synapomorphy. The evolutionary success of this elasmobranch group must be at least partly associated with this nutritional mode during pregnancy.

Fig. 14.9 A, 197 mm TL embryo above a 37 mm TL embryo from its stomach, left uterus, *Carcharias taurus*,14 August 1985; **B.** 198 mm TL embryo above a 46 mm TL embryo found dead within the same uterus having been extracted from its egg capsule by the 198 mm embryo, right uterus, *C. taurus*, 14 August 1985. Original.

Carcharias taurus embryos appear to be ravenous during this period consuming multiple egg capsules, at least up to 11-13 capsules per day. Unfertilized ova densely packed in ovoid egg capsules become the primary source of nutrition. During this period the embryo may consume over 19,000 ova, reaching 800-900 mm TL within 60 days. During late gestation embryos have swallowed so much yolk that their stomachs became greatly distended (Fig. 14.11). Cadenat (1956) found 1.5 kg of yolk (18.8 % total body weight) in a near-term *C. taurus* embryo weighing 8 kg. Some *C. taurus* embryos reach 1000 mm TL by 190 DAFI. Capsules are typically absent from uteri after 200 DAFI at which time the embryo reverts to the pre-parturition phase, developmental Stage VI (Fig. 14.13). Embryonic length and weight reach a plateau during the oophagous period as the last egg capsules are consumed (Fig. 14.12). It is interesting to note that the greatly extended yolk stomach has been documented in *Isurus*, *Lamna*, *Charcharodon*, and *Carcharias*, but not in *Alopias*. This may be due to lower ovulation rates and capsule formation/consumption rates in *Alopias* when compared to the other species. Mollett *et al.* (2000) found

Fig. 14.10 322 mm TL , 0.45 kg *Isurus oxyrinchus* embryo from female captured 7 December 1992 in the Florida Straits, revealing enlarged yolk stomach and flared gill openings, absence of embryophagy. Original.

that the condition factor of alopiid embryos remained constant throughout gestation further verifying that a yolk stomach does not develop in alopiids.

14.2.2.5 Stage VI. Pre-parturition phase. 900 - 1000 mm TL, 200 - 250/270 DAFI

The pre-parturition embryo digests consumed yolk that is packed in the distended stomach and spiral valve. Abdominal cavity distention declines to the point that the embryo takes on a very slender, fusiform appearance. This morphology undoubtedly aids passage through the cloaca, particularly when compared to the embryo birth length of 1.0 m.

Simultaneous to the decline in yolk consumption is an increase in the size of the embryo's liver (Fig. 14.13). The left and right lobes of the liver of the specimen from Sea World of Orlando measured 20.3 and 23.7 cm, respectively, with a total liver weight of 372 g (9% of total body weight). Cadenat (1956) found the liver of a near-term *Carcharias taurus* embryo to be relatively large, contributing 6.43% of the total body weight, in a 110 cm specimen. The larger liver in the near-term embryo compares favorably with the largest liver recorded in adults at 7.54% total body weight

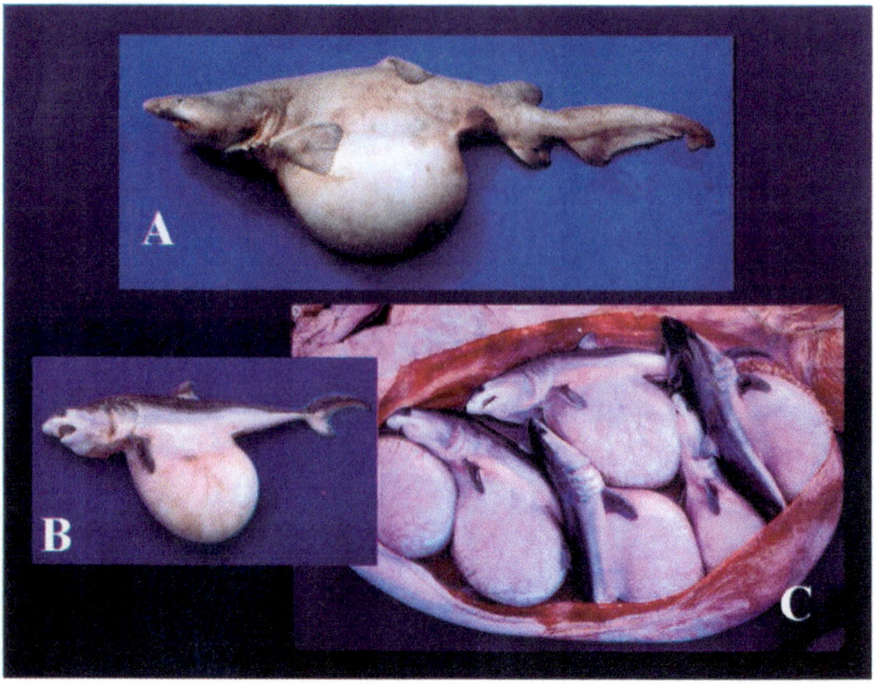

Fig. 14.11 A. Oophagous *Carcharias taurus* embryo, 750 mm TL w enlarged yolk stomach, 10-11 Oct. 1985; **B.** Oophagous 450 mm TL *I. oxyrinchus* embryo from 12 embryo litter, April 1985; **C.** Six of twelve oophagous *I. oxyrinchus* embryos, 430-492 mm TL, from a female captured April 1985. Original.

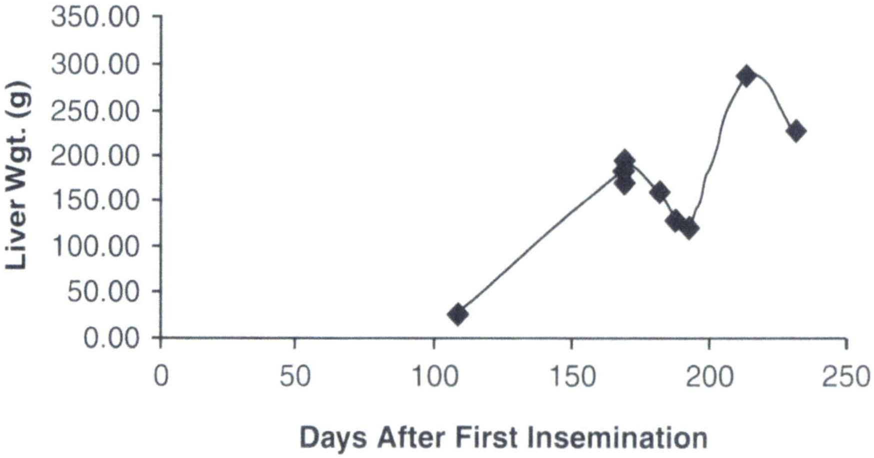

Fig. 14.12 Change in embryonic liver weight in *Carcharias taurus* embryos. Note change in embryonic liver growth rate when oophagy begins and ends 100-200 DAFI, and after all yolk capsules are consumed 200 DAFI when the embryo begins the pre-parturition embryonic phase. Original.

(Cadenat 1956). A plot of *C. taurus* embryonic liver growth during gestation shows a major increase in liver growth at around 200 DAFI the time when yolk capsules are no longer present in the uterus (Fig. 14.12).

A similar condition of large liver size and reduced yolk consumption has been observed in a near term oophagous embryo (97 cm TL) of *Isurus paucus* (Gilmore 1983), *Carcharodon carcharias* (16.5-18.6%, Francis 1996; Uchida *et al.* 1996), and in *I. oxyrinchus* (Mollet *et al.* 2000).

The embryonic length/weight plateau reached during the oophagous phase is maintained during the pre-parturition phase as the embryo digests yolk and stores energy in its liver (Fig. 14.13). Embryonic metabolism must be reduced as this stage may last 50-70 days. Embryonic length and weight appears to remain the same as embryos examined after 200 DAFI are typically at or above 900 mm TL with some individuals at the birth length of 1000-1100 mm TL as early as 190 DAFI. This stasis in embryonic morphology and metabolism may indicate some form of embryonic estivation. This period occurs during the winter, a period of reduced coastal seawater temperatures, thus possibly significantly reduced metabolism in the pregnant female as well as her embryos. The liver of pregnant near-term female *Carcharias taurus* reaches a minimum size at this time (2.88% total body weight, while it is normally 7- 8% the adult body weight (Cadenat 1956). Parturition occurs in late winter or early spring and may be triggered by increased parental activity associated with the arrival of males which are normally absent during the summer-fall gestation period, increasing ambient water temperatures, increasing adult metabolic rates and prey consumption.

The only observed births for *Carcharias taurus* captured in east Florida have been of captive specimens. Successful births occurred in the Marineland aquarium in NE Florida, 15 February 1959 and Sea World Shark Encounter aquaria of Orlando, 22 March 1981. In both cases at

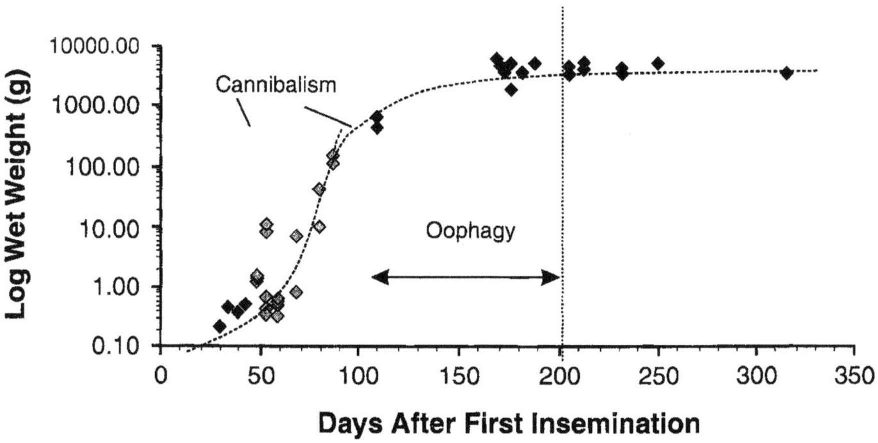

Fig. 14.13 Total length distribution for 105 *Carcharias taurus* embryos captured off the east central Florida coast, 1970 to 1987. Original.

least one neonate lived for more than 7 yrs in captivity indicating that these new born pups were sufficiently developed to survive after birth during February - March, at least 294 DAFI. It is possible that capture could have delayed embryonic development, but a range of 280 - 290 days for gestation seems reasonable for *C. taurus*. Mating may occur earlier than May as a 272 cm *C. taurus* female was captured 10 April 1946 off north Florida and kept in an aquarium for 11 mo and died on 9 March 1947. Her autopsy revealed two decomposing near term embryos 103-105 cm TL (6.1 and 6.4 kg) (Data from notes and letters provided by Robert Jenkins, F. G. Wood, and Stewart Springer 1976, also incorporated in Gilmore Springer 1948; *et al.* 1983; original photos of embryos and adult examined by authors). *C. taurus* mating behavior expeditions off Cape Hatteras 1987-1993 revealed parturition in March to April and male presence with scarred females April to July indicating a potential later parturition and mating period, potentially 30 days, for the most northern mating groups of *C. taurus* when compared to Florida populations. A maximum size of 1.2 m TL may be reached before birth according to Cadenat (1956). Birth in the wild takes place at depths <50 m on the continental shelf from Cape Canaveral, FL to Cape Hatteras, NC.

Isurus oxyrinchus is the only other lamnoid shark for which at least some data is available giving some indication of gestation periods. *I. oxyrinchus* gives birth to up 18 neonates each 50-60% the size of *Carcharias taurus* embryos. When the TL and date for all known northern hemisphere *I. oxyrinchus* are plotted it appears that their gestation period is extended past 12 mos, possibly 15 mos (450 DAFI) Fig. 14.14. The smaller *I. oxyrinchus* ova, 12.5% the mass of *C. taurus* ova, and smaller ovary in *I. oxyrinchus* may require a longer gestation period to produce multiple 60-70 cm TL embryos at parturition. Considering that *I. oxyrinchus* females must forage for prey in an oligotrophic pelagic environment relative to *C. taurus* females in a nearshore relatively eutrophic environment, prey limitations and higher metabolic rate may limit nutritional allocations to mako embryos when compared to sand tiger embryos.

14.2.3 Neonate Anatomy and Ecology

Neonate *Carcharias taurus* form large schools and migrate north entering temperate estuaries north of Florida during the spring and summer following parturition. There are no records of neonate *C. taurus* captures off east central Florida in spite of numerous adult captures over the past 50 yrs. We have records of twelve 965 - 1575 mm TL neonate *C. taurus* captured in the St. Marys River estuary of NE Florida during the summer of 1968 (13 July to 17 August). These sharks were migrating north from birthing grounds further south along the Florida coast. However, neonates are particularly common inshore from Cape Hatteras north to Cape Cod during spring and summer. These observations indicate that *C. taurus* young forage in temperate estuaries and nearshore continental shelves immediately after birth.

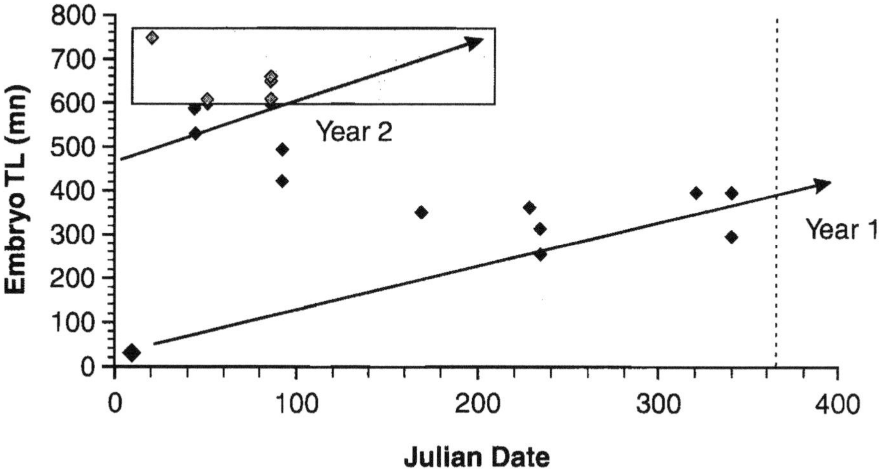

Fig. 14.14 Temporal distribution of total length range (only range end points are plotted for each litter) for all *Isurus oxyrinchus* embryos known from literature and specimens captured by the authors, limited to the northern hemisphere, Pacific and Atlantic Oceans. Shaded area represents potential mating and parturition period. Original.

Neonate *Isurus oxyrinchus* have been captured on long lines set for tuna and swordfish off east central Florida and in the Gulf of Mexico. These captures indicate that makos are pelagic immediately after birth. Mako neonate ecology, and potential predatory mortalities as well as foraging strategy differ significantly from that of *Carcharias taurus* and may have a significant impact on the reproductive strategy adopted by each of these species.

14.3 COMPARATIVE REPRODUCTIVE STRATEGIES IN LAMNOID SHARKS

The most successful reproductive strategy is one that produces as many fit progeny as necessary to insure species survival at the lowest cost in maternal resources. The individual strategies that accommodate the most successful reproductive scenario for the species depends on the niche occupied by the species. The individual niche dictates the limitations placed on the organism by its environment and interactions with other species. Lamnoid shark niche diversity is quite high for an order containing only 16 species in seven families. There are pelagic and mesopelagic planktivorous species, *Cetorhinus* and *Megachasma*; pelagic/neritic top predators, *Isurus oxyrinchus*, *Lamna nasus/ditropis* and *Alopias vulpinus/pelagicus*; mesopelagic predators, *Isurus paucus*, *Alopias superciliosus* and *Pseudocarcharias kamohari*; bathyal predators, *Mitsukorina* and *Odontaspis* spp; a coastal/pelagic predator, *Carcharodon carcharias*; and the only shallow inshore estuarine/coastal predator in the order, *Carcharias taurus*. This great lamnoid niche diversity insures a diversity of prey abundance and feeding strategies producing energetic resources for reproduction as well as a diversity of predatory environments for neonates after birth. All species appear to be oophagous which is the least complex method of nourishing large embryos within a modified oviduct (uterus), especially when compared to the complexities of placentation exhibited by the carcharhiniform sharks.

Two different oophagous strategies are now apparent in lamnoid sharks. The most common strategy is one that allows the development of multiple embryos 2-9 per uterus during relatively long gestation periods, 12 to possibly 15 months or more. These pregnancy periods indicate that species using this strategy would not mate and become impregnated every year, but would skip years. The relatively low number of pregnant females relative to the number of mature adult females captured indicate that this may be the case in most lamnoids studied, particularly in *Isurus oxyrinchus* and *Carcharodon carcharias*. Except for *C. carcharias* and *I. paucus* the species in this category typically produce young with body lengths (precaudal length, not total length) of 40 to 70 cm at birth. This category includes *I. oxyrinchus, Alopias* spp., *Lamna* spp., and *Pseudocarcharias kamohari*. Neonate size exceptions are observed in largest predaceous sharks within the order, *C. carcharias* and *I. paucus* which can produce multiple embryos up to 150 cm in length. Maximum fecundity occurs in *I. oxyrinchus,* 18 neonates and *C. carcharias,* with 14 neonates. Production of many large multiple embryos implies higher potential predatory mortality after birth in these two species.

The second and apparently the rarest oophagous strategy also incorporates embryophagy limiting the progeny to one large embryo per uterus. To date, this has only been documented in *Carcharias taurus* the only estuarine and shallow coastal shark within the order. The production of only two one meter neonates implies a lower neonate mortality risk. A one meter predatory neonate in estuarine and shallow coastal waters could be very competitive with most other predators encountered in these habitats, which are typically considerably smaller than neritic or oceanic predators. Since inshore estuarine and coastal environments are more likely to be eutrophic creating environments with an abundance of prey, maternal nourishment would likely be more predictable and constant than in the oligotrophic environment of the makos and threshers. Therefore, *C. taurus* would be more likely to produce larger ova at ovulation, from a relatively large ovary, mate consecutively over a longer period of time in a relatively protected environment and mate on an annual cycle, thus doubling the annual neonate production on a two year cycle matching *Alopias* and *Pseudocarcharias* production of four neonates every two years if multiple annual cycles are characteristic of these species. If sufficient predictable maternal food resources are available the lamnoid strategy should be to produce large embryos as quickly as possible, and as often as possible. The eutrophic environment of *C. taurus* insures abundant prey availability for maternal consumption and allows this species to devote considerable energy to reproduction allowing *C. taurus* embryos to grow more rapidly than those of other lamnoid sharks.

There is considerable evidence for annual reproduction in *Carcharias taurus* as, (1) all mature females captured south of Cape Hatteras from 1946 to 1995 have been pregnant during the May-January gestation period, 237 females examined (Springer 1948, Jacksonville Shark Club notes 1960 to 1975; F. G. Wood letters 1976, Dodrill 1977; Castro, pers. comm. South Carolina captures, Gilmore *et al.* 1983; Gilmore 1993); (2) individual marked female *C. taurus* observed at a specific mating site south of Cape Hatteras, NC were observed with mating scars during spring at this site on an annual basis for eight consecutive years, 1984-1992, by the same scuba divers and remotely operated vehicles and documented by still and video photography by the authors; (3) all pregnant females captured and examined at the same time of year over five decades off east Florida have been in the same embryonic developmental stage indicating a highly predictable reproductive pattern and population synchrony in reproductive periodicity. The embryonic development stage was very predictable for *C. taurus*, thus time of mating/insemination and location of mating.

There is no evidence for biannual reproduction in *Carcharias taurus*. The few aquarium observations of dead, aborted, still or live birthed *C. taurus* neonates and embryos indicate that aquarium observations of reproductive conditions are more likely to be misleading and exceptional when compared to observations made of populations under natural field conditions. The placement of a large wild predator in captivity, particularly a pregnant one which requires considerable food resources to support ravenous oophagous embryos, must place exceptional stress on maternal physiology and behavior, thus her embryos. Behavioral observations made in the wild and in public aquaria indicate that *C. taurus* may have a social hierarchy, particularly during mating periods, thus capture and placement in aquaria may place additional stress on the specimen based on social environmental changes.

The comparison of the *Carcharias taurus* reproductive strategy with that of *I. oxyrinchus* allows a life history strategy to be developed that supports the apparent differences in their reproductive strategies. They present two reproductive models to compare and examine more thoroughly as more specimens become available for study.

14.3.1 Functional Significance of Oophagy

Oophagy appears to be an universal embryonic nutritive source in lamnoid sharks. This strategy places several limits on maternal behavior and ecology. The pregnant female must reside where sufficient resources are available for efficient foraging, maintenance of maternal metabolism and yet produce large masses of yolk for embryonic consumption for several months at a time. In pelagic environments this may require migration with potential prey or to locations where prey aggregate at least seasonally. In the warm temperate coastal zone this will require maternal foraging where rich estuarine and coastal productivity will be most accessible. When these adult metabolic requirements are met, extended ovarian activity allows the nourishment of embryos throughout most of the gestation period. In lamnoid sharks this has evolved into the universal ovarian morphology designed for rapid ova production and maintenance of high ovulation rates. This method supercedes the need to place all embryonic nutritive sources in a single large ovum, as in scyliorhinids, orectolobids and most plesiomorphic elasmobranch groups. These latter reproductive modes require that the embryo forage for adequate resources immediately after hatching at a relatively small size. If hatching occurs in a highly predatory environment this strategy may not be effective in allowing neonate survival. Coastal and estuarine environments as well as oceanic pelagic environments are highly predatory environments. Undoubtedly there is selection for increasing litter size and/or larger embryos at birth in pelagic environments. Embryonic placentation in carcharhinoid sharks represents a alternative reproductive strategy to lamnoid oophagy. When compared to oophagy in lamnoids the most competitive reproductive strategy is placentation as observed in carcarhinoid sharks. The Cenozoic evolution of carcharhinoid sharks with their placental embryos undoubtedly had great influence on potential niche competition with lamnoid species and relative reproductive success. Both strategies produce large multiple precocious neonate predators. However the behavioral and metabolic pathways for embryonic support differ significantly and must have ecological and niche implications for the maternal adult.

14.3.2 Functional Significance of Cannibalism

Studies of *Carcharias taurus* have demonstrated that embryonic growth rate increases substantially when cannibalism occurs during gestation. At this point in development a relatively small embryo (55-60 mm TL) has hatched into a nourishing uterine environment filled with egg capsules. Some of these capsules contain moving metabolizing embryos smaller than the hatchling. Jaws and teeth are well developed and as the embryo reaches a size where it can cut through egg membranes it does so to capture and consume the embryos within. There is evidence that the first hatchling actually hunts and seeks those few capsules containing other embryos amidst numerous larger capsules present that contain unfertilized yolk. With the completion of cannibalism unfertilized yolk capsules are used as a food source, the oophagous stage. Embryonic mouth size relative to body size is greatest during this cannibalistic period, and declines in relative size as ontogeny progresses. However the relatively large size (335 mm TL) of the embryo after cannibalism is complete insures that the embryo can readily consume egg capsules. Cannibalism insures rapid growth for 50 days producing a large embryo ready to consume a backlog of ovoid egg capsules already within the uterus and an embryo capable of reaching 1,000 mm TL just 190-200 days after fertilization. If maternal food resources are available to support this great energy expenditure and allow the forfeiture of multiple embryos, due to low neonate mortality rates, then production of fewer large embryos more rapidly on an annual cycle appears to be a successful reproductive strategy, particularly in shallow coastal environments. Uterine cannibalism offers this selective advantage.

14.3.3 Reproductive Strategy and Phylogeny in Lamnoid Sharks

Even though *Carcharias taurus* appears to be an archaic lamnoid species the plesiomorphic reproductive strategy may likely be oophagy rather than embryophagy. Embryophagy may be a specialized derived condition associated with lamnoid invasion of shallow coastal estuaries and sublittoral zones. The potential for the *C. taurus* reproductive strategy to be specialized, possibly a derived character state, is evident when a variety of developmental characters are compared to other lamnoid species. *C. taurus* reproductive characters reveal a response to the need for very rapid embryo growth such as: (1) large ova; (2) exceptional ovulation rates; (3) endocoelomic yolk supplies in an "amphibian-like" early embryo; (4) 280-290 day gestation period; and an (5) annual reproductive cycle. It is likely that these characteristics evolved as a response to producing a very large predatory competitive embryo in a shallow high calorie environment while other lamnoids had to cope with higher neonate mortalities in a more oligotrophic environment. There is evidence that *C. taurus* may mate multiple times over several weeks, while it is more likely that *Isurus* spp. mate once, ephemerally and conserve multiple fertilized ova in the initial capsule formed. Mollett *et al.* (2000) predict a three year reproductive cycle for pelagic *I. oxyrinchus*, three times longer than that documented for sub-littoral *C. taurus*. Only with the capture of a sufficient number of reproductively active *Odontaspis ferox*, *O. noronhai*, *Mitsukurina owstoni* and *Megachasma pelagios*, will we finally know what may be the plesiomorphic reproductive mode in lamnoid sharks as these genera are considered among the most primitive living lamnoids.

14.4 ACKNOWLEDGEMENTS

We thank Tris Colkett and Eric Sander for furnishing the majority of *Carcharias taurus* and *Carcharodon carcharias* specimens examined in addition to pregnant specimens of *Isurus oxyrinchus*. The D & D Wholesale Fish House of Fort Pierce Florida furnished most of the *Isurus spp.* and *Alopias spp.* specimens examined. The Florida Shark Club of Jacksonville Florida furnished valuable *C. taurus* landing records. The original research for this chapter was supported by the Harbor Branch Oceanographic Institution and the Atlantic Foundation. Most lamnoid embryonic and reproductive organ specimens previously housed at the Harbor Branch Oceanographic Institution, the Harbor Branch Oceanographic Museum have been transferred to the Florida State Museum in Gainesville, curated by George Burgess and Robert Robins. Observations of *C. taurus* aggregations on the continental shelf off North Carolina, 1987-1993 were supported by the National Oceanic and Atmospheric Administration National Undersea Research Center Wilmington, North Carolina, The Audubon Society, Stan Waterman Productions and Jack McKenney Productions, Los Angeles California.

14.5 LITERATURE CITED

Branstetter, S. 1981. Biological notes on the sharks of the north central Gulf of Mexico. Contributions Marine Science, 24:13-34.

Cadenat, J. 1956. Note d'ichtyologie ouest-Africaine. XIV.-Remarques biologiques sur le requin-sable *Carcharias (Carcharias) taurus* Rafinesque 1810. Bulletin Institute French Africa Noire 18: 1249-1256.

Castro, J. I., E. Clark, K. Yano, and Nakaya, K. 1997. The gross anatomy of the female reproductive tract and associated organs of the Fukuoka megamouth shark (*Megachasma pelagios*). Pp. 115-119. In K. Yano, J. F. Morrissey, U. Y. Yabumototo and K. Nakaya (eds), *Biology of the Megamouth Shark.* Tokai University Press, Tokyo, Japan.

Dodrill, J. W. 1977. A hook and line survey of the sharks found within five hundred meters of shore along Melbourne Beach, Brevard County, Florida. M.S. Thesis, Florida Institute of Technology, Melbourne, Florida. 304 pp.

Francis, M. 1996. Chapter 15: Observations on a pregnant white shark with a review of reproductive biology. Pp. 157-172. In A. P. Klimley and D. G. Ainley, (eds), *Great White Sharks: The Biology of Carcharodon carcharias.* Academic Press, San Diego, CA.

Gilmore, R. G. 1983. Observations on the embryos of the longfin mako, *Isurus paucus*, and the bigeye thresher, *Alopias superciliosus.* Copeia, 1983: 375-382.

Gilmore, R. G., J. Dodrill and Linley, P. 1983. Reproduction and embryonic development of the sand tiger shark, *Odontaspis taurus* (Rafinesque). U.S. Fishery Bulletin, 81: 201-225.

Gilmore, R. G. 1993. Reproductive biology of lamnoid sharks. Environmental Biology Fish, 38: 95-114.

Gruber, S. H. and Compagno, L. J. V. 1981. Taxonomic status and biology of the bigeye thresher, *Alopias superciliosus.* U.S. Fishery Bulletin, 79: 617-640.

Gudger, E. W. 1940. The breeding habits, reproductive organs, and external embryonic development of *Chlamydoselachus* based on notes and drawings left by Bashford Dean. Pp. 521-646. In E. W. Gudger (ed.), *Bashford Dean memorial volume - archaic fishes.* Article 7, American Museum Natural History, NewYork, NY.

Hamlett, W. C. 1983. Maternal-fetal relations in elasmobranch fishes. Ph.D. Dissertation, Clemson Univ., Clemson, South Carolina. 228 pp.

Hamlett, W. C. and Koob, T. J. 1999. Chapter 15: Female reproductive system. Pp. 398-443. In W.C. Hamlett (ed.), *Sharks, skates and rays: The biology of Elasmobranch fishes*, Johns Hopkins University Press, Baltimore, MD.

Kent, B. W.1994. *Fossil sharks of the Chesapeake Bay region.* Egan Rees & Boyer, Inc., Columbia, Maryland. 146 pp.

Long, D. J. and Waggoner, B. M. 1996. Chapter 5. Evolutionary relationships of the white shark: a phylogeny of lamniform sharks based on dental morphology. Pp. 37-47. In A. P. Klimley and D. G. Ainley, (eds), *Great White Sharks: The Biology of Carcharodon carcharias*. Academic Press, San Diego, CA.

Martin, A. P. 1996. Chapter 6: Systematics of the lamnidae and the origination time of *Carcharodon carcharias* inferred from the comparative analysis of mitochondrial DNA sequences. Pp. 49-53. In A. P. Klimley and D. G. Ainley, (eds), *Great White Sharks: The Biology of Carcharodon carcharias*. Academic Press, San Diego, CA.

Martin, A. P. and Naylor, G. J. P. 1997. Independent origins of filter-feeding in megamouth and basking sharks (order lamniformes) inferred from phylogenetic analysis of cytochrome b gene sequences. Pp. 39-50. In K. Yano, J. F. Morrissey, U. Y. Yabumototo and K. Nakaya (eds), *Biology of the Megamouth Shark*. Tokai University Press, Tokyo, Japan.

Matthews, L. H. 1950. Reproduction in the basking shark, *Cetorhinus maximus* (Gunnerus). Philosophical Transactions Royal Society London, 234b: 247-316.

Mollet, H. F., G. Cliff, Pratt, Jr., H. L., and Stevens, J. D. 2000. Reproductive biology of the female shortfin mako *Isurus oxyrinchus* Rafinesque, 1810 with comments on the embryonic development of lamnoids. Fishery Bulletin, 98: 299-318.

Moreno, J. A. and Morón, J. 1992. Reproductive biology of the bigeye thresher shark, *Alopias superciliosus* (Lowe, 1839). Australian Journal Marine Freshwater Research, 43: 77-86.

Morrissey, J. F., Dunn, K. A. and Mule, F. 1997. The phylogenetic position of *Megachasma pelagios* inferred from mtDNA sequence data. Pp. 33-37. In K. Yano, J. F. Morrissey, U. Y. Yabumototo and K. Nakaya (eds), *Biology of the Megamouth Shark*. Tokai University Press, Tokyo, Japan.

Pratt, H. L., Jr. 1988. Elasmobranch gonad structure: a description and survey. Copeia, 1988: 719-729.

Pratt, H. L., Jr. 1993. The storage of spermatozoa in the oviducal glands of western North Atlantic sharks. Environmental Biology Fish, 38: 139-149.

Pratt, H.L., Jr. and Tanaka, S. 1994. Sperm storage in male elasmobranchs: a description and survey. Journal Morphology, 219:297-304

Smith, B. G. 1942. The heterodontid sharks: Their natural history and the external development of *Heterodontus (Cestracion) japonicus* based on notes and drawings by Bashford Dean. Pp. 651-770. In E. W. Gudger (ed.), *Bashford Dean memorial volume archaic fishes*, Article 8,. American Museum Natural History, N.Y.

Springer, S. 1948. Oviphagous embryos of the sand shark, *Odontaspis taurus*. Copeia, 1948: 153-157.

Stevens, J. D. 1983. Observations on reproduction in the shortfin mako. Copeia, 1983: 126-130.

Tanaka, S, and Yano, K. 1997. Histological observations on the reproductive organs of a female megamouth shark, *Megachasma pelagios*, from Hakata Bay, Japan. Pp. 121-129. In K. Yano, J. F. Morrissey, U. Y. Yabumototo and K. Nakaya (eds), *Biology of the Megamouth Shark*. Tokai Univeersity Press, Tokyo, Japan.

TeWinkel, L. E.1950. Notes on ovulation, ova, and early development in the smooth dogfish, *Mustelus canis*. Biological Bulletin (Woods Hole), 99: 474-486.

TeWinkel, L. E. 1963. Notes on the smooth dogfish, *Mustelus canis*, during the first three months of gestation. II. Structural modifications of yolk-sacs and yolk-stalks correlated with increasing absorptive function. Journal Experimental Zoology, 152: 123-137.

Uchida, S., Toda, M., Teshima, K., and Yano, K. 1996. Chapter 14: Pregnant white sharks and full-term embryos from Japan. Pp. 139-155. In A. P. Klimley and D. G.

Ainley (eds), *Great White Sharks: The Biology of Carcharodon carcharias*. Academic Press, San Diego, CA.

Wourms, J. P. 1977. Reproduction and development in chondrichthyan fishes. American Zoologist, 17: 379-410.

CHAPTER 15

Placentatrophy in Sharks

William C. Hamlett[1], Carolyn J. P. Jones[2] and Luana R. Paulesu[3]

15.1 INTRODUCTION

Placental sharks sequentially utilize three modes of embryonic/fetal nutrition to nourish their developing young viz. 1) lecithotrophy, 2) histotrophy and 3) placentatrophy. Preimplantation shark embryos begin their ontogenetic development by reliance on yolk. In vitellogenesis, energy-rich precursors of hepatic origin are stored in oocytes prior to ovulation. Following fertilization, embryogenesis and fetal development proceed at the expense of the yolk stores sequestered in the yolk sac, hence lecithotrophy (Hamlett and Wourms 1984). Yolk is made available to the developing young by two means. First, yolk is partially solubilized in the yolk syncytium of the yolk sac and subsequently modulated through the yolk sac endoderm. Yolk metabolites are then transported across the vitelline endothelium of yolk sac vessels to reach the fetal circulation (Hamlett 1987; Hamlett *et al.* 1987). Second, yolk granules are physically transported to the fetal gut by ciliated cells lining the ductus vitellointestinalis. The ductus is an endodermally lined conduit that extends from the yolk sac to connect to the fetal gut where digestion and absorption occur. As yolk stores in the yolk sac are being utilized and prior to development of the placenta, nutrient substances are supplied to the fetus by secretory activity of the uterus in the form of histotroph, hence histotrophy. Histotroph is then ingested and absorbed. In placentatrophy, the distal aspect of the yolk sac becomes modified to serve as a yolk sac placenta.

In the ontogenetic transition from lecithotrophy, to histotrophy, to placentatrophy, there is considerable overlap in the modes and a

[1]Department Anatomy and Cell Biology, Indiana University School of Medicine, Notre Dame, IN 46530, USA

[2]Academic Unit of Obstetrics and Gynaecology, University of Manchester, St. Mary's Hospital, Manchester M13 0JH, United Kingdom

[3]Departimento de Physiologia, Università di Siena, Siena, Italia

coincident shift in the function of the uterus. Some degree of histotrophy occurs throughout gestation and one may view different degrees of histotrophy in the three stages of gestation. During the lecithotrophic phase histotrophy is limited to mucous secretion to primarily serve as a lubricant. In the second phase there is a distinct change in the amount and type of secretions. During this phase, molecules of uterine origin accumulate in periembryonic fluids. One may describe placentatrophy as a highly modified form of histotrophy because there is no published evidence for hemotrophic transfer from mother to fetus; all placental transfer originates from uterine secretory activity. Uterine epithelium at the placental attachment site overlies a rich vascular tree and this area serves as the maternal component of the placenta. Adjacent areas of the uterus are termed paraplacental sites and continue to secrete mucus. Essential adaptations of the uterus include: enhanced secretory function, expansion to accommodate the embryos, respiration, osmoregulation, waste disposal, and protection of the young.

Evolution of placentation involves modifications of existing maternal and fetal membranes. The ontogenetic transformation of the shark oviduct into a functional uterus for the retention and metabolic support of developing young is regulated by endocrines (Chapter 9 of this volume). In the developing fetus, the yolk sac and stalk display terminal differentiation into a regionally specialized placenta and an umbilical cord respectively. Accessory vascular structures of the umbilical cord, termed appendiculae, are present in some species where they may serve a nutrient absorptive function as well as contribute secretions to periembryonic fluids (Southwell and Prasad 1919; Mahadevan 1949; Hamlett 1993; Hamlett *et al.* 1993a, b, c).

15.2 BACKGROUND

Early 19th century descriptions of the selachian yolk sac placenta include those of Ranzi (1932, 1934); Cate-Hodemaker (1938); Calzoni (1936) and Mahadevan (1940). Gilbert and Schnerlitzauer (1966) and Schlernitzauer and Gilbert (1966) provided brief reports and Teshima added to the data with an extensive series of papers (1972, 1973, 1974, 1975, 1978). The first transmission electron microscopic studies are those of Hamlett and Wourms (1984), Hamlett *et al.* (1985a, b, c) and Otake and Mizue (1985, 1986). These were followed by studies of Hamlett (1986, 1987, 1989, 1990a, b, 1993, 1998), Hamlett *et al.* (1993a, b, c) and Fishelson and Baranes (1998).

15.3 UTERINE COMPARTMENTALIZATION

Greatly distended abdomens characterize female placental sharks near term. This is a result of the enormous size of the gravid uteri (Fig. 15.1A). Uteri are thin and embryos can be seen through the walls without dissection. In early preimplantation stages individual fertilized eggs are enclosed by a tertiary egg envelope formed by two sheets fused at the

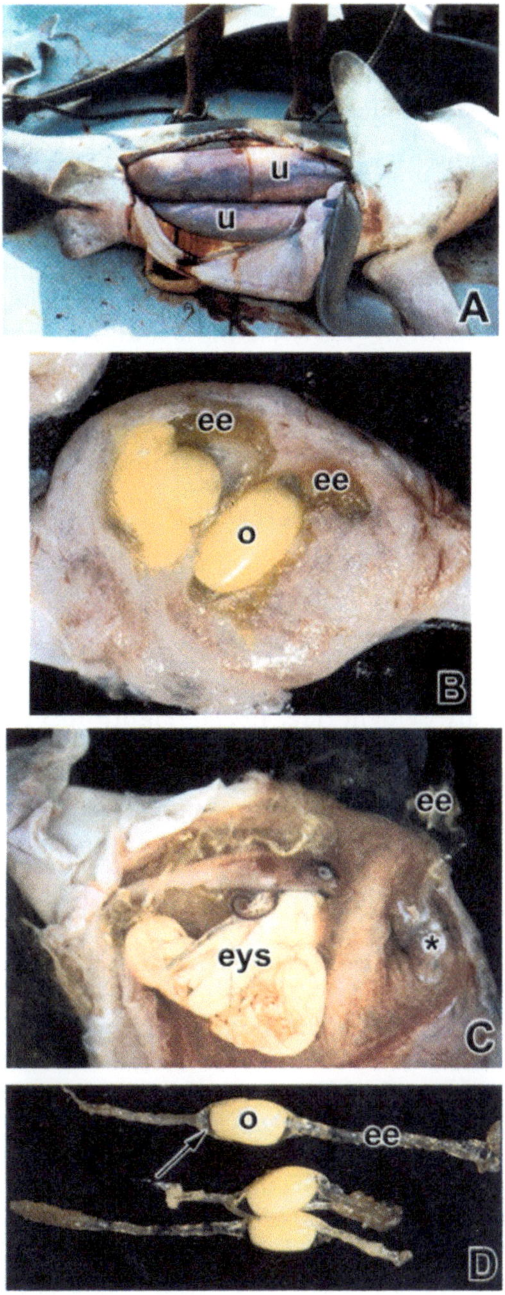

Fig. 15.1 A. Dilated uteri (u) of *Carcharhinus plumbeus* containing term fetuses. **B**. Uterus prior to uterine compartment formation containing two ova (o) surrounded by the tertiary egg envelope (ee) with the majority of the envelope plaited. **C.** Preimplantation embryo with external yolk sac (eys) in uterine compartment. Excess plaited egg envelope (ee) occupies the egg envelope reservoir (asterisk). **D.** Egg jelly (arrow) surrounds ova (o) enclosed by the egg envelope (ee). Original.

edges. The egg is situated at the posterior end of the envelope and 90% of the excess envelope is folded or plaited anterior to the egg (Fig. 15.1B). Initially eggs are free in the common uterine chamber but in the first few months of gestation dorsal and ventral uterine flaps approach each other to form a separate uterine compartment for each embryo (Fig. 15.1C). Excess egg envelope is sequestered in an egg envelope reservoir. Fertilized eggs pass through the oviducal gland (Hamlett *et al.* 1998; Chapter 10 of this volume) where they are invested with egg jelly and an amber colored, pliable and transparent tertiary egg envelope (Fig. 15.1D) (Baranes and Wendling 1981; Hamlett and Wourms 1984). The egg envelope is of sufficient size to accommodate the term embryo (Fig. 15.2A, B) and in *Carcharhinus plumbeus* the length of the extended envelope is 100-130 cm while the length of the compartment is initially only 12-14 cm long. As the embryo grows the uterine chamber enlarges and is oriented diagonally with the anterior end of each embryo pointing toward the oviducal gland and the yolk sac placenta (Fig. 15.2C) positioned at the posterior of each chamber. As the embryo grows, excess envelope is withdrawn from the egg envelope reservoir which becomes progressively smaller until it closes and disappears. The term embryo is enclosed by the tertiary egg envelope and bathed in periembryonic fluid (Fig. 15.2B). Following parturition, distinct raised vascular uterine scars, also called placental scars, are evident (Fig. 15.2D). Figure 15.3 illustrates stages in the formation of uterine compartments and the egg envelope reservoir.

15.4 DIVERSITY OF PLACENTA MORPHOLOGY

Despite their common origin from yolk sac and uterus to form the utero-placental complex, there is morphological diversity in placentae. In some species such as *Carcharhinus plumbeus* (Hamlett *et al.* 1985a, b, c), *Carcharhinus acronotus* (Hamlett 1990a, b), *Rhizoprionodon terraenovae* (Hamlett 1993, Hamlett *et al.* 1993a, b), the placenta is specialized into two segments: a smooth, sparsely vascularized proximal portion that has characteristics of a steroid-producing tissue and a more richly vascularized distal portion that is a nutrient-transporting tissue. In other species including *Mustelus canis* (Hamlett *et al.* 2002, Hamlett unpublished) and *Iago omanensis* (Fishelson and Baranes 1998) there is no distinct proximal part of the placenta. In *Scoliodon laticaudus* (Southwell and Prasad 1919, Thillayampalam 1928, Mahadevan 1940, Setna and Sardangdhar 1948, Teshima *et al.* 1978) the uterus forms a unique structure termed a trophonematous cup not present in any other species. In most species the tertiary egg envelope produced by the oviducal gland (Chapter 10 of this volume) persists throughout gestation and becomes a component of the maternal-fetal interface. In *Scoliodon* and *Prionace*, however this is not the case.

In sharks, as in most chondrichthyans, external yolk sac contents are gradually depleted during gestation and in most species the yolk stalk

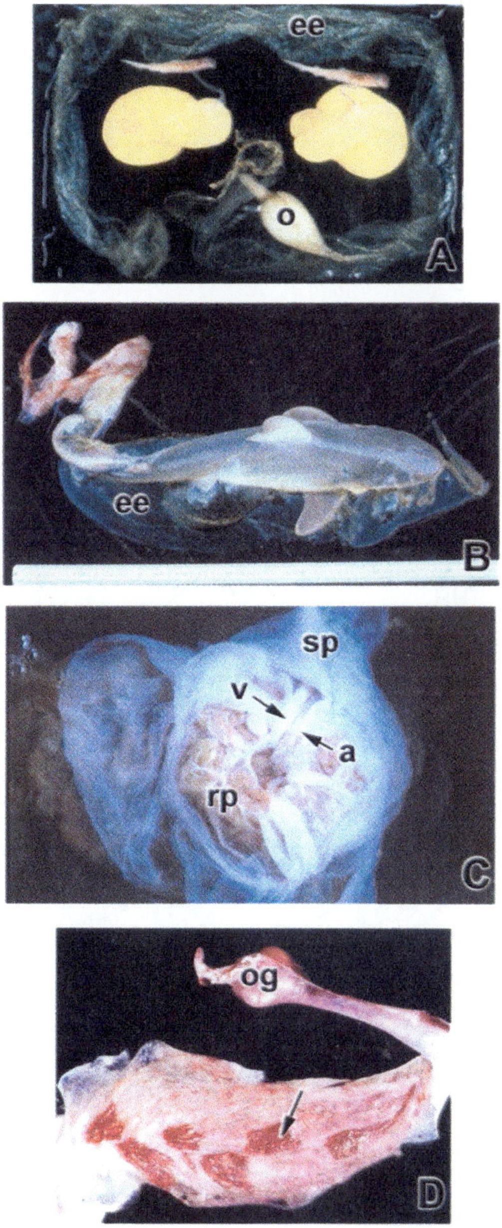

Fig. 15.2 A. *Negaprion brevirostris.* An egg envelope (ee) is shown unfolded. It is of sufficient size to accommodate the term fetus. One unfertilized ovum (o) is also surrounded by egg envelope. **B**. Term fetus *Carcharhinus plumbeus* fills the egg envelope (ee) which also contains periembryonic fluids. **C**. Term placenta of *C. plumbeus* with the smooth proximal portion of the placenta (sp) dissected and retracted to show the distal rugose (rp) portion of the yolk sac that forms the fetal portion of the placenta. No yolk remains inside the yolk sac and the umbilical artery (a) and vein (v) are visible. Original. **D**. Postpartum uterus of *Mustelus canis* shows red, elevated uterine scars (arrow) also referred to as placental scars. Oviducal gland = og. Original.

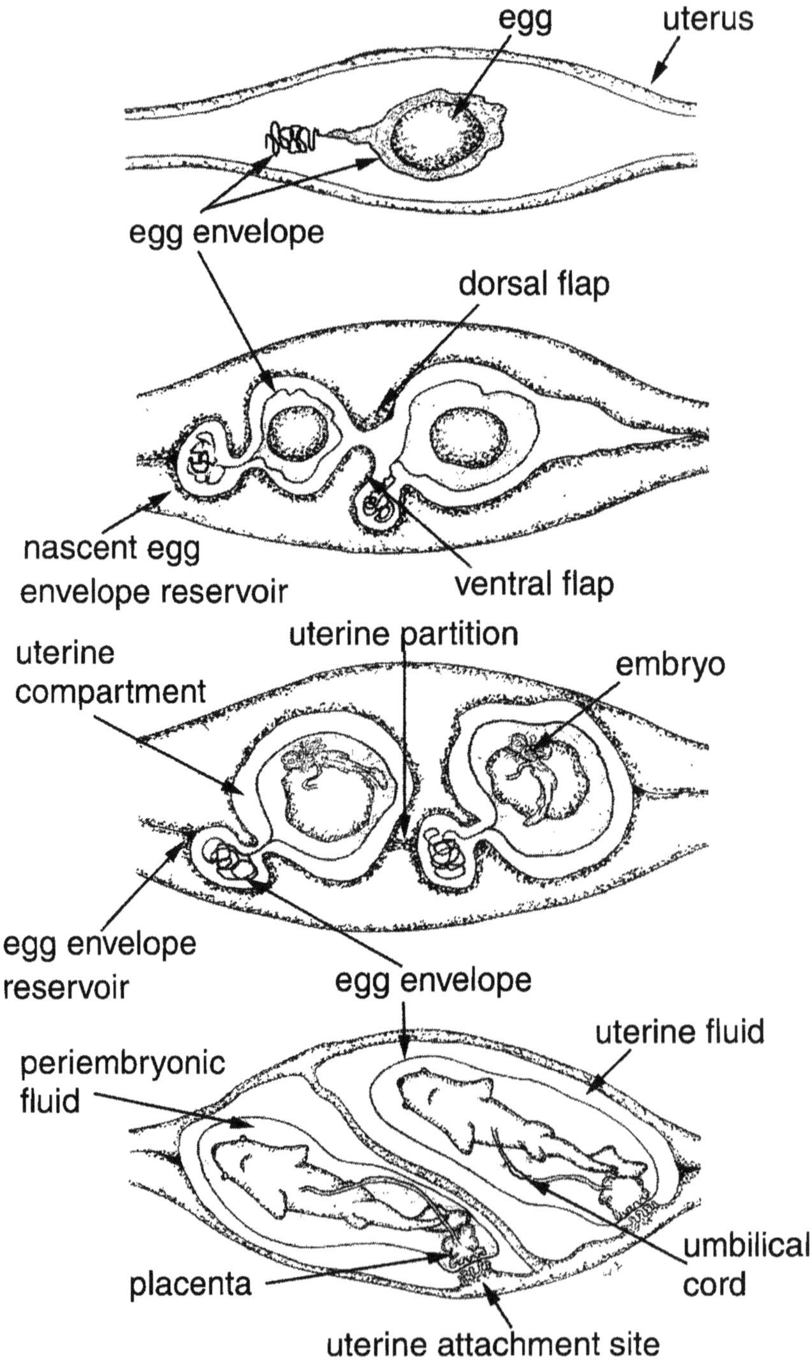

Fig. 15.3 Diagrams showing formation of uterine compartments and egg envelope reservoir. Original.

and sac are retracted and withdrawn such that all that remains is a transient yolk sac scar on the abdomen. In placental sharks the empty yolk sac persists to form the fetal portion of the yolk sac placenta and the yolk stalk elongates to form the umbilical cord. The general gross morphology of placentae in various groups shows some degree of diversity. Based on gross morphology, Mahadevan (1940) and Setna and Sragangdhar (1948) described three types of placentae in several Indian species. They termed the placenta type of *Scoliodon laticaudus* as columnar because the yolk sac portion of the placenta attaches to a trophonematous uterine stalk. Two other non-stalked placenta types are the discoidal type in which the placental portion is disk-like, oval, lobular and richly vascularized. They also recognize an entire type of placenta in which the corrugated yolk sac fits into a pocket in the uterine wall. Compagno (1988) reported the placenta type in *Leptocharias smithii* as globular where the fetal portion of the placenta is a smooth globe that attaches to a small circular area of the uterus.

In the triakid *Mustelus canis* the yolk sac is lobate and vascularized over its complete surface (Figs 15.4A, B). In several carcharhinid species including *Carcharhinus falciformis* (Gilbert and Schlernitzauer 1966), *Prionace glauca* (Otake and Mizue 1985, 1986), *Carcharhinus plumbeus* (Fig. 15.2C; 15.4C) (Hamlett *et al.* 1985a, b, c), *Carcharhinus acronotus* (Hamlett 1990) and *Rhizoprionodon terraenovae* (Fig. 15.4D) (Hamlett 1993; Hamlett *et al.* 1993a, b) a distinction between the relatively avascular smooth proximal portion of the yolk sac placenta and the richly vascularized, lobate distal portion of the yolk sac is noteworthy. Most reports have focused on the distal aspect of the yolk sac but two studies (Hamlett *et al.* 1985b; Hamlett *et al.* 1993b) have used ultrastructural tools to describe the fine structure of the proximal portion as well.

Another difference encountered is in the type of umbilical cord seen in various species. In most placental species including *Mustelus canis* (Fig. 15.4B) and *Carcharinus plumbeus* (Fig. 15.4C) the umbilical cord is smooth. In other species such as *Rhizoprionodon terraenovae* (Fig. 15.4D) the umbilical cord is decorated by vascularized appendages termed appendiculae. Appendiculae will be discussed in section 15.13 of this chapter.

15.5 LECITHOTROPHY: PREIMPLANTATION YOLK SAC

15.5.1 Ductus Vitellointestinalis

The yolk stalk of chondrichthyans contains three tubular structures, viz. (1) vitelline artery, (2) vitelline vein and (3) ductus vitellointestinalis (Fig. 15.5A). The extraembryonic coelom is also present. In placental species the vitelline artery and vein are later termed the umbilical vessels after the placenta is established. The ductus conveys yolk platelets to the embryonic alimentary canal for digestion and absorption. In *Mustelus canis* (TeWinkel 1963), *Rhizoprionodon terraenovae* (Hamlett *et al.* 1993b)

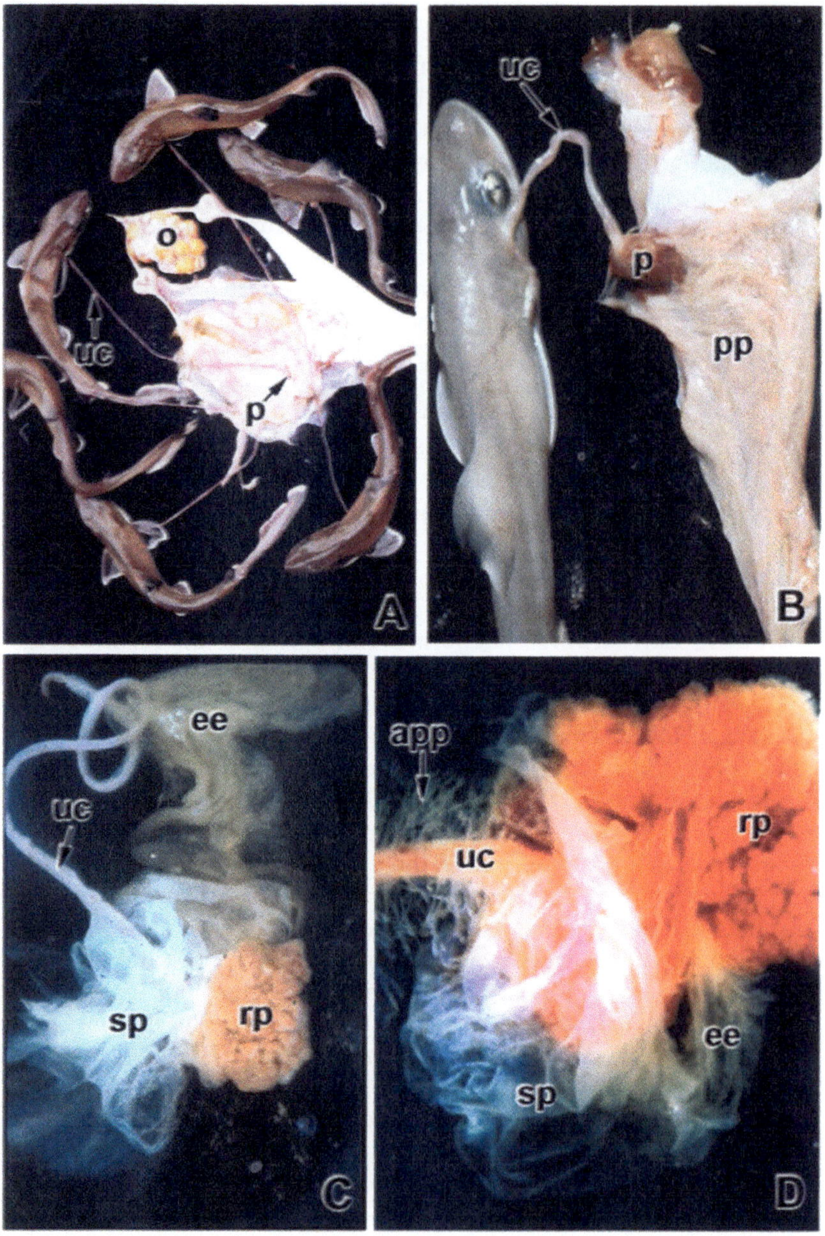

Fig. 15.4. A. Uterus of *Mustelus canis* containing 6 fetuses. Each placenta (p) is connected to the umbilical cord (uc). Ovary = o. **B**. Term *M. canis* fetus with umbilical cord (uc) connecting to the fetal portion of the placenta (p). Paraplacental uterine areas = pp. **C**. Term *Carcharhinus plumbeus* placenta showing differentiation of umbilical cord (uc), smooth proximal portion of the fetal placenta (sp) and the distal highly vascularized rugose portion (rp) of the fetal portion of the placenta. Egg envelope = ee. **D.** Term placenta of *Rhizoprionodon terraenovae* with umbilical cord (uc) adorned with appendiculae (app) and smooth proximal portion of the fetal placenta (sp) and the distal highly vascularized rugose portion (rp) of the fetal portion of the placenta. Egg envelope = ee. Original.

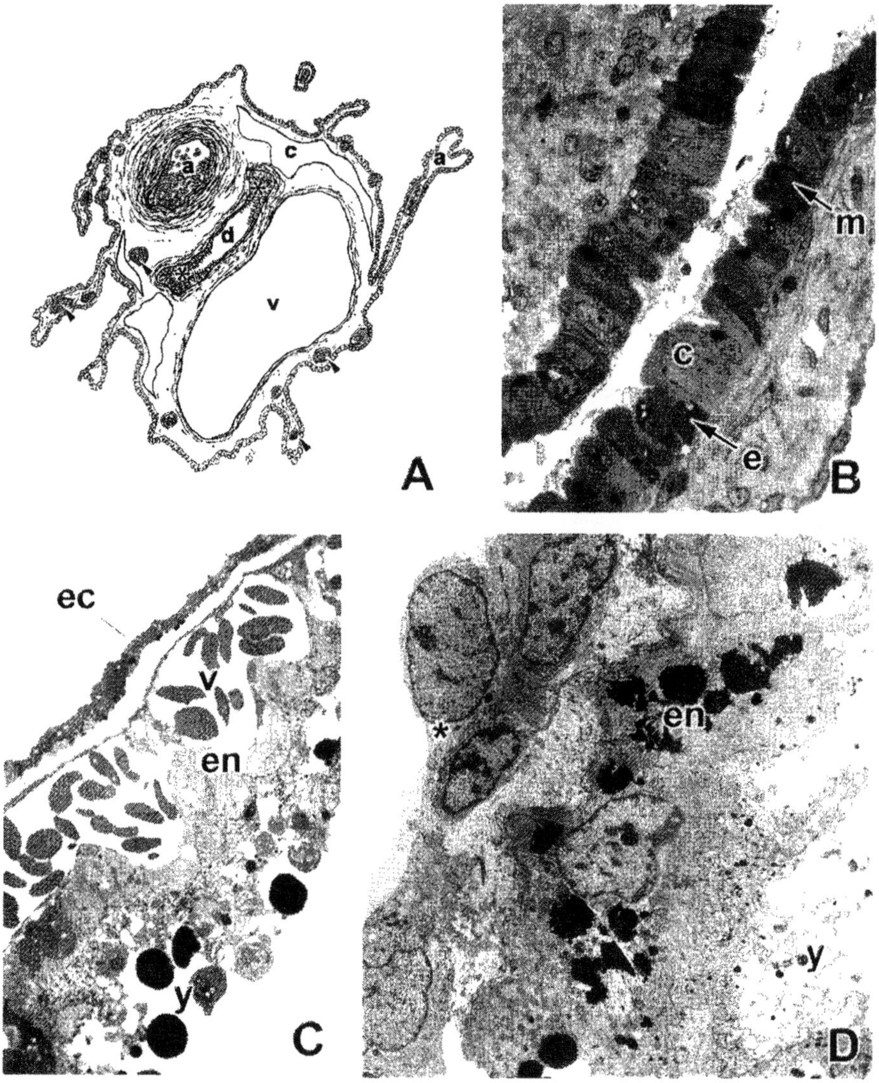

Fig. 15.5 A. Diagram of cross section of umbilical cord showing arteries (arrowheads), appendiculae (a), ductus vitellointestinalis (d), umbilical artery (a), umbilical vein (v) and extraembryonic coelom (c). From Hamlett, W. C. 1993. Environmental Biology of Fishes 38: 253-267, Fig. 4. **B.** Light micrograph of ductus vitellointestinalis showing microvillar (m), ciliated (c) and enteroendocrine (e) cells. X600. **C.** Light micrograph of preimplantation yolk sac showing ectoderm (ec), mesodermal vessels (v), endoderm (en) and yolk syncytium (y). X600. **D.** Transmission electron micrograph of preimplantation yolk sac showing yolk syncytium (y), endoderm (en), and endothelium (asterisk). X5,412. From W. C. Hamlett, W. C., Schwartz, F. J. and DiDio, L. J. A. 1987. Cell and Tissue Research 247: 275-285, Fig. 1.

and other placental sharks there is a marked expansion of the yolk sac and an increase in length and diameter of the yolk stalk. Hamlett *et al.* (1993b) used light microscopy (Fig. 15.5B) and transmission electron microscopy to examine the ductus vitellointestinalis of *R. terraenovae*. They found three types of epithelial cells viz. ciliated, microvillar and enteroendocrine. Ciliated cells also have sparse microvilli and are the tallest of the three cell types. Shorter microvillar cells have abundant glycogen and large lipid stores. Enteroendocrine cells are either pyramidal or round and lodged between the basal portions of surrounding cells. They contain basal secretory vesicles consistent with those of other vertebrate enteroendocrine cells where they exert paracrine regulation of local segments of the alimentary canal. Ciliated cells convey yolk to the embryo gut and microvillar cells may absorb some yolk metabolites that have been previously hydrolyzed in the yolk syncytium although this has yet to be experimentally verified. Teshima (1981) is of the opinion that unspecified cells of the ductus absorb yolk although electron microscopy was not employed in his study.

15.5.2 Yolk Syncytial-endoderm Complex

Hamlett *et al.* (1987) employed light and transmission electron microscopy to describe the yolk syncytial-endoderm complex of the preimplantation yolk sac of *Rhizoprionodon terraenovae* (Figs 15.5C, D). The yolk syncytium is bounded by a membrane anchored to the plasmalemma of adjacent endoderm cells by desmosomes. Enlarged nuclei, rough endoplasmic reticulum, Golgi complexes, mitochondria and other cellular organelles populate the syncytium. Microtubules and filamentous elements are also observed free in the syncytium. Yolk is present as pleomorphic droplets, the profiles of which are generally spherical but may be vesicular, especially at the periphery of large yolk droplets. Occasionally, large yolk droplets have a paracrystalline configuration. Small yolk droplets are modulated through the Golgi complex of the yolk syncytium where acid hydrolases are added. Small yolk droplets released from the maturing face of the Golgi are sequestered in membrane-limited packets (Fig. 15.6A). The membrane of the packets fuses with the membrane enveloping the yolk syncytium and the yolk droplets are released into the yolk syncytial-endoderm interspace (Fig. 15.6A). Subsequently, yolk droplets are endocytosed by the endoderm. Yolk droplets disperse and fuse to form the large irregular yolk inclusions of the endoderm. Yolk metabolites are then transported through the yolk sac endothelium. Yolk sac endoderm thus mediates the transfer of metabolites from the yolk mass to the extraembryonic circulation.

15.5.3 External Gill Filaments

During the lecithotrophic phase the uterus simultaneously produces mucoid secretions, hence a degree of histotrophy occurs throughout gestation (Hamlett 1989). Hamlett *et al.* (1985d) investigated the ability of

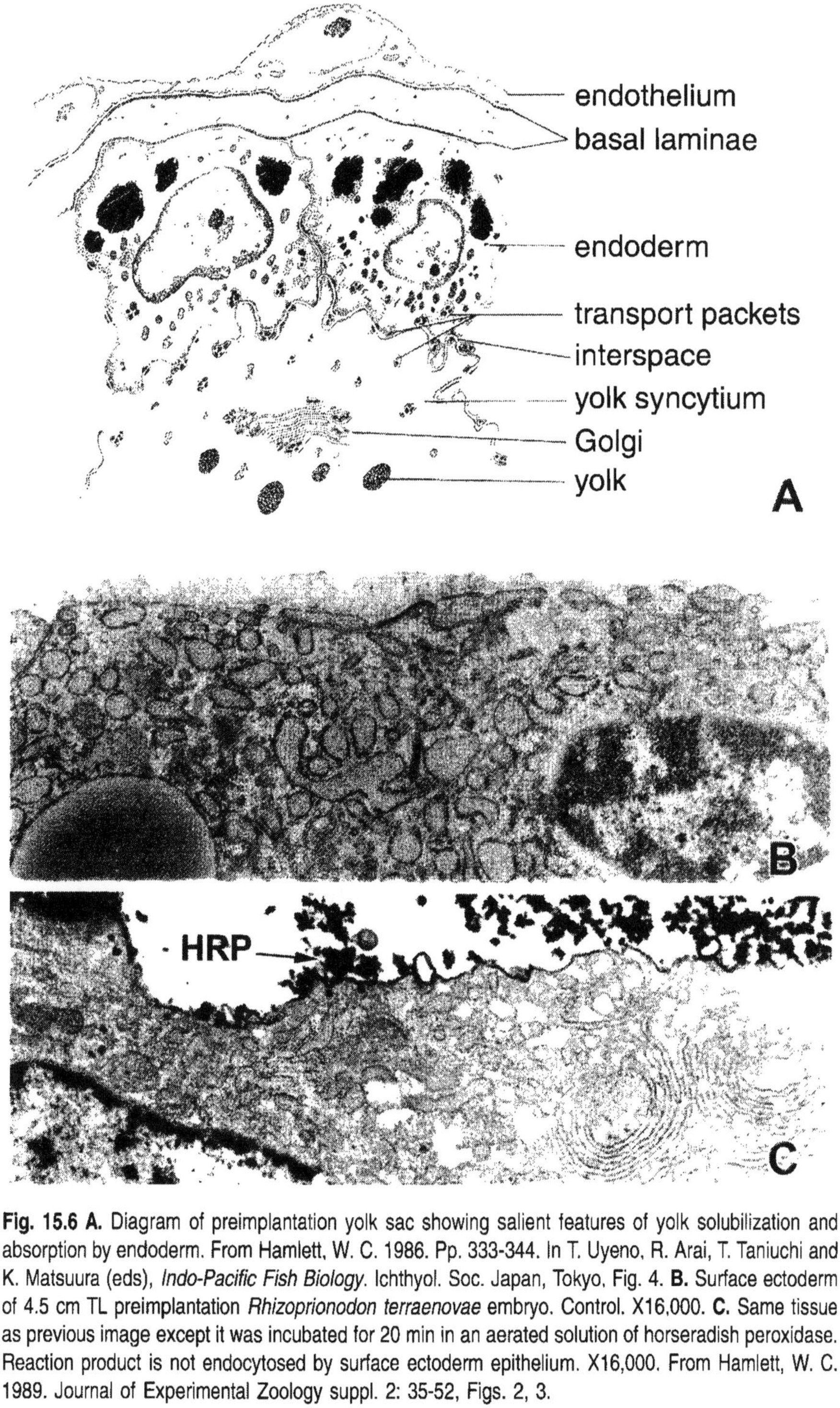

Fig. 15.6 A. Diagram of preimplantation yolk sac showing salient features of yolk solubilization and absorption by endoderm. From Hamlett, W. C. 1986. Pp. 333-344. In T. Uyeno, R. Arai, T. Taniuchi and K. Matsuura (eds), *Indo-Pacific Fish Biology*. Ichthyol. Soc. Japan, Tokyo, Fig. 4. **B.** Surface ectoderm of 4.5 cm TL preimplantation *Rhizoprionodon terraenovae* embryo. Control. X16,000. **C.** Same tissue as previous image except it was incubated for 20 min in an aerated solution of horseradish peroxidase. Reaction product is not endocytosed by surface ectoderm epithelium. X16,000. From Hamlett, W. C. 1989. Journal of Experimental Zoology suppl. 2: 35-52, Figs. 2, 3.

preimplantation stage *Rhizoprionodon terraenovae* embryo external gill filaments to absorb the macromolecular protein tracer horseradish peroxidase (HRP). External gill filaments are purely transient embryonic structures and are retracted as definitive permanent gills form. Embryos examined were 4.5 cm in total length (TL) and had no placenta. After 10 min exposure of gills *in vitro* to an aerated solution of HRP, reaction product was detected in smooth walled and tubular vesicles in the gill epithelium. Reaction product was also detected in vesicles within vessel endothelium suggesting external gill filaments are capable of endocytosis of periembryonic fluids prior to establishment of a placenta.

15.5.4 Preimplantation Yolk Sac Ectoderm

In a related experiment, Hamlett (1989) exposed the preimplantation yolk sac of 4.5 cm TL embryos of *Rhizoprionodon terraenovae* to an aerated solution of HRP under the same conditions as the gill experiment. After 20 min incubation, no reaction product was endocytosed by the yolk sac ectodermal epithelium (Fig. 15.6B, C). This suggests that at the preimplantation stage external gill filaments, in addition to respiratory functions, can absorb periembryonic fluids but the yolk sac ectodermal epithelium does not.

15.6 HISTOTROPHY

Hamlett (1989) indicated that during the period after yolk depletion and before establishment of a placental connection, embryos are nourished by uterine secretions or histotroph. He also noted that this phase corresponds with uterine compartment formation and the increased surface area may provide for increased histotroph production. He also noted that after establishment of the placental connection, paraplacental sites produce mucus which may be nutritive and/or serve as a lubricant (Hamlett 1989, Hamlett *et al.* 1993b). The increase in surface area in concert with depletion of external yolk sac contents heralds the transition from the initial lecithotrophic phase to the intermediate paraplacental histotrophic phase where nutrients of uterine origin are present in periembryonic fluids.

Mustelus canis embryos rely on yolk in the external yolk sac during the first 12 weeks of development (TeWinkel 1963). Further, after 14 weeks the distal portion of the yolk sac, egg envelope and uterine attachment sites are intimately associated (Graham 1967) to form a functional placenta. Graham (1967) suggested that paraplacental transfer commenced at 12 weeks. He reported that 90 min after injection of tritiated glucose into a *M. canis* female harboring 12 week embryos, ^{3}H (tritium) was present in the intracapsular fluid and embryos' stomach, umbilical cord and spiral intestine. However, tritium lability rendered Graham's results inconclusive as it was not clear if the tritium transferred to the embryo was still associated with injected glucose or its metabolites. Subsequently Graham and co-workers (1995) modified his original experiment and provided evidence for passage of maternally derived metabolites (enriched with stable non-radioactive ^{13}C atoms) to periembryonic fluids. They injected females containing embryos with no placental connection after 12 weeks gestation with 3-^{13}C-alanine. Periembryonic fluid was analyzed and found to contain ^{13}C-enriched signals from alanine C-3 and lactate C-3 carbons. They interpreted the results as indicating that matrotrophy via the paraplacental route commences after 12 weeks gestation. However, ^{13}C-enriched alanine was not detected in embryonic blood plasma and its uptake by embryos was not confirmed. In a related experiment, ^{13}C-1-glucose was injected in animals with near term fetuses and no ^{13}C-enriched glucose carbons were found in either periembryonic fluids or embryonic blood. This suggested to them that the mother utilizes glucose but does not transfer it to periembryonic fluid. These experiments demonstrate that the paraplacental matrotrophic nutrient route is established after 12 weeks of gestation in *M. canis*.

In *Scoliodon laticaudus* Setna and Sardangdhar (1948) concluded that copious uterine secretions are produced throughout gestation and the alimentary tract of the embryo is functional during this period. They further stated that nutrition in this species is far more histotrophic than hemotrophic. This refutes the speculative comments of Wourms (1993) and reinforces the importance of histotrophy even in placental species. Refer to the section on *Scoliodon* later in this chapter.

Okano *et al.* (1981) examined ultrastructural characteristics of the intestine of intrauterine embryos of *Mustelus manazo* (non-placental) and *Mustelus griseus* (placental). They observed columnar cells and goblet cells. Columnar cells had ultrastructural characteristics including microvilli, pinocytotic vesicles, multivesicular bodies and various other vesicles consistent with absorption. The conclusion is that embryos of both placental and non-placental species are imbibing and absorbing uterine fluids.

15.7 PLACENTATROPHY

As previously indicated, the preimplantation stage in placental sharks is characterized by the establishment of several compartments within the female reproductive tract. The trilaminar yolk sac consisting of ectoderm, mesoderm and endoderm delimit and enclose the yolk. The egg envelope surrounding the embryo creates a periembryonic compartment containing periembryonic fluids and separates it from the uterine compartment. Uterine epithelium secretes histotroph into the uterine compartment, some components of which may be transported across the egg envelope to the periembryonic compartment (Fig. 15.7A).

Depletion of yolk stores and intimate association of the distal aspect of the yolk sac and focal uterine attachment sites heralds the placental phase (Fig. 15.7B). Noteworthy changes are the expansion of the yolk sac and elongation of the yolk stalk (Te Winkel 1963). Two uterine areas are defined. The uterine attachment site is a vascular corrugated area that abuts the distal aspect of the yolk sac and paraplacental sites that continue to produce mucus. In species that reproduce annually such as *Rhizoprionodon terraenovae* and *M. canis* ovarian eggs are yolked as embryos grow such that at parturition the female has a set of fully yolked eggs to ovulate for the next reproductive cycle.

15.8 ULTRASTRUCTURE OF THE FETOMATERNAL INTERFACE

15.8.1 Uterus

In term *Rhizoprionodon terraenovae* (Hamlett 1993; Hamlett *et al.* 1993a) the uterine epithelium is loosely apposed to the egg envelope (Figs 15.8A, B). Gaps occur between the uterine surface and the egg envelope over the majority of the interface. Uterine epithelium is stratified squamous with two cell layers and scattered patches of microvilli on the free surface. Surface epithelium has scant microvilli and prominent merocrine secretory activity (Figs 15.8B, C). Classic tripartite junctional complexes of zonula occludens, zonula adherens and macula adherens junctions attach contiguous surface cells. Punctate desmosomes connect adjacent borders of neighboring cells and intercellular folds occur between contiguous surface cells and cells of the subjacent layer. Nuclei of surface uterine cells are generally fusiform with relatively smooth contours. Peripheral heterochromatin and a single prominent nucleolus are evident.

The cytoplasm of surface cells contains free ribosomes, mitochondria, a well developed rough endoplasmic reticulum frequently exhibiting dilated cisternae and a moderately distinct Golgi complex. Secretory vesicles containing homogeneously flocculent material are evident (Figs 15.8B), and those approaching the cell apex contain material that is morphologically identical to material present in the uterine lumen (Fig. 15.8A). Neighboring secretory vesicles may fuse to form a secretory conduit (Fig. 15.8C) to the uterine lumen. Immediately beneath the epithelium is a fairly distinct continuous basal lamina. The underlying vascular elements are in close proximity to the epithelium and the endothelium is continuous with undulating surface folds and marginal folds. The nucleus of the endothelial cells is flattened and protrudes modestly into the vascular lumen. Smooth walled caveolae occur on both the adluminal and abluminal surfaces. Deep to the vascular elements, a diffuse zone occurs that contains bundles of smooth muscle and collagen embedded in a matrix of undetermined character. Areas of the uterus not involved in transport are termed paraplacental uterine sites and are mucous in nature.

In near term *Carcharhinus plumbeus* (Hamlett *et al.* 1985c) the uterine epithelium is low columnar and shows no erosion. Cells have a plethora of rough endoplasmic reticulum with prominently dilated cisternae.

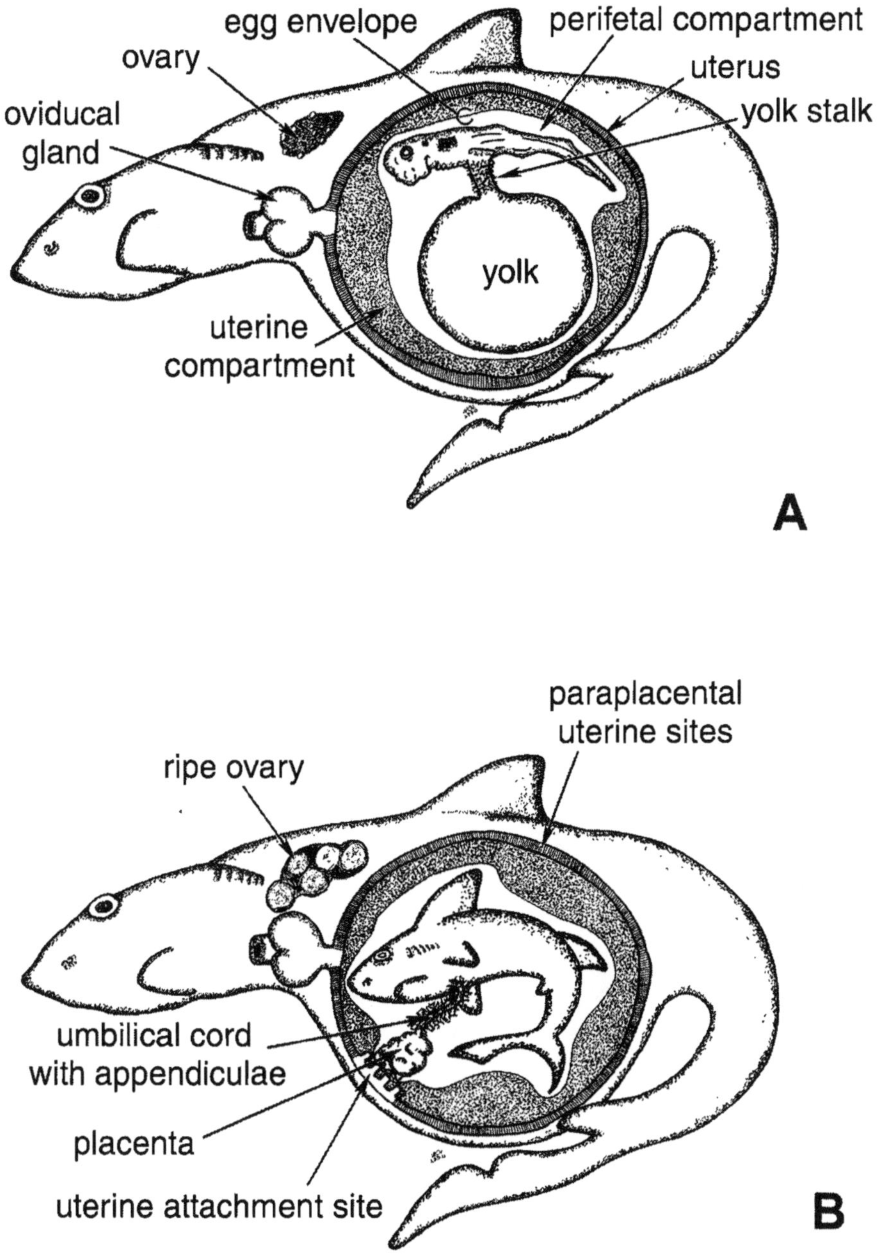

Fig. 15.7 A. During early stages of gestation, the ovary contains small, developing oocytes while the uterus harbors yolk dependent embryos. **B.** At term, the yolk stalk is transformed into an umbilical cord with appendiculae and the yolk sac contributes to the functional placenta. From W. C. Hamlett. 1993. Ontogeny of the umbilical cord and placenta in the Atlantic sharpnose shark, *Rhizoprionodon terraenovae*. Environmental Biology of Fishes 38: 253-267, Figs. 1, 8.

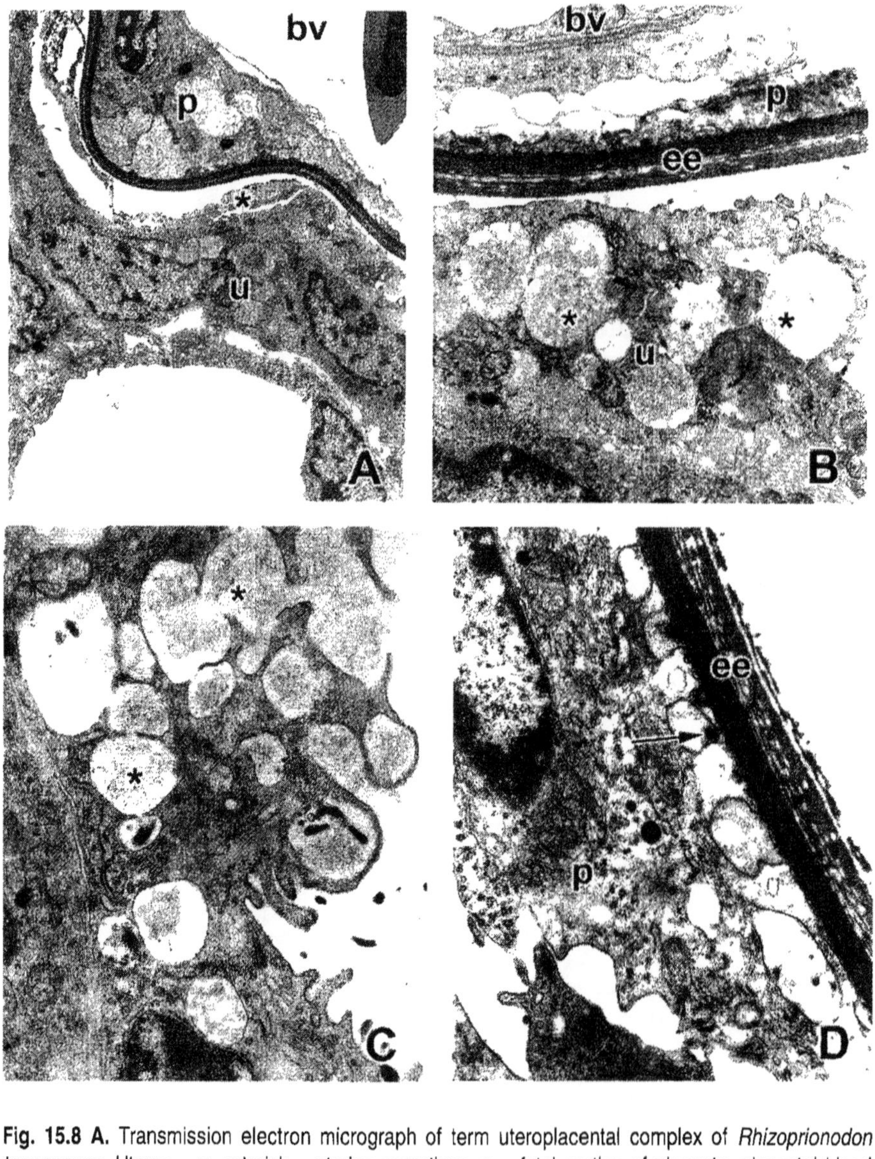

Fig. 15.8 A. Transmission electron micrograph of term uteroplacental complex of *Rhizoprionodon terraenovae*. Uterus = u, asterisk = uterine secretions, p = fetal portion of placenta, placental blood vessel = bv. 6,000X. **B.** Transmission electron micrograph of term uteroplacental complex of *R. terraenovae*. Uterus = u, asterisks = uterine secretory vesicles, egg envelope = ee, placenta = p, placental blood vessel endothelium = bv. X18,000. **C.** Transmission electron micrograph of term uterus of *R. terraenovae*. Uterine secretory vesicles often fuse during exocytosis. X20,000. **D.** Transmission electron micrograph of term uterus of *R. terraenovae*. The egg envelope (ee) has delaminations on the uterine side and is more compact on the placental side. Electron dense uterine secretions (arrow) are endocytosed by the placental epithelium (p). X36,000. Original.

There are few secretory vesicles but this might be cyclical in nature. Secretion may have just occurred and the rough endoplasmic reticulum and Golgi are in the process of forming new secretory vesicles.

15.8.2 Egg Envelope

The egg envelope intervenes between the uterus and yolk sac (Figs 15.8A, B, D; 15.9A, B, C) in all species thus far investigated except for *Prionace glauca* (Otake and Mizue 1985, 1986), *Iago omanensis* (Fishelson and Baranes 1998) and *Scoliodon laticaudus* (Setna and Sardangdhar 1948). In *Carcharhinus plumbeus* (Hamlett *et al.* 1985a), *Rhizoprionodon terraenovae* (Hamlett 1993; Hamlett *et al.* 1993) and *Carcharhinus acronotus* (Hamlett 1990) a physical gap separates the uterine endometrium from the egg envelope. In contrast the placental yolk sac epithelium makes contact with the egg envelope (Figs 15.8D; 15.9A). The egg envelope is compactly homogeneous on the placental surface. At high magnifications, when the plane of section is appropriate, axial periodicities are visible perpendicular to the envelope (Figs 15.9B, C). Slits or delaminations occur in the envelope consistently on the uterine surface (Fig. 15.8B).

15.8.3 Distal Yolk Sac

The distal portion of the yolk sac forms a continuous, intimate apposition with the egg envelope throughout its course. The plasmalemma and the egg envelope are parallel except where the cell membrane forms invaginations (Figs 15.8D; 15.9A). Placental epithelium is stratified squamous of two to three cell layers. Non-membrane bound electron dense material, frequently forming small spheres or globules, is present in the interspace between the egg envelope and the apical plasmalemma and in invaginations of the epithelium as well as in apical vesicles (Fig. 15.8D; 15.9A). Surface epithelial cells are laterally joined at their free surfaces by zonula occludens, zonula adherens and macula adherens junctions. Desmosomes attach the surface layer to the underlying components of the epithelium and intercellular folds occur between the cell margins.

Dense bodies with a regular whorled substructure characterize the surface epithelial cells. The inclusions are surrounded by a membrane and exhibit various configurations ranging from lamellar whorls to circular densities. Whorled configurations range in size from a few subunits to extensive lamellar arrays. The whorls may nearly fill a dense body or account for only a modest component of it, the remainder being present as homogeneous areas of moderate electron density. The overall contour of the dense bodies is generally spherical to ellipsoidal but irregularly shaped assemblages also occur. Dense bodies appear to be lysosomes (Fig. 15.9A). Other features of these cells include a prominent Golgi complex, mitochondria, fibrils and a modestly developed rough endoplasmic reticulum. More deeply situated epithelial cells have an extensive system of smooth walled caveolae on both their abluminal and

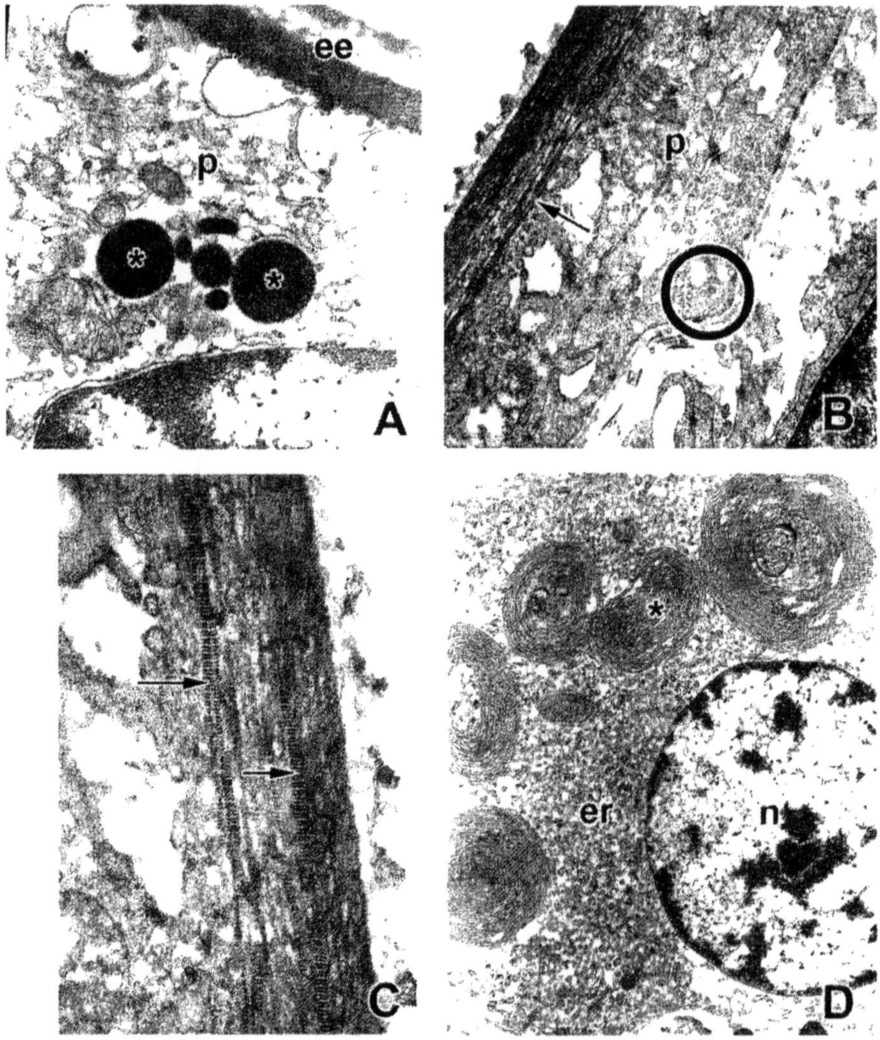

Fig. 15.9. A. Transmission electron micrograph of term placenta (p) and egg envelope (ee) in *Rhizoprionodon terraenovae* showing lysosomes (asterisks). X48,000. **B.** Transmission electron micrograph of term placenta (p) and egg envelope in *R. terraenovae* showing striations in the egg envelope (arrow) and pinocytotic vesicles (circle) in the placenta. X54,000. **C.** Transmission electron micrograph of term placenta and egg envelope in *R. terraenovae* showing striated fibers of the envelope (arrows). X72,000. **D.** Transmission electron micrograph of term smooth portion of the placenta in *R. terraenovae* showing smooth endoplasmic reticulum (er), nucleus (n) and whorled lamellar structures (asterisk). X24,000. Original.

adluminal surfaces (Fig. 15.9B) as well as in the endothelium. Marginal folds are present as are many fenestrae.

Hamlett *et al.* (1985a) noted similarity in ultrastructural appearance of non-membrane bound inclusions in the distal portion of the term placenta of the sandbar shark, *Carcharhinus plumbeus* with teleost (Walzer and Schonenberger 1979a, b) and amphibian (Ward, 1980) yolk protein precursors. It was suggested that the inclusions may represent residual bodies derived from partial lysosomal digestion of yolk. Residual bodies are frequently characterized by membrane fragments, myelin figures or whorled inclusions (de Duve and Wattiaux 1966). Another interpretation was also put forth (Hamlett *et al*. 1985a; Hamlett 1989). As yolk stores in the yolk sac diminish with the transition to histotrophy and placentatrophy, the maternal liver produces yolk protein precursors secreted into the blood where they are taken up by the new crop of developing oocytes for the next reproductive cycle. The same yolk precursors could then be transported across the uterus and fetal placenta to nourish the fetuses during the latter stages of gestation. To date this suggestion has not been verified.

Based on all the available data it appears that placental transport at the maternal-fetal interface is based on secretory activity of the uterine attachment sites and paraplacental endometrium, reinforcing the concept of histotrophy throughout gestation and placentatrophy as an extreme of that condition. In all cases thus far reported placental transfer is histotrophic, not hemotrophic. Fig. 15.10 summarizes the salient ultrastructural features of the utero-placental interface.

15.8.4 Proximal Yolk Sac

The epithelium of the proximal portion of the yolk sac, in species that have it, consists of one to two layers of cuboidal cells. Surface cells have microvilli and a round nucleus with clumps of heterochromatin dispersed in the nucleoplasm. Cell margins are generally straight and surface cells are joined by zonula occludens, zonula adherens and macula adherens junctions. The cytoplasm is characterized by a massive smooth endoplasmic reticulum, Golgi complexes and lamellar configurations (Fig. 15.9D). Ultrastructural characteristics of the smooth, proximal portion of the shark placenta suggest that it could be secreting steroid hormones. Although the human placenta produces steroid hormones, it lacks the complete repertoire of enzymes necessary for synthesis of all the hormones it produces. An exchange between the fetus and placenta by the feto-placental unit is responsible for the final hormone production (Van Tienhoven 1983). It is feasible that similar mechanisms might be operating in the shark placenta. Whether the shark placenta functions as an endocrine organ awaits investigation, however, preliminary studies employing radioimmunoassay procedures have identified measurable levels of some steroids in the placenta of the blacknose shark, *Carcharhinus acronotus* (Hamlett 1990).

15.9 FUNCTIONAL STUDIES

Otake (1990) performed two experiments to investigate maternal-fetal transfer in *Mustelus griseus*. In the first he injected 10-15 ml 0.8% trypan

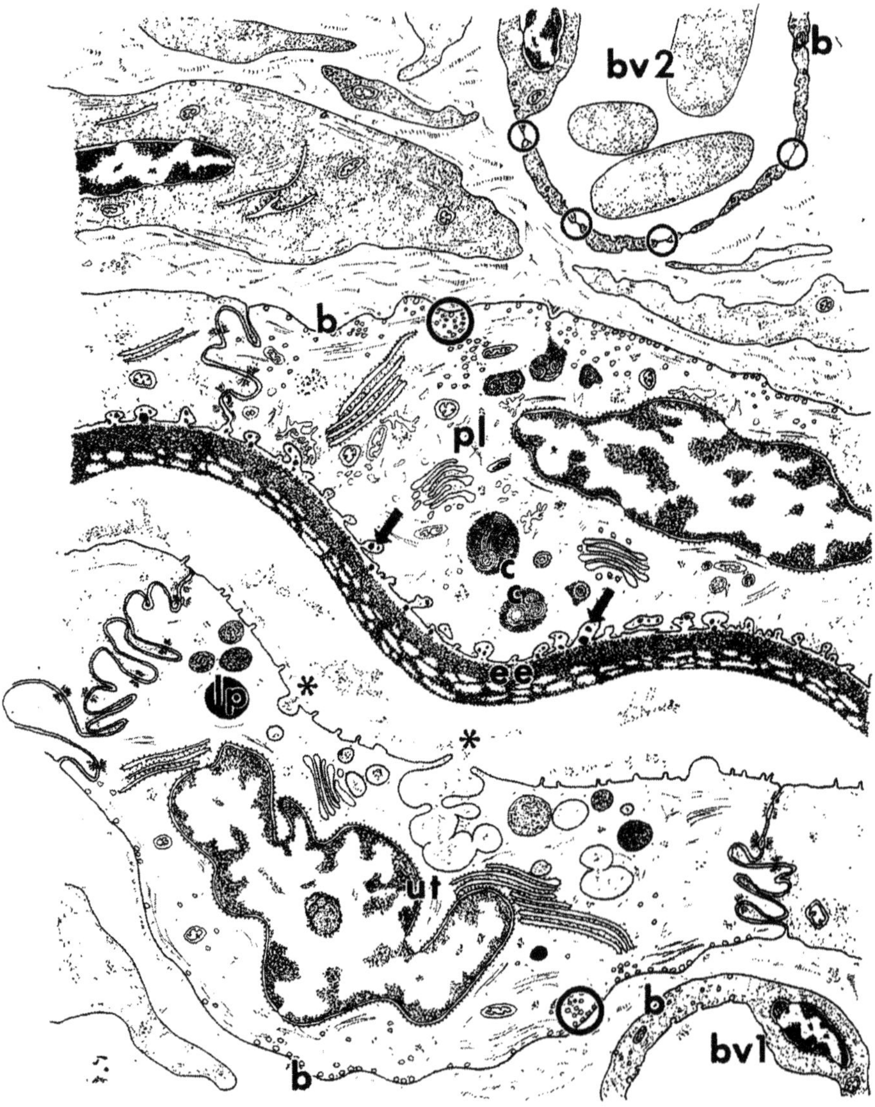

Fig. 15.10 At term the placenta is characterized by squamous uterine epithelium (ut) that has basal pinocytotic vesicles (large circles), lipid (lp) and secretory vesicles that release their contents to the uterine lumen (asterisks). Subjacent capillaries (bv1) are continuous. The egg envelope (ee) separates the fetal and maternal portions of the placenta. Electron dense vesicles (arrows) of uterine origin are endocytosed by the distal portion of the fetal placenta (pl). The epithelium has basal pinocytotic vesicles (large circles) and crystalline inclusions (c) that are interpreted to be lysosomes. Subjacent capillaries (bv2) are fenestrated (small circles). Basal lamina = b. From Hamlett, W. C. 1993. Environmental Biology of Fishes 38: 253-267, Fig. 12.

blue (MW 961) in physiological saline into maternal vessels of 3 pregnant females containing near term fetuses (20.4-27.5 cm TL). Size at birth is reported to be 28 cm (Compagno 1984). The females died one, two and four days after injections started. After 3 days muscles and internal organs (which organs not stated) were strongly stained. He also stated that "the mother is stained completely blue". Neither the embryo nor the embryonic component of the placenta was stained. It was not indicated if the maternal endometrium was stained. He concluded that trypan blue is not transported through placental tissue and uterine secretions. We urge caution in interpreting these results. First the fact that the mother was "stained completely blue" suggests that the volume of dye used was excessive. Whereas trypan blue is a smallish protein the chronic administration and the volume may have overwhelmed the cellular transport apparatus. As all egg envelopes thus far examined have been shown to have lamellae with spaces between them on the maternal side and to be more compacted on the fetal side, trypan blue may have clogged the pathway. Second, as all animals died without being sacrificed we interpret this as indicating the animals were injured by the experiment and failure to detect dye in the fetal placenta may be the result of cell injury. This experiment does not warrant the conclusion that trypan blue is not transported by uterine secretion. Uterine secretion in fact is the expected route of nutrient delivery via histotrophy and exogenous trypan blue injected into the maternal vasculature is unrelated to *de novo* secretion by the endometrium which is a completely different process.

In the second experiment, Otake (1990) injected a 10-15 ml solution of horseradish peroxidase (HRP) (type unspecified) into maternal vessels of six females carrying near term fetuses (24.5-30.8 cm TL). Animals were killed 10, 40, 60 min, 3 hrs and 24 hrs after injection. He detected reaction product in vesicles in the maternal portion of the placenta, egg envelope and surface invaginations of epithelial cells of the yolk sac portion of the placenta 40 min, 3 hrs and 24 hrs after administration. No reaction product was observed in embryonic epithelial cells or in capillaries under the epithelium. He concluded that transport of macromolecular weight organic materials the size of HRP (MW40,000-44,000) is minimal and that the embryonic placenta cannot endocytose this marker.

These results do not agree with conclusions reached by Hamlett *et al.* (1985a) in *Carcharhinus plumbeus*. Fresh whole placentae were immersed in an aerated solution of 1.0% trypan blue in 0.1 M phosphate buffer for 10 min and examined by light microscopy or small pieces of placentae were incubated in an aerated solution of 10.0 mg HRP (Sigma Type IV) per 50.0 ml phosphate buffer for 20 min. The results of the trypan blue experiment showed intense staining of the distal portion of the yolk sac placenta but scant staining of the smooth, proximal area. HRP was endocytosed into a canalicular system in the distal placenta and small transport vesicles budded off in the basal portion of the cells. No reaction product was seen in the endothelium. These results indicate that molecules the size of HRP can be endocytosed by the distal placenta. No pulse chase methods involving progressively longer chase times were employed which may account for failure to detect reaction product in the endothelium.

Lombardi and Files (1993) performed permeability studies on isolated pieces of egg envelope from *Mustelus canis*. They employed a Ussing chamber and subsequent analysis of molecules that traversed the egg envelope. They used solutions of insulin, lysozyme or bovine albumin for protein permeability; fast green fcf and rose bengal for dye permeability as well as glucose and urea. Their results reveal that the egg envelope is permeable to fast green fcf (808 Da) and rose bengal (1,000 Da) but impermeable to insulin (5,750 Da), lysozyme (14,300 Da) and bovine albumin (66,000 Da). Glucose and urea both pass. They conclude that maternal-embryonic nutrition involves passage of low (<5 kDa) molecular weight components. A major flaw in this study is a failure to realize that the egg envelope is asymmetrical. In *R. terraenovae* (Hamlett 1993; Hamlett *et al.* 1993a), *Carcharhinus plumbeus* (Hamlett *et al.* 1985a), *Carcharhinus acronotus* (Hamlett 1990) and *Mustelus canis* (Hamlett unpublished) the egg envelope has delaminations on the uterine side and the lamellae are compacted on the fetal side. Staining with ruthenium red in *C. plumbeus* showed the dye stained the maternal surface of the egg envelope and the interstices between lamellae. The asymmetry of the egg envelope causes concern about conclusions made by Lombardi and Files (1993). To be accepted as valid it is necessary to know which side of the egg envelope represents the maternal or fetal side. Transport phenomena reported must be questioned because in no case was the sidedness of the egg envelope known. It would be useful for these experiments to be repeated with proper regard to the asymmetry of the egg envelope.

Due to the conflicting results of these studies, conclusive experiments are necessary to resolve the question of transfer from the maternal vascular system to the embryonic vasculature. At present, it is concluded that all placental transfer is via uterine secretory activity.

15.10 ULTRASTRUCTURE OF THE UTEROPLACENTAL INTERFACE LACKING AN INTERVENING EGG ENVELOPE

Most shark placentae incorporate the intact egg envelope in the uteroplacental complex but in three species the egg envelope degenerates and is not a constituent of the interface, viz. *Scoliodon laticaudus* (Setna and Sardangdhar 1948; Otake 1990), *Prionace glauca* (Otake and Mizue 1985, 1986; Otake 1990) and *Iago omanensis* (Fishelson and Baranes 1998).

15.10.1 Placentation in *Prionace glauca*

Otake and Mizue (1986) described the ultrastructure of the endometrium of pregnant *Prionace glauca* containing near term fetuses. They differentiated between uterine attachment sites and paraplacental areas.

They reported that epithelium of the uterine attachment site had many supranuclear granules that discharge their contents to the interspace between maternal and embryonic epithelia. Lateral cell surfaces were smooth and intercellular spaces not dilated. These ultrastructural characteristics suggest an actively secreting epithelium. Paraplacental uterine areas were characterized by dilated intercellular spaces, basal mitochondria and lack of secretory structures although isolated mucus-secreting cells occurred. This led them to conclude paraplacental endometrium functions in water solute transport and gas exchange. The yolk sac placenta has both a proximal smooth portion to the yolk sac placenta and a distal rugose portion that abuts the endometrium.

Otake and Mizue (1985) and Otake (1990) reported that the epithelium of the embryonic portion of the placenta consists of two layers: giant cells, 50-70 μm in height and 75-100 μm in width and a flattened cell layer less than 5 μm in height. The giant cells were covered by microvilli except for the part closely attached to the maternal epithelium. Tubular invaginations and coated vesicles suggested absorption which is in keeping with histotrophic secretion at the placental endometrium.

15.10.2 Placentation in *Iago omanensis*

Fishelson and Barenes (1998) described placentation in *Iago omanensis*. They describe three methods of nourishment. The first stage is lecithotrophic where yolk stores are digested in the yolk sac endoderm and the fetal gut. The second is described as "mixed food provision" Placentation is said to commence at the beginning of embryogenesis. In embryos 10–12 mm TL the endometrium-yolk sac interface is firm but no blood vessels are present. Only fragments of the egg envelope persist. The maternal part of the placenta is circular, 6mm in diameter. Endometrial glands form, elongate and branch in the submucosa. The yolk sac forms "trophic villi" that extend into the yolk mass and contact the endometrium. In this second phase they reported yolk utilization and hemotrophic placental transfer although no evidence for this was presented to verify nutrients arose from maternal blood and not by endometrial secretion. They also described large cells as detaching from the endometrium and being located at the yolk sac-endometrial interface. They do not report the fate of these cells. Interestingly, they then go on to indicate it is not appropriate to speak of a yolk sac placenta in this animal because only a small area of the yolk sac and uterus are involved in placentation. In the third phase the yolk sac detaches from the endometrium and the embryo "swims free" in the uterus. During this phase uterine milk begins to form and the trophic villi of the yolk sac absorb it and the embryos ingest the milk as well as shed endometrial cells although no morphological evidence was presented. They mention "various particles of food material" present in the stomach of embryos but do not elaborate. Upon consideration of the data reported we feel that there certainly is an initial period of lecithotrophy as in all chondrichthyans. The degree to which the embryos rely on histotrophy seems to be great in this species. No data was reported on secretory activity of paraplacental uterine sites so we cannot determine exactly when histotroph secretion commences but it seems logical to conclude the uterus is secretory throughout. We do not think it justified to claim detached endometrial cells as constituting a nutrient source unless this phenomenon can be verified across the entire endometrium and not exclusively at the placenta site. The claim that trophic villi of the yolk sac absorb histotroph needs to be experimentally verified.

15.10.3 Placentation in *Scoliodon laticaudus*

Placentation in the spadenose shark, *Scoliodon laticaudus* was initially described by Setna and Sardangdhar (1948) with subsequent observations by Teshima *et al.* (1978) and Otake (1990). Egg diameter in this species is 1 mm and no ductus vitellointestinalis exists. The small yolk store must therefore be absorbed by the yolk sac endoderm as in other chondrichthyans (Hamlett *et al.* 1987). Diagonal uterine compartments (Fig. 15.11A, C) containing nutritive secretions form when embryos are 0.8mm TL. The compartments contain a single embryo initially surrounded by a transient tertiary egg envelope. In these early stages of gestation nutrition is primarily histotrophic. Implantation commences when embryos are 3 mm TL. At this stage appendiculae are short and stout (Fig. 15.11B) and the yolk sac has not yet differentiated into a placenta. By the time embryos are 22 mm TL appendiculae are long and vascular and the trophonematous complex has formed consisting of a fetal and endometrial component. The endometrial component consists of a trophonematous stalk, bulb, cord and cup that contacts the fetal yolk sac (Fig. 15.11C, D). Nutrition at this point is reported to be hemotrophic. Mahavedan (1940) and Setna and Sarangdhar (1948) report that the epithelium of the trophonematous cup is intact. Teshima *et al.* (1978) disagree and report that the cup epithelium is not intact and that maternal blood bathes the yolk sac epithelium at their point of contact. Otake (1990) used transmission electron microscopy and concurs with Teshima *et al.* (1978). In late gestation the placenta becomes "totally disorganized and very little nutrition, if at all, reaches the embryo through that organ" (Setna and Sarangdhar 1948). At this time uterine secretions continue until parturition. The authors conclude by saying that profuse quantities of uterine secretions are available to the embryos throughout gestation and the embryonic alimentary canal is functional and nutrition is "far more histotrophic than hemotrophic and that comparatively a very small quantity of nutrition is derived directly from the maternal blood vascular system". They believe the placenta is a primitive type in spite of its highly specialized structure.

Wourms (1993) examined uterine eggs and a single 90 mm embryo taken from a fish market and fixed in formalin. On the basis of scanning electron microscopy of these tissues, and extrapolation of weight gain

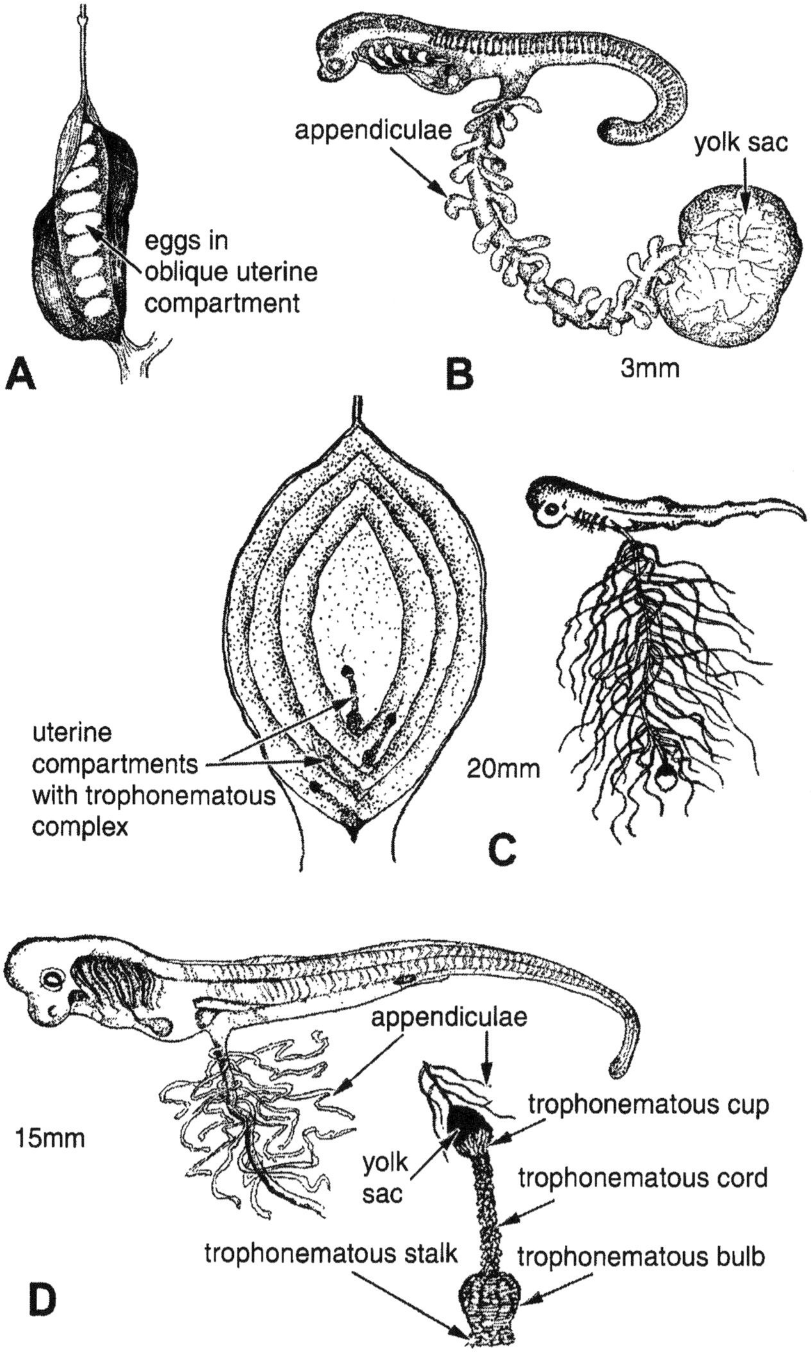

Fig. 15.11 A-D. Development of uterine compartments, appendiculae and the trophonematous complex in *Scoliodon laticaudus* is shown. Modified from Setna, S. B. and Sarangdhar, P. N. 1948. Records of the Indian Museum 46. 25-43, Figs 2, 4, 5, 6.

during gestation, he concluded that the placenta of *Scoliodon laticaudus* is the "most advanced of any shark and rivals that of mammalian placentae". This conclusion does not appear warranted. The speculative weight gain data will not be repeated here as it is totally unreliable and should not be perpetuated in the literature (see Chapter **13** of this volume). Wourms also concludes that placental transfer is hemotrophic although no evidence for this has ever been published. Indeed this is contrary to the above stated conclusion of Setna and Sardangdhar (1948) where they conclude histotrophy is the primary nutrient mode in this species with the placenta contributing very little. He also speculates that the uterine fluid (none has ever been analyzed) is low in organic content and arises from blood serum transported across the paraplacental uterine epithelium. He continues by discounting previous authors' conclusions and states that imbibition of uterine fluid is probably not very efficient. He also examined appendiculae and noted that the surface cells were devoid of microvilli and concluded they are not absorptive. Careful examination of the micrographs presented by Wourms (1993) clearly shows significant fixation artifacts. Post mortem degeneration is also suspected as the samples were not fresh and were fixed in formalin. The majority of the conclusions presented must be questioned.

15.11 INTERLEUKIN (IL-1) SYSTEM IN THE UTEROPLACENTAL INTERFACE

Although viviparity is advantageous for the nourishment and growth of offspring it involves immunological risk for the semi-allographic fetus. Prolonged exposure of embryonic tissues bearing paternal antigens to maternal uterine tissues can cause rejection of the embryo (Bainbridge 2000). Among immunological mechanisms to protect the embryo the local release of immunoregulatory peptides (cytokines) appears to play a primary role (Saito 2000). Since chondrichthyes are the oldest extant jawed vertebrates and the oldest line to have placentae, Cateni *et al.* (2003) employed immunohistochemistry to investigate the tissue expression of the cytokine interleukin-1 (IL-1) α, IL-1 β and its specific membrane receptor, IL-1 receptor type 1 (IL-1R tI) in term placentae, umbilical cord and paraplacental uterine sites of *Mustelus canis*. Immunohistochemistry for IL-1 α IL-1 β and the receptor reveals leukocytes of both the mother and fetus to be positive, as well as the apical aspect of paraplacental cells and apical vesicles in umbilical cord epithelium. Yolk sac endoderm is also positive with all the stains while ectoderm is positive only for IL-1 α. Immunoreactivity in uterine epithelium was obtained for IL-1 α and the receptor. The egg envelope is always negative. The finding of IL-1 in the yolk sac placenta of a chondrichthyan, in chorioallantoic placentae in squamate reptiles (Paulesu *et al.* 1995) and various types of mammalian placentae (Paulesu 1997), suggests that this molecule plays a critical role in interaction between maternal and fetal tissues in placental viviparity.

Moreover, this suggests that the role of interleukins in vertebrate placental evolution is independent of the germ layer derivative of the cells involved in placentation. In light of the recent finding of the IL-1β gene in a cartilaginous fish and the high level of conservation of proteins implicated in IL-1 action, Cateni *et al.* (2003) conclude the IL-1 system is a key mediator of materno-fetal interaction since the oldest extant placental vertebrates (Bird *et al.* 2002).

15.12 GLYCOSYLATION OF THE FETOMATERNAL INTERFACE

The evolution of the yolk sac placenta within chondrichthyan species differs from that of many other genera in that the egg envelope is preserved. This filmy membrane acts as a barrier between fetal and maternal epithelia and, as such, removes the necessity of developing an interface such as is found in other types of epitheliochorial placentation, where specific glycan-glycan interactions form an important prerequisite for successful implantation (Whyte and Robson 1984; Dantzer 1985; Bowen and Burghardt 2000).

In recent years, the importance of glycans in fetomaternal relationships has become increasingly appreciated, and many studies of mammalian placentation have suggested that successful implantation and maintenance of pregnancy is dependent on matching "glycotypes" of the mother and embryo (Surani and Fishel 1980; Jones *et al.* 1997). Part of this mechanism has no doubt been established to avoid inappropriate hybridization, and when embryos from one species have been planted in the uterus of another, the pregnancy invariably fails (Allen 1982; Allen *et al.* 1987) unless the blastocyst is manipulated so that the trophectoderm consists of cells of the host species (Fehilly *et al.* 1984). Jones and Hamlett (2004) have recently examined glycosylation at the interface of the yolk sac placenta and uterus of *Mustelus canis* using lectin histochemistry. Lectins are proteins or glycoproteins of non-immune origin that bind to glycans by more than one binding site, and as such are agglutinins (Sharon and Lis 1989). Most lectins used in biomedical research have had their binding properties very well characterized, so that their use with appropriate revealing systems enables the glycan composition of tissues to be ascertained. Examination of this interface with a panel of lectins and an avidin-biotin revealing system (Fig.15.12A-H) has shown clearly that each layer of the yolk sac placenta has a distinctive pattern of glycosylation. The ectoderm, elongated cells with a microvillous surface abutting the egg envelope, contained terminal β-galactosyl residues, which were mainly distributed on the basal surface (Fig. 15.12A), suggesting that they were associated primarily with the microvilli. This ectodermal cell surface must be instrumental in ingesting nutrients for transfer to the fetal circulation, and intracellular vesicles have been observed at the ultrastructural level. Also present in this cell layer were granules containing nonreducing terminal α-D-mannosyl residues, which may have been localized in lysosomes and so related to substance breakdown prior to transit to the embryo, as these termini are common in lysosomal structures (Litynska and Przybylo 1998). No bisected bi/tri-antennary or non-bisected tri/tetra-antennary complex N-glycans were present. This is a feature of some term mammalian trophoblasts such as the swine (Jones *et al.* 2004) and alpaca (Jones *et al.* 2002); the latter residue is also absent in the camel trophoblast and is present only in occasional secretory cells in the viviparous lizard *Chalcides chalcides* placenta (Jones *et al.* 2003), though fucosylated structures were plentiful, in both α1,2 (Fig. 15.12C) and α1,6 linkage. In this species, there were no non-bisected bi/tri-antennary complex N-glycans either (Fig.15.12D). Strong staining was shown by a variety of lectins binding to N-acetyl lactosamine/glucosamine oligomers (Fig. 15.12H), which is a feature of many types of placenta examined to date. The ectoderm also contained occasional granules expressing terminal N-acetyl galactosamine residues (Fig. 15.12G), which are also widespread in mammalian placentae. Sialic acid residues in α2,3 but not in α2,6 linkage were evident, the absence of the latter being a feature of many mammalian placentae and also of *Chalcides.*

The endodermal cells contained remnants of yolk which have been shown to be PAS-positive (Hamlett, unpublished). These cells expressed many classes of both N- and O-linked, though tri/tetraantennary non-bisected N-glycan and fucose in α1,2 linkage were sparse. The surface of the endoderm was composed of heavily glycosylated microvilli which bore subterminal β-galactosyl and N-acetyl galactosaminyl residues. Fucose in α1,6 linkage and N-acetyl glucosamine/lactosamine residues were again plentiful, especially on the cell surface (Fig. 15.12F).

The chemical composition of the multilaminate egg envelope is not known. Lectin histochemistry showed most staining with lectins binding to fucosylated structures and N-acetyl lactosamine (Fig. 15.12C,H). Structures containing the sequence Galβ1,3GalNAcα1- and the presence of some sialic acid was indicated. In mammalian epitheliochorial placentae, there is diffusion of nutrients and gases through layers of trophoblast and maternal uterine epithelium, with basal laminae also traversed, whereas in *Mustelus,* any transfer of material must take place via the egg envelope, which appears to be inert and contains no cellular components to facilitate transfer. Particulate matter seen on the outside and between separate lamellae may have been in transit across the membrane, as similar material was sometimes found on the inner face of the envelope, where the microvillous surface of the ectoderm is situated. Vesicles have also been seen within the ectoderm at the ultrastructural level, hence the egg envelope is regarded as semipermeable

The distal face of the egg envelope lies close to the apical surface of the uterine epithelium, which has been shown to bear well-developed microvilli which would be important in the transfer of nutrients to the fetus. Similar structures have been described in the uterus of the gravid sandbar shark *Carcharhinus plumbeus* (Hamlett *et al.* 1985c) and *Rhizoprionodon terraenovae* (Hamlett and Hysell 1998). Lectin binding studies have shown these somewhat attenuated cells to be heavily glycosylated, with abundant fucosyl termini (Fig. 15.12C) and N-acetyl lactosamine and/or glucosamine oligomers (Fig. 15.12H), possibly reflecting a biosynthetic role. Some β-galactosyl (Fig. 15.12B) and a few N-acetyl galactosamine residues were present (Fig. 15.12G), though neither terminal nor subterminal β-galactosyl residues. The lack of bisected bi/tri-antennary and non-bisected tri/tetra-antennary complex N-glycan in this tissue reflects a common finding in term mammalian uteri, the former being absent in, for example, alpaca and swine (Jones *et al.* 2002, 2004) and the latter in most epitheliochorial uteri studied by us to date (Jones *et al.* 1997, 2002, 2004a) excepting that of the horse (Jones *et al.* 1999). The viviparous lizard *Chalcides chalcides*, which also has an epitheliochorial placenta, showed a similar dearth of non-bisected tri/tetra-antennary complex N- glycan in the uterine epithelium (Jones *et al.* 2003) and a gradual decrease over pregnancy in the expression of bisected bi/tri-antennary structures, suggesting that such classes of complex N-glycan frequently have a restricted distribution in these tissues.

The *Mustelus* uterus is unusual in having, as well as the thin apical cells, a subpopulation of rather globular cells which express large amounts of N-acetyl glucosamine oligomers (Fig. 15.12F) and α2,3-linked sialic acid. Their sporadic nature on the surface of the uterus made their function unclear although the richly glycosylated, vesicular contents suggest a secretory function similar to that of the goblet cell, and they might be the source of the globular deposits occasionally seen lying against the undersurface of the egg envelope.

Under this apical layer was another layer of cells with a different pattern of glycosylation. This basal layer showed a selective distribution of terminal N-acetylgalactosamine residues (Fig. 15.12E), as well as other glycans bound by several lectins. That the difference in staining could be due to a maturation process cannot be ruled out, however, and this basal layer may be the progenitor of the surface cells. β-galactosyl residues subterminal to sialic acid were prevalent throughout this layer, as opposed to terminal β-galactose which had only a sparse distribution (Fig. 15.12A), and the presence of terminal sialic acid in α1,3 linkage was also found. There were also considerable amounts of N-acetyl lactosamine and/or glucosamine oligomers indicated (Fig. 15.12H). The presence of many secretory products has been described in the uterine epithelium of the sandbar shark, *C. plumbeus*, which were shown to be ruthenium red positive (Hamlett *et al.* 1993c) suggesting the presence of acid mucopoly saccharides as shown here.

The placental and uterine vasculature showed very similar patterns of glycosylation, which also resembled that of their therian counterparts, being heavily glycosylated, with sialic acid in a variety of linkages, sub-terminal galactose and N-acetyl lactosamine, and N-acetyl lactosamine/

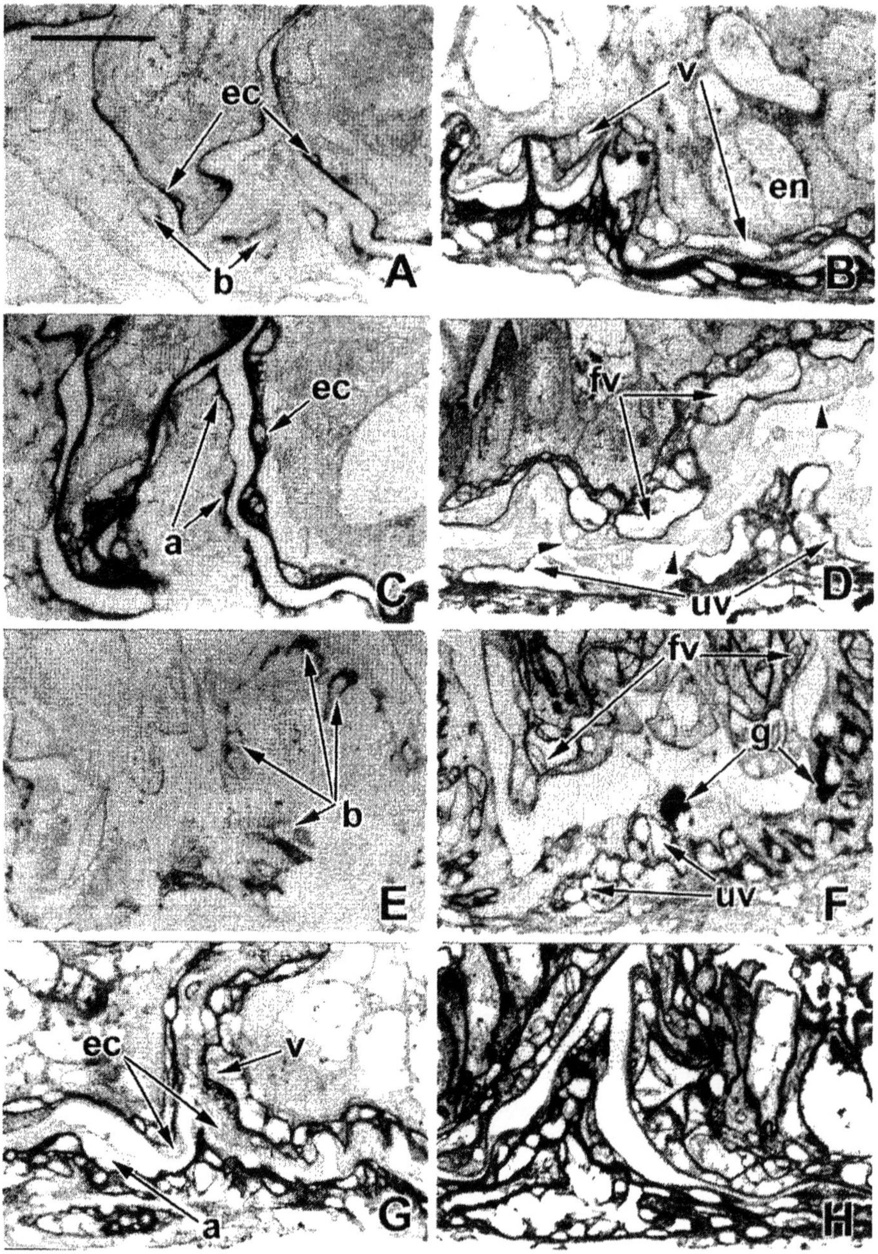

Fig. 15.12 A. Lectin histochemistry of the yolk sac placenta and uterine attachment site of *Mustelus canis.* Terminal ß-galactosyl residues in the ectoderm (ec) of the yolk sac placenta and occasional basal cells (b) of the uterine epithelium are shown by the binding of *Arachis hypogaea* agglutinin. Scale bar for A to H: 50 µm. **B.** Both apical and basal cells of the uterine epithelium express α-galactosyl residues with *Bandeiraea simplicifolia*-1B_4, as well as blood vessel walls in both uterus and

(Fig. 15.12 Contd...)

glucosamine oligomers (Fig. 15.12F and H) (Jones *et al.*1997). N-acetyl galactosamine was also present to some extent, as well as complex N-glycan with variable degrees of antennation (Fig. 15.12D, G). It was recently suggested that one function of this thick glycocalyx might be to act as an anti-adhesive barrier, to prevent attachment and sub-sequent migration of leukocytes across the endothelial cell barrier to the trophoblast (Jones *et al.* 2004b). However, in this case, the egg envelope would act as an additional obstacle.

Although generally classified as an epitheliochorial placenta, the yolk sac placenta of *Mustelus* is not really analogous to its mammalian counterparts as it lacks the close interdigitation between fetal trophoblast and maternal uterine epithelium. However, it is capable of expressing all classes of glycan which, in mammalian epitheliochorial placentae, are essential for successful attachment and adhesion between fetal and maternal tissues in the establishment of pregnancy. The egg envelope lies close to, but does not actually make contact with, the underlying uterine epithelium. The means by which it remains in such close proximity to the uterine epithelium is not clear as there is no direct contact between the two tissues, however, the irregularity of the uterine surface may be a factor. It is therefore remarkable that efficient diffusion of substances takes place, to enable the growth of an average of six young to reach a considerable size at birth.

15.13 UMBILICAL CORD AND APPENDICULAE

The term appendiculae was introduced by Alcock (1890) and Leuckart (1836) pictured appendiculae in a hammerhead embryo. Southwell and Prasad (1919), Thillayampalam (1928), Mahadevan (1940) and Setna and Sarangdhar (1948) described appendiculae and placentae from several species of *Scoliodon* from India. Appendiculae have also been shown in

---

Fig. 15.12 Contd. ...

placenta (v). The endoderm (en) is also moderately stained. **C.** Fucosyl residues in α1,2-linkage are present in both the placental ectoderm (ec) and uterine apical cells (a) after staining with *Anguilla anguilla*, and the egg envelope is also stained. **D.** Non-bisected bi-triantennary complex N-glycan, shown by staining with *Pisum sativum* lectin, is absent from the placental ectoderm, egg envelope (arrowheads) and uterine epithelial cells but can be found in the endoderm cells, fetal (fv) and uterine (uv) blood vessels and surrounding connective tissue. **E.** Only uterine epithelial basal cells (b) expressing terminal N-acetylgalactosamine are bound by the lectin from the snail *Helix pomatia*. No placental structures bind. **F.** Globular cells (g) and basal cells of the uterine epithelium are stained with the lectin from *Lycopersicon esculentum* which binds to N-acetylglucosamine oligomers, as well as fetal (fv) and uterine vessels (uv) and the surface of endoderm cells. **G.** The soybean lectin from *Glycine max* stains terminal N-acetyl galactosamine residues found in blood vessels (v) and connective tissue and basal cells of the uterine epithelium. Ectoderm (ec) and uterine apical cells (a) are weakly stained. **H** Most structures in both placenta and uterine wall express N-acetyl glucosamine oligomers bound by *Datura stramonium* lectin, and the egg envelope is also well stained. Original.

Sphyrna tiburo (Schlernitzauer and Gilbert, 1966) and *S. lewini* (Gilbert, 1982). The structure of the umbilical cord has received much less consideration (Mahadevan 1940; Budker 1949, 1953, 1958).

In early stages of development of the Atlantic sharpnose shark *Rhizoprionodon terraenovae* (Hamlett 1993; Hamlett *et* al. 1993b), the yolk stalk is a smooth, cylindrical structure. By the time the embryo is 25 mm TL, modest longitudinal folds form on the surface of the yolk stalk alternating with punctate depressions. These are the first indications of appendiculae. By the time the embryo has grown to 4 cm TL, appendiculae are present as rounded, longitudinal protuberances, and by 6 cm TL, the yolk stalk is adorned with a bushy covering of cylindrical and modestly branched appendiculae. Scanning electron microscopy revealed the surface epithelium to be columnar and a profusion of small blood vessels populate the connective tissue core in close proximity to the surface cells. The epithelium frequently displays surface bulges and foliose cell apices. Closer examination reveals two distinct cell types, viz. one with elongate microvilli and one with low relief microvilli. The yolk stalk of 10 cm TL embryos is covered with a dense mat of branched appendiculae. The bases of appendiculae are rounded but flatten distally as they branch. The yolk stalk of 18 cm TL embryos shows appendiculae basically characteristic of the term fetus.

In *Rhizoprionodon terraenovae*, the epithelial ectoderm of the somatopleure forms richly vascularized appendiculae (Fig. 15.13A) (Hamlett 1993; Hamlett *et al.* 1993b). Scanning electron microscopy reveals that the base and shaft of appendiculae are flattened while the distal portion may be expanded to form one to three lobes. The surface of appendiculae are again composed of two distinct cell types, the most plentiful being microvillar cells. The second cell type in this species contains prominent granules. These cells are much larger than the former and are partially submerged below the surface, except for the cell apex. These cells undergo secretory cycles ending in expulsion of their contents.

The presence of elongate microvilli establishes the possibility that these appendicular cells may serve as a nutrient absorptive site for uterine fluid bathing the embryos (Figs 15.13B, D). Hamlett (unpublished) incubated fresh umbilical cord appendiculae in an aerated solution of horseradish peroxidase for 10 min. He found HRP reaction product in endocytic vesicles of the microvillar cells but not the granulated cells.

The granulated cells present a more puzzling question. Clearly these cells are synthesizing and secreting material (Figs 15.13B, C). The nature of this material is currently unknown. It is known, however, that the granulated cells are formed early in embryonic development and, consequently, may play a role throughout gestation as well as near parturition (Hamlett, unpublished). The elaborated material may be utilized by the fetus directly or exert an influence by way of the perifetal environment. The nature and composition of the secretion await biochemical characterization.

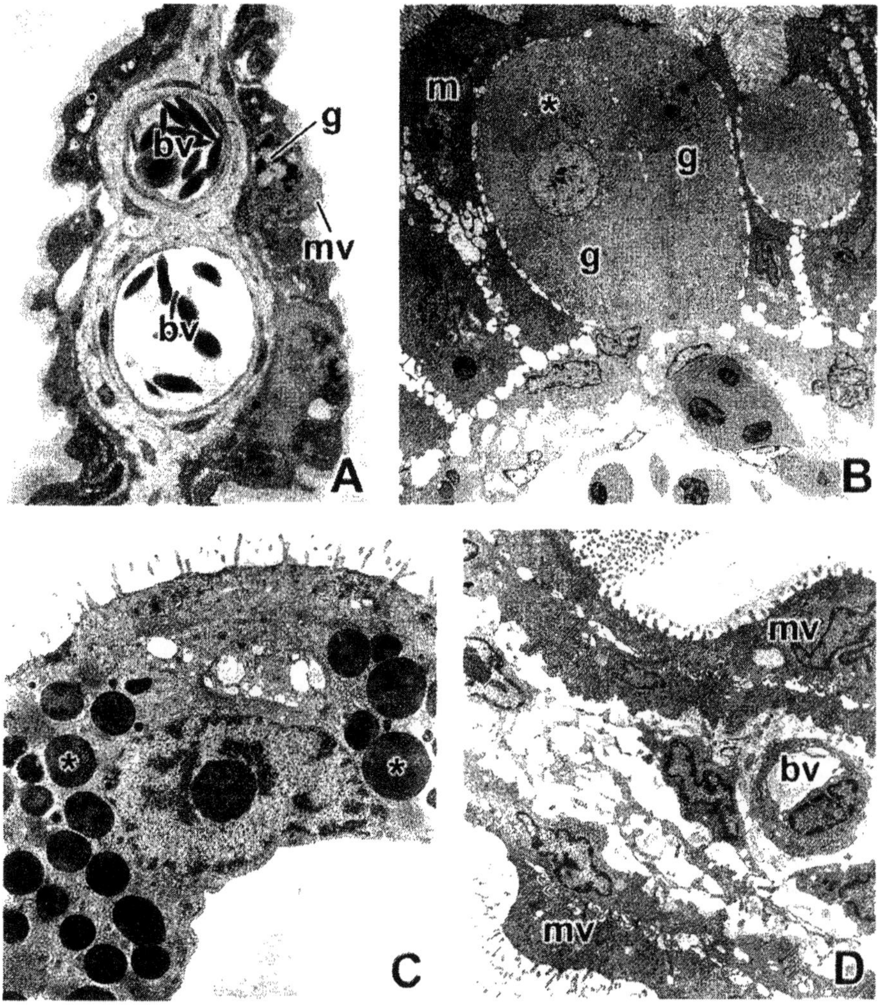

Fig. 15.13. A. Light micrograph of appendiculae of term *Rhizoprionodon terraenovae* showing large blood vessels (bv), microvillar cells (mv) and granulated secretory cells (g). X600. **B**. Transmission electron micrograph of appendiculae of term *R. terraenovae* showing granulated cells (g) that are synthesizing new secretory vesicles (asterisk) and microvillar cells (m). X1,340. **C**. Transmission electron micrograph of appendiculae of term *R. terraenovae* showing granulated cells with abundant secretory vesicles (asterisks). X5,200. **D**. Transmission electron micrograph of appendiculae of term *R. terraenovae* showing blood vessels (bv) at the center and microvillar cells (mv). X2,950. Original.

The granulated cells of the appendiculae bear striking ultrastructural similarity to the mitochondria-rich cells of the gills of the Atlantic hagfish, *Myxine glutinosa* (Bartels 1985; Bartels and Welsh 1986). Hagfish serum has high concentrations of sodium and potassium and is almost isosmotic with sea water (Alt *et al.* 1980). The mitochondria-rich cells are ovoid and their surface has scanty microvilli and is partially covered by neighboring cells. The mitochondria are numerous and occupy a basal location. A system of smooth-surfaced tubules occurs near the Golgi complex. The tubular system communicates with the intercellular space. These cells contain carbonic anhydrase and show sodium-potassium-ATPase activity and are actively involved in the regulation of the internal milieu of *M. glutinosa*. The tubular system in the granulated cells of the appendiculae does not communicate with the intercellular space. It is possible that granulated cells in *Rhizoprionodon terraenovae* may be involved in the regulation of the osmolarity of the embryonic fluids. Verification of this proposition awaits further biochemical and physiological studies.

Structure of the appendicular epithelium also closely resembles another tissue involved in active transport, the vertebrate gall bladder (Kaye *et al.* 1966). In gall bladders transporting fluid, the intercellular spaces are distended. No dilations occur, however, when Na^+ or Cl^- is omitted from the bathing media. The lateral and basal cell membranes remain attached via desmosomes even when the intercellular spaces are dilated. The same structural organization occurs in appendiculae.

In the gall bladder, the mucosa resorbs water and ions from bile and active transport of solute (Na^+) across the lateral cell membranes increases the concentration of solute in the intercellular space. Because of the osmotic gradient, water moves into and through the cell to the intercellular space, causing its dilation. Development of hydrostatic pressure in the spaces drives the solution across the basal lamina into the submucosa (Rose 1981).

15.14 CONCLUSIONS

Placentation occurs in relatively few species of sharks. Initially embryos rely on yolk stores (lecithotrophy) and as yolk is depleted the yolk stalk elongates and the yolk sac expands. Some species develop appendiculae on the umbilical cord that may absorb periembryonic fluids and also contribute secretions to the fluids. As yolk is being digested and before a placenta forms the uterus produces histotroph (histotrophy) The yolk sac then undergoes developmental modifications to form the fetal component of the placenta (placentatrophy). Complimentary sites on the uterus constitute the maternal component of the placenta. Another feature of shark placental species is development of uterine compartments. This conveys benefits including more surface area for gas exchange and histotrophy. This chapter points out the often neglected role of histotrophy during gestation in placental sharks. It also suggests placentation as a special case of histotrophy because the only documented types of placental transfer involve maternal secretion at the placental site rather than blood to blood transfer and calls into doubt whether any shark placenta is actually hemotrophic. Histotrophy throughout placental gestation calls for future studies of uptake in the embryonic alimentary canal as has been done in embryonic rays (Hamlett *et al.* 1996a, b). There is still much to be done to answer many questions regarding placentation in sharks.

15.15 ACKNOWLEDGEMENTS

This work was supported by funds provided by Indiana University School of Medicine (WCH).

15.16 LITERATURE CITED

Alcock, A. 1890. Observations on the gestation of some sharks and rays. Journal of the Asiatic Society of Bengal 59 (2): 51-56.

Allen, W. R. 1982. Immunological aspects of the endometrial cup reaction and the effects of xenogeneic pregnancy in horses and donkeys. Journal of Reproduction and Fertility Supplement 31: 57-94.

Allen, W. R., Kydd, J. H., Boyle, M. S. and Antczak, D. F. 1987. Extraspecific donkey-in-horse pregnancy as a model of early fetal death. Journal of Reproduction and Fertility Supplement 35: 197-209.

Alt, J. M., H. Stolte, G. M. Eisenbach, F. Walvig. 1980. Renal electrolyte and fluid excretion in the Atlantic hagfish *Myxine glutinosa*. Journal of Experimental Biolology 91:323-330.

Bainbridge, D. R. 2000. Evolution of mammalian pregnancy in the presence of the maternal immune system. Review of Reproduction 5: 67-74.

Baranes, A., and J. Wendling (1981) The early stages of development in *Carcharhinus plumbeus*. Journal of Fish Biology 18: 159-175.

Bartels, H. 1985. Assemblies of linear arrays of particles in the apical plasma membrane of mitochondria-rich cells in the gill epithelium of the Atlantic hagfish (*Myxine glutinosa*). Anatomical Record 211: 229-238.

Bartels, H. and U. Welsch. 1986. Mitochondria-rich cells in gill epithelium of cyclostomes, a thin section and freeze fracture study. Pp. 58-72. In T. Uyeno, R. Arai. T. Taniuchi and K. Matsuura (eds), *Proceedings of the Second International Conference on Indo-Pacific Fishes*, Ichthyological Society of Japan.

Bird, S., Wang, T., Zou, J., Cunningham, C and Secombes, C.J. 2002. The first cytokine sequence within cartilaginous fish: IL-1 beta in the small spotted catshark (*Scyliorhinus canicaula*). Journal of Immunology 168: 3329-3340.

Bowen, J. A. and Burghardt, R. C. 2000. Cellular mechanisms of implantation in domestics farm animals, Seminars in Cell and Developmental Biology 11: 93-104.

Budker, P. 1949. Note preliminaire sur le placenta et le cordon ombilical de trois selaciens vivipares de la Côte Occidentale d'Afrique C.R. XIII Congress International Zoology, Paris, 337-338 pp.

Budker, P. 1953. Sur le cordon ombilical des squales vivipares. Bulletin of the Museum of Natural History, Paris, 25: 541-545.

Budker, P. 1958. La viviparité chez les selaciens. Pp. 1755-1790. In Grasse, P. P. (ed.), *Traité de Zoologie*. Vol. 13, Part 2. Masson et Cie, Paris.

Calzoni, M. 1936. Ricerche sulla placenta del *Carcharias glaucus*. Pubbliccazioni della Stazione Zoologica di Napoli 15: 169-174.

Cate-Hoedemaker, N. J. Ten (1938) Die Beziehungen zwischen Muttertier and Embryo bei *Mustelus laevis* Risso and *Mustelus vulgaris* Day. Archives Neerlandaises de Zoologie 3: 89-96.

Cateni, C., Paulesu, L., Bigliardi, E., and Hamlett, W. C. 2003. The interleukin 1 (IL-1) system in the uteroplacental complex of a cartilaginous fish, the smoothhound shark, *Mustelus canis*. Reproductive Biology and Endocrinology 2003, peer reviewed electronic journal 1: 25.

Compagno, L. J. V. 1988. *Sharks of the order Carcharhiniformes*. Princeton, New Jersey: Princeton University Press. 486 pp.

Dantzer, V. 1985. Electron microscopy of the initial stages of placentation in the pig. Anatomy and Embryology 172: 281-293.

De Duve, C., and Wattiaux, R. 1966. Functions of lysosomes. Annual Review of Physiology 28: 435-492.

Fehilly, C. B., Willadsen, S. M. and Tucker, E. M. 1984. Interspecific chimaerism between sheep and goat. Nature 307: 634-636.

Fishelson, L. and Baranes, A. 1998. Observations on the Oman shark, *lago omanensis* (Triakidae), with emphasis on the morphological and cytological changes of the oviduct and yolk sac during gestation. Journal of Morphology 236: 151-165.

Gilbert, P. W. 1982. Patterns of shark reproduction. Oceanus 24: 30-39.

Gilbert, P. W. and Schlernitzauer, D. A. (1966) The placenta and gravid uterus of *Carcharhinus falciformis*. Copeia 1966: 451-457.

Graham, C. R. 1967. Nutrient transfer from mother to fetus and placental formation in *Mustelus* canis. Dissertation. University of Delaware, 55 pp.

Graham, C. R., Jr., Bond, C., Chacko, V. P., and Lombardi, J. 1995. NMR studies of glucose and alanine utilization and maternal-embryonic nutrient transfer in the smooth dogfish, *Mustelus canis*. Comparative Biochemistry and Physiology 111A: 199-207.

Gudger, E. W. 1915. Utero-gestation in the sharpnosed shark, *Scoliodon terraenovae*. Science N.S. 14 (1055): 439.

Hamlett, W. C. 1986. Prenatal nutrient absorptive structures in selachians. Pp. 333-344. In T. Uyeno, R. Arai, T. Taniuchi and K. Matsuura (eds), *Indo-Pacific Fish Biology*. Ichthyological Society of Japan, Tokyo.

Hamlett, W. C. 1987. Comparative morphology of the elasmobranch placental barrier. Archives de Biologie (Bruxelles) 98: 135-162.

Hamlett, W. C. 1989. Evolution and morphogenesis of the placenta in sharks. Journal of Experimental Zoology, Supplement 2: 35-52.

Hamlett, W. C. 1990a. Elasmobranch species as models for studies of placental viviparity and its endocrine regulation. Journal of Experimental Zoology Supplement 4: 129-131.

Hamlett, W. C. 1990b. Electron microscopic investigations of the fetal membranes and uterus in the blacknose shark, *Carcharhinus acronotus*. Anatomical Record 226: 40a.

Hamlett, W. C. 1993a. Ontogeny of the umbilical cord and placenta in the Atlantic sharpnose shark, *Rhizoprionodon terraenovae*. Environmental Biology of Fishes 38: 253-267.

Hamlett, W. C. and Wourms, J. P. 1984. Ultrastructure of the pre-implantation shark yolk sac. Tissue and Cell 16: 613-625.

Hamlett, W. C., Wourms, J. P. and Hudson, J. S. 1985a. Ultrastructure of the full term shark yolk sac placenta. I. Morphology and cellular transport at the fetal attachment site. Journal of Ultrastructure Research 91: 192-206.

Hamlett, W. C., Wourms, J. P. and Hudson, J. S. 1985b. Ultrastructure of the full term shark yolk sac placenta. II. The smooth, proximal segment. Journal of Ultrastructure Research 91: 207-220.

Hamlett, W. C., Wourms, J. P. and Hudson, J. S. 1985c. Ultrastructure of the full term shark yolk sac placenta. III. The maternal attachment site. Journal of Ultrastructure Research 91: 221-231.

Hamlett, W. C., Allen, D. J., Stribling, M. D., Schwartz, F.J . and DiDio, L. J. A. 1985d. Permeability of embryonic shark external gill filaments. Electron microscopic observations using horseradish peroxidase as a macromolecular tracer. Journal of Submicroscopic Cytology 17: 31-40.

Hamlett, W. C. 1986. Prenatal nutrient absorptive structures in selachians. In: *Indo-Pacific Fish Biology*. Pp. 333-343, In T. Uyeno, R. Arai. T. Taniuchi and K. Matsuura (eds), *Proceedings of the Second International Conference on Indo-Pacific Fishes*, Ichthyology Society of Japan.

Hamlett, W. C. 1987. Comparative morphology of the elasmobranch placental barrier. Archives of Biology (Bruxelles). 98: 135-162.

Hamlett, W. C. 1998. Placenta and placental analogs in elasmobranchs. Pp. 197-206. In E. Knobil and J. D. Neill (eds), *Encyclopedia of Reproduction*. Vol. 3. Academic Press,

Hamlett, W. C., Schwartz, F. J. and DiDio, L. J. A. 1987. Subcellular organization of the yolk syncytial-endoderm complex in the preimplantation yolk sac of the shark, *Rhizoprionodon terraenovae*. Cell and Tissue Research 247: 275-285.

Hamlett, W. C. and Hysell, M. K. 1998. Uterine specializations in elasmobranchs. Journal of Experimental Zoology 282: 438-459.

Hamlett, W. C., Ferri, A. G. and Miglino, M. A. 1992. Modes of reproduction in the elasmobranchs of Brazil. Pp. 3-18. In W. C. Hamlett (ed.), *Reproductive Biology of South American Vertebrates*. Springer Verlag, New York.

Hamlett, W. C., Eulitt, A. M., Jarrell, R. L. and Kelly, M. A. 1993. Uterogestation and placentation is elasmobranchs. Journal of Experimental Zoology 266: 347-367.

Hamlett, W. C., Miglino, M. A. and DiDio, L. J. A. 1993b. Subcellular organization of the placenta in the Atlantic sharpnose shark, *Rhizoprionodon terraenovae*. Journal of Submicroscopic Cytolology and Pathology 25: 535-545.

Hamlett, W. C., Miglino, M. A., Federman, D. J., Schafer, P. and DiDio, L. J. A. 1993c. Fine structure of the term umbilical cord in the Atlantic sharpnose shark, *Rhizoprionodon terraenovae*. Journal of Submicroscopic Cytolology and Pathology 25: 547-557.

Hamlett, W. C., Knight, D. P., Koob, T., Jezior, M., Luong, L., Rozycki, T., Brunette, N. and Hysell, M. 1998. Survey of oviducal gland structure and function in elasmobranchs. Journal of Experimental Zool.ogy 282: 399-420.

Hamlett, W. C. and Koob, T. 1999. Female reproductive system. Pp. 398-443. In W. C. Hamlett (ed.), *Sharks, Skates and Rays: Biology of Elasmobranch Fishes*. The Johns Hopkins University Press, Baltimore, Maryland.

Hamlett, W. C., Musick, J. A., Hysell, C. K. and Sever, D. M. 2002b. Uterine epithelial-sperm interaction, endometrial cycle and sperm storage in the terminal zone of the oviducal gland in the placental smoothhound, *Mustelus canis*. Journal of Experimental Zoology 292: 129-144.

Jones, C.J.P., Abd-Elnaeim, M., Bevilacqua, E., Oliveira, L.V. and Leiser, R. 2002. Comparison of uteroplacental glycosylation in the camel (*Camelus dromedarius*) and alpaca (*Lama pacos*). Reproduction 123: 115-126.

Jones, C. J. P. and Hamlett, W. C. 2004. Structure and glycosylation of the term yolk sac placenta and uterine attachment site in the viviparous shark *Mustelus canis*. Placenta 25: 820-828.

Jones, C. J. P., Cateni, C., Guarino, F. M. and Paulesu, L. R. 2003. Glycosylation of the materno-foetal interface in the pregnant viviparous placentotrophic lizard *Chalcides chalcides*: A lectin histochemical study. Placenta 24: 489-500.

Jones, C. J. P., Dantzer, V., Leiser, R., Krebs, C. and Stoddart, R. W. 1997. Localisation of glycans in the placenta: A comparative study of epitheliochorial, endotheliochorial, and haemomonochorial placentation. Microscopy Research and Technique 38: 100-114.

Jones, C. J. P., Santos, T. C., Abd-Elnaeim, M., Dantzer, V. and Miglino, M. A. 2004a. Placental glycosylation in peccary species and its relation to that of swine and dromedary. Placenta 25: 649-657.

Jones, C. J. P., Wooding, F. B. P., Dantzer, V., Leiser, R. and Stoddart, R. W. 1999. A lectin binding analysis of glycosylation patterns during development of the equine placenta. Placenta 20: 45-57.

Jones, C. J. P., Wooding, F. B. P., Mathias, S. S. and Allen, W. R. 2004b. Fetomaternal glycosylation of early placentation events in the African elephant *Loxodonta africana*. Placenta 25: 308-320

Kaye, G., Wheeler, H., Whitlock, R. and Lane, N. 1966. Fluid transport in the rabbit gallbladder. Journal of Cell Biology 30: 237-268.

Leuckart, F. S. 1836. Untersuchungen ueber ausseren Kiemen der Embryonen von Rochen und Hayden. 3: (embryo von hemmerfisch, *Zygaena tiburo*, 22-24 pp., pl. 3) Stuttgart.

Litynska, A. and Przybylo, M. 1998. Does glycosylation of lysosomal proteins show age-related changes in rat liver? Mechanisms of Ageing and Development 102: 33-43.

Lombardi, J., and Files, T. 1993. Egg capsule structure and permeability in the viviparous shark, *Mustelus canis*. Journal of Experimental Zoology 267: 76-85.

Lombardi, J., Jones, K. B., Garrity, C. A. and Files, T. 1993. Chemical composition of uterine fluid in four species of viviparous sharks (*Squalus acanthias, Carcharhinus plumbeus, Mustelus canis* and *Rhizoprionodon terraenovae*). Comparative Biochemistry and Physiology 105A: 91-102.

Mahadevan, G. 1940. Preliminary observations on the structure of the uterus and the placenta of a few Indian elasmobranchs. Proceedings of the Indian Natural Science Academy (Ser. B), 11: 2-47.

Mossman, H. W. 1987. *Vertebrate Fetal Membranes*, Rutgers University Press, New Brunswick, New Jersey. pp. 383.

Okano, S., Otake, S., Teshima, K. and Mizue, K. 1981. Studies on sharks. XX. Epithelial cells of the intestine in *Mustelus manazo* and *M. griseus*. Bulletin of the Faculty of Fisheries Nagasaki University 51: 23-28.

Otake, T. 1990. Classification of reproductive modes in sharks with comments on female reproductive tissues and structures. Pp. 518. In H. L. Pratt, S. H. Gruber, and T. Taniuchi, (eds), *Elasmobranchs as Living Resources: Advances in the Biology, Ecology, Systematics, and the Status of the Fisheries*, NOAA Tech. Report NMFS 90.

Otake, T. K. Mizue, K. 1985. The fine structure of the placenta of the blue shark, *Prionace glauca*. Japanese Journal of Ichthyology 32: 52-59.

Otake, T. and Mizue, K. 1986. The fine structure of the intra-uterine epithelium during late gestation in the blue shark, *Prionace glauca*. Japanese Journal of Ichthyology 33: 162-167.

Paulesu, L. 1997. Cytokines in mammalian reproduction and speculation about their possible involvement in nonmammalian viviparity. Microscopy Research Technique 38: 188-194.

Paulseu, L., Romagnoli, R., Marchetti, M., Cintorino, M., Ghiara, P., Guarino, F. M. and Ghira, G. 1995. Cytokines in the viviparous reproduction of squamate reptiles: Interleukin 1α (IL-1α) and IL-1β in placental structures of a skink. Placenta 16: 193-205.

Ranzi, S. 1932. Le basi fisio-morfologiche dello sviluppo embrionale dei Selaci. Parte I. Pubbliccazioni della Stazione Zoologica di Napoli 13: 209-290.

Ranzi, S. 1934. Le basi fisio-morfologiche dello sviluppo embrionale dei Selaci. Parti II e III. Pubbliccazioni della Stazione Zoologica di Napoli 13: 331-437.

Rose, R. C. 1981. Absorptive functions of the gallbladder. Pp. 1021-1033. In L. R. Johnson (ed.), *Physiology of the Gastrointestinal Tract*, Vol. 2. Raven Press, New York, NY.

Saito, S. 2000. Cytokine network at the feto-maternal interface. Journal of Reproduction and Immunology 47: 87-103.

Schlafke, S. and Enders, A. C. 1975. Cellular basis of interaction between trophoblast and uterus at implantation. Biology of Reproduction 12: 41-65.

Schlernitzauer, D. A. and Gilbert, P. W. 1966. Placentation and associated aspects of gestation in the bonnethead shark, *Sphyrna tiburo*. Journal of Morphology 120: 219-232.

Setna, S. B. and Sarangdhar, P. N. 1948. Description, bionomics and development of *Scoliodon sorrakow* (Cuvier). Records of the Indian Museum (Calcutta) 46: 25-53.

Sharon, N. and Lis, H. 1989. *Lectins*. Chapman and Hall, London. 454 pp.

Southwell, T. and B. Prashad. 1919. Notes from the Bengal fisheries laboratory, No. 6: Embryological and developmental studies of Indian fishes. Records of the Indian Museum 16: 215-240.

Surani, M. A. H. and Fishel, S. B. 1980. Blastocyst-uterine interactions at implantation. Progress in Reproductive Biology 7: 14-27.

Te Winkel, L. E. 1963. Notes on the smooth dogfish, *Mustelus canis*, during the first three months of gestation. II. Structural modifications of yolk-sacs and yolk-stalks correlated with increasing absorptive function. Journal of Experimental Zoology 152: 123-137.

Teshima, K., Ahmad, M. and Mizue, K. 1978. Studies on sharks XIV reproduction in the Telok Anson shark collected from Perak River, Malaysia. Japanese Journal of Icthyology 25: 181-189.

Teshima, K., Mizue, K. and Koga, S. 1974. Studies on sharks. VII. Reproduction in female *Mustelus griseus*. Journal of Shimonoseki University of Fisheries. 22: 199-206.

Teshima, K. and Mizue, K. 1972. Studies on sharks. I. Reproduction in the female sumitsuki shark *Carcharhinus dussumieri*. Marine Biology 14: 222-231.

Teshima, K. and Koga, S. 1973. Studies on sharks. V. Taxonomic characteristics of reproductive organs of Japanese *Mustelus*. Marine Biology 23: 337-341.

Teshima, K. (1975) Studies on sharks. VIII. Placentation in *Mustelus griseus*. Japanese Journal of Icthyology 22: 7-12

Teshima, K. 1981. Studies on the reproduction of Japanese smooth dogfishes, *Mustelus manazo* and *M. griseus*. Journal of the Shimonoseki University of Fisheries 29: 113-199.

Thillayampalam, E. M. 1928. *Scoliodon*, the common shark of the Indian Seas. Pp. 98-103. In K. N. Bahl, (ed.), *The Indian Zoological Memoirs on Indian Animal Types*. Lucknow, India.

Van Tienhoven, A. 1983. *Reproductive Physiology of Vertebrates*, 2nd edition, Ithaca, New York: Cornell University Press. 491 pp.

Walzer, C. and Schonenberger, N. 1979a. Ultrastructure and cytochemistry of the yolk syncytial layer in the alevin of trout (*Salmo fario trutta* L.) after hatching. I. The vitellolysis zone. Cell and Tissue Research. 196: 59-73.

Walzer, C. and Schonenberger, N. 1979b. Ultrastructure and cytochemistry of the yolk syncytial layer in the alevin of trout (*Salmo fario trutta* L. and *Salmo gairdneri* R.) after hatching. II. The cytoplasmic zone. Cell and Tissue Research. 196: 75-93.

Ward, R. T. 1980. The origin of protein and fatty yolk in *Rana pipiens*. V. Unusual paracrystalline configurations within the yolk precursor complex. Journal of Morphology. 165: 255-260.

Whyte, A. and Robson, T. 1984. Saccharides localised by fluorescent lectins on trophectoderm and endometrium prior to implantation in pigs, sheep and equids. Placenta 5: 533-540.

Widakowich, V. 1906. Uber Bau und Funktion des Nidamentalorgans von *Scyllium canicula*. Zeitschrift f¨ur Wissenschaftliche Zoologie 80: 1-21.

Widakowich, V. 1907. Uber eine Verschlussvorrichtung im Eileiter von *Squalus acanthias*. Zoologische Anzeiger 31: 636-643.

Wourms, J. P. 1993. Maximization of evolutionary trends for placental viviparity in the spadenose shark, *Scoliodon laticaudus*. Environmental Biology of Fishes. 38: 269-294.

CHAPTER 16

Checklist of Living Chondrichthyes

Leonard J. V. Compagno

16.1 INTRODUCTION

This chapter consists of an annotated checklist of living cartilaginous fishes that includes only described valid species but omits most synonyms and undescribed species. A generalized listing of geographic distribution and habitat is included for each genus. It is revised over previous published checklists as of June 11, 2004. Names of doubtful species are preceeded by a question mark; valid species that are doubtfully placed in agences have a question mark at the end of the generic name.

16.2 ANNOTATED CHECKLIST OF LIVING CHONDRICHTHYES

CLASS CHONDRICHTHYES. CARTILAGINOUS FISHES.

SUBCLASS ELASMOBRANCHII: SHARKS AND RAYS

ORDER HEXANCHIFORMES. COW AND FRILLED SHARKS.

FAMILY CHLAMYDOSELACHIDAE GARMAN, 1884. FRILLED SHARKS.

Genus *Chlamydoselachus* Garman, 1884. Frilled sharks. Wide-ranging but spotty in the Western North Atlantic, Eastern Atlantic, Southwestern Indian Ocean, Western Pacific, and Eastern Pacific, continental and insular slopes, occasionally on the shelves. Probably two species, including an undescribed one off southern Africa.

Chlamydoselachus anguineus Garman, 1884. Frilled shark.

Shark Research Center, Iziko-South African Museum, P.O. Box 61, Cape Town 8000, South Africa

FAMILY HEXANCHIDAE GRAY, 1851. COW SHARKS.

Genus *Heptranchias* Rafinesque, 1810. Sharpnose sevengill sharks. Sometimes placed in a separate family Heptranchiidae. Western North Atlantic, Eastern Atlantic and Mediterranean, Indo-West Pacific, and Western South Pacific, upper slopes and lower shelves. A single species.

Heptranchias perlo (Bonnaterre, 1788). Sharpnose sevengill shark.

Genus *Hexanchus* Rafinesque, 1810. Sixgill sharks. Circumglobal in temperate and tropical seas, continental and insular shelves and slopes, sea mounts. Two species.

Hexanchus griseus (Bonnaterre, 1788). Bluntnose sixgill shark.

Hexanchus nakamurai Teng, 1962. Bigeye sixgill shark. This species includes *H. vitulus* Springer & Waller, 1969 as a junior synonym.

Genus *Notorynchus* Ayres, 1855. Broadnose sevengill sharks. Sometimes placed in a separate family Notorynchidae. Temperate seas in the South Atlantic, Western Indian Ocean, and Pacific Ocean, continental shelves. A single species.

Notorynchus cepedianus (Peron, 1807). Broadnose sevengill shark.

ORDER SQUALIFORMES. DOGFISH SHARKS.

FAMILY ECHINORHINIDAE GILL, 1862. BRAMBLE SHARKS.

Genus *Echinorhinus* Blainville, 1816. Bramble sharks. Temperate seas in the Atlantic, Western Indian Ocean, and Pacific Ocean, continental and insular slopes and shelves, sea mounts. Two species.

Echinorhinus brucus (Bonnaterre, 1788). Bramble shark.

Echinorhinus cookei Pietschmann, 1928. Prickly shark.

FAMILY SQUALIDAE BLAINVILLE, 1816. DOGFISH SHARKS.

Genus *Cirrhigaleus* Tanaka, 1912. Roughskin dogfish. Wide-ranging in the Southwestern Indian Ocean, West-Central Pacific, and Western North Atlantic. Two species.

Cirrhigaleus asper (Merrett, 1973). Roughskin spurdog.

Cirrhigaleus barbifer Tanaka, 1912. Mandarin dogfish.

Genus *Squalus* Linnaeus, 1758. Spurdogs or spiny dogfishes. Circumglobal in temperate and tropical seas, continental and insular shelves and slopes, sea mounts. Eight valid species and at least six undescribed species:

Squalus acanthias Linnaeus, 1758. Piked dogfish.

Squalus blainvillei (Risso, 1826). Longnose spurdog.

Squalus cubensis Howell-Rivero, 1936. Cuban dogfish.

Squalus japonicus Ishikawa, 1908. Japanese spurdog.

Squalus megalops (Macleay, 1881). Shortnose spurdog. *Squalus brevirostris* Tanaka, 1912, and *Squalus probatovi* Myagkov & Kondyurin, 1986 are tentatively ranked as synonyms. *S. megalops* may be a species complex of several species.

Squalus melanurus Fourmanoir, 1979. Blacktail spurdog.

Squalus mitsukurii Jordan & Snyder, *in* Jordan & Fowler, 1903. Shortspine spurdog.

Squalus rancureli Fourmanoir, 1978. Cyrano spurdog.

FAMILY CENTROPHORIDAE BLEEKER, 1859. GULPER SHARKS.

Genus *Centrophorus* Müller & Henle, 1837. Gulper sharks. Virtually circumglobal, except for the Eastern Pacific and possibly the Western South Atlantic, continental and insular slopes and outer shelves. Genus *Pseudocentrophorus* Chu, Meng, & Liu, 1981 is a synonym of this genus. At least 10 species, some possibly species-complexes, plus possibly one or two undescribed species.

Centrophorus acus Garman, 1906. Needle dogfish.

Centrophorus atromarginatus Garman, 1913. Dwarf gulper shark.

Centrophorus granulosus (Bloch & Schneider, 1801). Gulper shark. *Squalus uyato* Rafinesque, 1810 is often used for this and other species of *Centrophorus*, but the name was apparently based on a species of *Squalus*.

Centrophorus harrissoni McCulloch, 1915. Longnose gulper shark.

Centrophorus isodon (Chu, Meng, & Liu, 1981). Blackfin gulper shark.

Centrophorus lusitanicus Bocage & Capello, 1864. Lowfin gulper shark.

Centrophorus moluccensis Bleeker, 1860. Smallfin gulper shark.

Centrophorus niaukang Teng, 1959. Taiwan gulper shark.

Centrophorus squamosus (Bonnaterre, 1788). Leafscale gulper shark.

Centrophorus tesselatus Garman, 1906. Mosaic gulper shark.

Genus *Deania* Jordan & Snyder, 1902. Birdbeak dogfishes. Western North Atlantic, Eastern Atlantic, Western Indian Ocean, Western Pacific, continental and insular slopes. At least four species, *D. calcea* possibly a species-complex.

Deania calcea (Lowe, 1839). Birdbeak dogfish.

Deania hystricosum (Garman, 1906). Rough longnose dogfish.

Deania profundorum (Smith & Radcliffe, 1912). Arrowhead dogfish.

Deania quadrispinosum (McCulloch, 1915). Longsnout dogfish.

FAMILY ETMOPTERIDAE FOWLER, 1934. LANTERN SHARKS.

Genus *Aculeola* de Buen, 1959. Hooktooth dogfishes. Eastern South Pacific, continental slopes. A single species.

Aculeola nigra de Buen, 1959. Hooktooth dogfish.

Genus *Centroscyllium* Müller & Henle, 1841. Combtooth dogfishes. Virtually circumglobal on continental and insular slopes. At least seven species.

Centroscyllium excelsum Shirai & Nakaya, 1990. Highfin dogfish.

Centroscyllium fabricii (Reinhardt, 1825). Black dogfish.

Centroscyllium granulatum Günther, 1887. Granular dogfish.

Centroscyllium kamoharai Abe, 1966. Bareskin dogfish.

Centroscyllium nigrum Garman, 1899. Combtooth dogfish.

Centroscyllium ornatum (Alcock, 1889). Ornate dogfish.

Centroscyllium ritteri Jordan & Fowler, 1903. Whitefin dogfish.

Genus *Etmopterus* Rafinesque, 1810. Lantern sharks. Circumglobal on the outer continental and insular slopes, with a few species oceanic. 31 valid species, plus three undescribed species.

Etmopterus baxteri Garrick, 1957. Giant lanternshark. Placed in synonymy of *E. granulosus* by some writers but apparently distinct. *E. tasmaniensis* Myagkov & Pavlov, 1986 is a possible synonym of this species or *E. unicolor*.

Etmopterus bigelowi Shirai & Tachikawa, 1993. Blurred smooth lanternshark.

Etmopterus brachyurus Smith & Radcliffe, 1912. Shorttail lanternshark.

Etmopterus bullisi Bigelow & Schroeder, 1957. Lined lanternshark.

Etmopterus carteri Springer & Burgess, 1985. Cylindrical lanternshark.

Etmopterus caudistigmus Last, Burgess & Seret, 2002. Tailspot lanternshark.

Etmopterus decacuspidatus Chan, 1966. Combtooth lanternshark.

Etmopterus dianthus Last, Burgess & Seret, 2002. Pink lanternshark.

Etmopterus dislineatus Last, Burgess & Seret, 2002. Lined lanternshark.

Etmopterus evansi Last, Burgess & Seret, 2002. Blackmouth lanternshark.

Etmopterus fusus Last, Burgess & Seret, 2002. Pygmy lanternshark.

Etmopterus gracilispinis Krefft, 1968. Broadband lanternshark.

Etmopterus granulosus (Günther, 1880). Southern lanternshark.

Etmopterus hillianus (Poey, 1861). Caribbean lanternshark.

Etmopterus litvinovi Parin & Kotlyar, in Kotlyar, 1990. Smalleye lanternshark.

Etmopterus lucifer Jordan & Snyder, 1902. Blackbelly lanternshark.

Etmopterus molleri (Whitley, 1939). Slendertail lanternshark. *E. schmidti* Dolganov, 1986 is a possible junior synonym of this species.

Etmopterus perryi Springer & Burgess, 1985. Dwarf lanternshark.

Etmopterus polli Bigelow, Schroeder, & Springer, 1953. African lanternshark.

Etmopterus princeps Collett, 1904. Great lanternshark.

Etmopterus pseudosqualiolus Last, Burgess & Seret, 2002. False lanternshark.

Etmopterus pusillus (Lowe, 1839). Smooth lanternshark.

Etmopterus pycnolepis Kotlyar, 1990. Densescale lanternshark.

Etmopterus robinsi Schofield & Burgess, 1997. West Indian lanternshark.

Etmopterus schultzi Bigelow, Schroeder & Springer, 1953. Fringefin lanternshark.

Etmopterus sentosus Bass, D'Aubrey & Kistnasamy, 1976. Thorny lanternshark.

Etmopterus spinax (Linnaeus, 1758). Velvet belly

Etmopterus splendidus Yano, 1988. Splendid lanternshark.

Etmopterus unicolor (Engelhardt, 1912). Brown or bristled lanternshark. *Etmopterus compagnoi* Fricke & Koch, 1990 is tentatively ranked as a junior synonym of this species.

Etmopterus villosus Gilbert, 1905. Hawaiian lanternshark.

Etmopterus virens Bigelow, Schroeder & Springer, 1953. Green lanternshark.

Genus *Miroscyllium* Shirai & Nakaya, 1990. Rasptooth dogfish. Western North Pacific, continental slopes. A single species.

Miroscyllium sheikoi (Dolganov, 1986). Rasptooth dogfish.

Genus *Trigonognathus* Mochizuki & Ohe, 1990. Viper dogfish. Western North Pacific, Japan, Central Pacific, Hawaiian Islands, continental slopes. A single species.

Trigonognathus kabeyai Mochizuki & Ohe, 1990. Viper dogfish or viper shark.

FAMILY SOMNIOSIDAE JORDAN, 1888. SLEEPER SHARKS.

Genus *Centroscymnus* Bocage & Capello, 1864. Velvet dogfishes. Western Atlantic, Eastern Atlantic, Southeastern and Central Indian Ocean, Western Pacific, Eastern South Pacific, continental and insular slopes. Two species as restricted here.

Centroscymnus coelolepis Bocage & Capello, 1864. Portugese dogfish.

Centroscymnus owstoni Garman, 1906. Roughskin dogfish. *Centroscymnus cryptacanthus* Regan, 1906 is a junior synonym of this species.

Genus *Centroselachus* Garman, 1913. Longnose velvet dogfishes. Eastern Atlantic, Southwestern Indian Ocean, Western Pacific, Eastern South Pacific, continental and insular slopes. A single species, formerly placed in *Centroscymnus*.

Centroselachus crepidater (Bocage & Capello, 1864). Longnose velvet dogfish.

Genus *Proscymnodon* Fowler, 1934. Largespine velvet dogfishes. Western South Atlantic, Southwestern Indian Ocean, Western South Pacific, continental and insular slopes. *Proscymnodon* was formerly a subgenus of *Scymnodon*. Two species, formerly placed in *Centroscymnus* or *Scymnodon*.

Proscymnodon macracanthus (Regan, 1906). Largespine velvet dogfish.

Proscymnodon plunketi (Waite, 1909). Plunket shark.

Genus *Scymnodalatias* Garrick, 1956. Spineless velvet dogfishes. North-Central Atlantic, Southeastern Indian Ocean, South Pacific, continental and insular slopes, semioceanic or oceanic Four species.

Scymnodalatias albicauda Taniuchi & Garrick, 1986. Whitetail dogfish.

Scymnodalatias garricki Kukuyev & Konovalenko, 1988. Azores dogfish.

Scymnodalatias oligodon Kukuyev & Konovalenko, 1988. Sparsetooth dogfish.

Scymnodalatias sherwoodi (Archey, 1921). Sherwood dogfish.

Genus *Scymnodon* Bocage & Capello, 1864. Knifetooth dogfishes. Eastern North Atlantic, Western South Pacific, slopes and oceanic. A single species.

Scymnodon ringens Bocage & Capello, 1864. Knifetooth dogfish.

Genus *Somniosus* Le Sueur, 1818. Sleeper sharks. Wide-ranging in Atlantic, Indian, Pacific, and Arctic Oceans, also off Subantarctic islands, continental and insular shelves (high latitudes) and slopes, also on sea mounts. Five species, falling in two well-marked subgenera, with an additional undescribed long-nosed species of uncertain subgeneric placement off Portugal.

Subgenus *Rhinoscymnus* Gill, 1862:

Somniosus (Rhinoscymnus) longus (Tanaka, 1912). Frog shark.

Somniosus (Rhinoscymnus) rostratus (Risso, 1810). Little sleeper shark.

Subgenus *Somniosus* Le Sueur, 1818:

Somniosus (Somniosus) antarcticus Whitley, 1939. Southern sleeper shark.

Somniosus (Somniosus) microcephalus (Bloch & Schneider, 1801). Greenland shark.

Somniosus (Somniosus) pacificus Bigelow & Schroeder, 1944. Pacific sleeper shark.

Genus *Zameus* Jordan & Fowler, 1903. Velvet dogfishes. Western and Eastern Atlantic, Indian Ocean, Western and Central Pacific, slopes and oceanic. Two species, formerly placed in *Scymnodon.*

Zameus ichiharai (Yano & Tanaka, 1984). Japanese velvet dogfish.

Zameus squamulosus (Günther, 1877). Velvet dogfish.

FAMILY OXYNOTIDAE GILL, 1872. ROUGH SHARKS.

Genus *Oxynotus* Rafinesque, 1810. Roughsharks or flatiron sharks. Western North and Eastern Atlantic and Mediterranean, possibly Southwestern Indian Ocean, Western Pacific, continental and insular shelves and slopes. Five species.

Oxynotus bruniensis (Ogilby, 1893). Prickly dogfish.

Oxynotus caribbaeus Cervigon, 1961. Caribbean roughshark.

Oxynotus centrina (Linnaeus, 1758). Angular roughshark. *Oxynotus shubnikovi* Myagkov, 1986 is a probable synonym of this species.

Oxynotus japonicus Yano & Murofushi, 1985. Japanese roughshark.

Oxynotus paradoxus Frade, 1929. Sailfin roughshark.

FAMILY DALATIIDAE GRAY, 1851. KITEFIN SHARKS.

Genus *Dalatias* Rafinesque, 1810. Kitefin sharks. Wide-ranging in most warm-temperate and tropical seas, absent in the Eastern Pacific, demersal on the continental and insular slopes. A single species.

Dalatias licha (Bonnaterre, 1788). Kitefin shark.

Genus *Euprotomicroides* Hulley & Penrith, 1966. Taillight sharks. South Atlantic, continental slopes and oceanic. A single species.

Euprotomicroides zantedeschia Hulley & Penrith, 1966. Taillight shark.

Genus *Euprotomicrus* Gill, 1865. Pygmy sharks. Circumglobal in temperate seas, oceanic. A single species.

Euprotomicrus bispinatus (Quoy & Gaimard, 1824). Pygmy shark. A single species.

Genus *Heteroscymnoides* Fowler, 1934. Longnose pygmy sharks. Eastern South Atlantic, Southwestern Indian Ocean, Eastern South Pacific, oceanic. A single species.

Heteroscymnoides marleyi Fowler, 1934. Longnose pygmy shark.

Genus *Isistius* Gill, 1865. Cookiecutter sharks. Circumglobal in temperate and tropical seas, oceanic. Perhaps three species.

Isistius brasiliensis (Quoy & Gaimard, 1824). Cookiecutter shark.
Isistius labialis Meng, Chu, & Li, 1985. South China cookiecutter shark.
Isistius plutodus Garrick & Springer, 1964. Largetooth cookiecutter shark.

Genus *Mollisquama* Dolganov, 1984. Pocket sharks. Eastern South Pacific on submarine ridges west of northern Chile, oceanic or demersal on slopes. A single species.

Mollisquama parini Dolganov, 1984. Pocket shark.

Genus *Squaliolus* Smith & Radcliffe, *in* Smith, 1912. Spined pygmy sharks. Western Atlantic, Eastern North Atlantic, Southwestern and Eastern Indian Ocean, Western Pacific, oceanic and on continental and insular shelves and slopes. Two species.

Squaliolus aliae Teng, 1959. Smalleye pigmy shark.
Squaliolus laticaudus Smith & Radcliffe, 1912. Spined pygmy shark.

ORDER PRISTIOPHORIFORMES. SAWSHARKS.

FAMILY PRISTIOPHORIDAE BLEEKER, 1859. SAWSHARKS.

Genus *Pliotrema* Regan, 1906. Sixgill sawsharks. Southern Africa and Madagascar, demersal on continental shelves and slopes. A single species.

Pliotrema warreni Regan, 1906. Sixgill sawshark.

Genus *Pristiophorus* Müller & Henle, 1837. Fivegill sawsharks. Western Indian Ocean, Western Pacific, Western North Atlantic, demersal on continental and insular shelves and slopes. Four described species, with four additional undescribed species from the Indo-West Pacific.

Pristiophorus cirratus (Latham, 1794). Longnose sawshark.
Pristiophorus japonicus Günther, 1870. Japanese sawshark.
Pristiophorus nudipinnis Günther, 1870. Shortnose sawshark.
Pristiophorus schroederi Springer & Bullis, 1960. Bahamas sawshark.

ORDER SQUATINIFORMES. ANGEL SHARKS.

FAMILY SQUATINIDAE BONAPARTE, 1838. ANGEL SHARKS.

Genus *Squatina* Dumeril, 1806. Angel sharks. Wide ranging on the continental shelves and upper slopes of most seas, mostly in temperate waters but penetrating the tropics. 15 species, plus at least three undescribed species from the Indo-West Pacific, possibly one from the Western North Atlantic, and one from the Eastern Pacific.

Squatina aculeata Dumeril, *in* Cuvier, 1817. Sawback angelshark.

Squatina africana Regan, 1908. African angelshark.

Squatina argentina (Marini, 1930). Argentine angelshark.

Squatina australis Regan, 1906. Australian angelshark.

Squatina californica Ayres, 1859. Pacific angelshark.

Squatina dumeril Lesueur, 1818. Sand devil.

Squatina formosa Shen & Ting, 1972. Taiwan angelshark.

Squatina guggenheim Marini, 1936. Hidden angelshark. *Squatina occulta* Vooren & da Silva, 1991 is a junior synonym. Formerly termed angular angelshark (see *S. punctata*).

Squatina japonica Bleeker, 1858. Japanese angelshark.

Squatina nebulosa Regan, 1906. Clouded angelshark.

Squatina oculata Bonaparte, 1840. Smoothback angelshark.

Squatina punctata Marini, 1936. Angular angelshark

Squatina squatina (Linnaeus, 1758). Angelshark.

Squatina tergocellata McCulloch, 1914. Ornate angelshark.

Squatina tergocellatoides Chen, 1963. Ocellated angelshark.

ORDER HETERODONTIFORMES. BULLHEAD SHARKS.

FAMILY HETERODONTIDAE GRAY, 1851. BULLHEAD SHARKS.

Genus *Heterodontus* Blainville, 1816. Bullhead sharks. Indo-Pacific, demersal on continental and insular shelves and uppermost slopes. Eight species, with an additional undescribed species in the Arabian Sea.

Heterodontus francisci (Girard, 1854). Horn shark.

Heterodontus galeatus (Günther, 1870). Crested bullhead shark.

Heterodontus japonicus (Maclay & Macleay, 1884). Japanese bullhead shark.

Heterodontus mexicanus Taylor & Castro-Aguirre, 1972. Mexican hornshark.

Heterodontus portusjacksoni (Meyer, 1793). Port Jackson shark.

Heterodontus quoyi (Freminville, 1840). Galapagos bullhead shark.

Heterodontus ramalheira (Smith, 1949). Whitespotted bullhead shark.

Heterodontus zebra (Gray, 1831). Zebra bullhead shark.

ORDER ORECTOLOBIFORMES. CARPET SHARKS.

FAMILY PARASCYLLIIDAE GILL, 1862. COLLARED CARPETSHARKS.

Genus *Cirrhoscyllium* Smith & Radcliffe, 1913. Barbeled carpetsharks. Western North Pacific, demersal on the continental and insular shelves. Three species.

Cirrhoscyllium expolitum Smith & Radcliffe, 1913. Barbelthroat carpetshark.

Cirrhoscyllium formosanum Teng, 1959. Taiwan saddled carpetshark.

Cirrhoscyllium japonicum Kamohara, 1943. Saddled carpetshark.

Genus *Parascyllium* Gill, 1862. Collared carpetsharks. Australia, continental shelves and upper slopes. Four species.

Parascyllium collare Ramsay & Ogilby, 1888. Collared carpetshark.
Parascyllium ferrugineum McCulloch, 1911. Rusty carpetshark.
Parascyllium variolatum (Dumeril, 1853). Necklace carpetshark.
Parascyllium sparsimaculata Goto & Last, 2002. Ginger carpetshark.

FAMILY BRACHAELURIDAE APPLEGATE, 1974. BLIND SHARKS.

Genus *Brachaelurus* Ogilby, 1907. Blind sharks. Continental shelf of eastern Australia. A single species.
Brachaelurus waddi (Bloch & Schneider, 1801). Blind shark.

Genus *Heteroscyllium* Regan, 1908. Bluegray carpetsharks. Continental shelf of Queensland, Australia. A single species.
Heteroscyllium colcloughi (Ogilby, 1907). Bluegray carpetshark.

FAMILY ORECTOLOBIDAE GILL, 1896. WOBBEGONGS. Arrangement of genera tentative.

Genus *Eucrossorhinus* Regan, 1908. Bearded or tasselled wobbegongs. Australia and Indo-Australian Archipelago, demersal on continental shelves. A single species.
Eucrossorhinus dasypogon (Bleeker, 1867). Tasselled wobbegong.

Genus *Orectolobus* Bonaparte, 1834. Wobbegongs. Western Pacific, demersal on continental shelves. Four valid species, and probably several undescribed species.
Orectolobus japonicus Regan, 1906. Japanese wobbegong.
Orectolobus maculatus (Bonnaterre, 1788). Spotted wobbegong.
Orectolobus ornatus (de Vis, 1883). Ornate wobbegong.
Orectolobus wardi Whitley, 1939. Northern wobbegong.

Genus *Sutorectus* Whitley, 1939. Cobbler wobbegongs. Demersal on continental shelves of Australia. A single species.
Sutorectus tentaculatus (Peters, 1864). Cobbler wobbegong.

FAMILY HEMISCYLLIIDAE GILL, 1862. LONGTAILED CARPETSHARKS.

Genus *Chiloscyllium* Müller & Henle, 1837. Bamboo sharks. Indo-West Pacific, demersal on continental shelves. Seven species, possibly including species complexes under *C. plagiosum* and *C. punctatum*.
Chiloscyllium arabicum Gubanov, *in* Gubanov & Schleib, 1980. Arabian carpetshark.
Chiloscyllium burmensis Dingerkus & DeFino, 1983. Burmese bambooshark.
Chiloscyllium griseum Müller & Henle, 1838. Gray bambooshark.
Chiloscyllium hasselti Bleeker, 1852. Indonesian bambooshark.
Chiloscyllium indicum (Gmelin, 1789). Slender bambooshark.
Chiloscyllium plagiosum (Bennett, 1830). Whitespotted bambooshark.
Chiloscyllium punctatum Müller & Henle, 1838. Brownbanded bambooshark.

Genus *Hemiscyllium* Müller & Henle, 1838. Epaulette sharks. Australia and Indo-Australian Archipelago, also Western Indian Ocean, demersal on continental shelves. Five species, and possibly at least one additional species.

Hemiscyllium freycineti (Quoy & Gaimard, 1824). Indonesian speckled carpetshark.

Hemiscyllium hallstromi Whitley, 1967. Papuan epaulette shark.

Hemiscyllium ocellatum (Bonnaterre, 1788). Epaulette shark.

Hemiscyllium strahani Whitley, 1967. Hooded carpetshark.

Hemiscyllium trispeculare Richardson, 1843. Speckeled carpetshark.

FAMILY GINGLYMOSTOMATIDAE GILL, 1862. NURSE SHARKS.

Genus *Ginglymostoma* Müller & Henle, 1837. Nurse sharks. Eastern Pacific, Atlantic, demersal on warm-temperate and tropical continental shelves. A single species.

Ginglymostoma cirratum (Bonnaterre, 1788). Nurse shark.

Genus *Pseudoginglymostoma* Dingerkus, 1986. Shorttail nurse sharks. Western Indian Ocean, demersal on tropical East Africa, continental shelves. Familial placement provisional. A single species.

Pseudoginglymostoma brevicaudatum (Günther, *in* Playfair & Günther, 1866). Shorttail nurse shark.

Genus *Nebrius* Rüppell, 1837. Tawny nurse sharks. Indo-West Pacific, demersal on tropical continental shelves. A single species.

Nebrius ferrugineus (Lesson, 1830). Tawny nurse shark. Commonly referred to *N. concolor* Rüppell, 1837 but this is a junior synonym.

FAMILY STEGOSTOMATIDAE GILL, 1862.

Genus *Stegostoma* Müller & Henle, 1837. Zebra sharks. Indo-West Pacific, demersal on continental shelves. A single species.

Stegostoma fasciatum (Hermann, 1783). Zebra shark.

FAMILY RHINCODONTIDAE MÜLLER & HENLE, 1839.

Genus *Rhincodon* Smith, 1829. Whale sharks. Circumglobal in all warm-temperate and tropical seas, littoral on shelves and oceanic in the epipelagic zone. A single species.

Rhincodon typus Smith, 1828. Whale shark.

ORDER LAMNIFORMES. MACKEREL SHARKS.

FAMILY MITSUKURINIDAE JORDAN, 1898. GOBLIN SHARKS.

Genus *Mitsukurina* Jordan, 1898. Goblin sharks. Atlantic, Indian Ocean, Western and Eastern Pacific, continental and insular slopes, demersal and semipelagic. A single species.

Mitsukurina owstoni Jordan, 1898. Goblin shark.

FAMILY ODONTASPIDIDAE MÜLLER & HENLE, 1839. SAND TIGER SHARKS.

Genus *Carcharias* Rafinesque, 1810. Largetooth sand tigers. Common junior synonyms include *Eugomphodus* Gill, 1862, *Synodontaspis* White,

1931, and *Paradontaspis* White, 1931. Atlantic, Indo-West Pacific, littoral on the continental and insular shelves. A single species.

Carcharias taurus Rafinesque, 1810. Sand tiger, spotted raggedtooth, or gray nurse shark. *Carcharias tricuspidatus* Day, 1878 is included as a tentative synonym of this species.

Genus *Odontaspis* Agassiz, 1838. Smalltooth sand tigers. Virtually circumglobal in warm-temperate and tropical seas, littoral on the outer shelves and epibenthic on slopes, seamounts, and ridges, also oceanic in the epipelagic and possibly mesopelagic zone. Two species.

Odontaspis ferox (Risso, 1810). Smalltooth sand tiger or bumpytail raggedtooth.

Odontaspis noronhai (Maul, 1955). Bigeye sand tiger.

FAMILY PSEUDOCARCHARIIDAE COMPAGNO, 1973. CROCODILE SHARKS.

Genus *Pseudocarcharias* Cadenat, 1963. Crocodile sharks. Virtually circumtropical, oceanic in the epipelagic and possibly mesopelagic zone. A single species.

Pseudocarcharias kamoharai (Matsubara, 1936). Crocodile shark.

FAMILY MEGACHASMIDAE TAYLOR, COMPAGNO & STRUHSAKER, 1983. MEGAMOUTH SHARKS.

Genus *Megachasma* Taylor, Compagno & Struhsaker, 1983. Megamouth sharks. Atlantic and Indo-Pacific, virtually circumtropical, oceanic in the epipelagic zone and littoral on continental shelves. A single species.

Megachasma pelagios Taylor, Compagno & Struhsaker, 1983. Megamouth shark.

FAMILY ALOPIIDAE BONAPARTE, 1838. THRESHER SHARKS.

Genus *Alopias* Rafinesque, 1810. Thresher sharks. Circumglobal in warm-temperate and tropical seas, oceanic, continental shelves and upper slopes. Three valid species; a fourth species suggested on molecular data for the Eastern Pacific, but this could not be verified.

Alopias pelagicus Nakamura, 1935. Pelagic thresher.

Alopias superciliosus (Lowe, 1839). Bigeye thresher.

Alopias vulpinus (Bonnaterre, 1788). Thresher shark.

FAMILY CETORHINIDAE GILL, 1862. BASKING SHARKS.

Genus *Cetorhinus* Blainville, 1816. Basking sharks. Circumglobal in temperate seas, littoral on the continental and insular shelves and ranging into the epipelagic zone. A single species.

Cetorhinus maximus (Gunnerus, 1765). Basking shark.

FAMILY LAMNIDAE MÜLLER & HENLE, 1838. MACKEREL SHARKS.

Genus *Carcharodon* Smith, *in* Müller & Henle, 1838. White sharks. Circumglobal in all seas but most commonly recorded in temperate seas, littoral on the continental and insular shelves and extending down the slopes; also oceanic and epipelagic and possibly mesopelagic. A single species.

Carcharodon carcharias (Linnaeus, 1758). Great white shark.

Genus *Isurus* Rafinesque, 1810. Makos. Circumglobal in all warm seas, littoral on the continental and insular shelves, mostly oceanic and epipelagic and possibly mesopelagic. Two species.

Isurus oxyrinchus Rafinesque, 1810. Shortfin mako.

Isurus paucus Guitart Manday, 1966. Longfin mako.

Genus *Lamna* Cuvier, 1816. Mackerel sharks. Wide-ranging in temperate to cold seas, littoral on continental and insular shelves, also oceanic and epipelagic. Two species.

Lamna ditropis Hubbs & Follett, 1947. Salmon shark.

Lamna nasus (Bonnaterre, 1788). Porbeagle shark.

ORDER CARCHARHINIFORMES. GROUND SHARKS.

FAMILY SCYLIORHINIDAE GILL, 1862. CAT SHARKS.

Genus *Apristurus* Garman, 1913. Ghost or demon catsharks. Circumglobal, demersal on continental and insular slopes and occasionally shelves. 31 valid species, plus about 13 undescribed species.

Apristurus albisoma Nakaya & Seret, 1999. White-bodied catshark.

Apristurus aphyodes Nakaya & Stehmann, 1998. White ghost catshark.

Apristurus brunneus (Gilbert, 1892). Brown catshark.

Apristurus canutus Springer & Heemstra, *in* Springer, 1979. Hoary catshark.

Apristurus exsanguis Sato, Nakaya & Stewart, 1999. Flaccid catshark.

Apristurus fedorovi Dolganov, 1985. Stout catshark.

Apristurus gibbosus Meng, Chu & Li, 1985. Humpback catshark.

Apristurus herklotsi (Fowler, 1934). Longfin catshark.

Apristurus indicus (Brauer, 1906). Smallbelly catshark.

Apristurus internatus Deng, Xiong & Zhan, 1988. Shortnose demon catshark.

Apristurus investigatoris (Misra, 1962). Broadnose catshark.

Apristurus japonicus Nakaya, 1975. Japanese catshark.

Apristurus kampae Taylor, 1972. Longnose catshark.

Apristurus laurussoni (Saemundsson, 1922). Iceland catshark. Junior synonyms include *A. atlanticus* (Koefoed, 1932) and *A. maderensis* Cadenat & Maul, 1966.

Apristurus longicephalus Nakaya, 1975. Longhead catshark.

Apristurus macrorhynchus (Tanaka, 1909). Flathead catshark.

Apristurus macrostomus Meng, Chu, & Li, 1985. Broadmouth catshark.

Apristurus manis (Springer, 1979). Ghost catshark.

Apristurus microps (Gilchrist, 1922). Smalleye catshark.

Apristurus micropterygeus Meng, Chu & Li, *in* Chu, Meng, & Li, 1986. Smalldorsal catshark.

Apristurus nasutus de Buen, 1959. Largenose catshark.

Apristurus parvipinnis Springer & Heemstra, *in* Springer, 1979. Smallfin catshark.

Apristurus pinguis Deng, Xiong, & Zhan, 1983. Fat catshark.

Apristurus platyrhynchus (Tanaka, 1909). Spatulasnout catshark. Junior synonyms include *A. acanutus* Chu, Meng, & Li, 1985 and *A. verweyi* (Fowler, 1934).

Apristurus profundorum (Goode & Bean, 1896). Deepwater catshark.

Apristurus riveri Bigelow & Schroeder, 1944. Broadgill catshark.

Apristurus saldanha (Barnard, 1925). Saldanha catshark.

Apristurus sibogae (Weber, 1913). Pale catshark.

Apristurus sinensis Chu & Hu, *in* Chu, Meng, Hu, & Li, 1981. South China catshark.

Apristurus spongiceps (Gilbert, 1895). Spongehead catshark.

Apristurus stenseni (Springer, 1979). Panama ghost catshark.

Genus *Asymbolus* Whitley, 1939. Australian spotted catsharks. Australia and New Caledonia, demersal on continental shelves and uppermost slopes. Eight species and at least one undescribed species.

Asymbolus analis (Ogilby, 1895). Grey spotted catshark.

Asymbolus funebris Compagno, Stevens & Last, *in* Last, 1999. Blotched catshark.

Asymbolus occiduus Last, Gomon & Gledhill, *in* Last, 1999. Western spotted catshark.

Asymbolus pallidus Last, Gomon & Gledhill, *in* Last, 1999. Pale spotted catshark.

Asymbolus parvus Compagno, Stevens & Last, *in* Last, 1999. Dwarf catshark.

Asymbolus rubiginosus Last, Gomon & Gledhill, *in* Last, 1999. Orange spotted catshark.

Asymbolus submaculatus Compagno, Stevens & Last, *in* Last, 1999. Variegated catshark.

Asymbolus vincenti (Zietz, 1908). Gulf catshark.

Genus *Atelomycterus* Garman, 1913. Coral catsharks. Northern Indian Ocean and Western Pacific, demersal on continental shelves. Three species, two of which may be species complexes.

Atelomycterus fasciatus Compagno & Stevens, 1993. Banded sand catshark.

Atelomycterus macleayi Whitley, 1939. Australian marbled catshark.

Atelomycterus marmoratus (Bennett, 1830). Coral catshark.

Genus *Aulohalaelurus* Fowler, 1934. Blackspotted catsharks. Australia and New Caledonia, demersal on continental and insular shelves. Two species.

Aulohalaelurus kanakorum Seret, 1990. New Caledonia catshark.

Aulohalaelurus labiosus (Waite, 1905). Blackspotted catshark.

Genus *Bythaelurus* Compagno, 1988. Dusky catsharks. Indo-West Pacific, Southeastern Pacific, demersal on continental and insular shelves and slopes. Formerly a subgenus of *Halaelurus* Gill, 1862. Six species, one dubious species, and at least two undescribed species.

?*Bythaelurus alcocki* (Garman, 1913). Arabian catshark. Generic placement uncertain.

Bythaelurus canescens (Günther, 1878). Dusky catshark.

Bythaelurus clevai (Seret, 1987). Broadhead catshark.

Bythaelurus dawsoni (Springer, 1971). New Zealand catshark.

Bythaelurus hispidus (Alcock, 1891). Bristly catshark.

Bythaelurus immaculatus (Chu & Meng, *in* Chu, Meng, Hu, & Li, 1982). Spotless catshark.

Bythaelurus lutarius (Springer & D'Aubrey, 1972). Mud catshark.

Genus *Cephaloscyllium* Gill, 1862. Swell sharks. Indo-Pacific, demersal on continental and insular shelves and upper slopes. Seven species, with possibly 11 undescribed species in addition to the following.

Cephaloscyllium fasciatum Chan, 1966. Reticulated swellshark.

Cephaloscyllium isabellum (Bonnaterre, 1788). Draughtsboard shark.

Cephaloscyllium laticeps (Dumeril, 1853). Australian swellshark.

Cephaloscyllium silasi (Talwar, 1974). Indian swellshark.

Cephaloscyllium sufflans (Regan, 1921). Balloon shark.

Cephaloscyllium umbratile Jordan & Fowler, 1903. Japanese swellshark.

Cephaloscyllium ventriosum (Garman, 1880). Swellshark.

Genus *Cephalurus* Bigelow & Schroeder, 1941. Lollipop catsharks. Eastern Pacific, demersal on continental slopes. One described species (Eastern North Pacific), possibly one additional species (Eastern South Pacific):

Cephalurus cephalus (Gilbert, 1892). Lollipop catshark.

Genus *Galeus* Rafinesque, 1810. Sawtail catsharks. Virtually circumglobal in temperate to tropical seas, demersal on continental shelves and slopes. 16 valid species, and at least three undescribed species.

Galeus antillensis Springer, 1979. Antilles catshark.

Galeus arae (Nichols, 1927). Roughtail catshark.

Galeus atlanticus (Vaillant, 1888). Atlantic sawtail catshark.

Galeus boardmani (Whitley, 1928). Australian sawtail catshark.

Galeus eastmani (Jordan & Snyder, 1904). Gecko catshark.

Galeus gracilis Compagno & Stevens, 1993. Slender sawtail catshark.

Galeus longirostris Tachikawa & Taniuchi, 1987. Longnose sawtail catshark.

Galeus melastomus Rafinesque, 1810. Blackmouth catshark.

Galeus mincaronei Soto, 2001. Southern sawtail catshark.

Galeus murinus (Collett, 1904). Mouse catshark.

Galeus nipponensis Nakaya, 1975. Broadfin sawtail catshark.

Galeus piperatus Springer & Wagner, 1966. Peppered catshark.

Galeus polli Cadenat, 1959. African sawtail catshark.

Galeus sauteri (Jordan & Richardson, 1909). Blacktip sawtail catshark.

Galeus schultzi Springer, 1979. Dwarf sawtail catshark.

Galeus springeri Konstantinou & Cozzi, 1998. Springer's sawtail catshark.

Genus *Halaelurus* Gill, 1862. Tiger catsharks. Indo-West Pacific, demersal on continental and insular shelves and slopes. Five species, *H. boesemani* possibly a species complex.

Halaelurus boesemani Springer & D'Aubrey, 1972. Speckeled catshark.

Halaelurus buergeri (Müller & Henle, 1838). Darkspot, blackspotted, or Nagasaki catshark.

Halaelurus lineatus Bass, D'Aubrey & Kistnasamy, 1975. Lined catshark.

Halaelurus natalensis (Regan, 1904). Tiger catshark.

Halaelurus quagga (Alcock, 1899). Quagga catshark.

Genus *Haploblepharus* Garman, 1913. Shysharks or happies. Southern Africa, demersal on continental shelves. Three described species, and one undescribed.

Haploblepharus edwardsii (Voigt, *in* Cuvier, 1832). Puffadder shyshark.

Haploblepharus fuscus Smith, 1950. Brown shyshark.

Haploblepharus pictus (Müller & Henle, 1838). Dark shyshark.

Genus *Holohalaelurus* Fowler, 1934. Izak catsharks. Southern and East Africa, demersal on outer continental shelves and upper slopes. Three described species, and a single new species.

Holohalaelurus melanostigma (Norman, 1939). Crying izak catshark.

Holohalaelurus punctatus (Gilchrist & Thompson, 1914). Small-spotted izak catshark.

Holohalaelurus regani (Gilchrist, 1922). Izak catshark.

Genus *Parmaturus* Garman, 1906. Filetail catsharks. Western Pacific, Eastern North Pacific, Western North Atlantic, demersal on continental and insular slopes. Five described species and three undescribed species.

Parmaturus campechiensis Springer, 1979. Campeche catshark.

Parmaturus macmillani Hardy, 1985. New Zealand filetail

Parmaturus melanobranchius (Chan, 1966). Blackgill catshark.

Parmaturus pilosus Garman, 1906. Salamander shark.

Parmaturus xaniurus (Gilbert, 1892). Filetail catshark.

Genus *Pentanchus* Smith & Radcliffe, *in* Smith, 1912. Onefin catsharks. Philippines, demersal on insular slopes, a single species.

Pentanchus profundicolus Smith & Radcliffe, 1912. Onefin catshark.

Genus *Poroderma* Smith, 1837. Barbeled catsharks. Temperate southern Africa, demersal on continental shelves. Two species.

Poroderma africanum (Gmelin, 1789). Striped catshark or pyjama shark.

Poroderma pantherinum (Smith, *in* Müller & Henle, 1838). Leopard catshark. *P. marleyi* Fowler, 1934 is a junior synonym.

Genus *Schroederichthys* Springer, 1966. Narrowtail catsharks. Eastern South Pacific and Western South Atlantic, South America, demersal on continental shelves and upper slopes. Five species.

Schroederichthys bivius (Smith, *in* Müller & Henle, 1838). Narrowmouth catshark.

Schroederichthys chilensis (Guichenot, *in* Gay, 1848). Redspotted catshark.

Schroederichthys maculatus Springer, 1966. Narrowtail catshark.

Schroederichthys saurisqualus Soto, 2001. Lizard catshark.

Schroederichthys tenuis Springer, 1966. Slender catshark.

Genus *Scyliorhinus* Blainville, 1816. Spotted catsharks. Atlantic, Southwestern Indian Ocean, Western Pacific, demersal on inner shelves to upper slopes. Fifteen species, plus at least one undescribed.

Scyliorhinus besnardi Springer & Sadowsky, 1970. Polkadot catshark.

Scyliorhinus boa Goode & Bean, 1896. Boa catshark.

Scyliorhinus canicula (Linnaeus, 1758). Smallspotted catshark.

Scyliorhinus capensis (Smith, *in* Müller & Henle, 1838). Yellowspotted catshark.

Scyliorhinus cervigoni Maurin & Bonnet, 1970. West African catshark.

Scyliorhinus comoroensis Compagno, 1989. Comoro catshark.

Scyliorhinus garmani (Fowler, 1934). Brownspotted catshark.

Scyliorhinus haeckelii (Ribeiro, 1907). Freckled catshark.

Scyliorhinus hesperius Springer, 1966. Whitesaddled catshark.

Scyliorhinus meadi Springer, 1966. Blotched catshark.

Scyliorhinus retifer (Garman, 1881). Chain catshark.

Scyliorhinus stellaris (Linnaeus, 1758). Nursehound

Scyliorhinus tokubee Shirai, Hagiwara & Nakaya, 1992. Izu catshark.

Scyliorhinus torazame (Tanaka, 1908). Cloudy catshark.

Scyliorhinus torrei Howell-Rivero, 1936. Dwarf catshark.

FAMILY PROSCYLLIIDAE FOWLER, 1941. FINBACK CATSHARKS.

Genus *Ctenacis* Compagno, 1973. Harlequin catsharks. East Africa, Northwestern Indian Ocean, demersal on continental slopes. A single species.

Ctenacis fehlmanni (Springer, 1968). Harlequin catshark.

Genus *Eridacnis* Smith, 1913. Ribbontail catsharks. Western North Atlantic, Indo-West Pacific, demersal on upper continental slopes. Three species.

Eridacnis barbouri (Bigelow & Schroeder, 1944). Cuban ribbontail catshark.

Eridacnis radcliffei Smith, 1913. Pygmy ribbontail catshark.

Eridacnis sinuans (Smith, 1957). African ribbontail catshark.

Genus *Proscyllium* Hilgendorf, 1904 Graceful catsharks. Western Pacific, demersal on continental shelves and uppermost slopes. One species, plus a new species to be described from the Indian Ocean.

Proscyllium habereri Hilgendorf, 1904. Graceful catshark. including *Calliscyllium venustum* Tanaka, 1913.

FAMILY PSEUDOTRIAKIDAE GILL, 1893. FALSE CATSHARKS. In addition to the two genera ascribed here to the family, there is an undescribed genus and species from the continental slopes of the northern Indian Ocean.

Genus *Gollum* Compagno, 1973. Slender smoothhounds or gollumsharks. Western Pacific, demersal on upper slopes. A single species described from New Zealand, plus two additional new species from elsewhere in the Western Pacific.

Gollum attenuatus (Garrick, 1954). Slender smoothhound.

Genus *Pseudotriakis* Capello, 1868 False catsharks. Wide-ranging but spotty in the North Atlantic, Southwestern Indian Ocean and West-Central Pacific, demersal on the continental and insular slopes. Apparently one species.

Pseudotriakis microdon Capello, 1868. False catshark. *P. acrales* Jordan & Snyder, 1904 is a junior synonym.

FAMILY LEPTOCHARIIDAE GRAY, 1851. BARBELED HOUNDSHARKS.

Genus *Leptocharias* Smith, in Müller & Henle, 1838. Barbeled houndsharks. Eastern Atlantic, demersal and littoral on the continental shelves. A single species.

Leptocharias smithii (Müller & Henle, 1839). Barbeled houndshark.

FAMILY TRIAKIDAE GRAY, 1851. HOUNDSHARKS.

Genus *Furgaleus* Whitley, 1951. Whiskery sharks. Australia, demersal and littoral on the continental shelves. A single species.

Furgaleus macki (Whitley, 1943). Whiskery shark.

Genus *Galeorhinus* Blainville, 1816. Tope, soupfin, school, and vitamin sharks. Eastern Atlantic and Southwestern Indian Ocean, Western South Pacific, Eastern Pacific, Western South Atlantic, littoral on the continental and insular shelves and epibenthic on the uppermost slopes. A single species.

Galeorhinus galeus (Linnaeus, 1758). Tope shark.

Genus *Gogolia* Compagno, 1973. Sailback houndsharks. Western South Pacific, Papua-New Guinea, littoral or demersal on the insular shelf. A single species.

Gogolia filewoodi Compagno, 1973. Sailback houndshark.

Genus *Hemitriakis* Herre, 1923. Whitefin topesharks. Southeastern Indian Ocean and Western Pacific, littoral or demersal on the continental and insular shelves and uppermost slopes. Five species, plus at least one undescribed species.

Hemitriakis abdita Compagno & Stevens, 1993. Deepwater sicklefin houndshark.

Hemitriakis complicofasciata Takahashi & Nakaya, 2004. Ocellate topeshark.

Hemitriakis japanica (Müller & Henle, 1839). Japanese topeshark.

Hemitriakis falcata Compagno & Stevens, 1993. Sicklefin houndshark.

Hemitriakis leucoperiptera Herre, 1923. Whitefin topeshark.

Genus *Hypogaleus* Smith, 1957. Blacktip topesharks or pencil sharks. Indo-West Pacific, littoral on the continental and insular shelves and uppermost slopes. A single species.

Hypogaleus hyugaensis (Miyosi, 1939). Blacktip topeshark.

Genus *Iago* Compagno & Springer, 1971. Bigeye houndsharks. Indo-West Pacific, demersal and littoral on the continental and insular shelves and demersal on the upper slopes. Two species, plus one or two undescribed species.

Iago garricki Fourmanoir, 1979. Longnose houndshark.
Iago omanensis (Norman, 1939). Bigeye houndshark.

Genus *Mustelus* Linck, 1790. Smoothhounds. Circumglobal in all temperate and tropical seas, demersal and littoral continental and insular shelves and uppermost slopes. 22 species, plus at least four undescribed species.

Mustelus antarcticus Günther, 1870. Gummy shark.
Mustelus asterias Cloquet, 1821. Starry smoothhound.
Mustelus californicus Gill, 1864. Gray smoothhound.
Mustelus canis (Mitchell, 1815). Dusky smoothhound.
Mustelus dorsalis Gill, 1864. Sharpnose smoothhound.
Mustelus fasciatus (Garman, 1913). Striped smoothhound.
Mustelus griseus Pitschmann, 1908. Spotless smoothhound.
Mustelus henlei (Gill, 1863). Brown smoothhound.
Mustelus higmani Springer & Lowe, 1963. Smalleye smoothhound.
Mustelus lenticulatus Phillipps, 1932. Spotted estuary smoothhound or rig.
Mustelus lunulatus Jordan & Gilbert, 1883. Sicklefin smoothhound.
Mustelus manazo Bleeker, 1854. Starspotted smoothhound.
Mustelus mento Cope, 1877. Speckled smoothhound.
Mustelus minicanis Heemstra, 1997. Venezuelan dwarf smoothhound.
Mustelus mosis Hemprich & Ehrenberg, 1899. Arabian, hardnose, or Moses smoothhound.
Mustelus mustelus (Linnaeus, 1758). Smoothhound.
Mustelus norrisi Springer, 1940. Narrowfin or Florida smoothhound.
Mustelus palumbes Smith, 1957. Whitespot smoothhound.
Mustelus punctulatus Risso, 1826. Blackspot smoothhound.
Mustelus schmitti Springer, 1940. Narrownose smoothhound.
Mustelus sinusmexicanus Heemstra, 1997. Gulf of Mexico smoothhound.
Mustelus whitneyi Chirichigno, 1973. Humpback smoothhound.

Genus *Scylliogaleus* Boulenger, 1902. Flapnose houndsharks. Southwestern Indian Ocean, South Africa, demersal and littoral on the continental shelf. One species.

Scylliogaleus quecketti Boulenger, 1902. Flapnose houndshark.

Genus *Triakis* Müller & Henle, 1838. Leopard sharks. Western North Pacific, Eastern South Atlantic and Southwestern Indian Ocean (southern

Africa), Eastern Pacific, demersal and littoral on continental and insular shelves. Five species, in two subgenera (sometimes ranked as genera).

Subgenus *Cazon* de Buen, 1959:

Triakis (Cazon) acutipinna Kato, 1968. Sharpfin houndshark.

Triakis (Cazon) maculata Kner & Steindachner, 1866. Spotted houndshark.

Triakis (Cazon) megalopterus (Smith, 1849). Spotted gully shark.

Subgenus *Triakis* Müller & Henle, 1838:

Triakis (Triakis) scyllium Müller & Henle, 1839. Banded houndshark.

Triakis (Triakis) semifasciata Girard, 1854. Leopard shark.

FAMILY HEMIGALEIDAE HASSE, 1879. WEASEL SHARKS.

Genus *Chaenogaleus* Gill, 1862. Hooktooth sharks. Northern Indian Ocean and Western Pacific, littoral on continental shelves. A single species.

Chaenogaleus macrostoma (Bleeker, 1852). Hooktooth shark.

Genus *Hemigaleus* Bleeker, 1852. Sicklefin weasel sharks. Indo-West Pacific, littoral on continental shelves. A single described species, and possibly one undescribed species in Australian waters.

Hemigaleus microstoma Bleeker, 1852. Sicklefin weasel shark.

Genus *Hemipristis* Agassiz, 1843. Snaggletooth sharks. Indo-West Pacific from South Africa to Australia, littoral on continental shelves. A single species.

Hemipristis elongatus (Klunzinger, 1871). Snaggletooth shark.

Genus *Paragaleus* Budker, 1935. Sharpnose weasel sharks. Eastern tropical Atlantic, possibly Western North Atlantic, Indo-West Pacific, continental shelves. Four species.

Paragaleus leucolomatus Compagno & Smale, 1985. Whitetip weasel shark.

Paragaleus pectoralis (Garman, 1906). Atlantic weasel shark.

Paragaleus randalli Compagno, Krupp, & Carpenter, 1996. Slender weasel shark.

Paragaleus tengi (Chen, 1963). Straighttooth weasel shark.

FAMILY CARCHARHINIDAE JORDAN & EVERMANN, 1896. REQUIEM SHARKS.

Genus *Carcharhinus* Blainville, 1816. Gray sharks. Circumglobal in all warm-temperate and tropical seas, continental and insular shelves and uppermost slopes, littoral to oceanic or semioceanic, fresh water tropical and warm-temperate rivers and lakes. 30 species, plus one undescribed species.

Carcharhinus acronotus (Poey, 1860). Blacknose shark.

Carcharhinus albimarginatus (Rüppell, 1837). Silvertip shark.

Carcharhinus altimus (Springer, 1950). Bignose shark.

Carcharhinus amblyrhynchoides (Whitley, 1934). Graceful shark.

Carcharhinus amblyrhynchos (Bleeker, 1856). Gray reef shark. *Carcharhinus wheeleri* Garrick, 1982 is a junior synonym.

Carcharhinus amboinensis (Müller & Henle, 1839). Pigeye or Java shark.

Carcharhinus borneensis (Bleeker, 1859). Borneo shark.

Carcharhinus brachyurus (Günther, 1870). Bronze whaler.

Carcharhinus brevipinna (Müller & Henle, 1839). Spinner shark.

Carcharhinus cautus (Whitley, 1945). Nervous shark.

Carcharhinus dussumieri (Valenciennes, *in* Müller & Henle, 1839). Whitecheek shark.

Carcharhinus falciformis (Bibron, *in* Müller & Henle, 1839). Silky shark.

Carcharhinus fitzroyensis (Whitley, 1943). Creek whaler.

Carcharhinus galapagensis (Snodgrass & Heller, 1905). Galapagos shark.

Carcharhinus hemiodon (Valenciennes, *in* Müller & Henle, 1839). Pondicherry shark.

Carcharhinus isodon (Valenciennes, *in* Müller & Henle, 1839). Finetooth shark.

Carcharhinus leiodon Garrick, 1985. Smoothtooth blacktip

Carcharhinus leucas (Valenciennes, *in* Müller & Henle, 1839). Bull shark.

Carcharhinus limbatus (Valenciennes, *in* Müller & Henle, 1839). Blacktip shark.

Carcharhinus longimanus (Poey, 1861). Oceanic whitetip shark.

Carcharhinus macloti (Müller & Henle, 1839). Hardnose shark.

Carcharhinus melanopterus (Quoy & Gaimard, 1824). Blacktip reef shark.

Carcharhinus obscurus (Lesueur, 1818). Dusky shark.

Carcharhinus perezi (Poey, 1876). Caribbean reef shark.

Carcharhinus plumbeus (Nardo, 1827). Sandbar shark.

Carcharhinus porosus (Ranzani, 1839). Smalltail shark.

Carcharhinus sealei (Pietschmann, 1916). Blackspot shark.

Carcharhinus signatus (Poey, 1868). Night shark.

Carcharhinus sorrah (Valenciennes, *in* Müller & Henle, 1839). Spottail shark.

Carcharhinus tilsoni (Whitley, 1950). Australian blacktip shark.

Genus *Galeocerdo* Müller & Henle, 1837. Circumglobal in warm-temperate and tropical seas, continental and insular shelves, littoral to semioceanic. A single species generally recognized, possibly two species though this is uncertain.

Galeocerdo cuvier (Peron & Lesueur, *in* Lesueur, 1822). Tiger shark.

Genus *Glyphis* Agassiz, 1843. River sharks. Northern Indian Ocean and Western South Pacific, littoral on the continental shelves and in tropical rivers. Three species, plus two or three undescribed species.

Glyphis gangeticus (Müller & Henle, 1839). Ganges shark.

Glyphis glyphis (Müller & Henle, 1839). Speartooth shark.

Glyphis siamensis (Steindachner, 1896). Irrawaddy river shark.

Genus *Isogomphodon* Gill, 1862. Tropical Atlantic coast of South America, littoral on continental shelves. A single species.

Isogomphodon oxyrhynchus (Müller & Henle, 1839). Daggernose shark.

Genus *Lamiopsis* Gill, 1862. Northern Indian Ocean and Western Pacific, littoral on continental shelves. One species.

Lamiopsis temmincki (Müller & Henle, 1839). Broadfin shark.

Genus *Loxodon* Müller & Henle, 1839. Indo-West Pacific, littoral on continental shelves. A single species.

Loxodon macrorhinus Müller & Henle, 1839. Sliteye shark.

Genus *Nasolamia* Compagno & Garrick, 1983. Eastern Pacific, littoral on continental shelves. A single species.

Nasolamia velox (Gilbert, *in* Jordan & Evermann, 1898). Whitenose shark.

Genus *Negaprion* Whitley, 1940. Lemon sharks. Circumtropical, littoral on continental and insular shelves. Two or three species.

Negaprion acutidens (Rüppell, 1837). Sharptooth lemon shark.

Negaprion brevirostris (Poey, 1868). Lemon shark. *N. fronto* (Jordan & Gilbert, 1882) from the Eastern Pacific is considered a junior synonym here, but could be distinct.

Genus *Prionace* Cantor, 1849. Circumglobal in all temperate and tropical seas, oceanic but penetrating the littoral of narrow continental shelves. A single species.

Prionace glauca (Linnaeus, 1758). Blue shark.

Genus *Rhizoprionodon* Whitley, 1929. Sharpnose sharks. Circumtropical and littoral on continental shelves, may penetrate mouths of rivers but not found far up rivers. Seven species.

Rhizoprionodon acutus (Rüppell, 1837). Milk shark.

Rhizoprionodon lalandei (Valenciennes, *in* Müller & Henle, 1839). Brazilian sharpnose shark.

Rhizoprionodon longurio (Jordan & Gilbert, 1882). Pacific sharpnose shark.

Rhizoprionodon oligolinx Springer, 1964. Gray sharpnose shark.

Rhizoprionodon porosus (Poey, 1861). Caribbean sharpnose shark.

Rhizoprionodon taylori (Ogilby, 1915). Australian sharpnose shark.

Rhizoprionodon terraenovae (Richardson, 1836). Atlantic sharpnose shark.

Genus *Scoliodon* Müller & Henle, 1837. Spadenose sharks. Indo-West Pacific, littoral on continental shelves. One species.

Scoliodon laticaudus Müller & Henle, 1838. Spadenose shark.

Genus *Triaenodon* Müller & Henle, 1837. Indo-Pacific, littoral and demersal on continental and insular shelves, commonly on coral reefs. One species. *Triaenodon obtusus* Day, 1878 was placed in this genus but is based on a fetus of *Carcharhinus amboinensis*.

Triaenodon obesus (Rüppell, 1837). Whitetip reef shark.

FAMILY SPHYRNIDAE GILL, 1872. HAMMERHEAD SHARKS.

Genus *Eusphyra* Gill, 1862. Northern Indian Ocean and Western Pacific, littoral on continental shelves.

Eusphyra blochii (Cuvier, 1817). Winghead shark.

Genus *Sphyrna* Rafinesque, 1810. Hammerhead and bonnethead sharks. Circumglobal in warm temperate and tropical seas, shelves and adjacent epipelagic zone. Seven species, arranged in three subgenera; possibly one additional species of large hammerhead (subgenus *Sphyrna*).

Subgenus *Mesozygaena* Compagno, 1988:

Sphyrna (Mesozygaena) corona Springer, 1940. Mallethead shark.

Sphyrna (Mesozygaena) media Springer, 1940. Scoophead shark.

Sphyrna (Mesozygaena) tudes (Valenciennes, 1822). Smalleye hammerhead.

Subgenus *Platysqualus* Swainson, 1839:

Sphyrna (Platysqualus) tiburo (Linnaeus, 1758). Bonnethead shark.

Subgenus *Sphyrna* Rafinesque, 1810:

Sphyrna (Sphyrna) lewini (Griffith & Smith, *in* Cuvier, Griffith & Smith, 1834). Scalloped hammerhead. *S. couardi* Cadenat, 1950 is tentatively placed here as a junior synonym.

Sphyrna (Sphyrna) mokarran (Rüppell, 1837). Great hammerhead.

Sphyrna (Sphyrna) zygaena (Linnaeus, 1758). Smooth hammerhead.

ORDER RAJIFORMES. RAYS (BATOIDS).

SUBORDER PRISTOIDEI. SAWFISHES.

FAMILY PRISTIDAE BONAPARTE, 1838. SAWFISHES.

Genus *Anoxypristis* White & Moy-Thomas, 1941. Knifetooth sawfishes. Northern Indian Ocean and Western Pacific, demersal and littoral on continental shelves and possibly fresh water. A single species.

Anoxypristis cuspidata (Latham, 1794). Knifetooth, pointed, or narrow sawfish.

Genus *Pristis* Linck, 1790. Narrowtooth sawfishes. Circumtropical, demersal and littoral on continental shelves and fresh water. Possibly four to six species based on distribution and morphology; molecular studies are being conducted to resolve species problems.

Pristis clavata Garman, 1906. Dwarf or Queensland sawfish.

Pristis microdon Latham, 1794. Greattooth or freshwater sawfish.

Pristis pectinata Latham, 1794. Smalltooth or wide sawfish.

Pristis perotteti Valenciennes, *in* Müller & Henle, 1841. Largetooth sawfish.

Pristis pristis (Linnaeus, 1758). Common sawfish.

Pristis zijsron Bleeker, 1851. Green sawfish.

SUBORDER RHINOIDEI. SHARKRAYS.

FAMILY RHINIDAE MÜLLER & HENLE, 1841. SHARKRAYS.

Genus *Rhina* Bloch & Schneider, 1801. Sharkrays. Indo-West Pacific, demersal and littoral on continental shelves. One species.

Rhina ancylostoma Bloch & Schneider, 1801. Bowmouth guitarfish or sharkray.

SUBORDER RHYNCHOBATOIDEI. WEDGEFISHES.

FAMILY RHYNCHOBATIDAE GARMAN, 1913. WEDGEFISHES.

Genus *Rhynchobatus* Müller & Henle, 1837. Wedgefishes. Eastern Atlantic and Indo-West Pacific, demersal and littoral on continental shelves. Four valid species plus at least two undescribed species.

Rhynchobatus australiae Whitley, 1939. Whitespotted shovelnose ray.

Rhynchobatus djiddensis (Forsskael, 1775). Whitespotted wedgefish or giant guitarfish.

Rhynchobatus laevis (Bloch & Schneider, 1801). Smoothnose wedgefish.

Rhynchobatus luebberti Ehrenbaum, 1914. African or spikenose wedgefish.

SUBORDER RHINOBATOIDEI. GUITARFISHES.

FAMILY RHINOBATIDAE MÜLLER & HENLE, 1837. GUITARFISHES.

Genus *Aptychotrema* Norman, 1926. Australasian shovelnose rays. West-central and Western South Pacific, continental shelves. Three species plus one undescribed species.

Aptychotrema rostrata (Shaw & Nodder, 1794). Eastern shovelnose ray.

Aptychotrema timorensis Last, 2004. Spotted shovelnose ray.

Aptychotrema vincentiana (Haake, 1885). Southern shovelnose ray.

Genus *Rhinobatos* Linck, 1790. Guitarfishes. Circumglobal in warm-temperate and tropical seas, demersal on continental shelves and uppermost slopes. Classification transitional, subgeneric arrangement provisional, two of the subgenera recognized here are probably to be elevated to genera in the FAO Catalog of World Batoids. Thirty-five valid species plus one dubious species and at least five undescribed species.

Subgenus *Acroteriobatus* Giltay, 1928:

Flapnose guitarfishes. Eastern Atlantic and Western Indian Ocean. Seven species, plus at least one undescribed species.

Rhinobatos (Acroteriobatus) annulatus Smith, *in* Müller & Henle, 1841. Lesser guitarfish.

Rhinobatos (Acroteriobatus) blochii Müller & Henle, 1841. Bluntnose guitarfish or fiddlefish.

Rhinobatos (Acroteriobatus) leucospilus Norman, 1926. Greyspot guitarfish.

Rhinobatos (Acroteriobatus) ocellatus Norman, 1926. Speckled guitarfish.
Rhinobatos (Acroteriobatus) salalah Randall & Compagno, 1995. Salalah guitarfish.
Rhinobatos (Acroteriobatus) variegatus Nair & Lal Mohan, 1973. Stripenose guitarfish.
Rhinobatos (Acroteriobatus) zanzibarensis Norman, 1926. Zanzibar guitarfish.

Subgenus *Glaucostegus* Bonaparte, 1846:

Spiny guitarfishes. Eastern Atlantic and Indo-West Pacific. Junior synonyms are subgenera *Platypornax* Whitley, 1939 and *Scobatus* Whitley, 1939. Eight valid species plus one dubious species.
Rhinobatos (Glaucostegus) cemiculus St. Hilaire, 1817. Blackchin guitarfish.
Rhinobatos (Glaucostegus) granulatus Cuvier, 1829. Sharpnose guitarfish.
Rhinobatos (Glaucostegus) halavi (Forsskael, 1775). Halavi guitarfish.
Rhinobatos (Glaucostegus) microphthalmus Teng, 1959. Smalleyed guitarfish.
Rhinobatos (Glaucostegus) obtusus Müller & Henle, 1841. Widenose guitarfish.
Rhinobatos (Glaucostegus) petiti Chabanaud, 1929. Madagascar guitarfish.
?*Rhinobatos (Glaucostegus) spinosus* Günther, 1870? Spiny guitarfish (possibly young of some other species).
Rhinobatos (Glaucostegus) thouin (Anonymous, 1798). Clubnose guitarfish.
Rhinobatos (Glaucostegus) typus Bennett, 1830. Giant shovelnose ray.

Subgenus *Rhinobatos* Linck, 1790:

Guitarfishes. Cosmopolitan in warm-temperate to tropical seas. Twenty species plus four undescribed species. This subgenus has two geographic components which may be separate taxa:

***Rhinobatos* Eastern Hemisphere.** Twelve species plus four undescribed species.
Rhinobatos albomaculatus Norman, 1930. Whitespotted guitarfish.
Rhinobatos annandalei Norman, 1926. Bengal guitarfish.
Rhinobatos formosensis Norman, 1926. Taiwan guitarfish.
Rhinobatos holcorhynchus Norman, 1922. Slender guitarfish.
Rhinobatos hynnicephalus Richardson, 1846. Ringstraked guitarfish.
Rhinobatos irvinei Norman, 1931. Spineback guitarfish.
Rhinobatos lionotus Norman, 1926. Smoothback guitarfish.
Rhinobatos nudidorsalis Last, 2004. Nakedback guitarfish.
Rhinobatos punctifer Compagno & Randall, 1987. Spotted guitarfish.
Rhinobatos rhinobatos (Linnaeus, 1758). Common guitarfish or violinfish.
Rhinobatos sainsburyi Last, 2004. Goldeneye shovelnose ray.
Rhinobatos schlegelii Müller & Henle, 1841. Brown guitarfish.

***Rhinobatos* Western Hemisphere.** Eight species.
Rhinobatos glaucostigmus Jordan & Gilbert, 1884. Slatyspotted guitarfish.
Rhinobatos horkelii Müller & Henle, 1841. Brazilian guitarfish.
Rhinobatos lentiginosus Garman, 1880. Freckled or Atlantic guitarfish.
Rhinobatos leucorhynchus Günther, 1866. Whitenose guitarfish.
Rhinobatos percellens (Walbaum, 1792). Southern guitarfish.
Rhinobatos planiceps Garman, 1880. Flathead guitarfish.

Rhinobatos prahli Acero & Franke, 1995. Gorgona guitarfish.
Rhinobatos productus Girard, 1854. Shovelnose guitarfish.

Genus *Trygonorrhina* Müller & Henle, 1838. Fiddler rays. Australia, demersal on continental shelves. One valid species, one dubious species, and one undescribed species.

Trygonorrhina fasciata Müller & Henle, 1841. Southern fiddler ray. Junior synonym *T. guanerius* Whitley, 1932.

?*Trygonorrhina melaleuca* Scott, 1954. Magpie fiddler ray. Possibly a piebald morph of *T. fasciata*?

Genus *Zapteryx* Jordan & Gilbert, 1880. Banded guitarfishes. Western South Atlantic (South America) and Eastern North Pacific, demersal on continental shelves. Three species.

Zapteryx brevirostris (Müller & Henle, 1841). Shortnose guitarfish.
Zapteryx exasperata (Jordan & Gilbert, 1880). Banded guitarfish.
Zapteryx xyster Jordan & Evermann, 1896. Southern banded guitarfish.

SUBORDER PLATYRHINOIDEI. THORNBACKS AND FANRAYS.

FAMILY PLATYRHINIDAE JORDAN, 1923. THORNBACKS AND FANRAYS.

Genus *Platyrhina* Müller & Henle, 1838. Fanrays. Western North Pacific, demersal on continental shelves. Two species.

Platyrhina limboonkengi Tang, 1933. Amoy fanray.
Platyrhina sinensis (Bloch & Schneider, 1801). Fanray.

Genus *Platyrhinoidis* Garman, 1881. Thornbacks. Eastern North Pacific, demersal on continental shelves. One species.

Platyrhinoidis triseriata (Jordan & Gilbert, 1881). Thornback.

SUBORDER ZANOBATOIDEI. PANRAYS OR STRIPED THORNBACKS.

FAMILY ZANOBATIDAE JORDAN, 1923. PANRAYS

Genus *Zanobatus* Garman, 1913. Panrays. Tropical West Africa, Eastern Atlantic, India. One valid species, one dubious species, plus one undescribed species. Status of Indian representative uncertain.

?*Zanobatus atlantica* (Chabanaud, 1928). Atlantic panray.
Zanobatus schoenleinii (Müller & Henle, 1841). Striped panray.

SUBORDER TORPEDINOIDEI. ELECTRIC RAYS.

FAMILY NARCINIDAE GILL, 1862. NUMBFISHES.

Genus *Benthobatis* Alcock, 1898. Blindrays. Western Atlantic, Northern Indian Ocean, Western North Pacific, Eastern South Pacific, continental and insular slopes. Four species, plus possibly one undescribed species.

Benthobatis kreffti Rincón, Stehmann & Vooren, 2001. South Atlantic blindray.

Benthobatis marcida Bean & Weed, 1909. Pale or deepsea blindray.

Benthobatis moresbyi Alcock, 1898. Dark blindray.

Benthobatis yangi de Carvalho, Compagno, & Ebert. 2003. Narrow blindray.

Genus *Diplobatis* Bigelow & Schroeder, 1948. Painted electric rays. Tropical Eastern Pacific, Western Atlantic, demersal on continental shelves. Four species, plus three subspecies.

Diplobatis columbiensis Fechhelm & McEachran, 1984. Colombian painted electric ray.

Diplobatis guamachensis Martin, 1957. Venezuelan painted electric ray.

Diplobatis ommata (Jordan & Gilbert, *in* Jordan & Bollman, 1889). Target ray.

Diplobatis pictus Palmer, 1950. Painted electric ray.

Genus *Discopyge* Heckel, *in* Tschudii, 1846. Apron rays. Eastern South Pacific and Western South Atlantic (South America), continental shelves. A single species.

Discopyge tschudii Heckel, *in* Tschudi, 1844. Apron ray.

Genus *Narcine* Henle, 1834. Numbfishes. Circumglobal in warm-temperate and tropical seas, continental and insular shelves and upper-most slopes. 17 species, plus about eight undescribed species.

Narcine atzi de Carvalho & Randall, 2003. Oman numbfish.

Narcine brasiliensis (Olfers, 1831). Lesser electric ray.

Narcine bancroftii (Griffith, 1834). Caribbean electric ray

Narcine brevilabiata Bessednov, 1966. Shortlip electric ray.

Narcine brunnea Annandale, 1909. Brown electric ray.

Narcine entemedor Jordan & Starks, 1895. Cortez electric ray.

Narcine insolita de Carvalho, Seret & Compagno, 2002. Madagascar electric ray

Narcine lasti de Carvalho & Seret, 2002. Western numbfish

Narcine leoparda de Carvalho, 2001. Leopard numbfish.

Narcine lingula Richardson, 1840. Rough electric ray.

Narcine maculata (Shaw, 1804). Darkspotted electric ray.

Narcine oculifera de Carvalho, Compagno & Mee, 2002. Bigeye electric ray.

Narcine prodorsalis Bessednov, 1966. Tonkin electric ray.

Narcine rierai (Lloris & Rucabado, 1991). Mozambique electric ray. (formerly in *Heteronarce*).

Narcine tasmaniensis Richardson, 1840. Tasmanian numbfish.

Narcine timlei (Bloch & Schneider, 1801). Blackspotted electric ray.

Narcine vermiculatus Breder, 1926. Vermiculated electric ray.

Narcine westralensis McKay, 1966. Banded numbfish.

FAMILY NARKIDAE FOWLER, 1934. SLEEPER RAYS.

Genus *Crassinarke* Takagi, 1951. Sleeper torpedos. Western North Pacific, Japan and South China Sea, demersal on continental shelf. Validity of genus and single species questionable.

Crassinarke dormitor Takagi, 1951. Sleeper torpedo.

Genus *Heteronarce* Regan, 1921. Soft sleeper rays. Western and Northern Indian Ocean, demersal on continental and insular shelves and upper slopes. Four species. An undescribed species from South Africa was formerly and tentatively ascribed to this genus by the writer, but apparently represents a new genus of Narkidae with two dorsal fins.

Heteronarce bentuvai (Baranes & Randall, 1989). Elat electric ray.
Heteronarce garmani Regan, 1921. Natal electric ray.
Heteronarce mollis (Lloyd, 1907). Soft electric ray.
Heteronarce prabhui Talwar, 1981. Quilon electric ray.

Genus *Narke* Kaup, 1826. Onefin sleeper rays. Indo-West Pacific, demersal on continental and insular shelves and upper slopes. Three species, plus possibly two undescribed species.

Narke capensis (Gmelin, 1789). Cape numbfish or onefin electric ray.
Narke dipterygia (Bloch & Schneider, 1801). Spottail electric ray.
Narke japonica (Temminck & Schlegel, 1850). Japanese spotted torpedo.

Genus *Temera* Gray, 1831. Finless sleeper rays. Southeast Asia, demersal on continental shelves. A single species.

Temera hardwickii Gray, 1831. Finless sleeper ray.

Genus *Typhlonarke* Waite, 1909. Legged torpedos. New Zealand, demersal on insular shelves and uppermost slopes. Two species.

Typhlonarke aysoni (Hamilton, 1902). Blind legged torpedo.
Typhlonarke tarakea Phillipps, 1929. Slender legged torpedo.

FAMILY HYPNIDAE GILL, 1862. COFFIN RAYS.

Genus *Hypnos* Dumeril, 1852. Coffin rays. Australia, demersal on continental shelf and uppermost slope. One species.

Hypnos monopterygius (Shaw & Nodder, 1795). Coffin ray.

FAMILY TORPEDINIDAE BONAPARTE, 1838. TORPEDO RAYS.

Genus *Torpedo* Houttuyn, 1764. Torpedos. Circumglobal, demersal and benthopelagic on continental and insular shelves and slopes. Two subgenera, 21 valid species, two dubious species, plus possibly three undescribed species.

Subgenus *Tetronarce* Gill, 1862:

Semipelagic torpedos. 10 valid species plus possibly three undescribed species.

Torpedo (Tetronarce) californica Ayres, 1855. Pacific torpedo.
Torpedo (Tetronarce) fairchildi Hutton, 1872. New Zealand torpedo.

Torpedo (Tetronarce) macneilli (Whitley, 1932). Australian torpedo.

Torpedo (Tetronarce) microdiscus Parin & Kotlyar, 1985. Smalldisk torpedo.

Torpedo (Tetronarce) nobiliana Bonaparte, 1835. Great, Atlantic, or black torpedo.

Torpedo (Tetronarce) peruana Chirichigno, 1963. Peruvian torpedo.

Torpedo (Tetronarce) puelcha Lahille, 1928. Argentine torpedo.

Torpedo (Tetronarce) semipelagica Parin & Kotlyar, 1985. Semipelagic torpedo.

Torpedo (Tetronarce) tokionis (Tanaka, 1908)? Trapezoid torpedo.

Torpedo (Tetronarce) tremens de Buen, 1959? Chilean torpedo.

Subgenus *Torpedo* Houttuyn, 1764:

Demersal torpedos. 11 valid species, two dubious species, plus possibly five undescribed species.

Torpedo (Torpedo) adenensis de Carvalho, Stehmann & Manilo, 2002. Aden torpedo.

Torpedo (Torpedo) alexandrinsis Mazhar, 1987. Alexandrine torpedo.

Torpedo (Torpedo) andersoni Bullis, 1962. Florida torpedo.

Torpedo (Torpedo) bauchotae Cadenat, Capape & Desoutter, 1978. Rosette torpedo.

Torpedo (Torpedo) fuscomaculata Peters, 1855. Blackspotted torpedo.

Torpedo (Torpedo) mackayana Metzelaar, 1919. Ringed torpedo.

Torpedo (Torpedo) marmorata Risso, 1810. Spotted or marbled torpedo.

Torpedo (Torpedo) panthera Olfers, 1831. Leopard torpedo.

?*Torpedo (Torpedo) polleni* Bleeker, 1866. Reunion torpedo.

Torpedo (Torpedo) sinuspersici Olfers, 1831. Gulf torpedo.

Torpedo (Torpedo) suissi Steindachner, 1898. Red Sea torpedo.

Torpedo (Torpedo) torpedo (Linnaeus, 1758). Ocellate or common torpedo.

?*Torpedo (Torpedo) zugmayeri* Engelhardt, 1912. Baluchistan torpedo.

SUBORDER RAJOIDEI. SKATES.

FAMILY ARHYNCHOBATIDAE FOWLER, 1934. SOFTNOSE SKATES. An alternative arrangement places this taxon in the family Rajidae.

Genus *Arhynchobatis* Waite, 1909. Longtailed skates. New Zealand, demersal on insular shelves. One species.

Arhynchobatis asperrimus Waite, 1909. Longtailed skate.

Genus *Atlantoraja* Menni, 1972. La Plata skates. Western South Atlantic coast of South America, demersal on continental shelves. Three species, plus possibly one undescribed species.

Atlantoraja castelnaui (Ribeiro, 1907). Spotback skate.

Atlantoraja cyclophora (Regan, 1903). Eyespot skate.

Atlantoraja platana (Günther, 1880). La Plata skate.

Genus *Bathyraja* Ishiyama, 1958. Softnose skates. Circumglobal, most diverse in high latitudes, continental and insular shelves and slopes. Forty-three valid species, plus at least two undescribed species.

Bathyraja abyssicola (Gilbert, 1896). Deepsea skate.
Bathyraja aguja (Kendall & Radcliffe, 1912). Aguja skate.
Bathyraja aleutica (Gilbert, 1895). Aleutian skate.
Bathyraja andriashevi Dolganov, 1985. Little-eyed skate.
Bathyraja bergi Dolganov, 1985. Bottom skate.
Bathyraja brachyurops (Fowler, 1910). Broadnose skate.
Bathyraja caeluronigricans Ishiyama & Ishihara, 1977. Purpleblack skate.
Bathyraja cousseaui de Astarloa & Mabragña, 2004. Cousseau's skate.
Bathyraja diplotaenia (Ishiyama, 1950). Duskypink skate.
Bathyraja eatonii (Günther, 1876). Eaton's skate.
Bathyraja fedorovi Dolganov, 1985. Cinnamon skate.
Bathyraja griseocauda (Norman, 1937). Graytail skate.
Bathyraja hesperafricana Stehmann, 1995. West African skate
Bathyraja irrasa Hureau & Ozouf-Costaz, 1980. Kerguelen sandpaper skate.
Bathyraja isotrachys (Günther, 1877). Raspback skate.
Bathyraja kincaidi (Garman, 1908). Sandpaper skate.
Bathyraja lindbergi Ishiyama & Ishihara, 1977. Commander skate.
Bathyraja longicauda (de Buen, 1959). Slimtail skate.
Bathyraja maccaini Springer, 1972. McCain's skate.
Bathyraja maculata Ishiyama & Ishihara, 1977. Whiteblotched skate.
Bathyraja mariposa Stevenson, Hoff, Orr & McEachran, 2004. Mariposa skate.
Bathyraja matsubarai (Ishiyama, 1952). Duskypurple skate.
Bathyraja meridionalis Stehmann, 1987. Darkbelly skate.
Bathyraja microtrachys (Osburn & Nichols, 1917). Finespined skate.
Bathyraja minispinosa Ishiyama & Ishihara, 1977. Smallthorn skate.
Bathyraja notoroensis Ishiyama & Ishihara, 1977. Notoro skate.
Bathyraja pallida (Forster, 1967). Pallid skate.
Bathyraja papilonifera Stehmann, 1985. Butterfly skate.
Bathyraja parmifera (Bean, 1881). Alaska skate.
Bathyraja peruana McEachran & Miyake, 1984. Peruvian skate.
Bathyraja richardsoni (Garrick, 1961). Richardson's skate.
Bathyraja scaphiops (Norman, 1937). Cuphead skate.
Bathyraja schroederi (Krefft, 1968). Whitemouth skate.
Bathyraja shuntovi Dolganov, 1985. Narrownose skate.
Bathyraja simoterus (Ishiyama, 1967). Hokkaido skate.
Bathyraja smirnovi (Soldatov & Lindberg, 1913). Golden skate.
Bathyraja smithii (Müller & Henle, 1841). African softnose skate.
Bathyraja spinicauda (Jensen, 1914). Spinetail or spinytail skate.
Bathyraja spinosissima (Beebe & Tee-Van, 1941). Pacific white skate.
Bathyraja trachouros (Ishiyama, 1958). Eremo skate.
Bathyraja trachura (Gilbert, 1892). Roughtail skate

Bathyraja tzinovskii Dolganov, 1985. Creamback skate.
Bathyraja violacea (Suvorov, 1935). Okhotsk skate.

Genus *Irolita* Whitley, 1931. Round skates. Australia, demersal on shelves and uppermost slopes. One valid species, and one undescribed species.
Irolita waitei (McCulloch, 1911). Southern round skate.

Genus *Notoraja* Ishiyama, 1958. Velvet skates. Western Pacific, slopes. Six species, plus five undescribed species.
Notoraja asperula (Garrick & Paul, 1974). Prickly deepsea skate.
Notoraja laxipella (Yearsley & Last, 1992). Eastern looseskin skate.
Notoraja ochroderma McEachran & Last, 1994. Pale skate.
Notoraja spinifera (Garrick & Paul, 1974). Spiny deepsea skate.
Notoraja subtilispinosa Stehmann, 1985. Velvet skate.
Notoraja tobitukai (Hiyama, 1940). Leadhued skate.

Genus *Pavoraja* Whitley, 1939. Peacock skates. Western South Pacific, demersal on shelves and slopes. Two species, plus four or five undescribed species.
Pavoraja alleni McEachran & Fechhelm, 1982. Allens skate.
Pavoraja nitida (Günther, 1880). Peacock skate.

Genus *Psammobatis* Günther, 1870. Sand skates. Eastern South Pacific and Western South Atlantic coasts of South America, shelves. Eight species.
Psammobatis bergi Marini, 1932. Blotched sandskate.
Psammobatis extenta Garman, 1913. Zipper sandskate.
Psammobatis lentiginosa McEachran, 1983. Freckled sandskate.
Psammobatis parvacauda McEachran, 1983. Smalltail sandskate.
Psammobatis normani McEachran, 1983. Shortfin sandskate.
Psammobatis rudis Günther, 1870. Smallthorn sandskate.
Psammobatis rutrum Jordan, 1890. Spade sandskate.
Psammobatis scobina (Philippi, 1857). Raspthorn sandskate.

Genus *Pseudoraja* Bigelow & Schroeder, 1954. Fanfin skates. Western North Atlantic, demersal on continental slopes. A single species.
Pseudoraja fischeri Bigelow & Schroeder, 1954. Fanfin skate.

Genus *Rhinoraja* Ishiyama, 1952. Jointnose skates. North Pacific, Western South Atlantic coast of South America, Antarctic, demersal on the continental and insular shelves and slopes. This may not be a monophyletic genus, as it contains species that are morphologically diverse and which differ from *Bathyraja* primarily in having a basal joint in the rostral cartilage. 13 species.
Rhinoraja albomaculata (Norman, 1937). Whitedotted skate.
Rhinoraja interrupta (Gill & Townsend, 1897). Bering skate.
Rhinoraja kujiensis (Tanaka, 1916). Dapplebellied softnose skate.
Rhinoraja longi Raschi & McEachran, 1991. Aleutian dotted skate
Rhinoraja longicauda Ishiyama, 1952. Whitebellied softnose skate.

Rhinoraja macloviana (Norman, 1937). Patagonian skate.
Rhinoraja magellanica (Philippi, 1902 or Steindachner, 1903). Magellan skate.
Rhinoraja multispinis (Norman, 1937). Multispine skate.
Rhinoraja murrayi (Günther, 1880). Murray's skate.
Rhinoraja obtusa (Gill & Townsend, 1897). Blunt skate.
Rhinoraja odai Ishiyama, 1952. Oda's skate.
Rhinoraja rosispinis (Gill & Townsend, 1897). Flathead skate.
Rhinoraja taranetzi Dolganov, 1985. Mudskate.

Genus *Rioraja* Whitley, 1939. Rio skates. Western South Atlantic coast of South America, demersal on continental shelves. A single species.
Rioraja agassizi (Müller & Henle, 1841). Rio skate.

Genus *Sympterygia* Müller & Henle, 1837. Fanskates. Eastern South Pacific and Western South Atlantic coasts of South America, demersal on continental shelves. Four species.
Sympterygia acuta Garman, 1877. Bignose fanskate.
Sympterygia bonapartei Müller & Henle, 1841. Smallnose fanskate.
Sympterygia brevicaudata Cope, 1877. Shorttail fanskate.
Sympterygia lima (Poeppig, 1835). Filetail fanskate.

FAMILY RAJIDAE BLAINVILLE, 1816. SKATES.

Genus *Amblyraja* Malm, 1877. Stout skates. Circumglobal, most records in higher latitudes and Northern Hemisphere but also in deep water in the tropics and in the Southern Hemisphere, demersal on shelves and slopes. Ten species, plus one or two undescribed species.
Amblyraja badia (Garman, 1899). Broad skate.
Amblyraja doellojuradoi (Pozzi, 1935). Southern thorny skate.
Amblyraja frerichsi (Krefft, 1968). Thickbody skate.
Amblyraja georgiana (Norman, 1938). Antarctic starry skate.
Amblyraja hyperborea (Collette, 1879). Arctic skate.
Amblyraja jenseni (Bigelow & Schroeder, 1950). Jensen's skate.
Amblyraja radiata (Donovan, 1808). Thorny skate.
Amblyraja reversa (Lloyd, 1906). Reversed skate.
Amblyraja robertsi (Hulley, 1970). Bigmouth skate.
Amblyraja taaf (Meisner, 1987). Whiteleg skate.

Genus *Breviraja* Bigelow & Schroeder, 1948. Lightnose skates. Western and Eastern Atlantic, demersal on the continental slopes. Six species, plus one undescribed species in Eastern Atlantic.
Breviraja claramaculata McEachran & Matheson, 1985. Brightspot skate.
Breviraja colesi Bigelow & Schroeder, 1948. Lightnose skate.
Breviraja marklei McEachran & Miyake, 1987. Nova Scotia skate.
Breviraja mouldi McEachran & Matheson, 1995. Blacknose skate.
Breviraja nigriventralis McEachran & Matheson, 1985. Blackbelly skate.
Breviraja spinosa Bigelow & Schroeder, 1950. Spinose skate.

Genus *Dactylobatus* Bean & Weed, 1909. Skilletskates. Western North Atlantic, demersal on the continental slopes. Two species.

Dactylobatus armatus Bean & Weed, 1909. Skilletskate.

Dactylobatus clarki (Bigelow & Schroeder, 1958). Hookskate.

Genus *Dipturus* Rafinesque, 1810. Longnosed skates. Virtually circumglobal in cool-temperate to tropical seas, demersal on the shelves and slopes. About 31 species, with several questionably placed in this genus, and at least 15 undescribed species.

Dipturus batis (Linnaeus, 1758). Gray skate.

Dipturus bullisi (Bigelow & Schroeder, 1962). Tortugas skate.

Dipturus campbelli (Wallace, 1967). Blackspot skate.

Dipturus? chilensis (Guichenot, 1848). Yellownose skate.

Dipturus crosnieri (Seret, 1989). Madagascar skate.

Dipturus diehli Soto & Mincarone, 2001. Thornytail skate.

Dipturus doutrei (Cadenat, 1960). Javalin skate.

Dipturus ecuadoriensis (Beebe & Tee-Van, 1941). Ecuador skate.

Dipturus garricki (Bigelow & Schroeder, 1958). San Blas skate.

Dipturus gigas (Ishiyama, 1958). Giant skate.

Dipturus? gudgeri (Whitley, 1940). Bight skate.

Dipturus? innominatus (Garrick & Paul, 1974). New Zealand smooth skate.

Dipturus johannisdavesi (Alcock, 1899). Travancore skate.

Dipturus kwangtungensis (Chu, 1960). Kwangtung skate.

Dipturus lanceorostrata (Wallace, 1967). Rattail skate.

Dipturus laevis (Mitchill, 1817). Barndoor skate.

Dipturus leptocauda (Krefft & Stehmann, 1974). Thintail skate.

Dipturus? linteus (Fries, 1838). Sailskate or sailray.

Dipturus macrocauda (Ishiyama, 1955). Bigtail skate.

Dipturus mennii Gomes & Paragó, 2001. South Brazilian skate.

Dipturus? nasutus (Banks *in* Müller & Henle, 1841). New Zealand rough skate.

Dipturus nidarosiensis (Collett, 1880). Norwegian skate.

Dipturus olseni (Bigelow & Schroeder, 1951). Spreadfin skate.

Dipturus oregoni (Bigelow & Schroeder, 1958). Hooktail skate.

Dipturus oxyrhynchus (Linnaeus, 1758). Sharpnose skate

Dipturus pullopunctata (Smith, 1964). Slime skate.

Dipturus springeri (Wallace, 1967). Roughbelly skate.

Dipturus stenorhynchus (Wallace, 1967). Prownose skate.

Dipturus teevani (Bigelow & Schroeder, 1951). Caribbean skate.

Dipturus tengu (Jordan & Fowler, 1903). Acutenose or tengu skate.

Dipturus trachyderma (Krefft & Stehmann, 1974). Roughskin skate.

Genus *Fenestraja* McEachran & Compagno, 1982. Pluto skates. Western North Atlantic, Western Indian Ocean (Madagascar), Western Pacific (Celebes), demersal on continental and insular slopes. Eight species.

Fenestraja atripinna (Bigelow & Schroeder, 1950). Blackfin pygmy skate.

Fenestraja cubensis (Bigelow & Schroeder, 1950). Cuban pygmy skate.
Fenestraja ishiyamai (Bigelow & Schroeder, 1962). Plain pygmy skate.
Fenestraja maceachrani (Seret, 1989). Madagascar pygmy skate.
Fenestraja mamillidens (Alcock, 1889). Prickly skate.
Fenestraja plutonia (Garman, 1881). Pluto skate.
Fenestraja sibogae (Weber, 1913). Siboga pygmy skate.
Fenestraja sinusmexicanus (Bigelow & Schroeder, 1950). Gulf of Mexico pygmy skate.

Genus *Gurgesiella* de Buen, 1959. Finless skates. Western South Atlantic, Eastern South Pacific, demersal on continental and insular slopes. Three species.

Gurgesiella atlantica (Bigelow & Schroeder, 1962). Atlantic pygmy skate.
Gurgesiella dorsalifera McEachran & Compagno, 1980. Onefin skate.
Gurgesiella furvescens de Buen, 1959. Dusky finless skate.

Genus *Leucoraja* Malm, 1877. Rough skates. Western North Atlantic, Eastern Atlantic, Mediterranean and southwestern Indian Ocean, Australia, shelves and slopes. 12 species, plus two undescribed species.

Leucoraja circularis (Couch, 1838). Sandy skate or ray.
Leucoraja compagnoi (Stehmann, 1995). Tigertail skate.
Leucoraja erinacea (Mitchill, 1825). Little skate.
Leucoraja fullonica (Linnaeus, 1758). Shagreen skate or ray.
Leucoraja garmani (Whitley, 1939). Rosette skate.
Leucoraja lentiginosa (Bigelow & Schroeder, 1951). Freckled skate.
Leucoraja leucosticta (Stehmann, 1971). Whitedappled skate.
Leucoraja melitensis (Clark, 1926). Maltese skate or ray.
Leucoraja naevus (Müller & Henle, 1841). Cuckoo skate or ray.
Leucoraja ocellata (Mitchill, 1815). Winter skate.
Leucoraja wallacei (Hulley, 1970). Yellowspot or blancmange skate.
Leucoraja yucatanensis (Bigelow & Schroeder, 1950). Yucatan skate.

Genus *Malacoraja* Stehmann, 1970. Soft skates. Eastern South Pacific, Western Atlantic, Eastern North Atlantic, slopes. Three species.

Malacoraja kreffti (Stehmann, 1978). Krefft's skate or ray.
Malacoraja senta (Garman, 1885). Smooth skate.
Malacoraja spinacidermis (Barnard, 1923). Prickled skate or ray, roughskin skate. Provisionally including the Northern Hemisphere *Raja mollis* Bigelow & Schroeder, 1950 as a junior synonym.

Genus *Neoraja* McEachran & Compagno, 1982. Pygmy skates. Western North Atlantic, Eastern Atlantic and Northern Indian Ocean, demersal on continental slopes. Four species.

Neoraja africana Stehmann & Seret, 1983. West African pygmy skate
Neoraja caerulea (Stehmann, 1976). Blue pygmy skate.
Neoraja carolinensis McEachran & Stehmann, 1984. Carolina pygmy skate.
Neoraja stehmanni (Hulley, 1972). South African pygmy skate.

Genus *Okamejei* Ishiyama, 1958. Spiny rasp skates. Indian Ocean and Western Pacific, demersal on shelves and slopes. About 14 species, plus two undescribed species.

Okamejei acutispina (Ishiyama, 1958). Sharpspine skate.
Okamejei australis (Macleay, 1884). Sydney skate.
Okamejei boesemani (Ishihara, 1987). Black sand skate.
Okamejei cerva (Whitley, 1939). White-spotted skate.
Okamejei heemstrai (McEachran & Fechhelm, 1982). East African skate.
Okamejei hollandi (Jordan & Richardson, 1909). Yellow-spotted skate.
Okamejei kenojei (Müller & Henle, 1841). Spiny rasp, swarthy, or ocellate spot skate.
Okamejei? koreana (Jeong & Nakabo, 1997). Korean skate.
Okamejei lemprieri (Richardson, 1846). Australian thornback skate.
Okamejei meerdervoorti (Bleeker, 1860). Bigeye skate.
Okamejei philipi (Lloyd, 1906). Aden ringed skate.
Okamejei pita (Fricke & Al-Hussar, 1995). Pita skate.
Okamejei powelli (Alcock, 1898). Indian ringed skate.
Okamejei schmidti (Ishiyama, 1958). Browneye skate.

Genus *Raja* Linnaeus, 1758. Ocellate skates. Eastern Atlantic and Mediterranean, Southwestern Indian Ocean. Demersal on continental and insular shelves and slopes. 12 valid species, two dubious species, plus one undescribed species.

?*Raja africana* Capape, 1977. African skate or ray.
Raja asterias Delaroche, 1809. Atlantic starry skate.
Raja brachyura Lafont, 1873. Blonde skate or ray.
Raja clavata Linnaeus, 1758. Thornback skate or ray.
Raja herwigi Krefft, 1965. Cape Verde skate.
Raja maderensis Lowe, 1841. Madeira skate or ray.
Raja microocellata Montagu, 1818. Smalleyed skate or ray, painted skate.
Raja miraletus Linnaeus, 1758. Brown or twineye skate or ray.
Raja montagui Fowler, 1910. Spotted skate or ray
Raja polystigma Regan, 1923. Speckled skate or ray.
Raja radula Delaroche, 1809. Rough skate or ray.
?*Raja rondeleti* Bougis, 1959. Rondelet's skate or ray.
Raja straeleni Poll, 1951. Biscuit skate.
Raja undulata Lacepede, 1802. Undulate skate or ray.

Genus *Rajella* Stehmann, 1970. Gray skates. Atlantic and southwestern Indian Ocean, demersal on shelves and slopes. 14 valid species, plus one dubious species and four undescribed species.

Rajella annandalei (Weber, 1913). Indonesian round skate.
Rajella barnardi (Norman, 1935). Bigthorn skate.
Rajella bathyphila (Holt & Byrne, 1908). Deepwater skate or ray.
Rajella bigelowi (Stehmann, 1978). Bigelow's skate or ray.
Rajella caudaspinosa (von Bonde & Swart, 1923). Munchkin skate.

Rajella dissimilis (Hulley, 1970). Ghost skate.
Rajella fuliginea (Bigelow & Schroeder, 1954). Sooty skate.
Rajella fyllae (Luetken, 1888). Round skate or ray.
Rajella kukujevi (Dolganov, 1985). Mid-Atlantic skate.
Rajella leopardus (von Bonde & Swart, 1923). Leopard skate.
Rajella nigerrima (de Buen, 1960). Blackish skate.
Rajella purpuriventralis (Bigelow & Schroeder, 1962). Purplebelly skate.
Rajella ravidula (Hulley, 1970). Smoothback skate.
Rajella sadowskyii (Krefft & Stehmann, 1974). Brazilian skate.
Rajella? alia (Garman, 1899). Blake skate.

Genus *Rostroraja* Hulley, 1972. Bottlenose skates. Eastern Atlantic and southwestern Indian Ocean, demersal on shelves and slopes. A single species.
Rostroraja alba (Lacepede, 1803). White, bottlenose, or spearnose skate.

Appendix: Species representing undescribed genus-group taxa, nominally retained in *Raja*.

1. Undescribed genus for the 'North Pacific Assemblage' of McEachran & Dunn, 1998, including *Dipturus*-like giant species and *Okamejei*-like species:
Raja binoculata Girard, 1854. Big skate.
Raja cortezensis McEachran & Miyake, 1988. Cortez skate.
Raja inornata Jordan & Gilbert, 1880. California skate.
Raja 'pulchra' Liu, 1932. Mottled skate. Junior homonym of *Raja pulchra* Schafhaeutl, 1863, for fossil dermal tubercles from the Eocene of Bavaria.
Raja rhina Jordan & Gilbert, 1880. Longnose skate.
Raja stellulata Jordan & Gilbert, 1880. Pacific starry skate.

2. Undescribed genus for the 'Amphi-American Assemblage' of McEachran & Dunn (1998), including mostly *Raja*-like species from the Western Atlantic and Eastern Pacific:
Raja ackleyi Garman, 1881. Ocellate skate.
Raja bahamensis Bigelow & Schroeder, 1965. Bahama skate.
Raja cervigoni Bigelow & Schroeder, 1964. Venezuela skate.
Raja eglanteria Bosc, 1802. Clearnose skate.
Raja equatorialis Jordan & Bollman, 1890. Equatorial skate.
Raja texana Chandler, 1921. Roundel skate.
Raja velezi Chirichigno, 1973. Rasptail skate.

3. Australian species, with generic and familial placement provisional, including two named species and at least three undescribed taxa.
Raja polyommata Ogilby, 1910. Argus skate.
Raja whitleyi Iredale, 1938. Melbourne skate.

FAMILY ANACANTHOBATIDAE VON BONDE & SWART, 1924.
LEGSKATES. An alternative arrangement places this taxon in the family Rajidae.

Genus *Anacanthobatis* von Bonde & Swart, 1924. Smooth legskates. Western North Atlantic, southwestern Indian Ocean, Western Pacific, demersal on continental and insular slopes. 10 described species, and at least five undescribed species.

Anacanthobatis americanus Bigelow & Schroeder, 1962. American legskate.
Anacanthobatis borneensis Chan, 1965. Borneo legskate.
Anacanthobatis donghaiensis (Deng, Xiong, & Zhan, 1983). East China legskate.
Anacanthobatis folirostris (Bigelow & Schroeder, 1951). Leafnose legskate.
Anacanthobatis longirostris Bigelow & Schroeder, 1962. Longnose legskate.
Anacanthobatis marmoratus (von Bonde & Swart, 1924). Spotted legskate.
Anacanthobatis melanosoma (Chan, 1965). Blackbodied legskate.
Anacanthobatis nanhaiensis (Meng & Li, 1981). South China legskate.
Anacanthobatis ori (Wallace, 1967). Black legskate.
Anacanthobatis stenosoma (Li & Hu, 1982). Narrow legskate.

Genus *Cruriraja* Bigelow & Schroeder, 1948. Rough legskates. Western North Atlantic, Eastern South Atlantic and Western Indian Ocean, demersal on continental and insular shelves and slopes. Eight species.

Cruriraja andamanica (Lloyd, 1909). Andaman legskate.
Cruriraja atlantis Bigelow & Schroeder, 1948. Atlantic legskate.
Cruriraja cadenati Bigelow & Schroeder, 1962. Broadfoot legskate.
Cruriraja durbanensis (von Bonde & Swart, 1924). Smoothnose legskate.
Cruriraja parcomaculata (von Bonde & Swart, 1924). Roughnose legskate.
Cruriraja poeyi Bigelow & Schroeder, 1948. Cuban legskate.
Cruriraja rugosa Bigelow & Schroeder, 1958. Rough legskate.
Cruriraja triangularis Smith, 1964. Triangular legskate.

SUBORDER MYLIOBATOIDEI STINGRAYS.

FAMILY PLESIOBATIDAE NISHIDA, 1990. GIANT STINGAREES.

Genus *Plesiobatis* Nishida, 1990. Giant stingaree or deepwater stingray. Wide-ranging in Indo-West Pacific from South Africa to Hawaii, demersal on continental and insular slopes. Formerly included in the genus *Urotrygon* and the family Urolophidae, but placed in its own genus and family by Nishida, 1990.

Plesiobatis daviesi (Wallace, 1967). Giant stingaree.

FAMILY HEXATRYGONIDAE HEEMSTRA & SMITH, 1980. SIXGILL STINGRAYS.

Genus *Hexatrygon* Heemstra & Smith, 1980. Sixgill stingray. Wide-ranging in Eastern South Atlantic and Indo-West Pacific from South Africa to Hawaii, demersal on continental and insular slopes but occasionally caught on shelves. Several species described but probably only one variable species.

Hexatrygon bickelli Heemstra & Smith, 1980. Sixgill stingray. Provisionally including *H. longirostrum* (Chu & Meng, 1981), *H. yangi*

Shen & Liu, 1984, *H. taiwanensis* Shen, 1986, and *H. brevirostra* Shen, 1986 as junior synonyms.

FAMILY UROTRYGONIDAE McEACHRAN, DUNN & MIYAKE, 1996. AMERICAN ROUND STINGRAYS.

Genus *Urobatis* Garman, 1913. Shorttail round stingrays. Western North Atlantic and Eastern Pacific, demersal on continental shelves. Six species.

Urobatis concentricus Osburn & Nichols, 1916. Bullseye stingray.
Urobatis halleri (Cooper, 1863). Round stingray.
Urobatis jamaicensis (Cuvier, 1817). Yellow stingray.
Urobatis maculatus Garman, 1913. Cortez round stingray.
Urobatis marmoratus (Philippi, 1893). Chilean round stingray.
Urobatis tumbesensis (Chirichigno & McEachran, 1979). Tumbes round stingray.

Genus *Urotrygon* Gill, 1864. Longtail round stingrays. Eastern Pacific and Western Atlantic, demersal on continental shelves. 10 species.

Urotrygon aspidura (Jordan & Gilbert, 1882). Roughtail round stingray.
Urotrygon chilensis (Günther, 1871). Thorny round stingray.
Urotrygon cimar López S. & Bussing 1998. Costa Rica round stingray.
Urotrygon microphthalmum Delsman, 1941. Smalleyed round stingray.
Urotrygon munda Gill, 1863. Shortfin round stingray.
Urotrygon nana Miyake & McEachran, 1988. Dwarf round stingray.
Urotrygon reticulata Miyake & McEachran, 1988. Reticulate round stingray.
Urotrygon rogersi (Jordan & Starks, 1895). Lined round stingray.
Urotrygon simulatrix Miyake & McEachran, 1988. Stellate round stingray.
Urotrygon venezuelae Schultz, 1949. Venezuela round stingray.

FAMILY UROLOPHIDAE MÜLLER & HENLE, 1841. STINGAREES.

Genus *Trygonoptera* Müller & Henle, 1841. Shovelnose stingarees. Australia, demersal on continental shelves to slope edge. Four species, plus two undescribed species.

Trygonoptera mucosa (Whitley, 1939). Western shovelnose stingaree.
Trygonoptera ovalis Last & Gomon, 1987. Striped stingaree.
Trygonoptera personalis Last & Gomon, 1987. Masked stingaree.
Trygonoptera testacea Banks, *in* Müller & Henle, 1841. Common stingaree.

Genus *Urolophus* Müller & Henle, 1837. Stingarees. Western Pacific, demersal on continental shelves and upper slopes.20 species, plus at least two undescribed species.

Urolophus armatus Valencienes, *in* Müller & Henle, 1841. New Ireland stingaree.
Urolophus aurantiacus Müller & Henle, 1841. Sepia stingray.
Urolophus bucculentus Macleay, 1884. Sandyback stingaree.
Urolophus circularis McKay, 1966. Circular stingaree.
Urolophus cruciatus (Lacepede, 1804). Banded or crossback stingaree.
Urolophus deforgesi Séret & Last 2003. Chesterfield Island stingaree.

Urolophus expansus McCulloch, 1916. Wide stingaree.
Urolophus flavomosaicus Last & Gomon, 1987. Patchwork stingaree.
Urolophus gigas Scott, 1954. Spotted or Sinclair's stingaree.
Urolophus javanicus (Martens, 1864). Java stingaree.
Urolophus kaianus Günther, 1880. Kai stingaree.
Urolophus lobatus McKay, 1966. Lobed stingaree.
Urolophus mitosis Last & Gomon, 1987. Mitotic or blotched stingaree.
Urolophus neocaledoniensis Séret & Last 2003. New Caledonian stingaree.
Urolophus orarius Last & Gomon, 1987. Coastal stingaree.
Urolophus paucimaculatus Dixon, 1969. Sparsely-spotted, Dixons, or white-spotted stingaree.
Urolophus piperatus Séret & Last 2003. Peppered stingaree.
Urolophus sufflavus Whitley, 1929. Yellowback stingaree.
Urolophus viridis McCulloch, 1916. Greenback stingaree.
Urolophus westraliensis Last & Gomon, 1987. Brown stingaree.

FAMILY POTAMOTRYGONIDAE GARMAN, 1877. RIVER STINGRAYS.

Genus *Paratrygon* Dumeril, 1865. Discusrays. Rivers of northern Bolivia, eastern Peru, and northern Brazil. Demersal, one species; a second undescribed species may fall in this genus or a new one.
Paratrygon aireba (Müller & Henle, 1841). Discusray.

Genus *Plesiotrygon* Rosa, Castello, & Thorson, 1987. Longtailed river stingrays. Upper and Mid-Amazon River and tributaries in Ecuador and Brazil. Demersal. A single species.
Plesiotrygon iwamae Rosa, Castello, & Thorson, 1987. Longtailed river stingray.

Genus *Potamotrygon* Garman, 1877. Shorttailed river stingrays. Rivers of Colombia, Venezuela, Bolivia, Guyana, French Guiana, Surinam, Peru, Brazil, Argentina, Uruguay, and Paraguay. Demersal. 18 species, plus three or more undescribed species.
Potamotrygon brachyura (Günther, 1880). Shorttailed river stingray.
Potamotrygon castexi Castello & Yagolkowski, 1969. Vermiculate river stingray.
Potamotrygon constellata (Vaillant, 1880). Thorny river stingray.
Potamotrygon dumerilii (Castelnau, 1855). Anglespot river stingray.
Potamotrygon falkneri Castex & Maciel, 1963. Largespot river stingray.
Potamotrygon henlei (Castelnau, 1855). Bigtooth river stingray.
Potamotrygon histrix (Müller & Henle, *in* Orbigny, 1834). Porcupine river stingray.
Potamotrygon humerosa Garman, 1913. Roughback river stingray.
Potamotrygon leopoldi Castex & Castello, 1970. Whiteblotched river stingray.
Potamotrygon magdalenae (Valenciennes, *in* Dumeril, 1865). Magdalena river stingray.
Potamotrygon motoro (Natterer, in Müller & Henle, 1841). Ocellate river stingray.

Potamotrygon ocellata (Engelhardt, 1912). Redblotched river stingray.
Potamotrygon orbignyi (Castelnau, 1855). Smoothback river stingray.
Potamotrygon schroederi Fernandez Yepez, 1957. Rosette river stingray.
Potamotrygon schuemacheri Castex, 1964. Parana river stingray.
Potamotrygon scobina Garman, 1913. Raspy river stingray.
Potamotrygon signata Garman, 1913. Parnaiba river stingray.
Potamotrygon yepezi Castex & Castello, 1970. Maracaibo river stingray.

FAMILY DASYATIDAE JORDAN, 1888. WHIPTAIL STINGRAYS.
Classification under revision; members of genus *Taeniura* and American species of *Himantura* may belong in the family Potamotrygonidae.

Genus *Dasyatis* Rafinesque, 1810. Fintail stingrays. Circumglobal in all temperate and tropical seas, demersal on the continental and insular shelves and uppermost slopes, also in tropical and warm-temperate rivers and lakes. 38 valid species, one dubious species, and possibly five undescribed species.

Dasyatis acutirostra Nishida & Nakaya, 1988. Sharpnose stingray.
Dasyatis akajei (Müller & Henle, 1841). Red stingray.
Dasyatis americana Hildebrand & Schroeder, 1928. Southern stingray.
Dasyatis annotata Last, 1987. Plain maskray.
Dasyatis bennetti (Müller & Henle, 1841). Bennett's cowtail or frilltailed stingray.
Dasyatis brevicaudata (Hutton, 1875). Shorttail or smooth stingray.
Dasyatis centroura (Mitchill, 1815). Roughtail stingray.
Dasyatis chrysonota (Smith, 1828). Blue or marbled stingray.
Dasyatis colarensis Santos, Gomes & Charvet-Almeida, 2004. Colares stingray.
Dasyatis dipterura Jordan & Gilbert, 1880. Diamond stingray. *Dasyatis brevis* (Garman, 1880) a junior synonym.
Dasyatis fluviorum Ogilby, 1908. Estuary stingray.
Dasyatis garouaensis (Stauch & Blanc, 1962). Smooth freshwater stingray, Niger stingray.
Dasyatis geijskesi Boeseman, 1948. Wingfin stingray.
Dasyatis gigantea (Lindberg, 1930). Giant stumptail stingray.
Dasyatis guttata (Bloch & Schneider, 1801). Longnose stingray.
Dasyatis hypostigma Santos & de Carvalho, 2004. Groovebelly stingray.
Dasyatis izuensis Nishida & Nakaya, 1988. Izu stingray.
Dasyatis kuhlii (Müller & Henle, 1841). Bluespotted stingray or maskray.
Dasyatis laevigata Chu, 1960. Yantai stingray.
Dasyatis laosensis Roberts & Karnasuta, 1987. Mekong freshwater stingray.
Dasyatis lata (Garman, 1880). Brown stingray.
Dasyatis leylandi Last, 1987. Painted maskray.
Dasyatis longa (Garman, 1880). Longtail stingray.
Dasyatis margarita (Günther, 1870). Daisy stingray.
Dasyatis margaritella Compagno & Roberts, 1984. Pearl stingray.

Dasyatis matsubarai Miyosi, 1939. Pitted stingray.
Dasyatis microps (Annandale, 1908). Thickspine giant stingray.
Dasyatis multispinosa (Tokarev, 1959). Multispine giant stingray.
Dasyatis navarrae (Steindachner, 1892). Blackish stingray.
Dasyatis pastinaca (Linnaeus, 1758). Common stingray.
Dasyatis rudis (Günther, 1870). Smalltooth stingray.
Dasyatis sabina (Lesueur, 1824). Atlantic stingray.
Dasyatis sayi (Lesueur, 1817). Bluntnose stingray.
Dasyatis sinensis (Steindachner, 1892). Chinese stingray.
Dasyatis thetidis Ogilby, *in* Waite, 1899. Thorntail or black stingray.
?*Dasyatis tortonesei* Capape, 1977. Tortonese's stingray.
Dasyatis ushiei Jordan & Hubbs, 1925. Cow stingray.
Dasyatis zugei (Müller & Henle, 1841). Pale-edged stingray.

Genus *Himantura* Müller & Henle, 1837. Whiprays. Tropical Indo-West Pacific. Demersal on the continental and insular shelves and in tropical rivers and lakes. 21 valid species, one doubtful species, at least three undescribed species, and two species in the Western Hemisphere that probably belong to a separate genus (see below).

Himantura alcocki (Annandale, 1909). Palespot whipray.

Himantura bleekeri (Blyth, 1860). Whiptail stingray.

Himantura chaophraya Monkolprasit & Roberts, 1990. Giant freshwater stingray or whipray.

Himantura fai Jordan & Seale, 1906. Pink whipray.

Himantura fava (Annandale, 1909). Ocellate whipray.

?*Himantura fluviatilis* (?Hamilton-Buchanan, 1822, or Annandale, 1910). Ganges stingray.

Himantura gerrardi (Gray, 1851). Sharpnose stingray, Bluntnose whiptail ray or whipray, banded whiptail ray (possibly a species complex).

Himantura granulata (Macleay, 1883). Mangrove whipray.

Himantura imbricata (Bloch & Schneider, 1801). Scaly stingray or whipray.

Himantura jenkinsii (Annandale, 1909). Pointed-nose stingray, dragon stingray, or golden whipray. *H. draco* Compagno & Heemstra, 1984 is a junior synonym of this species.

Himantura krempfi (Chabanaud, 1923). Marbled whipray.

Himantura marginata (Blyth, 1860). Blackedge whipray.

Himantura microphthalma (Chen, 1948). Smalleye whipray.

Himantura oxyrhyncha (Sauvage, 1878). Longnosed marbled whipray.

Himantura pastinacoides (Bleeker, 1852). Round whipray.

Himantura pareh (Bleeker, 1852). Pareh whipray.

Himantura signifer Compagno & Roberts, 1982. White-edge freshwater whipray.

Himantura toshi Whitley, 1939. Blackspotted whipray or coachwhip ray.

Himantura uarnacoides (Bleeker, 1852). Whitenose whipray.

Himantura uarnak (Forsskael, 1775). Honeycomb or leopard stingray or reticulate whipray (probably a species complex).

Himantura undulata (Bleeker, 1852). Undulated whipray.
Himantura walga (Müller & Henle, 1841). Dwarf whipray.

Western Hemisphere *Himantura*. Probably a separate genus from the otherwise Eastern Hemisphere *Himantura*, and assigned in some classifications to the Potamotrygonidae. Tropical Atlantic and Eastern Pacific, demersal on the continental and insular shelves, one species preferring brackish water. Two species.
Himantura pacifica (Beebe & Tee-Van, 1941). Pacific whipray.
Himantura schmardae (Werner, 1904). Chupare stingray.

Genus *Pastinachus* Rüppell, 1829. Feathertail stingrays. Tropical Indo-West Pacific, demersal on the continental shelves, also tropical rivers and lakes. Often placed in genus *Hypolophus* Müller & Henle, 1837 or in *Dasyatis*. One wide-ranging species that may be a species complex, and at least two additional species that may be undescribed.
Pastinachus sephen (Forsskael, 1775). Feathertail or cowtail stingray. *Dasyatis gruveli* Chabanaud, 1923 a junior synonym.

Genus *Pteroplatytrygon* Fowler, 1910. Pelagic stingrays. Circumglobal in tropical and temparate seas, oceanic and epipelagic. One species, often placed in genus *Dasyatis*.
Pteroplatytrygon violacea (Bonaparte, 1832). Pelagic stingray.

Genus *Taeniura* Müller & Henle, 1837. Ribbontail stingrays. Tropical and warm-temperate Eastern Atlantic, Mediterranean, Indo-West Pacific, and Eastern Pacific, demersal on the continental and insular shelves and uppermost slopes. An alternate classification places this genus in the Potamotrygonidae. Three species.
Taeniura grabata (Geoffroy St. Hilaire, 1817). Round fantail stingray.
Taeniura lymma (Forsskael, 1775). Ribbontailed stingray, Bluespotted ribbontail or fantail ray.
Taeniura meyeni Müller & Henle, 1841. Fantail stingray, round ribbontail ray, speckled stingray (including *T. melanospilos* Bleeker, 1853).

Genus *Urogymnus* Müller & Henle, 1837. Porcupine rays. Tropical Eastern Atlantic and Indo-West Pacific, demersal on continental shelves and tropical rivers and lakes of West Africa, demersal. Two species, possibly three.
Urogymnus asperrimus (Bloch & Schneider, 1801). Porcupine ray. Possibly including *U. africanus* (Bloch & Schneider, 1801).
Urogymnus ukpam (Smith, 1863). Pincushion ray or thorny freshwater stingray.

FAMILY GYMNURIDAE FOWLER, 1934. BUTTERFLY RAYS.

Genus *Aetoplatea* Valenciennes, *in* Müller & Henle, 1841. Fintail butterfly rays. Northern Indian Ocean and Western Pacific, continental shelves. Genus possibly not distinct from *Gymnura*. Two species.
Aetoplatea tentaculata Valenciennes, *in* Müller & Henle, 1841. Tentacled butterfly ray.
Aetoplatea zonura Bleeker, 1852. Zonetail butterfly ray.

Genus *Gymnura* Kuhl *in* van Hasselt, 1823. Butterfly rays. Circumglobal in temperate and tropical seas, demersal and epibenthic on the continental shelves. Nine species, plus one dubious species.

Gymnura afuerae (Hildebrand, 1946). Peruvian butterfly ray.

Gymnura altavela (Linnaeus, 1758). Spiny butterfly ray.

Gymnura australis (Ramsay & Ogilby, 1885). Australian butterfly ray.

Gymnura bimaculata (Norman, 1925). Twinspot butterfly ray.

?*Gymnura hirundo* (Lowe, 1843). Madeira butterfly ray.

Gymnura japonica (Schlegal, 1850). Japanese butterfly ray.

Gymnura marmorata (Cooper, 1863). California butterfly ray. *G. crebripunctata* (Peters, 1869) is a junior synonym of this species.

Gymnura micrura (Bloch & Schneider, 1801). Smooth butterfly ray.

Gymnura natalensis (Gilchrist & Thompson, 1911). Diamond ray or backwater butterfly ray.

Gymnura poecilura (Shaw, 1804). Longtail butterfly ray.

FAMILY MYLIOBATIDAE BONAPARTE, 1838. EAGLE RAYS.

Genus *Aetobatus* Blainville, 1816. Bonnetrays. Circumglobal in all warm-temperate and tropical seas, demersal and littoral on continental and insular shelves and possibly semioceanic. At least three species.

Aetobatus flagellum (Bloch & Schneider, 1801). Longheaded eagle ray.

Aetobatus narinari (Euphrasen, 1790). Spotted eagle ray or bonnetray. Possibly a species complex. Tentatively including *Myliobatis punctatus* Maclay & Macleay, 1885 as a junior synonym, formerly placed in *Pteromylaeus*.

?*Aetobatus guttatus* (Shaw, 1804). Indian eagle ray.

Genus *Aetomylaeus* Garman, 1908. Smoothtail eagle rays. Indo-West Pacific, demersal and littoral on continental and insular shelves and possibly semioceanic. Four species.

Aetomylaeus maculatus (Gray, 1832). Mottled eagle ray.

Aetomylaeus milvus (Valenciennes, *in* Müller & Henle, 1841). Ocellate eagle ray or vulturine ray.

Aetomylaeus nichofii (Bloch & Schneider, 1801). Banded or Nieuhof's eagle ray.

Aetomylaeus vespertilio (Bleeker, 1852). Ornate or reticulate eagle ray.

Genus *Myliobatis* Cuvier, 1817. Eagle rays. Circumglobal in temperate and tropical seas, with most diversity in temperate waters, demersal and littoral on continental shelves. 11 apparently valid species, plus one dubious species.

Myliobatis aquila (Linnaeus, 1758). Common eagle ray or bullray.

Myliobatis australis Macleay, 1881. Southern eagle ray.

Myliobatis californicus Gill, 1865. Bat ray.

Myliobatis chilensis Philippi, 1892?. Chilean eagle ray.

Myliobatis freminvillii Lesueur, 1824. Bullnose ray.

Myliobatis goodei Garman, 1885 . Southern eagle ray.

Myliobatis hamlyni Ogilby, 1911. Purple eagle ray.

Myliobatis longirostris Applegate & Fitch, 1964. Longnose eagle ray

Myliobatis peruanus Garman, 1913. Peruvian eagle ray.
?*Myliobatis rhombus* Basilewsky, 1855. Rhombic eagle ray.
Myliobatis tenuicaudatus Hector, 1877. New Zealand eagle ray.
Myliobatis tobijei Bleeker, 1854. Kite ray.

Genus *Pteromylaeus* Garman, 1913. Bullrays. Eastern North Pacific, Eastern Atlantic and Mediterranean, southwestern Indian Ocean, demersal and littoral on continental shelves. *Myliobatis punctatus* Maclay & Macleay, 1885 was formerly placed in this genus but apparently is a species of *Aetobatus*.

Pteromylaeus asperrimus (Jordan & Evermann, 1898). Roughskin bullray.
Pteromylaeus bovinus (Geoffroy St. Hilaire, 1817). Bullray or duckbill ray.

FAMILY RHINOPTERIDAE JORDAN & EVERMANN, 1896. COWNOSE RAYS.

Genus *Rhinoptera* Kuhl in Cuvier, 1829. Cownose or flapnose rays. Circumglobal in tropical and warm temperate seas, demersal and littoral on continental shelves. Possibly seven valid species and four dubious species.

Rhinoptera adspersa Valenciennes, *in* Müller & Henle, 1841. Rough cownose ray.

Rhinoptera bonasus (Mitchill, 1815). Cownosed ray.

Rhinoptera brasiliensis Müller & Henle, 1841. Brazilian cownose ray.

?*Rhinoptera hainanensis* Chu, 1960? Hainan cownose ray.

Rhinoptera javanica Müller & Henle, 1841. Javanese cownose ray or flapnose ray.

?*Rhinoptera jayakari* Boulenger, 1895. Oman cownose ray.

Rhinoptera marginata (Geoffroy St. Hilaire, 1817). Lusitanian cownose ray.

Rhinoptera neglecta Ogilby, 1912. Australian cownose ray.

?*Rhinoptera peli* Bleeker, 1863. African cownose ray.

?*Rhinoptera sewelli* Misra, 1947. Indian cownose ray.

Rhinoptera steindachneri Evermann & Jenkins, 1891. Hawkray or Pacific cownose ray.

FAMILY MOBULIDAE GILL, 1893. DEVIL RAYS.

Genus *Manta* Bancroft, 1828. Mantas. Virtually circumglobal in all warm seas, littoral and epipelagic. Genus *Indomanta* Whitley 1939 synonymized with this genus. Possibly one species, but this is uncertain and the genus is in need of a world-wide review.

Manta birostris (Walbaum 1792). Manta.. Manta. The most commonly cited synonyms are *M. birostris* (Donndorff, 1798), *M. hamiltoni* (Newman, 1849) from the Eastern Pacific, *M. alfredi* (Krefft, 1868) from the Western South Pacific, and *Indomanta tombazii*Whitley 1936 from the northwestern Indian Ocean.

Genus *Mobula* Rafinesque, 1810. Devilrays. Wide-ranging in all warm seas, mostly littoral in inshore waters but some species oceanic or semioceanic in the epipelagic zone. Nine species.

Mobula eregoodootenkee (Bleeker, 1859). Longfin devilray or oxray. Possibly a junior synonym of *M. diabolus* (Shaw, 1804) which has been synonymized with *M. mobular* but may refer to this species.

Mobula hypostoma (Bancroft, 1831). Atlantic devilray. Including *Ceratobatis robertsi* Boulenger, 1897

Mobula japanica (Müller & Henle, 1841). Spinetail devilray. Possibly = *M. mobular*

Mobula kuhlii (Valenciennes, *in* Müller & Henle, 1841). Shortfin devilray.

Mobula mobular (Bonnaterre, 1788). Giant devilray or devil ray.

Mobula munkiana Di Sciara, 1988. Pygmy devilray.

Mobula rochebrunei (Vaillant, 1879). Lesser Guinean devilray.

Mobula tarapacana (Philippi, 1892). Sicklefin devilray.

Mobula thurstoni (Lloyd, 1908). Bentfin or smoothtail devilray. *M. eregoodoo* (Cantor, 1849) may be a senior synonym of this species.

SUBCLASS HOLOCEPHALII. CHIMAERAS.

ORDER CHIMAERIFORMES. MODERN CHIMAERAS.

FAMILY CALLORHYNCHIDAE GARMAN, 1901. ELEPHANT FISHES. The emended form Callorhinchidae Fowler, 1941 is used by some authors to conform with the genus *Callorhinchus*.

Genus *Callorhinchus* Lacépède, 1798. Elephantfishes. Confined to the Southern Hemisphere, off South America, southern Africa, Australia and New Zealand. Demersal on the continental and insular shelves and uppermost slopes. Three species.

Callorhinchus callorynchus (Linnaeus, 1758). American elephantfish or chickenfish.

Callorhinchus capensis Dumeril, 1865. Cape elephantfish or St. Joseph.

Callorhinchus milii Bory de St. Vincent, 1823. Elephant fish.

FAMILY RHINOCHIMAERIDAE GARMAN, 1901. LONGNOSE CHIMAERAS. Senior synonym of this taxon is Subfamily Harriottinae Gill, 1893, but this is not used by most authors in favor of Rhinochimaeridae.

Genus *Harriotta* Goode & Bean, 1895. Narrownose chimaeras. Wide-ranging in most seas. Demersal on continental and insular slopes. Two species.

Harriotta haeckeli Karrer, 1972. Smallspine spookfish.

Harriotta raleighana Goode & Bean, 1895. Narrownose chimaera, bentnose rabbitfish, bigspine spookfish, or longnose chimaera.

Genus *Neoharriotta* Bigelow & Schroeder, 1950. Sicklefin chimaeras. Western Atlantic (Caribbean Sea), Eastern Atlantic, northwestern Indian Ocean. Demersal on continental and insular slopes. Three species.

Neoharriotta carri Bullis & Carpenter, 1966. Dwarf sicklefin chimaera.

Neoharriotta pinnata (Schnakenbeck, 1931). Sicklefin chimaera.

Neoharriotta pumila Didier & Stehmann, 1996. Arabian sicklefin chimaera.

Genus *Rhinochimaera* Garman, 1901. Knifetooth chimaeras. Wide-ranging in the Western North Atlantic, Eastern Atlantic, and Indo-Pacific. Demersal on continental and insular slopes. Three species.

Rhinochimaera africana Compagno, Stehmann & Ebert, 1990. Paddlenose chimaera or spookfish.

Rhinochimaera atlantica Holt & Byrne, 1909. Spearnose chimaera or straightnose rabbitfish.

Rhinochimaera pacifica (Mitsukuri, 1895). Pacific spookfish or knifenose chimaera.

FAMILY CHIMAERIDAE RAFINESQUE, 1815. SHORTNOSE CHIMAERAS.

Genus *Chimaera* Linnaeus, 1758. Rabbitfishes. Wide-ranging in the Western North and Eastern Atlantic, and Indo-West Pacific. Demersal on continental and insular shelves and slopes. Seven valid species, plus four or five undescribed species.

Chimaera cubana Howell-Rivero, 1936. Cuban chimaera.

Chimaera jordani Tanaka, 1905. Jordan's chimaera.

Chimaera lignaria Didier 2002. Giant chimaera.

Chimaera monstrosa Linnaeus, 1758. Rabbitfish.

Chimaera owstoni Tanaka, 1905. Owston's chimaera.

Chimaera panthera Didier, 1998. Leopard chimaera.

Chimaera phantasma Jordan & Snyder, 1900. Silver chimaera. *C. pseudomonstrosa* Fang & Wang, 1932 is a tentative junior synonym of this species.

Genus *Hydrolagus* Gill, 1863. Ratfishes. Circumglobal in cold-temperate to tropical seas. Demersal on continental and insular shelves and slopes. 15 valid species, plus possibly seven undescribed species.

Hydrolagus affinis (Capello, 1867). Atlantic chimaera or smalleyed rabbitfish.

Hydrolagus africanus (Gilchrist, 1922). African chimaera.

Hydrolagus alberti Bigelow & Schroeder, 1951. Gulf chimaera.

Hydrolagus barbouri (Garman, 1908). Ninespot chimaera.

Hydrolagus bemisi Didier 2002. Pale ghostshark.

Hydrolagus colliei (Lay & Bennett, 1839). Spotted ratfish. *H. media* (Garman, 1911) a junior synonym of this species.

Hydrolagus lemures (Whitley, 1939). Blackfin ghostshark.

Hydrolagus macrophthalmus de Buen, 1959. Bigeye chimaera.

Hydrolagus matallanasi Soto & Vooren, 2004. Striped rabbitfish.

Hydrolagus mirabilis (Collett, 1904). Large-eyed rabbitfish or spectral chimaera.

Hydrolagus mitsukurii (Dean in Jordan & Snyder, 1904). Mitsukurii's chimaera. *H. deani* (Smith & Radcliffe, 1912) from Philippines is probably a synonym of this species.

Hydrolagus novaezealandiae (Fowler, 1910). New Zealand ghostshark.

Hydrolagus ogilbyi (Waite, 1898). Ogilby's ghostshark. *Hydrolagus waitei* Fowler 1907 is probably a junior synonym of this species.

Hydrolagus pallidus Hardy & Stehmann, 1990. Pale chimaera.

Hydrolagus purpurescens (Gilbert, 1905). Purple chimaera. *H. eidolon* (Jordan & Hubbs, 1925) is a probable synonym of this species.

Hydrolagus trolli Didier & Séret 2002. Pointynosed blue chimaera.

Index